Solving Physics Problems

Solving Physics Problems

Exploring New Thinking Paradigms

Yipeng Gu

Jenny Stanford Publishing

Published by

Jenny Stanford Publishing Pte. Ltd.
101 Thomson Road
#06-01, United Square
Singapore 307591

Email: editorial@jennystanford.com
Web: www.jennystanford.com

British Library Cataloguing-in-Publication Data
A catalogue record for this book is available from the British Library.

Solving Physics Problems: Exploring New Thinking Paradigms

ISBN 978-981-4877-41-1 (Hardcover)
ISBN 978-1-003-16254-4 (eBook)

Contents

Preface

Since 2003, the author has been a physics competition coach in Changchun City, Jilin Province, teaching basic college physics courses and special competition courses to students participating in Physics Olympiad at all levels. High-level physics competition requires students to master physics theories similar to junior college students, but the requirement for thinking ability definitely exceeds the general investigation capacity of college students. High-level students can solve extremely complex problems with limited and basic knowledge, which manifests their excellent thinking ability and at the same time drives educators to contemplate the balance between knowledge cultivation and thinking mode cultivation. From a broader perspective, the author realized that the ability to tackle in a higher level route the specific physics problems with physics theories is the basic requirement for all physics researchers, from beginners to physicists. Unfortunately, the means of obtaining this ability are extremely limited. Physics exercises and examinations, as means of evaluation, have acted as lame "teachers" in the long history of physics. However, there is no denying the reality that the compilation of exercises aiming at examination-oriented guidance cannot meet the requirement of helping people to acquire this ability perfectly. Therefore, it is imperative to develop a set of teaching materials with visionary perspective, clear epistemology and methodology, which can cultivate people to acquire this basic ability in a real sense. Based on this philosophy, the idea of writing this book sprouts, grows and yields fruit. The guiding ideology and specific contents of the book are briefly introduced as follows:

1. **Main Contents**
   (a) Throughout the history of scientific development, natural science with nature as its research object (hereinafter referred to as science) and philosophy of science with science (or the ideology to construct science) as its research object are always intertwined and back-feeding,

to develop in a mutually reinforcing manner. However, the latter often lags behind the former. Modern science, represented by classical physics, emerged around the 16th century and developed rapidly under the impetus of a large number of intellectual elites in the world. Until the 20th century, it had experienced several revolutionary developments to become modern physics. By contrast, the philosophy of science, which had sprouted in Aristotle's time, has not been well established until the 20th century (marked by scientific realism and logical positivism). Knowledge should not be diminished by its lateness; after all, science divorced from the guidance of philosophy of science is no different from runaway horses that sometimes run very fast but still cannot avoid an uncertain future. This is especially true for physics, the foundation of natural science. Physicists have found that both "constructing physics theory" and "obtaining an understanding of reality based on physics theory" demand a clear epistemology and methodology, otherwise, they will suffer a lack of fundamental thinking, and will eventually be forced to become scientific cognitive theorists due to the unavoidable needs of their research topics. Those who deliberately avoid philosophy of science often replace it with an immature philosophy unconsciously. As a belated test, the author has developed the thinking paradigm for constructing the theoretical system of physics and the thinking paradigm for addressing specific physics problems by adopting physics theories respectively, with the latter included in this book (called ***MLQ*-(*ST*) *C paradigm***). The application of this thinking paradigm is fully demonstrated by the study of a large number of physics problems. The clarification of thinking paradigm has cleared away the obstacles of people's underlying thinking and made the mathematical thinking and physical thinking built on it clear and smooth.

(b) The reality of physics is often presented in the form of physics problems and taken as the study object of physics theory. Apart from the universal approach mentioned above, the unique academic value of the problems themselves can be excavated to different degrees according to the research depth of researchers. Some problems, owing to their fundamentality, can enable researchers to obtain a deeper or even a brand-new understanding of the existing basic theories. Some problems can directly provide important clues for basic theoretical research or applied research due to their novelty. For those new problems with strong generality and their conclusions, they can be directly elevated into the framework of basic theory. In the long-term teaching process, the author found that many physics problems proposed are quite original and worthy of study, but unfortunately the searchable studies (solutions) has not made progress with the times in terms of logical clarity, selection of mathematical tools and depth of physical theories. In view of people's demand for high-quality physics problems and their studies (solutions), the author has conducted the following work: (1) The author carried out deep excavation on some classical physics problems, made meaningful promotion, and conducted a detailed study; (2) the existing solutions to many physics problems have been systematically optimized and deepened to be more profound and rigorous; (3) different from the default classical solutions to some physics problems repeatedly copied by many similar books, brand-new solutions are provided; (4) the author revealed and corrected some widely spread erroneous solutions that are not easy to be found; (5) some original and enlightening physics problems have been created, and the solutions have been provided. The solutions to all problems are formalized under the guidance of the "***MLQ-(ST)C paradigm***" proposed by the author, and exquisite schematic diagrams are drawn. The classification of problems and solutions is introduced as follows:

<table>
<tr><th colspan="4">Classification of problems and solutions (a, A)</th></tr>
<tr><th colspan="2">Classification of problems and corresponding notes</th><th colspan="2">Classification of solutions and corresponding notes</th></tr>
<tr><td>Problems originally proposed by the author.</td><td>a</td><td rowspan="2">Author's original solutions.</td><td rowspan="2">A</td></tr>
<tr><td>The author's important promotion or adaptation of famous historical problems, high-quality problems and high-level examination problems.</td><td>b</td></tr>
<tr><td>High-level examinations and high-quality problems found in individual materials (for such problems, the author generally makes some textual adjustment and content expansion or reduction according to the overall style of the book, which does not affect the main topic meaning).</td><td>c</td><td>The author's deepening or preciseness of the solutions given by predecessors; The correction on some wrong solutions which have been widely spread.</td><td>B</td></tr>
<tr><td>Famous problems with a relatively long history, well-known classical models, popular examples or exercises existing in many textbooks, reference books and literatures.</td><td>d</td><td>The common solutions have already existed, given by predecessors.</td><td>C</td></tr>
</table>

2. The Value Orientation of the Book

(a) A physics exercise must be a physics problem; vice versa, any physics problem can also appear as a physics exercise. Whether it is regarded as a physics exercise to be solved or as a physics problem to be further studied is entirely up to people's subjective attitude, but these two realms differ in a large extent. This book regards all the included physics problems as serious topics, controls macro thinking with optimal logic, and deals with all the thinking details with meticulous standards to ensure perfection. This book aims to return all physics problems to more meaningful study values, instead of being confined to the mediocre value of strengthening students' learning or examining learning effects.

(b) The deeper intention of this book is to make the physics problems included in this book serve as a carrier for readers to study the thinking paradigm. Therefore, the physics problems selected do not involve too broad or too deep theories (tentatively limited to classical physics dominated by classical mechanics, and later version may be expanded to special relativity). Albert Einstein once stated: **If a young man has trained his muscles and physical endurance by gymnastics and walking, he will later be fitted for every physical work. This is also analogous to the training of the mind and the exercising of the mental and manual skill. Thus, the wit was not wrong who defined education in this way: "Education is that which remains, if one has forgotten everything he learned in school."** Many of physics problems will slowly be "outdated" with the development of the times, and will be naturally and even voluntarily forgotten by people because of their declining importance. However, a set of excellent epistemology or methodology possesses eternal vitality. Although the thinking paradigm provided in this book is not the only effective logic system that physicists can obtain through universal meditation on philosophy of science, it is expected to be as vital as seeds, taking root, sprouting and blossoming in different fields of physics.

Introduction: The MLQ-(ST)C Paradigm

I.1 The Deductive Study of Specific Physical Reality Based on Known Physical Theory

The purpose of this book is to establish a set of orderly thinking paradigms for the implementation of a kind of study defined in the following ways:

Definition (**the deductive study of specific physical reality based on known physical theory**): For a specific physical reality, within the deconstruction ability of known physical theories, certain conditions are given including but not limited to: (1) values, or conditions satisfied by values, of certain related physical quantities, such as value of constants, initial condition and boundary condition of variables, etc., (2) the mathematical relations between certain related physical quantities, (3) constraints. To determine the evolution of certain related physical quantities in the specified space-time range.

The establishment of and the solution to this problem are based on the understanding of the following two themes of philosophy of science: the construction of physical theory and the deconstruction of physical reality by using physical theory (here, the term "deconstruction" is defined as "the recognition of reality based on theory"). The discussion on these two themes of philosophy of

Solving Physics Problems: Exploring New Thinking Paradigms
Yipeng Gu

ISBN 978-981-4877-41-1 (Hardcover), 978-1-003-16254-4 (eBook)
www.jennystanford.com

science is a grand topic, which is suitable for a systematic and in-depth discussion with an independent monograph, as the basis of the specific problem studied in this book, only an overview is given here:

The construction of physical theory

The thinking process in which the thinking subject simulates the observable (directly or indirectly) physical reality—including "the physical objects and the patterns of their regular phenomena" into the theoretical objects is called **the construction of physical theory**. The thinking subject can conceptualize physical objects and objectively describe the patterns of their regular phenomena by using the specific standardized natural language (natural language with strict regimentation) and the specific variables (or constants) with units and the mathematical relations between them. The two kinds of theoretical objects obtained though this method are called physical model and physical law respectively. As a theoretical object, the specific variable (or constant) with units used in this method is called physical quantity. That is to say, the observable (directly or indirectly) physical reality can be simulated by the thinking subject into physical theories with physical models, physical law, and physical quantity as the core elements.

If we can construct a physical theory that can do true simulation or approximate simulation to a physical reality in the objective sense, then the physical meaning of this physical theory is fully endowed by the physical reality described by them, in which the physical meaning of a physical model or a physical law directly corresponds to the physical reality described by it, the physical meaning of a physical quantity is in different depths reflected in all the physical laws satisfied by various physical models constructed with it, and also comes from the reality it describes [for example, the physical meaning of gravitational mass can only be reflected in the law of gravitation constructed with it and its derived laws; the physical meaning of inertial mass can only be reflected in the basic dynamic equation constructed with it and its derived laws]. It can be considered that a physical quantity can be regarded as a characteristic of the physical models corresponding to the physical laws constructed with it. We can define a new physical quantity in a completely formal way without thinking about any physical law, if it cannot bear the function of expressing a certain physical law, it is regarded as a physical quantity without clear physical meaning.

The deconstruction of the physical reality based on the physical theory

The thinking process of the thinking subject to realize the recognition (cognition) to the physical reality (physical object or its certain phenomenon) to be studied by matching the information of the physical reality to be studied with the known (constructed) physical theory, is called **the deconstruction of the physical reality based on the physical theory**.

I.2 The Thinking Paradigm for Implementing "the Deductive Study of Specific Physical Reality Based on Known Physical Theory"

We can further establish the thinking paradigm to solve arbitrary specific physics problems based on the thought of "**the construction of physical theory**" and "**the deconstruction of physical reality by using physical theory**" proposed above.

I.2.1 The Thinking Objects Involved in the Implementation of "the Deductive Study of Specific Physical Reality Based on Known Physical Theory"

Definition (**space-time framework**): The theoretical tool for describing space and time (including physical space and reference frame with coordinate system) is called **space-time framework**.

Definition (**object of study**): The physical object provided by the original problem and newly constructed necessary physical object (for auxiliary study) that can be deconstructed by a known physical model, or the physical object that has been deconstructed by the physical model in the original problem (directly named by the physical model) is called **object of study**.

Definition (**state/process to be studied**): An arbitrary state or the state satisfying specific conditions in which the *object of study* is at (defined by the time or space configuration under the constructed *space-time framework*) is called **state to be studied**. An arbitrary process or the process satisfying specific conditions experienced by the *object of study* (defined by the initial and final states) is called **process to be studied**.

Definition (**physical fact to be considered**): The physical laws satisfied by the *object of study* at the *state to be studied* (or in the *process to be studied*) or the formal definition of physical quantities related is called **physical fact to be considered**.

Definition (**quantity to be introduced**): The physical quantity or mathematical quantity that needs to be used to specifically represent the *physical fact to be considered*, but not be given in the original problem, is called **quantity to be introduced**.

Definition (**relevant given conditions**): Various conditions related to *physical fact to be considered* that are given by the original problem are called **related given conditions**.

I.2.2 MLQ-(ST)C Paradigm

Definition: The thinking norms determined by the following thinking steps is called the ***MLQ-(ST)C paradigm***—the thinking paradigm for implementing "the deductive study of specific physical reality based on known physical theory". In the analysis of a specific physics problem, a sequence formed by these thinking steps is called a **thinking module** that satisfies this thinking paradigm:

Step 1. Setting the *space-time framework*.

[Taking the introduction of space-time background as the starting point of a set of thinking steps, to establish a strict *space-time framework* for subsequent thinking steps. It can effectively avoid the confusion of space-time background in the same set of thinking steps.]

Step 2. On the basis of necessary deconstruction of the physical objects in the original problem by using physical models, selecting the *object of study*, then noting the physical model used to deconstruct the *object of study*.

[Selecting the *object of study* in a formal and strict way can effectively avoid the confusion of subsequent thinking steps caused by unclear physical models.]

Step 3. For the selected *object of study*, selecting the *state/process to be studied.*

Step 4. Selecting *physical fact to be considered* from the physical facts satisfied by the *object of study* at the *state to be studied* (or in the *process to be studied*) (which is a fuzzy retrieval process), then noting the *relevant given conditions* for the fact to hold in the original problem.

[The selection of physical facts is mainly based on forward reasoning, giving consideration to backward reasoning. Forward reasoning enables the subject of thinking to solve problems in a wide enough perspective, and fully releases the flexibility and diversity of thinking. On the basis of solving the original problems, more valuable information can often be revealed. And the moderate backward reasoning is the moderate restriction of thinking direction, which can avoid too many invalid trial and error.]

Step 5. Determining the *quantity to be introduced* according to the selected *physical fact to be considered* (accomplished in thinking), then concretely expressing the selected *physical fact to be considered*, then annotating each *quantity to be introduced*.

[The determination of the *quantity to be introduced* is strictly based on the determination of *physical fact to be considered*, rather than the introduction of arbitrary physical quantities without purpose. In this way, on the one hand, it inherits the basic philosophical cognition that the physical significance of physical quantity is reflected in physical laws, On the other hand, it avoids the possible negative interference to thinking caused by the introduction of physical quantities for unknown reasons.]

Step 6. Expressing the *relevant given conditions* that need to be expressed independently.

The above steps are expressed based on standardized natural language or mathematical form as required.

If more than one *object of study* is studied in the same *space-time framework*, or the same *object of study* corresponds to more than one *state/process to be studied*, or the same *state/process to be studied* corresponds to more than one *physical fact to be considered*, these related thinking modules can be reduced to a tree structure (it is recommended to use this method for the problem that need to be simplified in form due to too much information).

According to its characteristics, the thinking paradigm can be named ***MLQ-(ST)C paradigm***, in which

M represents physical models,
L represents physical laws,
Q represents physical quantities,
ST represents space-time, and
C represents conditions.

Among them, the first three items ***MLQ*** represent the three core elements of physics theory and constitute the main body of physical thinking, and the latter two items (***ST***)***C*** constitute the main background of physical thinking.

The complete implementation of "*the deductive study of specific physical reality based on known physical theory*" can be composed of a series of thinking modules that satisfy the ***MLQ-(ST)C paradigm***, necessary mathematical facts, and mathematical deduction.

Considering the limitation of space, the use of this thinking paradigm will be appropriately simplified in this book according to different situations, mainly in the following situations:

1. The physical problems in this book are temporarily limited to the classical physics category. The physical space used for the solution is Euclidean space by default and no formal introduction will be made.
2. When the reference frame used for analysis is the default reference frame for the original problem to express information, the introduction of the reference frame will be omitted.
3. As a demonstration, the annotation of the physical model used to deconstruct the *object of study* is mainly expressed in some problems in Chapter 2, Dynamics, and basically omitted in other chapters.
4. For individual simple physical facts, when the *object of study* and *state/process to be studied* are very easy to be recognized

by the readers, the expression of these two thinking steps can be omitted to directly express the physical facts.

5. Omit the repetitive annotations, such as the model annotations of the same *object of study* and the conditional annotations of the physical facts under the same conditions.
6. When the coordinate system is clear, omit the annotations of the coordinates of the specific *object of study* (when the *object of study* is not unique, the coordinates will be distinguished by the indicative corner marks).
7. Omit the standardized introduction of some interaction theories familiar to readers.

Here is an example of using the ***MLQ-(ST)C paradigm*** to solve specific problems. We use the symbols "***M***", "***L***", "***Q***", "***ST***", and "***C***" to represent the thinking objects in this paradigm; the details are as follows:

"YO-YO" eliminating rotation

In order to keep the satellite's symmetry axis stable in the specified direction during its orbiting the planet, one of the simplest methods is to let the satellite rotate around its own symmetry axis all the time when orbiting, but sometimes in order to change the satellite's direction, it is also required to slow down or eliminate the rotation of the satellite. A method that can be realized is called the "YO-YO" eliminating rotation method, whose principle is shown in Fig. 1.

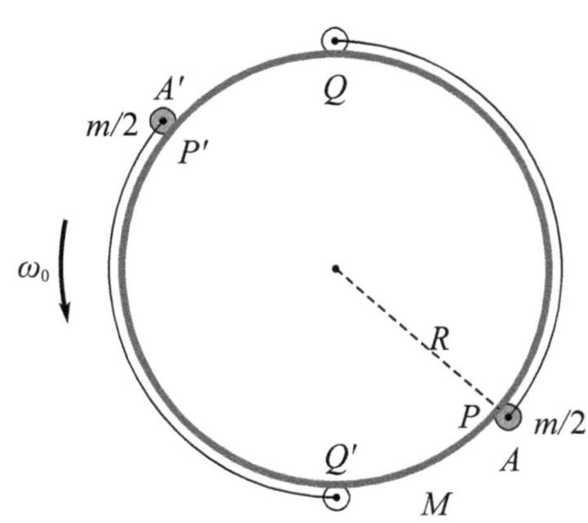

Figure 1

Let the satellite be a homogeneous thin-walled cylinder of sectional radius R and mass M, whose cross section is shown in the figure. There are two equal length light rigid ropes, one end of which are respectively fixed at Points Q and Q' on the outer surface of the cylinder (Points Q and $Q'$are symmetrical with respect to the central axis of the cylinder), and the other end are respectively tied with small balls A and $A'$of equal mass $m/2$. Under normal circumstances, the two ropes are twined on the outer surface of the cylinder, and the two balls are respectively locked at Points P and P' on the outer surface of the cylinder with pins (Points P and P' are symmetrical with respect to the central axis of the cylinder and are co-circular with Points Q and Q'), forming a whole with the cylinder. If you want the satellite to slow down or stop rotating, you can instantly remove the pins to release the balls, so that they can leave the cylinder surface. In the whole process, the segment of the rope that has left the cylinder surface is straightened all the time. It is required that the cylinder just stops rotating when the tangent points between the two ropes and the cylinder just reach Points Q and Q'. At this time point, the two ropes and the cylinder shall be separated immediately to remove the interaction between them. Given the initial angular velocity of cylinder rotation is $\boldsymbol{\omega}_0$.

(1) Find the length of each rope that has left the cylinder surface when the angular velocity of the satellite decreases to ω, and find the total length of each rope according to this.
(2) Find the time taken for the satellite's eliminating rotation.
(3) Find the tension in each rope at an arbitrary time.

Solution to Problem (1)

ST: The time point when the two balls are released is defined as time zero.

The translational reference frame that is at rest with respect to the central axis of the cylinder is called the S reference frame, which can be regarded as an inertial reference frame. The rotating reference frame with the central axis of the cylinder as the rotation axis fixed to the cylinder is called the S'reference frame.

As shown in Fig. 2, the midpoint of PP' is denoted by O, the tangent points of the two ropes and the cylinder at arbitrary Time

t [the points on the ropes] are denoted by T and T' respectively, the segments of the ropes that have left the cylinder at arbitrary Time t are denoted by Rope AT and Rope AT' respectively. The unit vector in the direction of Point O pointing to Point P is denoted by $\boldsymbol{i}$, the unit vector perpendicular to the plane of the figure outward is denoted by $\boldsymbol{k}$, $\boldsymbol{k} \times \boldsymbol{i}$ is denoted by $\boldsymbol{j}$. ***ST***: Establishing the Cartesian coordinate system with the frame $(O, \boldsymbol{i}, \boldsymbol{j}, \boldsymbol{k})$ fixed to the S' reference frame, let x, y, z be the Cartesian coordinates. Introducing the tangential unit vector $\boldsymbol{\tau}$ and normal unit vector $\boldsymbol{n}$ of Rope AT.

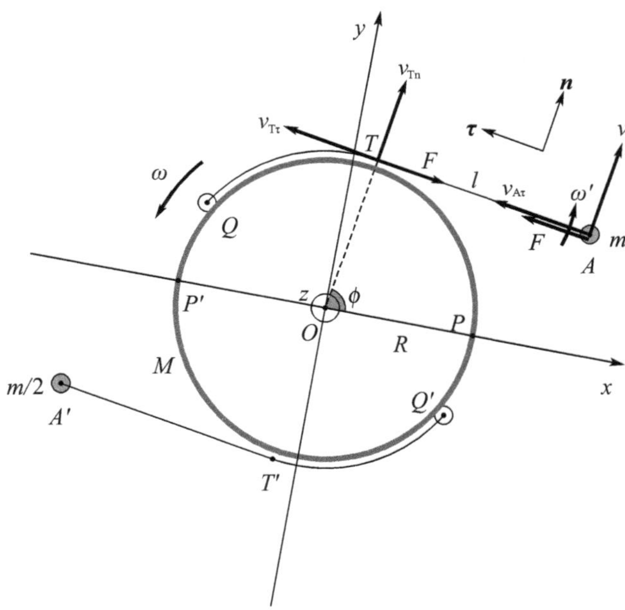

Figure 2

The motions of balls A and A' are symmetrical; therefore only one of them needs to be studied.

M: Taking *Ball A* as the object of study. ***ST***: Studying *the state at arbitrary Time t*. ***L***: Considering ***theorem for composition of velocities of a particle*** [taking the S reference frame as the stationary reference frame, and the translational reference frame that is at rest with respect to Point T as the moving reference frame] (the $\boldsymbol{\tau}$-direction and $\boldsymbol{n}$-direction component equations):

$$\textbf{L}: \quad \begin{cases} v_{A\tau} = v_{T\tau} + v_{AT\tau} & \text{(1a)} \\ v_{An} = v_{Tn} + v_{ATn} & \text{(1b)} \end{cases}$$

Q: $v_{A\tau}$ and v_{An}—the $\boldsymbol{\tau}$-direction and $\boldsymbol{n}$-direction components of the velocity of ball *A* at Time *t*.

$v_{T\tau}$ and v_{Tn}—the $\boldsymbol{\tau}$-direction and $\boldsymbol{n}$-direction components of the velocity of Point *T* at Time *t*.

$v_{AT\tau}$ and v_{ATn}—the $\boldsymbol{\tau}$-direction and $\boldsymbol{n}$-direction components of the velocity of ball *A* relative to Point *T* at Time *t*.

L: Considering ***the components form of the velocity of a particle in polar coordinate system*** [***C***: the constraint condition: the cylinder is a rigid body; the rope length remains unchanged and the rope is tightly twined on the cylinder]:

$$\textbf{L}: \quad \begin{cases} v_{T\tau} = \omega R & \text{(2a)} \\ v_{Tn} = 0 & \text{(2b)} \end{cases}$$

Q: ω—the ***k***-direction component of the angular velocity of the cylinder at arbitrary Time *t*.

L: Considering ***the components form of the velocity of a particle in polar coordinate system*** [***C***: the constraint condition: the rope length remains unchanged]:

$$\textbf{L}: \quad \begin{cases} v_{AT\tau} = 0 & \text{(3a)} \\ v_{ATn} = \omega' l & \text{(3b)} \end{cases}$$

Q: ω'—the ***k***-direction component of the angular velocity of Rope *AT* at arbitrary Time *t*.

l—the length of Rope *AT* at arbitrary Time *t*.

M: Taking *Rope AT* as the object of study. ***ST***: Studying *the state at arbitrary Time t*. ***L***: Considering ***theorem for the composition of angular velocities*** [taking the *S* reference frame as the stationary reference frame, and the *S′* reference frame as the moving reference frame] (the ***k***-direction component equation):

$$\textbf{L}: \quad \omega' = \omega + \dot{\phi} \tag{4}$$

Q: ϕ—the angle between *OT* and *OP* at Time *t*.

C: Considering the constraint condition: the rope is tightly twined on the cylinder:

C: $$l = R\phi \tag{5}$$

Combining (1) to (5),

$$\Rightarrow \begin{cases} v_{A\tau} = \omega R & \text{(6a)} \\ v_{An} = (\omega + \dot{\phi})l & \text{(6b)} \end{cases}$$

M: Taking *the system composed of the cylinder, the two ropes and the two balls* as the object of study. ***ST***: Studying *the process from time zero to arbitrary Time t*. ***L***: Considering (a) ***law of conservation of mechanical energy*** [***C***: in the *S* reference frame, the work done by the external force and non-conservative internal force acting on the system is zero; therefore the mechanical energy of the system is conserved], (b) ***law of conservation of angular momentum*** [***C***: in the *S* reference frame, taking the central axis of the cylinder as the rotation axis, the resultant external torque acting on the system is zero; therefore the angular momentum of the system is conserved] (the ***k***-direction component equation), (c) ***the form of kinetic energy of a rigid body performing fixed-axis rotation*** [integrating into (a)], (d) ***the form of angular momentum of a rigid body performing fixed-axis rotation*** [integrating into (b)]:

$$\textbf{\textit{L}}: \begin{cases} \frac{1}{2}\left[I + 2\left(\frac{1}{2}m\right)R^2\right]\omega_0^2 = \frac{1}{2}I\omega^2 + 2\cdot\frac{1}{2}\left(\frac{1}{2}m\right)(v_{A\tau}^2 + v_{An}^2) & \text{(7a)} \\ \left[I + 2\left(\frac{1}{2}m\right)R^2\right]\omega_0 = I\omega + 2\cdot\left(\frac{1}{2}m\right)(Rv_{A\tau} + lv_{An}) & \text{(7b)} \end{cases}$$

Q: *I*—the moment of inertia of the cylinder about its central axis.

L: Considering ***the form of moment of inertia of a homogeneous hollow cylinder about its central axis***:

L: $$I = MR^2 \tag{8}$$

(6) and (8) → (7),

$$\Rightarrow \begin{cases} \frac{1}{2}MR^2\omega_0^2 + \frac{1}{2}mR^2\omega_0^2 = \frac{1}{2}MR^2\omega^2 + \frac{1}{2}mR^2\left[\omega^2 + (\omega + \dot{\phi})^2\phi^2\right] & \text{(9a)} \\ MR^2\omega_0 + mR^2\omega_0 = MR^2\omega + mR^2\left[\omega + (\omega + \dot{\phi})\phi^2\right] & \text{(9b)} \end{cases}$$

$$\Rightarrow \begin{cases} \omega_0^2 - \omega^2 = \frac{m}{M+m}(\omega + \dot{\phi})^2\phi^2 & \text{(10a)} \\ \omega_0 - \omega = \frac{m}{M+m}(\omega + \dot{\phi})\phi^2 & \text{(10b)} \end{cases}$$

$$\Rightarrow \qquad \omega_0 + \omega = \omega + \dot{\phi}$$

$$\Rightarrow \qquad \dot{\phi} = \omega_0 \tag{11}$$

(11) → (10b),

$$\Rightarrow \qquad \phi = \sqrt{\frac{M+m}{m}\frac{\omega_0 - \omega}{\omega_0 + \omega}} \tag{12}$$

Q: Defining the constant k:

$$k = \frac{M+m}{m} \tag{13}$$

$$\Rightarrow \qquad \phi = \sqrt{k\frac{\omega_0 - \omega}{\omega_0 + \omega}} \tag{14}$$

(14) → (5),

$$\Rightarrow \qquad l = R\sqrt{k\frac{\omega_0 - \omega}{\omega_0 + \omega}} \tag{15}$$

Q: Defining L—the total length of the rope.

$$\Rightarrow \qquad L = l\big|_{\omega=0} = R\sqrt{k} \tag{16}$$

Solution to Problem (2)

From (11),

$$\Rightarrow \qquad \phi = \omega_0 t \tag{17}$$

(17) → (14),

$$\Rightarrow \qquad \omega = \frac{k - \omega_0^2 t^2}{k + \omega_0^2 t^2}\omega_0 \tag{18}$$

Q: Defining τ—the time taken for the satellite's eliminating rotation.

$$\Rightarrow \qquad \tau = t\big|_{\omega=0} = \frac{\sqrt{k}}{\omega_0} \tag{19}$$

Solution to Problem (3)

M: Taking *the cylinder* as the object of study. ***ST***: Studying *the state at arbitrary Time t*. ***L***: Considering the ***law of rotation*** [in the S reference frame, taking its central axis as the rotation axis] (the ***k***-direction component equation):

$$-2FR = I\dot{\omega} \tag{20}$$

Q: F—the magnitude of the tension in the two ropes at Time t.

From (18),

$$\Rightarrow \qquad \dot{\omega} = -\frac{4k\omega_0^3 t}{(k+\omega_0^2 t^2)^2} \tag{21}$$

(8) and (21) $\rightarrow$ (20),

$$\Rightarrow \qquad F = MR\frac{2k\omega_0^3 t}{(k+\omega_0^2 t^2)^2} \tag{22}$$

Another method to get the velocity of ball *A*:

M: Taking *Ball A* as the object of study. ***ST***: Studying *the state at arbitrary Time t*. ***L***: Considering ***theorem for composition of velocities of a particle*** [taking the S reference frame as the stationary reference frame, and the S' reference frame as the moving reference frame]:

$$\textbf{\textit{L}}: \qquad \boldsymbol{v}_A = \left(\frac{\mathrm{d}}{\mathrm{d}t}\right)' \boldsymbol{r}' + \boldsymbol{\omega}\times\boldsymbol{r}' \tag{23}$$

Q:

$\boldsymbol{v}_A$—the velocity of ball A at Time t.
$\boldsymbol{r}'$—the position vector of ball A in the S' reference frame at Time t with Point O as the reference point.
$\boldsymbol{\omega}$—the angular velocity of the cylinder at Time t.

Considering the following geometric relation in Fig. 2:

$$\boldsymbol{r}' = (R\cos\phi + l\sin\phi)\boldsymbol{i} + (R\sin\phi - l\cos\phi)\boldsymbol{j} \tag{24}$$

(5) $\rightarrow$ (24),

$$\Rightarrow \qquad \boldsymbol{r}' = R(\cos\phi + \phi\sin\phi)\boldsymbol{i} + R(\sin\phi - \phi\cos\phi)\boldsymbol{j} \tag{25}$$

Known:

$$\boldsymbol{\omega} = \omega\boldsymbol{k} \tag{26}$$

(25) and (26) $\rightarrow$ (23),

$$\begin{aligned}\Rightarrow \boldsymbol{v}_A &= R(-\dot{\phi}\sin\phi + \dot{\phi}\sin\phi + \phi\dot{\phi}\cos\phi)\boldsymbol{i} + \\ &\quad R(\dot{\phi}\cos\phi - \dot{\phi}\cos\phi + \phi\dot{\phi}\sin\phi)\boldsymbol{j} + \\ &\quad R(\omega\cos\phi + \omega\phi\sin\phi)\boldsymbol{j} - R(\omega\sin\phi - \omega\phi\cos\phi)\boldsymbol{i} \\ &= R\left[\phi(\omega+\dot{\phi})\cos\phi - \omega\sin\phi\right]\boldsymbol{i} + R\left[\omega\cos\phi + \phi(\omega+\dot{\phi})\sin\phi\right]\boldsymbol{j}\end{aligned} \tag{27}$$

Considering the transformation of $\boldsymbol{v}_A$ from Base $(\boldsymbol{i}, \boldsymbol{j})$ to Base $(\boldsymbol{n}, \boldsymbol{\tau})$:

$$\Rightarrow \qquad \boldsymbol{v}_A = \begin{pmatrix} \cos\phi & \sin\phi \\ -\sin\phi & \cos\phi \end{pmatrix} \begin{pmatrix} R[\phi(\omega+\dot{\phi})\cos\phi - \omega\sin\phi] \\ R[\omega\cos\phi + \phi(\omega+\dot{\phi})\sin\phi] \end{pmatrix}$$

$$= \begin{pmatrix} R\phi(\omega+\dot{\phi}) \\ R\omega \end{pmatrix} \tag{28}$$

Chapter 1

Kinematics

1.1 Particle Motion and Rigid Body Motion

Classical mechanics studies all problems on the basis of an absolute space-time view. The description of a particle's space position and all the kinematics quantities defined by its space position and time depends on the reference system which is established by the fixed connection of the artificially selected reference object and the coordinate frame. The component relations of any kinematics quantity in two coordinate frames which are stationary to each other can be described by coordinates' transformation or base vectors' transformation, which can be regarded as a pure mathematical problem. It is a physical fact that the value relation of any kinematics quantity in two different reference systems is described by Galileo transformation.

The basic kinematic characteristics of a particle are described by the function relation between its coordinates and time (its coordinates as the function of time), that is, the kinematics equations. The physical facts of dynamics can provide us with the differential equations of a particle's kinematics quantities with respect to time; it is a pure mathematical problem to solve the kinematics equations on this basis. As a typical kinematics characteristic of a particle,

Solving Physics Problems: Exploring New Thinking Paradigms
Yipeng Gu

ISBN 978-981-4877-41-1 (Hardcover), 978-1-003-16254-4 (eBook)
www.jennystanford.com

the trajectory of the particle is described by the relation between its coordinates, the information of a particle's trajectory equation is contained in its kinematics equations, for trajectory equation, kinematics equations can be regarded as its system of equations with time as the parameter.

Based on the principle of vector superposition, a complex motion can be regarded as the superposition of simple motions; therefore, the research on the motion with simple constraints on the kinematics quantities or trajectory is of vital importance. These simple kinematics models can develop many specific problems worthy of study.

The motion of a rigid body can be described jointly by the motion of a selected base point on the rigid body and the rotation of the rigid body about that base point, based on composition of kinematics quantities. The general form of rotation of the rigid body about the base point is called fixed-point rotation, which can be described by Euler kinematics equations. A typical degradation of fixed-point rotation is fixed-axis rotation, whose description becomes relatively simple.

In this section, we will study some interesting and inspiring particle motion models and rigid body motion models.

1.1.1 The Round-Trip Motion of an Airplane in the Wind (d, *A*)

An aircraft can fly with the highest speed v relative to the ground when there is no wind, and the longest distance that its fuel can support its flight is R (including flight out and flight back). Now the aircraft is flying in the wind with the velocity of the magnitude u and the direction α degrees east by north, while the actual direction of its flight is β degrees east by north, given that $v > u$. In this case, find the longest distance that the aircraft can fly (including flying out and flying back).

Solution

The time point when the plane departs is defined as Time zero.

As shown in Figs. 1 and 2, the starting point of the aircraft is denoted by O, the unit vector in the east direction is denoted by $\boldsymbol{i}$, and the unit vector in the north direction is denoted by $\boldsymbol{j}$. Establishing the

Cartesian coordinate system with the frame $(O, \boldsymbol{i}, \boldsymbol{j})$, let x, y be the Cartesian coordinates.

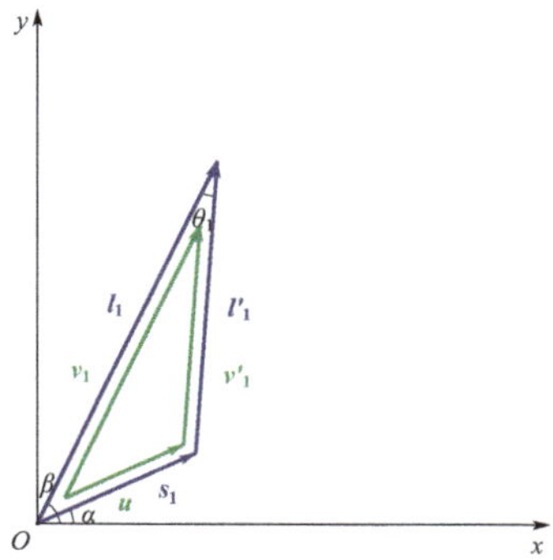

Figure 1

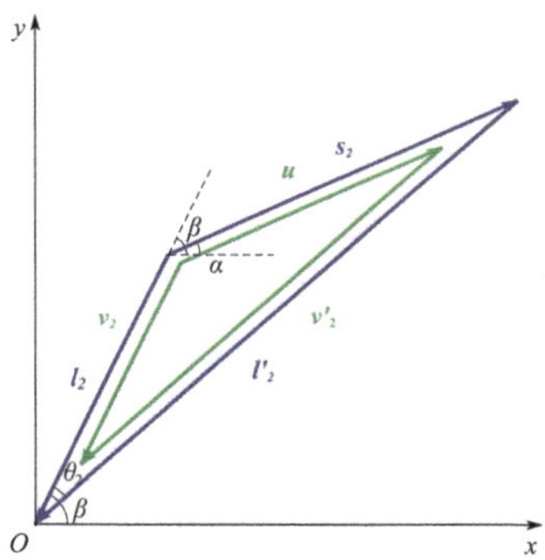

Figure 2

Taking *the aircraft (can be regarded as a particle)* as the object of study. Studying (A) *the departure process.* Considering ***theorem for composition of displacements of a particle and theorem for composition of velocities of a particle*** [taking the ground as the stationary reference frame, and the translational reference frame that is at rest with respect to the wind as the moving reference frame]:

$$\begin{cases} \boldsymbol{l}_1 = \boldsymbol{s}_1 + \boldsymbol{l}_1' & \text{(1a)} \\ \boldsymbol{v}_1 = \boldsymbol{u} + \boldsymbol{v}_1' & \text{(1b)} \end{cases}$$

$\boldsymbol{l}_1$—the displacement of the aircraft relative to the ground in the whole departure process.

$\boldsymbol{s}_1$—the displacement of the wind in the whole departure process.

$\boldsymbol{l'}_1$—the displacement of the aircraft relative to the wind in the whole departure process.

$\boldsymbol{v}_1$—the velocity of the aircraft relative to the ground at arbitrary Time t in the departure process.

$\boldsymbol{u}$—the wind velocity at arbitrary Time t in the departure process.

$\boldsymbol{v'}_1$—the velocity of the aircraft relative to the wind at arbitrary Time t in the departure process.

Studying (B) *the returning process.* Considering ***theorem for composition of displacements of a particle and theorem for composition of velocities of a particle*** [taking the ground as the stationary reference frame, and the translational reference frame that is at rest with respect to the wind as the moving reference frame]:

$$\begin{cases} \boldsymbol{l}_2 = \boldsymbol{s}_2 + \boldsymbol{l}_2' & \text{(2a)} \\ \boldsymbol{v}_2 = \boldsymbol{u} + \boldsymbol{v}_2' & \text{(2b)} \end{cases}$$

$\boldsymbol{l}_2$—the displacement of the aircraft relative to the ground in the whole returning process.

$\boldsymbol{s}_2$—the displacement of the wind in the whole returning process.

$\boldsymbol{l'}_2$—the displacement of the aircraft relative to the wind in the whole returning process.

$\boldsymbol{v}_2$—the velocity of the aircraft relative to the ground at arbitrary Time t in the returning process.

$\boldsymbol{v'}_2$—the velocity of the aircraft relative to the wind at arbitrary Time t in the returning process.

Known:

$$|\boldsymbol{v}_1'| = |\boldsymbol{v}_2'| = v \qquad \text{(3a, b)}$$

The unit vector of the displacement direction of the aircraft in the departure process is denoted by $\boldsymbol{e}$.

Defining l—the flight distance of the aircraft relative to the ground in the whole process. Considering that the aircraft returns back along the original route, we obtain:

$$\boldsymbol{l}_1 = -\boldsymbol{l}_2 = \frac{l}{2}\boldsymbol{e} \tag{4a, b}$$

Because the motion of the wind, the aircraft relative to the wind and the aircraft relative to the ground are all uniform motion; therefore in the same process the vectorgraph of the composition of its displacements is similar to the vectorgraph of the composition of its velocities. According to this, we can draw the vector diagrams corresponding to Eqs. (1) and (2) (see Figs. 1 and 2):

Considering that both two triangles in Fig. 1 satisfy sine theorem:

$$\Rightarrow \begin{cases} \dfrac{v}{\sin(\beta-\alpha)} = \dfrac{u}{\sin\theta_1} & \text{(5a)} \\ \dfrac{l_1'}{\sin(\beta-\alpha)} = \dfrac{l/2}{\sin(\beta-\alpha+\theta_1)} & \text{(5b)} \end{cases}$$

θ_1—the angle between $\boldsymbol{l}_1$ and $\boldsymbol{l}'_1$.

Considering that both two triangles in Fig. 2 satisfy sine theorem:

$$\Rightarrow \begin{cases} \dfrac{v}{\sin(\beta-\alpha)} = \dfrac{u}{\sin\theta_2} & \text{(6a)} \\ \dfrac{l_2'}{\sin(\beta-\alpha)} = \dfrac{l/2}{\sin(\beta-\alpha-\theta_2)} & \text{(6b)} \end{cases}$$

θ_2—the angle between $\boldsymbol{l}_2$ and $\boldsymbol{l}'_2$.

Combining (5) and (6),

$$\Rightarrow \begin{cases} l_1' = \dfrac{v}{\sqrt{v^2-u^2\sin^2(\beta-\alpha)}+u\cos(\beta-\alpha)}\dfrac{l}{2} & \text{(7a)} \\ l_2' = \dfrac{v}{\sqrt{v^2-u^2\sin^2(\beta-\alpha)}-u\cos(\beta-\alpha)}\dfrac{l}{2} & \text{(7b)} \end{cases}$$

Considering that under the same fuel consumption, the longest distance that the aircraft can fly relative to the wind is equal to the longest distance that the aircraft can fly relative to the ground when there is no wind:

$$l_1' + l_2' = R \tag{8}$$

$$\Rightarrow \quad l = \frac{v^2-u^2}{v\sqrt{v^2-u^2\sin^2(\beta-\alpha)}}R \tag{9}$$

$$\Rightarrow \qquad l < R \tag{10}$$

The further study

From (5a) and (6a),

$$\Rightarrow \qquad \theta_1 = \theta_2 = \arcsin\left[\frac{u}{v}\sin(\beta-\alpha)\right] \tag{11}$$

Combining (9) and (11),

$$\Rightarrow \qquad \begin{cases} l_{max}\Big|_{\beta=\alpha+\frac{\pi}{2},\alpha+\frac{3\pi}{2}\left(\theta_1=\theta_2=\arcsin\frac{u}{v}\right)} = \sqrt{1-\frac{u^2}{v^2}}R & \text{(12a)} \\ l_{min}\Big|_{\beta=\alpha,\alpha+\pi(\theta_1=\theta_2=0)} = \left(1-\frac{u^2}{v^2}\right)R & \text{(12b)} \end{cases}$$

1.1.2 The Crawling of a Caterpillar on an Elastic Rope (*a, A*)

As shown in Fig. 1, a linear coordinate system is established on a horizontal ground. There is a light elastic rope of natural length L and uniform elastic property, one end of which is fixed at the origin O, the whole is placed along the positive x-axis. Initially, a little caterpillar of negligible mass statically stands on the elastic rope at the position with the distance of L/α from its fixed end, $\alpha > 1$. Pull the free end of the elastic rope to make it move at the velocity $v_0\boldsymbol{i}$, at the same time, the caterpillar starts to crawl towards the free end of the rope, at arbitrary time, the velocity of the caterpillar relative to the point on the elastic rope in contact with it is $(x/T_0)\,\boldsymbol{i}$, where x is the coordinate of the caterpillar at that time, $T_0 > 0$.

(1) Find the time taken for the caterpillar to crawl from the initial position to the free end of the rope.

(2) Find the coordinate and velocity of the caterpillar at arbitrary time.

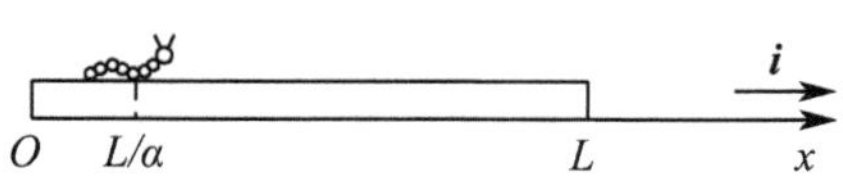

Figure 1

Solution to Problem (1)

The time point that the caterpillar starts to crawl is defined as Time zero.

The free end of the elastic rope is denoted by *A*, the caterpillar is denoted by *P*. At arbitrary Time *t*, the point on the elastic rope in contact with the caterpillar is denoted by *P* (*t*).

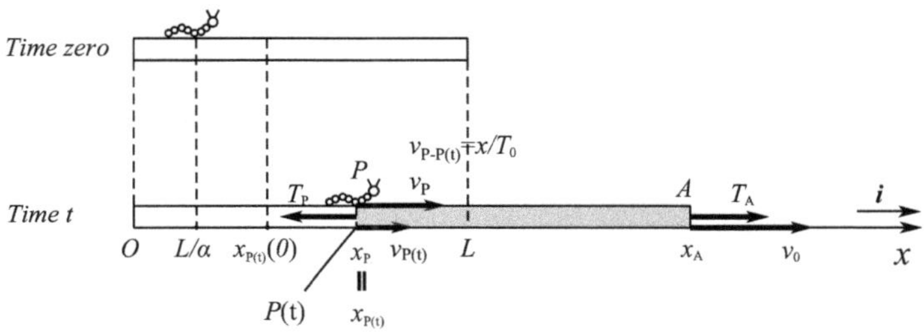

Figure 2

Taking *the whole elastic rope and the segment of the elastic rope between Point A and Point P (t) at arbitrary Time t* as the object of study. Studying *the state at Time t*. Considering ***Hooke's law***:

$$\begin{cases} T_A = SY\dfrac{x_A - L}{L} & \text{(1a)} \\ T_{P(t)} = SY\dfrac{x_{P(t)} - x_{P(t)}(0)}{x_{P(t)}(0)} & \text{(1b)} \end{cases}$$

S—the cross-sectional area of the elastic rope.

Y—the Young's modulus of the elastic rope.

T_A and $T_{P(t)}$—the magnitude of the tension in the elastic rope at Point *A* and Point *P* (*t*) at Time *t*.

$x_{P(t)}(0)$—the coordinate of Point *P* (*t*) at Time zero.

Taking *the segment of the elastic rope between Point A and Point P (t) at arbitrary Time t (can be regarded as a system of particles)* as the object of study. Studying *the state at Time t*. Considering ***Newton's Second Law*** [the mass of the elastic rope is zero; therefore, the resultant external force acting on it is zero all the time] (the ***i***-direction component equation):

$$T_A - T_{P(t)} = 0 \tag{2}$$

(1) → (2), [In this book, the arrow symbols "→" are used to represent the substitution operation between equations, for example: "(1) → (2)" represent "substituting Eq. (1) into Eq. (2)".]

$$\Rightarrow \qquad \frac{x_{P(t)}}{x_A} = \frac{x_{P(t)}(0)}{L} = \text{const}$$

$$\Rightarrow \qquad \frac{x_{P(t)}}{x_A} = \frac{x_{P(t)} + \mathrm{d}x_{P(t)}}{x_A + \mathrm{d}x_A} \tag{3}$$

$$\Rightarrow \qquad \frac{x_{P(t)}}{x_A} = \frac{\mathrm{d}x_{P(t)}}{\mathrm{d}x_A}$$

Considering the definition of ***velocity***:

$$\begin{cases} v_{P(t)} = \dfrac{\mathrm{d}x_{P(t)}}{\mathrm{d}t} & \text{(4a)} \\[2ex] v_0 = \dfrac{\mathrm{d}x_A}{\mathrm{d}t} & \text{(4b)} \end{cases}$$

$v_{P(t)}$—the ***i***-direction component of the velocity of Point $P(t)$ at Time t.

From (4b),

$$\Rightarrow \qquad x_A = L + v_0 t \tag{5}$$

Combining (3), (4) and (5),

$$\Rightarrow \qquad v_{P(t)} = \frac{x_{P(t)}}{L + v_0 t} v_0 \tag{6}$$

Taking *the caterpillar (can be regarded as a particle)* as the object of study. Studying *the state at arbitrary Time t*. Considering ***theorem for composition of velocities of a particle*** [taking the ground as the stationary reference frame, and the translational reference frame that is at rest with respect to Point P (t) as the moving reference frame] (the ***i***-direction component equation):

$$v_P = v_{P(t)} + v_{P-P(t)} \tag{7}$$

v_P—the ***i***-direction component of the velocity of the caterpillar at Time t.

$v_{P\text{-}P(t)}$—the ***i***-direction component of the velocity of the caterpillar relative to Point P (t) at Time t.

Known:

$$v_{P-P(t)} = \frac{x_P}{T_0} \tag{8}$$

$$x_{P(t)} = x_P \tag{9}$$

Combining (6) to (9),

$$\Rightarrow \qquad v_P = \frac{x_P}{L+v_0 t} v_0 + \frac{x_P}{T_0} \tag{10}$$

Considering the definition of ***velocity***:

$$v_P = \frac{\mathrm{d}x_P}{\mathrm{d}t} \tag{11}$$

(11) → (10),

$$\Rightarrow \qquad \frac{\mathrm{d}x_P}{x_P} = \left(\frac{v_0}{L+v_0 t} + \frac{1}{T_0} \right) \mathrm{d}t \tag{12}$$

$$\Rightarrow \qquad \int_{L/\alpha}^{x_A(T)} \frac{\mathrm{d}x_P}{x_P} = \int_0^T \left(\frac{v_0}{L+v_0 t} + \frac{1}{T_0} \right) \mathrm{d}t \tag{13}$$

T—the Time when the caterpillar reaches Point A.

$x_A\,(T)$—the coordinate of Point A at Time T.

(5) → (13),

$$\Rightarrow \qquad \int_{L/\alpha}^{L+v_0 T} \frac{\mathrm{d}x_P}{x_P} = \int_0^T \left(\frac{v_0}{L+v_0 t} + \frac{1}{T_0} \right) \mathrm{d}t \tag{14}$$

$$\Rightarrow \qquad T = T_0 \ln \alpha$$

Solution to Problem (2)

From (12),

$$\Rightarrow \qquad \int_{L/\alpha}^{x_P} \frac{\mathrm{d}x_P}{x_P} = \int_0^t \left(\frac{v_0}{L+v_0 t} + \frac{1}{T_0} \right) \mathrm{d}t$$

$$\Rightarrow \qquad \ln \frac{x_P}{L/\alpha} = \ln \frac{L+v_0 t}{L} + \frac{t}{T_0}$$

$$\Rightarrow \qquad x_P = \frac{L+v_0 t}{\alpha} e^{t/T_0} \tag{15}$$

(15) → (11),

$$\Rightarrow \qquad v_P = \frac{L + v_0(t + T_0)}{\alpha T_0} e^{t/T_0} \tag{16}$$

1.1.3 Intercepting a Bomb (d, *C*)

1. A bomber flies at a constant velocity in the horizontal right direction with the speed of v_1, with the height of h from the ground (see Fig. 1). In order to make the bomb released freely from the bomber can hit a target on the ground in front of it, find the release position of the bomb relative to the target.

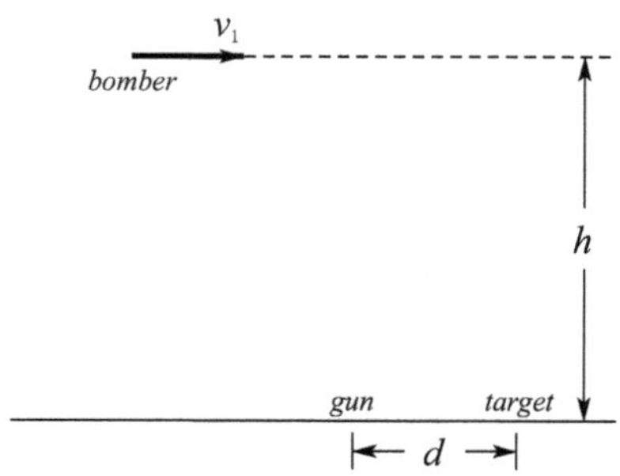

Figure 1

2. On the ground (in the vertical plane where the bomber and the bombing target are located) there is an anti-aircraft gun at the distance d from the bombing target (on its left side), which will fire an intercepting shell once the bomber releases the bomb. The shell is required to hit the bomb in the air, determine the minimum firing speed and corresponding firing direction of the shell.

Solution to Problem 1

The time point when the bomb was released is defines as Time zero.

As shown in Fig. 2, the position where the bomb was released, the position of the bombing target and the position of the anti-aircraft gun are denoted by M, N and P respectively, the positions of the bomb and the shell at arbitrary Time t are denoted by A and B, and the projection of Point M on the ground is denoted by O. Establishing the Cartesian coordinate system with the frame $(O, \boldsymbol{i}, \boldsymbol{j})$, let x, y be the Cartesian coordinates.

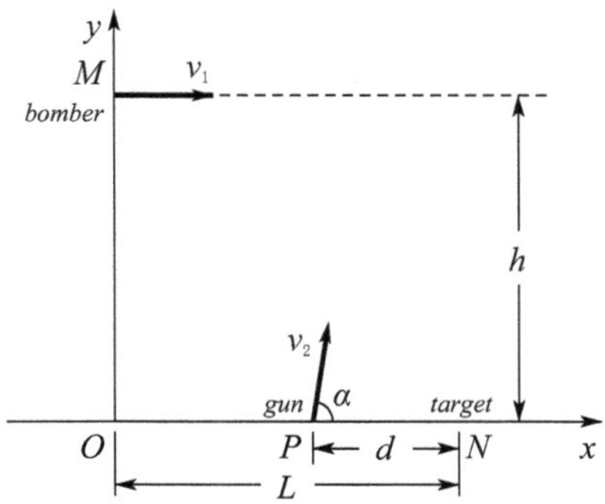

Figure 2

Considering ***the displacement-time relation of uniformly variable motion*** (the component equations in the $(\boldsymbol{i}, \boldsymbol{j})$-directions):

$$\begin{cases} L = v_1 T & \text{(1a)} \\ T = \sqrt{\dfrac{2h}{g}} & \text{(1b)} \end{cases}$$

L—the distance from the target to Point O.

T—the Time when the bomb hit the target.

$\Rightarrow$

$$L = v_1\sqrt{\frac{2h}{g}} \tag{2}$$

Solution [1] to Problem 2

Considering ***the displacement-time relation of uniformly variable motion*** (the component equations in the $(\boldsymbol{i}, \boldsymbol{j})$-directions), for the bomb and the shell respectively:

$$\begin{cases} x_A = v_1 t & \text{(3a)} \\ y_A = h - \dfrac{1}{2}gt^2 & \text{(3b)} \end{cases}$$

$$\begin{cases} x_B = L - d + (v_2 \cos\alpha)t & \text{(4a)} \\ y_B = (v_2 \sin\alpha)t - \dfrac{1}{2}gt^2 & \text{(4b)} \end{cases}$$

v_2—the firing speed of the shell.

α—the angle between the firing velocity of the shell and $\boldsymbol{i}$.

The trajectory of the bomb and the shell are shown in Fig. 3.

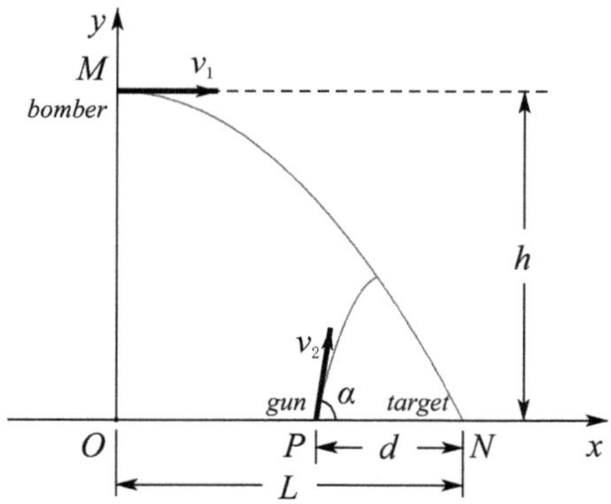

Figure 3

The condition that the shell can hit the bomb is as follows:

$$\begin{cases} x_A = x_B & (5a) \\ y_B = y_B & (5b) \end{cases}$$

$$\Rightarrow \quad \frac{L-d}{h}\sin\alpha = \frac{v_1}{v_2} - \cos\alpha \tag{6}$$

Defining the constants J and K:

$$\begin{cases} J = \dfrac{L-d}{h} & (7a) \\ K = \dfrac{v_1}{v_2} & (7b) \end{cases}$$

$$\Rightarrow \quad J\sin\alpha = K - \cos\alpha$$

$$\Rightarrow \quad (1+J^2)\cos^2\alpha - 2K\cos\alpha + K^2 - J^2 = 0 \tag{8}$$

Equation (8) is a one-variable quadratic equation with $\cos\alpha$ as the variable, the necessary condition for it to have real number solution is that its discriminant is non-negative:

$$\Delta = 4K^2 - 4(1+J^2)(K^2 - J^2) \geq 0 \tag{9}$$

$$\Rightarrow \qquad 1+J^2 \geq K^2 \tag{10}$$

(7) → (10),

$$\Rightarrow \qquad v_2 \geq \frac{h}{\sqrt{h^2+(L-d)^2}} v_1 \tag{11}$$

$$\Rightarrow \qquad v_{2(\min)} = \frac{h}{\sqrt{h^2+(L-d)^2}} v_1 \tag{12}$$

$$\Rightarrow \qquad \cos\alpha\Big|_{v_2=v_{2(\min)}} = \frac{K\big|_{v_2=v_{2(\min)}}}{1+J^2} = \frac{h}{\sqrt{h^2+(L-d)^2}} \tag{13}$$

$$\Rightarrow \qquad \alpha\Big|_{v_2=v_{2(\min)}} = \arccos\frac{h}{\sqrt{h^2+(L-d)^2}} \tag{14}$$

Solution [2] to Problem 2

The translational reference frame with the acceleration of $\boldsymbol{g}$ (the free-falling reference frame) is denoted by Reference frame S'. The following studies this problem in Reference frame S'. In Reference frame S', the bomb and the shell both perform the uniform linear motion (see Fig. 4), the point where they meet is denoted by Q, and the projection of Point N on the line MQ is denoted by N'.

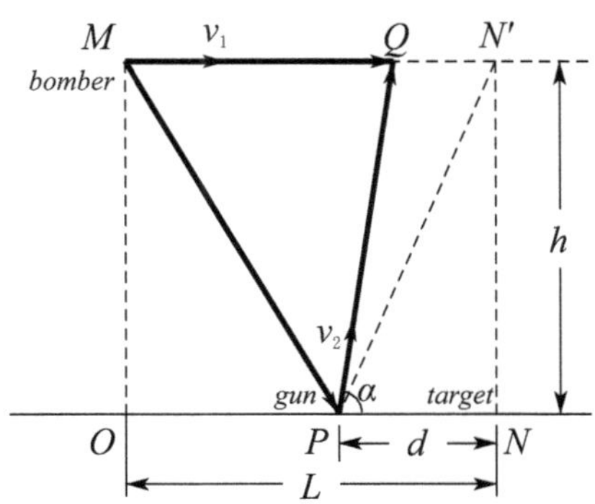

Figure 4

Considering the principle of vector superposition:

$$\overrightarrow{MQ} = \overrightarrow{MP} + \overrightarrow{PQ} \tag{15}$$

Considering that ΔMPQ satisfies sine theorem (see Fig. 4):

$$\frac{\overline{MQ}}{\sin\angle MPQ}=\frac{\overline{PQ}}{\sin\angle QMP} \tag{16}$$

Considering ***the displacement-time relation of uniform linear motion***, for the bomb and the shell respectively:

$$\begin{cases}\overline{MQ}=v_1 t_{meet} & (17a)\\ \overline{PQ}=v_2 t_{meet} & (17b)\end{cases}$$

t_{meet}—the Time when the bomb and the shell meet.

$$\Rightarrow \qquad v_2=\frac{\sin\angle QMP}{\sin\angle MPQ}v_1=\frac{\sin\angle OPM}{\sin\angle MPQ}v_1 \tag{18}$$

Considering the following geometric relations in Fig. 4:

$$\angle OPM=\arcsin\frac{h}{\sqrt{h^2+(L-d)^2}} \tag{19}$$

$$\angle N'PN=\arcsin\frac{h}{\sqrt{h^2+d^2}} \tag{20}$$

$$\alpha=\pi-(\angle OPM+\angle MPQ) \tag{21}$$

The shell should hit the bomb before the bomb hits the target; therefore Point Q should be on the left side of Point N', from the geometric relation, we can know that:

$$\alpha>\angle N'PN \tag{22}$$

$$\Rightarrow \qquad \angle MPQ<\pi-(\angle OPM+\angle N'PN) \tag{23}$$

Situation 1

$$\angle OPM+\angle N'PN<\frac{\pi}{2} \tag{24}$$

From (18) and (19),

$$\Rightarrow v_{2(\min)}=\left.\frac{\sin\angle OPM}{\sin\angle MPQ}v_1\right|_{\angle MPQ=\frac{\pi}{2}}=v_1\sin\angle OPM=\frac{h}{\sqrt{h^2+(L-d)^2}}v_1 \tag{25}$$

From (19) and (21),

$$\Rightarrow \qquad \alpha\big|_{v_2=v_{2(\min)}} = \frac{\pi}{2} - \angle OPM = \arccos\frac{h}{\sqrt{h^2+(L-d)^2}} \tag{26}$$

Situation 2

$$\angle OPM + \angle N'PN \geq \frac{\pi}{2} \tag{27}$$

From (23) and (27),

$$\Rightarrow \qquad (\angle MPQ)_{\max} \rightarrow \pi - (\angle OPM + \angle N'PN) \tag{28}$$

From (18), (19) and (28),

$$\Rightarrow \quad v_{2(\min)} \rightarrow \frac{\sin\angle OPM}{\sin\angle MPQ}v_1\bigg|_{\angle MPQ=\pi-(\angle OPM+\angle N'PN)}$$

$$= \frac{\sin\angle OPM}{\sin(\angle OPM+\angle N'PN)}v_1$$

$$= \frac{\sin\angle OPM}{\sin\angle OPM\cos\angle N'PN+\cos\angle OPM\sin\angle N'PN}v_1$$

$$= \frac{v_1}{\cos\angle N'PN+\cot\angle OPM\sin\angle N'PN}$$

$$= \frac{v_1}{\dfrac{d}{\sqrt{h^2+d^2}}+\dfrac{L-d}{h}\dfrac{h}{\sqrt{h^2+d^2}}}$$

$$= \frac{\sqrt{h^2+d^2}}{L}v_1 \tag{29}$$

(28) $\rightarrow$ (21),

$$\Rightarrow \qquad \alpha\big|_{v_2=v_{2(\min)}} \rightarrow \angle N'PN = \arcsin\frac{h}{\sqrt{h^2+d^2}} \tag{30}$$

Therefore, in this situation, the minimum firing speed and the corresponding firing angle of the shell will be infinitely close to the above limit, that is, the shell will intercept the bomb when it is infinitely close to the target.

1.1.4 Fountain Spraying Model (d, *C*)

There is a small hemispherical fountain nozzle on a horizontal ground, with many spray holes of area S_0 densely distributed on it. Water is sprayed from these holes at the same speed v_0. Suppose that the area of the whole nozzle is much smaller than that of the landing area of the water sprayed from it.

(1) Find the envelope surface of the space area that the water sprayed from the nozzle can reach.
(2) In order to make the water fall on the ground evenly, determine the distribution law of the spray holes.

Solution to Problem (1)

The radius of the nozzle is denoted by r_0, the center of the nozzle is denoted by O (see Fig. 1), taking an arbitrary unit vector in the horizontal direction and denoting it by $\boldsymbol{i}$, and the unit vector in the vertical upward direction is denoted by $\boldsymbol{k}$, $\boldsymbol{k} \times \boldsymbol{i}$ is denoted by $\boldsymbol{j}$, establishing the Cartesian coordinate system with the frame (O, $\boldsymbol{i}$, $\boldsymbol{j}$, $\boldsymbol{k}$), let x, y, z be the Cartesian coordinates. Taking the origin O as the pole, the $\boldsymbol{k}$-direction as the zenith direction, the $\boldsymbol{i}$-direction as the reference direction on the reference plane, establishing the spherical coordinate system, let r, φ, θ be the spherical coordinates.

For the water sprayed from the spray hole at Point (r_0, φ, θ), considering ***the displacement-time relation of uniformly variable motion***:

$$\begin{cases} r = v_0 \cos\theta t & \text{(1a)} \\ z = v_0 \sin\theta t - \dfrac{1}{2} g t^2 & \text{(1b)} \end{cases}$$

$$\Rightarrow \qquad z = -\frac{g}{2v_0^2}(1+\tan^2\theta)r^2 + r\tan\theta \qquad (2)$$

Equation (2) is the trajectory equation of the water sprayed form the spray hole at Point (r_0, φ, θ), which shows that the trajectory of the water is a parabola.

Taking r as the parameter and θ as the variable, the function $z(\theta)$ describes the distribution of the intersection points of all the water's trajectory and the cylinder of sectional radius r. The enveloping surface of the space area that the water sprayed form the nozzle can reach is given by traversing the highest intersection points on all the cylinders, that is, finding the envelope surface is equivalent to finding the maximum value of z as a function of θ. The necessary condition for z to take the extreme value is as follows:

$$\frac{\mathrm{d}z}{\mathrm{d}\theta}=0 \tag{3}$$

$$\Rightarrow \qquad \tan\theta=\frac{v_0^2}{gr} \tag{4}$$

$$\Rightarrow \qquad \left.\frac{\mathrm{d}^2z}{\mathrm{d}\theta^2}\right|_{\tan\theta=v_0^2/gr}=-\frac{gr^2}{v_0^2}\left(\frac{v_0^4}{g^2r^2}+1\right)^2<0 \tag{5}$$

$$\Rightarrow \qquad Z_{r\max}=z\big|_{\tan\theta=v_0^2/gr}=-\frac{g}{2v_0^2}r^2+\frac{v_0^2}{2g} \tag{6}$$

$Z_{r\max}$—when r is regarded as a constant, the maximum value of z as a function of θ.

Taking r as the variable, Eq. (6) is the enveloping surface equation of the space area that the water sprayed form the nozzle can reach. It is a rotating paraboloid with the z-axis as the symmetry axis. The schematic diagram is as follows (see Fig. 1):

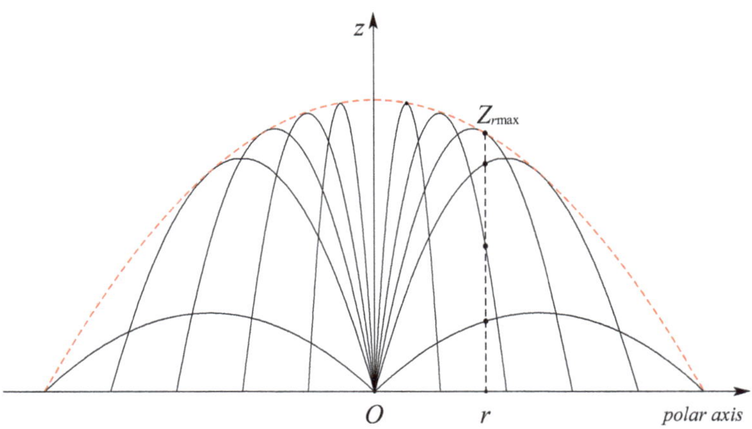

Figure 1

Solution to Problem (2)

Defining R_θ—the polar radius of the landing points of the water sprayed from the spray holes at the polar angle θ.

$$\Rightarrow \qquad R_\theta = r\big|_{z=0} = \frac{2v_0^2}{g}\frac{\tan\theta}{1+\tan^2\theta} = \frac{2v_0^2}{g}\sin\theta\cos\theta = \frac{v_0^2}{g}\sin 2\theta \tag{7}$$

Considering ***flow conservation***:

$$\mathrm{d}Q = \mathrm{d}Q' \tag{8}$$

$\mathrm{d}Q$—the volume rate of flow of the water sprayed from the spray holes on the annulus between θ and $\theta+\mathrm{d}\theta$ on the nozzle.

$\mathrm{d}Q'$—the volume rate of flow of the water falling on the annulus between R_θ and $R_{\theta+\mathrm{d}\theta}$ on the ground.

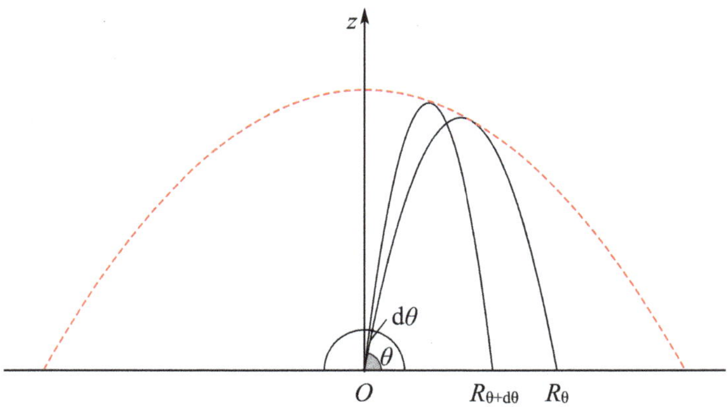

Figure 2

Considering the definition of ***volume rate of flow***:

$$\begin{cases} \mathrm{d}Q = v_0 \cdot S_0 n_\theta \mathrm{d}\Omega & \text{(9a)} \\ \mathrm{d}Q' = k\mathrm{d}S & \text{(9b)} \end{cases}$$

n_θ—the number of the spray holes in the unit solid angle at the polar angle θ on the nozzle.

k—the volume rate of flow of the water falling on the per unit area at the polar radius R_θ on the ground.

$\mathrm{d}\Omega$—the solid angle of the annulus between θ and $\theta+\mathrm{d}\theta$.

dS—the area of the annulus between R_θ and $R_{\theta+d\theta}$ on the ground.

Considering the following geometric relations in Fig. 2, then place (7) in to obtain:

$$\begin{cases} d\Omega = 2\pi \sin\theta d\theta & (10a) \\ dS = 2\pi R_\theta dR_\theta = \dfrac{2\pi v_0^4}{g^2}\sin 4\theta & (10b) \end{cases}$$

Combining (8) to (10),

$$\Rightarrow \qquad n_\theta = \frac{k v_0^3}{S_0 g^2}\frac{\sin 4\theta}{\sin\theta} \tag{11}$$

Another solution of the enveloping surface

Rewriting Eq. (2) as a one-variable quadratic equation with $\tan\theta$ as the variable:

$$\Rightarrow \qquad \frac{gr^2}{2v_0^2}\tan^2\theta - r\tan\theta + \left(z + \frac{gr^2}{2v_0^2}\right) = 0 \tag{12}$$

The necessary condition for $\tan\theta$ to have real number solution is that the discriminant is non-negative:

$$\Delta = r^2 - 4\frac{gr^2}{2v_0^2}\left(z + \frac{gr^2}{2v_0^2}\right) \geq 0 \tag{13}$$

$$\Rightarrow \qquad z \leq -\frac{g}{2v_0^2}r^2 + \frac{v_0^2}{2g} \tag{14}$$

1.1.5 Noncoplanar Maneuver Model (d, *C*)

Consider the following noncoplanar maneuver model: a spacecraft orbiting around the earth transfers from one elliptical orbit to another elliptical orbit by impulse maneuver (jetting gaseous-fuel to change its velocity in an instantaneous process). Given that the speed of the spacecraft at the time point before and after the maneuver is v_1 and v_2 respectively, the flight path angle (the angle between the velocity direction and the angle direction) is γ_1 and γ_2 respectively, and the dihedral angle between the two elliptical orbits is δ.

(1) Find the magnitude of the velocity increment of the spacecraft obtained through the impulse maneuver.

(2) If the two elliptical orbits are coplanar (i.e., δ is zero, it is called coplanar maneuver), find the magnitude of the velocity increment of the spacecraft obtained through the impulse maneuver.

(3) If the impulse maneuver position of the spacecraft is exactly the perigee or apogee of the two elliptical orbits (i.e., γ_1 and γ_2 are both zero), find the magnitude of the velocity increment of it obtained through the impulse maneuver, and compare it with the magnitude of the velocity increment obtained through the following two indirect impulse maneuvers: (i) first change the velocity direction, then change the velocity magnitude, (ii) first change the velocity magnitude, then change the velocity direction.

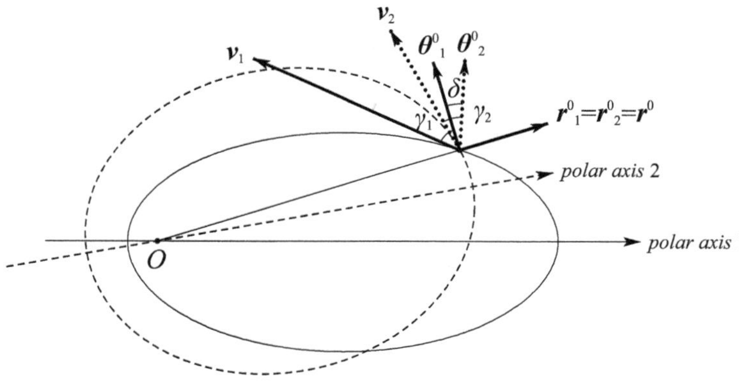

Figure 1

Solution to Problem (1)

The elliptical orbits of the spacecraft before and after the impulse maneuver are called Orbit 1 and Orbit 2 respectively.

Defining $\boldsymbol{v}_1$ and $\boldsymbol{v}_2$—the velocity of the spacecraft at the time point before and after the impulse maneuver.

The orthogonal decomposition of $\boldsymbol{v}_1$ and $\boldsymbol{v}_2$ respectively in the polar coordinate systems 1 and 2 are as follows:

$$\begin{cases} \boldsymbol{v}_1 = v_{1r}\boldsymbol{r}^0 + v_{1\theta_1}\boldsymbol{\theta}_1^0 & \text{(1a)} \\ \boldsymbol{v}_2 = v_{2r}\boldsymbol{r}^0 + v_{2\theta_2}\boldsymbol{\theta}_2^0 & \text{(1b)} \end{cases}$$

$(v_{1r}, v_{1\theta_1})$—the components of $\boldsymbol{v}_1$ in the polar coordinate system 1.

$(v_{2r}, v_{2\theta_2})$—the components of $\boldsymbol{v}_2$ in the polar coordinate system 2.

Defining $\Delta\boldsymbol{v}$—the velocity increment of the spacecraft obtained through the impulse maneuver.

$$\Rightarrow \qquad \Delta\boldsymbol{v} = \boldsymbol{v}_2 - \boldsymbol{v}_1 = (v_{2r} - v_{1r})\boldsymbol{r}^0 + v_{2\theta_2}\boldsymbol{\theta}_2^0 - v_{1\theta_1}\boldsymbol{\theta}_1^0 \tag{2}$$

$$\Rightarrow (\Delta v)^2 = (v_{2r} - v_{1r})^2 + v_{1\theta_1}^2 + v_{2\theta_2}^2 +$$
$$(v_{2r} - v_{1r})v_{2\theta_2}\boldsymbol{r}^0 \cdot \boldsymbol{\theta}_2^0 - (v_{2r} - v_{1r})v_{1\theta_1}\boldsymbol{r}^0 \cdot \boldsymbol{\theta}_1^0 - 2v_{1\theta_1}v_{2\theta_2}\boldsymbol{\theta}_1^0 \cdot \boldsymbol{\theta}_2^0 \tag{3}$$

Considering (a) the orthogonality relation of the base vectors of polar coordinate system, (b) the defining of dihedral angle:

$$\begin{cases} \boldsymbol{r}^0 \cdot \boldsymbol{\theta}_2^0 = \boldsymbol{r}^0 \cdot \boldsymbol{\theta}_1^0 = 0 & \text{(4a)} \\ \boldsymbol{\theta}_1^0 \cdot \boldsymbol{\theta}_2^0 = \cos\delta & \text{(4b)} \end{cases}$$

$$\Rightarrow \qquad \Delta v = \sqrt{(v_{2r} - v_{1r})^2 + v_{1\theta_1}^2 + v_{2\theta_2}^2 - 2v_{1\theta_1}v_{2\theta_2}\cos\delta} \tag{5}$$

Expressing $(v_{1r}, v_{1\theta_1})$ and $(v_{2r}, v_{2\theta_2})$ by the flight path angle γ_1 and γ_2:

$$\begin{cases} v_{1r} = -v_1 \sin\gamma_1 & \text{(6a)} \\ v_{1\theta_1} = v_1 \cos\gamma_1 & \text{(6b)} \end{cases}$$

$$\begin{cases} v_{2r} = -v_2 \sin\gamma_2 & \text{(7a)} \\ v_{2\theta_2} = v_2 \cos\gamma_2 & \text{(7b)} \end{cases}$$

$$\Rightarrow \qquad \Delta v = \sqrt{\begin{aligned}&(v_2\sin\gamma_2 - v_1\sin\gamma_1)^2 + v_1^2\cos^2\gamma_1 + v_2^2\cos^2\gamma_2 - \\ &2v_1v_2\cos\gamma_1\cos\gamma_2\cos\delta\end{aligned}}$$

$$= \sqrt{v_1^2 + v_2^2 - 2v_1v_2\sin\gamma_1\sin\gamma_2 - 2v_1v_2\cos\gamma_1\cos\gamma_2\cos\delta}$$

$$= \sqrt{v_1^2 + v_2^2 - 2v_1v_2\begin{bmatrix}(\cos\gamma_1\cos\gamma_2 + \sin\gamma_1\sin\gamma_2) - \\ \cos\gamma_1\cos\gamma(1-\cos\delta)\end{bmatrix}}$$

$$= \sqrt{v_1^2 + v_2^2 - 2v_1v_2\left[\mathrm{co}(\gamma_2 - \gamma_1) - \cos\gamma_1\cos\gamma_2(1-\cos\delta)\right]}$$

$$= \sqrt{v_1^2 + v_2^2 - 2v_1v_2\left[\cos(\gamma_2 - \gamma_1) - 2\cos\gamma_1\cos\gamma_2\sin^2\frac{\delta}{2}\right]} \tag{8}$$

Solution to Problem (2)

From (8), we obtain the magnitude of the velocity increment of the spacecraft obtained through the coplanar maneuver:

$$\Rightarrow \qquad \Delta v\big|_{\delta=0} = \sqrt{v_1^2 + v_2^2 - 2v_1v_2\cos(\gamma_2 - \gamma_1)} \tag{9}$$

Solution to Problem (3)

Known:

$$v_{1r} = v_{2r} = 0 \tag{10}$$

From (6) and (7),

$$\Rightarrow \qquad \begin{cases} v_1 = v_{1\theta_1} \\ v_2 = v_{1\theta_2} \end{cases} \tag{11}$$

Defining Δv_a—the magnitude of the velocity increment of the spacecraft obtained through the impulse maneuver when the impulse maneuver position is exactly the perigee or apogee of the two elliptical orbits (i.e., γ_1 and γ_2 are both zero).

(10) and (11) $\rightarrow$ (5),

$$\begin{aligned} \Rightarrow \qquad \Delta v_a &= \Delta v\big|_{\substack{v_{1r}=v_{2r}=0 \\ v_1=v_{1\theta_1}, v_2=v_{1\theta_2}}} \\ &= \sqrt{v_1^2 + v_2^2 - 2v_1v_2\cos\delta} \\ &= \sqrt{v_1^2 + v_2^2 - 2v_1v_2\left(1 - 2\sin^2\frac{\delta}{2}\right)} \\ &= \sqrt{(v_2 - v_1)^2 + 4v_1v_2\sin^2\frac{\delta}{2}} \end{aligned} \tag{12}$$

The vector diagram corresponding to the equation above is shown in Fig. 2a.

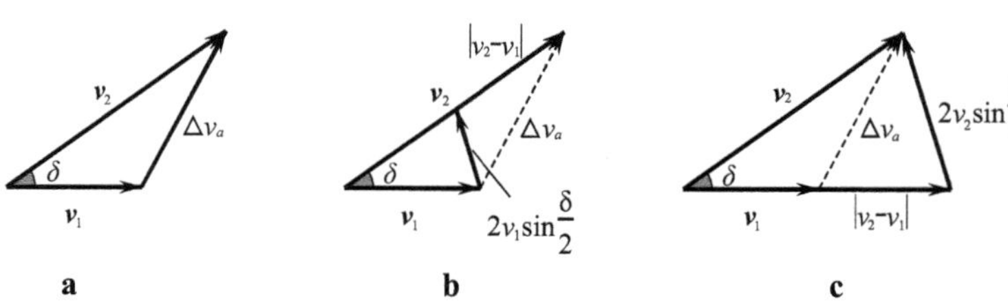

Figure 2a–c

Defining Δv_{I}—the magnitude of the velocity increment obtained through the indirect impulse maneuvers I (first changes the velocity direction, then changes the velocity magnitude).

Considering the following geometric relation in Fig. 2b:

$$\Delta v_{\text{I}} = 2v_1 \sin\frac{\delta}{2} + |v_2 - v_1| \tag{13}$$

Comparing Δv_{I} and Δv_a:

$$\Rightarrow \quad (\Delta v_{\text{I}})^2 - (\Delta v_a)^2 = 4v_1 \sin\frac{\delta}{2}\left[(v_1 - v_2)\sin\frac{\delta}{2} + |v_2 - v_1|\right] > 0 \tag{14}$$

Defining Δv_{II}—the magnitude of the velocity increment obtained through the indirect impulse maneuvers II (first changes the velocity magnitude, then changes the velocity direction).

Considering the following geometric relation in Fig. 2c:

$$\Delta v_{\text{II}} = |v_2 - v_1| + 2v_2 \sin\frac{\delta}{2} \tag{15}$$

Comparing Δv_{II} and Δv_a:

$$\Rightarrow \quad (\Delta v_{\text{II}})^2 - (\Delta v_a)^2 = 4v_2 \sin\frac{\delta}{2}\left[(v_2 - v_1)\sin\frac{\delta}{2} + |v_2 - v_1|\right] > 0 \tag{16}$$

That is, the magnitude of the velocity increment obtained by these two indirect impulse maneuvers is greater than that obtained by the direct impulse maneuver. Of course, the magnitude relations of these three quantities can be more easily seen directly from the vector diagrams.

1.1.6 Determining the Speed of Light by Observing the Motion of Jupiter's Moon (*b*, *A*)[1]

The speed of light can be determined by observing the orbital period of Jupiter's moon around Jupiter (measuring on the earth the time interval between the twice successive appearances of Jupiter's moon from the Jupiter's sun shadow). One observation is as follows: once the sun, the earth and Jupiter are collinear is defined as Time zero (the earth lies between the sun and Jupiter), the moon appears from the Jupiter's sun shadow at a certain Time *t*, from Time *t*, two successive observed values of the orbital period of Jupiter's moon

[1]This problem is adapted from a test question of the 1th APhO (Asian Physics Olympiads) in 2000.

are T and $T + \Delta T$ respectively. Determine the speed of light according to the above observation results.

The orbital period of the earth and Jupiter around the sun are respectively denoted by T_E and T_J, T_E, $T_J >> T$. The distances from the earth and Jupiter to the sun are respectively denoted by R_E and R_J, $R_E << R_J$.

Solution [1]

The line segment between the sun and the earth is denoted by SE, and the line segment between the sun and Jupiter is denoted by SJ.

The translational reference frame that is at rest with respect to the sun is denoted by Reference frame S, and the rotating reference frame with the sun as the center and rotating with SJ is denoted by Reference frame SJ.

As shown in Fig. 1, the position of the sun is denoted by O, the unit vector perpendicular to the plane of the figure outward is denoted by $\boldsymbol{k}$. Taking the origin O as the pole, the direction of the sun pointing to Jupiter as the positive direction of the polar axis, the right-hand helix direction with regard to the $\boldsymbol{k}$-direction as the positive direction of polar angle, establishing the polar coordinate system fixed to Reference frame SJ, let r, θ be the polar coordinates.

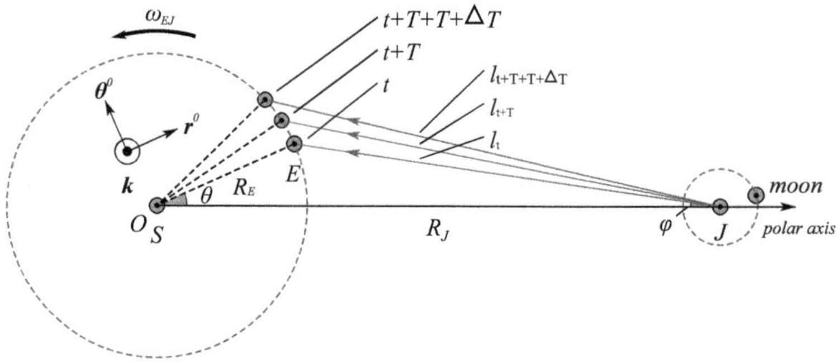

Figure 1

Considering the definition of ***angular velocity***:

$$\begin{cases} \omega_E = \dfrac{2\pi}{T_E} & \text{(1a)} \\ \omega_J = \dfrac{2\pi}{T_J} & \text{(1b)} \end{cases}$$

ω_E and ω_J—in Reference frame S, the $\boldsymbol{k}$-direction component of the angular velocity of revolution of the earth and Jupiter.

Taking SE as the object of study. Studying *the state at arbitrary Time t'*. Considering ***theorem for composition of angular velocities*** [taking Reference frame S as the stationary reference frame and Reference frame SJ as the moving reference frame]:

$$\omega_E = \omega_J + \omega_{EJ} \tag{2}$$

ω_{EJ}—the $\boldsymbol{k}$-direction component of the angular velocity of SE in Reference frame SJ.

$$\Rightarrow \qquad \omega_{EJ} = 2\pi\left(\frac{1}{T_E} - \frac{1}{T_J}\right) \tag{3}$$

Considering ***the angular displacement-time relation of uniform rotation***:

$$\theta = \omega_{EJ} t \tag{4}$$

θ—the polar angle of the earth at Time t.

Defining l_t—the distance between the earth and Jupiter at Time t. Considering the following geometric relation in Fig. 1, then place (4) in to obtain:

$$\begin{aligned} l_t &= \left(R_J^2 + R_E^2 - 2R_J R_E \cos\theta\right)^{1/2} \\ &= \left(R_J^2 + R_E^2 - 2R_J R_E \cos\omega_{EJ} t\right)^{1/2} \\ &= R_J\left[1 + \left(\frac{R_E}{R_J}\right)^2 - 2\frac{R_E}{R_J}\cos\omega_{EJ} t\right]^{1/2} \\ &\approx R_J\left(1 - 2\frac{R_E}{R_J}\cos\omega_{EJ} t\right)^{1/2} \\ &\approx R_J\left(1 - \frac{R_E}{R_J}\cos\omega_{EJ} t\right) \\ &= R_J - R_E \cos\omega_{EJ} t \end{aligned} \tag{5}$$

This operation has considered R_E/R_J as a small quantity and taken a low order approximation.

Considering ***the displacement-time relation of uniform linear motion***:

$$\begin{cases} \tau + \dfrac{l_t}{c} = t & \text{(6a)} \\ (\tau + T_0) + \dfrac{l_{t+T}}{c} = t + T & \text{(6b)} \end{cases}$$

τ—the emitting Time of the light of the Jupiter's moon whose receiving Time is t on the earth.

T_0—the orbital period of Jupiter's moon.

l_{t+T}—the distance between the earth and Jupiter at Time $t+T$.

$$\Rightarrow \qquad T = T_0 + \frac{l_{t+T} - l_t}{c} \tag{7}$$

Considering that the orbital period of Jupiter's moon is a small quantity, the receiving Time of the light emitted by the Jupiter's moon can be regarded as a continuous variable, introducing Δl—the differential of l_t as the continuous function of t, then place (5) in to obtain:

$$\Rightarrow \qquad \Delta l = l_{t+T} - l_t \approx \frac{\mathrm{d}l}{\mathrm{d}t} T = (\omega_{EJ} R_E \sin \omega_{EJ} t) T \tag{8}$$

$$\Rightarrow \qquad T_0 = \left(1 - \frac{\omega_{EJ} R_E}{c} \sin \omega_{EJ} t\right) T \tag{9}$$

Replacing t and T in Eq. (9) with $t + T$ and $T + \Delta T$, respectively:

$$\Rightarrow \qquad T_0 = \left[1 - \frac{\omega_{EJ} R_E}{c} \sin \omega_{EJ} (t+T)\right](T + \Delta T) \tag{10}$$

Combining (9) and (10):

$$\Rightarrow \qquad c = \omega_{EJ} R_E \frac{(T + \Delta T) \sin \omega_{EJ} (t+T) - T \sin \omega_{EJ} t}{\Delta T} \tag{11}$$

Solution [2]

Considering that T_0 is a small quantity, Jupiter's moon can be regarded as a wave source with $1/T_0$ as the pulse frequency, the array formed by the light emitted by the Jupiter's moon twice successively appearing from the Jupiter's sun shadow can be regarded as a pulse of Jupiter's moon.

From Time zero, the *i*-th pulse of Jupiter's moon received by the earth is denoted by Pulse *i*.

Taking *Pulse i* as the object of study. Studying *the process of its moving from Jupiter's moon to the earth*. Considering ***Doppler effect***:

$$\frac{1}{T(t_i)}=\left[1+\frac{v_E'}{c}\cos\alpha(t_i)\right]\frac{1}{T_0} \tag{12}$$

t_i—the receiving Time of Pulse *i* on the earth.

$T(t_i)$—the period of Pulse *i* measured on the earth.

v'_E—the speed of the earth in Reference frame *SJ*.

$\alpha(t_i)$—the supplementary angle of the angle between the propagation direction of Pulse *i* and the velocity direction of the earth.

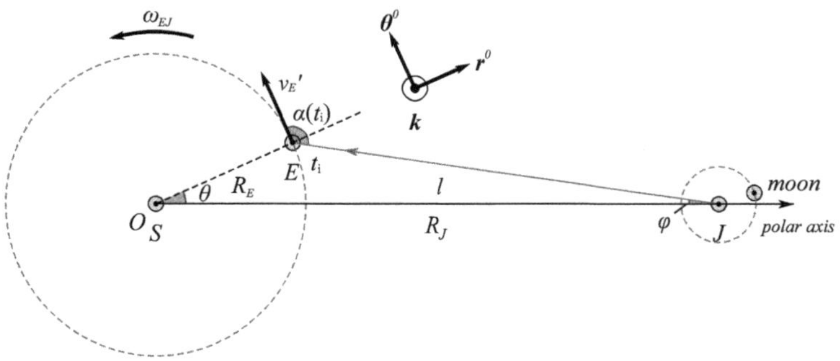

Figure 2

Considering ***the transverse component form of velocity***:

$$v_E' = \omega_{EJ} R_E \tag{13}$$

Considering the following geometric relation in Fig. 2, then place (4) in to obtain:

$$\alpha(t_i)=(\theta+\varphi)+\frac{\pi}{2}=(\omega_{EJ}t_i+\varphi)+\frac{\pi}{2} \tag{14}$$

φ—the supplementary angle of the angle between the propagation direction of Pulse i and the positive direction of the polar axis.
Considering that φ is a small angle:

$$\Rightarrow \qquad \alpha(t_i) \approx \omega_{EJ} t_i + \frac{\pi}{2} \tag{15}$$

(13) and (15) $\rightarrow$ (12):

$$\Rightarrow \qquad T(t_i) = \frac{T_0}{1 - \dfrac{\omega_E R_E}{c} \sin \omega_{EJ} t_i} \tag{16}$$

$$\Rightarrow \qquad \begin{cases} T = T(t_i)|_{t_i = t} = \dfrac{T_0}{1 - \dfrac{\omega_E R_E}{c} \sin \omega_{EJ} t} & \text{(17a)} \\ T + \Delta T = T(t_i)|_{t_i = t+T} = \dfrac{T_0}{1 - \dfrac{\omega_E R_E}{c} \sin \omega_{EJ} (t+T)} & \text{(17b)} \end{cases}$$

Combining (17a) and (17b) (eliminating T_0):

$$\Rightarrow \qquad c = \omega_E R_E \frac{(T + \Delta T) \sin \omega_{EJ} (t+T) - T \sin \omega_{EJ} t}{\Delta T} \tag{18}$$

1.1.7 The Landing of a Rotating Homogeneous Rod (*a*, *A*)

As shown in Fig. 1, consider a homogeneous rod of length $2l$, initially, it is in the vertical direction, with its center of mass C static at h above a horizontal ground, and with the rotating angular velocity of the magnitude ω_0 and the direction anticlockwise. Determine the condition for the rod to be just in the horizontal direction when it lands on the ground.

Solution

The initial time is defined as Time zero.

As shown in Fig. 2, the position of the center of mass C at Time zero is denoted by O, establishing the Cartesian coordinate system with the frame $(O, \mathbf{i}, \mathbf{j}, \mathbf{k})$, let x, y, z be the Cartesian coordinates.

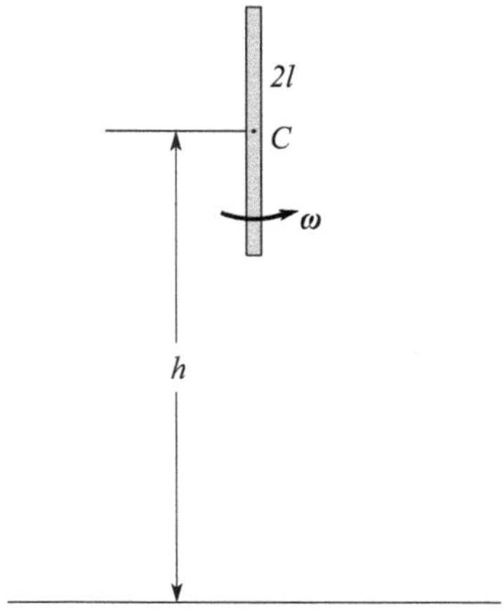

Figure 1

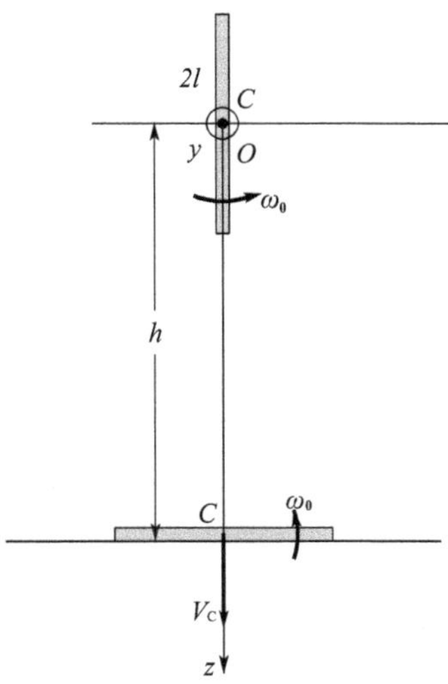

Figure 2

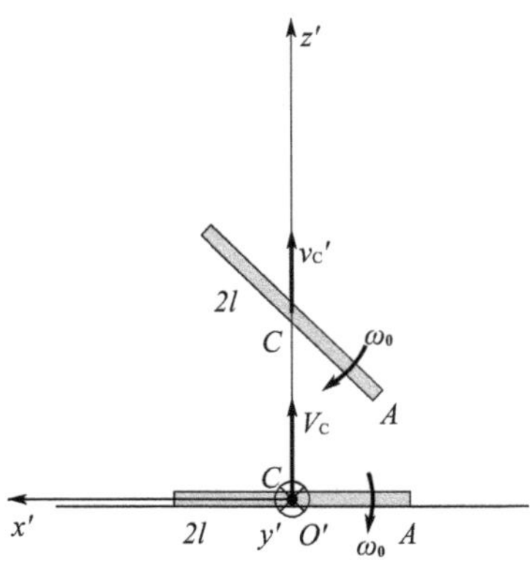

Figure 3

Taking *the rod (can be regarded as a rigid body)* as the object of study. Studying *the state at arbitrary Time t in its falling process.* Considering (a) ***theorem of motion of center of mass*** (the $\boldsymbol{k}$-direction component equation), (b) ***law of rotation*** [in its center-of-mass reference frame, taking its central axis parallel to the y-axis as the rotation axis] (the ***j***-direction component equation), (c) ***in the center-of-mass reference frame, the resultant torque of the inertia force acting on system of particles is zero***:

$$\begin{cases} mg = ma & (1a) \\ 0 = I\beta & (1b) \end{cases}$$

m—the mass of the rod.

I—the moment of inertia of the rod about its central axis parallel to the y-axis.

a—the ***k***-direction component of the center-of-mass acceleration of the rod at Time t in the falling process.

β—the ***j***-direction component of the rotating angular acceleration of the rod at Time t in the falling process.

$$\begin{cases} a = g & (2a) \\ \beta = 0 & (2b) \end{cases}$$

That is, the center of mass of the rod moves at a uniformly variable velocity and the rod rotates at a uniform angular velocity in the falling process.

Considering ***the angular displacement-time relation of uniform rotation***:

$$\Rightarrow \quad \theta = \omega_0 t \quad (3)$$

θ—the angle of rotation of the rod in the anticlockwise direction [the finite angular displacement is not a vector] at arbitrary Time t in the falling process.

One condition for the rod to be just in the horizontal direction when it lands on the ground is as follows (hereinafter referred to as Condition 1):

$$\theta\Big|_{t=t_C} = \left(k + \frac{1}{2}\right)\pi \qquad k = 0,1,2\cdots\cdots \quad (4)$$

t_C—the Time when the rod lands on the ground.

$$\Rightarrow \quad t_C = \left(k + \frac{1}{2}\right)\frac{\pi}{\omega_0} \quad (5)$$

Considering ***the displacement-time relation of uniformly variable motion***:

$$\Rightarrow \qquad h=\frac{1}{2}gt_C^2=\frac{1}{2}\left(k+\frac{1}{2}\right)^2\frac{\pi^2 g}{\omega_0^2} \qquad (6)$$

The right endpoint of the rod when it lands on the ground is denoted by A. Another condition for the rod to be just in the horizontal direction when it lands on the ground is that Point A does not contact the ground during the quarter rotation period before is lands on the ground (hereinafter referred to as Condition 2).

Defining T—the rotation period of the rod:

$$\Rightarrow \qquad T=\frac{2\pi}{\omega_0} \qquad (7)$$

Considering ***the velocity-time relation of uniformly variable motion***:

$$\Rightarrow \qquad V_C=gt_C=\left(k+\frac{1}{2}\right)\frac{\pi g}{\omega_0} \qquad (8)$$

V_C—the $\boldsymbol{k}$-direction component of the center-of-mass velocity of the rod at the time point when it lands on the ground.

In order to simplify the study of Condition 2, we can study the inverse motion process of the rod—the motion of the rod from the ground at the initial center-of-mass velocity $-V_C\boldsymbol{k}$ and the initial rotating angular velocity $-\omega_0\boldsymbol{j}$. By analogy with the study of the real motion of the rod, it can be seen that the inverse motion process of the rod is also the superposition of a uniformly variable motion of its center of mass and a uniform rotation.

The projection of Point O on the ground is denoted by O', establishing the Cartesian coordinate system with the frame (O', $\boldsymbol{i}'$, $\boldsymbol{j}'$, $\boldsymbol{k}'$), where $\boldsymbol{i}' = -\boldsymbol{i}$, $\boldsymbol{j}' = -\boldsymbol{j}$, $\boldsymbol{k}' = -\boldsymbol{k}$, let x', y', z' be the Cartesian coordinates.

For the inverse motion process of the rod, the time point when it is on the ground is defined as Time (inverse) zero.

Considering ***the displacement-time relation of uniformly variable motion***:

$$\Rightarrow \qquad z_C'=V_C t'-\frac{1}{2}gt'^2=\left(k+\frac{1}{2}\right)\frac{g\pi}{\omega_0}t'-\frac{1}{2}gt'^2 \qquad (9)$$

z'_C—the z'-coordinate of the center of mass of the rod at arbitrary Time t' in the inverse motion process.

Taking *Point A* as the object of study. Studying *the state at arbitrary Time t' in the inverse motion process*. Considering ***theorem for composition of displacements of a particle*** [taking the ground as the stationary reference frame and the center-of-mass reference frame as the moving reference frame] (the ***k'***-direction component equation), then place (9) in to obtain:

$$\Rightarrow \quad z'_A = z'_C - l\sin(\omega_0 t') = \left(k+\frac{1}{2}\right)\frac{g\pi}{\omega_0}t' - \frac{1}{2}gt'^2 - l\sin(\omega_0 t') \qquad (10)$$

z'_A—the z'-coordinate of Point A at arbitrary Time t' in the inverse motion process.

Condition 2 can be expressed as follows:

$$z'_A\Big|_{t'\le\frac{T}{4}} > 0 \qquad (11)$$

$$\Rightarrow \quad \left(k+\frac{1}{2}\right)\frac{\pi g}{\omega_0}t' - \frac{1}{2}gt'^2 - l\sin(\omega_0 t') > 0 \quad \text{when } t' \le \frac{T}{4}$$

$$\Rightarrow \quad k > \frac{\omega_0 t'}{2\pi} + \frac{\omega_0^2 l}{\pi g}\frac{\sin(\omega_0 t')}{\omega_0 t'} - \frac{1}{2} \quad \text{when } t' \le \frac{T}{4} \qquad (12)$$

There is no analytic solution to this inequality.

The following let's discuss an unnecessary but sufficient condition of Condition 2 (a stronger condition than Condition 2, hereinafter referred to as Condition 3): at arbitrary Time t' in the first quarter rotation period in the inverse motion process, the ***k'***-direction component of the velocity of Point A is greater than zero.

Considering ***the velocity-time relation of uniformly variable motion***:

$$\Rightarrow \quad v'_C = V_C - gt' = \left(k+\frac{1}{2}\right)\frac{\pi g}{\omega_0} - gt' \qquad (13)$$

v'_C—the ***k'***-direction component of the center-of-mass velocity of the rod at arbitrary Time t' in the inverse motion process.

Taking *Point A* as the object of study. Studying *the state at arbitrary Time t' in its inverse motion process*. Considering ***theorem***

for composition of velocities of a particle [taking the ground as the stationary reference frame, and the center-of-mass reference frame of the rod as the moving reference frame] (the ***k′***-direction component equation):

$$\Rightarrow \quad v'_A = v'_C - \omega_0 l\cos(\omega_0 t') = \left(k+\frac{1}{2}\right)\frac{\pi g}{\omega_0} - gt' - \omega_0 l\cos(\omega_0 t') \quad (14)$$

v'_A—the ***k′***-direction component of the velocity of Point A at arbitrary Time t' in the inverse motion process.

Condition 3 can be expressed as follows:

$$v'_A\big|_{t'\le\frac{T}{4}} > 0 \quad (15)$$

$$\Rightarrow \quad k > \frac{\omega_0}{\pi g}\left[gt' + \omega_0 l\cos(\omega_0 t')\right] - \frac{1}{2} \quad \text{when } t' \le \frac{T}{4} \quad (16)$$

Defining the function $f(t')$:

$$f(t') = \frac{\omega_0}{\pi g}\left[gt' + \omega_0 l\cos(\omega_0 t')\right] - \frac{1}{2} \quad (17)$$

$$\Rightarrow \quad k > f(t')_{\max}\big|_{t'\le\frac{T}{4}} \quad (18)$$

From (17),

$$\Rightarrow \quad \frac{df(t')}{dt'} = \frac{\omega_0}{\pi g}\left[g - \omega_0^2 l\sin(\omega_0 t')\right] \quad (19)$$

Situation 1, under the following condition:

$$\omega_0^2 l < g \quad (20)$$

$$\Rightarrow \quad \frac{df(t')}{dt'} > 0 \quad (21)$$

Therefore $f(t')$ is a monotonically increasing function.

$$\Rightarrow \quad f(t')_{\max}\big|_{t'\le\frac{T}{4}} = f\left(\frac{T}{4}\right) = 0 \quad (22)$$

(22) → (18),

$$\Rightarrow \quad k > 0 \quad (23)$$

$$\Rightarrow \quad k = 1, 2, 3 \ldots\ldots \quad (24)$$

That is, k can take all positive integers in this situation.

Equations (6) and (24) are the sufficient conditions for the rod to be just in the horizontal direction when it lands on the ground in Situation 1.

Situation 2, under the following condition:

$$\omega_0^2 l \geq g \tag{25}$$

The necessary condition for $f(t')$ to take the maximum value is as follows:

$$\left.\frac{\mathrm{d}f(t')}{\mathrm{d}t'}\right|_{t'=\tau} = 0 \tag{26}$$

τ—the value of t' when $f(t')$ takes the extremum.

(19) $\rightarrow$ (26):

$$\Rightarrow \quad g - \omega_0^2 l \sin(\omega_0 \tau) = 0 \tag{27}$$

$$\Rightarrow \quad \tau = \frac{1}{\omega_0} \arcsin \frac{g}{\omega_0^2 l} \tag{28}$$

$$\Rightarrow \quad \left.\frac{\mathrm{d}^2 f(t')}{\mathrm{d}t'^2}\right|_{t'=\tau} = -\frac{\omega_0^4 l}{\pi g}\sqrt{1-\left(\frac{g}{\omega_0^2 l}\right)^2} < 0 \tag{29}$$

Therefore $f(t')$ takes the maximum value when $t' = \tau$.

(28) $\rightarrow$ (17),

$$\Rightarrow \quad f(t')_{\max}\Big|_{t' \leq \frac{T}{4}} = f(\tau) = \frac{1}{\pi}\left[\arcsin\frac{g}{\omega_0^2 l} + \sqrt{\left(\frac{\omega_0^2 l}{g}\right)^2 - 1}\right] - \frac{1}{2} \tag{30}$$

(30) $\rightarrow$ (18),

$$\Rightarrow \quad k > \frac{1}{\pi}\left[\arcsin\frac{g}{\omega_0^2 l} + \sqrt{\left(\frac{\omega_0^2 l}{g}\right)^2 - 1}\right] - \frac{1}{2} \tag{31}$$

$$\Rightarrow \quad k_{\min} = \text{Ceiling}\left\{\frac{1}{\pi}\left[\arcsin\frac{g}{\omega_0^2 l} + \sqrt{\left(\frac{\omega_0^2 l}{g}\right)^2 - 1}\right] - \frac{1}{2}\right\} \tag{32}$$

where function Ceiling (x) represents the minimum integer not smaller than the variable x.

$$\Rightarrow \qquad k = k_{\min}, k_{\min}+1, k_{\min}+2\cdots\cdots \qquad (33)$$

Equations (6) and (33) are the sufficient conditions for the rod to be just in the horizontal direction when it lands on the ground in Situation 2.

1.1.8 The Reflection of a Beam of Light by a Variable Angular Velocity Rotating Plane Mirror (*a*, *A*)

As shown in Fig. 1, consider a horizontal ground intersects a vertical wall, taking an arbitrary point on the intersecting line of the two and denoting it by O, establishing the Cartesian coordinate system with the frame $(O, \boldsymbol{i}, \boldsymbol{j}, \boldsymbol{k})$, let x, y, z be the Cartesian coordinates. A plane mirror rotates around the z-axis, a point light source (denoted by A) moves along the x-axis and emits light in the $\boldsymbol{j}$-direction continually. The light reflected by the plane mirror (the dynamic reflection point is denoted by B) propagates toward the wall (the dynamic intersection point is denoted by C). Time zero is arbitrarily set, at Time t, the length of OA is denoted by l, the $\boldsymbol{i}$-direction component of the velocity and acceleration of Point A are denoted by v_A and a_A, the angle between the plane mirror and the wall is denoted by θ, the $\boldsymbol{k}$-direction component of the angular velocity and angular acceleration of the plane mirror are denoted by ω and β. Find the velocity and acceleration of Point C at Time t.

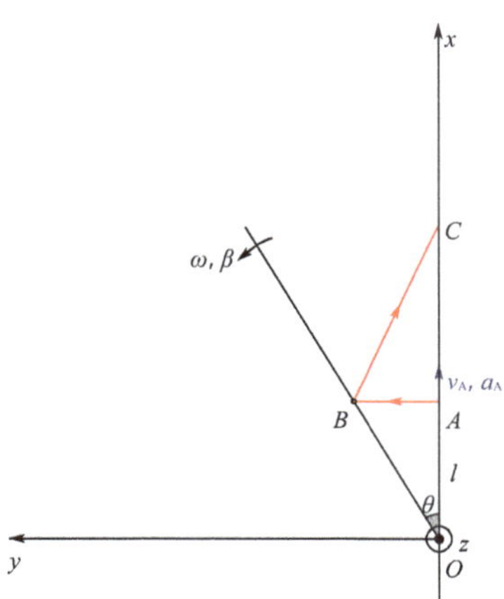

Figure 1

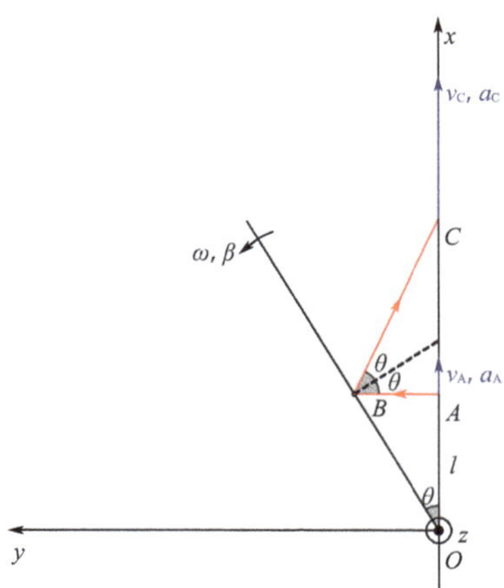

Figure 2

Solution [1]

As shown in Fig. 2, considering the definition of ***velocity***:

$$v_C = \frac{\mathrm{d}\overline{OC}}{\mathrm{d}t} \tag{1}$$

v_C—the ***i***-direction component of the velocity of Point *C* at Time *t*.

Considering the following geometric relation in Fig. 2:

$$\overline{OC} = \overline{OA} + \overline{AC} = \overline{OA} + \overline{AB}\tan 2\theta = \overline{OA} + \overline{OA}\tan\theta\tan 2\theta = \frac{l}{\cos 2\theta} \tag{2}$$

$$\Rightarrow \qquad v_C = \frac{1}{\cos 2\theta}\cdot\frac{\mathrm{d}l}{\mathrm{d}t} + l\cdot\frac{\tan 2\theta}{\cos 2\theta}\cdot 2\frac{\mathrm{d}\theta}{\mathrm{d}t} \tag{3}$$

Considering the definition of ***velocity*** and ***angular velocity***:

$$v_A = \frac{\mathrm{d}l}{\mathrm{d}t} \tag{4}$$

$$\omega = \frac{\mathrm{d}\theta}{\mathrm{d}t} \tag{5}$$

(4) and (5) $\rightarrow$ (3):

$$\Rightarrow \qquad v_C = v_A\frac{1}{\cos 2\theta} + 2\omega l\frac{\tan 2\theta}{\cos 2\theta} \tag{6}$$

Considering the definition of ***acceleration***:

$$a_C = \frac{\mathrm{d}v_C}{\mathrm{d}t} \tag{7}$$

a_C—the ***i***-direction component of the acceleration of Point *C* at Time *t*.

$$\Rightarrow \qquad a_C = \frac{1}{\cos 2\theta}\frac{\mathrm{d}v_A}{\mathrm{d}t} + v_A\frac{\sin 2\theta}{(\cos 2\theta)^2}\cdot 2\frac{\mathrm{d}\theta}{\mathrm{d}t} + 2l\frac{\tan 2\theta}{\cos 2\theta}\frac{\mathrm{d}\omega}{\mathrm{d}t} + \tag{8}$$

$$2\omega l\frac{\sec^2 2\theta + \tan 2\theta\tan 2\theta}{\cos 2\theta}\cdot 2\frac{\mathrm{d}\theta}{\mathrm{d}t} + 2\omega\frac{\tan 2\theta}{\cos 2\theta}\frac{\mathrm{d}l}{\mathrm{d}t}$$

Considering the definition of ***acceleration*** and ***angular acceleration***:

$$a_A = \frac{\mathrm{d}v_A}{\mathrm{d}t} \tag{9}$$

$$\beta = \frac{\mathrm{d}\omega}{\mathrm{d}t} \tag{10}$$

(9) and (10) → (8),

$$\Rightarrow \quad a_C = a_A\frac{1}{\cos 2\theta} + v_A\frac{\sin 2\theta}{(\cos 2\theta)^2}2\omega + 2\beta l\frac{\tan 2\theta}{\cos 2\theta} +$$

$$4\frac{\sec^2 2\theta + \tan 2\theta \tan 2\theta}{\cos 2\theta}\omega^2 l + 2\omega\frac{\tan 2\theta}{\cos 2\theta}v_A$$

$$= \frac{1}{\cos 2\theta}\left[a_A + 2\beta l \tan 2\theta + 4(1 + 2\tan^2 2\theta)\omega^2 l + 4\omega v_A \tan 2\theta\right] \tag{11}$$

Solution [2]

Taking the origin O as the pole, the positive x-axis as the polar axis, the right-hand helix direction with regard to the $\boldsymbol{k}$-direction as the positive direction of polar angle, establishing the polar coordinate system, let r, θ be the polar coordinates (see Fig. 3). Introducing the mirror image of the wall relative to the mirror, the mirror image of Point A is denoted by A'.

Considering the following geometric relations in Fig. 3:

$$\overline{OA} = \overline{OA'} = l \tag{12}$$

$$\overline{CA'} = l \tan 2\theta \tag{13}$$

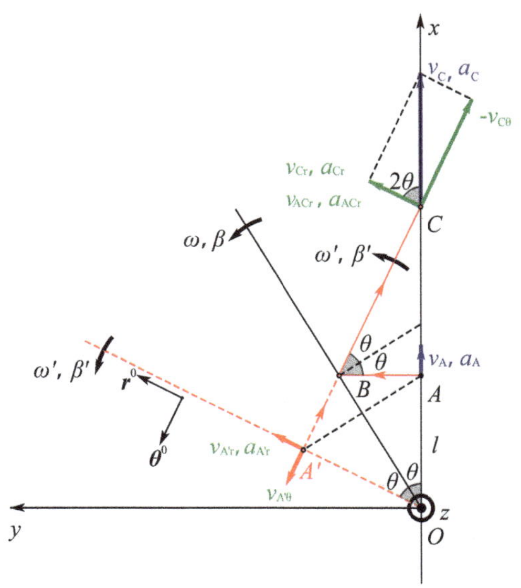

Figure 3

Considering the definition of ***angular velocity*** and ***angular acceleration***:

$$\begin{cases} \omega' = \dfrac{\mathrm{d}(2\theta)}{\mathrm{d}t} & (14a) \\ \beta' = \dfrac{\mathrm{d}\omega'}{\mathrm{d}t} = \dfrac{\mathrm{d}^2(2\theta)}{\mathrm{d}t^2} & (14b) \end{cases}$$

ω' and β'—the $\boldsymbol{k}$-direction component of the angular velocity and angular acceleration of the mirror image of the wall.

(5) → (14a), (10) → (14b),

$$\Rightarrow \quad \begin{cases} \omega' = 2\omega & (15a) \\ \beta' = 2\beta & (15b) \end{cases}$$

Taking *Point C* as the object of study. Studying *the state at Time t*. Considering ***theorem for composition of velocities of a particle*** [taking the ground as the stationary reference frame, and the translational reference frame that is at rest with respect to Point A' as the moving reference frame] (the $\boldsymbol{r}^0$-direction component equation):

$$v_{Cr} = v_{A'r} + v_{CA'r} \tag{16}$$

v_{Cr}, $v_{A'r}$, $v_{CA'r}$—the $\boldsymbol{r}^0$-direction component of the velocity of Point C, Point A', and Point C relative to Point A' at Time t.

$$v_{Cr} = v_C \cos 2\theta \tag{17}$$

Considering the spatial symmetry, we can obtain:

$$v_{A'r} = v_A \tag{18}$$

Considering ***the transverse component form of velocity***, then place (13) and (15a) in to obtain:

$$v_{CA'r} = \omega' \overline{CA'} = 2\omega l \tan 2\theta \tag{19}$$

(17) to (19) → (16),

$$\Rightarrow \quad v_C \cos 2\theta = v_A + 2\omega l \tan 2\theta$$

$$\Rightarrow \quad v_C = v_A \frac{1}{\cos 2\theta} + 2\omega l \frac{\tan 2\theta}{\cos 2\theta} \tag{20}$$

Taking *Point C* as the object of study. Studying *the state at Time t*. Considering ***theorem for composition of accelerations of a particle*** [taking the ground as the stationary reference frame, and the translational reference frame that is at rest with respect to Point A' as the moving reference frame] (the $\boldsymbol{r}^0$-direction component equation):

$$a_{Cr} = a_{A'r} + a_{CA'r} \tag{21}$$

a_{Cr}, $a_{A'r}$, $a_{CA'r}$—the $\boldsymbol{r}^0$-direction component of the acceleration of Point C, Point A', and Point C relative to Point A' at Time t.

$$a_{Cr} = a_C \cos 2\theta \tag{22}$$

Considering ***the radial component form of acceleration***, then place (12) and (15a) in to obtain:

$$a_{A'r} = \frac{\mathrm{d}^2 \overline{OA'}}{\mathrm{d}t^2} - \omega'^2 \overline{OA'} = \frac{\mathrm{d}^2 l}{\mathrm{d}t^2} - 4\omega^2 l \tag{23}$$

Considering the definition of ***acceleration***:

$$\Rightarrow \qquad a_A = \frac{\mathrm{d}^2 l}{\mathrm{d}t^2} \tag{24}$$

(24) $\rightarrow$ (23),

$$\Rightarrow \qquad a_{A'r} = a_A - 4\omega^2 l \tag{25}$$

Considering ***the transverse component form of acceleration***:

$$a_{CA'r} = \beta' \overline{CA'} + 2\omega' \left[(-v_{C\theta}) - (-v_{A'\theta}) \right] \tag{26}$$

$v_{C\theta}$ and $v_{A'\theta}$—the $\boldsymbol{\theta}^0$-direction component of the velocity of Point C and Point A' at Time t.

From (20),

$$-v_{C\theta} = v_C \sin 2\theta = v_A \tan 2\theta + 2\omega l \tan^2 2\theta \tag{27}$$

Considering ***the transverse component form of velocity***, then place (12) and (15a) in to obtain:

$$v_{A'\theta} = \omega' \overline{OA'} = 2\omega l \tag{28}$$

(13), (27) and (28) $\rightarrow$ (26),

$$\Rightarrow \qquad a_{CA'r} = 2\beta l \tan 2\theta + 8(1 + \tan^2 2\theta)\omega^2 l + 4\omega v_A \tan 2\theta \tag{29}$$

(22), (25) and (29) $\rightarrow$ (21),

$$\Rightarrow \quad a_C = \frac{1}{\cos 2\theta} \left[a_A + 2\beta l \tan 2\theta + 4(1 + 2\tan^2 2\theta)\omega^2 l + 4\omega v_A \tan 2\theta \right] \tag{30}$$

Solution [3]

As shown in Fig. 4, the image of Point C is denoted by C'.

Considering the following geometric relation in Fig. 4:

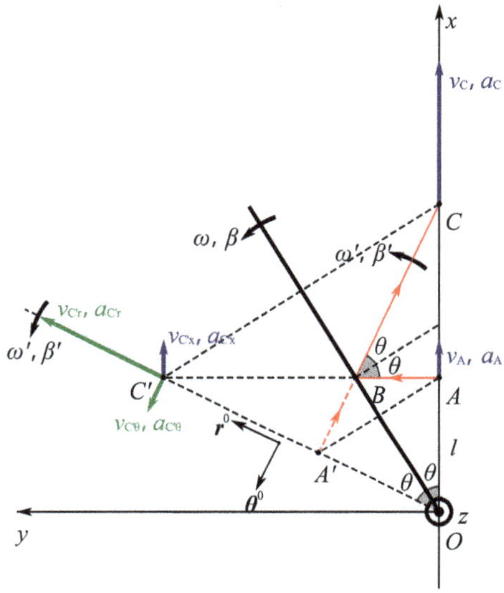

Figure 4

$$\overline{OC} = \overline{OC'} = \frac{l}{\cos 2\theta} \tag{31}$$

Considering the spatial symmetry, we can obtain:

$$v_C = v_{C'r} \tag{32}$$

$v_{C'r}$—the $\boldsymbol{r}^0$-direction component of the velocity of Point C' at Time t.

Considering ***the transverse component form of velocity*** (AC' does not rotate), then place (31) in to obtain:

$$v_{C'x} = v_A \tag{33}$$

$$v_{C'\theta} = \omega'\overline{OC'} = \frac{2\omega l}{\cos 2\theta} \tag{34}$$

$v_{C'\theta}$ and $v_{C'x}$—the $\boldsymbol{r}^0$-direction and $\boldsymbol{i}$-direction component of the velocity of Point C' at Time t.

Considering the orthogonality of the base ($\boldsymbol{r}^0$, $\boldsymbol{\theta}^0$):

$$v_{C'x} = v_{C'r}\cos 2\theta - v_{C'\theta}\sin 2\theta \tag{35}$$

Combining (32) to (35),

$$\Rightarrow \qquad v_C = v_A \frac{1}{\cos 2\theta} + 2\omega l \frac{\tan 2\theta}{\cos 2\theta} \tag{36}$$

Considering ***the component of the acceleration of a particle in polar coordinate system*** (AC' does not rotate), then place (15), (31) and (32) in to obtain:

$$a_{C'x} = a_A \tag{37}$$

$$\begin{cases} a_{C'r} = \dfrac{\mathrm{d}^2 \overline{OC'}}{\mathrm{d}t^2} - \omega'^2 \overline{OC'} = \dfrac{\mathrm{d}^2 \overline{OC}}{\mathrm{d}t^2} - \dfrac{4\omega^2 l}{\cos 2\theta} & \text{(38a)} \\ a_{C'\theta} = \beta' \overline{OC'} + 2\omega' v_{C'r} = \dfrac{2\beta l}{\cos 2\theta} + 4\omega v_C & \text{(38b)} \end{cases}$$

Considering the definition of ***acceleration***:

$$a_C = \frac{\mathrm{d}^2 \overline{OC}}{\mathrm{d}t^2} \tag{39}$$

Considering the orthogonality of the base ($\boldsymbol{r}^0$, $\boldsymbol{\theta}^0$):

$$a_{C'x} = a_{C'r} \cos 2\theta - a_{C'\theta} \sin 2\theta \tag{40}$$

Combining (36) to (40),

$$\Rightarrow \; a_C = \frac{1}{\cos 2\theta} \left[a_A + 2\beta l \tan 2\theta + 4(1 + 2\tan^2 2\theta)\omega^2 l + 4\omega v_A \tan 2\theta \right] \tag{41}$$

Solution [4]

Considering the following geometric relations in Fig. 5:

$$\overline{OB} = \frac{l}{\cos\theta} \tag{42}$$

$$\overline{BC} = \frac{l \tan\theta}{\cos 2\theta} \tag{43}$$

Taking *Point C* as the object of study. Studying *the state at Time t*. Considering ***theorem for composition of velocities of a particle*** [taking the ground as the stationary reference frame, and the translational reference frame that is at rest with respect to Point *B* as the moving reference frame] (the $\boldsymbol{r}^0$-direction component equation):

$$v_{Cr} = v_{Br} + v_{CBr} \tag{44}$$

v_{Br} and v_{CBr}—the $\boldsymbol{r}^0$-direction component of the velocity of Point B and Point C relative to Point B at Time t.

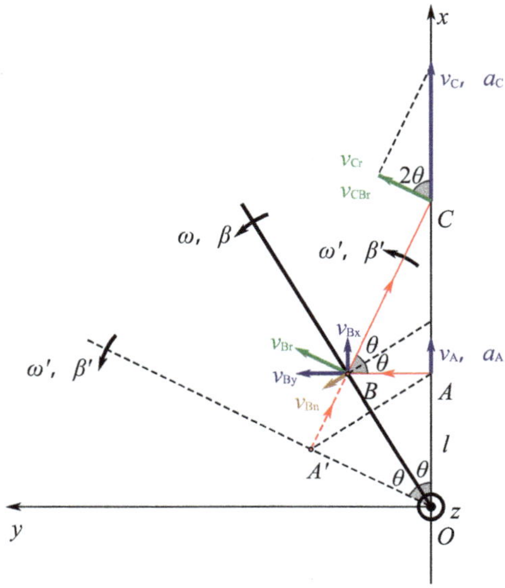

Figure 5

Considering ***the transverse component form of velocity*** (AB does not rotate), then place (42) in to obtain:

$$\begin{cases} v_{Bx} = v_A & (45) \\ v_{Bn} = \omega \overline{OB} = \dfrac{\omega l}{\cos\theta} & (46) \end{cases}$$

v_{Bx} and v_{Bn}—the $\boldsymbol{i}$-direction and $\boldsymbol{n}$-direction component of the velocity of Point B at Time t.

Considering the orthogonality of the base $(\boldsymbol{i}, \boldsymbol{j})$:

$$v_{Bn} = v_{By}\cos\theta - v_{Bx}\sin\theta \tag{47}$$

v_{By}—the $\boldsymbol{j}$-direction component of the velocity of Point B at Time t.

Combining (45) to (47),

$$\Rightarrow \qquad v_{By} = v_A \tan\theta + \omega l \frac{1}{\cos^2\theta} \tag{48}$$

Considering the orthogonality of the base $(\boldsymbol{i}, \boldsymbol{j})$:

$$v_{Br} = v_{Bx}\cos 2\theta + v_{By}\sin 2\theta \tag{49}$$

(45) and (48) → (49),

$$\Rightarrow \qquad v_{Br} = v_A(\tan\theta\sin 2\theta + \cos 2\theta) + \omega l \frac{\sin 2\theta}{\cos^2\theta} \tag{50}$$

Considering ***the transverse component form of velocity***, then place (15a) and (43) in to obtain:

$$v_{CBr} = \omega' \overline{BC} = 2\omega \frac{l\tan\theta}{\cos 2\theta} \tag{51}$$

(17), (50) and (51) → (44),

$$\Rightarrow \qquad v_C \cos 2\theta = (v_{Bx}\sin 2\theta + v_{By}\cos 2\theta) + 2\omega \cdot \frac{l\tan\theta}{\cos 2\theta}$$

$$\Rightarrow \qquad v_C = v_A \frac{1}{\cos 2\theta} + 2\omega l \frac{\tan 2\theta}{\cos 2\theta} \tag{52}$$

1.1.9 Rolling Curve (d, *C*)

(1) Consider a rigid ring of radius R performing the pure rolling on a horizontal ground horizontally to the right (see Fig. 1). Determine the trajectory of an arbitrary point on the ring.

(2) Time zero is arbitrarily set, at arbitrary Time t, the components of the rotating angular velocity and angular acceleration of the ring in the direction perpendicular to the plane of the figure inward are denoted by ω and β respectively. Determine the velocity and acceleration of an arbitrary point on the ring at Time t.

(3) Determine the radius of curvature at an arbitrary point on the trajectory of an arbitrary point on the ring.

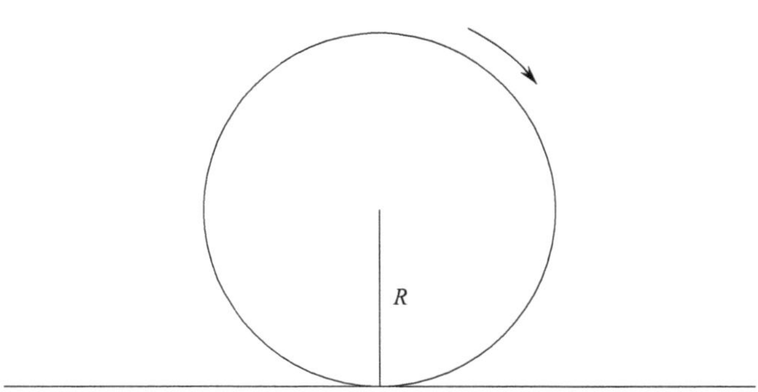

Figure 1

Solution to Problem (1)

As shown in Fig. 2, the center of the ring is denoted by O, the points on the ring in contact with the ground at Time zero and arbitrary Time t are denoted by A and P respectively, $\angle POA$ is denoted by φ, the point on the ground in contact with the ring at Time zero is denoted by O', the unit vector in the horizontal right direction is denoted by $\boldsymbol{i}$, and the unit vector in the vertical upward direction is denoted by $\boldsymbol{j}$. Establishing the Cartesian coordinate system with the frame $(O', \boldsymbol{i}, \boldsymbol{j})$, let x, y be the Cartesian coordinates.

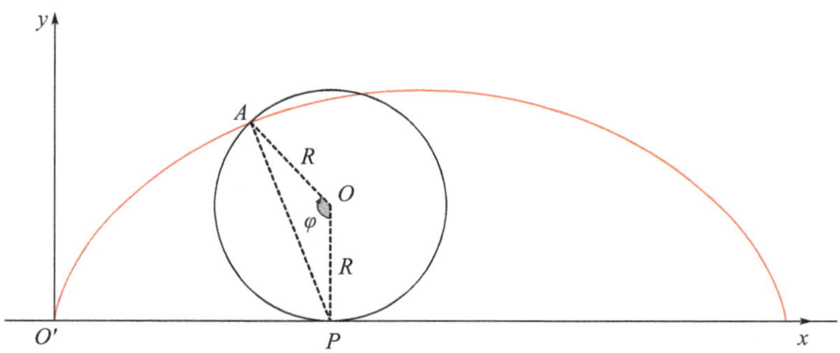

Figure 2

Considering the constraint condition: there is no relative sliding between the ring and the ground, it can be seen that the x-coordinate of Point P is equal to the arc length of the ring has rotated.

$$x_p = \overset{\frown}{AP} = R\varphi \tag{1}$$

Considering the following geometric relations in Fig. 2:

$$\Rightarrow \quad \begin{cases} x_A = R\varphi - R\sin\varphi & \text{(2a)} \\ y_A = R - R\cos\varphi & \text{(2b)} \end{cases}$$

$$\Rightarrow \quad x_A = R\arccos\left(1 - \frac{y_A}{R}\right) - \sqrt{y_A(2R - y_A)} \tag{3}$$

Equation (3) is the trajectory equation of Point P.

The curve that satisfies Eq. (3) is called the rolling curve (or cycloid) [Eq. (2) is the parameter equation of the rolling curve with φ as the parameter], which is denoted by the Curve Γ in this problem.

Solution to Problem (2)

As shown in Fig. 3, taking *Point O* as the object of study. Studying *the state at arbitrary Time t*. Considering ***theorem for composition of velocities of a particle*** [taking the ground as the stationary reference frame, and the translational reference frame that is at rest with respect to Point *P* as the moving reference frame] (the component equations in the (***i***, ***j***)-directions).

$$\begin{cases} v_{Ox} = v_{Px} + v_{OPx} & \text{(4a)} \\ v_{Oy} = v_{Py} + v_{OPy} & \text{(4b)} \end{cases}$$

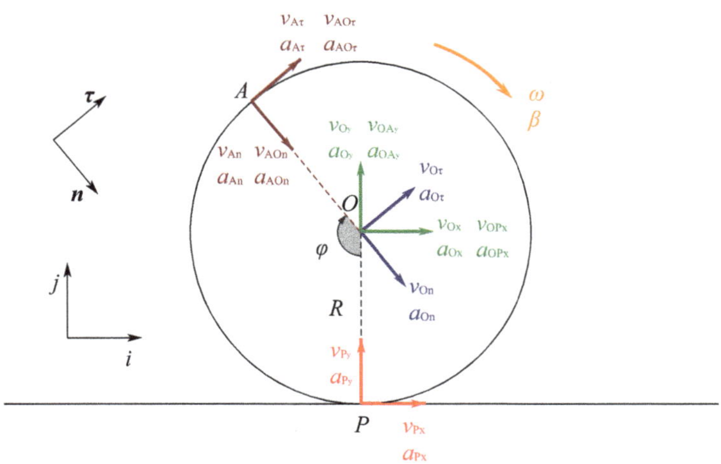

Figure 3

v_{Ox} and v_{Oy}—the components of the velocity of Point *O* in the (***i***, ***j***)-directions at Time *t*.

v_{Px} and v_{Py}—the components of the velocity of Point *P* in the (***i***, ***j***)-directions at Time *t*.

v_{OPx} and v_{OPy}—the components of the velocity of Point *O* relative to Point *P* in the (***i***, ***j***)-directions at Time *t*.

Considering the constraint condition: the ring is in contact with the ground all the time and there is no relative sliding between the ring and the ground, it can be seen that the velocity of Point *P* at Time *t* is zero.

$$\begin{cases} v_{Px} = 0 & (5a) \\ v_{Py} = 0 & (5b) \end{cases}$$

Considering ***the components form of the velocity of a particle in polar coordinate system*** (the constraint condition: the ring is a rigid body, that is the distance between Point O and Point P remains unchanged):

$$\begin{cases} v_{OPx} = \omega R & (6a) \\ v_{OPy} = 0 & (6b) \end{cases}$$

(5) and (6) $\rightarrow$ (4),

$$\Rightarrow \quad \begin{cases} v_{Ox} = \omega R & (7a) \\ v_{Oy} = 0 & (7b) \end{cases}$$

Considering the definition of ***acceleration***:

$$\begin{cases} a_{Ox} = \dfrac{\mathrm{d}v_{Ox}}{\mathrm{d}t} & (8a) \\ a_{Oy} = \dfrac{\mathrm{d}v_{Oy}}{\mathrm{d}t} & (8b) \end{cases}$$

a_{Ox} and a_{Oy}—the components of the acceleration of Point O in the $(\boldsymbol{i}, \boldsymbol{j})$-directions at Time t.

Considering the definition of ***angular acceleration***:

$$\beta = \frac{\mathrm{d}\omega}{\mathrm{d}t} \tag{9}$$

Combining (7) to (9),

$$\Rightarrow \quad \begin{cases} a_{Ox} = \beta R \\ a_{Oy} = 0 \end{cases} \tag{10}$$

Introducing the tangent direction $\boldsymbol{\tau}$ and normal direction $\boldsymbol{n}$ of the ring at Point A at Time t (see Fig. 3).

Taking *Point A* as the object of study. Studying *the state at arbitrary Time t*. Considering ***theorem for composition of velocities of a particle and theorem for composition of accelerations of a particle*** [taking the ground as the stationary reference frame, and the translational reference frame that is at rest with respect to Point O as the moving reference frame] (the $\boldsymbol{\tau}$-direction and $\boldsymbol{n}$-direction component equations):

$$\begin{cases} v_{A\tau} = v_{O\tau} + v_{AO\tau} & (11a) \\ v_{An} = v_{On} + v_{AOn} & (11b) \end{cases}$$

$$\begin{cases} a_{A\tau} = a_{O\tau} + a_{AO\tau} & (12a) \\ a_{An} = a_{On} + a_{AOn} & (12b) \end{cases}$$

$v_{A\tau}$ and v_{An}—the components of the velocity of Point A in the $(\boldsymbol{\tau}, \boldsymbol{n})$-directions at Time t.

$v_{O\tau}$ and v_{On}—the components of the velocity of Point O in the $(\boldsymbol{\tau}, \boldsymbol{n})$-directions at Time t.

$v_{AO\tau}$ and v_{AOn}—the components of the velocity of Point A relative to Point O in the $(\boldsymbol{\tau}, \boldsymbol{n})$-directions at Time t.

$a_{A\tau}$ and a_{An}—the components of the acceleration of Point A in the $(\boldsymbol{\tau}, \boldsymbol{n})$-directions at Time t.

$a_{O\tau}$ and a_{On}—the components of the acceleration of Point O in the $(\boldsymbol{\tau}, \boldsymbol{n})$-directions at Time t.

$a_{AO\tau}$ and a_{AOn}—the components of the acceleration of Point A relative to Point O in the $(\boldsymbol{\tau}, \boldsymbol{n})$-directions at Time t.

Considering the orthogonality of the base $(\boldsymbol{i}, \boldsymbol{j})$, from (7),

$$\Rightarrow \quad \begin{cases} v_{O\tau} = -v_{Ox}\cos\varphi + v_{Oy}\sin\varphi = -\omega R\cos\varphi & (13a) \\ v_{On} = v_{Ox}\sin\varphi + v_{Oy}\cos\varphi = \omega R\sin\varphi & (13b) \end{cases}$$

From (10),

$$\Rightarrow \quad \begin{cases} a_{O\tau} = -a_{Ox}\cos\varphi + a_{Oy}\sin\varphi = -\beta R\cos\varphi & (14a) \\ a_{On} = a_{Ox}\sin\varphi + a_{Oy}\cos\varphi = \beta R\sin\varphi & (14b) \end{cases}$$

Considering ***the components form of the velocity and acceleration of a particle in polar coordinate system*** (the constraint condition: the ring is a rigid body, that is the distance between Point A and Point O remains unchanged):

$$\begin{cases} v_{AO\tau} = \omega R & (15a) \\ v_{AOn} = 0 & (15b) \end{cases}$$

$$\begin{cases} a_{AO\tau} = \beta R & (16a) \\ a_{AOn} = \omega^2 R & (16b) \end{cases}$$

(13) and (15) → (11),

$$\Rightarrow \quad \begin{cases} v_{A\tau} = \omega R(1-\cos\varphi) & (17a) \\ v_{An} = \omega R\sin\theta & (17b) \end{cases}$$

(14) and (16) → (12),

$$\Rightarrow \quad \begin{cases} a_{A\tau} = \beta R(1-\cos\varphi) & (18a) \\ a_{An} = \beta R\sin\varphi + \omega^2 R & (18b) \end{cases}$$

Considering the degenerate form of Eqs. (18a) and (18b) when $\varphi = 0$ (Point A is in contact with the ground):

$$\Rightarrow \quad \begin{cases} a_{A\tau}\big|_{\varphi=0} = 0 & (19a) \\ a_{An}\big|_{\varphi=0} = \omega^2 R & (19b) \end{cases}$$

It can be seen from Eq. (19) that the point on the ring in contact with the ground has a vertical upward acceleration component, which is not easy to be seen intuitively.

Solution [1] to Problem (3)

For simplicity, the following studies the situation of $\beta = 0$.

Taking *Point A* as the object of study. Studying *the state at Time t*. Considering ***theorem for composition of velocities of a particle*** [taking the ground as the stationary reference frame, and the translational reference frame that is at rest with respect to Point O as the moving reference frame]:

$$\boldsymbol{v}_A = \boldsymbol{v}_O + \boldsymbol{v}_{AO} \qquad (20)$$

$\boldsymbol{v}_A$—the velocity of Point A at Time t.

$\boldsymbol{v}_O$—the velocity of Point O at Time t.

$\boldsymbol{v}_{AO}$—the velocity of Point A relative to Point O at Time t.

It can be seen from Eq. (7) that $\boldsymbol{v}_O$ has only the component in the $\boldsymbol{i}$-direction, and from Eq. (15) that $\boldsymbol{v}_{AO}$ has only the component in the $\boldsymbol{\tau}$-direction, according to this, we can obtain the vector diagram corresponding to Eq. (20) (see Fig. 4):

Combining (7) and (15),

$$\Rightarrow \quad v_O = v_{AO} = \omega R \qquad (21)$$

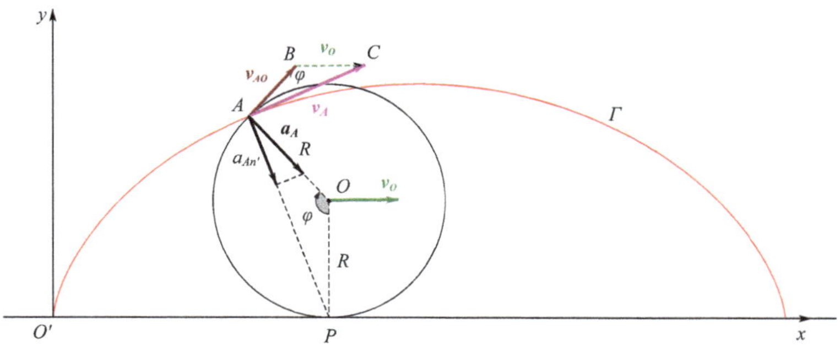

Figure 4

Considering the following geometric relation in Fig. 4:

$$\Rightarrow \qquad \angle ABC = \varphi \qquad (22)$$

$$\Rightarrow \qquad v_A = 2\omega R \sin\frac{\varphi}{2} \qquad (23)$$

It can be seen from the geometric relation in Fig. 4 that $\Delta AOP \sim \Delta ABC$, It can be further known that the normal of Curve Γ at Point *A* passes through Point *P*.

From (18),

$$\Rightarrow \qquad \boldsymbol{a}_A = \omega^2 R \boldsymbol{n} \qquad (24)$$

$$\Rightarrow \qquad a_{An'} = a_A \sin\frac{\varphi}{2} = \omega^2 R \sin\frac{\varphi}{2} \qquad (25)$$

$a_{An'}$—the normal direction component of the acceleration of Point *A*. Considering ***the normal component form of acceleration***:

$$a_{An'} = \frac{v_A^2}{\rho} \qquad (26)$$

ρ—the radius of curvature of Curve Γ at Point *A*.

Combining (23), (25) and (26),

$$\Rightarrow \qquad \rho = 4R\sin\frac{\varphi}{2} \qquad (27)$$

Solution [2] to Problem (3)

Considering the form of the radius of curvature at an arbitrary point on the plane curve by parameter equations:

$$\Rightarrow \qquad \rho = \frac{\left\{[x'(\varphi)]^2 + [y'(\varphi)]^2\right\}^{3/2}}{|x'(\varphi)y''(\varphi) - x''(\varphi)y'(\varphi)|} \tag{28}$$

$(2) \rightarrow (28)$,

$$\Rightarrow \qquad \rho = 4R\sin\frac{\varphi}{2} \tag{29}$$

1.1.10 The Regular Precession of a Ring (d, *A*)

As shown in Fig. 1, there is a circular ring of radius r performing the pure rolling on a horizontal ground with the self-rotating angular velocity of the magnitude ω and the direction as shown in the figure, the angle between the ring and the vertical direction is θ all the time, the herpolhode of the ring on the ground is a circle of radius R. The centers of circle of the ring and its herpolhode are denoted by O_1 and O_2 respectively. Find the precession angular velocity of the ring.

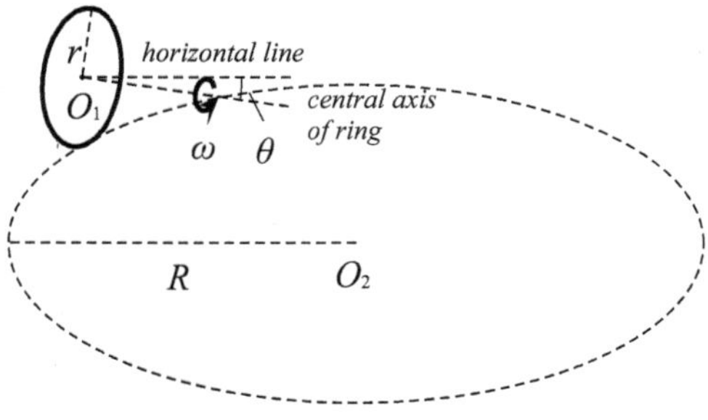

Figure 1

Solution

Time zero is arbitrarily set.

As shown in Fig. 2, the center of the circle trajectory of Point O_1 is denoted by O_3, the intersection point of the central axis of the ring and the central axis of the herpolhode of the ring on the ground is denoted by O. At arbitrary Time t, the point on the ring in contact with the ground is denoted by A. Establishing the body coordinate system with the frame $(O, \boldsymbol{i}, \boldsymbol{j}, \boldsymbol{k})$ (the coordinate system of principal axis of the ring), let *x*, *y*, *z* be the Cartesian coordinates, and establishing the space coordinate system with the frame $(O, \boldsymbol{i}', \boldsymbol{j}', \boldsymbol{k}')$, let *X*, *Y*, *Z* be the Cartesian coordinates. The Time when $\boldsymbol{j}$ coincides with $\boldsymbol{j}'$ is denoted by Time *T*.

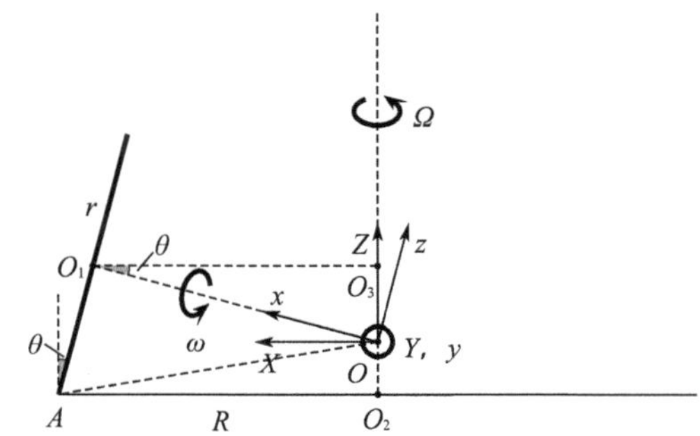

Figure 2

Defining $\boldsymbol{\omega}$—the angular velocity of the ring at Time *T*.

Defining $(\omega_x, \omega_y, \omega_z)$—the components of the angular velocity of the ring in the body coordinate system $(O, \boldsymbol{i}, \boldsymbol{j}, \boldsymbol{k})$ at Time *T*.

$$\boldsymbol{\omega} = \omega_x \boldsymbol{i} + \omega_y \boldsymbol{j} + \omega_z \boldsymbol{k} \tag{1}$$

Considering ***Euler kinematics equations***:

$$\begin{cases} \omega_x = -\omega + \Omega \sin\theta & \text{(2a)} \\ \omega_y = 0 & \text{(2b)} \\ \omega_z = \Omega \cos\theta & \text{(2c)} \end{cases}$$

Ω—the component of the precession angular velocity of the ring in the vertical up direction at Time *T*.

Taking *Point A* as the object of study. Studying *the state at Time T*. Considering ***theorem for composition of velocities of a particle*** [taking the ground as the stationary reference frame, and the translational reference frame that is at rest with respect to Point O_1 as the moving reference frame]:

$$\boldsymbol{v}_A = \boldsymbol{v}_{O_1} + \boldsymbol{v}_{AO_1} \tag{3}$$

$\boldsymbol{v}_A$—the velocity of Point *A* at Time *T*.

$\boldsymbol{v}_{O_1}$—the velocity of Point O_1 at Time *T*.

$\boldsymbol{v}_{AO_1}$—the velocity of Point *A* relative to Point O_1 at Time *T*.

Considering the definition of ***velocity*** and Poisson formula, then place (1) and (2) in to obtain:

$$\Rightarrow \begin{cases} \boldsymbol{v}_{O_1} = \dfrac{\mathrm{d}\overrightarrow{O_3O_1}}{\mathrm{d}t} = \Omega\boldsymbol{k}' \times \overrightarrow{O_3O_1} = \Omega\boldsymbol{k}' \times (R - r\sin\theta)\boldsymbol{i}' & \text{(4a)} \\ \quad = \Omega(R - r\sin\theta)\boldsymbol{j}' = \Omega(R - r\sin\theta)\boldsymbol{j} & \\ \boldsymbol{v}_{AO_1} = \dfrac{\mathrm{d}\overrightarrow{O_1A}}{\mathrm{d}t} = \boldsymbol{\omega} \times \overrightarrow{O_1A} = (\omega_x\boldsymbol{i} + \omega_y\boldsymbol{j} + \omega_z\boldsymbol{k}) \times r(-\boldsymbol{k}) & \\ \quad = (-\omega r + \Omega r\sin\theta)\boldsymbol{j} & \text{(4b)} \end{cases}$$

(4) → (3),

$$\boldsymbol{v}_A = (-\omega r + \Omega R)\,\boldsymbol{j} \tag{5}$$

The condition for the ring to be in contact with the ground all the time without relative slipping is as follows:

$$\boldsymbol{v}_A = 0 \tag{6}$$

$$\Rightarrow \qquad \Omega = \frac{r}{R}\omega \tag{7}$$

Another method to get $\boldsymbol{v}_A$

Considering the definition of ***velocity*** and Poisson formula, then place (1) and (2) in to obtain:

$$\Rightarrow \boldsymbol{v}_A = \frac{\mathrm{d}\overrightarrow{OA}}{\mathrm{d}t} = \boldsymbol{\omega} \times \overrightarrow{OA} = (\omega_x\boldsymbol{i} + \omega_y\boldsymbol{j} + \omega_z\boldsymbol{k}) \times \left[\frac{R - r\sin\theta}{\cos\theta}\boldsymbol{i} + r(-\boldsymbol{k})\right] \tag{8}$$

$$= (-\omega r + \Omega R)\boldsymbol{j}$$

1.1.11 Cylinder-String-Ball Model 1 (d, *C*)

As shown in Fig. 1, there is a light rigid string all tightly twined on a fixed cylinder of sectional radius R (twined perpendicular to the central axis of the cylinder, and the center is denoted by O), the free end of which is tied with a small ball. The ball is subjected to an instantaneous impact and gets an initial velocity of the magnitude v_0 and the direction radial outward. Given that there is no relative sliding between the string and the cylinder.

(1) Find the trajectory equation and kinematics equation of the ball.
(2) Find the acceleration of the ball at arbitrary time.

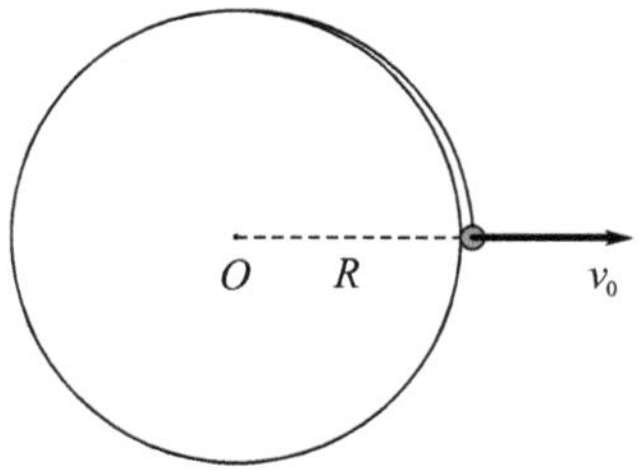

Figure 1

Solution [1] to Problem (1)

The time point when the ball is impacted is defined as Time zero.

As shown in Fig. 2, the position of the ball at Time zero is denoted by P_0, the point on the string in contact with the cylinder at arbitrary Time t is denoted by P. Introducing the orthogonal unit vectors $\boldsymbol{i}, \boldsymbol{j}, \boldsymbol{k}$. Establishing the Cartesian coordinate system with the frame $(O, \boldsymbol{i}, \boldsymbol{j})$, let x, y be the Cartesian coordinates. Taking the origin O as the pole, the positive x-axis as the polar axis, the right-hand helix direction with regard to the $\boldsymbol{k}$-direction as the positive direction of polar angle, establishing the polar coordinate system, let r, φ be the polar coordinates. Introducing the tangential unit vector $\boldsymbol{e}_1$ and normal unit vector $\boldsymbol{e}_2$ of the segment of the string between the ball and the cylinder at arbitrary Time t.

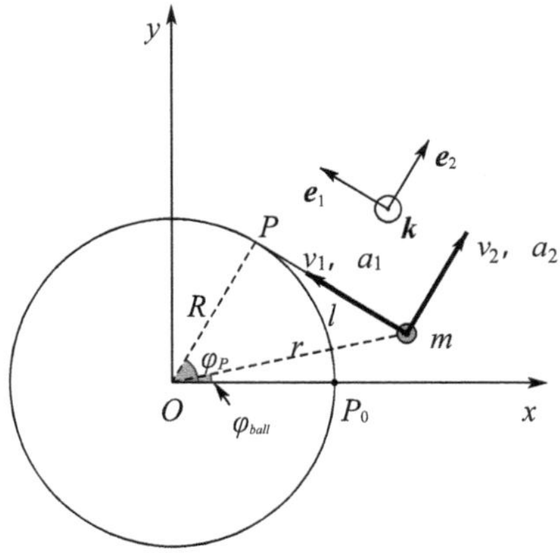

Figure 2

Taking *the ball (can be regarded as a particle)* as the object of study. Studying *the process from Time zero to arbitrary Time t.* Considering ***law of conservation of mechanical energy*** [the tension in the string acting on the ball is perpendicular to its velocity all the time, no work is done; therefore the mechanical energy of the ball is conserved]:

$$\frac{1}{2}mv_0^2 = \frac{1}{2}m|\boldsymbol{v}|^2 \tag{1}$$

m—the mass of the ball.

$\boldsymbol{v}$—the velocity of the ball at arbitrary Time t.

$$\Rightarrow \qquad |\boldsymbol{v}| = v_0 \tag{2}$$

For the motion of the ball relative to Point P at Time t (at Time t, the velocity of Point P is zero, but the acceleration is not zero), considering ***the components form of the velocity of a particle in polar coordinate system***, [the constraint condition: the string length remains unchanged, and the segment of the string between the ball and the cylinder is tangent to the cylinder all the time]:

$$\begin{cases} v_1 = 0 & \text{(3a)} \\ v_2 = l\dfrac{\mathrm{d}\varphi_P}{\mathrm{d}t} & \text{(3b)} \end{cases}$$

v_1 and v_2—the $\boldsymbol{e}_1$-direction and $\boldsymbol{e}_2$-direction component of the velocity of the ball at Time t.

l—the length of the segment of the string between the ball and the cylinder at Time t.

$$\Rightarrow \qquad \boldsymbol{v} = v_1\boldsymbol{e}_1 + v_2\boldsymbol{e}_2 = l\frac{\mathrm{d}\varphi_P}{\mathrm{d}t}\boldsymbol{e}_2 \tag{4}$$

Considering the constraint condition: the string is tightly twined on the cylinder and there is no relative sliding between them:

$$l = R\varphi_P \tag{5}$$

Combining (2), (4) and (5) [φ_P increases with time]:

$$\Rightarrow \qquad \varphi_P\mathrm{d}\varphi_P = \frac{v_0}{R}\mathrm{d}t \tag{6}$$

$$\Rightarrow \qquad \int_0^{\varphi_P} \varphi_P\mathrm{d}\varphi_P = \frac{v_0}{R}\int_0^t \mathrm{d}t$$

$$\Rightarrow \qquad \frac{1}{2}\varphi_P^2 = \frac{v_0}{R}t$$

$$\Rightarrow \qquad \varphi_P = \sqrt{\frac{2v_0 t}{R}} \qquad (7)$$

(7) → (5):

$$\Rightarrow \qquad l = \sqrt{2Rv_0 t} \qquad (8)$$

Considering the following geometric relations in Fig. 2:

$$\begin{cases} r = \sqrt{R^2 + l^2} & (9a) \\ \tan(\varphi_P - \varphi_{ball}) = \dfrac{l}{R} & (9b) \end{cases}$$

Combining (5) and (9): we obtain the trajectory equation of the ball in the polar coordinate system:

$$\Rightarrow \qquad \varphi_{ball} = \sqrt{\left(\frac{r}{R}\right)^2 - 1} - \arctan\sqrt{\left(\frac{r}{R}\right)^2 - 1} \qquad (10)$$

(5) and (8) → (9a) and (9b), we obtain the kinematics equations of the ball in the polar coordinate system:

$$\Rightarrow \qquad \begin{cases} r = R\sqrt{1 + \dfrac{2v_0 t}{R}} & (11a) \\ \varphi_{ball} = \sqrt{\dfrac{2v_0 t}{R}} - \arctan\sqrt{\dfrac{2v_0 t}{R}} & (11b) \end{cases}$$

Solution [2] to Problem (1)

Considering the following geometric relations in Fig. 2:

$$\begin{cases} x = R\cos\varphi_P + l\sin\varphi_P & (12a) \\ y = R\sin\varphi_P - l\cos\varphi_P & (12b) \end{cases}$$

(5) → (12), we obtain the trajectory equations of the ball in the Cartesian coordinate system (with φ_P as the parameter):

$$\Rightarrow \qquad \begin{cases} x = R(\cos\varphi_P + \varphi_P \sin\varphi_P) & (13a) \\ y = R(\sin\varphi_P - \varphi_P \cos\varphi_P) & (13b) \end{cases}$$

(7) → (13), we obtain the kinematics equations of the ball in the Cartesian coordinate system:

$$\Rightarrow \quad \begin{cases} x = R\left(\cos\sqrt{\dfrac{2v_0t}{R}} + \sqrt{\dfrac{2v_0t}{R}}\sin\sqrt{\dfrac{2v_0t}{R}}\right) & (14a) \\ y = R\left(\sin\sqrt{\dfrac{2v_0t}{R}} - \sqrt{\dfrac{2v_0t}{R}}\cos\sqrt{\dfrac{2v_0t}{R}}\right) & (14b) \end{cases}$$

Solution [1] to Problem (2)

Defining $\boldsymbol{a}$—the acceleration of the ball at arbitrary Time t. Considering the definition of ***acceleration***:

$$\boldsymbol{a} = \frac{\mathrm{d}\boldsymbol{v}}{\mathrm{d}t} = \frac{v\mathrm{d}\varphi_P}{\mathrm{d}t}\boldsymbol{e}_1 \quad (15)$$

(2) and (7) → (15):

$$\Rightarrow \quad \boldsymbol{a} = \frac{v_0\sqrt{v_0}}{\sqrt{2Rt}}\boldsymbol{e}_1 \quad (16)$$

Solution [2] to Problem (2)

Defining a_1 and a_2—the $\boldsymbol{e}_1$-direction and $\boldsymbol{e}_2$-direction component of the acceleration of the ball at Time t. Considering ***the components form of the acceleration of a particle in natural coordinate system***:

$$\begin{cases} a_1 = \dfrac{v^2}{l} & (17a) \\ a_2 = \dfrac{\mathrm{d}v}{\mathrm{d}t} = 0 & (17b) \end{cases}$$

$$\Rightarrow \quad \boldsymbol{a} = a_1\boldsymbol{e}_1 + a_2\boldsymbol{e}_2 = \frac{v^2}{l}\boldsymbol{e}_1 \quad (18)$$

(2) and (8) → (18):

$$\Rightarrow \quad \boldsymbol{a} = \frac{v_0\sqrt{v_0}}{\sqrt{2Rt}}\boldsymbol{e}_1 \quad (19)$$

1.1.12 Cylinder-String-Ball Model 2 (*a*, *A*)

As shown in Fig. 1, on a horizontal ground there is a cylinder of sectional radius R with a light rigid string tightly twined on it (twined perpendicular to the central axis of the cylinder, and the center is denoted by O), and a particle is tied to the free end of the string. The cylinder performs purely rolling on the ground in the direction away from the particle (horizontally to the right in the figure) under the action of the external force, given that there is no relative sliding between the string and the cylinder, and the string is in a straightened state all the time.

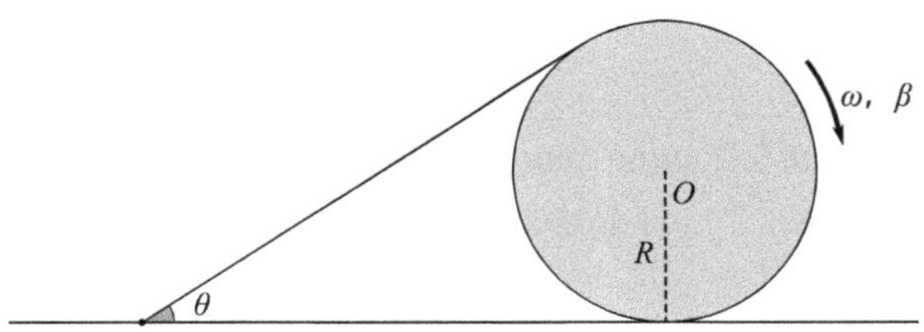

Figure 1

(1) Time zero is arbitrarily set, at arbitrary Time t, the clockwise direction component of the rotating angular velocity and angular acceleration of the cylinder are denoted by ω and β respectively, and the angle between the string and the ground is denoted by θ. Find the rotating angular velocity and angular acceleration of the segment of the string between the cylinder and the particle at arbitrary time.

(2) If the cylinder rotates at a uniform angular velocity of the clockwise direction component ω_0, find the time taken for the angle between the string and the ground to change from θ_0 to θ_1.

Solution to Problem (1)

As shown in Fig. 2, at arbitrary Time t, the point on the cylinder in contact with the string is denoted by A, the point on the string in contact with the cylinder is denoted by B, and the position of the particle is denoted by C. Introducing the tangential unit vector $\boldsymbol{\tau}$ and the normal unit vector $\boldsymbol{n}$ of the string at Point B.

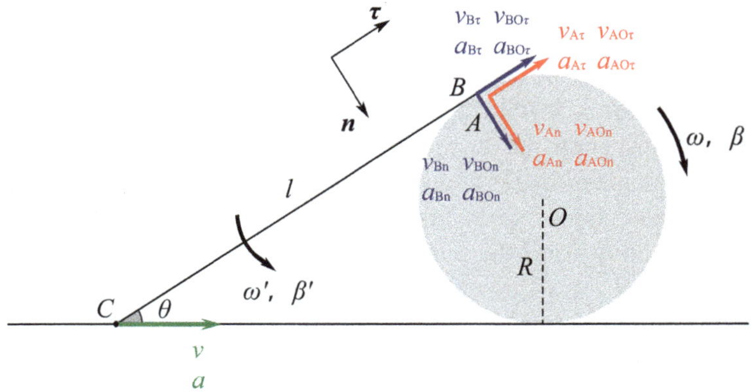

Figure 2

Considering the following geometric relation in Fig. 2:

$$l\sin\theta = R(1+\cos\theta) \tag{1}$$

l—the length of BC.

$$\Rightarrow \quad l = R(\cot\theta + \csc\theta) \tag{2}$$

Taking *Point B* as the object of study. Studying *the state at arbitrary Time t*. Considering ***theorem for composition of velocities of a particle and theorem for composition of accelerations of a particle*** [taking the ground as the stationary reference frame, and the translational reference frame that is at rest with respect to Point *A* as the moving reference frame] (the $\boldsymbol{\tau}$-direction and $\boldsymbol{n}$-direction component equations):

$$\begin{cases} v_{B\tau} = v_{A\tau} + v_{BA\tau} & (3a) \\ v_{Bn} = v_{An} + v_{BAn} & (3b) \end{cases}$$

$$\begin{cases} a_{B\tau} = a_{A\tau} + a_{BA\tau} & (4a) \\ a_{Bn} = a_{An} + a_{BAn} & (4b) \end{cases}$$

($v_{B\tau}$, v_{Bn}) and ($a_{B\tau}$, a_{Bn})—the $\boldsymbol{\tau}$-direction and $\boldsymbol{n}$-direction components of the velocity and acceleration of Point B at Time t.

($v_{A\tau}$, v_{An}) and ($a_{A\tau}$, a_{An})—the $\boldsymbol{\tau}$-direction and $\boldsymbol{n}$-direction components of the velocity and acceleration of Point A at Time t.

($v_{BA\tau}$, v_{BAn}) and ($a_{BA\tau}$, a_{BAn})—the $\boldsymbol{\tau}$-direction and $\boldsymbol{n}$-direction components of the velocity and acceleration of Point B relative to Point A at Time t.

From the conclusion of the previous problem, Section 1.1.9, "Rolling curve", we can obtain:

$$\begin{cases} v_{A\tau} = \omega R(1+\cos\theta) & (5a) \\ v_{An} = \omega R\sin\theta & (5b) \end{cases}$$

$$\begin{cases} a_{A\tau} = \beta R(1+\cos\theta) & (6a) \\ a_{An} = \beta R\sin\theta + \omega^2 R & (6b) \end{cases}$$

From the conclusion of the previous problem Section 1.1.9, "Rolling curve", we can obtain [the constraint condition: the string is in contact with the cylinder without relative sliding all the time]:

$$\begin{cases} v_{BA\tau} = 0 & (7a) \\ v_{BAn} = 0 & (7b) \end{cases}$$

$$\begin{cases} a_{BA\tau} = 0 & (8a) \\ a_{BAn} = -(\omega+\omega')^2 R & (8b) \end{cases}$$

ω'—the anticlockwise direction component of the rotating angular velocity of BC at arbitrary Time t.

(5) and (7) $\rightarrow$ (3),

$$\Rightarrow \quad \begin{cases} v_{B\tau} = \omega R(1+\cos\theta) & (9a) \\ v_{Bn} = \omega R\sin\theta & (9b) \end{cases}$$

(6) and (8) $\rightarrow$ (4),

$$\Rightarrow \quad \begin{cases} a_{B\tau} = \beta R(1+\cos\theta) & (10a) \\ a_{Bn} = \beta R\sin\theta + \omega^2 R - (\omega+\omega')^2 R & (10b) \end{cases}$$

Taking *the particle* as the object of study. Studying *the state at arbitrary Time t*. Considering ***theorem for composition of velocities of a particle and theorem for composition of accelerations of a particle*** [taking the ground as the stationary reference frame, and the translational reference frame that is at rest with respect to Point *B* as the moving reference frame] (the $\boldsymbol{\tau}$-direction and $\boldsymbol{n}$-direction component equations):

$$\begin{cases} v\cos\theta = v_{B\tau} + v_{CB\tau} & (11a) \\ v\sin\theta = v_{Bn} + v_{CBn} & (11b) \end{cases}$$

$$\begin{cases} a\cos\theta = a_{B\tau} + a_{CB\tau} & (12a) \\ a\sin\theta = a_{Bn} + a_{CBn} & (12b) \end{cases}$$

v and a—the component of the velocity and acceleration of the particle in the horizontal right direction at arbitrary Time t.

($v_{CB\tau}$, v_{CBn}) and ($a_{CB\tau}$, a_{CBn})—the $\boldsymbol{\tau}$-direction and $\boldsymbol{n}$-direction components of the velocity and acceleration of the particle relative to Point B at arbitrary Time t.

Considering ***the components form of the velocity and acceleration of a particle in polar coordinate system*** [the constraint condition: the string length remains unchanged]:

$$\begin{cases} v_{CB\tau} = 0 & (13a) \\ v_{CBn} = \omega' l & (13b) \end{cases}$$

$$\begin{cases} a_{CB\tau} = \omega'^2 l & (14a) \\ a_{CBn} = \beta' l & (14b) \end{cases}$$

β'—the anticlockwise direction component of the angular acceleration of BC at arbitrary Time t.

(9) and (13) $\rightarrow$ (11),

$$\Rightarrow \quad \begin{cases} v\cos\theta = \omega R(1+\cos\theta) & (15a) \\ v\sin\theta = \omega R\sin\theta + \omega' l & (15b) \end{cases}$$

(10) and (14) $\rightarrow$ (12),

$$\Rightarrow \quad \begin{cases} a\cos\theta = \beta R(1+\cos\theta) + \omega'^2 l & (16a) \\ a\sin\theta = \beta R\sin\theta + \omega^2 R - (\omega+\omega')^2 R + \beta' l & (16b) \end{cases}$$

From (15a),

$$\Rightarrow \quad v = \omega R(1+\sec\theta) \quad (17)$$

(2) and (17) $\rightarrow$ (15b),

$$\Rightarrow \quad \omega' = \omega\frac{\sec\theta\sin\theta}{\cot\theta+\csc\theta} = \omega(\sec\theta - 1) \quad (18)$$

(2) and (18) $\rightarrow$ (16a),

$$\Rightarrow \quad a = \beta R(1+\sec\theta) + \omega^2 R\frac{\tan^3\theta}{1+\cos\theta} \quad (19)$$

(2), (18) and (19) $\rightarrow$ (16b),

$$\Rightarrow \quad \beta' = \beta\frac{\sin\theta\sec\theta}{\cot\theta+\csc\theta} + \omega^2\frac{\left(\dfrac{\tan\theta}{1+\cos\theta}\sin\theta+1\right)\tan^2\theta}{\cot\theta+\csc\theta}$$

$$= \beta(\sec\theta-1) + \omega^2\left(\frac{\tan\theta}{1+\cos\theta}\sin\theta+1\right)\tan\theta(\sec\theta-1)$$

$$= \beta(\sec\theta-1) + \omega^2\tan\theta\sec\theta(\sec\theta-1) \tag{20}$$

Solution to Problem (2)

Considering the definition of ***angular velocity***:

$$\omega' = \frac{\mathrm{d}\theta}{\mathrm{d}t} \tag{21}$$

$(18) \rightarrow (21)$,

$$\Rightarrow \qquad \left.\frac{\mathrm{d}\theta}{\mathrm{d}t}\right|_{\omega=\omega_0} = \omega_0(\sec\theta-1) \tag{22}$$

$$\Rightarrow \qquad \mathrm{d}t = \frac{1}{\omega_0}\frac{\mathrm{d}\theta}{\sec\theta-1}$$

$$\Rightarrow \quad T = \int_{t_0}^{t_0+T}\mathrm{d}t = \frac{1}{\omega_0}\int_{\theta_0}^{\theta_1}\frac{\mathrm{d}\theta}{\sec\theta-1} = \frac{1}{\omega_0}\int_{\theta_0}^{\theta_1}\frac{\sec\theta+1}{\sec^2\theta-1}\mathrm{d}\theta$$

$$= \frac{1}{\omega_0}\int_{\theta_0}^{\theta_1}\frac{\sec\theta+1}{\tan^2\theta}\mathrm{d}\theta = \frac{1}{\omega_0}\left(\int_{\theta_0}^{\theta_1}\frac{\cos\theta}{\sin^2\theta}\mathrm{d}\theta + \int_{\theta_0}^{\theta_1}\cot^2\theta\mathrm{d}\theta\right)$$

$$= \frac{1}{\omega_0}\left[\int_{\theta_0}^{\theta_1}\frac{1}{\sin^2\theta}\mathrm{d}(\sin\theta) + \int_{\theta_0}^{\theta_1}(\csc^2\theta-1)\mathrm{d}\theta\right]$$

$$= \frac{1}{\omega_0}\left[\left(\frac{1}{\sin\theta_0} - \frac{1}{\sin\theta_1}\right) + (\cot\theta_0 - \cot\theta_1) + (\theta_0-\theta_1)\right] \tag{23}$$

t_0—the Time when the angle between the string and the ground is θ_0.

T—the time taken for the angle between the string and the ground to change from θ_0 to θ_1.

1.2 Chase Motion

Chasing problem is a kind of motion constraint problem of two-body model, which is vivid and interesting. A certain chasing strategy can be summed up as the constraints of various kinematic quantities of the two objects involved in chasing. The integrable chasing problem of kinematics equation or trajectory equation is worthy of studying. In this section, we will study some interesting, seemingly complex but integrable chasing problems.

1.2.1 Collinear and Equal Speed Chasing (d, *B*)

A wolf runs at a uniform speed along a circle of radius R, and a hound chases the wolf from the center of the circle at the same speed. Given that the wolf, the hound, and the center of the circle are collinear all the time. Find the trajectory of the hound.

Solution

The time point when the hound starts to chase the wolf is defined as Time zero.

As shown in Fig. 1, the center of the wolf's circular trajectory is denoted by O, the positions of the hound and the wolf at arbitrary Time t is denoted by A and B respectively. Taking Point O as the pole, the direction of the hound pointing to the wolf at Time zero as the positive direction of the polar axis, the rotation direction of OB as the positive direction of polar angle, establishing the polar coordinate system, let r, θ be the Polar coordinates.

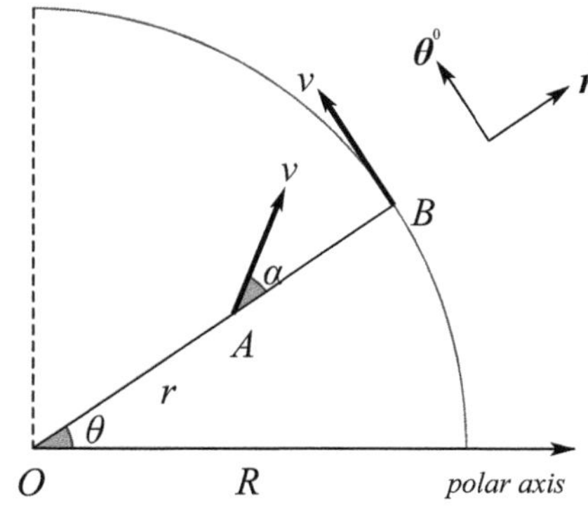

Figure 1

Respectively for the motion of the wolf and the hound, considering ***the components form of velocity in polar coordinate system***:

$$v = R\frac{\mathrm{d}\theta}{\mathrm{d}t} \tag{1}$$

$$\begin{cases} v\sin\alpha = r\dfrac{\mathrm{d}\theta}{\mathrm{d}t} & \text{(2a)} \\ v\cos\alpha = \dfrac{\mathrm{d}r}{\mathrm{d}t} & \text{(2b)} \end{cases}$$

v—the constant speed of the hound and the wolf.

α—the angle between the velocity of the hound and $\boldsymbol{r}^0$ at arbitrary Time t.

Combining (1) and (2) [eliminating v and α],

$$\Rightarrow \qquad \frac{\mathrm{d}r}{\sqrt{R^2 - r^2}} = \mathrm{d}\theta \tag{3}$$

$$\Rightarrow \qquad \int_0^r \frac{\mathrm{d}r}{\sqrt{R^2 - r^2}} = \int_0^\theta \mathrm{d}\theta \tag{4}$$

Defining the variable φ:

$$r = R\sin\varphi \tag{5}$$

$$\Rightarrow \qquad \mathrm{d}r = R\cos\varphi\mathrm{d}\varphi \tag{6}$$

$$\Rightarrow \qquad \int_0^{\arcsin\frac{r}{R}} \mathrm{d}\varphi = \int_0^\theta \mathrm{d}\theta$$

$$\Rightarrow \qquad \arcsin\frac{r}{R} = \theta$$

$$\Rightarrow \qquad r = R\sin\theta \tag{7}$$

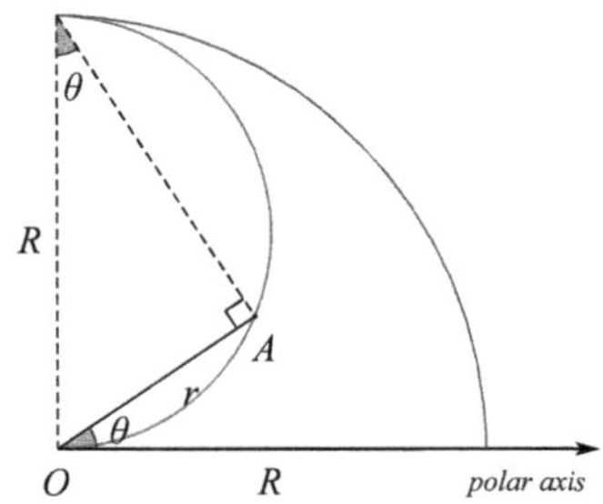

Figure 2

That is, the trajectory of the hound is the circle of diameter R (see Fig. 2), the hound will catch up with the wolf at Point $(R, \pi/2)$.

Another method to get Eq. (3)

Figure 3 is the schematic diagram of the instantaneous process from arbitrary Time t to Time $t + \Delta t$. Considering that $\Delta OA_tA_{t+\Delta t}$ satisfies sine theorem (see Fig. 3):

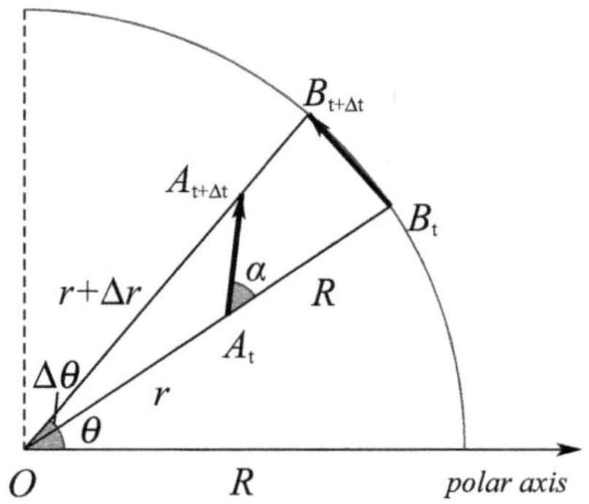

Figure 3

$$\frac{\overline{A_tA_{t+\Delta t}}}{\sin\Delta\theta}=\frac{r}{\sin(\alpha-\Delta\theta)}=\frac{r+\Delta r}{\sin\alpha} \tag{8a, b}$$

$$\overline{B_tB_{t+\Delta t}}\approx\widehat{B_tB_{t+\Delta t}}=R\Delta\theta \tag{9a, b}$$

It can be seen from the fact that the speed of the hound and wolf are equal:

$$\overline{A_tA_{t+\Delta t}}\approx\overline{B_tB_{t+\Delta t}} \tag{10}$$

$$\Rightarrow\qquad \frac{R\Delta\theta}{\Delta\theta}=\frac{r}{\sin(\alpha-\Delta\theta)}=\frac{r+\Delta r}{\sin\alpha} \tag{11a, b}$$

$\Delta\theta$ is a small quantity, taking the following first-order approximations:

$$\begin{cases}\sin\Delta\theta\approx\Delta\theta & \text{(12a)}\\ \sin(\alpha-\Delta\theta)=\sin\alpha\cos\Delta\theta-\cos\alpha\sin\Delta\theta\approx\sin\alpha-\Delta\theta\cos\alpha & \text{(12b)}\end{cases}$$

$(12)\rightarrow(11)$,

$$\Rightarrow\qquad \frac{R\Delta\theta}{\Delta\theta}=\frac{r+\Delta r}{\sin\alpha}=\frac{\Delta r}{\Delta\theta\cos\alpha} \tag{13a, b}$$

The operation above has been simplified by using the partition ratio theorem.

From Eqs. (13a, b) [eliminating α by combining the two equations],

$$\Rightarrow \qquad \frac{\Delta r}{\sqrt{R^2 - r^2}} = \Delta\theta \qquad (14)$$

When Δt is extremely small, Δr and $\Delta\theta$ can be regarded as the differential of r and θ respectively:

$$\Rightarrow \qquad \frac{\mathrm{d}r}{\sqrt{R^2 - r^2}} = \mathrm{d}\theta \qquad (15)$$

Another method to get Eq. (14)

Considering that $\Delta OA_tA_{t+\Delta t}$ satisfies cosine theorem (see Fig. 3):

$$\begin{aligned}\left(\overline{A_tA_{t+\Delta t}}\right)^2 &= r^2 + (r+\Delta r)^2 - 2r(r+\Delta r)\cos\Delta\theta \\ &\approx 2r^2 + 2r\Delta r + (\Delta r)^2 - 2r(r+\Delta r)\left[1 - \frac{1}{2}(\Delta\theta)^2\right] \\ &\approx (\Delta r)^2 + (r\Delta\theta)^2 \qquad (16)\end{aligned}$$

The operation above has considered $\Delta\theta$ as a small quantity and taken the second-order approximation.

(9) and (10) $\rightarrow$ (16),

$$\Rightarrow \qquad \frac{\Delta r}{\sqrt{R^2 - r^2}} = \Delta\theta \qquad (17)$$

1.2.2 N-Regular Polygon Chasing Game (d, *ABC*)

There are N children each standing at one vertex of a N-regular polygon with the circumscribed radius r_0. At some point, each child starts to chase one adjacent child at the equal speed v, the velocity direction is always aim at the chased target. Given that the rotating direction of the line linking any two adjacent children is anticlockwise, thereupon that all children's positions form a N-regular polygon all the time.

(1) Find the trajectory equation and the kinematics equation of each child.

(2) Find the acceleration of each child when the radius of the circumcircle of the N-regular polygon takes an arbitrary value.
(3) Find the running distance of each child in the whole chasing process.

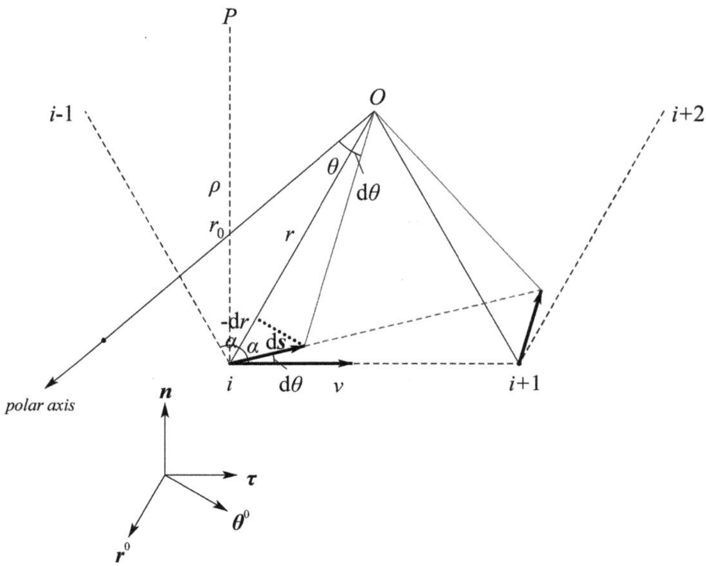

Figure 1

Solution to Problem (1) (type C)

The time point when the N children start to move is defined as Time zero.

Considering the spatial symmetry, it can be seen that the center of the N-regular polygon formed by all children remains stationary, which is denoted by O.

The vertex angle of the N-regular polygon is denoted by 2α. Sorting the children with positive integers anticlockwise (serial number 1 is arbitrarily set). According to the spatial symmetry, the motions of all children are similar. The following studies an arbitrary child.

Taking Point O as the pole, the direction of Point O pointing No. i child at Time zero as the positive direction of the polar axis, anticlockwise direction as the positive direction of polar angle, establishing the polar coordinate system, let r, θ be the polar coordinates, in addition, introducing the natural frame ($\boldsymbol{n}$, $\boldsymbol{\tau}$) of No. i child (see Fig. 1).

Considering ***the components form of velocity in polar coordinate system***:

$$\begin{cases} v_r = -v\cos\alpha = \dfrac{\mathrm{d}r}{\mathrm{d}t} & \text{(1a)} \\ v_\theta = v\sin\alpha = r\dfrac{\mathrm{d}\theta}{\mathrm{d}t} & \text{(1b)} \end{cases}$$

(v_r, v_θ)—the components of the velocity of No. i child in the ($\boldsymbol{r}^0$, $\boldsymbol{\theta}^0$)-directions at arbitrary Time t.

$$\Rightarrow \quad \frac{\mathrm{d}r}{r} = -\cot\alpha\,\mathrm{d}\theta \tag{2}$$

$$\Rightarrow \quad \int_{r_0}^{r}\frac{\mathrm{d}r}{r} = -\cot\alpha\int_0^{\theta}\mathrm{d}\theta$$

$$\Rightarrow \quad \ln\frac{r}{r_0} = -\theta\cot\alpha$$

$$\Rightarrow \quad r = r_0\exp(-\theta\cot\alpha) \tag{3}$$

Equation (3) is the trajectory equation of No. i child in the polar coordinate system.

From (1a),

$$\Rightarrow \quad \mathrm{d}r = -v\cos\alpha\,\mathrm{d}t \tag{4}$$

$$\Rightarrow \quad \int_{r_0}^{r}\mathrm{d}r = -v\cos\alpha\int_0^{t}\mathrm{d}t$$

$$\Rightarrow \quad r = r_0 - vt\cos\alpha \tag{5}$$

(5) → (3),

$$\Rightarrow \quad \theta = -\ln\left(1 - \frac{vt}{r_0}\cos\alpha\right)\tan\alpha \tag{6}$$

Equations (5) and (6) are the kinematics equations of No. i child in the polar coordinate system.

Another method to get Eq. (2) (type C)

Considering the following geometric relation in Fig. 1:

$$-\mathrm{d}r \approx r\mathrm{d}\theta\cdot\cot\alpha \tag{7}$$

$$\Rightarrow \qquad \frac{\mathrm{d}r}{r} = -\cot\alpha \mathrm{d}\theta \tag{8}$$

Figure 2 is the schematic diagram of the all the children's trajectories in the special case of N=6.

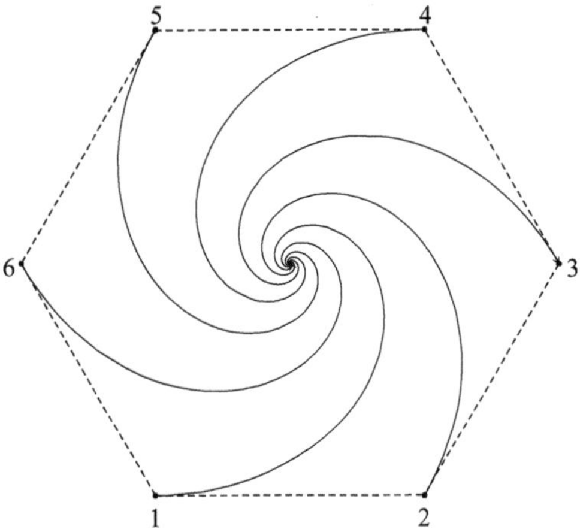

Figure 2

Solution [1] to Problem (2) (type B)

Defining $\boldsymbol{a}$—the acceleration of No. i child at Time t. Considering the definition of ***acceleration*** (expanded in the child's natural frame) [the constraint condition: the child moves at a constant speed]:

$$\boldsymbol{a} = \frac{\mathrm{d}\boldsymbol{v}}{\mathrm{d}t} = \frac{\mathrm{d}v}{\mathrm{d}t}\boldsymbol{\tau} + \frac{v\mathrm{d}\theta}{\mathrm{d}t}\boldsymbol{n} = \frac{v\mathrm{d}\theta}{\mathrm{d}t}\boldsymbol{n} \tag{9}$$

Defining d$\boldsymbol{s}$—the displacement of No. i child in dt time interval at Time t. Considering the following geometric relation in Fig. 1:

$$\Rightarrow \qquad |\mathrm{d}\boldsymbol{s}|\sin\alpha \approx \left(r - |\mathrm{d}\boldsymbol{s}|\cos\alpha\right)\mathrm{d}\theta \approx r\mathrm{d}\theta \tag{10}$$

$$\Rightarrow \qquad \boldsymbol{a} = \frac{v\sin\alpha}{r}\frac{|\mathrm{d}\boldsymbol{s}|}{\mathrm{d}t}\boldsymbol{n} \tag{11}$$

Considering the definition of ***velocity***:

$$v = \frac{|\mathrm{d}\boldsymbol{s}|}{\mathrm{d}t} \tag{12}$$

$$\Rightarrow \qquad \boldsymbol{a} = \frac{v^2 \sin\alpha}{r}\boldsymbol{n} \tag{13}$$

Solution [2] to Problem (2) (type A)

Considering ***the components form of acceleration in polar coordinate system***:

$$\begin{cases} a_r = \dfrac{\mathrm{d}^2 r}{\mathrm{d}t^2} - r\left(\dfrac{\mathrm{d}\theta}{\mathrm{d}t}\right)^2 & \text{(14a)} \\ a_\theta = r\dfrac{\mathrm{d}^2\theta}{\mathrm{d}t^2} + 2\dfrac{\mathrm{d}r}{\mathrm{d}t}\dfrac{\mathrm{d}\theta}{\mathrm{d}t} & \text{(14b)} \end{cases}$$

(a_r, a_θ)—the components of the acceleration of No. i child in the $(\boldsymbol{r}^0, \boldsymbol{\theta}^0)$-directions at arbitrary Time t.

(1), (5) and (6) $\rightarrow$ (14):

$$\Rightarrow \begin{cases} a_r = -\dfrac{(v\sin\alpha)^2}{r} & \text{(15a)} \\ a_\theta = r\dfrac{\mathrm{d}}{\mathrm{d}t}\left(\dfrac{v\sin\alpha}{r}\right) + 2(-v\cos\alpha)\dfrac{v\sin\alpha}{r} \\ \quad = r\left(-\dfrac{v\sin\alpha}{r^2}\dfrac{\mathrm{d}r}{\mathrm{d}t} + \dfrac{\sin\alpha}{r}\dfrac{\mathrm{d}v}{\mathrm{d}t}\right) - \dfrac{v^2\sin 2\alpha}{r} \\ \quad = r\left[-\dfrac{v\sin\alpha}{r^2}(-v\cos\alpha)\right] - \dfrac{v^2\sin 2\alpha}{r} \\ \quad = \dfrac{v^2\sin 2\alpha}{2r} - \dfrac{v^2\sin 2\alpha}{r} = -\dfrac{v^2\sin 2\alpha}{2r} & \text{(15b)} \end{cases}$$

Considering the orthogonality of the base $(\boldsymbol{r}^0, \boldsymbol{\theta}^0)$:

$$\Rightarrow \begin{cases} a_\tau = -a_r\cos\alpha + a_\theta\sin\alpha = 0 & \text{(16a)} \\ a_n = -a_r\sin\alpha - a_\theta\cos\alpha = \dfrac{v^2\sin\alpha}{r} & \text{(16b)} \end{cases}$$

(a_τ, a_n)—the components of the acceleration of No. i child in nature coordinate system at arbitrary Time t.

Solution [3] to Problem (2) (type C)

Considering ***the components form of acceleration in nature coordinate system*** [the child moves at a constant speed]:

$$\begin{cases} a_\tau = 0 & (17a) \\ a_n = \dfrac{v^2}{\rho} & (17b) \end{cases}$$

ρ—the radius of curvature of the trajectory of No. i child at the position where the child is at arbitrary Time t.

Considering the definition of ***radius of curvature***:

$$\rho = \frac{|\mathrm{d}s|}{\mathrm{d}\theta} \tag{18}$$

(10) → (18),

$$\Rightarrow \qquad \rho = \frac{r}{\sin\alpha} \tag{19}$$

$$\Rightarrow \qquad a_n = \frac{v^2 \sin\alpha}{r} \tag{20}$$

Another method to get Eq. (19) (type A)

Considering the form of the radius of curvature at an arbitrary point on the plane curve in polar coordinate system:

$$\rho = \frac{\left[r^2 + \left(\dfrac{\mathrm{d}r}{\mathrm{d}\theta}\right)^2\right]^{3/2}}{\left|r^2 + 2\left(\dfrac{\mathrm{d}r}{\mathrm{d}\theta}\right)^2 - r\dfrac{\mathrm{d}^2 r}{\mathrm{d}\theta^2}\right|} \tag{21}$$

(3) → (21),

$$\Rightarrow \qquad \rho = r_0 \exp(-\theta\cot\alpha)(1+\cot^2\alpha)^{1/2} = \frac{r}{\sin\alpha} \tag{22}$$

Solution to Problem (3) (type C)

According to the spatial symmetry, all children will eventually meet at Point O.

Considering ***the arc length formula of a curved element in polar coordinate system***:

$$|\mathrm{d}s| = \sqrt{(\mathrm{d}r)^2 + (r\mathrm{d}\theta)^2} \tag{23}$$

(3) → (23),

$$\Rightarrow \quad |d\boldsymbol{s}| = \sqrt{r_0^2 \cot^2\alpha \exp^2(-\theta\cot\alpha) + r_0^2 \exp^2(-\theta\cot\alpha)}\,d\theta$$

$$= r_0 \exp(-\theta\cot\alpha)\sqrt{\cot^2\alpha + 1}\,d\theta$$

$$= r_0 \frac{\exp(-\theta\cot\alpha)}{\sin\alpha} d\theta \tag{24}$$

From (3),

$$\Rightarrow \quad \theta\big|_{r\to 0} \to \infty \tag{25}$$

Defining s—the running distance of No. i child in the whole process of chasing No. $(i + 1)$ child.

$$\Rightarrow \quad s = \int |d\boldsymbol{s}| = \frac{r_0}{\sin\alpha}\int_0^\infty \exp(-\theta\cot\alpha)d\theta$$

$$= -\frac{r_0}{\cos\alpha}\exp(-\theta\cot\alpha)\Big|_0^\infty = \frac{r_0}{\cos\alpha} \tag{26}$$

1.2.3 Orientational and Constant Speed Chasing (d, *C*)

A Cartesian coordinate system with the frame $(O, \boldsymbol{i}, \boldsymbol{j})$ is established in a vertical plane (see Fig. 1). An aircraft leaves from the origin O and flies at the constant velocity $(0, v_1)$ along the y-axis. When the aircraft just leaves the origin O, a tracker is launched from Point $(x_0, 0)$ to track the aircraft, the tracking strategy is as follows: the velocity magnitude of the tracker is always v_2 $(v_2 > v_1)$, and the direction always points to the aircraft.

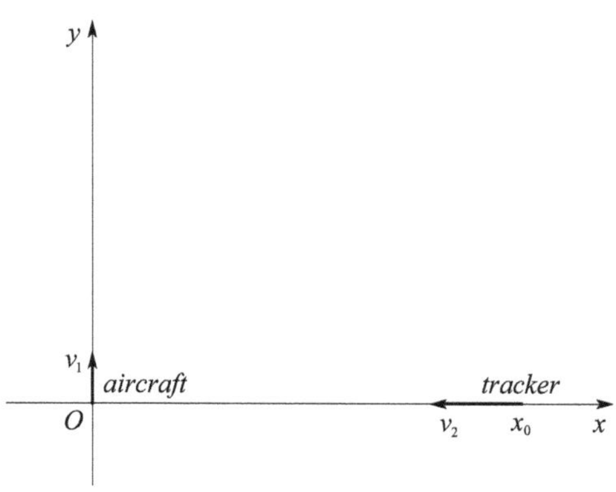

Figure 1

(1) Find the trajectory equation of the tracker.
(2) Find the time taken for the tracker to catch up with the aircraft.
(3) Find the acceleration of the tracker at arbitrary time.

Solution to Problem (1)

The time point when the aircraft leaves from the origin O is defined as Time zero.

As shown in Fig. 2, the positions of the aircraft and the tracker at arbitrary Time t are denoted by A and B respectively (for simplicity, in the following, the corner mark of the coordinates of Point B will be omitted). Establishing the polar coordinate system fixed to the tracker, let r, θ be the polar coordinates.

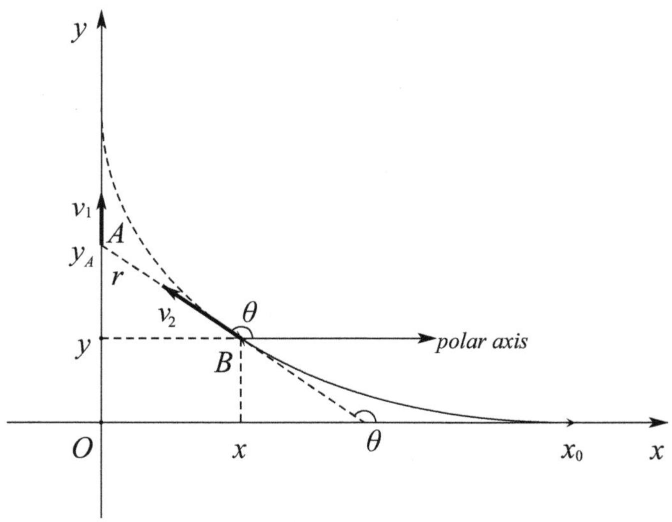

Figure 2

Considering the definition of ***velocity***:

$$\begin{cases} v_1 = \dfrac{\mathrm{d}s_1}{\mathrm{d}t} = \dfrac{\mathrm{d}y_A}{\mathrm{d}t} & \text{(1a)} \\ v_2 = \dfrac{\mathrm{d}s_2}{\mathrm{d}t} & \text{(1b)} \end{cases}$$

$\mathrm{d}s_1$ and $\mathrm{d}s_2$—the displacement magnitude of the aircraft and the tracker in $\mathrm{d}t$ time interval at arbitrary Time t.

Considering ***the arc length formula of the curved element in Cartesian coordinate system***:

$$ds_2 = \sqrt{1+\left(\frac{dy}{dx}\right)^2}\,dx \tag{2}$$

$\Rightarrow$
$$dy_A = \frac{v_1}{v_2}\sqrt{1+\left(\frac{dy}{dx}\right)^2}\,dx \tag{3}$$

Considering the following geometric relation in Fig. 2:

$$\tan\theta = -\frac{y_A - y}{x} \tag{4}$$

Considering ***the geometric meaning of derivative***:

$$\frac{dy}{dx} = \tan\theta \tag{5}$$

$\Rightarrow$
$$y_A = y - x\frac{dy}{dx} \tag{6}$$

$\Rightarrow$
$$dy_A = dy - \frac{dy}{dx}dx - \frac{d^2y}{dx^2}xdx = -\frac{d^2y}{dx^2}xdx \tag{7}$$

Combining (3) and (7):

$\Rightarrow$
$$\frac{v_1}{v_2}\sqrt{1+\left(\frac{dy}{dx}\right)^2} = -\frac{d^2y}{dx^2}x \tag{8}$$

Defining the integration variable u and the constant α:

$$u = \frac{dy}{dx} \tag{9a}$$

$$\alpha = \frac{v_1}{v_2} \tag{9b}$$

$\Rightarrow$
$$\alpha\sqrt{1+u^2} = -\frac{du}{dx}x \tag{10}$$

$\Rightarrow$
$$\frac{du}{\sqrt{1+u^2}} = -\alpha\frac{dx}{x}$$

$\Rightarrow$
$$\int_0^u \frac{du}{\sqrt{1+u^2}} = -\alpha\int_{x_0}^x \frac{dx}{x}$$

$$\Rightarrow \qquad \ln\left(u+\sqrt{1+u^2}\right)=-\alpha\ln\frac{x}{x_0}$$

$$\Rightarrow \qquad \operatorname{arcsinh} u=-\alpha\ln\frac{x}{x_0}=\ln\left(\frac{x_0}{x}\right)^{\alpha}$$

$$\Rightarrow u=-\sinh\left[\ln\left(\frac{x_0}{x}\right)^{\alpha}\right]=\frac{1}{2}\left\{\exp\left[-\ln\left(\frac{x_0}{x}\right)^{\alpha}\right]-\exp\left[\ln\left(\frac{x_0}{x}\right)^{\alpha}\right]\right\}$$

$$=\frac{1}{2}\left[\left(\frac{x}{x_0}\right)^{\alpha}-\left(\frac{x}{x_0}\right)^{-\alpha}\right] \tag{11}$$

$$\Rightarrow \qquad dy=\frac{1}{2}\left[\left(\frac{x}{x_0}\right)^{\alpha}-\left(\frac{x}{x_0}\right)^{-\alpha}\right]dx \tag{12}$$

$$\Rightarrow \qquad y=\int_0^y dy=\frac{1}{2}\int_{x_0}^{x}\left[\left(\frac{x}{x_0}\right)^{\alpha}-\left(\frac{x}{x_0}\right)^{-\alpha}\right]dx$$

$$=\frac{1}{2}x_0\int_{x_0}^{x}\left[\left(\frac{x}{x_0}\right)^{\alpha}-\left(\frac{x}{x_0}\right)^{-\alpha}\right]d\left(\frac{x}{x_0}\right)$$

$$=\frac{1}{2}x_0\left\{\frac{1}{1+\alpha}\left[\left(\frac{x}{x_0}\right)^{1+\alpha}-1\right]-\frac{1}{1-\alpha}\left[\left(\frac{x}{x_0}\right)^{1-\alpha}-1\right]\right\} \tag{13}$$

Equation (13) is the trajectory equation of the tracker.

Solution [1] to Problem (2)

From (13),

$$\Rightarrow \qquad y\big|_{x=0}=x_0\frac{\alpha}{1-\alpha^2} \tag{14}$$

Defining T—the time taken for the tracker to catch up with the aircraft. Considering ***the displacement-time relation of uniform motion***:

$$\Rightarrow \qquad T=\frac{y\big|_{x=0}}{v_1}=\frac{x_0}{v_1}\frac{\alpha}{1-\alpha^2}=x_0\frac{v_2}{v_2^2-v_1^2} \tag{15}$$

Solution [2] to Problem (2)

For the motion of the aircraft relative to the tracker, considering ***the radial component form of velocity***:

$$v_1 \sin\theta - v_2 = \frac{\mathrm{d}r}{\mathrm{d}t} \tag{16}$$

$$\Rightarrow \quad \mathrm{d}r = (v_1 \sin\theta - v_2)\mathrm{d}t$$

$$\Rightarrow \quad \int_{x_0}^{0} \mathrm{d}r = \int_0^T (v_1 \sin\theta - v_2)\mathrm{d}t$$

$$\Rightarrow \quad -x_0 = v_1 \int_0^T \sin\theta \mathrm{d}t - v_2 T \tag{17}$$

Defining the variable y':

$$y' = y_A - y \tag{18}$$

Considering the definition of ***velocity***:

$$v_1 - v_2 \sin\theta = \frac{\mathrm{d}y'}{\mathrm{d}t} \tag{19}$$

$$\Rightarrow \quad \mathrm{d}y' = (v_1 - v_2 \sin\theta)\mathrm{d}t$$

$$\Rightarrow \quad \int_0^0 \mathrm{d}y' = \int_0^T (v_1 - v_2 \sin\theta)\mathrm{d}t$$

$$\Rightarrow \quad v_1 T - v_2 \int_0^T \sin\theta \mathrm{d}t = 0 \tag{20}$$

Combining (17) and (20),

$$\Rightarrow \quad T = \frac{v_2}{v_2^2 - v_1^2} x_0 \tag{21}$$

Solution to Problem (3)

Introducing the natural frame ($\boldsymbol{\tau}, \boldsymbol{n}$) of the tracker (see Fig. 3).

Considering ***the normal component form of acceleration***:

$$a_n = \frac{v_2^2}{\rho} \tag{22}$$

a_n—the $\boldsymbol{n}$-direction component of the acceleration of the tracker at arbitrary Time t.

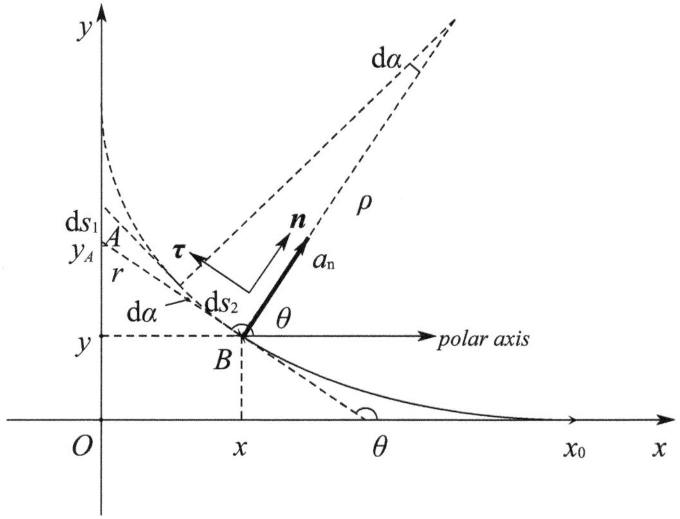

Figure 3

ρ—the radius of curvature of the tracker's trajectory at the position where the tracker is at arbitrary Time t.

Considering the following geometric relation in Fig. 3:

$$d\alpha = \frac{ds_2}{\rho} = \frac{ds_1 \cdot \sin\theta}{r} \tag{23}$$

(1) → (23):

⇒
$$\rho = r\frac{v_2}{v_1 \sin\theta} \tag{24}$$

(24) → (22):

⇒
$$a_n = \frac{v_1 v_2}{r}\sin\theta \tag{25}$$

1.2.4 Orientational and Fixed-Distance Chasing (*a*, *A*)

In the previous problem Section 1.2.3, "Orientational and constant speed chasing", The constant velocity of the aircraft is re-denoted as $(0, v_0)$, the chasing strategy is changed as follows: at arbitrary time, the velocity direction of the tracker points to the aircraft, and the distance between the tracker and the aircraft remains constant.

(1) Find the kinematics equation and trajectory equation of the tracker.

(2) Find the velocity and acceleration of the tracker at arbitrary time.

Solution to Problem (1)

The definition of Time zero and the establishment of the coordinate systems are the same as the previous problem in Section 1.2.3 (see Fig. 1).

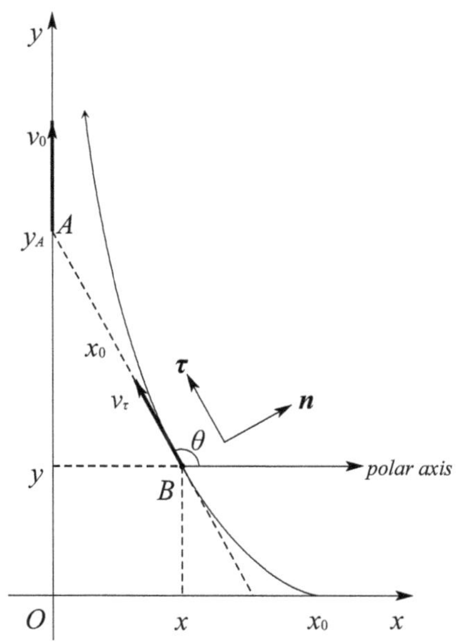

Figure 1

Considering the following geometric relations in Fig. 1:

$$\begin{cases} x = -x_0 \cos\theta & \text{(1a)} \\ y = y_A - x_0 \sin\theta & \text{(1b)} \end{cases}$$

Considering ***the displacement-time relation of uniform linear motion***:

$$y_A = v_0 t \tag{2}$$

For the motion of the aircraft relative to the tracker, considering ***the transverse component form of velocity*** [the constraint condition: the velocity direction of the tracker points to the aircraft all the time]:

$$v_0 \cos\theta = x_0 \frac{\mathrm{d}\theta}{\mathrm{d}t} \tag{3}$$

$$\Rightarrow \qquad \sec\theta \mathrm{d}\theta = -\frac{v_0}{x_0}\mathrm{d}t$$

$$\Rightarrow \qquad \int_{\pi}^{\theta} \sec\theta \mathrm{d}\theta = -\frac{v_0}{x_0}\int_0^t \mathrm{d}t$$

$$\Rightarrow \qquad \ln|\sec\theta + \tan\theta| = -\frac{v_0 t}{x_0}$$

$$\Rightarrow \qquad \operatorname{arccosh}(\sec\theta) = -\frac{v_0 t}{x_0}$$

$$\Rightarrow \qquad \sec\theta = \cosh\left(-\frac{v_0 t}{x_0}\right) = \cosh\left(\frac{v_0 t}{x_0}\right) \tag{4}$$

(2) and (4) $\rightarrow$ (1), we obtain the kinematics equations and the trajectory equation of the tracker:

$$\Rightarrow \qquad \begin{cases} x = -x_0 \operatorname{sech}\left(\dfrac{v_0 t}{x_0}\right) & \text{(5a)} \\ y = v_0 t - x_0 \tanh\left(\dfrac{v_0 t}{x_0}\right) & \text{(5b)} \end{cases}$$

$$\Rightarrow \qquad y = x_0\left[\operatorname{arcsech}\left(\frac{x}{x_0}\right) - \tanh\left[\operatorname{arcsech}\left(\frac{x}{x_0}\right)\right]\right]$$

$$= x_0\left[\ln\left(\frac{x_0}{x} + \sqrt{\left(\frac{x_0}{x}\right)^2 - 1}\right) - \sqrt{1 - \left(\frac{x}{x_0}\right)^2}\right] \tag{6}$$

Another solution of the trajectory equation of the tracker: Considering the following geometric relations in Fig. 1:

$$x^2 + (y_A - y)^2 = x_0^2 \tag{7}$$

$$\tan\theta = -\frac{y_A - y}{x} \tag{8}$$

Considering the geometric meaning of derivative:

$$\frac{\mathrm{d}y}{\mathrm{d}x} = \tan\theta \tag{9}$$

Combining (7), (8) and (9),

$$\Rightarrow \qquad \mathrm{d}y = -\sqrt{\left(\frac{x_0}{x}\right)^2 - 1}\,\mathrm{d}x \tag{10}$$

$$\Rightarrow \quad y = \int_0^y \mathrm{d}y = -\int_{x_0}^{x_B} \sqrt{\left(\frac{x_0}{x}\right)^2 - 1}\,\mathrm{d}x = x_0 \int_{\pi}^{\pi - \arccos\frac{x}{x_0}} (\sec\theta - \cos\theta)\mathrm{d}\theta$$

$$= x_0 \left(\ln\left|\sec\theta + \tan\theta\right| - \sin\theta\right)\Big|_{\pi}^{\pi - \arccos\frac{x}{x_0}}$$

$$= x_0 \left[\ln\left(\frac{x_0}{x} + \sqrt{\left(\frac{x_0}{x}\right)^2 - 1}\right) - \sqrt{1 - \left(\frac{x}{x_0}\right)^2}\right]$$

$$= x_0 \left(\operatorname{arccosh}\frac{x_0}{x} - \sqrt{1 - \left(\frac{x}{x_0}\right)^2}\right) \tag{11}$$

Solution [1] to Problem (2)

For the motion of the aircraft relative to the tracker, considering ***the radial component form of velocity*** [the constraint condition: the distance between the tracker and the aircraft remains constant]:

$$v_0 \sin\theta - v_\tau = 0 \tag{12}$$

v_τ—the $\boldsymbol{\tau}$-direction component of the velocity of the tracker at arbitrary Time t.

Combining (4) and (12),

$$\Rightarrow \qquad v_\tau = v_0 \tanh\left(\frac{v_0 t}{x_0}\right) \tag{13}$$

For the motion of the aircraft relative to the tracker, considering ***components form of the acceleration in polar coordinate system*** [the constraint condition: the distance between the tracker and the aircraft remains constant]:

$$\begin{cases} 0 - a_\tau = -x_0 \left(\dfrac{\mathrm{d}\theta}{\mathrm{d}t}\right)^2 & (14a) \\ 0 - (-a_n) = x_0 \dfrac{\mathrm{d}^2\theta}{\mathrm{d}t^2} & (14b) \end{cases}$$

(a_τ, a_n)—the components of the acceleration of the tracker in the natural coordinate system at arbitrary Time t.

From (4),

$$\Rightarrow \qquad \sec^2\theta \mathrm{d}\theta = -\frac{v_0}{x_0}\cosh\left(\frac{v_0 t}{x_0}\right)\mathrm{d}t \tag{15}$$

$$\Rightarrow \qquad \frac{\mathrm{d}\theta}{\mathrm{d}t} = -\frac{v_0}{x_0}\frac{\cosh\left(\frac{v_0 t}{x_0}\right)}{1+\sinh^2\left(\frac{v_0 t}{x_0}\right)} = -\frac{v_0}{x_0}\frac{1}{\cosh\left(\frac{v_0 t}{x_0}\right)}$$

$$= -\frac{v_0}{x_0}\operatorname{sech}\left(\frac{v_0 t}{x_0}\right) \tag{16}$$

$$\Rightarrow \qquad \frac{\mathrm{d}^2\theta}{\mathrm{d}t^2} = \left(\frac{v_0}{x_0}\right)^2 \operatorname{sech}\left(\frac{v_0 t}{x_0}\right)\tanh\left(\frac{v_0 t}{x_0}\right) \tag{17}$$

$$\Rightarrow \qquad \begin{cases} a_\tau = \dfrac{v_0^2}{x_0}\operatorname{sech}^2\left(\dfrac{v_0 t}{x_0}\right) & (18a) \\ a_n = \dfrac{v_0^2}{x_0}\operatorname{sech}\left(\dfrac{v_0 t}{x_0}\right)\tanh\left(\dfrac{v_0 t}{x_0}\right) & (18b) \end{cases}$$

Solution [2] to Problem (2)

Considering ***the normal component form of acceleration***:

$$a_n = \frac{v_\tau^2}{\rho} \tag{19}$$

ρ—the radius of curvature of the tracker's trajectory at the position where the tracker is at arbitrary Time t.

Considering the form of the radius of curvature at an arbitrary point on the plane curve in Cartesian coordinate system:

$$\rho = \frac{\left[1+\left(\frac{\mathrm{d}y}{\mathrm{d}x}\right)^2\right]^{3/2}}{\left|\frac{\mathrm{d}^2 y}{\mathrm{d}x^2}\right|} \tag{20}$$

From (6),

$$\Rightarrow \qquad \frac{\mathrm{d}y}{\mathrm{d}x} = -\sinh\left(\frac{v_0 t}{x_0}\right) \tag{21}$$

$$\Rightarrow \qquad \frac{\mathrm{d}^2 y}{\mathrm{d}x^2} = -\frac{v_0}{x_0}\cosh\left(\frac{v_0 t}{x_0}\right)\frac{\mathrm{d}t}{\mathrm{d}x} = \frac{\cosh^2\left(\frac{v_0 t}{x_0}\right)}{\tanh\left(\frac{v_0 t}{x_0}\right)x_0} \tag{22}$$

$$\Rightarrow \qquad \rho = x_0 \sinh\left(\frac{v_0 t}{x_0}\right) = -x_0 \tan\theta \tag{23}$$

(13) and (23) $\rightarrow$ (19),

$$\Rightarrow \qquad a_n = \frac{v_0^2}{x_0}\operatorname{sech}\left(\frac{v_0 t}{x_0}\right)\tanh\left(\frac{v_0 t}{x_0}\right) \tag{24}$$

Another solution to ρ

Considering the form of the radius of curvature at an arbitrary point on the plane curve by parameter equations:

$$\rho = \frac{\left[\left(\frac{\mathrm{d}x}{\mathrm{d}\theta}\right)^2 + \left(\frac{\mathrm{d}y}{\mathrm{d}\theta}\right)^2\right]^{3/2}}{\left|\frac{\mathrm{d}x}{\mathrm{d}\theta}\frac{\mathrm{d}^2 y}{\mathrm{d}\theta^2} - \frac{\mathrm{d}^2 x}{\mathrm{d}\theta^2}\frac{\mathrm{d}y}{\mathrm{d}\theta}\right|} \tag{25}$$

(4) $\rightarrow$ (5),

$$\begin{cases} x = -x_0\cos\theta & \text{(26a)} \\ y = x_0\left[\ln[-(\sec\theta + \tan\theta)] - \sin\theta\right] & \text{(26b)} \end{cases}$$

$$\Rightarrow \begin{cases} \dfrac{dx}{d\theta} = x_0 \sin\theta & (27a) \\ \dfrac{dy}{d\theta} = x_0 \dfrac{\sec\theta\tan\theta + \tan^2\theta - \sin\theta}{\sec\theta + \tan\theta} = x_0 \tan\theta \sin\theta & (27b) \end{cases}$$

$$\Rightarrow \begin{cases} \dfrac{d^2x}{d\theta^2} = x_0 \cos\theta & (28a) \\ \dfrac{d^2y}{d\theta^2} = x_0 \tan\theta(\sec\theta + \cos\theta) & (28b) \end{cases}$$

$$\Rightarrow \qquad \rho = -x_0 \tan\theta \qquad (29)$$

1.2.5 The Motion of a Ship across a Flowing River to the Target Wharf on Opposite Bank (d, *C*)

As shown in Fig. 1, there is a river with two parallel banks, flowing at a constant velocity $\boldsymbol{v}_0$. A small boat starts from Point *A* on the bank and is ready to reach the target wharf (denoted by Point *O*) on the opposite bank. The speed of the boat relative to the river water is v' all the time, and the direction of the boat velocity points to Point *O* all the time. Given $\overline{AO} = r_0$, $\angle AOP = \varphi_0$. Find the trajectory equation of the boat.

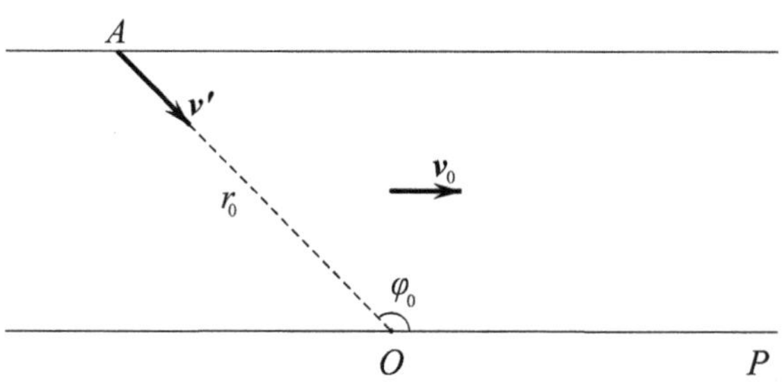

Figure 1

Solution

The departure time of the boat is defined as Time zero.

As shown in Fig. 2, establishing the polar coordinate system, let r, φ be the polar coordinates.

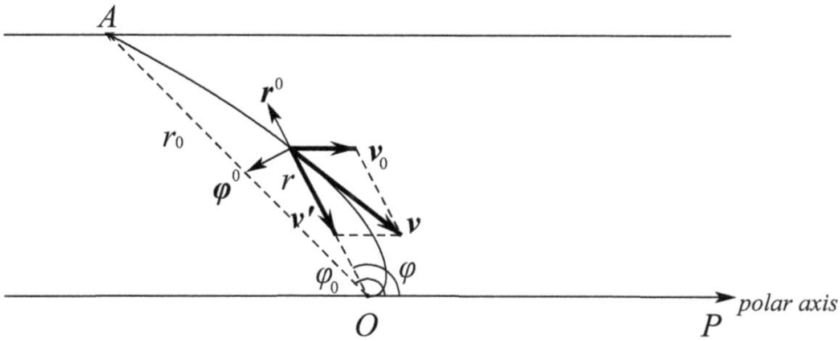

Figure 2

Taking *the boat (can be regarded as a particle)* as the object of study. Studying *the state at arbitrary Time t*. Considering ***theorem for composition of velocities of a particle*** [taking the ground as the stationary reference frame, and the river as the moving reference frame] (the $\boldsymbol{r}^0$-direction and φ^0-direction component equations):

$$\begin{cases} v_r = v_0 \cos\varphi - v' & (1a) \\ v_\varphi = -v_0 \sin\varphi & (1b) \end{cases}$$

(v_r, v_φ)—the components of the boat velocity in the $(\boldsymbol{r}^0, \varphi^0)$-directions at Time t.

Considering ***the component of the velocity of a particle in polar coordinate system***:

$$\begin{cases} v_r = \dot{r} & (2a) \\ v_\varphi = r\dot{\varphi} & (2b) \end{cases}$$

$$\Rightarrow \qquad \frac{v_0 \cos\varphi - v'}{-v_0 \sin\varphi} = \frac{\dot{r}}{r\dot{\varphi}} = \frac{\dfrac{\mathrm{d}r}{\mathrm{d}t}}{r\dfrac{\mathrm{d}\varphi}{\mathrm{d}t}} = \frac{\mathrm{d}r}{r\mathrm{d}\varphi} \qquad (3)$$

$$\Rightarrow \qquad \frac{\mathrm{d}r}{r} = \left(\frac{v'}{v_0}\csc\varphi - \cot\varphi\right)\mathrm{d}\varphi \qquad (4)$$

$$\Rightarrow \qquad \int_{r_0}^{r} \frac{\mathrm{d}r}{r} = \int_{\varphi_0}^{\varphi} \left(\frac{v'}{v_0}\csc\varphi - \cot\varphi\right)\mathrm{d}\varphi$$

$$\Rightarrow \qquad \ln\frac{r}{r_0} = \ln\left[\left(\frac{\tan\frac{\varphi}{2}}{\tan\frac{\varphi_\varphi}{2}}\right)^{\frac{v'}{v_0}} \frac{\sin\varphi_0}{\sin\varphi}\right]$$

$$\Rightarrow \qquad r = r_0\left(\frac{\tan\frac{\varphi}{2}}{\tan\frac{\varphi_0}{2}}\right)^{\frac{v'}{v_0}} \frac{\sin\varphi_0}{\sin\varphi} \tag{5}$$

1.3 The Most Time-Saving Path Problem

The most time-saving path problem, which is to solve the most time-saving path of a particle from the starting point to the end point under the specific constraints of the kinematics quantities, is a valuable kinematics problem. This kind of problems can be regarded as a special case of optimal control problem in the field of kinematics. The brachistochrone problem is the most classical and influential model. The variational method is a general method to deal with the most time-saving path problem. For some simple special cases, some more elementary and effective mathematical tools can be preferred, but they are not without profundity. The exploration on fruitful research methods for this kind of problems is interesting and inspiring.

1.3.1 The Most Time-Saving Ship Chasing Path (d, *C*)

As shown in Fig. 1, there is a straight lakeshore *MN*. A ship starts from the Point *A* on the lakeshore and travels at a constant velocity in the lake in the direction of angle $\boldsymbol{\alpha}$ (an acute angle) with $\overrightarrow{MN}$. A person starts from Point *A* at the same time as the ship, first walks along the lakeshore with the speed v_1 and then enters the water to swim with the speed v_2 to chase the ship, $v_1 > v_2$. To ensure that this person can catch up with the ship, what is the maximum speed of the ship?

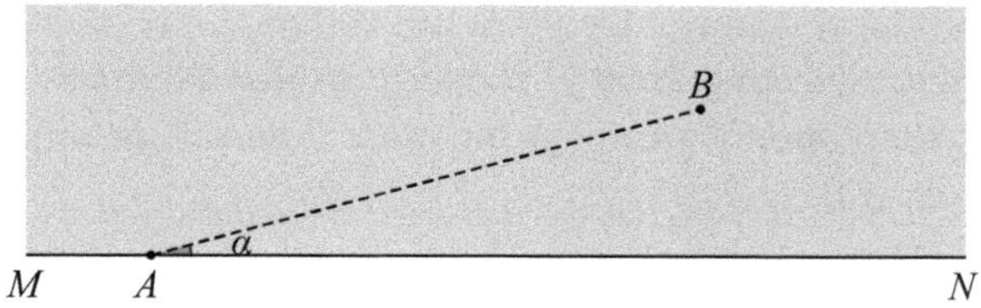

Figure 1

Solution [1]

The position where the person meets the ship is denoted by B. There are infinite paths for the person to chase the ship, such as $A \to C \to B$, $A \to D \to B$, $A \to E \to B$, etc. (see Fig. 2), where C, D, E, etc. are all possible water entry points, and the one with the shortest time taken in these paths corresponds to the maximum allowable ship speed.

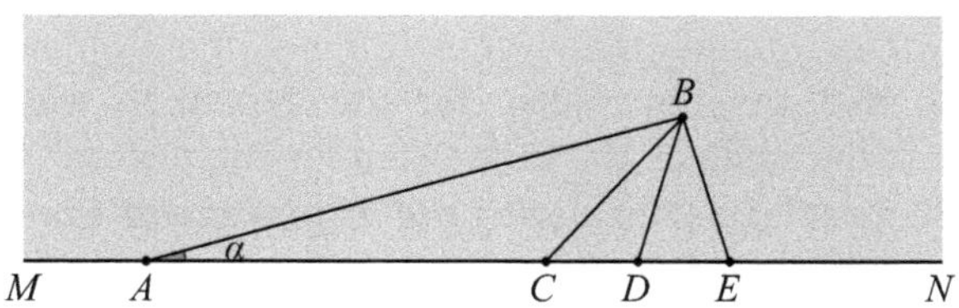

Figure 2

In this chapter, an isochronous ship-chasing scheme is conceived, in which the actual motion process of the person is equivalent to the motion process of swimming from start to finish in the same time period: Supposing that the land area also turns into a lake water area. Making a ray AP in this area, and draw perpendiculars CC', DD', EE' etc. from the actual water entry Points C, D, E etc. to AP. Taking Points C', D', E' etc. as the hypothetical starting point, and the corresponding hypothetical ship chasing paths are $C \to C' \to B$, $D \to D' \to B$, $E \to E' \to B$, etc. (see Fig. 3), whose time taken are equal to the corresponding actual ship chasing paths respectively (that is, the time taken for the person to walk on the lakeshore is the same as the time taken for the person to swim in the hypothetical lake). The shortest of these hypothetical paths takes the shortest time, obviously, when Point B, the actual water entry point and the hypothetical water entry point are collinear, the path length is the shortest. Selecting Point D properly so that Points B, D, D' are collinear, thereupon we can get the most time-saving ship chasing path $A \rightarrow D \rightarrow B$.

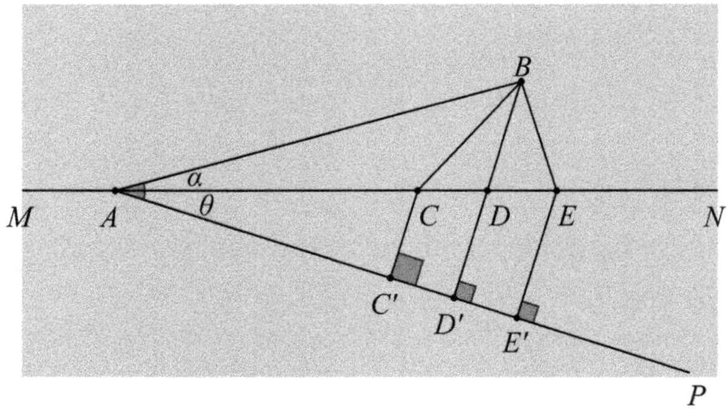

Figure 3

Known that the time taken for the person to walk on the lakeshore is the same as the time taken for the person to swim in the hypothetical lake. Considering ***the displacement-time relation of uniform linear motion***:

$$t_{\text{lakeshore}} = \frac{AD}{v_1} = \frac{D'D}{v_2} \tag{1}$$

$t_{\text{lakeshore}}$—the time taken for the person to walk from Point A to Point D along the lakeshore.

Considering that $\Delta AD'D$ satisfies sine theorem (see Fig. 3):

$$\sin\theta = \frac{D'D}{AD} \tag{2}$$

θ—the angle between $\overrightarrow{AP}$ and $\overrightarrow{MN}$.

$\Rightarrow$

$$\sin\theta = \frac{v_2}{v_1} \tag{3}$$

Known that the motion time of the ship and the person is equal in the whole process. Considering ***the displacement-time relation of uniform linear motion***:

$$t_{\min} = \frac{AB}{u_{\max}} = \frac{D'B}{v_2} \tag{4}$$

$t_{\min}$—the shortest time it takes for the person to catch up with the ship.

$u_{\max}$—the maximum allowable ship speed for the person to catch up with the ship successfully.

Considering that $\Delta AD'B$ satisfies sine theorem (see Fig. 3):

$$\sin(\alpha+\theta)=\frac{D'B}{AB} \tag{5}$$

Combining (3), (4) and (5),

$$\Rightarrow \qquad u_{\max}=\frac{v_2}{\sin(\alpha+\theta)}=\frac{v_2}{\sin\alpha\cos\theta+\cos\alpha\sin\theta}$$

$$=\frac{v_1v_2}{\sqrt{v_1^2-v_2^2}\sin\alpha+v_2\cos\alpha} \tag{6}$$

Solution [2]

As shown in Fig. 4, the position where the person meets the ship is denoted by *B*, in all the optional ship chasing paths, the water entry point corresponding to the most time-saving path is denoted by *D*, the time taken for the person to catch up with the ship when entering water at Point *D* shall be the minimum value of that when entering water at all points in the neighborhood of Point *D*. That is, for an arbitrary point (denoted by *C*) on the left side of Point *D*, there must be a point (denoted by *E*) on the right side corresponding to it, so that the time taken to catch up with the ship is equal when entering water at Point *C* and Point *E*.

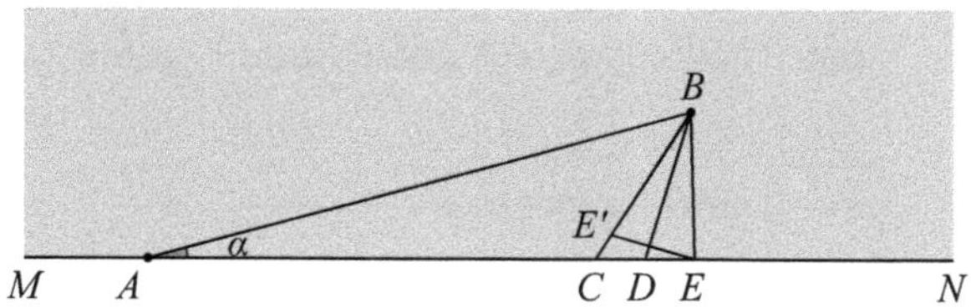

Figure 4

Taking a point on *BC* (denoted by *E'*), let *BE'*=*BE*, the walking time of the person from Point *C* to Point *E* along the lakeshore is equal to the swimming time of the person from Point *C* to Point *E'* in the lake. When Point *C* approaches Point *D* infinitely, Point *E* must approach Point D at the same time. From the geometric relation in Fig. 4, it can be seen that $\angle BCN$, $\angle BDN$ and $\angle BEN$ will tend to be equal, and $\angle CE'E$ will approximate a right angle.

Considering ***the displacement-time relation of uniform linear motion***:

$$\Delta t = \frac{CE'}{v_2} = \frac{CE}{v_1} \tag{7}$$

Δt—the time taken for the person to swim from Point C to Point E' in the lake.

Considering that $\Delta CEE'$ and ΔADB satisfy sine theorem (see Fig. 4), then place (7) in to obtain:

$$\cos\angle BDE \approx \cos\angle BCE = \frac{CE'}{CE} = \frac{v_2}{v_1} \tag{8}$$

$$\frac{AB}{\sin\angle BDE} = \frac{DB}{\sin\alpha} = \frac{AD}{\sin(\angle BDE - \alpha)} \tag{9}$$

Known that the motion time of the ship and the person is equal in the whole process. Considering ***the displacement-time relation of uniform linear motion***:

$$t_{\min} = \frac{AB}{u_{\max}} = \frac{AD}{v_1} + \frac{DB}{v_2} \tag{10}$$

Combining (8), (9) and (10),

$$\Rightarrow \quad u_{\max} = \frac{v_1 v_2 \sin\angle BDE}{v_2 \sin(\angle BDE - \alpha) + v_1 \sin\alpha}$$

$$= \frac{v_1 v_2 \sin\angle BDE}{v_2(\sin\angle BDE \cos\alpha - \cos\angle BDE \sin\alpha) + v_1 \sin\alpha}$$

$$= \frac{v_1 v_2}{\sqrt{v_1^2 - v_2^2}\,\sin\alpha + v_2 \cos\alpha} \tag{11}$$

Solution [3]

As shown in Fig. 5, in a given time, the lake area that the person can reach with an arbitrary possible water entry point is a semicircle with the water entry point as the center; therefore in the given time, the envelope of the semicircular lake areas corresponding to all possible water entry points is the boundary of the lake area that the person can reach with all possible water entry points (hereinafter referred to as motion boundary). The envelope is composed of two line segments and an arc (denoted by *FG*, $\overset{\frown}{GH}$, *HI* respectively). If the ship does not reach the motion boundary within the given time, the person can definitely catch up with the ship by properly selecting the water entry point; therefore, in the given time, the speed at which the ship can just reach the motion boundary is the maximum allowable speed. The position where the person meets the ship just at the motion boundary is denoted by *B*.

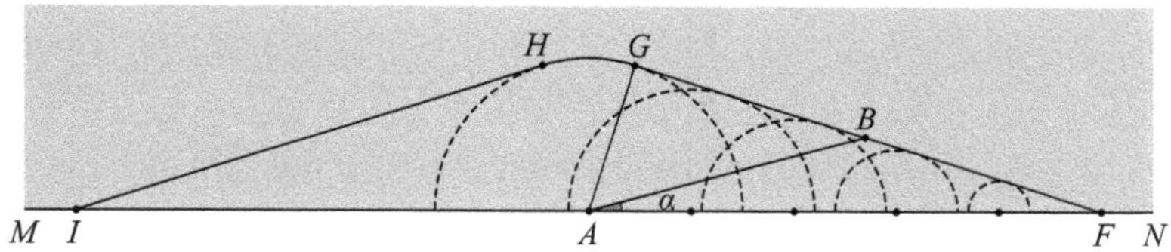

Figure 5

Known that the motion time of the ship and the person is equal in the whole process (that is, the time taken for the ship to reach Point *B* is the same as the time taken for the person to reach an arbitrary point on the motion boundary). Considering ***the displacement-time relation of uniform linear motion***:

$$t_{\min} = \frac{AB}{u_{\max}} = \frac{AF}{v_1} = \frac{AG}{v_2} \tag{12}$$

Considering that ΔAFG satisfies sine theorem (see Fig. 5), then place (12) in to obtain:

$$\Rightarrow \qquad \sin\angle AFB = \frac{AG}{AF} = \frac{v_2}{v_1} \tag{13}$$

Considering that ΔAFB satisfies sine theorem (see Fig. 5):

$$\frac{AB}{\sin\angle AFB} = \frac{AF}{\sin(\alpha + \angle AFB)} \tag{14}$$

Combining (12), (13) and (14),

$$\Rightarrow \quad u_{\min} = v_1 \frac{\sin\angle AFB}{\sin(\alpha + \angle AFB)} = v_1 \frac{\sin\angle AFB}{\sin\alpha\cos\angle AFB + \cos\alpha\sin\angle AFB}$$

$$= \frac{v_1 v_2}{\sqrt{v_1^2 - v_2^2}\sin\alpha + v_2\cos\alpha} \tag{15}$$

Solution [4]

The time point when the ship starts is defined as Time zero.

Taking *the person (can be regarded as a particle)* as the object of study. Studying *the states at arbitrary Time t_1 in the process of his/her walking along the lakeshore and at arbitrary Time t_2 in the process of his/her swimming in the lake respectively*. Considering ***theorem for composition of velocities of a particle*** [taking the ship as the stationary reference frame and the lakeshore as the moving reference frame]:

$$\begin{cases} \boldsymbol{v}_1' = \boldsymbol{v}_1 + (-\boldsymbol{u}) & (16a) \\ \boldsymbol{v}_2' = \boldsymbol{v}_2 + (-\boldsymbol{u}) & (16b) \end{cases}$$

$\boldsymbol{v}'_1$ and $\boldsymbol{v}'_2$—the velocity of the person relative to the ship at Time t_1 in the process of his/her walking along the lakeshore and at Time t_2 in the process of his/her swimming in the lake.

We can draw the schematic diagrams corresponding to Eqs. (16a) and (16b) (see Fig. 6a), the two triangles are denoted by ΔI and ΔII respectively.

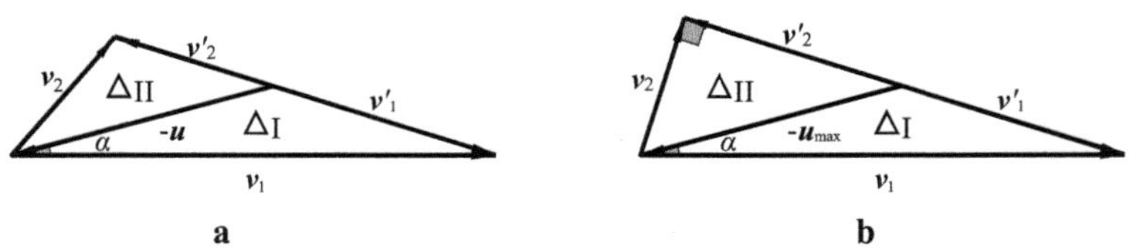

Figure 6a, b

The condition for the person to catch up with the ship successfully is that $\boldsymbol{v}'_1$ and $\boldsymbol{v}'_2$ are in opposite directions.

Considering that ΔII satisfies sine theorem (see Fig. 6a):

$$\frac{u}{\sin\langle \boldsymbol{v}_2, \boldsymbol{v}_2' \rangle} = \frac{v_2}{\sin\langle -\boldsymbol{u}, \boldsymbol{v}_2' \rangle} \tag{17}$$

When the magnitudes of $\boldsymbol{v}_2$ and $<-\boldsymbol{u}, \boldsymbol{v}'_2>$ are determined, u can take the maximum value when $<\boldsymbol{v}_2, \boldsymbol{v}'_2> = \pi/2$, that is, $\boldsymbol{v}_2$ is perpendicular to $\boldsymbol{v}'_2$ (see Fig. 6b).

$$\Rightarrow \quad u_{\max} = u\Big|_{\langle \boldsymbol{v}_2, \boldsymbol{v}_2' \rangle = \frac{\pi}{2}} = \frac{v_2}{\sin\langle -\boldsymbol{u}_{\max}, \boldsymbol{v}_2' \rangle} \sin\frac{\pi}{2} = \frac{v_2}{\sin\langle -\boldsymbol{u}_{\max}, \boldsymbol{v}_2' \rangle} \tag{18}$$

Considering that ΔI satisfies sine theorem (see Fig. 6b):

$$\frac{u_{\max}}{\sin\left(\langle -\boldsymbol{u}_{\max},\boldsymbol{v}_2'\rangle-\alpha\right)}=\frac{v_1}{\sin\langle -\boldsymbol{u}_{\max},\boldsymbol{v}_1'\rangle}=\frac{v_1}{\sin\langle -\boldsymbol{u}_{\max},\boldsymbol{v}_2'\rangle} \tag{19}$$

Combining (18) and (19),

$$\Rightarrow \qquad \sin\left(\langle -\boldsymbol{u},\boldsymbol{v}_2'\rangle-\alpha\right)=\frac{v_2}{v_1}$$

$$\Rightarrow \qquad \langle -\boldsymbol{u},\boldsymbol{v}_2'\rangle=\alpha+\arcsin\frac{v_2}{v_1} \tag{20}$$

(20) → (18),

$$\Rightarrow \qquad u_{\max}=\frac{v_1v_2}{\sqrt{v_1^1-v_2^2}\sin\alpha+v_2\cos\alpha} \tag{21}$$

Solution [5]

The position where the person meets the ship is denoted by *B*, an arbitrary optional water entry point is denoted by *D* (see Fig. 7).

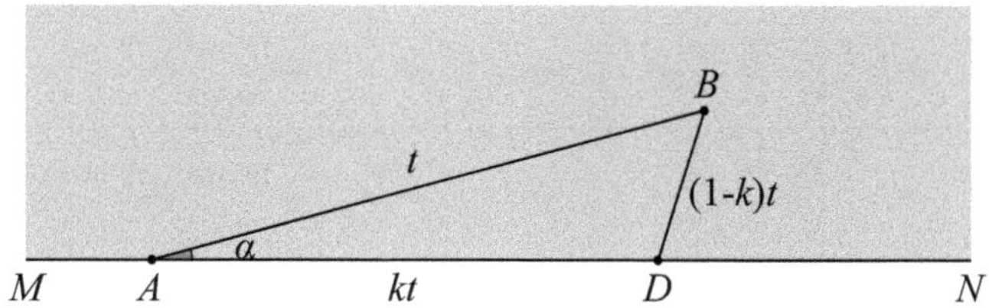

Figure 7

Considering that ΔADB satisfies cosine theorem (see Fig. 7):

$$DB^2=AB^2+AD^2-2AB\cdot AD\cos\alpha \tag{22}$$

Known that the motion time of the ship and the person is equal in the whole process. Considering ***the displacement-time relation of uniform linear motion***:

$$\begin{cases} t=\dfrac{AB}{u} & (23a)\\ kt=\dfrac{AD}{v_1} & (23b)\\ (1-k)t=\dfrac{DB}{v_2} & (23c)\end{cases}$$

t—the time taken of the whole process.

k—the proportionality coefficient that is greater than zero and less than 1.

u—the speed of the ship.

$$\Rightarrow \qquad [(1-k)v_2t]^2 = (ut)^2 + (kv_1t)^2 - 2(ut)(kv_1t)\cos\alpha \tag{24}$$

Rewriting Eq. (24) as a one-variable quadratic equation with k as the variable and u as the parameter:

$$\Rightarrow \qquad \left(v_1^2 - v_2^2\right)k^2 + 2\left(v_2^2 - uv_1\cos\alpha\right)k + \left(u^2 - v_2^2\right) = 0 \tag{25}$$

The necessary condition for k to have real number solution is that the discriminant is non-negative:

$$\Delta = 4\left(v_2^2 - uv_1\cos\alpha\right)^2 - 4\left(v_1^2 - v_2^2\right)\left(u^2 - v_2^2\right) \geq 0 \tag{26}$$

$$\Rightarrow \qquad \left(v_1^2\sin^2\alpha - v_2^2\right)u^2 + 2v_1v_2^2\cos\alpha \cdot u - v_1^2v_2^2 \leq 0 \tag{27}$$

The inequality above is a one-variable quadratic inequality with u as the variable, and its solution is as follows:

$$\Rightarrow u \leq \frac{v_1v_2\left(\sqrt{v_1^2 - v_2^2}\sin\alpha - v_2\cos\alpha\right)}{v_1^2\sin^2\alpha - v_2^2} = \frac{v_1v_2}{\sqrt{v_1^2 - v_2^2}\sin\alpha + v_2\cos\alpha} \tag{28}$$

1.3.2 Brachistochrone (d, *C*)

As shown in Fig. 1, a Cartesian coordinate system with the frame (O, ***i***, ***j***) is established in a vertical plane, where ***i*** is the horizontal right direction and ***j*** is the vertical down direction. There is a smooth slideway in the O-xy plane. The coordinates of its endpoints A and B are (0, 0) and (X, Y), respectively, X, $Y > 0$. Given that the shape of the slideway is a smooth curve, in order to minimize the time taken for a particle free released from Point A to slide along the slideway to Point B, determine the curve equation of the slideway.

Solution [1]

The time point when the particle starts to move is defined as Time zero.

Taking *the particle* as the object of study. Studying *the process from Time zero to arbitrary Time t.* Considering ***kinetic energy***

theorem [the work done by the normal contact force exerted on the particle by the slideway is zero]:

$$mgy = \frac{1}{2}mv^2 \tag{1}$$

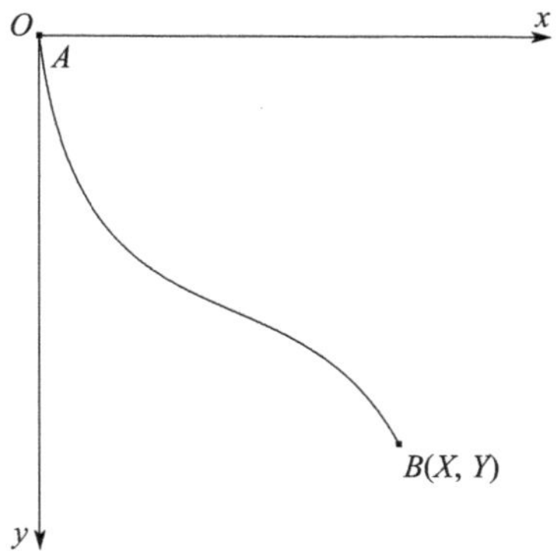

Figure 1

m—the mass of the particle.

v—the speed of the particle at Time t.

$$\Rightarrow \qquad v = \sqrt{2gy} \tag{2}$$

Considering the definition of ***velocity***:

$$v = \frac{\mathrm{d}s}{\mathrm{d}t} \tag{3}$$

ds—the displacement magnitude of the particle in dt time interval at arbitrary Time t.

The displacement magnitude of the particle in the instantaneous process at arbitrary Time is equal to the arc length of the curve element of its motion path. Considering ***the arc length formula of the curved element***:

$$\mathrm{d}s = \sqrt{1 + y'^2}\,\mathrm{d}x \tag{4}$$

Combining (2) to (4),

$$\Rightarrow \qquad \mathrm{d}t = \frac{\sqrt{1 + y'^2}}{\sqrt{2gy}}\mathrm{d}x \tag{5}$$

$$\Rightarrow \qquad T\left[y(x)\right]=\int_0^T \mathrm{d}t=\int_A^B \frac{\sqrt{1+y'^2}}{\sqrt{2gy}}\mathrm{d}x \tag{6}$$

T—the time taken for the particle to move from Point A to Point B. Its value is determined by the form of the function $y\,(x)$, that is, it is the functional of the function $y\,(x)$, which is denoted by $T[y(x)]$.

Defining the integral function $F(x, y, y')$:

$$F(x,y,y')=\frac{\sqrt{1+y'^2}}{\sqrt{2gy}} \tag{7}$$

$$\Rightarrow \qquad T\left[y(x)\right]=\int_A^B F(x,y,y')\mathrm{d}x \tag{8}$$

The condition for functional to take extremum is that its first variation is zero:

$$\delta T\left[y(x)\right]=0 \tag{9}$$

Considering ***Euler equation***:

$$\Rightarrow \qquad \frac{\partial F}{\partial y}-\frac{\mathrm{d}}{\mathrm{d}x}\left(\frac{\partial F}{\partial y'}\right)=0 \tag{10}$$

(7) $\rightarrow$ (10),

$$\Rightarrow \qquad -\frac{\sqrt{1+y'^2}}{2\sqrt{2g}\,y^{3/2}}-\frac{\mathrm{d}}{\mathrm{d}x}\left(\frac{y'}{\sqrt{2gy}\sqrt{1+y'^2}}\right)=0 \tag{11}$$

Taking the following differential transformations:

$$\begin{aligned}\frac{\mathrm{d}}{\mathrm{d}x}&=\frac{\partial}{\partial x}+\frac{\mathrm{d}y}{\mathrm{d}x}\frac{\partial}{\partial y}+\frac{\mathrm{d}y'}{\mathrm{d}x}\frac{\partial}{\partial y'}=\frac{\partial}{\partial x}+\frac{\mathrm{d}y}{\mathrm{d}x}\frac{\partial}{\partial y}+\frac{\mathrm{d}y'}{\mathrm{d}y}\frac{\mathrm{d}y}{\mathrm{d}x}\frac{\partial}{\partial y'}\\ &=\frac{\partial}{\partial x}+y'\left(\frac{\partial}{\partial y}+\frac{\mathrm{d}y'}{\mathrm{d}y}\frac{\partial}{\partial y'}\right)=\frac{\partial}{\partial x}+y'\frac{\mathrm{d}}{\mathrm{d}y}\end{aligned} \tag{12}$$

The function acted by $\frac{\mathrm{d}}{\mathrm{d}x}$ is not the explicit function of x.

$$\Rightarrow \qquad \frac{\mathrm{d}}{\mathrm{d}x}=y'\frac{\mathrm{d}}{\mathrm{d}y} \tag{13}$$

$$\Rightarrow \qquad \frac{\sqrt{1+y'^2}}{y^{3/2}}+2y'\frac{\mathrm{d}}{\mathrm{d}y}\left(\frac{y'}{\sqrt{y}\sqrt{1+y'^2}}\right)=0 \tag{14}$$

$$\Rightarrow \qquad \frac{dy}{y} + \frac{2y'dy'}{1+y'^2} = 0$$

$$\Rightarrow \qquad \int \frac{dy}{y} + \int \frac{d(1+y'^2)}{1+y'^2} = 0$$

$$\Rightarrow \qquad \ln y(1+y'^2) = C_0 \tag{15}$$

C_0—the integral constant.

Defining the constant C_1:

$$2C_1 = e^{C_0} \tag{16}$$

$$\Rightarrow \qquad y' = \sqrt{\frac{2C_1}{y} - 1} \tag{17}$$

$$\Rightarrow \qquad dx = \frac{\sqrt{y}dy}{\sqrt{2C_1 - y}} \tag{18}$$

$$\Rightarrow x = \int dx = \int \frac{\sqrt{y}dy}{\sqrt{2C_1 - y}} = -\sqrt{2C_1 y - y^2} + C_1 \arccos\left(1 - \frac{y}{C_1}\right) + C_2 \tag{19}$$

C_2—the integral constant.

Defining the variable θ:

$$\theta = \arccos\left(1 - \frac{y}{C_1}\right) \tag{20}$$

$$\Rightarrow \qquad \begin{cases} x = C_1(\theta - \sin\theta) + C_2 & (21a) \\ y = C_1(1 - \cos\theta) & (21b) \end{cases}$$

Considering the boundary conditions:

$$\begin{cases} x|_{y=0} = 0 & (22a) \\ x|_{y=Y} = X & (22b) \end{cases}$$

$$\Rightarrow \qquad \begin{cases} C_2 = 0 & (23a) \\ X = C_1 \arccos\left(1 - \dfrac{Y}{C_1}\right) - \sqrt{2C_1 Y - Y^2} & (23b) \end{cases}$$

Equation (23b) is the transcendental equation satisfied by C_1, the solution C_1 (X, Y) is denoted by k:

$$k = C_1(X, Y) \tag{24}$$

$$\Rightarrow \quad \begin{cases} x = k(\theta - \sin\theta) & (25a) \\ y = k(1 - \cos\theta) & (25b) \end{cases}$$

The system of equations above are the trajectory equations of the most time-saving path, it can be seen that it's a rolling curve (see Fig. 2).

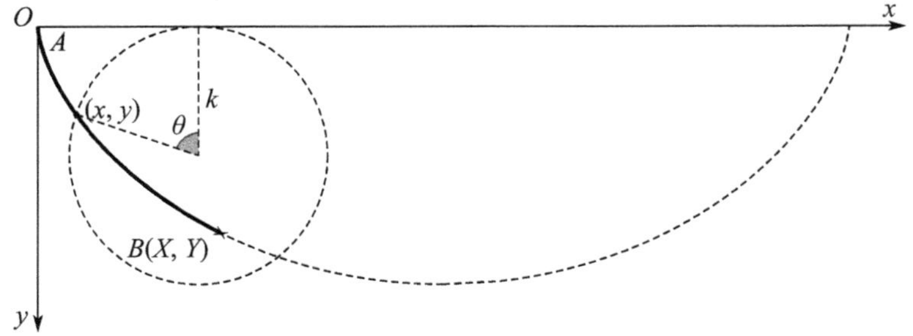

Figure 2

(25) → (6), we obtain the minimum time taken for the particle to slide from Point A along the slideway to Point B:

$$\Rightarrow T_{\min} = \int_A^B \frac{\sqrt{\left(\frac{dx}{d\theta}\right)^2 + \left(\frac{dy}{d\theta}\right)^2}\, d\theta}{\sqrt{2gy}} = \int_0^{\arccos\left(1-\frac{Y}{k}\right)} \frac{k\sqrt{2(1-\cos\theta)}d\theta}{\sqrt{2gk(1-\cos\theta)}}$$

$$= \sqrt{\frac{k}{g}} \int_0^{\arccos\left(1-\frac{Y}{k}\right)} d\theta = \sqrt{\frac{k}{g}} \arccos\left(1 - \frac{Y}{k}\right) \quad (26)$$

Solution 2

As shown in Fig. 3, the most time-saving path is denoted by L, which is divided into a series of path infinitesimal elements, and sorted by positive integer from top to bottom, the No. i path infinitesimal element is denoted by L_i, the connection point of Path L_{i-1} and Path L_i is denoted by P_i. The shape of each path infinitesimal element can be regarded as a line segment, and the motion of the particle on each path infinitesimal element can be regarded as a uniform linear motion. From the fact that the whole path is the most time-saving path, it can be seen that for any positive integer i, the combination path of Path L_i and Path L_{i+1} is the most time-saving path from Point

P_i to Point P_{i+2}. By analogy with the propagation of light, it can be seen that the motion of the particle in Path L_i and Path L_{i+1} should satisfy ***law of refraction***:

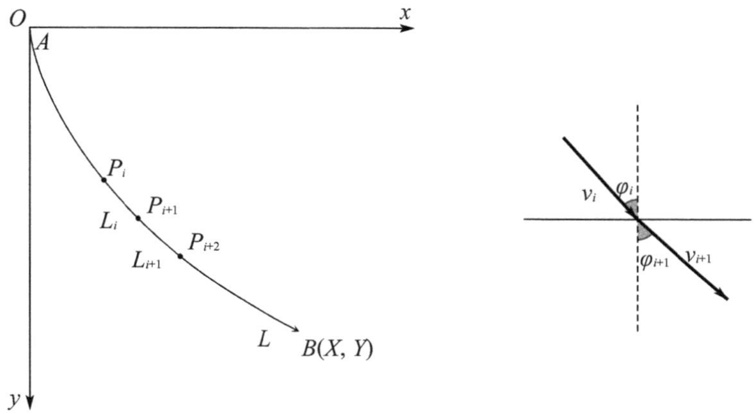

Figure 3

$$\frac{\sin\varphi_i}{v_i} = \frac{\sin\varphi_{i+1}}{v_{i+1}} \tag{27}$$

v_i—the speed of the particle moving in the Path L_i.

φ_i—the angle between Path L_i and the vertical direction.

Defining φ—the angle between the tangent of an arbitrary point of the most time-saving path L and the vertical direction.

$$\Rightarrow \qquad \frac{\sin\varphi}{v} = C \tag{28}$$

(3) → (28),

$$\Rightarrow \qquad \frac{\sin\varphi}{\sqrt{2gy}} = C \tag{29}$$

Considering the geometric meaning of derivative:

$$y' = \cot\varphi \tag{30}$$

Combining (29) and (30),

$$\Rightarrow \qquad y' = \sqrt{\frac{1}{2gC^2 y} - 1} \tag{31}$$

The equation above has the same form as Eq. (17).

Chapter 2

Dynamics

2.1 Simple Problems of Particle Dynamics

In classical mechanics, the law of dynamics of a particle can be described by Newton's law. The establishment of the basic dynamic equation of a particle is based on the selection of reference frame (the reference object and the coordinate frame attached to it), the force analysis based on various interaction theories, and the necessary kinematic analysis. For the non-inertial reference frame, the introduction of inertial force is a necessary means to keep the form of basic dynamic equation unchanged.

In this book, all statics problems are regarded as the degradation of dynamics problems and treated in a unified way.

In this section, we will study some simple particle dynamics problems.

2.1.1 The Motion of a Projectile under Damping Force 1 (d, *C*)

Consider a particle moving in the air, which is subject to the air drag, the magnitude of the air drag is directly proportional to the speed of the particle, the proportionality coefficient is k, and the direction of the air drag is opposite to the velocity direction of the particle.

Solving Physics Problems: Exploring New Thinking Paradigms
Yipeng Gu

ISBN 978-981-4877-41-1 (Hardcover), 978-1-003-16254-4 (eBook)
www.jennystanford.com

Given that the particle's initial velocity magnitude is v_0, and the angle between the initial velocity direction and the horizontal right direction is θ_0.

(1) Find the kinematics equation and trajectory equation of the particle and take the low order approximation for the trajectory equation when k is a small quantity.

(2) Find the position of the highest point of the trajectory of the particle and take the low order approximation when k is a small quantity.

Solution to Problem (1)

The initial time is denoted by Time zero.

As shown in Fig. 1, the initial position of the particle is denoted by O, the unit vector in the horizontal right direction is denoted by $\boldsymbol{i}$, and the unit vector in the vertical upward direction is denoted by $\boldsymbol{j}$. Establishing the Cartesian coordinate system with the frame $(O, \boldsymbol{i}, \boldsymbol{j})$, let x, y be the Cartesian coordinates.

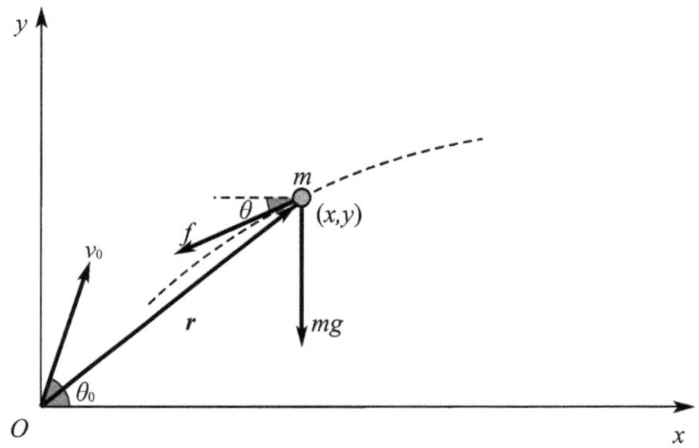

Figure 1

Taking *the particle* as the object of study. Studying *the state at arbitrary Time t*. Considering ***Newton's second law*** (the component equations in the $(\boldsymbol{i}, \boldsymbol{j})$-directions):

$$\begin{cases} f_x = m\ddot{x} & \text{(1a)} \\ -mg + f_y = m\ddot{y} & \text{(1b)} \end{cases}$$

m—the mass of the particle.

(f_x, f_y)—the components of the air drag acting on the particle in the $(\boldsymbol{i}, \boldsymbol{j})$-directions at Time t.

Known:

$$\boldsymbol{f} = -k\dot{\boldsymbol{r}} \tag{2}$$

$\boldsymbol{r}$—the position vector of the particle at arbitrary Time t.

$\boldsymbol{f}$—the air drag acting on the particle at arbitrary Time t.

$$\Rightarrow \quad \begin{cases} f_x = -k\dot{x} & \text{(3a)} \\ f_y = -k\dot{y} & \text{(3b)} \end{cases}$$

$(3) \rightarrow (1)$,

$$\Rightarrow \quad \begin{cases} -k\dot{x} = m\ddot{x} & \text{(4a)} \\ -mg - k\dot{y} = m\ddot{y} & \text{(4b)} \end{cases}$$

$$\Rightarrow \quad \begin{cases} \dfrac{\mathrm{d}\dot{x}}{\dot{x}} = -\dfrac{k}{m}\mathrm{d}t \\ \dfrac{\mathrm{d}\dot{y}}{\dfrac{mg}{k} + \dot{y}} = -\dfrac{k}{m}\mathrm{d}t \end{cases} \Rightarrow \begin{cases} \displaystyle\int_{v_0 \cos\theta_0}^{\dot{x}} \frac{\mathrm{d}\dot{x}}{\dot{x}} = -\frac{k}{m}\int_0^t \mathrm{d}t \\ \displaystyle\int_{v_0 \sin\theta_0}^{\dot{y}} \frac{\mathrm{d}\dot{y}}{\dfrac{mg}{k} + \dot{y}} = -\frac{k}{m}\int_0^t \mathrm{d}t \end{cases}$$

$$\Rightarrow \quad \begin{cases} \dot{x} = (v_0 \cos\theta_0) e^{-\frac{k}{m}t} \\ \dot{y} = \left(v_0 \sin\theta_0 + \dfrac{mg}{k}\right) e^{-\frac{k}{m}t} - \dfrac{mg}{k} \end{cases}$$

$$\Rightarrow \quad \begin{cases} \mathrm{d}x = (v_0 \cos\theta_0) e^{-\frac{k}{m}t}\,\mathrm{d}t \\ \mathrm{d}y = \left[\left(v_0 \sin\theta_0 + \dfrac{mg}{k}\right) e^{-\frac{k}{m}t} - \dfrac{mg}{k}\right]\mathrm{d}t \end{cases}$$

$$\Rightarrow \quad \begin{cases} x = \displaystyle\int_0^x \mathrm{d}x = \int_0^t (v_0 \cos\theta_0) e^{-\frac{k}{m}t}\,\mathrm{d}t \\ \quad = \dfrac{m}{k}(v_0 \cos\theta_0)\left(1 - e^{-\frac{k}{m}t}\right) & \text{(5a)} \\ y = \displaystyle\int_0^y \mathrm{d}y = \int_0^t \left[\left(v_0 \sin\theta_0 + \frac{mg}{k}\right) e^{-\frac{k}{m}t} - \frac{mg}{k}\right]\mathrm{d}t \\ \quad = \dfrac{m}{k}\left(v_0 \sin\theta_0 + \dfrac{mg}{k}\right)\left(1 - e^{-\frac{k}{m}t}\right) - \dfrac{mg}{k}t & \text{(5b)} \end{cases}$$

$$\Rightarrow \quad y = \left(\tan\theta_0 + \frac{mg}{kv_0\cos\theta_0}\right)x + \frac{m^2 g}{k^2}\ln\left(1 - \frac{k}{mv_0\cos\theta_0}x\right) \qquad (6)$$

Equations (5a) and (5b) are the kinematics equations of the particle, Eq. (6) is the trajectory equation of the particle.

Considering the Maclaurin expansion of $\ln(1 + x)$:

$$\Rightarrow \quad \ln\left(1 - \frac{k}{mv_0\cos\theta_0}x\right)\Bigg|_{k\ll 1}$$

$$= -\frac{k}{mv_0\cos\theta_0}x - \frac{1}{2}\left(\frac{k}{mv_0\cos\theta_0}\right)^2 x^2 - \frac{1}{3}\left(\frac{k}{mv_0\cos\theta_0}\right)^3 x^3 + \cdots \qquad (7)$$

$$\Rightarrow \; y\big|_{k\ll 1} = x\tan\theta_0 - \frac{g}{2(v_0\cos\theta_0)^2}x^2 - \frac{gk}{3m(v_0\cos\theta_0)^3}x^3 + O(k) \qquad (8)$$

It can be seen that if all items containing k are removed, the trajectory of the particle will degenerate to a parabola, which is just the trajectory of the particle without air drag. We can draw the schematic diagram of the trajectory of the particle (see Fig. 2).

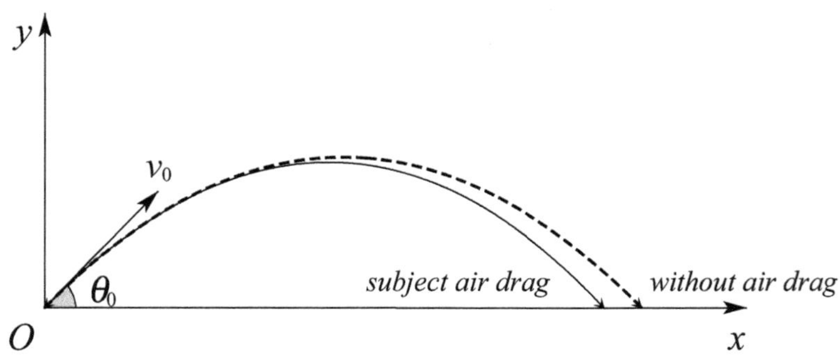

Figure 2

Solution to Problem (2)

From (6),

$$\Rightarrow \qquad \frac{\mathrm{d}y}{\mathrm{d}x} = \tan\theta_0 - \frac{mg}{kv_0\cos\theta_0}\left(\frac{1}{\dfrac{k}{mv_0\cos\theta_0} - x} - 1\right)^{-1} \qquad (9)$$

$$\Rightarrow \qquad \frac{d^2y}{dx^2} = -\frac{g}{\left(v_0\cos\theta_0 - \frac{k}{m}x\right)^2} < 0 \tag{10}$$

The necessary condition for y to take the maximum value is as follows:

$$\left.\frac{dy}{dx}\right|_{x=x_0} = 0 \tag{11}$$

x_0—the value of x when y takes the extremum.

$$\Rightarrow \qquad x_0 = \frac{v_0^2\sin\theta_0\cos\theta_0}{g\left(1+\frac{kv_0\sin\theta_0}{mg}\right)} \tag{12}$$

(12) → (6),

$$\Rightarrow \qquad y_{\max}\Big|_{x=x_0} = \frac{mv_0\sin\theta_0}{k} - \frac{m^2g}{k^2}\ln\left(1+\frac{kv_0\sin\theta_0}{mg}\right) \tag{13}$$

Considering the Maclaurin expansion of $\ln(1 + x)$:

$$\Rightarrow \qquad \ln\left(1+\frac{kv_0\sin\theta_0}{mg}\right)\Bigg|_{k\ll 1}$$

$$= \frac{kv_0\sin\theta_0}{mg} - \frac{1}{2}\left(\frac{kv_0\sin\theta_0}{mg}\right)^2 + \frac{1}{3}\left(\frac{kv_0\sin\theta_0}{mg}\right)^3 + \cdots \tag{14}$$

$$\Rightarrow \qquad y_{\max}\Big|_{k\ll 1} = \frac{(v_0\sin\theta_0)^2}{2g} - \frac{(v_0\sin\theta_0)^3}{3mg^2}k + O(k) \tag{15}$$

2.1.2 The Motion of a Projectile under Damping Force 2 (d, C)

There is a particle moving in the air, it is subject to the air drag, the magnitude of the air drag is directly proportional to the square of the speed of the particle, the proportionality coefficient is c, and the direction of the air drag is opposite to the velocity direction of the particle. Given that the magnitude of the initial velocity of the particle is v_0, and the angle between the direction of the initial velocity and the horizontal right direction is θ_0. Find the kinematics equation and trajectory equation of the particle when θ_0 is a small quantity, and take the low order approximation for the trajectory equation when c is a small quantity.

Solution [1]

The initial time is denoted by Time zero.

As shown in Fig. 1, the initial position of the particle is denoted by O, the unit vector in the horizontal right direction is denoted by $\boldsymbol{i}$, and the unit vector in the vertical upward direction is denoted by $\boldsymbol{j}$. Establishing the Cartesian coordinate system with the frame $(O, \boldsymbol{i}, \boldsymbol{j})$, let x, y be the Cartesian coordinates.

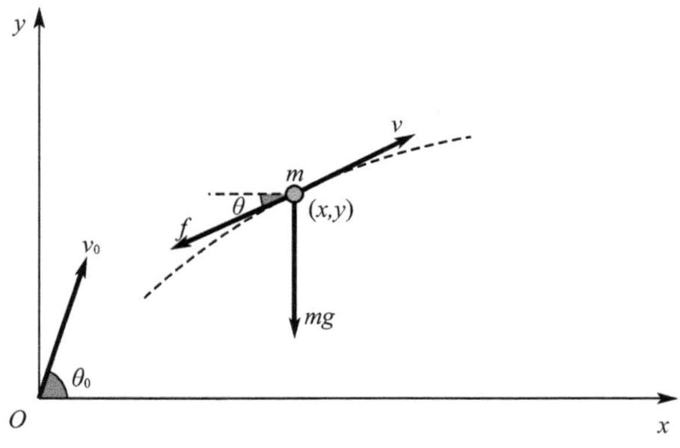

Figure 1

Taking *the particle* as the object of study. Studying *the state at arbitrary Time t*. Considering ***Newton's second law*** (the component equations in the $(\boldsymbol{i}, \boldsymbol{j})$-directions):

$$\begin{cases} -f\cos\theta = m\ddot{x} & (1a) \\ -mg - f\sin\theta = m\ddot{y} & (1b) \end{cases}$$

m—the mass of the particle.

θ—the angle between the velocity of the particle and the $\boldsymbol{i}$-direction at Time t.

f—the magnitude of the air drag acting on the particle at Time t.

Known:

$$f = cv^2 \tag{2}$$

v—the speed of the particle at arbitrary Time t.

$$\begin{cases} \dot{x} = v\cos\theta & (3a) \\ \dot{y} = v\sin\theta & (3b) \end{cases}$$

Defining the constant β:

$$\beta = \frac{c}{m} \tag{4}$$

(2), (3) and (4) $\rightarrow$ (1),

$$\Rightarrow \quad \begin{cases} -\beta v\dot{x} = \ddot{x} & (5a) \\ -g - \beta v\dot{y} = \ddot{y} & (5b) \end{cases}$$

Taking the following differential transformations:

$$\ddot{x} = \frac{\mathrm{d}}{\mathrm{d}t}\left(\frac{\mathrm{d}x}{\mathrm{d}t}\right) = \frac{\mathrm{d}}{\mathrm{d}x}\left(\frac{\mathrm{d}x}{\mathrm{d}t}\right)\frac{\mathrm{d}x}{\mathrm{d}t} = \frac{1}{2}\frac{\mathrm{d}\left(\frac{\mathrm{d}x}{\mathrm{d}t}\right)^2}{\mathrm{d}x} = \frac{1}{2}\frac{\mathrm{d}\dot{x}^2}{\mathrm{d}x} \tag{6}$$

Defining ds—the displacement magnitude of the particle in dt time interval at arbitrary Time t. Considering the definition of ***velocity***:

$$v = \frac{\mathrm{d}s}{\mathrm{d}t} = \frac{\mathrm{d}s}{\mathrm{d}x}\frac{\mathrm{d}x}{\mathrm{d}t} = \dot{x}\frac{\mathrm{d}s}{\mathrm{d}x} \tag{7}$$

(6) and (7) $\rightarrow$ (5a),

$$\Rightarrow \quad \frac{\mathrm{d}\dot{x}^2}{\dot{x}^2} = -2\beta\,\mathrm{d}s$$

$$\Rightarrow \quad \int_{v_0\cos\theta_0}^{\dot{x}} \frac{\mathrm{d}\dot{x}^2}{\dot{x}^2} = -2\beta\int_0^s \mathrm{d}s$$

$$\Rightarrow \quad \dot{x} = v_0\cos\theta_0 e^{-\beta s} \tag{8}$$

Taking the following differential transformations:

$$\dot{y} = \frac{\mathrm{d}y}{\mathrm{d}t} = \frac{\mathrm{d}y}{\mathrm{d}x}\frac{\mathrm{d}x}{\mathrm{d}t} = \dot{x}\frac{\mathrm{d}y}{\mathrm{d}x} \tag{9}$$

$$\ddot{y}=\frac{\mathrm{d}}{\mathrm{d}t}\left(\frac{\mathrm{d}y}{\mathrm{d}t}\right)=\frac{\mathrm{d}}{\mathrm{d}t}\left(\frac{\mathrm{d}y}{\mathrm{d}x}\frac{\mathrm{d}x}{\mathrm{d}t}\right)=\frac{\mathrm{d}}{\mathrm{d}t}\left(\frac{\mathrm{d}y}{\mathrm{d}x}\right)\frac{\mathrm{d}x}{\mathrm{d}t}+\frac{\mathrm{d}^2x}{\mathrm{d}t^2}\frac{\mathrm{d}y}{\mathrm{d}x}$$

$$=\frac{\mathrm{d}}{\mathrm{d}x}\left(\frac{\mathrm{d}y}{\mathrm{d}x}\right)\left(\frac{\mathrm{d}x}{\mathrm{d}t}\right)^2+\frac{\mathrm{d}^2x}{\mathrm{d}t^2}\frac{\mathrm{d}y}{\mathrm{d}x}=\frac{\mathrm{d}^2y}{\mathrm{d}x^2}\left(\frac{\mathrm{d}x}{\mathrm{d}t}\right)^2+\frac{\mathrm{d}^2x}{\mathrm{d}t^2}\frac{\mathrm{d}y}{\mathrm{d}x}$$

$$=\dot{x}^2\frac{\mathrm{d}^2y}{\mathrm{d}x^2}+\ddot{x}\frac{\mathrm{d}y}{\mathrm{d}x} \tag{10}$$

(9) and (10) $\rightarrow$ (5b),

$$\Rightarrow \qquad \dot{x}^2\frac{\mathrm{d}^2y}{\mathrm{d}x^2}+(\ddot{x}+\beta v\dot{x})\frac{\mathrm{d}y}{\mathrm{d}x}=-g \tag{11}$$

Combining (5a) and (11),

$$\Rightarrow \qquad \dot{x}^2\frac{\mathrm{d}^2y}{\mathrm{d}x^2}=-g \tag{12}$$

(8) $\rightarrow$ (12),

$$\Rightarrow \qquad \frac{\mathrm{d}^2y}{\mathrm{d}x^2}=-\frac{ge^{2\beta s}}{(v_0\cos\theta_0)^2} \tag{13}$$

When θ_0 is a small angle, the displacement of the particle in the ***j***-direction is a small quantity, we obtain:

$$s\approx x \tag{14}$$

(14) $\rightarrow$ (8),

$$\Rightarrow \qquad \dot{x}\approx v_0\cos\theta_0 e^{-\beta x}$$

$$\Rightarrow \qquad e^{\beta x}\mathrm{d}x=v_0\cos\theta_0\mathrm{d}t$$

$$\Rightarrow \qquad \int_0^x e^{\beta x}\mathrm{d}x=v_0\cos\theta_0\int_0^t\mathrm{d}t$$

$$\Rightarrow \qquad x=\frac{1}{\beta}\ln(1+\beta v_0\cos\theta_0 t) \tag{15}$$

Equation (15) is the kinematics equation of the particle when θ_0 is a small angle.

(14) $\rightarrow$ (13),

$$\Rightarrow \qquad \frac{\mathrm{d}^2y}{\mathrm{d}x^2}\approx-\frac{ge^{2\beta x}}{(v_0\cos\theta_0)^2} \tag{16}$$

Defining k—the slope of the tangent line at an arbitrary point on the trajectory of the particle:

$$k = \frac{\mathrm{d}y}{\mathrm{d}x} \tag{17}$$

$(17) \rightarrow (16)$,

$$\Rightarrow \qquad \mathrm{d}k = -\frac{g e^{2\beta x}}{(v_0 \cos\theta_0)^2}\mathrm{d}x$$

$$\Rightarrow \qquad \int_{\tan\theta_0}^{k} \mathrm{d}k = -\frac{g}{(v_0 \cos\theta_0)^2}\int_0^x e^{2\beta x}\mathrm{d}x$$

$$\Rightarrow \qquad k = \tan\theta_0 + \frac{g}{2\beta(v_0 \cos\theta_0)^2}(1 - e^{2\beta x}) \tag{18}$$

$(17) \rightarrow (18)$,

$$\Rightarrow \qquad \mathrm{d}y = \left[\tan\theta_0 + \frac{g}{2\beta(v_0 \cos\theta_0)^2}(1 - e^{2\beta x})\right]\mathrm{d}x$$

$$\Rightarrow \qquad y = \int_0^y \mathrm{d}y = \int_0^x \left[\tan\theta_0 + \frac{g}{2\beta(v_0 \cos\theta_0)^2}(1 - e^{2\beta x})\right]\mathrm{d}x$$

$$= \left[\tan\theta_0 + \frac{g}{2\beta(v_0 \cos\theta_0)^2}\right]x + \frac{g}{4\beta^2(v_0 \cos\theta_0)^2}(1 - e^{2\beta x}) \tag{19}$$

Considering the Maclaurin expansion of e^x:

$$\Rightarrow \qquad e^{2\beta x}\Big|_{\beta \ll 1} = 1 + 2\beta x + \frac{1}{2}(2\beta x)^2 + \frac{1}{6}(2\beta x)^3 + \cdots \tag{20}$$

$$\Rightarrow \quad y\big|_{\beta \ll 1} = x\tan\theta_0 - \frac{g}{2(v_0 \cos\theta_0)^2}x^2 - \frac{g\beta}{3(v_0 \cos\theta_0)^2}x^3 + O(\beta) \tag{21}$$

It can be seen that if all items containing β are removed, the trajectory of the particle will degenerate to a parabola, which is just the trajectory of the particle without air drag. We can draw the schematic diagram of the trajectory of the particle (see Fig. 2).

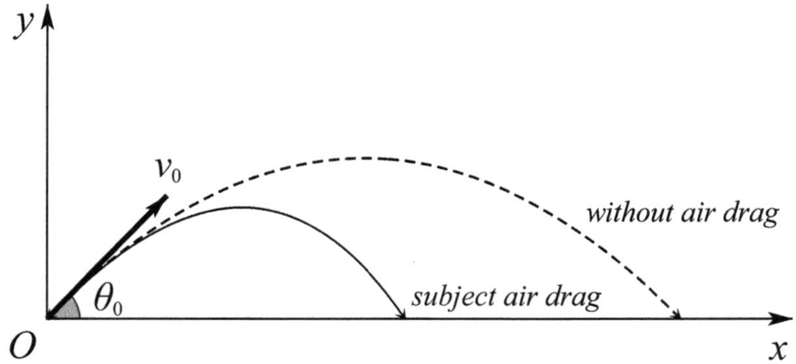

Figure 2

Solution [2]

As shown in Fig. 3, introducing the natural frame $(\boldsymbol{\tau}, \boldsymbol{n})$ of the particle.

Taking *the particle* as the object of study. Studying *the state at arbitrary Time t*. Considering ***Newton's second law*** (the component equations in the natural coordinate system):

$$\begin{cases} -f - mg\sin\theta = ma_\tau & \text{(22a)} \\ mg\cos\theta = ma_n & \text{(22b)} \end{cases}$$

(a_τ, a_n)—the components of the acceleration of the particle in the $(\boldsymbol{\tau}, \boldsymbol{n})$-directions at Time t.

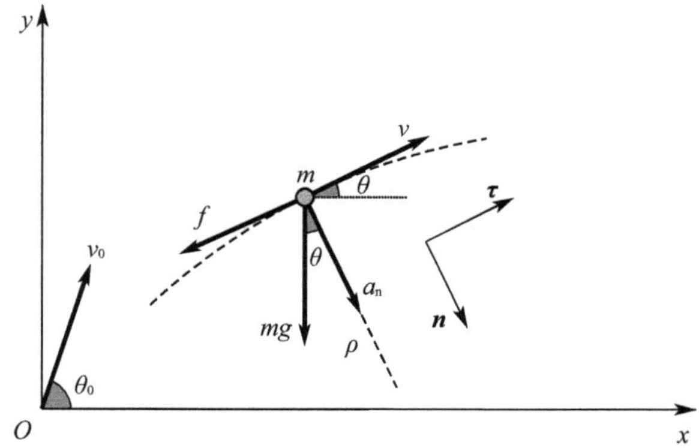

Figure 3

Considering ***the components form of acceleration in natural coordinate system***:

$$\begin{cases} a_\tau = \dfrac{\mathrm{d}v}{\mathrm{d}t} & (23a) \\ a_n = \dfrac{v^2}{\rho} & (23b) \end{cases}$$

ρ—the radius of curvature of the particle's trajectory at the position where the particle is at arbitrary Time t.

Considering the definition of ***radius of curvature***:

$$\rho = -\frac{\mathrm{d}s}{\mathrm{d}\theta} \qquad (24)$$

(23b) → (22b),

$$\rho = \frac{v^2}{g\cos\theta} \qquad (25)$$

Combining (24) and (25),

$$\Rightarrow \qquad \mathrm{d}s = -\frac{v^2}{g}\frac{\mathrm{d}\theta}{\cos\theta} \qquad (26)$$

(2), (4) and (23a) → (22a),

$$\Rightarrow \qquad -\beta v^2 - g\sin\theta = \frac{\mathrm{d}v}{\mathrm{d}t} \qquad (27)$$

(7) → (26),

$$\Rightarrow \qquad \mathrm{d}t = -\frac{v\mathrm{d}\theta}{g\cos\theta} \qquad (28)$$

Combining (27) and (28),

$$\Rightarrow \qquad \frac{\beta}{g}v^3\mathrm{d}\theta = \cos\theta\mathrm{d}v - v\sin\theta\mathrm{d}\theta = \mathrm{d}(v\cos\theta) \qquad (29)$$

$$\Rightarrow \qquad \frac{\mathrm{d}(v\cos\theta)}{(v\cos\theta)^3} = \frac{\beta}{g}\frac{\mathrm{d}\theta}{\cos^3\theta}$$

$$\Rightarrow \qquad \int_{v_0\cos\theta_0}^{v\cos\theta}\frac{\mathrm{d}(v\cos\theta)}{(v\cos\theta)^3} = \frac{\beta}{g}\int_{\theta_0}^{\theta}\frac{\mathrm{d}\theta}{\cos^3\theta}$$

$$\Rightarrow \quad -\frac{1}{v^2\cos^2\theta}+\frac{1}{v_0^2\cos^2\theta_0}$$

$$=\frac{2\beta}{g}\left[\frac{1}{2}\left(\frac{\sin\theta}{\cos^2\theta}+\ln\frac{1+\sin\theta}{\cos\theta}\right)-\frac{1}{2}\left(\frac{\sin\theta_0}{\cos^2\theta_0}+\ln\frac{1+\sin\theta_0}{\cos\theta_0}\right)\right] \tag{30}$$

Defining the function $F(\theta)$:

$$F(\theta)=\frac{1}{2}\left(\frac{\sin\theta}{\cos^2\theta}+\ln\frac{1+\sin\theta}{\cos\theta}\right) \tag{31}$$

$$\Rightarrow \quad v=v(\theta)=v_0\left[\frac{\cos^2\theta}{\cos^2\theta_0}-\frac{2\beta v_0^2\cos^2\theta}{g}\left[F(\theta)-F(\theta_0)\right]\right]^{-1/2} \tag{32}$$

(32) → (26), we obtain the differential equation satisfied by $s\ (\theta)$ and further obtain the parameter equation of the trajectory of the particle with θ as the parameter:

$$\mathrm{d}s=-\frac{v^2(\theta)\mathrm{d}\theta}{g\cos\theta}$$

$$\Rightarrow \quad \begin{cases} \mathrm{d}x=\mathrm{d}s\cos\theta=-\dfrac{v^2(\theta)\mathrm{d}\theta}{g} & \text{(33a)} \\ \mathrm{d}y=\mathrm{d}s\sin\theta=-\dfrac{v^2(\theta)\tan\theta\,\mathrm{d}\theta}{g} & \text{(33b)} \end{cases}$$

(32) → (28), we obtain the differential equation satisfied by $\theta\,(t)$:

$$\Rightarrow \quad \mathrm{d}t=-\frac{v(\theta)\mathrm{d}\theta}{g\cos\theta} \tag{34}$$

Equation (34) contains the function relation between θ and t. The differential equations satisfied by $x\ (t)$ and $y\ (t)$ can be obtained by combining this equation with Eqs. (32) and (33).

(3a) → (29),

$$\Rightarrow \quad \frac{\mathrm{d}\dot{x}}{\dot{x}}=\frac{\beta v^2}{g}\frac{\mathrm{d}\theta}{\cos\theta} \tag{35}$$

(26) → (35),

$$\Rightarrow \quad \frac{\mathrm{d}\dot{x}}{\dot{x}}=-\beta\mathrm{d}s$$

$$\Rightarrow \quad \int_{v_0\cos\theta_0}^{\dot{x}}\frac{\mathrm{d}\dot{x}}{\dot{x}}=-\beta\int_0^s\mathrm{d}s$$

$$\Rightarrow \qquad \ln \frac{\dot{x}}{v_0 \cos\theta_0} = -\beta s$$

$$\Rightarrow \qquad \dot{x} = v_0 \cos\theta_0 e^{-\beta s} \tag{36}$$

Considering the form of the radius of curvature at an arbitrary point on the plane curve in Cartesian coordinate system:

$$\rho = \frac{\left[1+\left(\frac{\mathrm{d}y}{\mathrm{d}x}\right)^2\right]^{3/2}}{\left|\frac{\mathrm{d}^2 y}{\mathrm{d}x^2}\right|} \tag{37}$$

Considering (a) the geometric meaning of derivative, (b) the trajectory of the particle is curved downward:

$$\frac{\mathrm{d}y}{\mathrm{d}x} = \tan\theta \tag{38}$$

$$\frac{\mathrm{d}^2 y}{\mathrm{d}x^2} < 0 \tag{39}$$

(38) $\rightarrow$ (37),

$$\Rightarrow \qquad \rho = -\frac{1}{\frac{\mathrm{d}^2 y}{\mathrm{d}x^2}\cos^3\theta} \tag{40}$$

Combining (25) and (40),

$$\Rightarrow \qquad \frac{\mathrm{d}^2 y}{\mathrm{d}x^2} = -\frac{g}{v^2\cos^2\theta} \tag{41}$$

Combining (3a), (36) and (41),

$$\Rightarrow \qquad \frac{\mathrm{d}^2 y}{\mathrm{d}x^2} = -\frac{g}{\dot{x}^2} = -\frac{g e^{2\beta s}}{(v_0\cos\theta_0)^2} \tag{42}$$

This result is the same as Eq. (13).

2.1.3 The Conical Pendulum in Uniformly Variable Velocity Reference Frame (d, *AB*)

Consider a conical pendulum hung in a static container rotating with the cone angle 2θ (see Fig. 1). Once the pendulum ball is in the leftmost position, the container starts a uniformly variable motion with the acceleration of the magnitude a and the direction horizontally to the right. Is it possible that the pendulum ball to start a new conical pendulum motion relative to the container from this time point?

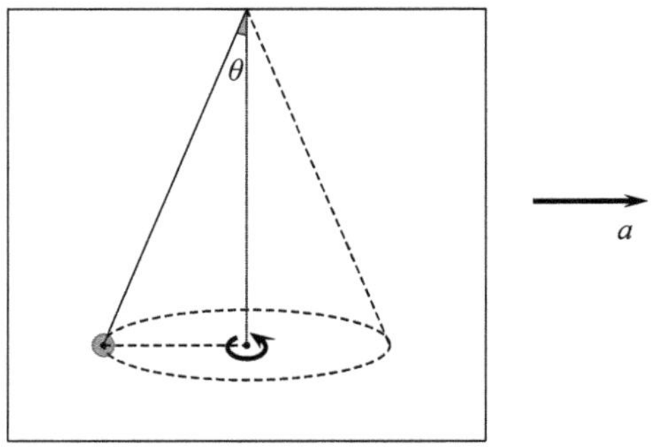

Figure 1

Solution [1] (type B)

Introducing the resultant force of the gravity and the inertia force acting on the pendulum ball in the container reference frame, which is regarded as the equivalent gravity, whose acceleration of gravity is denoted by $\boldsymbol{g}'$ (see Fig. 2):

$$m\boldsymbol{g}' = m\boldsymbol{g} + m\boldsymbol{a} \tag{1}$$

m—the mass of the pendulum ball.

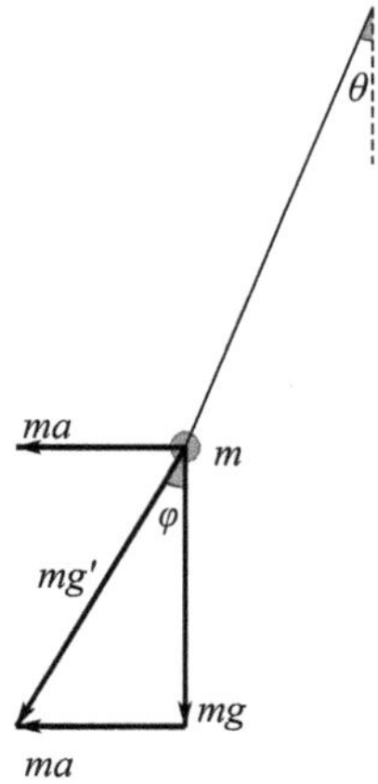

Figure 2

Considering the following geometric relations in Fig. 2:

$$\begin{cases} (mg')^2 = (mg)^2 + (ma)^2 \Rightarrow g' = \sqrt{g^2 + a^2} & \text{(2a)} \\ \tan\varphi = \dfrac{ma}{mg} = \dfrac{a}{g} & \text{(2b)} \end{cases}$$

φ—the angle between $\boldsymbol{g}'$ and $\boldsymbol{g}$.

It is assumed that the pendulum ball can start a new conical pendulum motion just when the container is accelerated, by analogy with the cone pendulum motion in the real gravity field, it can be seen that its axis should pass through the hanging point and be parallel to the equivalent gravity.

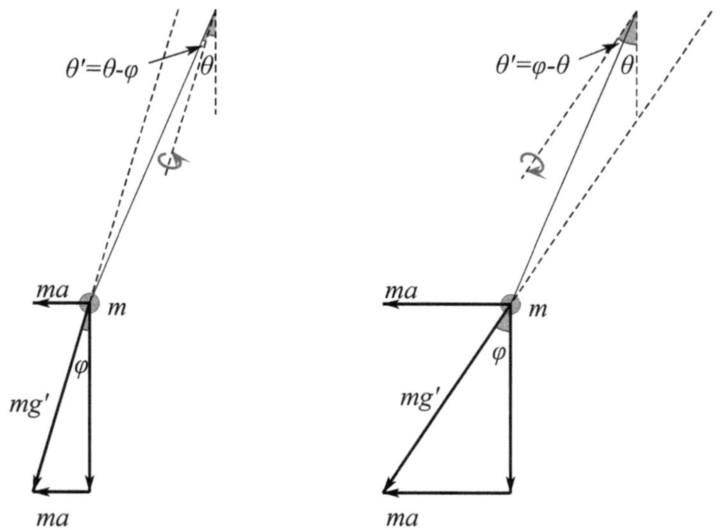

Figure 3

As shown in Fig. 3, the half cone angle θ' of the new cone pendulum motion is defined in the following two situations:

$$\theta' = \begin{cases} \theta - \varphi & \text{when } \theta > \varphi & \text{(3a)} \\ \varphi - \theta & \text{when } \theta < \varphi & \text{(3b)} \end{cases}$$

Taking *the pendulum ball (can be regarded as a particle)* as the object of study. Studying *the states at the time point before and after the container is accelerated, respectively*. Considering ***Newton's second law***:

$$\begin{cases} \boldsymbol{T} + m\boldsymbol{g} = m\boldsymbol{a}_m & \text{(4a)} \\ \boldsymbol{T}' + m\boldsymbol{g}' = m\boldsymbol{a}'_m & \text{(4b)} \end{cases}$$

$\boldsymbol{T}$ and $\boldsymbol{T}'$—the tension in the pendulum string acting on the pendulum ball at the time point before and after the container is accelerated.
$\boldsymbol{a}_m$ and $\boldsymbol{a}'_m$—the acceleration of the pendulum ball in the container reference frame at the time point before and after the container is accelerated.

Introducing ($\boldsymbol{\tau}$, $\boldsymbol{n}$) and ($\boldsymbol{\tau}'$, $\boldsymbol{n}'$)—the natural frames of the pendulum ball at the time point before and after the container is accelerated.

Considering ***the components form of acceleration in natural frame*** (according to the assumption, after the container is accelerated, the pendulum ball performs the uniform circular motion relative to the container; therefore, the component of the acceleration of the pendulum ball in the tangent direction of its trajectory is zero):

$$\begin{cases} \boldsymbol{a}_m = \dfrac{v^2}{l\sin\theta}\boldsymbol{n} & \text{(5a)} \\ \boldsymbol{a}'_m = \dfrac{v'^2}{l\sin\theta'}\boldsymbol{n}' & \text{(5b)} \end{cases}$$

l—the length of the pendulum string.
v and v'—the speed of the pendulum ball in the container reference frame at the time point before and after the container is accelerated.

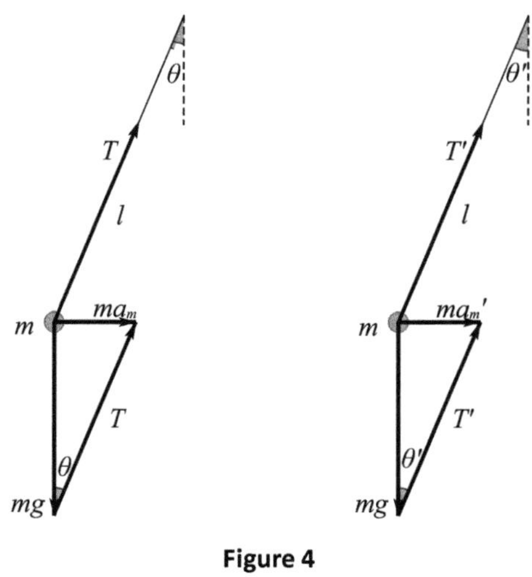

Figure 4

Taking *the pendulum ball* as the object of study. Studying *the instantaneous process of the container being accelerated.* Considering ***law of conservation of momentum*** [the gravity of the pendulum ball, the tension in the pendulum string and the inertia force acting on the pendulum ball are all finite quantity, acting on an extremely short process, their impulse are all infinitely small quantities; therefore, the momentum of the pendulum ball is approximately conserved] (the $\boldsymbol{\tau}$-direction component equation):

$$mv = mv' \tag{6}$$

$$\Rightarrow \qquad v = v' \tag{7}$$

From (4), (5) and (7), and considering the following geometric relations in Fig. 4:

$$\begin{cases} mg\tan\theta = m\dfrac{v^2}{l\sin\theta} & \text{(8a)} \\ mg'\tan\theta' = m\dfrac{v^2}{l\sin\theta'} & \text{(8b)} \end{cases}$$

Combining (2a), (3) and (8),

$$\Rightarrow \qquad g\frac{\sin^2\theta}{\cos\theta} = g'\frac{\sin^2\theta'}{\cos\theta'} = \sqrt{g^2+a^2}\,\frac{\sin^2[\pm(\theta-\varphi)]}{\cos[\pm(\theta-\varphi)]} \tag{9}$$

Taking the following transformation of trigonometric function, then place (2b) in to obtain:

$$\frac{\sin^2[\pm(\theta-\varphi)]}{\cos[\pm(\theta-\varphi)]} = \frac{\sin^2(\theta-\varphi)}{\cos(\theta-\varphi)} = \frac{(\sin\theta\cos\varphi-\cos\theta\sin\varphi)^2}{\cos\theta\cos\varphi+\sin\theta\sin\varphi}$$

$$= \frac{\left(\sin\theta\dfrac{1}{\sqrt{1+\tan^2\varphi}} - \cos\theta\dfrac{\tan\varphi}{\sqrt{1+\tan^2\varphi}}\right)^2}{\cos\theta\dfrac{1}{\sqrt{1+\tan^2\varphi}} + \sin\theta\dfrac{\tan\varphi}{\sqrt{1+\tan^2\varphi}}}$$

$$= \frac{\dfrac{1}{g^2+a^2}(g\sin\theta - a\cos\theta)^2}{\dfrac{1}{\sqrt{g^2+a^2}}(g\cos\theta + a\sin\theta)}$$

$$= \frac{(g\sin\theta - a\cos\theta)^2}{\sqrt{g^2+a^2}\,(g\cos\theta + a\sin\theta)} \tag{10}$$

$(10) \to (9)$,

$$\Rightarrow \qquad g\tan^2\theta = \frac{(g\tan\theta - a)^2}{g + a\tan\theta}$$

$$\Rightarrow \qquad a = g\tan\theta(\tan^2\theta + 2) \tag{11}$$

The trivial solution: $a = 0$ has been removed in the operation above.

$(11) \to (2b)$,

$$\Rightarrow \qquad \tan\varphi = \tan\theta(\tan^2\theta + 2)$$

$$\Rightarrow \qquad \varphi = \arctan[\tan\theta(\tan^2\theta + 2)] \tag{12}$$

$$\Rightarrow \qquad \tan\varphi > \tan\theta$$

$$\Rightarrow \qquad \varphi > \theta \tag{13}$$

From (3), (12) and (13),

$$\Rightarrow \qquad \theta' = \varphi - \theta = \arctan[\tan\theta(\tan^2\theta + 2)] - \theta \tag{14}$$

It can be seen that, the pendulum ball can start a new conical pendulum motion just when the container is accelerated according to Eq. (11), whose half cone angle is determined by Eq. (14).

Solution [2] (type A)

Considering the following geometric relation in Fig. 2:

$$g' = \frac{g}{\cos\varphi} \tag{15}$$

$(3) \to (15)$,

$$g' = \frac{g}{\cos(\theta \pm \theta')} \tag{16}$$

$(16) \to (9)$,

$$\Rightarrow \qquad g\frac{\sin^2\theta}{\cos\theta} = \frac{g}{\cos(\theta \pm \theta')}\frac{\sin^2\theta'}{\cos\theta'}$$

$$= g\frac{1}{\cos\theta\cos\theta' \mp \sin\theta\sin\theta'}\frac{\sin^2\theta'}{\cos\theta'}$$

$$= g\frac{\tan^2\theta'}{\cos\theta \mp \sin\theta\tan\theta'} \tag{17}$$

$$\Rightarrow \qquad \tan^2\theta' \pm \sin^2\theta\tan\theta\tan\theta' - \sin^2\theta = 0 \tag{18}$$

$$\Rightarrow \qquad \tan\theta' = \frac{\mp\sin^2\theta\tan\theta \pm \sqrt{\sin^4\theta\tan^2\theta + 4\sin^2\theta}}{2}$$

$$= \frac{\mp\dfrac{\tan^3\theta}{1+\tan^2\theta} \pm \sqrt{\dfrac{\tan^2\theta}{1+\tan^2\theta}\left(\dfrac{\tan^2\theta}{1+\tan^2\theta}\tan^2\theta + 4\right)}}{2}$$

$$= \frac{\mp\dfrac{\tan^3\theta}{1+\tan^2\theta} \pm \sqrt{\dfrac{\tan^2\theta}{1+\tan^2\theta}\dfrac{(\tan^2\theta+2)^2}{1+\tan^2\theta}}}{2}$$

$$= \frac{\tan\theta[\mp\tan^2\theta \pm (\tan^2\theta+2)]}{2(1+\tan^2\theta)}$$

$$= \begin{cases} \dfrac{\tan\theta}{1+\tan^2\theta} & \text{when } \theta < \varphi \\ \tan\theta & \text{when } \theta > \varphi \end{cases} \tag{19}$$

Above, for the equations corresponding to the two possible cases only positive solutions are taken, where $\tan\theta' = \tan\theta$ is the trivial solution corresponding to $\varphi = 0$ (i.e., $a = 0$).

$$\Rightarrow \qquad \tan\theta' = \frac{\tan\theta}{1+\tan^2\theta} = \sin^2\theta < \tan\theta \tag{20}$$

$$\Rightarrow \qquad \theta' = \arctan(\sin^2\theta) \tag{21}$$

It can be proved that Eq. (21) is equivalent to Eq. (14).

2.1.4 The Smooth Connection of Segmented Motions of a Small Ball Which Is Constrained by a String (d, *B*)

As shown in Fig. 1, there are two fixed nails located on a horizontal line, with the distance l, whose positions are denoted by Point O and Point A, respectively. There is a light rigid string of length $2l$, one end of which is fixed at the nail at Point O and the other end is tied with a small ball. Initially, the string is straightened to the right of Point

O. The ball starts to move with an initial velocity of the magnitude v_0 and the direction vertical downward due to an instantaneous impact. What is the minimum value of v_0 to ensure that the ball hit the nail at Point O?

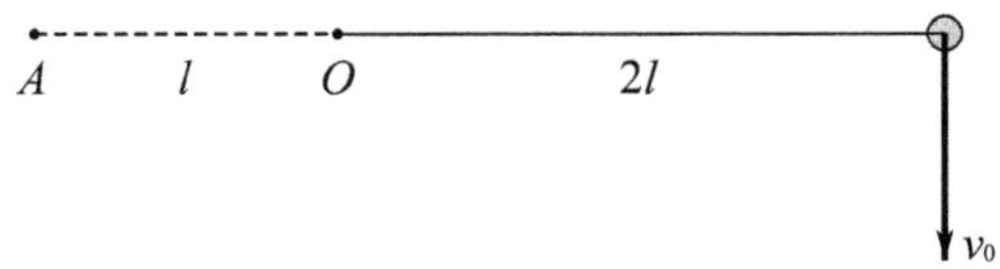

Figure 1

Solution

The time point when the ball starts to move is defined as Time zero.

As shown in Fig. 2, taking the origin O as the pole, the horizontal left direction as the positive direction of the polar axis, the clockwise direction as the positive direction of polar angle, establishing the polar coordinate system, let r, θ be the polar coordinates.

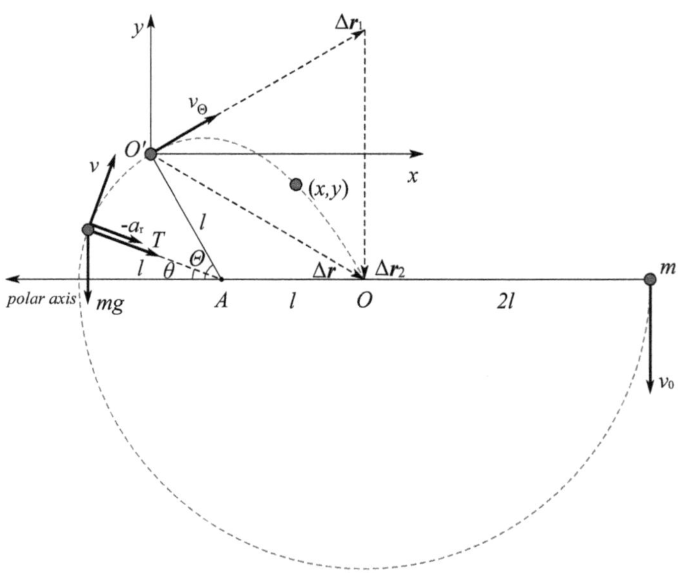

Figure 2

The process of the circular motion of the ball with Point O as the center before the string contacts the nail at Point A is denoted by

Process 1. The process of the circular motion of the ball with Point A as the center after the string contacts the nail at Point A is denoted by Process 2.

Taking *the system composed of the ball and the earth* as the object of study. Studying *the process from Time zero to arbitrary Time t in Process 2*. Consider ***law of conservation of mechanical energy*** [the work done by the external force on the system is zero (the direction of the tension in the string is perpendicular to the ball's velocity direction all the time; therefore no work is done), and the work done by the non-conservative internal force acting on the system is zero; therefore, the mechanical energy of the system is conserved] (taking Point O as the zero point of gravitational potential energy):

$$\frac{1}{2}mv_0^2 = \frac{1}{2}mv^2 + mgl\sin\theta \tag{1}$$

m—the mass of the ball.

v—the speed of the ball at Time t in Process 2.

Taking *the ball (can be regarded as a particle)* as the object of study. Studying *the state at arbitrary Time t in Process 2*. Considering ***Newton's second law*** (the $-\boldsymbol{r}^0$-direction component equation):

$$mg\sin\theta + T = m(-a_r) \tag{2}$$

T—the magnitude of the tension in the string at Time t in Process 2.

a_r—the $\boldsymbol{r}^0$-direction component of the acceleration of the ball at Time t in Process 2.

Considering ***the radial component form of acceleration*** (the constraint condition: the string length remains unchanged):

$$a_r = -\frac{v^2}{l} \tag{3}$$

Combining (1) to (3):

$$\Rightarrow \qquad T = m\frac{v_0^2}{l} - 3mg\sin\theta \tag{4}$$

$$\Rightarrow \qquad \frac{\mathrm{d}T}{\mathrm{d}\theta} = -3mg\cos\theta$$

$$\Rightarrow \qquad \left.\frac{\mathrm{d}T}{\mathrm{d}\theta}\right|_{\theta\in(0,\pi/2)} < 0 \tag{5}$$

Assumption I: when the ball moves to a certain point, the tension in the string will decrease to zero. Defining Θ—the polar angle of the ball when the tension in the string is zero:

$$T|_{\theta=\Theta} = 0 \tag{6}$$

Combining (1), (4) and (6),

$$\Rightarrow \qquad \sin\Theta = \frac{v_0^2}{3gl} = \frac{v^2}{gl} \tag{7a, b}$$

Assumption II: after the tension in the string decreases to zero, the string will never be straightened, and the ball will finally reach Point O and hit the nail there. This process is denoted by Process 3.

The motion of the ball in Process 3 can be regarded as the superposition of the uniform linear motion with the actual initial velocity and the free-falling motion with the initial velocity of zero. Considering ***the principle of vector superposition***:

$$\Delta \boldsymbol{r} = \Delta \boldsymbol{r}_1 + \Delta \boldsymbol{r}_2 \tag{8}$$

$\Delta \boldsymbol{r}$, $\Delta \boldsymbol{r}_1$ and $\Delta \boldsymbol{r}_2$—in Process 3, the ball's actual displacement, the displacement of the uniform linear motion with the actual initial velocity and the displacement of the free-falling motion with the initial velocity of zero.

Considering that the triangle of displacement vector of the ball satisfies sine theorem (see Fig. 2):

$$\frac{\Delta r_1}{\cos\frac{\Theta}{2}} = \frac{\Delta r_2}{\cos\frac{\Theta}{2}} = \frac{\Delta r}{\sin\Theta} \tag{9a, b}$$

$$\Delta r = 2l\cos\frac{\Theta}{2} \tag{10}$$

Considering ***the displacement-time relation of uniform linear motion and the displacement-time relation of uniformly variable motion***:

$$\Delta r_1 = v_\Theta \tau \tag{11}$$

$$\Delta r_2 = \frac{1}{2} g\tau^2 \tag{12}$$

τ—the time taken in Process 3.

(10) to (12) $\rightarrow$ (9),

$$\Rightarrow \qquad \frac{v_\Theta \tau}{\cos\frac{\Theta}{2}} = \frac{\frac{1}{2}g\tau^2}{\cos\frac{\Theta}{2}} = \frac{2l\cos\frac{\Theta}{2}}{\sin\Theta} \qquad \text{(13a, b)}$$

Combining (7) and (13):

$$\Rightarrow \qquad \begin{cases} v_0 = \sqrt{\dfrac{3\sqrt{3}}{2}gl} & \text{(14a)} \\ v_\Theta = \sqrt{\dfrac{\sqrt{3}}{2}gl} & \text{(14b)} \\ \sin\Theta = \dfrac{\sqrt{3}}{2} & \text{(14c)} \end{cases}$$

Proof of Assumption II

As shown in Fig. 2, Point (l, Θ) is denoted by O'. Establishing the Cartesian coordinate system with the frame (O', $\boldsymbol{i}$, $\boldsymbol{j}$), let x, y be the Cartesian coordinates.

Considering ***the displacement-time relations of uniformly variable motion***:

$$\begin{cases} x = v_\Theta \sin\Theta \cdot t & \text{(15a)} \\ y = v_\Theta \cos\Theta \cdot t - \dfrac{1}{2}gt^2 & \text{(15b)} \end{cases}$$

(14b) and (14c) $\rightarrow$ (15), we obtain the trajectory equation of the ball in Process 3:

$$\Rightarrow \qquad y = -\frac{4}{3\sqrt{3}l}x^2 + \frac{\sqrt{3}}{3}x \qquad (16)$$

Considering the following geometric relation in Fig. 2:

$$r = \sqrt{(x - x_O)^2 + (y - y_O)^2} \qquad (17)$$

r—the distance between the ball and Point O at arbitrary Time t in Process 3.

Considering the following geometric relations in Fig. 2:

$$\begin{cases} x_O = l\cos\Theta & (18a) \\ y_O = -l\sin\Theta & (18b) \end{cases}$$

Combining (14c), (17) and (18):

$$\Rightarrow \qquad r = \sqrt{\left(x - \frac{1}{2}l\right)^2 + \left(y + \frac{\sqrt{3}}{2}l\right)^2}$$

$$= \sqrt{\frac{16}{27l^2}x^4 - \frac{8}{9l}x^3 + l^2}$$

$$= \sqrt{x^3\left(\frac{16}{27l^2}x - \frac{8}{9l}\right) + l^2} \qquad (19)$$

Considering the following geometric relation in Fig. 2, then place (14c) in to obtain:

$$x \le l\cos\Theta + l = \frac{3}{2}l \qquad (20)$$

$(20) \to (19)$:

$$\Rightarrow \qquad r \le l \qquad (21)$$

Assumption II has been proved.

2.1.5 The Elliptic Pendulum Formed by a Small Ring Which is Sleeved on a Light Rigid String (*a*, *A*)

As shown in Fig. 1, there is a light rigid string of length L, the two ends of which are fixed at the Point O_1 and Point O_2 on the ceiling, respectively, $l<L$. A small ring of mass m is sleeved on the string. Pull the small ring to the leftmost point, then release the ring freely, ignore the friction between the small ring and the string. Determine the tension in the string when the small ring moves to an arbitrary position.

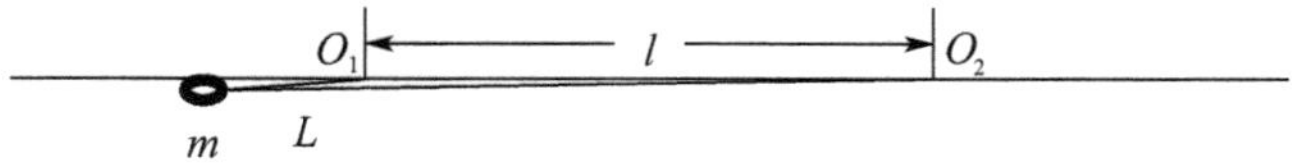

Figure 1

Solution

The time point when the ring is released is defined as Time zero.

As shown in Fig. 2, establishing the polar coordinate system, let r, θ be the polar coordinates. The position of the ring at arbitrary Time t is denoted by A.

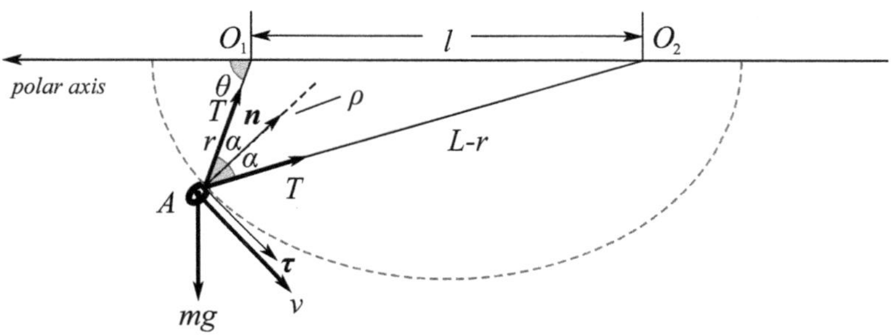

Figure 2

According to the definition of ellipse, it can be seen that the trajectory of the ring is an ellipse with Point O_1 and Point O_2 as the focuses, the trajectory equation is as follows:

$$r = \frac{p}{1 + e\cos\theta} \tag{1}$$

e and p—the eccentricity and semi latus rectum of the elliptic trajectory of the ring.

Defining a—the length of semi-major axis, b—the length of semi-minor axis, c—the focal distance. Considering the definitions of the ellipse parameters and relations between the ellipse parameters:

$$\begin{cases} a = \dfrac{1}{2}L & \text{(2a)} \\ c = \dfrac{1}{2}l & \text{(2b)} \end{cases}$$

$$\Rightarrow \qquad b = \sqrt{a^2 - c^2} = \frac{1}{2}\sqrt{L^2 - l^2} \tag{3}$$

$$\Rightarrow \qquad \begin{cases} p = \dfrac{b^2}{a} = \dfrac{1}{2}\dfrac{L^2 - l^2}{L} & \text{(4a)} \\ e = \dfrac{c}{a} = \dfrac{l}{L} & \text{(4b)} \end{cases}$$

Taking *the ring (can be regarded as a particle)* as the object of study. Studying *the process from Time zero to arbitrary Time t*. Considering ***theorem of kinetic energy*** [the work done by the tension in the string is zero]:

$$mgr\sin\theta = \frac{1}{2}mv^2 \tag{5}$$

v—the speed of the ring at Time t.

$$\Rightarrow \qquad v = \sqrt{2g\frac{p\sin\theta}{1+e\cos\theta}} \tag{6}$$

Introducing the natural frame $(\boldsymbol{\tau}, \boldsymbol{n})$ of the ring.

Considering the following geometric relation: the angle between the normal of an arbitrary point of an ellipse and the two focus chords passing through that point are equal. The angle between the segments of the string on both sides of the ring and $\boldsymbol{n}$ at arbitrary Time t is denoted by α.

Taking *the ring* as the object of study. Studying *the state at arbitrary Time t*. Considering ***Newton's second law*** (the $\boldsymbol{n}$-direction component equation):

$$2T\cos\alpha - mg\sin(\theta-\alpha) = m\frac{v^2}{\rho} \tag{7}$$

ρ—the radius of curvature of the ring's trajectory at the position where the ring is at Time t.

Considering the form of the radius of curvature at an arbitrary point on the plane curve in polar coordinate system:

$$\rho = \frac{\left[r^2 + \left(\frac{\mathrm{d}r}{\mathrm{d}\theta}\right)^2\right]^{3/2}}{\left|r^2 + 2\left(\frac{\mathrm{d}r}{\mathrm{d}\theta}\right)^2 - r\frac{\mathrm{d}^2r}{\mathrm{d}\theta^2}\right|} \tag{8}$$

(1) → (8),

$$\Rightarrow \qquad \rho = p\left[\frac{1+2e\cos\theta+e^2}{(1+e\cos\theta)^2}\right]^{3/2} \tag{9}$$

Considering that ΔO_1AO_2 satisfy cosine theorem (see Fig. 2):

$$\cos\angle O_1AO_2 = \frac{\overline{O_1A}^2 + \overline{O_2A}^2 - \overline{O_1O_2}^2}{2\overline{O_1A}\cdot\overline{O_2A}} \tag{10}$$

$\Rightarrow$
$$\cos 2\alpha = \frac{r^2 + (L-r)^2 - l^2}{2r(L-r)} \tag{11}$$

Considering the following trigonometric function transformations:

$\Rightarrow$
$$\begin{cases} \sin\alpha = \sqrt{\dfrac{1-\cos 2\alpha}{2}} = \dfrac{e\sin\theta}{\sqrt{1+2e\cos\theta+e^2}} & \text{(12a)} \\ \cos\alpha = \sqrt{\dfrac{1+\cos 2\alpha}{2}} = \dfrac{1+e\cos\theta}{\sqrt{1+2e\cos\theta+e^2}} & \text{(12b)} \end{cases}$$

(6), (9) and (12) $\rightarrow$ (7),

$$T = \frac{m}{2\cos\alpha}\left[\frac{v^2}{\rho} + g\sin(\theta-\alpha)\right]$$

$$= \frac{m}{2\cos\alpha}\left[\frac{v^2}{\rho} + g(\sin\theta\cos\alpha - \cos\theta\sin\alpha)\right]$$

$$= \frac{1}{2}mg\sin\theta\left[\frac{1}{1+e\cos\theta} + \frac{2(1+e\cos\theta)}{1+2e\cos\theta+e^2}\right] \tag{13}$$

2.1.6 The Equilibrium Problem of Two Small Balls Constrained Each Other by a Light Rigid String on a Rough Plane (d, *C*)

As shown in Fig. 1, two small balls 1 and 2 of equal mass are, respectively, tied at both ends of a light rigid string, the whole is placed on an inclined plane of inclination angle θ. Given the friction coefficients between the balls and the inclined plane are μ_1 (Ball 1) and μ_2 (Ball 2), respectively, $\mu_1 < \mu_2$, $\tan\theta = \sqrt{\mu_1\mu_2}$. When the string is straightened and the system is in mechanical equilibrium, the angle between the string and the line of the maximum slope on the inclined plane is denoted by α. Determine the maximum value of α.

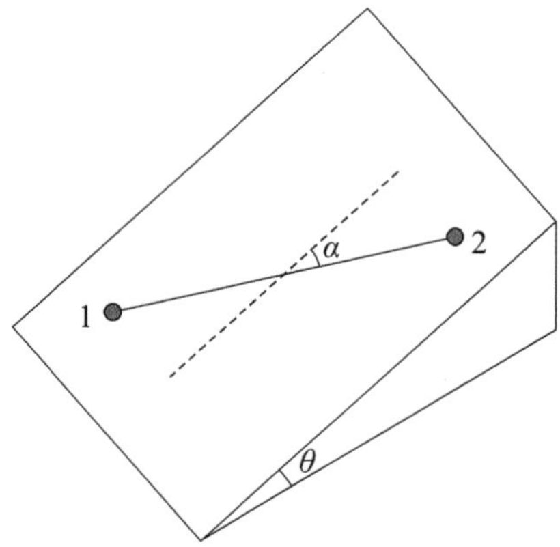

Figure 1

Solution

As shown in Fig. 2, the unit vector in the upward direction of the normal of the inclined plane is denoted by $\boldsymbol{n}$, the unit vector in the downward direction of the line of the maximum slope on the inclined plane is denoted by $\boldsymbol{e}_1$, when the system is in mechanical equilibrium, the unit vector of Ball 1 pointing to Ball 2 is denoted by $\boldsymbol{e}_2$.

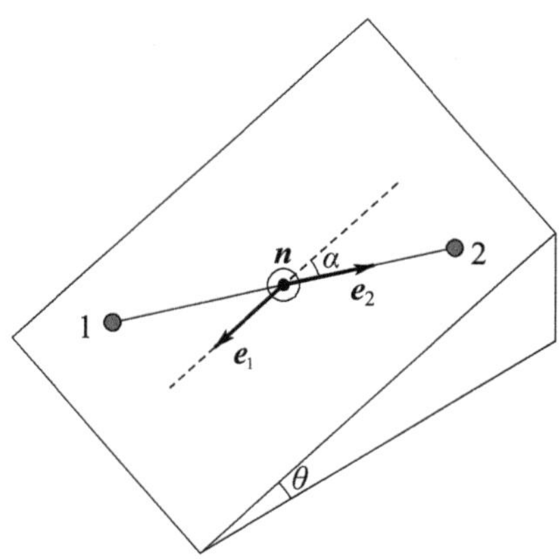

Figure 2

Taking *Ball 1 and Ball 2 (both can be regarded as a particle)* as the object of study, respectively. Studying *the state when the system is in mechanical equilibrium*. Considering ***Newton's second law*** (the $\boldsymbol{n}$-direction component equation):

$$\begin{cases} N_1 - mg\cos\theta = 0 & \text{(1a)} \\ N_2 - mg\cos\theta = 0 & \text{(1b)} \end{cases}$$

N_1 and N_2—the ***n***-direction component of the normal contact force exerted on Ball 1 and Ball 2 by the inclined plane when the system is in mechanical equilibrium.

Considering ***friction law***:

$$\begin{cases} f_{1\max} = \mu_1 N_1 & \text{(2a)} \\ f_{2\max} = \mu_2 N_2 & \text{(2b)} \end{cases}$$

$f_{1\max}$ and $f_{2\max}$—the maximum static friction force exerted on Ball 1 and Ball 2 by the inclined plane.

$$\Rightarrow \quad \begin{cases} f_{1\max} = \mu_1 mg\cos\theta & \text{(3a)} \\ f_{2\max} = \mu_2 mg\cos\theta & \text{(3b)} \end{cases}$$

Taking *Ball 1 and Ball 2* as the object of study, respectively. Studying *the state when the system is in mechanical equilibrium*. Considering ***Newton's second law*** (the projection equation in the inclined plane):

$$\begin{cases} mg\sin\theta\boldsymbol{e}_1 + \boldsymbol{T} + \boldsymbol{f}_1 = 0 & \text{(4a)} \\ mg\sin\theta\boldsymbol{e}_1 - \boldsymbol{T} + \boldsymbol{f}_2 = 0 & \text{(4b)} \end{cases}$$

T—the tension in the string acting on Ball 1 when the system is in mechanical equilibrium.

$\boldsymbol{f}_1$ and $\boldsymbol{f}_2$—the static friction force exerted on Ball 1 and Ball 2 by the inclined plane when the system is in mechanical equilibrium.

The schematic diagrams corresponding to the equations above are as follows (see Fig. 3):

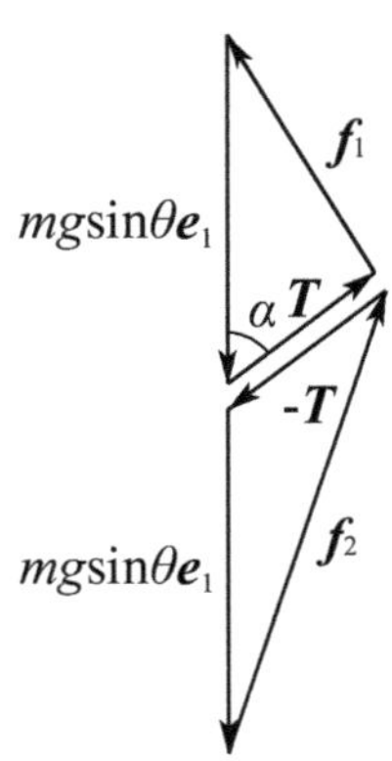

Figure 3

Situation 1

When α is taken as the maximum value, Ball 1 reaches the maximum relative motion trend relative to the inclined plane, that is, the friction exerted on Ball 1 by the inclined plane is the maximum static friction force. Considering the constraint of Ball 1 by the string, it can be seen that the maximum static friction force is perpendicular to the string:

$$\boldsymbol{f}_1 = \boldsymbol{f}_{1\max} \perp \boldsymbol{e}_2 \tag{5}$$

We can obtain the schematic diagram corresponding to this situation (see Fig. 4a).

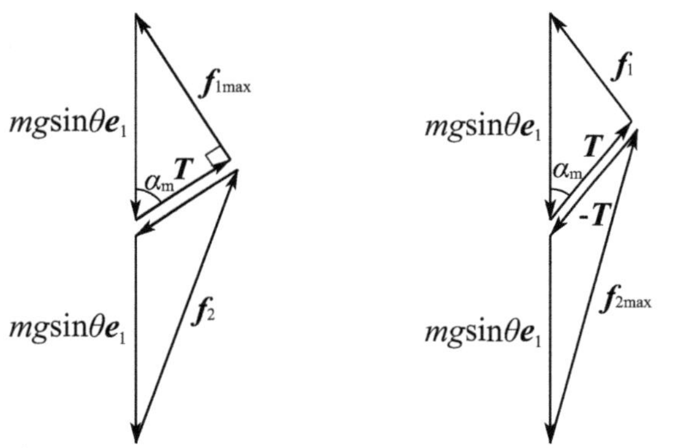

Figure 4a,b

Considering the following geometric relations in Fig. 4a:

$$\Rightarrow \qquad \begin{cases} f_{1\max} = mg\sin\theta\sin\alpha_m & \text{(6a)} \\ T = mg\sin\theta\cos\alpha_m & \text{(6b)} \end{cases}$$

Known:

$$\Rightarrow \qquad \tan\theta = \sqrt{\mu_1\mu_2} \tag{7}$$

Combining (2a), (6a) and (7),

$$\Rightarrow \qquad \sin\alpha_m = \sqrt{\frac{\mu_1}{\mu_2}} \tag{8}$$

Considering the following geometric relation in Fig. 4a:

$$\Rightarrow \qquad f_2^2 = (mg\sin\theta)^2 + T^2 + 2mg\sin\theta T\cos\alpha_m \tag{9}$$

(6b) → (9),

$$\Rightarrow \qquad f_2 = mg\sin\theta\sqrt{1+3\cos^2\alpha_m} \qquad (10)$$

Considering ***friction law***:

$$f_2 \le f_{2\max} \qquad (11)$$

(3b), (10) → (11),

$$\Rightarrow \qquad \tan^2\theta(1+3\cos^2\alpha_m) \le \mu_2^2 \qquad (12)$$

(7) and (8) → (12), we can obtain the condition for Situation 1 to hold:

$$\Rightarrow \qquad 3\left(\frac{\mu_1}{\mu_2}\right)^2 - 4\frac{\mu_1}{\mu_2} + 1 \ge 0 \qquad (13)$$

$$\Rightarrow \qquad \frac{\mu_1}{\mu_2} \ge 1 \quad \text{or} \quad \frac{\mu_1}{\mu_2} \le \frac{1}{3} \qquad (14)$$

Situation 2

Taking the complementary set of in Eq. (14):

$$\mu_1 < \mu_2 < 3\mu_1 \qquad (15)$$

Under these conditions, when α takes the maximum value, Ball 2 reaches the maximum relative motion trend relative to the inclined plane, that is, the friction exerted on Ball 2 by the inclined plane is the maximum static friction force:

$$\boldsymbol{f}_2 = \boldsymbol{f}_{2\max} \qquad (16)$$

We can obtain the schematic diagram corresponding to this situation (see Fig. 4b).

Considering the following geometric relations in Fig. 4b:

$$\begin{cases} f_1^2 = (mg\sin\theta)^2 + T^2 - 2mg\sin\theta T\cos\alpha_m & (17a) \\ f_{2\max}^2 = (mg\sin\theta)^2 + T^2 + 2mg\sin\theta T\cos\alpha_m & (17b) \end{cases}$$

Considering ***friction law***:

$$f_1 \le f_{1\max} \qquad (18)$$

Combining (3), (17) and (18),

$$\Rightarrow \qquad \begin{cases} (mg\sin\theta)^2 + T^2 - 2mg\sin\theta T\cos\alpha_m \le (\mu_1 mg\cos\theta)^2 & (19a) \\ (mg\sin\theta)^2 + T^2 + 2mg\sin\theta T\cos\alpha_m = (\mu_2 mg\cos\theta)^2 & (19b) \end{cases}$$

(19a) plus (19b),

$$\Rightarrow \qquad T \le mg\cos\theta\sqrt{\frac{1}{2}(\mu_1^2+\mu_2^2)-\tan^2\theta}$$

$$= mg\cos\theta\sqrt{\frac{1}{2}(\mu_1^2+\mu_2^2)-\mu_1\mu_2}$$

$$= \frac{\sqrt{2}}{2}(\mu_2-\mu_1)mg\cos\theta \qquad (20)$$

(19b) minus (19a),

$$\Rightarrow \qquad T \ge \frac{1}{4}mg(\mu_2^2-\mu_1^2)\frac{\cos\theta}{\tan\theta\cos\alpha_m} = \frac{1}{4}\frac{\mu_2^2-\mu_1^2}{\sqrt{\mu_1\mu_2}}\frac{mg\cos\theta}{\cos\alpha_m} \qquad (21)$$

Combining (20) and (21),

$$\Rightarrow \qquad \cos\alpha_m \ge \frac{\sqrt{2}}{4}\frac{\mu_1+\mu_2}{\sqrt{\mu_1\mu_2}} \qquad (22)$$

Inequalities (14) and (22) are the conditions for Situation 2 to hold.

2.1.7 The Quasi-Static Motion of a Small Charged Ball under Several Forces (*c*, *AC*)[1]

There is a horizontal table top with a small hole on it (see Fig. 1), the position of the hole is denoted by *O*, the unit vector in the vertical upward direction is denoted by $\boldsymbol{k}$, establishing the Cartesian coordinate system with the frame $(O, \boldsymbol{i}, \boldsymbol{j}, \boldsymbol{k})$, let x, y, z be the Cartesian coordinates. A small ball with negative electricity is statically placed at Point $(x_0, 0, 0)$ on the table top. A light rigid string, one end of which is tied with the ball, and the other end passes through the hole. Apply a uniform electric field in the space, whose electric field intensity is located in the *O-yz* plane, the angle with $\boldsymbol{k}$ is θ. Given the friction coefficient between the ball and the table top is μ, $\mu = \tan\theta$. Initially, the string is straightened, pull the free end of the string slowly to make the ball perform a quasi-static motion. Determine the trajectory of the ball.

[1]This problem is a test question of the 16th CphO (China Physics Olympiads) in 1999.

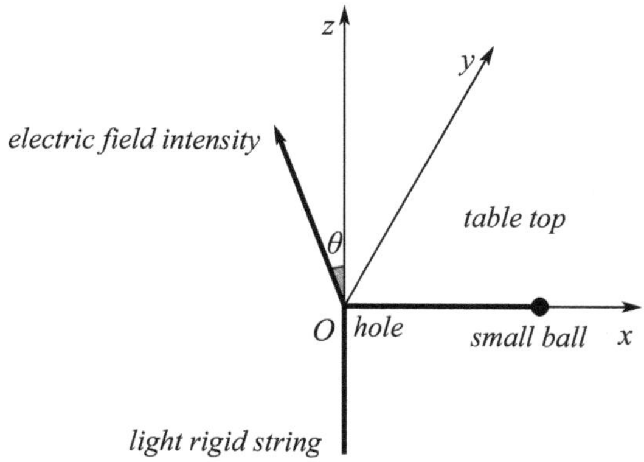

Figure 1

Solution [1] (type C)

The time point when the ball starts to move is defined as Time zero.

Taking *the ball (can be regarded as a particle)* as the object of study. Studying *the state at arbitrary Time t*. Considering ***Newton's second law*** (the component equations in the (***i***, ***j***, ***k***)-directions), (b) ***friction law***:

$$\begin{cases} -T\cos\alpha + f\cos\beta = 0 & \text{(1a)} \\ qE\sin\theta - T\sin\alpha - f\sin\beta = 0 & \text{(1b)} \\ N - qE\cos\theta = 0 & \text{(1c)} \end{cases}$$

$$f = \mu N \quad (2)$$

q—the magnitude of the electric quantity of the ball.

E—the magnitude of the electric field intensity.

T and f—the magnitude of the tension in the string and the sliding friction exerted on the ball by the table top at Time t.

N—the ***k***-direction component of the normal contact force exerted on the ball by the table top at Time t.

α—the angle between the tension in the string acting on the ball and the −***i***-direction at Time t.

β—the angle between the sliding friction acting on the ball and the ***i***-direction at Time t.

Known:

$$\mu = \tan\theta \tag{3}$$

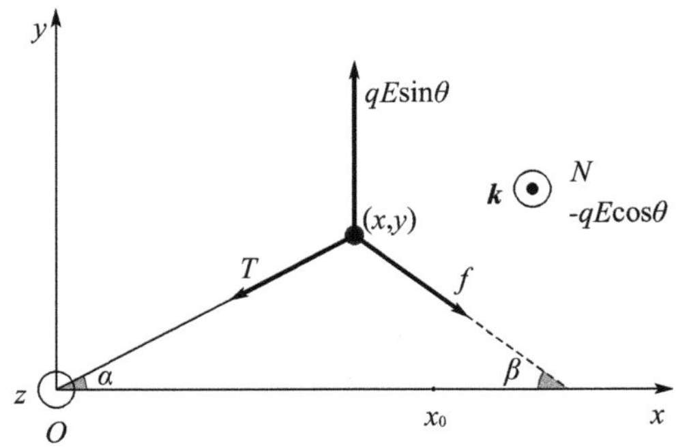

Figure 2

Combining (1), (2) and (3),

$$\Rightarrow \begin{cases} T = 2qE\sin\theta\sin\alpha & \text{(4a)} \\ N = qE\cos\theta & \text{(4b)} \\ f = qE\sin\theta & \text{(4c)} \\ \tan\beta = \cot 2\alpha & \text{(4d)} \end{cases}$$

Considering the following geometric relation in Fig. 2:

$$\tan\alpha = \frac{y}{x} \tag{5}$$

Taking the following trigonometric function transformation:

$$\cot 2\alpha = \frac{1-\tan^2\alpha}{2\tan\alpha} \tag{6}$$

Combining (4d), (5) and (6),

$$\Rightarrow \quad \tan\beta = \frac{x^2-y^2}{2xy} \tag{7}$$

The direction of the sliding friction acting on the ball is the tangent direction of its trajectory. Considering the geometric meaning of derivative:

$$\tan(\pi-\beta) = \frac{\mathrm{d}y}{\mathrm{d}x} \tag{8}$$

$$\Rightarrow \qquad \frac{\mathrm{d}y}{\mathrm{d}x}=\frac{y}{2x}-\frac{x}{2y} \tag{9}$$

Defining the integration variable u:

$$u=\frac{y}{x} \tag{10}$$

$$\Rightarrow \qquad y = ux$$

$$\Rightarrow \qquad \frac{\mathrm{d}y}{\mathrm{d}x}=u+x\frac{\mathrm{d}u}{\mathrm{d}x} \tag{11}$$

(10) and (11) $\rightarrow$ (9),

$$\Rightarrow \qquad u+x\frac{\mathrm{d}u}{\mathrm{d}x}=\frac{u}{2}-\frac{1}{2u} \tag{12}$$

$$\Rightarrow \qquad \frac{2u\mathrm{d}u}{u^2+1}=-\frac{\mathrm{d}x}{x}$$

$$\Rightarrow \qquad \int_0^{\frac{y}{x}}\frac{2u\mathrm{d}u}{u^2+1}=-\int_{x_0}^{x}\frac{\mathrm{d}x}{x}$$

$$\Rightarrow \qquad \ln\left[\left(\frac{y}{x}\right)^2+1\right]=\ln\frac{x_0}{x}$$

$$\Rightarrow \qquad \left(\frac{y}{x}\right)^2+1=\frac{x_0}{x}$$

$$\Rightarrow \qquad \left(x-\frac{x_0}{2}\right)^2+y^2=\left(\frac{x_0}{2}\right)^2 \tag{13}$$

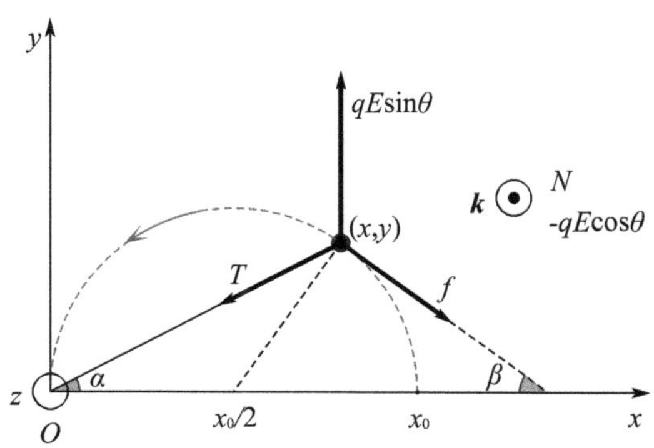

Figure 3

That is, the trajectory of the ball is the semicircle with Point $\left(\frac{x_0}{2}, 0\right)$ as the center and $\frac{x_0}{2}$ as the radius (see Fig. 3).

Solution [2] (type A)

As shown in Fig. 4, taking the origin O as the pole, the positive x-axis as the polar axis, the anticlockwise direction as the positive direction of polar angle, establishing the polar coordinate system, let r, α be the polar coordinates.

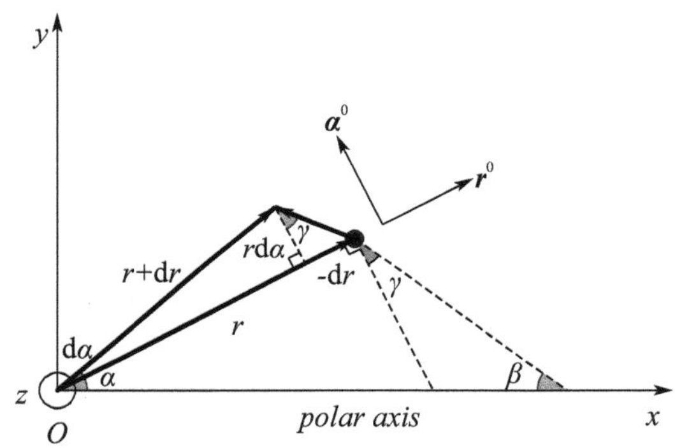

Figure 4

Studying the instantaneous process of the ball moving from Point (r, α) to Point $(r + \mathrm{d}r, \alpha + \mathrm{d}\alpha)$. Considering the following geometric relations in Fig. 4:

$$\tan\gamma = \frac{-\mathrm{d}r}{r\mathrm{d}\alpha} \tag{14}$$

$$\gamma \approx \pi - (\alpha + \beta) - \frac{\pi}{2} = \frac{\pi}{2} - (\alpha + \beta) \tag{15}$$

$$\Rightarrow \quad \tan\gamma = \cot(\alpha + \beta) = \frac{1 - \tan\beta\tan\alpha}{\tan\alpha + \tan\beta} \tag{16}$$

(4d) → (16),

$$\Rightarrow \quad \tan\gamma = \frac{1 - \cot 2\alpha\tan\alpha}{\tan\alpha + \cot 2\alpha} = \frac{\tan 2\alpha - \tan\alpha}{\tan 2\alpha\tan\alpha + 1} \tag{17}$$

Taking the following trigonometric function transformation:

$$\tan 2\alpha = \frac{2\tan\alpha}{1-\tan^2\alpha} \tag{18}$$

(18) → (17),

$$\Rightarrow \quad \tan\gamma = \tan\alpha$$

$$\Rightarrow \quad \gamma = \alpha \tag{19}$$

(19) → (14),

$$\Rightarrow \quad \frac{\mathrm{d}r}{r} = -\tan\alpha\,\mathrm{d}\alpha \tag{20}$$

$$\Rightarrow \quad \int_{x_0}^{r} \frac{\mathrm{d}r}{r} = -\int_0^{\alpha} \tan\alpha\,\mathrm{d}\alpha$$

$$\Rightarrow \quad \ln\frac{r}{x_0} = \ln\cos\alpha$$

$$\Rightarrow \quad r = x_0\cos\alpha \tag{21}$$

Equations (21) and (13) describe the same circular trajectory.

2.1.8 The Reciprocating Collisions of a Small Ball between Two Slowly Approaching Parallel Flat Boards (d, *AB*)

As shown in Fig. 1, two flat boards are fixed and parallel to each other, with the distance x_0, a small ball of mass m performs reciprocating motion between the two boards in the path perpendicular to them. Given that the collisions between the ball and the two boards are completely elastic, and the speed of the ball is v_0 all the time. The right board starts to move towards the left board at a constant speed u ($u << v_0$) under an external force.

(1) Find the speed of the ball when the right board moves to an arbitrary position.

(2) Find the work done by the external force acting on the right board in the process from the beginning till the right board moves to an arbitrary position.

(3) Find the external force acting on the right board when the right board moves to an arbitrary position.

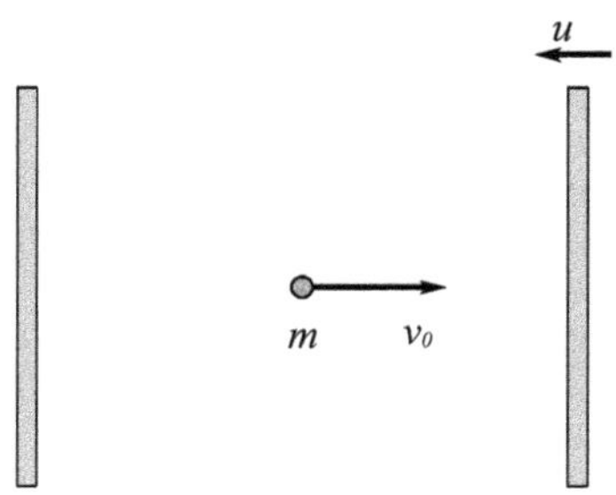

Figure 1

Solution to Problem (1) (type B)

The time point when the right board starts to move is defined as Time zero.

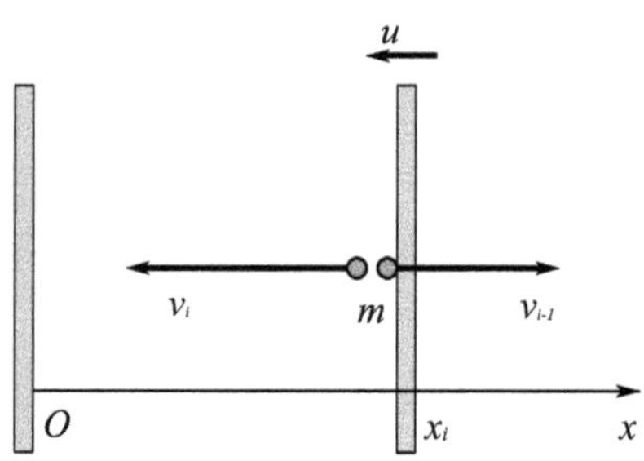

Figure 2

As shown in Fig. 2, taking an arbitrary point on the left board and denoting it by *O*, the unit vector in the horizontal right direction is denoted by ***i***. Establishing the linear coordinate system with Point *O* as the origin and the ***i***-direction as the positive direction of coordinate axis, let *x* be the coordinate.

Taking *the system composed of the ball and the right board (can be regarded as a particle and a rigid body, respectively)* as the object of study. Studying *the instantaneous process of the i-th collision between the two*. Considering ***Newton-collision-formula for completely elastic collision*** [for the ball, its collision with each board can be regarded as the completely elastic collision with the object of infinite mass]:

$$\frac{v_i - u}{u - (-v_{i-1})} = 1 \tag{1}$$

v_i—the speed of the ball after its i-th collision with the right board (i.e., before its $(i + 1)$-th collision with the right board).

$$\Rightarrow \quad v_i = v_{i-1} + 2u \tag{2}$$

Defining Δv_i—the speed increment that the ball obtained in its i-th collision with the right board:

$$\Rightarrow \quad \Delta v_i = v_i - v_{i-1} = 2u \tag{3}$$

Considering ***the displacement-time relation of uniform linear motion***:

$$\begin{cases} \Delta x_i = -u\Delta t_i & \text{(4a)} \\ x_i = v_i \Delta t_{i(\text{left})} & \text{(4b)} \\ x_{i+1} = (v_i + u)\Delta t_{i(\text{right})} & \text{(4c)} \end{cases}$$

x_i—the coordinate of the right board when the ball collides with it for the i-th time.

Δx_i—the displacement of the right board in the process from the i-th collision between it and the ball to the $(i + 1)$-th collision.

Δt_i, $\Delta t_{i(\text{left})}$, $\Delta t_{i(\text{right})}$—the total time of the process from the i-th collision between the ball and the right board to the $(i + 1)$-th collision, and the motion time of the ball to the left and right in this process.

Considering the following relations:

$$\begin{cases} \Delta x_i = x_{i+1} - x_i & \text{(5a)} \\ \Delta t_i = \Delta t_{i(\text{left})} + \Delta t_{i(\text{right})} & \text{(5a)} \end{cases}$$

Combining (4) and (5),

$$\Rightarrow \quad \Delta t_i = \frac{x_i}{v_i} + \frac{x_i - u\Delta t_i}{v_i + u} = \frac{x_i}{v_i} + \frac{x_i - u\Delta t_i}{v_i}\left(1 + \frac{u}{v_i}\right)^{-1} \approx \frac{2x_i - u\Delta t_i}{v_i} \tag{6}$$

$$\Rightarrow \quad \Delta t_i \approx \frac{2x_i}{v_i + u} = \frac{2x_i}{v_i}\left(1 + \frac{u}{v_i}\right)^{-1} \approx \frac{2x_i}{v_i} \tag{7}$$

The two operations above have considered u/v_i as a small quantity and taken the low order approximations.

(3) → (4a),

$$\Delta x_i = -\frac{1}{2}\Delta v_i \Delta t_i \tag{8}$$

Combining (7) and (8),

$$\Rightarrow \qquad \frac{\Delta x_i}{\Delta v_i} = -\frac{x_i}{v_i} \tag{9}$$

Δx_i and Δv_i are small quantities; therefore, it can be assumed that the coordinate of the right board and the speed of the ball change continuously with time. Defining x—the coordinate of the right board at arbitrary Time t, v—the velocity of the ball at arbitrary Time t, Δx_i and Δv_i can be regarded as the differentiation of x and v, respectively:

$$\Rightarrow \qquad \frac{\mathrm{d}x}{\mathrm{d}v} = -\frac{x}{v} \tag{10}$$

$$\Rightarrow \qquad \frac{\mathrm{d}v}{v} = -\frac{\mathrm{d}x}{x}$$

$$\Rightarrow \qquad \int_{v_0}^{v} \frac{\mathrm{d}v}{v} = -\int_{x_0}^{x} \frac{\mathrm{d}x}{x}$$

$$\Rightarrow \qquad \ln\frac{v}{v_0} = \ln\frac{x_0}{x}$$

$$\Rightarrow \qquad v = \frac{x_0}{x} v_0 \tag{11}$$

Solution to Problem (2) (type A)

Taking *the system composed of the ball and the right board* as the object of study. Studying *the instantaneous process at arbitrary Time t*. Considering ***theorem of kinetic energy***:

$$f\mathrm{d}x = \mathrm{d}\left(\frac{1}{2}mv\right)^2 \tag{12}$$

$$\Rightarrow \qquad f = mv\frac{\mathrm{d}v}{\mathrm{d}x} = -m\frac{(x_0 v_0)^2}{x^3} \tag{13}$$

f—the ***i***-direction component of the average external force acting on the right board in the instantaneous process at Time t.

Solution [1] of Problem (3) (type A)

Taking *the system composed of the ball and the right board* as the object of study. Studying *the process from Time zero to arbitrary Time t*. Considering ***theorem of kinetic energy***:

$$W=\frac{1}{2}mv^2-\frac{1}{2}mv_0^2 \tag{14}$$

W—the work done by the external force acting on the right board in the process from Time zero to Time t.

(11) → (14),

$$\Rightarrow \quad W=\frac{1}{2}mv_0^2\left(\frac{x_0^2}{x^2}-1\right) \tag{15}$$

Solution [2] of Problem (3) (type A)

Taking *the system composed of the ball and the right board* as the object of study. Studying *the instantaneous process of the i-th collision between the two*. Considering ***theorem of kinetic energy***:

$$W_i=\frac{1}{2}mv_i^2-\frac{1}{2}mv_{i-1}^2 \tag{16}$$

W_i—the work done by the external force acting on the right board in the i-th collision between the two.

(3) → (16),

$$\Rightarrow \quad W_i=\frac{1}{2}m(v_i+v_{i-1})(v_i-v_{i-1})=2mv_iu-2mu^2\approx 2mv_iu \tag{17}$$

Combining (4a) and (7):

$$\Rightarrow \quad u=-v_i\frac{\Delta x_i}{2x_i} \tag{18}$$

From (11),

$$\Rightarrow \quad v_i=\frac{x_0}{x_i}v_0 \tag{19}$$

$$\Rightarrow \quad u=-v_0x_0\frac{\Delta x_i}{2x_i^2} \tag{20}$$

(19) and (20) → (17),

$$\Rightarrow \qquad W_i = -mv_0^2 x_0^2 \frac{\Delta x_i}{x_i^3} \tag{21}$$

$$\Rightarrow W = \sum W_i = -mv_0^2 x_0^2 \sum \frac{\Delta x_i}{x_i^3} = -mv_0^2 x_0^2 \int_{x_0}^{x} \frac{\mathrm{d}x}{x^3} = \frac{1}{2} mv_0^2 \left(\frac{x_0^2}{x^2} - 1 \right) \tag{22}$$

Solution [3] of Problem (3) (type C)

Considering the definition of ***work***, then place (13) in to obtain:

$$\Rightarrow \qquad W = \int_{x_0}^{x} f\mathrm{d}x = -m(x_0 v_0)^2 \int_{x_0}^{x} \frac{\mathrm{d}x}{x^3} = \frac{1}{2} mv_0^2 \left(\frac{x_0^2}{x^2} - 1 \right) \tag{23}$$

2.2 Charged Particle in Electromagnetic Field

The dynamics problems of charged particles in electromagnetic field have important research value in view of its fundamentality.

As the components of electromagnetic tensor, the value of electric field intensity and magnetic field intensity depends on the selection of reference frame, which is beyond the research scope of classical physics. Therefore, in classical physics, the dynamics problems of charged particles in electromagnetic field can only be studied in a single reference frame.

Because of the complexity of the basic form of electromagnetic force and the variety of forms that real electromagnetic fields possibly have, the dynamic equation of charged particles in electromagnetic field often has complex mathematical form. In some simple electromagnetic fields (such as uniform electrostatic field, electrostatic field excited by point charge, uniform magnetostatic field, etc.), the motion of charged particles can be solved accurately. In more complex electromagnetic fields, the dynamic problems of charged particles that can be solved accurately or approximately is worthy of studying (such as in superposition field of uniform electrostatic field and uniform magnetostatic field, gradual change magnetostatic field, etc.), these problems will be studied in this section.

2.2.1 The Motion of a Charged Particle in an Orthogonal Electromagnetic Field (b, C)

An orthogonal electromagnetic field composed of a uniform electric field with the electric field intensity $(0, E, 0)$ and a uniform magnetic field with the magnetic induction intensity $(0, 0, B)$ is applied in space (see Fig. 1), in which a charged particle of mass m and electric charge q is moving, $q > 0$. Initially, the particle is at the origin O, and the initial velocity is $(v_0, 0, 0)$. Determine the kinematics equation of the particle and then draw the schematic diagrams of its trajectories for several special situations.

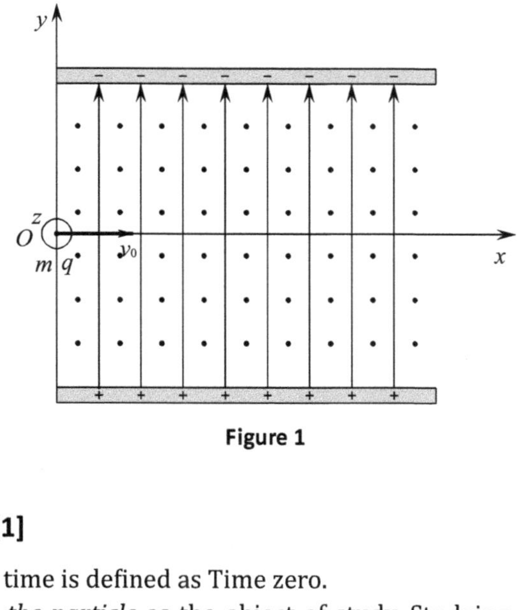

Figure 1

Solution [1]

The initial time is defined as Time zero.

Taking *the particle* as the object of study. Studying *the state at arbitrary Time t*. Considering ***Newton's second law*** (the ***i***-direction and ***j***-direction component equations):

$$\begin{cases} qB\dot{y} = m\ddot{x} & \text{(1a)} \\ qE - qB\dot{x} = m\ddot{y} & \text{(1b)} \end{cases}$$

$\Rightarrow$
$$qB\text{d}y = m\dot{x}$$

$\Rightarrow$
$$qBy = qB\int_0^y \text{d}y = m\int_{v_0}^{\dot{x}} \dot{x} = m(\dot{x} - v_0)$$

$$\Rightarrow \qquad \dot{x} = \frac{q}{m} By + v_0 \qquad (2)$$

$$\Rightarrow \qquad \ddot{y} + \frac{q^2}{m^2} B^2 y = -\frac{q}{m} B(v_0 - E/B) \qquad (3)$$

Equation (3) is the differential equation of the particle performing simple harmonic vibration parallel to the y-axis, and its angular frequency ω is:

$$\Rightarrow \qquad \omega = \frac{q}{m} B \qquad (4)$$

The general solution of Eq. (3) is:

$$\Rightarrow \qquad y = A\cos(\omega t + \varphi) - \frac{m}{q} \frac{v_0 - E/B}{B} \qquad (5)$$

A and φ—the amplitude and initial phase of the simple harmonic vibration.

$$\Rightarrow \qquad \dot{y} = -\omega A \sin(\omega t + \varphi) \qquad (6)$$

Considering the initial conditions:

$$\begin{cases} y|_{t=0} = 0 & (7a) \\ \dot{y}|_{t=0} = 0 & (7b) \end{cases}$$

(5) and (6) $\rightarrow$ (7),

$$\Rightarrow \qquad \begin{cases} A = \frac{m}{q} \frac{v_0 - E/B}{B} & (8a) \\ \varphi = 0 & (8b) \end{cases}$$

$$\Rightarrow \qquad y = \frac{m}{q} \frac{v_0 - E/B}{B} (\cos \omega t - 1) \qquad (9)$$

$$\Rightarrow \qquad \dot{x} = (v_0 - E/B)\cos \omega t + E/B$$

$$\Rightarrow \qquad \mathrm{d}x = [(v_0 - E/B)\cos \omega t + E/B]\mathrm{d}t$$

$$\Rightarrow \qquad x = \int_0^x \mathrm{d}x = \int_0^t \left[(v_0 - E/B)\cos \omega t + E/B\right]\mathrm{d}t$$

$$= \frac{v_0 - E/B}{\omega} \sin \omega t + (E/B)t$$

$$= \frac{m}{q}\frac{v_0 - E/B}{B}\sin\omega t + (E/B)t \tag{10}$$

That is, the projection motion of the particle on the *x*-axis is the superposition of a simple harmonic vibration and a uniform linear motion.

Solution [2]

Taking *the particle* as the object of study. Studying *the state at arbitrary Time t*. Considering ***Newton's second law***:

$$qE\boldsymbol{j} + q\boldsymbol{v} \times B\boldsymbol{k} = m\boldsymbol{a} \tag{11}$$

The motion of the particle can be decomposed into any two independent Sub-motions I and II:

$$\begin{cases} \boldsymbol{v} = \boldsymbol{v}_1 + \boldsymbol{v}_2 & \text{(12a)} \\ \boldsymbol{a} = \boldsymbol{a}_1 + \boldsymbol{a}_2 & \text{(12b)} \end{cases}$$

$\boldsymbol{v}_1$ and $\boldsymbol{v}_2$—the velocity of the particle's Sub-motion I and II at arbitrary Time *t*.

$\boldsymbol{a}_1$ and $\boldsymbol{a}_2$—the acceleration of the particle's Sub-motion I and II at arbitrary Time *t*.

$$\Rightarrow \quad qE\boldsymbol{j} + q(\boldsymbol{v}_1 + \boldsymbol{v}_2) \times B\boldsymbol{k} = m(\boldsymbol{a}_1 + \boldsymbol{a}_2) \tag{13}$$

Studying the following special case: the particle's Sub-motion I is the uniform linear motion under the action of the electric field force and the magnetic field force:

$$\boldsymbol{a}_1 = 0 \tag{14}$$

$$\boldsymbol{v}_1 = \boldsymbol{v}_{10} \tag{15}$$

$$qE\boldsymbol{j} + q\boldsymbol{v}_{10} \times B\boldsymbol{k} = 0 \tag{16}$$

Combining (13) to (16),

$$\Rightarrow \quad q\boldsymbol{v}_2 \times B\boldsymbol{k} = m\boldsymbol{a}_2 \tag{17}$$

It can be seen that the particle's Sub-motion II is the uniform circular motion under the action of the magnetic field force.

From (16),

$$\Rightarrow \quad \boldsymbol{v}_{10} = (E/B)\boldsymbol{i} \tag{18}$$

$$\Rightarrow \qquad \boldsymbol{v}_{20} = (v_0 - E/B)\boldsymbol{i} \tag{19}$$

$$\Rightarrow \qquad |\boldsymbol{v}_2| = |\boldsymbol{v}_{20}| = |v_0 - E/B| \tag{20}$$

Introducing the nature frame $(\boldsymbol{\tau}, \boldsymbol{n})$ of the particle's Sub-motion II. Considering ***the normal component form of acceleration***:

$$\boldsymbol{a}_2 = \frac{|\boldsymbol{v}_2|^2}{R}\boldsymbol{n} \tag{21}$$

R—the radius of the circular trajectory (Sub-motion II) of the particle.

$$\Rightarrow \qquad R = \frac{m}{q}\frac{|v_0 - E/B|}{B} \tag{22}$$

Defining $\boldsymbol{\omega}$—the angular velocity of the uniform circular motion (Sub-motion II) of the particle at arbitrary Time t. Considering the definition of ***angular velocity***:

$$\Rightarrow \qquad \boldsymbol{\omega} = -\frac{|\boldsymbol{v}_2|}{R}\boldsymbol{k} = -\frac{qB}{m}\boldsymbol{k} \tag{23}$$

Considering the following special situations:

Situation 1: $v_0 = E/B$

In this situation, the uniform circular motion does not exist, and the motion of the particle is the uniform linear motion, the kinematics equation is as follows:

$$x = (\boldsymbol{v}_{10} \cdot \boldsymbol{i})t = (E/B)t \tag{24}$$

Situation 2: $v_0 > E/B$

In this situation, the uniform circular motion is clockwise, and its center is below the x-axis. The kinematics equation of the particle is the superposition of the uniform linear motion and the uniform circular motion:

$$\Rightarrow \qquad \begin{cases} x = (\boldsymbol{v}_{10} \cdot \boldsymbol{i})t + R\sin\omega t = (E/B)t + R\sin\omega t & \text{(25a)} \\ y = (\boldsymbol{v}_{10} \cdot \boldsymbol{j})t - R + R\cos\omega t = -R + R\cos\omega t & \text{(25b)} \end{cases}$$

The minimum value of $\dot{x}$ in this situation is analyzed as follows:

$$\Rightarrow \qquad \dot{x}_{\min} = \boldsymbol{v}_{10} \cdot \boldsymbol{i} - |\boldsymbol{v}_{20}| = 2E/B - v_0 \begin{cases} <0 & v_0 > 2E/B \\ =0 & v_0 = 2E/B \\ >0 & v_0 < 2E/B \end{cases} \tag{26}$$

It can be seen that the sign of $\dot{x}$ in this situation is to be determined according to different sub situations. We can draw the schematic diagrams of the three sub situations (see Figs. 3–5).

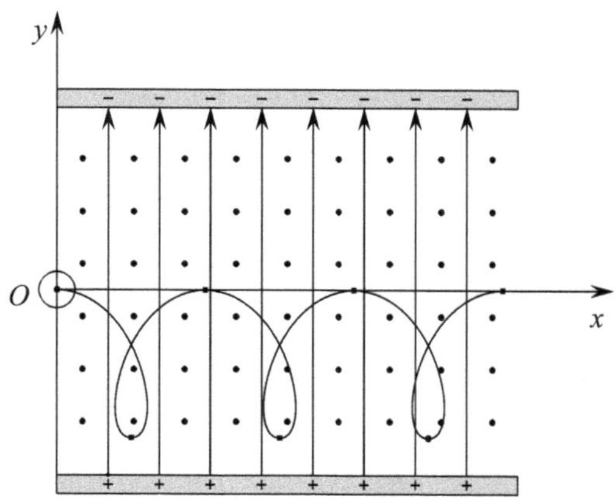

Figure 3 $v_0 > 2E/B$

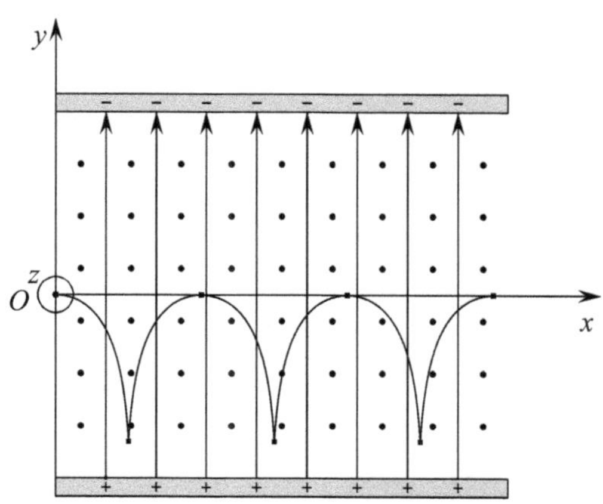

Figure 4 $v_0 = 2E/B$

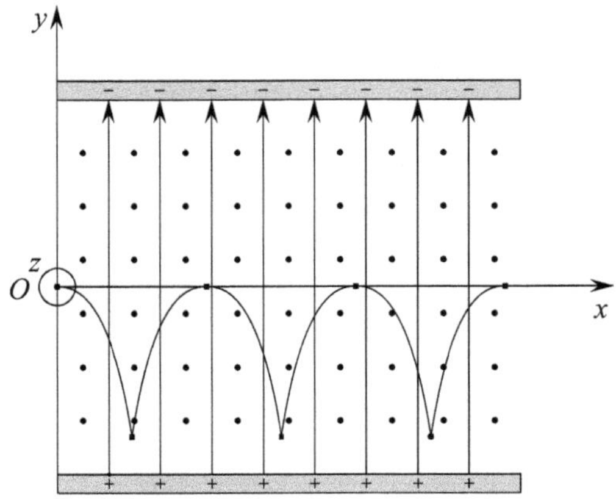

Figure 5 $E/B < v_0 < 2E/B$

Situation 3: $v_0 < E/B$

In this situation, the uniform circular motion is clockwise, and its center is above the x-axis. The kinematics equation of the particle is the superposition of the uniform linear motion and the uniform circular motion:

$$\Rightarrow \quad \begin{cases} x = (\boldsymbol{v}_{10} \cdot \boldsymbol{j})t - R\sin\omega t = (E/B)t - R\sin\omega t & (27a) \\ y = (\boldsymbol{v}_{10} \cdot \boldsymbol{j})t + R - R\cos\omega t = R - R\cos\omega t & (27b) \end{cases}$$

The minimum value of $\dot{x}$ in this situation is analyzed as follow:

$$\Rightarrow \quad \dot{x}_{\min} = (\boldsymbol{v}_{10} \cdot \boldsymbol{i}) - |\boldsymbol{v}_{20}| = v_0 \quad (28)$$

It can be seen that $\dot{x}$ is positive all the time in this situation. We can draw the schematic diagram of this situation (see Fig. 6):

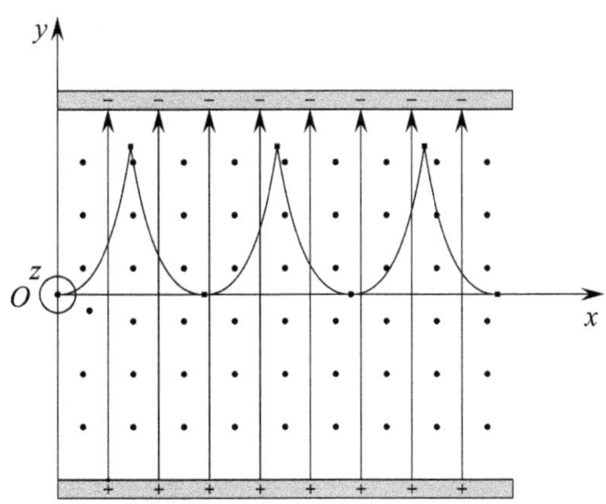

Figure 6 $v_0 < E/B$

2.2.2 The Charged Particle's Spiral Motion in a Gradual Magnetic Field and the Phenomenon of Magnetic Mirror (d, *C*)

Apply an axisymmetric non-uniform magnetic field in space, whose magnetic induction intensity components in the cylindrical coordinate system provided are $(B_r, 0, B_z)$ (see Fig. 1), $B_r << B_z$, and the magnitude range of the magnetic induction intensity is $B_1 < B < B_2$. A charged particle of mass m and charge q moves along a gradual spiral trajectory in the magnetic field and gradually enters the region with strong magnetic induction intensity from the region with weak magnetic induction intensity. Given that the magnetic induction intensity at the position where the particle is initially is B_1, the particle has the initial velocity of the magnitude v_0 and with the angle φ_0 from the $\boldsymbol{k}$-direction.

(1) Defining the magnetic moment of charged particle performing Larmor gyration in the magnetic field: $\boldsymbol{\mu} = iS\boldsymbol{n}$, where i is the magnitude of current intensity formed by the particle's Larmor gyration, S is the area of the particle's Larmor gyration trajectory, $\boldsymbol{n}$ is the unit vector of the right-hand helix direction of the particle's Larmor gyration. Please prove that the magnetic moment of the particle's Larmor gyration is an adiabatic invariant.

(2) Determine the condition that the particle can be reflected.

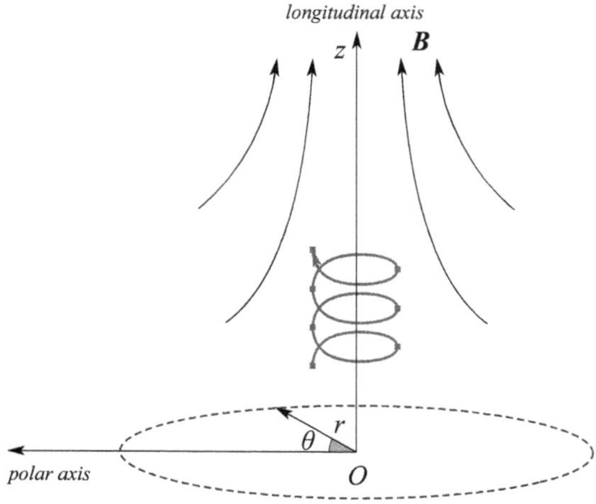

Figure 1

Solution to Problem (1)

Taking *the magnetic field* as the object of study. Studying *the state at arbitrary Time t*. Considering ***Gauss's law for magnetism*** [the closed surface is taken the cylindrical surface with the z-axis as the central axis, r as the radius (r is a small quantity), and whose top and bottom are located at z and z + dz, respectively] [It can be seen from the spatial symmetry that B_r is equal everywhere at the side of the cylinder. It can be seen from the fact that r is a small quantity that B_z is approximately equal everywhere at the top and bottom of the cylinder] (see Fig. 2):

$$(B_z + \mathrm{d}B_z)\cdot \pi r^2 - B_z \cdot \pi r^2 + B_r \cdot 2\pi r \mathrm{d}z = 0 \tag{1}$$

$$\Rightarrow \qquad B_r = -\frac{r}{2}\frac{\mathrm{d}B_z}{\mathrm{d}z} \tag{2}$$

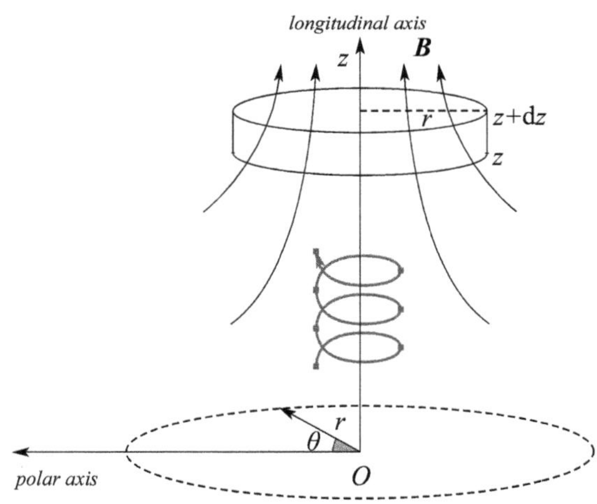

Figure 2

Taking *the particle* as the object of study. Studying *an arbitrary process from Time zero*. Considering ***law of conservation of mechanical energy*** [the work done by the magnetic force acting on the particle is zero]:

$$\frac{1}{2}mv_0^2 = \frac{1}{2}mv_z^2 + \frac{1}{2}mv_\perp^2 \tag{3}$$

v_z and $v_\perp$—the components of the velocity of the particle in the ***k***-direction and perpendicular to the ***k***-direction at arbitrary Time t.

Defining E_{kz} and $E_{k\perp}$:

$$\begin{cases} E_{kz} = \frac{1}{2}mv_z^2 & (4a) \\ E_{k\perp} = \frac{1}{2}mv_\perp^2 & (4b) \end{cases}$$

$$\Rightarrow \quad \frac{dE_{kz}}{dt} + \frac{dE_{k\perp}}{dt} = 0 \tag{5}$$

Taking *the particle* as the object of study. Studying *the state at arbitrary Time t.* Considering ***Newton's second law*** (the $\boldsymbol{r}^0$-direction component equation):

$$-qv_\perp B_z = ma_r \tag{6}$$

a_r—the $\boldsymbol{r}^0$-direction component of the acceleration of the particle at Time t.

Considering ***the radial component form of acceleration*** (the radius of the gyration trajectory of the particle changes slowly with time):

$$a_r \approx -\frac{v_\perp^2}{r} \qquad (7a)\ (7b)$$

r—the radius of the gyration trajectory of the particle.

$$\Rightarrow \quad r = \frac{mv_\perp}{qB_z} \approx \frac{mv_\perp}{qB} \tag{8}$$

Defining T—the gyration period of the particle at arbitrary Time t:

$$\Rightarrow \quad T = \frac{2\pi r}{v_\perp} = \frac{2\pi m}{qB_z} \approx \frac{2\pi m}{qB} \tag{9}$$

Taking *the particle* as the object of study. Studying *the state at arbitrary Time t.* Considering ***Newton's second law*** (the $\boldsymbol{k}$-direction component equation):

$$qv_\perp B_r = m\frac{dv_z}{dt} \tag{10}$$

From (2), (4), (8) and (10),

$$\Rightarrow \quad \frac{dE_{kz}}{dt} = mv_z\frac{dv_z}{dt} = qv_\perp v_z B_r = -\frac{1}{2}qv_\perp v_z r\frac{dB_z}{dz}$$

$$= -\frac{mv_\perp^2 v_z}{2B}\frac{\mathrm{d}B_z}{\mathrm{d}z} = -\frac{E_{k\perp}v_z}{B}\frac{\mathrm{d}B_z}{\mathrm{d}z} = -\frac{E_{k\perp}}{B}\frac{\mathrm{d}z}{\mathrm{d}t}\frac{\mathrm{d}B_z}{\mathrm{d}z}$$

$$= -\frac{E_{k\perp}}{B}\frac{\mathrm{d}B_z}{\mathrm{d}t} \approx -\frac{E_{k\perp}}{B}\frac{\mathrm{d}B}{\mathrm{d}t} \tag{11}$$

Considering the definitions of ***magnetic moment*** and ***current intensity***:

$$\mu = i \cdot \pi r^2 \tag{12}$$

$$i = \frac{q}{T} \tag{13}$$

$$\Rightarrow \quad \mu = \frac{mv_\perp^2}{2B} = \frac{E_{k\perp}}{B} \tag{14}$$

$(14) \to (11)$,

$$\Rightarrow \quad \frac{\mathrm{d}E_{kz}}{\mathrm{d}t} = -\mu\frac{\mathrm{d}B}{\mathrm{d}t} \tag{15}$$

From (14),

$$\Rightarrow \quad \frac{\mathrm{d}E_{k\perp}}{\mathrm{d}t} = \mu\frac{\mathrm{d}B}{\mathrm{d}t} + B\frac{\mathrm{d}\mu}{\mathrm{d}t} \tag{16}$$

(15) and $(16) \to (5)$,

$$\Rightarrow \quad \frac{\mathrm{d}\mu}{\mathrm{d}t} = 0 \tag{17}$$

$$\Rightarrow \quad \mu = \frac{mv_\perp^2}{2B} = \text{const} \tag{18}$$

QED.

Solution [1] to Problem (2)

$(4) \to (11)$,

$$\Rightarrow \quad -v_\perp^2\frac{\mathrm{d}B}{B} = \mathrm{d}v_z^2 \tag{19}$$

$v_z = 0$ when the particle is reflected.

$(3) \to (19)$,

$$\Rightarrow \quad \frac{\mathrm{d}v_z^2}{v_0^2 - v_z^2} = -\frac{\mathrm{d}B}{B} \tag{20}$$

$$\Rightarrow \qquad \int_{v_z|_{t=0}}^{0} \frac{\mathrm{d}v_z^2}{v_0^2 - v_z^2} = -\int_{B_1}^{B|_{v_z=0}} \frac{\mathrm{d}B_z}{B_z}$$

$$\Rightarrow \qquad \ln \frac{v_0^2}{v_0^2 - \left(v_z|_{t=0}\right)^2} = \ln \frac{B|_{v_z=0}}{B_1}$$

$$\Rightarrow \qquad \frac{v_0^2}{v_0^2 - \left(v_z|_{t=0}\right)^2} = \frac{B|_{v_z=0}}{B_1} \tag{21}$$

Considering the relation between ***principal velocity*** and ***component velocity***:

$$\frac{1}{2}mv_0^2 = \left.\left(\frac{1}{2}mv_\perp^2 + \frac{1}{2}mv_z^2\right)\right|_{t=0} \tag{22}$$

$$\Rightarrow \qquad \frac{v_\perp|_{t=0}}{v_0} = \sqrt{\frac{B_1}{B|_{v_z=0}}} \tag{23}$$

Known:

$$B|_{v_z=0} \le B_2 \tag{24}$$

$$\Rightarrow \qquad \frac{v_\perp|_{t=0}}{v_0} \ge \sqrt{\frac{B_1}{B_2}} \tag{25}$$

Defining φ_0—the angle between the particle's initial velocity and $\boldsymbol{k}$:

$$\sin\varphi_0 = \frac{v_\perp|_{t=0}}{v_0} \tag{26}$$

$$\Rightarrow \qquad \sin\varphi_0 \ge \sqrt{\frac{B_1}{B_2}} \tag{27}$$

$$\Rightarrow \qquad \varphi_0 \ge \arcsin\left(\sqrt{\frac{B_1}{B_2}}\right) \tag{28}$$

Solution [2] to Problem (2)

For the process from Time zero to the particle being reflected, Eq. (18) degenerates into the following form:

$$\frac{mv_\perp^2\big|_{t=0}}{2B_1} = \frac{mv_\perp^2\big|_{v_z=0}}{2B\big|_{v_z=0}} \tag{29}$$

(3) → (29),

$$\Rightarrow \qquad \frac{v_\perp^2\big|_{t=0}}{B_1} = \frac{v_0^2}{B\big|_{v_z=0}} \tag{30}$$

This result is the same as Eq. (23).

2.2.3 Magnetic Lens (d, *C*)

A conductor ring of radius R is fixed in space, whose central axis coincides with the z-axis of the cylindrical coordinate system provided (see Fig. 1), and in which there is the current with the current intensity of I, whose direction is the right-hand helix direction with regard to the $\boldsymbol{k}$-direction. Some high-speed charged particles of mass m and charge q near the z-axis move toward the ring from infinite distance at the initial velocity of the magnitude v_0 and the direction $\boldsymbol{k}$. Please prove that these particles will converge at a certain point on the z-axis approximately, and find the position of the convergence point.

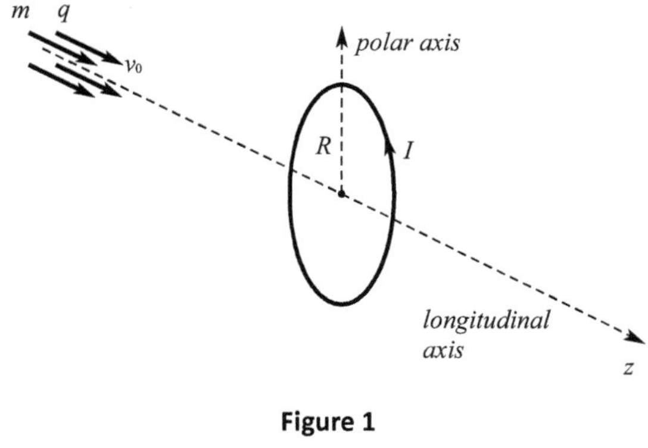

Figure 1

Solution

Taking *the current in the ring and the magnetic field it excites at an arbitrary point near the central axis of the ring* as the object of study. Studying *the state at arbitrary Time t*. Considering ***the form of magnetic induction intensity at an arbitrary point near the central axis of an current-carrying loop***:

$$\begin{cases} B_z = \dfrac{\mu_0}{2} \dfrac{I}{R(1+z^2/R^2)^{3/2}} & \text{(1a)} \\ B_r = -\dfrac{r}{2}\dfrac{\mathrm{d}B_z}{\mathrm{d}z} = -\dfrac{3\mu_0}{2}\dfrac{Izr}{R^3(1+z^2/R^2)^{5/2}} \ll B_z & \text{(1b)} \\ B_\theta = 0 & \text{(1c)} \end{cases}$$

(B_r, B_θ, B_z)—the components of the magnetic induction intensity at an arbitrary point near the central axis of the ring in the $(\boldsymbol{r}^0, \boldsymbol{\theta}^0, \boldsymbol{k})$-directions.

$$\Rightarrow \qquad \frac{B_z\big|_{z=2R}}{B_z\big|_{z=0}} = \left(\frac{1}{5}\right)^{3/2} \approx 0.09 \qquad (2)$$

That is, when $|z| > 2R$, B_z will decrease to less than 10% of its maximum value, thereupon it can be approximately considered that the charged particles are only affected by magnetic field force in the interval $|z| \le 2R$, which is called the main region of magnetic field.

For high-speed charged particles, it can be considered that the distance from them to the z-axis approximately remains unchanged during them passing through the main region of the magnetic field.

The particles with initial distance r to the z-axis are called particles (r).

Taking *an arbitrary particle (r)* as the object of study. Studying *the instantaneous process (the small time interval dt) at arbitrary Time t (the time point when the particle is at infinity is defined as Time zero)*. Considering ***theorem of angular momentum*** (the $\boldsymbol{k}$-direction component equation):

$$(qv_zB_r - qv_rB_z)r\mathrm{d}t = \mathrm{d}(mv_\theta r) \qquad (3)$$

$$\Rightarrow \qquad m\frac{\mathrm{d}}{\mathrm{d}t}(v_\theta r) = -q\left(\frac{1}{2}v_z r^2 \frac{\mathrm{d}B_z}{\mathrm{d}z} + rv_rB_z\right) \qquad (4)$$

(v_r, v_θ, v_z)—the components of the velocity of the particle (r) in the $(\boldsymbol{r}^0, \boldsymbol{\theta}^0, \boldsymbol{k})$-directions at Time t.

Considering ***the components form of velocity in cylindrical coordinate system***:

$$\begin{cases} v_r = \dfrac{\mathrm{d}r}{\mathrm{d}t} & \text{(5a)} \\ v_\theta = r\dfrac{\mathrm{d}\theta}{\mathrm{d}t} & \text{(5b)} \\ v_z = \dfrac{\mathrm{d}z}{\mathrm{d}t} & \text{(5c)} \end{cases}$$

$$\Rightarrow \qquad m\frac{\mathrm{d}}{\mathrm{d}t}\left(r^2\frac{\mathrm{d}\theta}{\mathrm{d}t}\right) = -q\left(\frac{1}{2}\frac{\mathrm{d}z}{\mathrm{d}t}r^2\frac{\mathrm{d}B_z}{\mathrm{d}z} + r\frac{\mathrm{d}r}{\mathrm{d}t}B_z\right)$$

$$= -\frac{1}{2}q\left(r^2\frac{\mathrm{d}B_z}{\mathrm{d}t} + \frac{\mathrm{d}r^2}{\mathrm{d}t}B_z\right)$$

$$= -\frac{1}{2}q\frac{\mathrm{d}}{\mathrm{d}t}(r^2B_z) \tag{6}$$

$$\Rightarrow \qquad \mathrm{d}\left(r^2\frac{\mathrm{d}\theta}{\mathrm{d}t}\right) = -\frac{q}{2m}\mathrm{d}(r^2B_z)$$

$$\Rightarrow \qquad r^2\frac{\mathrm{d}\theta}{\mathrm{d}t} = \int_0^{r^2\frac{\mathrm{d}\theta}{\mathrm{d}t}}\mathrm{d}\left(r^2\frac{\mathrm{d}\theta}{\mathrm{d}t}\right) = -\frac{q}{2m}\int_0^{r^2B_z}\mathrm{d}(r^2B_z) = -r^2\frac{qB_z}{2m}$$

$$\Rightarrow \qquad \frac{\mathrm{d}\theta}{\mathrm{d}t} = -\frac{qB_z}{2m} \tag{7}$$

Taking *an arbitrary particle (r)* as the object of study. Studying *the state at arbitrary Time t.* Considering ***Newton's second law*** (the $\boldsymbol{r}^0$-direction component equation):

$$qv_\theta B_z = ma_r \tag{8}$$

a_r—the $\boldsymbol{r}^0$-direction component of the acceleration of the particle (r) at Time t.

Considering ***the radial component form of acceleration***:

$$a_r = \frac{\mathrm{d}^2r}{\mathrm{d}t^2} - r\left(\frac{\mathrm{d}\theta}{\mathrm{d}t}\right)^2 \tag{9}$$

(7) → [(5b) and (9) → (8)],

$$\Rightarrow \qquad \frac{\mathrm{d}^2r}{\mathrm{d}t^2} = -\frac{q^2B_z^2r}{4m^2} \tag{10}$$

From (5) and (10),

$$\Rightarrow \qquad \frac{\mathrm{d}v_r}{\mathrm{d}t} = \frac{\mathrm{d}^2 r}{\mathrm{d}t^2} \qquad (11)$$

$$\Rightarrow \qquad \mathrm{d}v_r = -\frac{q^2 B_z^2 r}{4m^2}\mathrm{d}t = -\frac{q^2 B_z^2 r}{4m^2 v_z}\mathrm{d}z \approx -\frac{q^2 B_z^2 r}{4m^2 v_0}\mathrm{d}z \qquad (12)$$

$$\Rightarrow V_r = \int_0^{V_r} \mathrm{d}v_r = -\frac{q^2 r}{4m^2 v_0}\int_{-\infty}^{\infty} B_z^2 \mathrm{d}z = -\frac{\mu_0^2 q^2 I^2 r}{16m^2 v_0 R^2}\int_{-\infty}^{\infty}\frac{dz}{(1+z^2/R^2)^3}$$

(13)

V_r—the $\boldsymbol{r}^0$-direction component of the velocity of the particle (r) when it passes through the main region of the magnetic field.

The operation above has considered that r approximately remains unchanged during the particle (r) passing through the main region of the magnetic field.

Defining the definite integral I':

$$I' = \int_{-\infty}^{\infty}\frac{\mathrm{d}z}{(1+z^2/R^2)^3} \qquad (14)$$

Defining the variable φ:

$$z = R\tan\varphi \qquad (15)$$

$$\Rightarrow \qquad \mathrm{d}z = R\sec^2\varphi\,\mathrm{d}\varphi \qquad (16)$$

$$\Rightarrow \qquad I' = R\int_{-\pi/2}^{\pi/2}\cos^4\varphi\,\mathrm{d}\varphi = \frac{3\pi R}{8} \qquad (17)$$

$$\Rightarrow \qquad V_r = -\frac{3\pi\mu_0^2 q^2 I^2 r}{128m^2 v_0 R} \qquad (18)$$

Defining α—the deflection angle of the velocity of the particle (r) relative to the z-axis when it passes through the main region of the magnetic field. Δz—the $\boldsymbol{k}$-direction displacement of the particle (r) during the process from passing through the main region of the magnetic field to reaching the z-axis.

From the geometric relation, we can know that:

$$\tan\alpha = -\frac{V_r}{v_z} \approx -\frac{V_r}{v_0} \qquad (19)$$

$$\tan\alpha = \frac{r}{\Delta z} \tag{20}$$

$$\Rightarrow \qquad \Delta z = \frac{128 m^2 v_0^2 R}{3\pi \mu_0^2 q^2 I^2} \tag{21}$$

It can be seen that Δz is independent of r; therefore, all particles will converge at the same point on the z-axis. If the current-carrying ring is regarded as a "thin lens" focusing charged particles, then Δz can be regarded as its focal length.

2.2.4 The Magnetic Focusing Created by a Pair of Square Column Magnets (d, *B*)

As shown in Fig. 1, there are two square column magnets with the same shape, whose cross section are both square of side length L. Place the two magnets in the position in which the two poles are parallel and align with each other. The distance between the two magnets is much smaller than L. Establishing the Cartesian coordinate system with the frame $(O, \boldsymbol{i}, \boldsymbol{j}, \boldsymbol{k})$ as shown in the figure, let x, y, z be the Cartesian coordinates. The magnetic field between the two magnets is approximately uniform, with the magnetic induction intensity of the magnitude B_0 and the direction $\boldsymbol{i}$. There is a charged particle of mass m and charge q, $q>0$, whose initial coordinates is $(x_0, 0, 0)$ and initial velocity is $(0, 0, v_0)$, x_0 is much smaller than the distance between the two magnets, $v_0 >> \frac{q}{m} B_0 L$.

(1) Find the deflection angle of the particle in the $\boldsymbol{j}$-direction just after passing through the space area between the two poles of the magnets.
(2) Find the deflection angle of the particle in the $\boldsymbol{i}$-direction after passing through the space area between the two poles of the magnets and reaching the distance.
(3) If there is a bunch of particles that are the same as the above particle, which are continuously distributed in the interval $[-x_0, x_0]$ on the x-axis, and have the same initial velocity as the above particle, please prove that these particles will converge at a certain point on the z-axis approximately, and find the position of the convergence point.

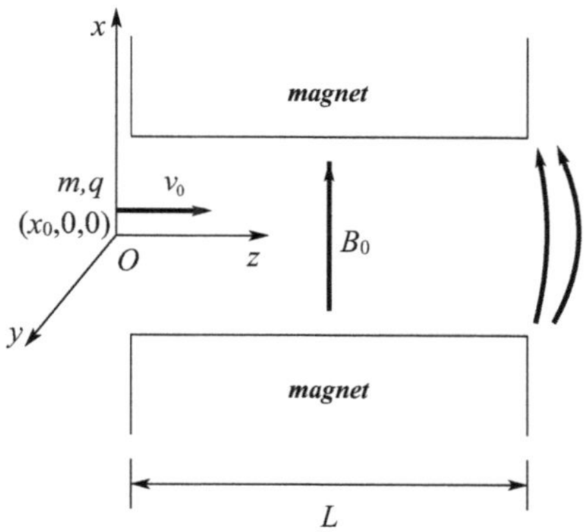

Figure 1

Solution to Problem (1)

The magnetic field region between the two poles of the magnets is called the main region of magnetic field.

The motion process of the particle in the main region of magnetic field is denoted by Process 1. The motion process of the particle after passing through the main region of magnetic field is denoted by Process 2.

Taking *the particle* as the object of study. Studying *the state at arbitrary Time t in Process 1*. Considering (a) ***Newton's second law*** (the component equation in the radial inward direction), (b) ***the normal component form of acceleration*** [integrating in (a)]:

$$qB_0v_0 = m\frac{v_0^2}{R} \tag{1}$$

R—the radius of the circular trajectory of the particle.

$$\Rightarrow \qquad R = \frac{mv_0}{qB_0} \tag{2}$$

Known:

$$mv_0 \gg qB_0L \tag{3}$$

$$\Rightarrow \qquad R \gg L \tag{4}$$

Defining θ_y—the deflection angle of the particle in the ***j***-direction just after passing through the main region of magnetic field. Considering the following geometric relation in Fig. 2:

$$\Rightarrow \qquad \sin\theta_y = \frac{L}{R} \ll 1 \tag{5}$$

That is, θ_y is a small quantity, taking the first-order approximation of $\sin\theta_y$:

$$\Rightarrow \qquad \theta_y \approx \sin\theta_y = \frac{L}{R} \tag{6}$$

Considering the following geometric relation in Fig. 2:

$$\Rightarrow \qquad \begin{cases} v_y\big|_{z=L} = v_0 \sin\theta_y \approx v_0\theta_y = \dfrac{L}{R} v_0 \ll v_0 & \text{(7a)} \\ v_z\big|_{z=L} = v_0 \cos\theta_y \approx v_0 & \text{(7b)} \end{cases}$$

The operations above have taken the first-order approximation of θ_y.

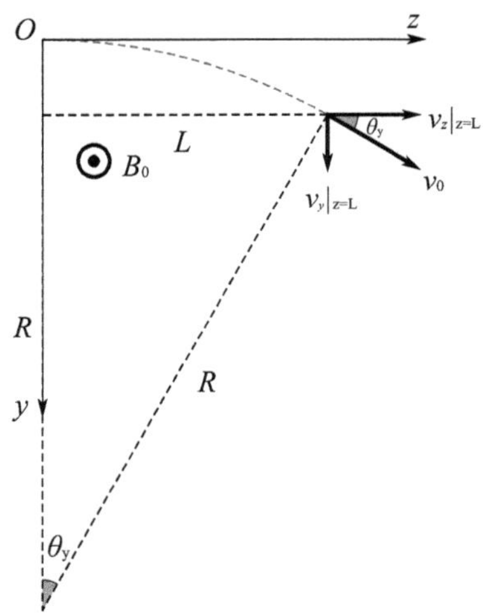

Figure 2

Solution to Problem (2)

Taking *the particle* as the object of study. Studying *the state at arbitrary Time t in Process 2*. Considering ***Newton's second law*** (the ***i***-direction component equation):

$$qv_yB_z = m\frac{dv_x}{dt} \tag{8}$$

B_z—the ***k***-direction component of the magnetic induction intensity at the position where the particle is at Time t in Process 2.

v_x and v_y—the ***i***-direction and ***j***-direction components of particle's velocity at Time *t* in Process 2.

Assumption I: after the particle passing through the main region of magnetic field, the ***j***-direction component of its acceleration is a small quantity at arbitrary Time *t*, that is, the ***j***-direction component of its velocity is approximately conserved.

From (7a),

$$v_y \approx v_y\big|_{z=L} = \frac{L}{R}v_0 \tag{9}$$

(9) → (8),

$$\Rightarrow \qquad \frac{dv_x}{dt} = \frac{B_z L v_0^2}{B_0 R^2} \tag{10}$$

Taking the following differential transformation and considering the definition of ***velocity***:

$$\frac{dv_x}{dt} = \frac{dv_x}{dz}\frac{dz}{dt} = \frac{v_z dv_x}{dz} \tag{11}$$

Taking *the particle* as the object of study. Studying *the state at arbitrary Time t in Process 2*. Considering ***Newton's second law*** (the ***k***-direction component equation):

$$qv_x B_y = m\frac{dv_z}{dt} \tag{12}$$

B_y—the ***j***-direction component of the magnetic induction intensity at the position where the particle is at Time *t* in Process 2.

v_z—the ***k***-direction component of particle's velocity at Time *t* in Process 2.

Assumption II: after the particle passing through the main region of magnetic field, the ***i***-direction component of its velocity is a small quantity at arbitrary Time *t* (much smaller than v_0):

$$v_x \ll v_0 \tag{13}$$

$$\Rightarrow \qquad \frac{dv_z}{dt} \approx 0 \tag{14}$$

Combining (7b) and (14),

$$\Rightarrow \qquad v_z \approx v_z\big|_{z=L} = v_0 \tag{15}$$

(15) → (11),

$$\Rightarrow \qquad \frac{dv_x}{dt} \approx \frac{v_0 dv_x}{dz} \tag{16}$$

Combining (10) and (16),

$$\Rightarrow \qquad dv_x = \frac{B_z L v_0}{B_0 R^2} dz \tag{17}$$

$$\Rightarrow \qquad v_x\Big|_{z\to\infty} = \int_0^{v_x|_{z\to\infty}} dv_x = \frac{L v_0}{B_0 R^2} \int_L^{\infty} B_z dz \tag{18}$$

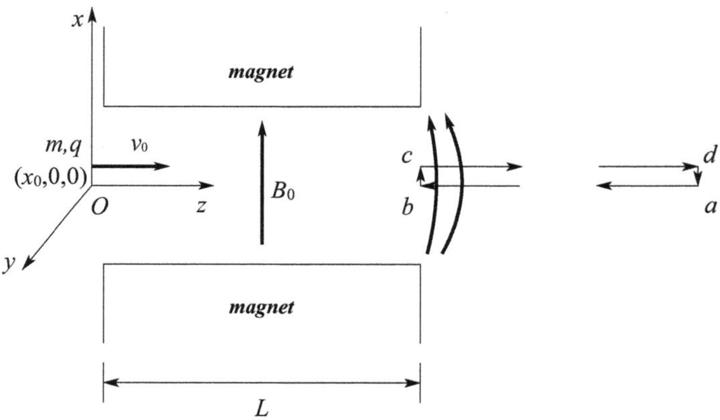

Figure 3

Taking *the magnetic field* as the object of study. Studying *the state at arbitrary Time t*. Considering ***Ampere circuital theorem*** [taking the rectangular *abcd* whose vertices are Point *a* (0, 0, ∞), Point *b* (0, 0, *L*), Point c (x_0, 0, *L*), Point *d* (x_0, 0, ∞) as the integral path] (taking the clockwise sense as the positive direction of curvilinear integral) (see Fig. 3):

$$B_0 x_0 + \int_L^{\infty} B_z dz = 0 \tag{19}$$

Combining (18) and (19):

$$\Rightarrow \qquad v_x\Big|_{z\to\infty} = -\frac{x_0 L}{R^2} v_0 \ll v_0 \tag{20}$$

That is, Assumption II is hold.

Known:

$$\Rightarrow \qquad x_0 \ll L \tag{21}$$

Defining θ_x—the deflection angle of the particle in the ***i***-direction after passing through the main region of magnetic field and reaching the distance. Considering the following geometric relation, then place (15) and (20) in to obtain:

$$\Rightarrow \qquad \tan\theta_x = \frac{v_x\big|_{z\to\infty}}{v_z\big|_{z\to\infty}} = -\frac{x_0 L}{R^2} \ll 1 \tag{22}$$

That is, θ_x is a small quantity, taking the first-order approximation of $\tan\theta_x$:

$$\Rightarrow \qquad \theta_x \approx \tan\theta_x = -\frac{x_0 L}{R^2} = -\frac{x_0}{L}\theta_y^2 \ll \theta_y \tag{23}$$

Taking *the particle* as the object of study. Studying *its motion process from initial position to infinity*. Considering ***law of conservation of mechanical energy*** [the work done by the magnetic force acting on the particle is zero]:

$$\frac{1}{2}mv_0^2 = \frac{1}{2}m\left[\left(v_x\big|_{z\to\infty}\right)^2 + \left(v_y\big|_{z\to\infty}\right)^2 + \left(v_z\big|_{z\to\infty}\right)^2\right] \tag{24}$$

(15) and (20) $\to$ (24):

$$\Rightarrow \qquad v_y\big|_{z\to\infty} = v_0\sqrt{\sin^2\theta_y - \left(\frac{x_0 L}{R^2}\right)^2} \tag{25}$$

(5) and (9) $\to$ (25),

$$\Rightarrow \qquad v_y\big|_{z\to\infty} = v_0\sqrt{\left(\frac{L}{R}\right)^2 - \left(\frac{x_0 L}{R^2}\right)^2}$$

$$= v_0\frac{L}{R}\sqrt{1-\left(\frac{x_0}{R}\right)^2} \approx v_0\frac{L}{R} = v_y\big|_{z=L} \tag{26}$$

That is, Assumption I is hold.

Solution to Problem (3)

Extending the result of Eq. (22) to the particle with the initial position of $(x, 0, 0)$:

$$\tan\theta_x = -\frac{xL}{R^2} \tag{27}$$

Defining Δz—the $\boldsymbol{k}$-direction displacement of the particle with the initial position of $(x, 0, 0)$ during the process from passing through the main region of magnetic field to reaching the z-axis. Considering the following geometrical relation [ignore the $\boldsymbol{j}$-direction displacement], then place (27) in to obtain:

$$\Rightarrow \qquad \Delta z = \left|\frac{x}{\tan\theta_x}\right| \approx \left|\frac{x}{\theta_x}\right| = \frac{R^2}{L} \gg L \tag{28}$$

It can be seen that Δz is independent of x, that is, the particles emitted in the interval $[-x_0, x_0]$ will converge at the same point on the z-axis.

Defining the focal length f:

$$\Rightarrow \qquad f = \Delta z = \frac{R^2}{L} \tag{29}$$

All results in this paper retain R determined by Eq. (2).

2.2.5 The Charged Particle's Motion Near an Infinitely Long Straight Current-Carrying Wire (a, *A*)

An infinitely long straight conducting wire coincides with the z-axis of the cylindrical coordinate system provided, electrified with the steady current of current intensity I, whose direction is the positive direction of the longitudinal axis. A point charge of mass m and electric quantity q can move freely around the conducting wire, its initial position is (r_0, θ_0, z_0) and initial velocity is $(v_{r0}, v_{\theta 0}, v_{z0})$.

(1) Determine the velocity of the point charge as the function of its polar radius.

(2) Determine the equation satisfied by the extreme value of polar radius of the point charge, then find its solution under the condition of $v_{\theta 0} = 0$, $v_{z0} > v_{r0}$.

Solution

The initial time is defined as Time zero.

Taking *the current in the conducting wire and the magnetic field it excites in space* as the object of study. Studying *the state at arbitrary Time t*. Considering ***Ampere circuital theorem*** [taking the circle with the wire as the central axis in the plane perpendicular to the wire as the integral path] (taking the right-hand helix direction with regard to the ***k***-direction as the positive direction of curvilinear integral):

$$B \cdot 2\pi r = \mu_0 I \tag{1}$$

B—the $\boldsymbol{\theta}^0$-direction component of the magnetic induction intensity at an arbitrary point on the circular loop.

$$\Rightarrow \qquad B = \frac{\mu_0}{2\pi}\frac{I}{r} \tag{2}$$

Taking *the point charge (can be regarded as a particle)* as the object of study. Studying *the instantaneous process (the small time interval dt) at arbitrary Time t*. Considering ***momentum theorem*** (the ***k***-direction component equation):

$$qv_r B\mathrm{d}t = \mathrm{d}(mv_z) \tag{3}$$

v_r and v_z—the $\boldsymbol{r}^0$-direction and the ***k***-direction component of the velocity of the point charge at Time *t*.

Considering ***the radial component form of velocity***:

$$v_r = \frac{\mathrm{d}r}{\mathrm{d}t} \tag{4}$$

(2) and (4) $\rightarrow$ (3),

$$\Rightarrow \qquad \frac{\mu_0}{2\pi}\frac{q}{m}I\frac{\mathrm{d}r}{r} = \mathrm{d}v_z \tag{5}$$

Defining the constant α:

$$\alpha = \frac{\mu_0}{2\pi}\frac{q}{m}I \tag{6}$$

$$\Rightarrow \qquad \alpha\frac{\mathrm{d}r}{r} = \mathrm{d}v_z$$

$$\Rightarrow \qquad \alpha\int_{r_0}^{r}\frac{\mathrm{d}r}{r} = \int_{v_{z0}}^{v_z}\mathrm{d}v_z$$

$$\Rightarrow \qquad v_z = v_{z0} + \alpha\ln\frac{r}{r_0} \tag{7}$$

Taking *the point charge* as the object of study. Studying *the process from Time zero to arbitrary Time t*. Considering ***law of conservation***

of angular momentum [taking the z-axis as the rotation axis, the $\boldsymbol{k}$-direction component of resultant torque acting on the point charge at arbitrary Time is zero (the $\boldsymbol{\theta}^0$-direction component of the magnetic field force acting on the point charge at arbitrary Time is zero); therefore, the $\boldsymbol{k}$-direction component of angular momentum of the point charge is conserved] (the $\boldsymbol{k}$-direction component equation):

$$mv_{\theta 0}r_0 = mv_\theta r \tag{8}$$

v_θ—the $\boldsymbol{\theta}^0$-direction component of the velocity of the point charge at Time t.

$$\Rightarrow \qquad v_\theta = v_{\theta 0}\frac{r_0}{r} \tag{9}$$

Taking *the point charge* as the object of study. Studying *the process from Time zero to arbitrary Time t*. Considering ***law of conservation of mechanical energy*** [the work done by the magnetic force acting on the point is zero; therefore, the mechanical energy of the point charge is conserved]:

$$\frac{1}{2}m(v_{r0}^2 + v_{\theta 0}^2 + v_{z0}^2) = \frac{1}{2}m(v_r^2 + v_\theta^2 + v_z^2) \tag{10}$$

(7) and (9) $\rightarrow$ (10),

$$\Rightarrow \qquad v_r = \sqrt{v_{r0}^2 + v_{\theta 0}^2\left(1 - \frac{r_0^2}{r^2}\right) - 2v_{z0}\alpha\ln\frac{r}{r_0} + \left(\alpha\ln\frac{r}{r_0}\right)^2} \tag{11}$$

The necessary condition for the polar radius of the point charge to take extreme value is as follows:

$$\frac{\mathrm{d}r}{\mathrm{d}t} = 0 \tag{12}$$

Combining (4), (11) and (12),

$$\Rightarrow \qquad v_{r0}^2 + v_{\theta 0}^2\left(1 - \frac{r_0^2}{r^2}\right) - 2v_{z0}\alpha\ln\frac{r}{r_0} + \left(\alpha\ln\frac{r}{r_0}\right)^2 = 0 \tag{13}$$

Under the condition of $v_{\theta 0} = 0$, the equation above degenerates to the following form:

$$\Rightarrow \qquad v_{r0}^2 - 2v_{z0}\alpha\ln\frac{r}{r_0} + \left(\alpha\ln\frac{r}{r_0}\right)^2 = 0 \tag{14}$$

Defining the variable V:

$$V = \alpha \ln \frac{r}{r_0} \tag{15}$$

$$\Rightarrow \quad V^2 - 2v_{z0}V + v_{r0}^2 = 0 \tag{16}$$

$$\Rightarrow \quad V_{1,2} = v_{z0}\left(1 \pm \sqrt{1 - \frac{v_{r0}^2}{v_{z0}^2}}\right)$$

$$\Rightarrow \quad \begin{cases} r_{\max} = r_0 \exp\left[\dfrac{1}{\alpha} v_{z0}\left(1 + \sqrt{1 - \dfrac{v_{r0}^2}{v_{z0}^2}}\right)\right] & \text{(17a)} \\ r_{\min} = r_0 \exp\left[\dfrac{1}{\alpha} v_{z0}\left(1 - \sqrt{1 - \dfrac{v_{r0}^2}{v_{z0}^2}}\right)\right] & \text{(17b)} \end{cases}$$

All the results above have retained α defined by Eq. (6).

2.2.6 The Charged Particle's Motion in the Cylindrical Symmetric Electrostatic Field and Axial Uniform Magnetic Field (c, *B*)[2]

A long cylindrical conductor of sectional radius a and a long hollow cylindrical conductor of inner radius b are coaxially placed, and it is vacuum between the two (see Fig. 1). The outer conductor has a positive potential U relative to the inner conductor (hereinafter referred to as the outer conductor as the anode and the inner conductor as the cathode). Apply a uniform magnetic field parallel to the central axis between the two conductors, the magnetic induction intensity of which is $\mathbf{B} = B\mathbf{k}$ in the cylindrical coordinate system provided in the figure. When the cathode is heated, electrons will escape from its surface at the initial velocity of $(v_{r0}, v_{\varphi 0}, v_{z0})$. Induced charges on the conductors are neglected. Determine the critical magnetic induction intensity that just makes the electrons to reach the anode.

Solution

Taking *an arbitrary electron escaping from the cathode* as the object of study. Studying *the instantaneous process (the small time interval*

[2]This problem is a test question of the 27th IPhO (International Physics Olympiad) in 1996.

dt) at arbitrary Time t (the time point when it escapes from the cathode is defined as Time zero). Considering ***theorem of angular momentum*** [taking the pole O as the reference point]:

$$\mathrm{d}\boldsymbol{L} = \boldsymbol{r} \times \left[(-e)\boldsymbol{E} + (-e)\boldsymbol{v} \times \boldsymbol{B}\right]\mathrm{d}t \tag{1}$$

$$\Rightarrow \qquad \mathrm{d}(L_r \boldsymbol{r}^0 + L_\varphi \boldsymbol{\varphi}^0 + L_z \boldsymbol{k})$$

$$= (r\boldsymbol{r}^0 + z\boldsymbol{k}) \times \left[(-e)E_r \boldsymbol{r}^0 + (-e)(v_r \boldsymbol{r}^0 + v_\varphi \boldsymbol{\varphi}^0 + v_z \boldsymbol{k}) \times B_z \boldsymbol{k}\right]\mathrm{d}t$$

(L_r, L_φ, L_z)—the components of the angular momentum of the electron relative to the pole O in the $(\boldsymbol{r}^0, \boldsymbol{\varphi}^0, \boldsymbol{k})$-directions at Time t.

(v_r, v_φ, v_z)—the components of the velocity of the electron in the $(\boldsymbol{r}^0, \boldsymbol{\varphi}^0, \boldsymbol{k})$-directions at Time t.

E_r—the $\boldsymbol{r}^0$-direction component of the electric field strength at the position where the electron is at Time t.

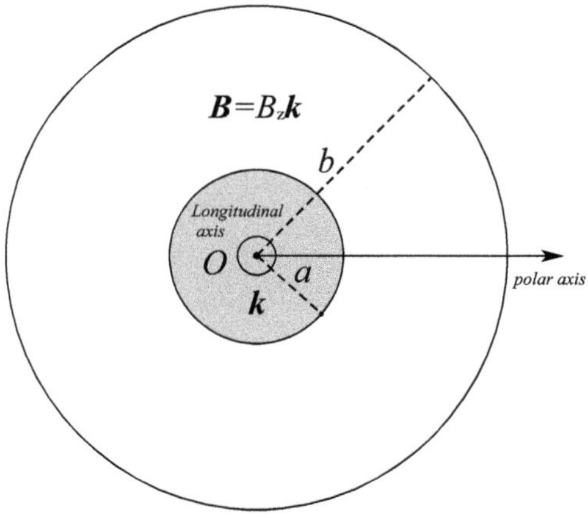

Figure 1

The projection equation of Eq. (1) in the $\boldsymbol{k}$-direction is as follows:

$$\Rightarrow \qquad \mathrm{d}(L_z \boldsymbol{k}) = -r\boldsymbol{r}^0 \times \left[e(v_r \boldsymbol{r}^0 + v_\varphi \boldsymbol{\varphi}^0) \times B_z \boldsymbol{k}\right]\mathrm{d}t$$

$$= -r\boldsymbol{r}^0 \times (-ev_r B_z \boldsymbol{\varphi}^0 + ev_\varphi B_z \boldsymbol{r}^0)\mathrm{d}t$$

$$= -r\boldsymbol{r}^0 \times (-ev_r B_z \boldsymbol{\varphi}^0)\mathrm{d}t$$

$$= ev_r B_z r\boldsymbol{k}\mathrm{d}t$$

$$\Rightarrow \qquad \mathrm{d}L_z = ev_r B_z r\mathrm{d}t \tag{2}$$

Considering ***the radial component form of velocity***:

$$v_r = \frac{\mathrm{d}r}{\mathrm{d}t} \tag{3}$$

$$\Rightarrow \qquad \mathrm{d}L_z = eB_z r\mathrm{d}r \tag{4}$$

$$\Rightarrow \qquad \int L_z = eB_z \int r\mathrm{d}r$$

$$\Rightarrow \qquad L_z - \frac{1}{2}eB_z r^2 = \text{const} \tag{5}$$

That is, the quantity defined by the expression $L_z - \frac{1}{2}eB_z r^2$ is conserved.

Considering the definition of ***angular momentum***, and extending Eq. (5) to the whole process of an arbitrary electron moving from the cathode to the anode:

$$\Rightarrow \qquad mv_{\varphi 0}a - \frac{1}{2}eB_{zC}a^2 = mv_{\varphi 1}b - \frac{1}{2}eB_{zC}b^2 \tag{6}$$

$v_{\varphi 1}$—the $\boldsymbol{\varphi}^0$-direction component of the velocity of the electron when it reaches the anode.

B_{zC}—the critical magnetic induction intensity B_z that just makes the electron to reach the anode.

Taking *an arbitrary electron escaping from the cathode* as the object of study. Studying *the process of its moving from the cathode to the anode*. Considering ***theorem of kinetic energy***:

$$eU = \frac{1}{2}m(v_{r1}^2 + v_{\varphi 1}^2 + v_{z1}^2) - \frac{1}{2}m(v_{r0}^2 + v_{\varphi 0}^2 + v_{z0}^2) \tag{7}$$

v_{r1} and v_{z1}—the $\boldsymbol{r}$-direction and $\boldsymbol{k}$-direction component of the velocity of the electron when it reaches the anode.

Taking *an arbitrary electron escaping from the cathode* as the object of study. Studying *the process of its moving from the cathode to the anode*. Considering ***law of conservation of momentum*** [the $\boldsymbol{k}$-direction component of the electric field force and magnetic field force acting on the electron are both zero; therefore, the $\boldsymbol{k}$-direction component of the momentum of the electron is conserved] (the $\boldsymbol{k}$-direction component equation):

$$m_e v_{z0} = m_e v_{z1} \tag{8}$$

$$\Rightarrow \qquad v_{z1} = v_{z0} \tag{9}$$

The condition that the electron can just reach the anode is as follows:

$$\Rightarrow \qquad v_{r1} = 0 \tag{10}$$

Combining (6), (7), (9) and (10),

$$\Rightarrow \qquad B_{zC} = \frac{2mb}{e(b^2 - a^2)}\left(\sqrt{v_{r0}^2 + v_{\varphi 0}^2 + \frac{2eU}{m}} - v_{\varphi 0}\frac{a}{b}\right) \tag{11}$$

We can draw the schematic diagram of the electron trajectory in the critical situation (see Fig. 2).

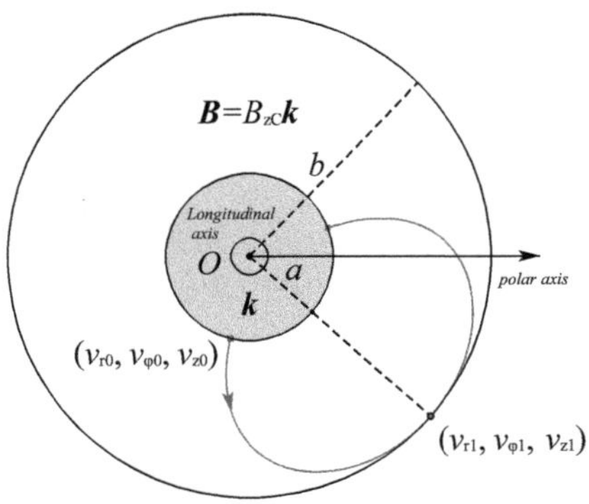

Figure 2

2.3 Particle in Rotating Reference Frame

Particles can exhibit rich dynamics phenomena in rotating reference frame, which is due to the complex inertial forces acting on the particles in the rotating reference frame. The inertia forces acting on a particle in a rotating reference frame can be divided into two categories: transport inertia force (determined entirely by transport motion) [It can also be subdivided into tangential transport inertia force and normal transport inertia force (also called inertial centrifugal force)] and Coriolis inertial force (determined by the interaction between transport motion and relative motion). It is difficult to solve the dynamic equations of particles in such a complex force field, except for taking the low angular speed approximation in a rotating reference frame which is rotating uniformly and slowly.

The Foucault pendulum problem is the most classical and famous problem in this model field, in this section, some typical dynamics problems of particles in rotating reference frame including Foucault pendulum problem will be studied.

2.3.1 The Target Practice on a Rotating Disk (d, *C*)

Consider a disk of radius R in a horizontal plane rotating around its central axis at the constant angular velocity $\boldsymbol{\omega}$. A shooter stays at the endpoint A of the diameter AB of the disk and is ready to fire a bullet in the horizontal plane to hit a target at the other endpoint B. Given the initial speed of the bullet relative to the disk is v_0, which is far greater than the speed of an arbitrary point on the disk, ignore the air resistance.

(1) Determine the aim direction of the shooter to hit the target.

(2) Determine the trajectory of the bullet in the disk reference frame.

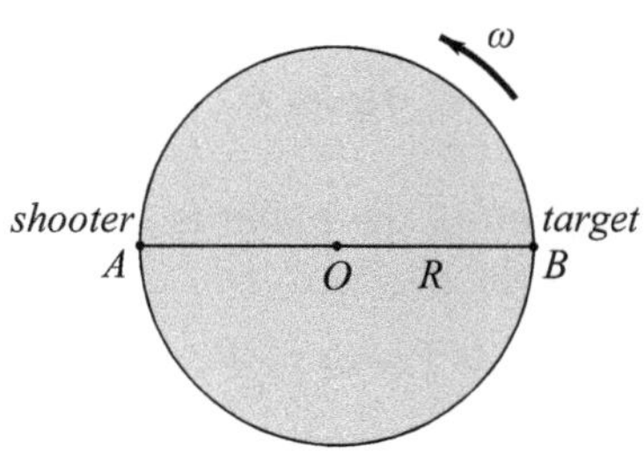

Figure 1

Solution [1]

The time point when the bullet is fired is defined as Time zero.

Establishing the Cartesian coordinate system with the frame $(O, \mathbf{i}, \mathbf{j}, \mathbf{k})$ fixed to the disk, let x, y, z be the Cartesian coordinates (see Fig. 2).

Taking *the bullet fired by the shooter (can be regarded as a particle)* as the object of study. Studying *the state at arbitrary Time t*. Considering (a) ***Newton's second law*** [in the disk reference frame], (b) ***inertial centrifugal force theorem*** [integrating into (a)], (c) ***Coriolis force theorem*** [integrating into (a)]:

$$-m\boldsymbol{\omega}\times(\boldsymbol{\omega}\times\boldsymbol{r}')-2m\boldsymbol{\omega}\times\boldsymbol{v}'=m\ddot{\boldsymbol{r}}' \tag{1}$$

$\boldsymbol{r}'$—the position vector of the bullet in the disk reference frame at Time t.

$\boldsymbol{v}'$—the velocity of the bullet in the disk reference frame at Time t.

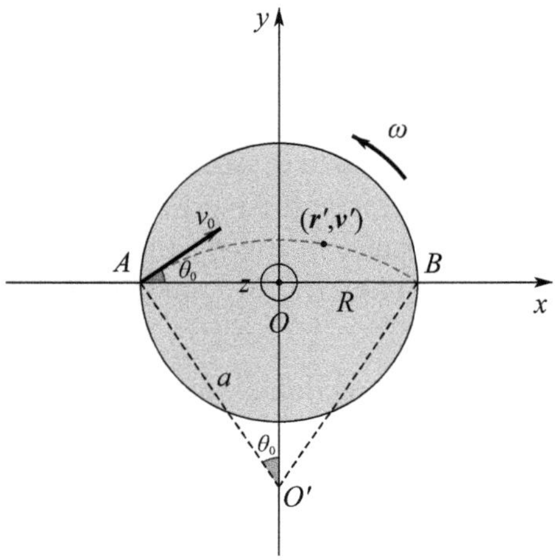

Figure 2

Known that the speed of the bullet is far faster than that of an arbitrary point on the disk, that is:

$$|\boldsymbol{v}'|\gg|\boldsymbol{\omega}\times\boldsymbol{r}'| \tag{2}$$

Therefore, the inertial centrifugal force can be ignored, only the Coriolis force needs to be considered:

$$\Rightarrow \qquad -2m\boldsymbol{\omega}\times\boldsymbol{v}'\approx m\ddot{\boldsymbol{r}}' \tag{3}$$

The components of each kinematics quantity of the bullet in the $(\boldsymbol{i},\boldsymbol{j},\boldsymbol{k})$-directions are as follows:

$$\begin{cases}\boldsymbol{\omega}=(0,0,\omega) & \text{(4a)}\\ \ddot{\boldsymbol{r}}'=(\ddot{x},\ddot{y},0) & \text{(4b)}\\ \boldsymbol{v}'=(\dot{x},\dot{y},0) & \text{(4c)}\end{cases}$$

$$\Rightarrow \qquad \boldsymbol{\omega}\times\boldsymbol{v}'=\begin{vmatrix}\boldsymbol{i} & \boldsymbol{j} & \boldsymbol{k}\\ 0 & 0 & \omega\\ \dot{x} & \dot{y} & 0\end{vmatrix}=-\omega\dot{y}\boldsymbol{i}+\omega\dot{x}\boldsymbol{j} \tag{5}$$

$$\Rightarrow \qquad \begin{cases} \ddot{x} = 2\omega\dot{y} & (6a) \\ \ddot{y} = -2\omega\dot{x} & (6b) \end{cases}$$

Considering the initial conditions:

$$\begin{cases} \dot{x}\big|_{x=-R,y=0} = v_0 \cos\theta_0 & (7a) \\ \dot{y}\big|_{x=-R,y=0} = v_0 \sin\theta_0 & (7b) \end{cases}$$

θ_0—the angle between the initial velocity of the bullet and the ***i***-direction.

From (6) and (7),

$$\Rightarrow \qquad \begin{cases} \dot{x} = 2\omega y + v_0 \cos\theta_0 & (8a) \\ \dot{y} = -2\omega(x+R) + v_0 \sin\theta_0 & (8b) \end{cases}$$

$$\Rightarrow \qquad \frac{dy}{dx} = \frac{-2\omega(x+R) + v_0 \sin\theta_0}{2\omega y + v_0 \cos\theta_0}$$

$$\Rightarrow \qquad \left[-2\omega(x+R) + v_0 \sin\theta_0\right] dx = (2\omega y + v_0 \cos\theta_0) dy$$

$$\Rightarrow \qquad \int_{-R}^{x} \left[-2\omega(x+R) + v_0 \sin\theta_0\right] dx = \int_0^y (2\omega y + v_0 \cos\theta_0)\, dy$$

$$\Rightarrow \qquad \left[x - \left(-R + \frac{v_0}{2\omega}\sin\theta_0\right)\right]^2 + \left[y - \left(-\frac{v_0}{2\omega}\cos\theta_0\right)\right]^2 = \left(\frac{v_0}{2\omega}\right)^2 \qquad (9)$$

That is, the trajectory of the bullet is a circle in the disk reference frame.

Considering the constraint condition: the trajectory of the bullet passes through Point B:

$$(x_B, y_B) = (R, 0) \qquad (10)$$

(10) $\rightarrow$ (9),

$$\Rightarrow \qquad \sin\theta_0 = \frac{2\omega R}{v_0} \qquad (11)$$

$$\Rightarrow \qquad \theta_0 = \arcsin\left(\frac{2\omega R}{v_0}\right) \qquad (12)$$

(11) $\rightarrow$ (9), we can obtain the trajectory equation of the bullet in the disk reference frame:

$$\Rightarrow \qquad x^2+\left[y+\sqrt{\left(\frac{v_0}{2\omega}\right)^2-R^2}\right]^2=\left(\frac{v_0}{2\omega}\right)^2 \tag{13}$$

The center of the circular trajectory of the bullet in the disk reference frame is denoted by O', defining a—the radius of the circle.

$$\Rightarrow \qquad \begin{cases} (x_{O'}, y_{O'})=\left(0,\ -\sqrt{\left(\frac{v_0}{2\omega}\right)^2-R^2}\right) & (14a) \\ a=\frac{v_0}{2\omega} & (14b) \end{cases}$$

Solution [2]

It can be seen from Eq. (3) that in the disk reference frame, the resultant force acting on the bullet is perpendicular to its velocity all the time; therefore the bullet performs the uniform circular motion.

Considering ***the normal component form of acceleration***:

$$|\ddot{\mathbf{r}}'|=\frac{v_0^2}{a} \tag{15}$$

(15) $\rightarrow$ (3),

$$\Rightarrow \qquad 2mv_0\omega=m\frac{v_0^2}{a} \tag{16}$$

$$\Rightarrow \qquad a=\frac{v_0}{2\omega} \tag{17}$$

Considering the constraint condition: the trajectory of the bullet passes through Point B. Considering the following geometric relation in Fig. 2:

$$\sin\theta_0=\frac{R}{a} \tag{18}$$

(17) $\rightarrow$ (18),

$$\Rightarrow \qquad \sin\theta_0=\frac{2\omega R}{v_0} \tag{19}$$

2.3.2 The Motion of a Small Ring Sleeved on a Big Rotating Ring (d, *C*)

A small ring (denoted by A) is sleeved on a smooth big ring which is on a horizontal ground (see Fig. 1). Under the external force, the big ring rotates around a fixed point on it (denoted by O) at a constant angular velocity of the magnitude ω and the direction vertical upward. The center of the big ring is denoted by C, the angle between $\overrightarrow{OC}$ and $\overrightarrow{CA}$ at arbitrary time is denoted by θ. Find the kinematics equation satisfied by θ.

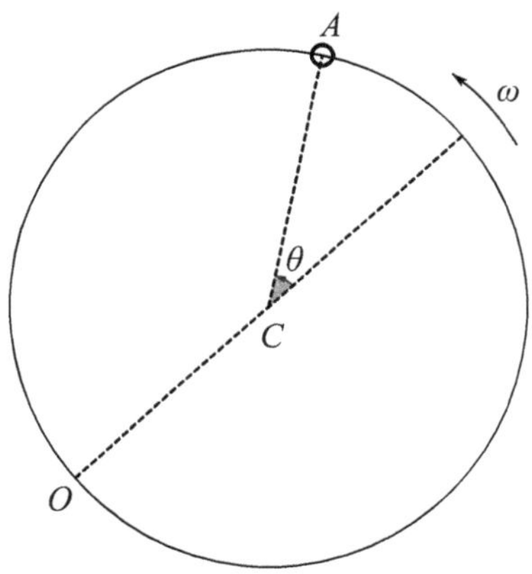

Figure 1

Solution [1]

Time zero is arbitrarily set.

As shown in Fig. 2, taking Point O as the pole, an arbitrary direction in the horizontal plane as the positive direction of the polar axis, anticlockwise rotation direction as the positive direction of polar angle, establishing the polar coordinate system, let r, φ be the polar coordinates. Introducing the tangent unit vector $\boldsymbol{\tau}$ and normal unit vector $\boldsymbol{n}$ of the big ring at Point A.

Taking *the small ring (can be regarded as a particle)* as the object of study. Studying *the state at arbitrary Time t.* Considering (a) ***Newton's second law*** (the component equations in the ($\boldsymbol{r}^0$, $\boldsymbol{\varphi}^0$)-directions), (b) ***the components form of the acceleration of a particle in polar coordinate system*** [integrating into (a)]:

$$\begin{cases} -N\cos\dfrac{\theta}{2} = m(\ddot{r} - r\dot{\varphi}^2) & \text{(1a)} \\ -N\sin\dfrac{\theta}{2} = m(r\ddot{\varphi} + 2\dot{r}\dot{\varphi}) & \text{(1b)} \end{cases}$$

m—the mass of the small ring.

N—the $\boldsymbol{n}$-direction component of the normal contact force exerted on the small ring by the big ring at Time t.

Considering the following geometric relations in Fig. 2:

$$r = 2R \cdot \cos\frac{\theta}{2} \tag{2}$$

$$\varphi = \varphi_C + \frac{\theta}{2} \tag{3}$$

R—the radius of the big ring.

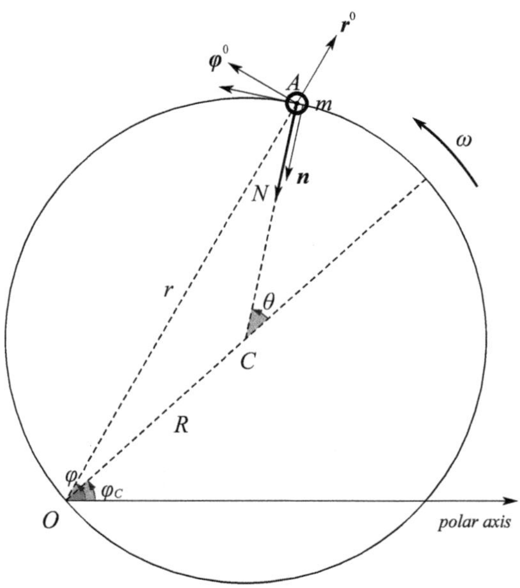

Figure 2

Considering the definition of ***angular velocity***:

$$\dot{\varphi}_C = \omega \tag{4}$$

From (2),

$$\Rightarrow \qquad \dot{r} = -R\dot{\theta}\sin\frac{\theta}{2} \tag{5}$$

$$\Rightarrow \qquad \ddot{r} = -R\ddot{\theta}\sin\frac{\theta}{2} - \frac{1}{2}R\dot{\theta}^2\cos\frac{\theta}{2} \tag{6}$$

Combining (3) and (4),

$$\Rightarrow \qquad \dot{\varphi} = \omega + \frac{\dot{\theta}}{2} \tag{7}$$

$$\Rightarrow \qquad \ddot{\varphi} = \frac{\ddot{\theta}}{2} \tag{8}$$

(5) to (8) → (1),

$$\Rightarrow \begin{cases} -N\cos\frac{\theta}{2} = m\left[-R\ddot{\theta}\sin\frac{\theta}{2} - \frac{1}{2}R\dot{\theta}^2\cos\frac{\theta}{2} - 2R\cos\frac{\theta}{2}\left(\omega + \frac{\dot{\theta}}{2}\right)^2\right] & \text{(9a)} \\ -N\sin\frac{\theta}{2} = m\left[R\ddot{\theta}\cos\frac{\theta}{2} - 2R\dot{\theta}\sin\frac{\theta}{2}\left(\omega + \frac{\dot{\theta}}{2}\right)\right] & \text{(9b)} \end{cases}$$

Combining (9a) and (9b), we obtain the kinematics equation satisfied by θ:

$$\Rightarrow \qquad \ddot{\theta} + \omega^2 \sin\theta = 0 \tag{10}$$

Solution [2]

Taking *the small ring* as the object of study. Studying *the state at arbitrary Time t*. Considering (a) ***Newton's second law*** [in the rotating reference frame with the pole O as the center and $\boldsymbol{\omega}$ as the angular velocity] (the $\boldsymbol{\tau}$-direction component equation), (b) ***inertial centrifugal force theorem*** [integrating into (a)] (c) ***the transverse component form of acceleration*** [integrating into (a)]:

$$-mr\dot{\varphi}_C^2 \sin\frac{\theta}{2} = mR\ddot{\theta} \tag{11}$$

(2) and (4) → (11), we obtain the kinematics equation satisfied by θ:

$$\Rightarrow \qquad \ddot{\theta} + \omega^2 \sin\theta = 0 \tag{12}$$

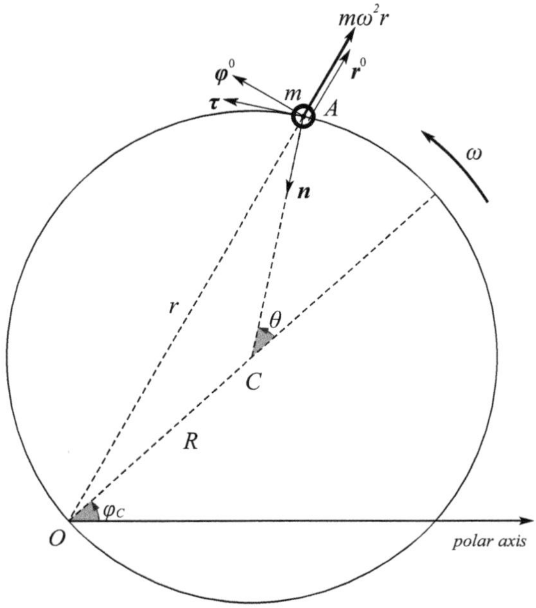

Figure 3

2.3.3 The Free Fall Considering Coriolis Force (d, *C*)

The rotating angular velocity of the earth is denoted by $\boldsymbol{\omega}$, $\omega = |\boldsymbol{\omega}|$ is a small quantity, the gravitational acceleration at latitude φ is denoted by $\boldsymbol{g}$ (pure gravitational field that doesn't take into account inertial forces). A particle falls freely from h above the ground at latitude φ, the air resistance can be ignored. Find the landing position of the particle (under the first-order approximation of ω).

Solution

The time point when the particle starts to fall is defined as Time zero.

The rotating reference frame fixed to the earth is denoted by Reference frame *S*. As shown in Fig. 1, the projection of the point of release on the earth is denoted by *O*, the unit vector tangent to the longitude line passing through Point *O* is denoted by $\boldsymbol{i}$, the unit vector tangent to the latitude line passing through Point *O* is denoted by $\boldsymbol{j}$, the unit vector in the vertical upward direction is denoted by $\boldsymbol{k}$. Establishing the Cartesian coordinate system with the frame $(O, \boldsymbol{i}, \boldsymbol{j}, \boldsymbol{k})$ fixed to Reference frame *S*, let x, y, z be the Cartesian coordinates.

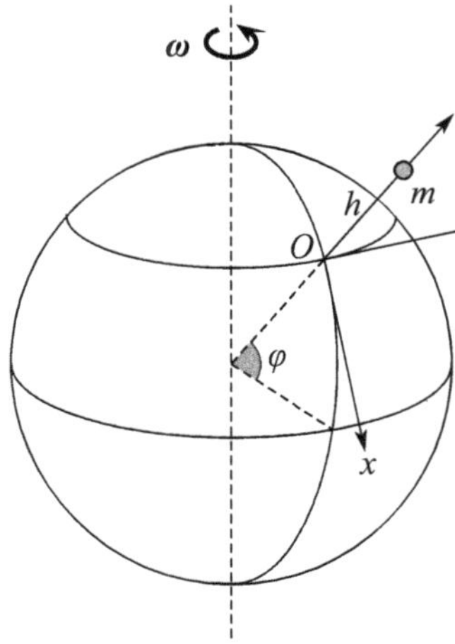

Figure 1

Taking *the particle* as the object of study. Studying *the state at arbitrary Time t*. Considering (a) ***Newton's second law***, (b) ***law of gravitation*** [integrating into (a)], (c) ***inertial centrifugal force theorem*** [integrating into (a)], (d) ***Coriolis force theorem*** [integrating into (a)]:

$$m\ddot{\boldsymbol{r}} = m\boldsymbol{g} - m\boldsymbol{\omega}\times(\boldsymbol{\omega}\times\boldsymbol{r}) - 2m\boldsymbol{\omega}\times\boldsymbol{v}'$$
$$\approx m\boldsymbol{g} - 2m\boldsymbol{\omega}\times\boldsymbol{v}' \quad (1)$$

m—the mass of the particle.

$\boldsymbol{v}'$—the velocity of the particle in Reference frame S at Time t.

The operation above has taken the first-order approximation of ω (the inertial centrifugal force is a second-order small quantity, which is ignored).

Known:

$$\begin{cases} \boldsymbol{\omega} = (-\omega\cos\varphi, 0, \omega\sin\varphi) & (2a) \\ \boldsymbol{g} = (0,0,-g) & (2b) \\ \ddot{\boldsymbol{r}} = (\ddot{x},\ddot{y},\ddot{z}) & (2c) \\ \boldsymbol{v}' = (\dot{x},\dot{y},\dot{z}) & (2d) \end{cases}$$

$$\Rightarrow \qquad \boldsymbol{\omega}\times\boldsymbol{v}' = \begin{vmatrix} \boldsymbol{i} & \boldsymbol{j} & \boldsymbol{k} \\ -\omega\cos\varphi & 0 & \omega\sin\varphi \\ \dot{x} & \dot{y} & \dot{z} \end{vmatrix}$$

$$= -\omega\dot{y}\sin\varphi\boldsymbol{i} + \omega(\dot{z}\cos\varphi + \dot{x}\sin\varphi)\boldsymbol{j} - \omega\dot{y}\cos\varphi\boldsymbol{k} \quad (3)$$

(2) and (3) $\rightarrow$ (1),

$$\Rightarrow \begin{cases} \ddot{x} = 2\omega\dot{y}\sin\varphi & (4a) \\ \ddot{y} = -2\omega(\dot{z}\cos\varphi + \dot{x}\sin\varphi) & (4b) \\ \ddot{z} = -g + 2\omega\dot{y}\cos\varphi & (4c) \end{cases}$$

Considering the initial conditions:

$$\begin{cases} x\big|_{t=0} = y\big|_{t=0} = 0, z\big|_{t=0} = h \\ \dot{x}\big|_{t=0} = \dot{y}\big|_{t=0} = \dot{z}\big|_{t=0} = 0 \end{cases} \quad (5a\text{–}f)$$

Integrating the system of equations (4) under the initial conditions:

$$\Rightarrow \begin{cases} \dot{x} = 2\omega y\sin\varphi & (6a) \\ \dot{y} = -2\omega[(z-h)\cos\varphi + x\sin\varphi] & (6b) \\ \dot{z} = -gt + 2\omega y\cos\varphi & (6c) \end{cases}$$

(6) $\rightarrow$ (4):

$$\Rightarrow \begin{cases} \ddot{x} = -4\omega^2\sin\varphi[(z-h)\cos\varphi + x\sin\varphi] \approx 0 & (7a) \\ \ddot{y} = (2\omega g\cos\varphi)t - 4\omega^2 y \approx (2\omega g\cos\varphi)t & (7b) \\ \ddot{z} = -g - 4\omega^2\cos\varphi[(z-h)\cos\varphi + x\sin\varphi] \approx -g & (7c) \end{cases}$$

Integrating the system of equations (7) under the initial conditions:

$$\Rightarrow \begin{cases} x = 0 & (8a) \\ y = \left(\frac{1}{3}\omega g\cos\varphi\right)t^3 & (8b) \\ z = h - \frac{1}{2}gt^2 & (8c) \end{cases}$$

Defining T—the landing Time of the particle.

$$\Rightarrow \quad T = t\big|_{z=0} = \sqrt{\frac{2h}{g}} \quad (9)$$

$$\Rightarrow \begin{cases} x\big|_{t=T} = 0 & (10a) \\ y\big|_{t=T} = \frac{1}{3}\omega g\left(\frac{2h}{g}\right)^{3/2}\cos\varphi & (10b) \end{cases}$$

2.3.4 Foucault Pendulum (d, *C*)

The rotating angular velocity of the earth is denoted by $\boldsymbol{\omega}$, $\omega = |\boldsymbol{\omega}|$ is a small quantity, and the gravitational acceleration at latitude φ is denoted by $\boldsymbol{g}$ (pure gravitational field that doesn't take into account inertial forces). Find the law of the motion of the small-angle simple pendulum at latitude φ (under the first-order approximation of ω). Given the length of the pendulum is l.

Solution

Time zero is arbitrarily set.

The rotating reference frame fixed to the earth is denoted by Reference frame S. As shown in Fig. 1, the projection of the hanging point of the simple pendulum on the earth is denoted by O, the unit vector tangent to the longitude line passing through Point O is denoted by $\boldsymbol{i}$, the unit vector tangent to the latitude line passing through Point O is denoted by $\boldsymbol{j}$, the unit vector in the vertical upward direction is denoted by $\boldsymbol{k}$. Establishing the Cartesian coordinate system with the frame $(O, \boldsymbol{i}, \boldsymbol{j}, \boldsymbol{k})$ fixed to Reference frame S, let x, y, z be the Cartesian coordinates.

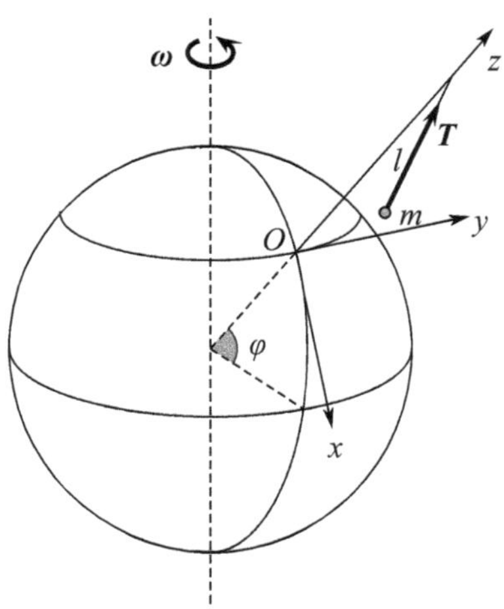

Figure 1

Taking *the pendulum ball (can be regarded as a particle)* as the object of study. Studying *the state at arbitrary Time t*. Considering (a) ***Newton's second law***, (b) ***law of gravitation*** [integrating into (a)], (c) ***inertial centrifugal force theorem*** [integrating into (a)], (d) ***Coriolis force theorem*** [integrating into (a)]:

$$\begin{aligned} m\ddot{\boldsymbol{r}} &= \boldsymbol{T} + m\boldsymbol{g} - m\boldsymbol{\omega}\times(\boldsymbol{\omega}\times\boldsymbol{r}) - 2m\boldsymbol{\omega}\times\boldsymbol{v}' \\ &\approx \boldsymbol{T} + m\boldsymbol{g} - 2m\boldsymbol{\omega}\times\boldsymbol{v}' \end{aligned} \tag{1}$$

m—the mass of the pendulum ball.

$\boldsymbol{v}'$—the velocity of the pendulum ball in Reference frame S at Time t.

$\boldsymbol{T}$—the tension in the string acting on the pendulum ball at Time t.

The operation above has taken the first-order approximation of ω (the inertial centrifugal force is a second-order small quantity, which is ignored).

Known:

$$\begin{cases} \boldsymbol{\omega} = (-\omega\cos\varphi, 0, \omega\sin\varphi) & \text{(2a)} \\ \boldsymbol{g} = (0,0,-g) & \text{(2b)} \\ \ddot{\boldsymbol{r}} = (\ddot{x},\ddot{y},\ddot{z}) & \text{(2c)} \\ \boldsymbol{v}' = (\dot{x},\dot{y},\dot{z}) & \text{(2d)} \\ \boldsymbol{T} = (T\cos\alpha, T\cos\beta, T\cos\gamma) & \text{(2e)} \end{cases}$$

(α, β, γ)—the direction angles of $\boldsymbol{T}$.

The direction cosines of $\boldsymbol{T}$ are as follows:

$$\begin{cases} \cos\alpha = -\dfrac{x}{l} & \text{(3a)} \\ \cos\beta = -\dfrac{y}{l} & \text{(3b)} \\ \cos\gamma \approx 1 - \dfrac{\gamma^2}{2} = \dfrac{l-z}{l} = 1 - \dfrac{z}{l} \approx 1 & \text{(3c)} \end{cases}$$

The operation above has considered γ as a small quantity and taken the first-order approximation (z is a second-order small quantity, which is ignored):

$$\Rightarrow \qquad \boldsymbol{\omega}\times\boldsymbol{v}' = \begin{vmatrix} \boldsymbol{i} & \boldsymbol{j} & \boldsymbol{k} \\ -\omega\cos\varphi & 0 & \omega\sin\varphi \\ \dot{x} & \dot{y} & \dot{z} \end{vmatrix}$$

$$= -\omega\dot{y}\sin\varphi\boldsymbol{i} + \omega(\dot{z}\cos\varphi + \dot{x}\sin\varphi)\boldsymbol{j} - \omega\dot{y}\cos\varphi\boldsymbol{k} \tag{4}$$

Combining (1) to (4),

$$\Rightarrow \quad \begin{cases} m\ddot{x} = -\dfrac{x}{l}T + 2m\omega\dot{y}\sin\varphi & (5a) \\ m\ddot{y} = -\dfrac{y}{l}T - 2m\omega(\dot{z}\cos\varphi + \dot{x}\sin\varphi) & (5b) \\ m\ddot{z} = T - mg + 2m\omega\dot{y}\cos\varphi & (5c) \end{cases}$$

z is a second-order small quantity, taking the following approximations:

$$\Rightarrow \quad \begin{cases} \dot{z} \approx 0 & (6a) \\ \ddot{z} \approx 0 & (6b) \end{cases}$$

(6) $\rightarrow$ (5c),

$$\Rightarrow \quad T = mg - 2m\omega\dot{y}\cos\varphi \approx mg \qquad (7)$$

$$\Rightarrow \quad \begin{cases} \ddot{x} - 2\omega\dot{y}\sin\varphi + \dfrac{g}{l}x = 0 & (8a) \\ \ddot{y} + 2\omega\dot{x}\sin\varphi + \dfrac{g}{l}x = 0 & (8b) \end{cases}$$

Defining the constant ω_0 and ω':

$$\begin{cases} \omega_0 = \sqrt{\dfrac{g}{l}} & (9a) \\ \omega' = \omega\sin\varphi & (9b) \end{cases}$$

$$\Rightarrow \quad \begin{cases} \ddot{x} - 2\omega'\dot{y} + \omega_0^2 x = 0 & (10a) \\ \ddot{y} + 2\omega'\dot{x} + \omega_0^2 y = 0 & (10b) \end{cases}$$

Multiply Eq. (10b) by i and plus Eq. (10a):

$$\Rightarrow \quad (\ddot{x} + i\ddot{y}) + 2i\omega'(\dot{x} + i\dot{y}) + \omega_0^2(x + iy) = 0 \qquad (11)$$

Defining the complex number Z:

$$Z = x + iy \qquad (12)$$

$$\Rightarrow \quad \ddot{Z} + 2i\omega'\dot{Z} + \omega_0^2 Z = 0 \qquad (13)$$

The general solution of Eq. (13) is as follows:

$$\Rightarrow\ Z = e^{-i\omega' t}\left(Ae^{i\omega_0 t} + Be^{-i\omega_0 t}\right) = Ae^{i(\omega_0-\omega')t} + Be^{-i(\omega_0+\omega')t}$$

$$= A\left[\cos(\omega_0 - \omega')t + i\sin(\omega_0 - \omega')t\right] + B\left[\cos(\omega_0 + \omega')t - i\sin(\omega_0 + \omega')t\right]$$

$$= \left[A\cos(\omega_0 - \omega')t + B\cos(\omega_0 + \omega')t\right] + i\left[A\sin(\omega_0 - \omega')t - B\sin(\omega_0 + \omega')t\right]$$

$$= \left[(A+B)\cos(\omega_0 t)\cos(\omega' t) + (A-B)\sin(\omega_0 t)\sin(\omega' t)\right] + i\left[(A-B)\sin(\omega_0 t)\cos(\omega' t) - (A+B)\cos(\omega_0 t)\sin(\omega' t)\right] \tag{14}$$

$$\Rightarrow\ \begin{cases} x = (A+B)\cos(\omega_0 t)\cos(\omega' t) + (A-B)\sin(\omega_0 t)\sin(\omega' t) & \text{(15a)} \\ y = (A-B)\sin(\omega_0 t)\cos(\omega' t) - (A+B)\cos(\omega_0 t)\sin(\omega' t) & \text{(15b)} \end{cases}$$

As shown in Fig. 2a, introducing the rotating reference frame with the angular velocity of $-\omega'\boldsymbol{k}$ relative to Reference frame S, which is denoted by Reference frame S', and establishing the Cartesian coordinate system with the frame $(O, \boldsymbol{i}', \boldsymbol{j}', \boldsymbol{k}')$ fixed to it, where $\boldsymbol{k}' = \boldsymbol{k}$, let x', y', z' be the Cartesian coordinates. The two coordinate systems coincide at Time zero.

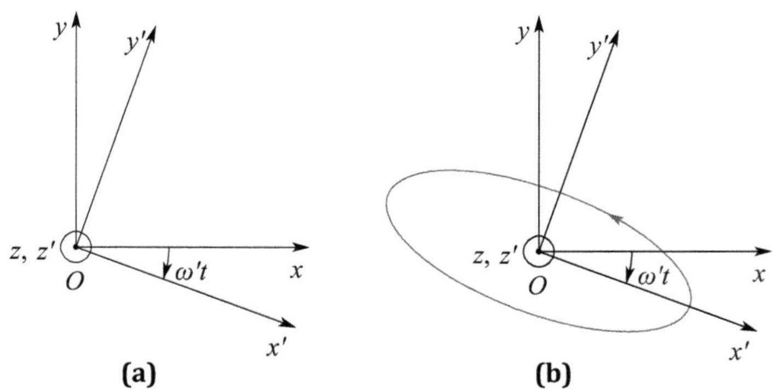

Figure 2a,b

Defining the complex number Z':

$$Z' = x' + iy' \tag{16}$$

The transformation relation between Z and Z' is as follows:

$$Z = Z'e^{-i\omega' t} \tag{17}$$

$$\Rightarrow \quad Z' = Ae^{i\omega_0 t} + Be^{-i\omega_0 t} = (A+B)\cos(\omega_0 t) + i(A-B)\sin(\omega_0 t) \qquad (18)$$

$$\Rightarrow \quad \begin{cases} x' = (A+B)\cos(\omega_0 t) & (19a) \\ y' = (A-B)\sin(\omega_0 t) & (19b) \end{cases}$$

It can be seen that:

(a) In the Reference frame S', the sub-motions of the pendulum ball in the $\boldsymbol{i}'$-direction and $\boldsymbol{j}'$-direction are simple harmonic vibrations with the angular frequency ω_0, that is, in the Reference frame S', the projection motion of the pendulum ball in the O-$x'y'$ plane is an elliptical motion with the angular velocity $\omega_0 \boldsymbol{k}$.

(b) The elliptical motion performs precession with the angular velocity $-\omega' \boldsymbol{k}$ relative to the Reference frame S.

The corresponding schematic diagram can be obtained (see Fig. 2b).

Defining T_0—the rotation period of the earth, T—the rotation period of Reference frame S' relative to Reference frame S:

$$\begin{cases} T_0 = \dfrac{2\pi}{\omega} & (20a) \\ T = \dfrac{2\pi}{\omega'} & (20b) \end{cases}$$

$$\Rightarrow \quad T = \frac{T_0}{\sin\varphi} \qquad (21)$$

Another way to get Eq. (19)

Considering the transformation of the coordinates of the pendulum ball from Base $(\boldsymbol{i}, \boldsymbol{j})$ to Base $(\boldsymbol{i}', \boldsymbol{j}')$, then place (15) in to obtain:

$$\Rightarrow \quad \begin{pmatrix} x' \\ y' \end{pmatrix} = \begin{pmatrix} \cos(\omega' t) & -\sin(\omega' t) \\ \sin(\omega' t) & \cos(\omega' t) \end{pmatrix} \begin{pmatrix} x \\ y \end{pmatrix} = \begin{pmatrix} (A+B)\cos(\omega_0 t) \\ (A-B)\sin(\omega_0 t) \end{pmatrix} \qquad \begin{matrix} (22a) \\ (22b) \end{matrix}$$

2.4 Simple Problems of Dynamics of System of Particles

The dynamic law of a system of particles at arbitrary time can be described jointly by the dynamic equations of all particles constituting the system of particles. This is the most general description method, but it is often not the easiest to solve mathematically. It can be proved that: taking a series of different subsystems (the number of which is required to be equal to the number of all particles, and their sum aggregate is the whole system) in the system of particles, the dynamic information of the system of particles can be fully described based on the dynamic equations of all these subsystems.

The dynamic information of an arbitrary evolution process of a system of particles can be described by the integral form of Newtonian dynamic equation (momentum equation, energy equation, and angular momentum equation). Depending on the keen insight to space translation invariance, time translation invariance and space rotation invariance, the system can be properly selected to satisfy different conservation conditions, and establishing the corresponding conservation equation, which can not only reflect the beauty of physics theory, but also provide great convenience for solving specific problems.

The center-of-mass reference frame of a system of particles, as a space frame with the maximum symmetry of mass distribution, in which the expression of its dynamic equation has the greatest simplicity, which is the preferred reference frame for studying many specific problems.

According to the continuity of particle distribution, the system of particles can be divided into two categories: discrete system of particles and continuous system of particles (called continuum). The continuum can be roughly divided into two categories: rigid continuum and elastic continuum. In this section, we will study the dynamics problems of some discrete system of particles and simple continuum.

2.4.1 N People Jumping Off the Flatcar (*c, C*)[3]

A railroad flatcar of mass M can roll without friction along a straight horizontal track (see Fig. 1). N people, each of mass m are initially standing on the flatcar that is at rest.

Figure 1

[3]This problem is a test question of CUSPEA (China-U.S. Physics Examination and Application) in 1983.

(1) The N people run to one end of the flatcar in unison to jump off the flatcar at the same relative velocity of magnitude u parallel to the track (at the same time). Find the final velocity of the flatcar.

(2) The N people run to one end of the flatcar one by one to jump off the flatcar at the same relative velocity of magnitude u parallel to the track. Find the final velocity of the flatcar.

(3) In which case, (1) or (2) does the flatcar attain the greater speed?

Solution to Case (1)

The unit vector of the velocity direction of the N people relative to the flatcar when they jump off the flatcar is denoted by $\boldsymbol{i}$.

Taking *the system composed of the flatcar and the N people (can be regarded as a system of particles)* as the object of study. Studying *the process of the N people jumping off the flatcar.* Considering ***law of conservation of momentum*** [the $\boldsymbol{i}$-direction component of the resultant external force acting on the system is zero; therefore, the $\boldsymbol{i}$-direction component of the momentum of the system is conserved] (the $\boldsymbol{i}$-direction component equation):

$$Mv + Nmv' = 0 \tag{1}$$

v and v'—the $\boldsymbol{i}$-direction component of the velocity of the flatcar and the N people at the time point after the N people jump off the flatcar.

Taking *the system composed of the N people (can be regarded as a particle)* as the object of study. Studying *the state at the time point after the N people jump off the flatcar.* Considering ***theorem for composition of velocities of a particle*** [taking the ground as the stationary reference frame, and the flatcar as the moving reference frame] (the $\boldsymbol{i}$-direction component equation):

$$v' = v + u \tag{2}$$

(2) → (1),

$$\Rightarrow \qquad v = -\frac{Nm}{M + Nm}u \tag{3}$$

Solution [1] to Case (2)

The No. n person jumping off the flatcar is denoted by No. n person.

Defining v_j—the ***i***-direction component of the velocity of the system composed of the flatcar and the remaining $N-j$ people on the flatcar at the time point after the No. j people jumps off the flatcar. v'_j—the ***i***-direction component of the velocity of No. j people at the time point after he/she jumps off the flatcar.

Taking *the system composed of No. 1 person, the flatcar and the remaining N−1 people on the flatcar* as the object of study. Studying *the process of No. 1 person jumping off the flatcar*. Considering ***law of conservation of momentum*** (the ***i***-direction component equation):

$$[M+(N-1)m]v_1+mv'_1=0 \tag{4}$$

Taking *No. 1 person* as the object of study. Studying *the state at the time point after he/she jumping off the flatcar*. Considering ***theorem for composition of velocities of a particle*** (the ***i***-direction component equation):

$$v'_1=v_1+u \tag{5}$$

(5) → (4),

$$\Rightarrow \qquad v_1=-\frac{m}{M+Nm}u \tag{6}$$

Taking *the system composed of No. 2 person, the flatcar and the remaining N−2 people on the flatcar* as the object of study. Studying *the process of No. 2 person jumping off the flatcar*. Considering ***law of conservation of momentum*** (the ***i***-direction component equation):

$$[M+(N-1)m]v_1=[M+(N-2)m]v_2+mv'_2 \tag{7}$$

Taking *No. 2 person* as the object of study. Studying *the state at the time point after he/she jumps off the flatcar*. Considering ***theorem for composition of velocities of a particle*** (the ***i***-direction component equation):

$$v'_2=v_2+u \tag{8}$$

Combining (6) to (8),

$$\Rightarrow \qquad v_2=-\left[\frac{m}{M+Nm}+\frac{m}{M+(N-1)m}\right]u \tag{9}$$

Based on the above results, using similar analysis, we can obtain:

$$\Rightarrow \qquad v_3=-\left[\frac{m}{M+Nm}+\frac{m}{M+(N-1)m}+\frac{m}{M+(N-2)m}\right]u \tag{10}$$

According to the law that v_1, v_2, v_3 satisfy, we can infer:

$$\Rightarrow \qquad v_j = -\left[\sum_{i=0}^{j-1} \frac{m}{M+(N-i)m}\right]u \tag{11}$$

$$\Rightarrow \qquad v_N = -\left[\sum_{i=0}^{N-1} \frac{m}{M+(N-i)m}\right]u \tag{12}$$

Solution [2] to Case (2)

Taking *the system composed of No. (i + 1) person, the flatcar and the remaining N − i − 1 people on the flatcar* as the object of study. Studying ***the process of No. (i + 1) person jumping off the flatcar***. Considering ***law of conservation of momentum*** (the ***i***-direction component equation):

$$[M+(N-i)m]v_i = [M+(N-i-1)m]v_{i+1} + mv'_{i+1} \tag{13}$$

Taking *No. (i + 1) person* as the object of study. Studying *the state at the time point after he/she jumps off the flatcar*. Considering ***theorem for composition of velocities of a particle*** (the ***i***-direction component equation):

$$v'_{i+1} = v_{i+1} + u \tag{14}$$

(14) → (13),

$$\Rightarrow \qquad v_i - v_{i+1} = \frac{m}{M+(N-i)m}u \tag{15}$$

$$\Rightarrow \qquad v_0 - v_j = \sum_{i=0}^{j-1}(v_i - v_{i+1}) = \left[\sum_{i=0}^{j-1} \frac{m}{M+(N-i)m}\right]u$$

$$\Rightarrow \qquad v_j = v_0 - \left[\sum_{i=0}^{j-1} \frac{m}{M+(N-i)m}\right]u \tag{16}$$

Known:

$$v_0 = 0 \tag{17}$$

$$\Rightarrow \qquad v_j = -\left[\sum_{i=0}^{j-1} \frac{m}{M+(N-i)m}\right]u \tag{18}$$

$$\Rightarrow \qquad v_N = -\left[\sum_{i=0}^{N-1} \frac{m}{M+(N-i)m}\right]u \tag{19}$$

2.4.2 Ellipse Pendulum Realized by the System Composed of Two Particles and a Light Rigid String (d, *C*)

A block of mass M can slide freely on a smooth straight rail in the horizontal direction. There is a light rigid string of length l, one end of which is tied to the block and the other end is tied to a small ball of mass m. Initially, the string is straightened to the left, and the block and the small ball are at rest. Release the block and the small ball at the same time.

(1) Find the tension in the string as a function of the angle between the string and vertical direction.
(2) Find the motion trajectory of the small ball.

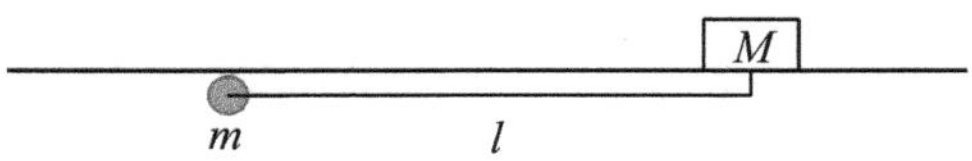

Figure 1

Solution to Problem (1)

The time point when the block and the small ball are released is defined as Time zero.

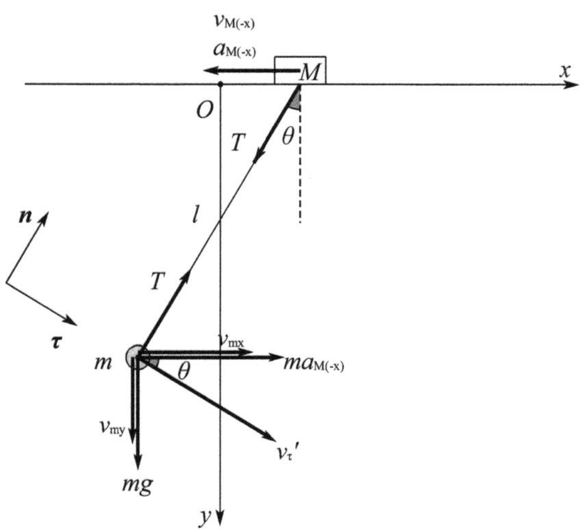

Figure 2

As shown in Fig. 2, the center of mass of the system composed of the block and the ball is denoted by C, whose projection on the straight rail at Time zero is denoted by O, the unit vector in the horizontal right direction is denoted by $\boldsymbol{i}$, the unit vector in the vertical downward direction is denoted by $\boldsymbol{j}$. Establishing the Cartesian coordinate system with the frame $(O, \boldsymbol{i}, \boldsymbol{j})$, let x, y be the Cartesian coordinates. Introducing the natural frame $(\boldsymbol{n}, \boldsymbol{\tau})$ of the ball in the block reference frame.

Taking *the system composed of the block and the ball* as the object of study. Studying *the process from Time zero to arbitrary Time t.* Considering ***law of conservation of momentum*** [the $\boldsymbol{i}$-direction component of the resultant external force acting on the system is zero (the tension in the string acting on the block and the ball have the same magnitude and opposite directions); therefore, the $\boldsymbol{i}$-direction component of the momentum of the system is conserved] (the $\boldsymbol{i}$-direction component equation); Taking *the system composed of the block, the ball and the earth* as the object of study. Studying *the process from Time zero to arbitrary Time t.* Considering ***law of conservation of mechanical energy*** [the work done by the external force and non-conservative internal force acting on the system is zero (the tension in the string acting on the block and the ball have the same magnitude and opposite directions, and there is no relative displacement between the block and the ball throughout); therefore, the mechanical energy of the system is conserved] (taking the position of the ball at the final state as the zero point of gravitational potential energy):

$$\begin{cases} -Mv_{M(-x)} + mv_{mx} = 0 & \text{(1a)} \\ mgl\cos\theta = \dfrac{1}{2}M_{Mx}v_{M(-x)}^2 + \dfrac{1}{2}m(v_{mx}^2 + v_{my}^2) & \text{(1b)} \end{cases}$$

θ—the angle between the string and the vertical direction at Time t.

$v_{M(-x)}$—the $-\boldsymbol{i}$-direction component of the velocity of the block at Time t.

v_{mx} and v_{my}—the $\boldsymbol{i}$-direction and the $\boldsymbol{j}$-direction component of the velocity of the ball at Time t.

Considering the constraint condition: the string length remains unchanged:

$$\Rightarrow \qquad v'_n = 0 \qquad (2)$$

v'_n—the ***n***-direction component of the velocity of the ball relative to the block at arbitrary Time *t*.

Taking *the ball* as the object of study. Studying *the state at arbitrary Time t*. Considering ***theorem for composition of velocities of a particle*** [taking the ground as the stationary reference frame, and the block reference frame as the moving reference frame]:

$$\begin{cases} v_{mx} = -v_{M(-x)} + v'_\tau \cos\theta & (3a) \\ v_{my} = v'_\tau \sin\theta & (3b) \end{cases}$$

v'_τ—the ***τ***-direction component of the velocity of the ball relative to the block at Time *t*.

(3) → (1),

$$\Rightarrow \qquad v'_\tau = \sqrt{\frac{2(M+m)\cos\theta}{M+m\sin^2\theta}gl} \qquad (4)$$

Taking *the block and the ball* as the object of study, respectively. Studying *the state at arbitrary Time t*. Considering ***Newton's second law*** [for the block: in the ground reference frame; for the ball: in the block reference frame] (for the block: the ***i***-direction component equation; for the ball: the ***n***-direction component equation), (b) ***inertial force theorem*** [integrating into (a)], (c) ***the radial component form of acceleration*** [the constraint condition: the string length remains unchanged] [integrating into (a)]:

$$\begin{cases} T\sin\theta = Ma_{M(-x)} & (5a) \\ T + ma_{M(-x)}\sin\theta - mg\cos\theta = m\dfrac{v'^2_\tau}{l} & (5b) \end{cases}$$

T—the magnitude of the tension in the string at Time *t*.

$a_{M(-x)}$—the −***i***-direction component of the acceleration of the block at Time *t*.

(4) → (5),

$$\Rightarrow \qquad T = \frac{Mm[3M + m(2+\sin^2\theta)]}{(M+m\sin^2\theta)^2}g\cos\theta \qquad (6)$$

Solution to Problem (2)

Taking *the system composed of the block and the ball* as the object of study. Studying *the process from Time zero to arbitrary Time t.* Considering (a) ***law of conservation of momentum*** (the ***i***-direction component equation), (b) ***the total momentum of the system of particles is equal to the center-of-mass momentum*** [integrating into (a)]:

$$(M+m)v_{Cx}=0 \tag{7}$$

v_{Cx}—the ***i***-direction component of the velocity of the center of mass C at Time t.

$$\Rightarrow \quad v_{Cx}=0 \tag{8}$$

That is, the projection of the center of mass C on the x-axis remains stationary.

$$\Rightarrow \quad x_{Cx}=\text{const}=0 \tag{9}$$

Considering the definition of ***center of mass***:

$$x_{Cx}=\frac{Mx_M+mx_m}{M+m} \tag{10}$$

$$\Rightarrow \quad Mx_M+mx_m=0 \tag{11}$$

Considering the following geometric relations in Fig. 2:

$$\begin{cases} x_M-x_m=l\sin\theta & \text{(12a)} \\ y_m=l\cos\theta & \text{(12b)} \end{cases}$$

Combining (11) and (12a):

$$\Rightarrow \quad x_m=-\frac{M}{M+m}l\sin\theta \tag{13}$$

Combining (12b) and (13):

$$\Rightarrow \quad \frac{x_m^2}{\left(\frac{M}{M+m}l\right)^2}+\frac{y_m^2}{l^2}=1 \tag{14}$$

That is, the trajectory of the ball in the ground reference frame is a semi-ellipse, the corresponding schematic diagram can be obtain (see Fig. 3):

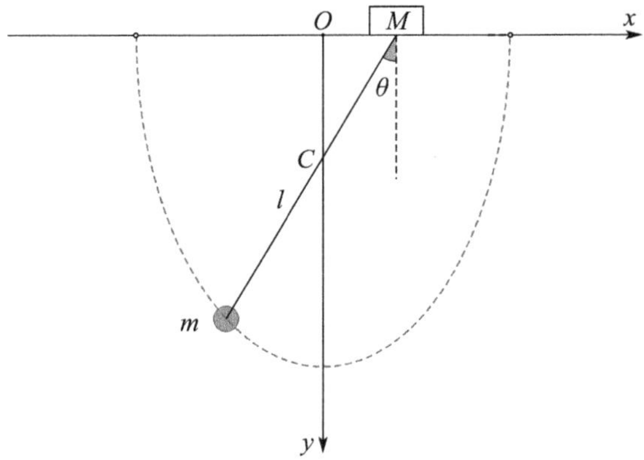

Figure 3

2.4.3 An Object Exploding into Three Parts (*c*, *C*)[4]

(1) A static block explodes into three blocks of equal mass (denoted by 1, 2, 3, respectively, all moving at low speed). The total energy released in the explosion is Q, and the kinetic energies of the three blocks are denoted by T_1, T_2 and T_3, respectively, there are many possibilities for their values. Consider an equilateral triangle of height Q, a explode event can be represented by a point (denoted by P) inside the triangle; T_1, T_2 and T_3 are the distances from this point to the three sides (see Fig. 1). Determine the location range of Point P which is physically permitted.

(2) If a static positronium decays into three photons whose total energy is Q, determine the position range of Point P in the above method.

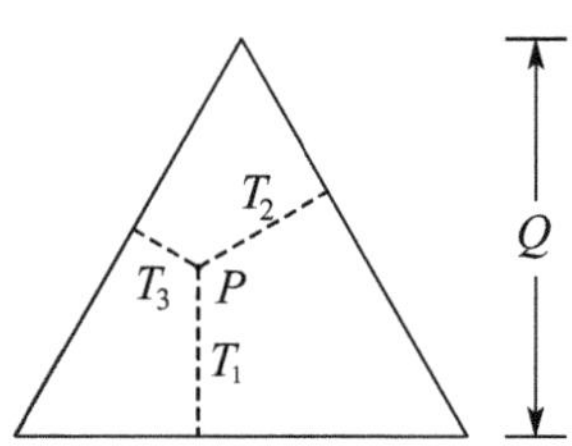

Figure 1

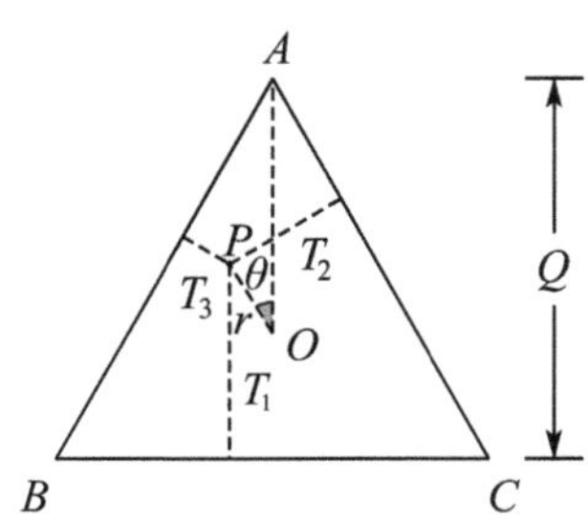

Figure 2

[4]This problem is a test question of CUSPEA (China-U.S. Physics Examination and Application) in 1987.

Solution

As shown in Fig. 2, the center of the equilateral triangle is denoted by O, and the three vertices are denoted by A, B and C, respectively.

Taking *the block/the positronium (can be regarded as a particle)* as the object of study. Studying *the process of its explosion*. Considering (a) ***law of conservation of energy***, (b) ***law of conservation of momentum*** [the resultant external force acting on the system is zero; therefore, the momentum of the system is conserved]:

$$\begin{cases} T_1 + T_2 + T_3 = Q & (1a) \\ \boldsymbol{p}_1 + \boldsymbol{p}_2 + \boldsymbol{p}_3 = 0 & (1b) \end{cases}$$

$\boldsymbol{p}_1$, $\boldsymbol{p}_2$, $\boldsymbol{p}_3$—the momentum of Block 1, 2 and 3 (for the positronium, Photon 1, 2 and 3) at the time point after the explosion (for the positronium, the decay).

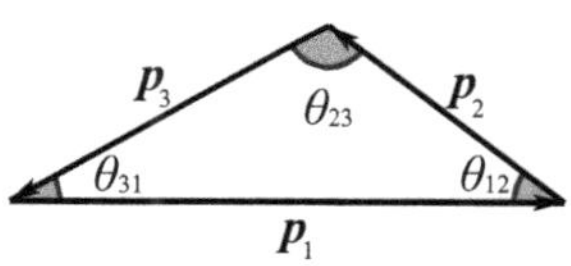

Figure 3

The vector diagram corresponding to Eq. (1b) is shown in Fig. 3. Considering the following geometric relations in Fig. 3:

$$\Rightarrow \quad p_i^2 = p_j^2 + p_k^2 - 2p_j p_k \cos\theta_{jk} \qquad i,j,k = 1,2,3 \quad i \neq j \neq k \tag{2}$$

θ_{jk}—the angle between $\boldsymbol{p}_j$ and $\boldsymbol{p}_k$.

$$\left|\cos\theta_{jk}\right| \leq 1 \tag{3}$$

$$\Rightarrow \quad \left(\frac{p_j^2 + p_k^2 - p_i^2}{2p_j p_k}\right)^2 \leq 1 \tag{4}$$

For Problem (1), considering ***the relation between the kinetic energy and the momentum under non-relativistic limit***:

$$T_i = \frac{p_i^2}{2m} \tag{5}$$

$$\Rightarrow \quad \frac{(T_j + T_k - T_i)^2}{4T_j T_k} \leq 1 \tag{6}$$

Defining r—the distance from Point P to Point O, θ—the angle between OP and OA. Considering the following geometric relations in Fig. 2:

$$\begin{cases} T_1 = \dfrac{Q}{3} + r\cos\theta & \text{(7a)} \\ T_2 = \dfrac{Q}{3} + r\cos\left(\dfrac{4\pi}{3} + \theta\right) & \text{(7b)} \\ T_3 = \dfrac{Q}{3} + r\cos\left(\dfrac{2\pi}{3} + \theta\right) & \text{(7c)} \end{cases}$$

$(7) \to (6)$:

$$\Rightarrow \quad r \le \frac{Q}{3} \tag{8}$$

It can be seen that the location range of Point P which is physically permitted is bounded by the inscribed circle of ΔABC (see Fig. 4).

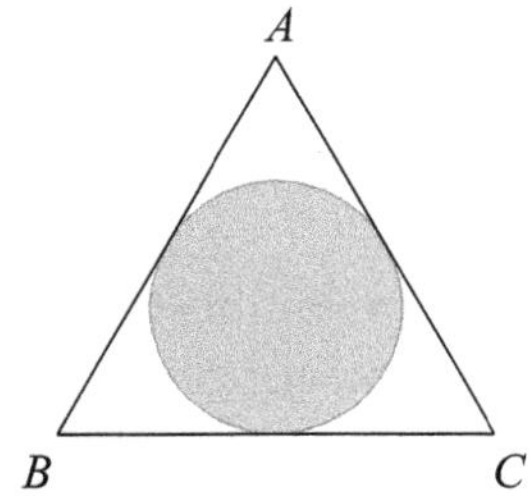

Figure 4

For Problem (2), considering ***the relation between the kinetic energy and the momentum of photon under extreme relativistic limit***:

$$T_i = cp_i \tag{9}$$

$(9) \to (4)$:

$$\Rightarrow \quad \left|\frac{T_j^2 + T_k^2 - T_i^2}{2T_jT_k}\right| \le 1 \tag{10}$$

$$\Rightarrow \quad T_j^2 + T_k^2 - T_i^2 \le 2T_jT_k$$

$$\Rightarrow \quad (T_j - T_k)^2 \le T_i^2$$

$$\Rightarrow \qquad T_j \le T_k + T_i \qquad (11)$$

(1a) → (11):

$$\Rightarrow \qquad 2T_j \le T_k + T_i + T_j = Q$$

$$\Rightarrow \qquad T_j \le \frac{Q}{2} \qquad (12)$$

It can be seen that the position range of Point P which is physically permitted is bounded by the equilateral triangle with the midpoints of the three sides of ΔABC as its vertices (see Fig. 5).

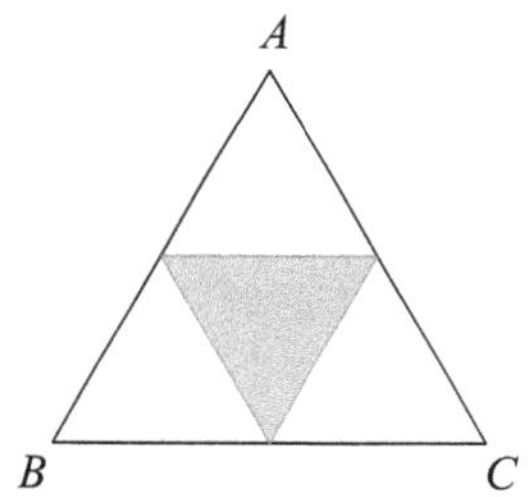

Figure 5

2.4.4 The Motion of a Small Ball along the Inner Wall of a Free, Homogeneous Ring (*d, B*)

A homogeneous ring of mass M and radius R is placed on a smooth ground, and a small ball of mass m can slide along the smooth inner wall of the ring without friction (see Fig. 1). Initially, the ring is static, its center O is at the origin A of the Cartesian coordinate system provided in the figure, and the ball is static at Point $(0, R)$. Give the ball an instant impact to make it get an initial velocity $v_0\boldsymbol{i}$.

(1) Determine the kinematics equation of the ball.
(2) Determine the radius of curvature at the inflexion points of the trajectory of the ball.

Solution to Problem (1)

The time point when the ball starts to move is defined as Time zero.

The center of mass of the system composed of the ring and the ball is denoted by C. Establishing the Cartesian coordinate system with the frame $(C, \boldsymbol{i}, \boldsymbol{j}, \boldsymbol{k})$ fixed to the center of mass C, let x', y', z' be the Cartesian coordinates (see Fig. 2).

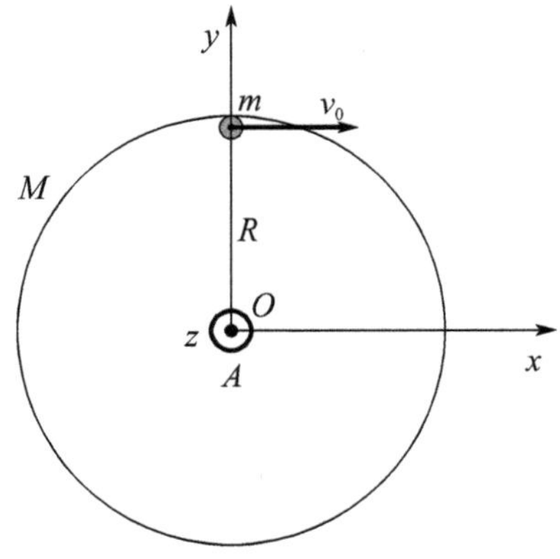

Figure 1

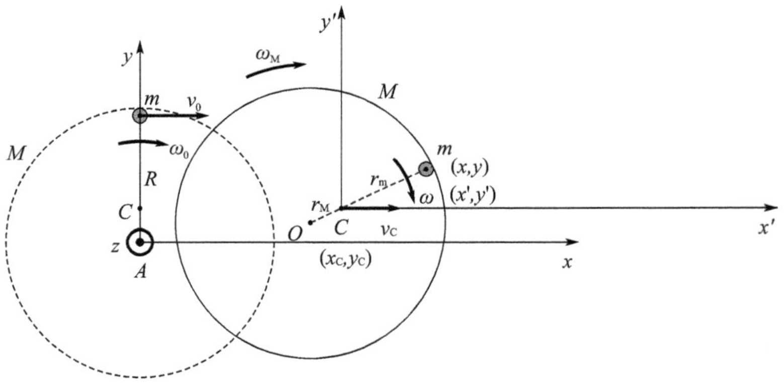

Figure 2

Taking *the ball (can be regarded as a particle)* as the object of study. Studying *the state at arbitrary Time t*. Considering ***Galilean transformation*** [taking the ground as the stationary reference frame, and the center-of-mass reference frame of the system as the moving reference frame] (the component equations in the $(\boldsymbol{i}, \boldsymbol{j})$-directions):

$$\begin{cases} x = x_C + x' & \text{(1a)} \\ y = y_C + y' & \text{(1b)} \end{cases}$$

Taking *the system composed of the ring and the ball (can be regarded as a system of particles)* as the object of study. Studying *the process from Time zero to arbitrary Time t*. Considering (a) ***law of conservation of momentum*** [the resultant external force acting on the system is zero; therefore, the momentum of the system is conserved] (the ***i***-direction component equation), (b) ***the total momentum of the system of particles is equal to the center-of-mass momentum*** [integrating into (a)]:

$$mv_0 = (M+m)v_C \tag{2}$$

$$\Rightarrow \qquad v_C = \frac{m}{M+m}v_0 \tag{3}$$

v_C—the ***i***-direction component of the center-of-mass velocity of the system at Time *t*.

Considering the definition of ***center of mass***:

$$\begin{cases} r_m = \dfrac{M}{M+m}R & \text{(4a)} \\ r_M = \dfrac{m}{M+m}R & \text{(4b)} \end{cases}$$

r_m and r_M—the distance from the ball and the center *O* to the center of mass *C* at arbitrary Time *t*.

$$\Rightarrow \qquad \begin{cases} x_C\big|_{t=0} = 0 & \text{(5a)} \\ y_C\big|_{t=0} = \dfrac{m}{M+m}R & \text{(5b)} \end{cases}$$

Considering ***the displacement-time relation of uniform linear motion*** (the component equations in the (***i***, ***j***)-directions):

$$\begin{cases} x_C = x_C\big|_{t=0} + v_C t & \text{(6a)} \\ y_C = y_C\big|_{t=0} & \text{(6b)} \end{cases}$$

(3) and (5) → (6), we obtain the kinematics equations of the center of mass *C* in the ground reference frame:

$$\Rightarrow \qquad \begin{cases} x_C = \dfrac{m}{M+m}v_0 t & \text{(7a)} \\ y_C = \dfrac{m}{M+m}R & \text{(7b)} \end{cases}$$

Taking *the system composed of the ring and the ball* as the object of study. Studying *the process from Time zero to arbitrary Time t*. Considering (a) ***law of conservation of angular momentum*** [in its center-of-mass reference frame, taking the center of mass C as the reference point, the resultant external torque acting on the system is zero; therefore, the angular momentum of the system is conserved] (the $-\boldsymbol{k}$-direction component equation), (b) ***theorem for the composition of angular momentums based on center-of-mass reference frame*** [integrating into (a)], (c) ***the form of angular momentum of a rigid body performing fixed-axis rotation*** [integrating into (b)]. Taking *the ring* [can be regarded as a rigid body] as the object of study. Studying *the process from Time zero to arbitrary Time t*. Considering (a) ***law of conservation of angular momentum*** [in its center-of-mass reference frame, taking its center of mass (the center O) as the reference point, the resultant external torque acting on the system is zero; therefore, the angular momentum of the system is conserved] (the $-\boldsymbol{k}$-direction component equation), (b) ***the form of angular momentum of a rigid body performing fixed-axis rotation*** [integrating into (a)]:

$$\begin{cases} M\omega_0 r_M^2 + m\omega_0 r_m^2 = (M\omega r_M^2 + I\omega_M) + m\omega r_m^2 & \text{(8a)} \\ 0 = I\omega_M & \text{(8b)} \end{cases}$$

ω_0 and ω—the $-\boldsymbol{k}$-direction component of the rotating angular velocity of the line segment linking the center O and the ball at Time zero and Time t.

ω_M—the $-\boldsymbol{k}$-direction component of the self-rotation angular velocity of the ring at Time t.

I—the moment of inertia of the ring about its central axis parallel to the z-axis.

$$\Rightarrow \qquad \omega_M = 0 \tag{9}$$

That is, the ring does not self-rotate.

Considering the definition of ***angular velocity***:

$$\omega_0 = \frac{v_0}{R} \tag{10}$$

$$\Rightarrow \qquad \omega = \frac{v_0}{R} \tag{11}$$

That is, the line segment linking the center O and the ball performs uniform angular velocity rotation.

Considering ***the kinematics equations of uniform angular velocity rotation***, we obtain the kinematics equations of the ball in the center-of-mass reference frame of the system:

$$\begin{cases} x' = r_m \sin \omega t & (12a) \\ y' = r_m \cos \omega t & (12b) \end{cases}$$

(7) and (12) → (1), we obtain the kinematics equations of the ball in the ground reference frame:

$$\Rightarrow \begin{cases} x = \dfrac{m}{M+m} v_0 t + \dfrac{M}{M+m} R \sin \omega t & (13a) \\ y = \dfrac{m}{M+m} R + \dfrac{M}{M+m} R \cos \omega t & (13b) \end{cases}$$

Solution to Problem (2)

Considering the definition of ***velocity***, then place (13) in to obtain:

$$\Rightarrow \begin{cases} v_x = \dfrac{\mathrm{d}x}{\mathrm{d}t} = \dfrac{m}{M+m} v_0 + \dfrac{M}{M+m} v_0 \cos \omega t & (14a) \\ v_y = \dfrac{\mathrm{d}y}{\mathrm{d}t} = -\dfrac{M}{M+m} v_0 \sin \omega t & (14b) \end{cases}$$

(v_x, v_y)—the components of the velocity of the ball in the $(\boldsymbol{i}, \boldsymbol{j})$-directions at arbitrary Time t.

Considering the definition of ***acceleration***, then place (14) in to obtain:

$$\Rightarrow \begin{cases} a_x = \dfrac{\mathrm{d}v_x}{\mathrm{d}t} = -\dfrac{M}{M+m} \dfrac{v_0^2}{R} \sin \omega t & (15a) \\ a_y = \dfrac{\mathrm{d}v_y}{\mathrm{d}t} = -\dfrac{M}{M+m} \dfrac{v_0^2}{R} \cos \omega t & (15b) \end{cases}$$

(a_x, a_y)—the components of the acceleration of the ball in the $(\boldsymbol{i}, \boldsymbol{j})$-directions at arbitrary Time t.

Taking the coordinate extreme points of ball's local trajectory in the ***i***-direction and ***j***-direction as the inflexion points of its trajectory.

Considering ***the normal component form of acceleration***:

$$a_n = \frac{|v|^2}{\rho} \qquad (16)$$

v—the speed of the ball at arbitrary Time t.

a_n—the normal direction component of the acceleration of the ball at arbitrary Time t.

ρ—the radius of curvature of the ball's trajectory at the position where the ball is at arbitrary Time t.

The necessary condition for the y-coordinate of the ball to take extreme value is as follows:

$$v_y = \frac{dy}{dt} = 0 \tag{17}$$

(14b) $\rightarrow$ (17),

$$\Rightarrow \qquad \sin \omega t = 0 \tag{18}$$

$$\Rightarrow \qquad \omega t = n\pi \qquad n \in \mathbf{N} \tag{19}$$

(19) $\rightarrow$ (14a) and (15b),

$$\Rightarrow \begin{cases} v\big|_{v_y=0} = v_x = \begin{cases} v_0 & \omega t = 2k\pi & \text{(20a)} \\ -\dfrac{M-m}{M+m} v_0 & \omega t = (2k+1)\pi & \text{(20b)} \end{cases} \\ a_n\big|_{v_y=0} = a_y = \begin{cases} -\dfrac{M}{M+m}\dfrac{v_0^2}{R} & \omega t = 2k\pi & \text{(21a)} \\ \dfrac{M}{M+m}\dfrac{v_0^2}{R} & \omega t = (2k+1)\pi & \text{(21b)} \end{cases} \end{cases}$$

(20) and (21) $\rightarrow$ (16),

$$\Rightarrow \qquad \rho\big|_{v_y=0} = \begin{cases} \dfrac{M+m}{M} R & \omega t = 2k\pi & \text{(22a)} \\ \dfrac{(M-m)^2}{M(M+m)} R & \omega t = (2k+1)\pi & \text{(22b)} \end{cases}$$

The necessary condition for the x-coordinate of the ball to take extreme value is as follows:

$$v_x = \frac{dx}{dt} = 0 \tag{23}$$

(14a) $\rightarrow$ (23),

$$\Rightarrow \qquad \cos \omega t = -\frac{m}{M} \tag{24}$$

$$\Rightarrow \qquad \omega t = (2k+1)\pi \pm \arccos\frac{m}{M} \qquad k \in \mathbf{N} \tag{25}$$

(25) → (14b) and (15a),

$$\Rightarrow \begin{cases} v\big|_{v_x=0} = v_y = \begin{cases} -\sqrt{\dfrac{M-m}{M+m}}v_0 & \omega t = (2k+1)\pi - \arccos\dfrac{m}{M} & \text{(26a)} \\ \sqrt{\dfrac{M-m}{M+m}}v_0 & \omega t = (2k+1)\pi + \arccos\dfrac{m}{M} & \text{(26b)} \end{cases} \\ a_n\big|_{v_x=0} = a_x = \begin{cases} -\sqrt{\dfrac{M-m}{M+m}}\dfrac{v_0^2}{R} & \omega t = (2k+1)\pi - \arccos\dfrac{m}{M} & \text{(27a)} \\ \sqrt{\dfrac{M-m}{M+m}}\dfrac{v_0^2}{R} & \omega t = (2k+1)\pi + \arccos\dfrac{m}{M} & \text{(27b)} \end{cases} \end{cases}$$

(26) and (27) → (16),

$$\Rightarrow \qquad \rho\big|_{v_x=0} = \sqrt{\frac{M-m}{M+m}}R \tag{28}$$

Comparing the results of all the above radii of curvature:

$$\frac{(M-m)^2}{M(M+m)}R < \sqrt{\frac{M-m}{M+m}}R < R < \frac{M+m}{M}R \tag{29}$$

$$\Rightarrow \qquad \rho\big|_{v_y=0,\ \omega t=(2k+1)\pi} < \rho\big|_{v_x=0} < R < \rho\big|_{v_y=0,\ \omega t=2k\pi} \tag{30}$$

According to the results above, we can draw the schematic diagram of the ball's trajectory (see Fig. 3):

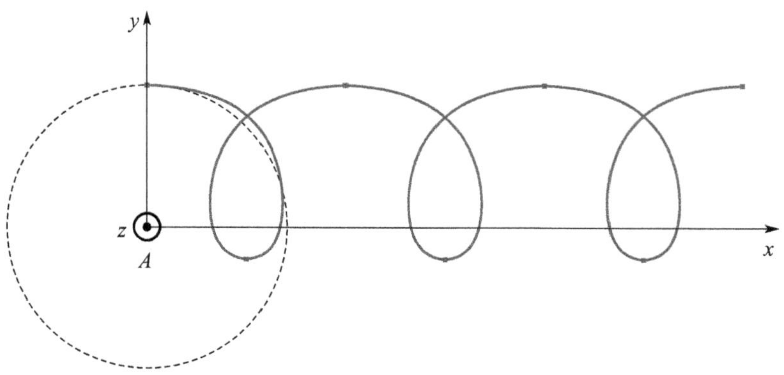

Figure 3

2.4.5 The Crawling of a Beetle on a Ring (*a*, *A*)

Consider a homogeneous ring of mass M and radius R statically placed on a horizontal ground, on which a light rod AB of length L is fixed (see Fig. 1). Initially, a beetle of mass m stands at Point A on the rod. The beetle starts to crawl along the rod toward Point B, and its speed relative to the point on the rod in contact with it is u. Find the center-of-mass velocity and the rotating angular velocity of the ring when the beetle starts to crawl.

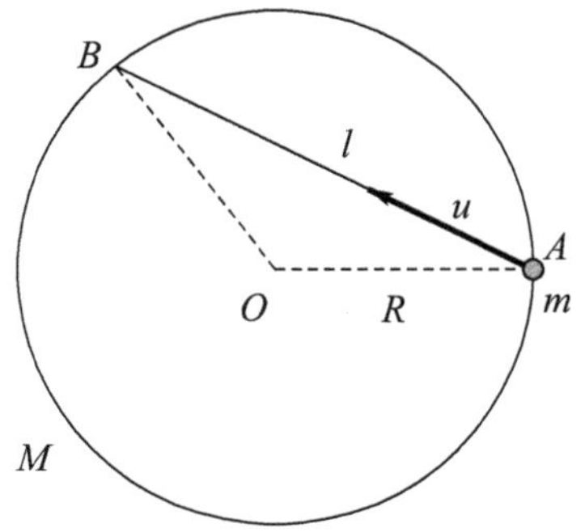

Figure 1

Solution

The time point when the beetle starts to move is defined as Time zero.

As shown in Fig. 2, the center of mass of the system composed of the ring and the beetle at Time zero is denoted by O', the unit vector that the center O pointing to Point A at Time zero is denoted by $\boldsymbol{i}$, the unit vector in the vertical downward direction is denoted by $\boldsymbol{k}$. Establishing the Cartesian coordinate system with the frame $(O', \boldsymbol{i}, \boldsymbol{j}, \boldsymbol{k})$, where $\boldsymbol{j} = \boldsymbol{k} \times \boldsymbol{i}$, let x, y, z be the Cartesian coordinates.

Defining θ—the angle between AO and AB. Considering the following geometric relation in Fig. 2:

$$\theta = \arccos\frac{l}{2R} \tag{1}$$

Taking *the system composed of the ring and the beetle (can be regarded as a system of particles)* as the object of study. Studying *the instantaneous process in which the beetle starts to crawl.* Considering ***law of conservation of momentum*** [the resultant external force acting on the system is zero; therefore, the momentum of the system is conserved] (the ***i***-direction and ***j***-direction component equations):

$$\begin{cases} Mv_{Mx} + mv_{mx} = 0 & \text{(2a)} \\ Mv_{My} + mv_{my} = 0 & \text{(2b)} \end{cases}$$

(v_{Mx}, v_{My}) and (v_{mx}, v_{my})—the ***i***-direction and ***j***-direction components of the velocity of the center *O* and the beetle at Time zero.

Taking *the beetle (can be regarded as a particle)* as the object of study. Studying *the state at Time zero.* Considering ***theorem for composition of velocities of a particle*** [taking the ground as the stationary reference frame, and the translational reference frame that is at rest with respect to Point *A* as the moving reference frame] (the ***i***-direction and ***j***-direction component equations):

$$\begin{cases} v_{mx} = v_{Ax} - u\cos\theta & \text{(3a)} \\ v_{mx} = v_{Ay} - u\sin\theta & \text{(3b)} \end{cases}$$

(v_{Ax}, v_{Ay})—the ***i***-direction and ***j***-direction components of the velocity of Point *A* at Time zero.

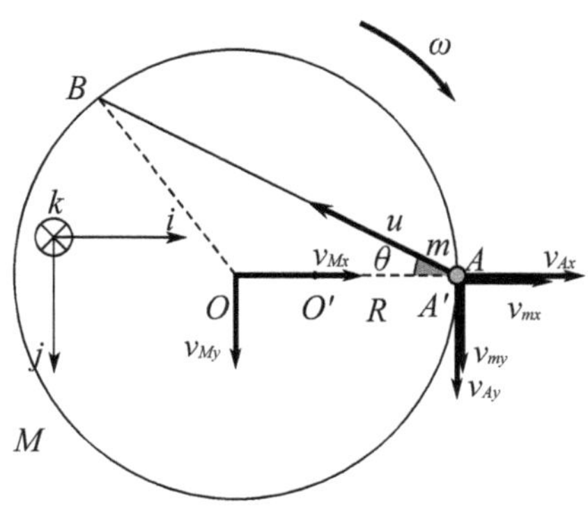

Figure 2

Taking *Point A* as the object of study. Studying *the state at Time zero*. Considering ***theorem for composition of velocities of a particle*** [taking the ground as the stationary reference frame, and the translational reference frame that is at rest with respect to Point *O* as the moving reference frame] (the ***i***-direction and ***j***-direction component equations):

$$\begin{cases} v_{Ax} = v_{Mx} & \text{(4a)} \\ v_{Ay} = v_{My} + \omega R & \text{(4b)} \end{cases}$$

ω—the $\boldsymbol{k}$-direction component of the rotating angular velocity of the ring at Time zero.

$$\Rightarrow \quad \begin{cases} Mv_{Mx} + m(v_{Mx} - u\cos\theta) = 0 & \text{(5a)} \\ Mv_{My} + m[(v_{My} + \omega R) - u\sin\theta] = 0 & \text{(5b)} \end{cases}$$

The point on the ground in contact with Point A at Time zero is denoted by Point A'.

Taking *the system composed of the ring and the beetle* as the object of study. Studying *the instantaneous process in which the beetle starts to crawl*. Considering (a) ***law of conservation of angular momentum*** [taking Point A' as the reference point, the resultant external torque acting on the system is zero; therefore, the angular momentum of the system is conserved] (the $\boldsymbol{k}$-direction component equation), (b) ***theorem for the composition of angular momentums based on center-of-mass reference frame*** [integrating into (a)]:

$$L_{O-A'} + L'_O = 0 \tag{6}$$

$L_{O\text{-}A'}$—the $\boldsymbol{k}$-direction component of the ring's orbital angular momentum at Time zero (taking Point A' as the reference point)

L'_O—the $\boldsymbol{k}$-direction component of the ring's spin angular momentum at Time zero.

Considering the definition of ***angular momentum***:

$$\begin{cases} L_{O-A'} = -Mv_{My}R & \text{(7a)} \\ L'_O = \sum \Delta M(\omega R)R = M\omega R^2 & \text{(7b)} \end{cases}$$

(7) → (6),

$$\Rightarrow \quad -Mv_{My}R + M\omega R^2 = 0 \tag{8}$$

Combining (5) and (8),

$$\Rightarrow \quad \begin{cases} v_{Mx} = \dfrac{m}{M+m} u\cos\theta & \text{(9a)} \\ v_{My} = \dfrac{m}{M+2m} u\sin\theta & \text{(9b)} \\ \omega = \dfrac{m}{M+2m} \dfrac{u\sin\theta}{R} & \text{(9c)} \end{cases}$$

The results retain θ determined by Eq. (1).

2.4.6 The Motion of Water in a Cuboid Water Cup with Uniformly Variable Velocity (*d*, *C*)

As shown in Fig. 1, there is a cuboid water cup placed on a horizontal table with water in it, which is at rest initially. The distance between the water surface and the cup edge is h. Under the action of the external force, the water cup starts to move at the acceleration of the magnitude a and the direction parallel to its edge of length l (horizontally to the right in the figure). Find the condition that the water will not overflow the cup.

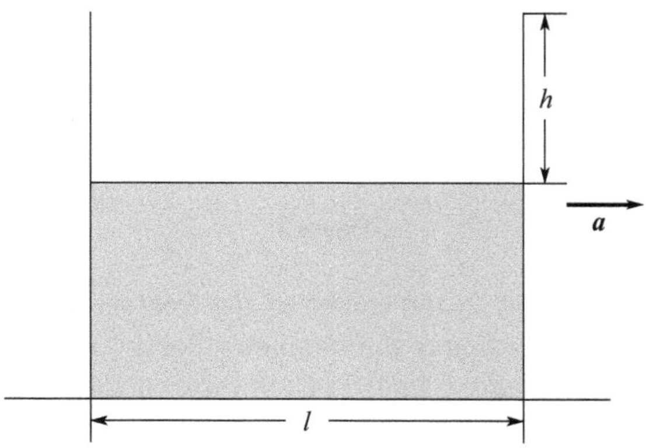

Figure 1

Solution

As shown in Fig. 2, introducing the resultant force of the gravity and the inertia force acting on the water infinitesimal element of unit mass in the water cup reference frame, which is regarded as the equivalent gravity, whose acceleration of gravity is denoted by $\boldsymbol{g}'$:

$$\boldsymbol{g}' = \boldsymbol{g} + \boldsymbol{a} \tag{1}$$

Considering the following geometric relations in Fig. 2:

$$\Rightarrow \quad \begin{cases} g' = \sqrt{g^2 + a^2} & \text{(2a)} \\ \tan\theta = \dfrac{ma}{mg} = \dfrac{a}{g} & \text{(2b)} \end{cases}$$

θ—the angle between $\boldsymbol{g}'$ and $\boldsymbol{g}$.

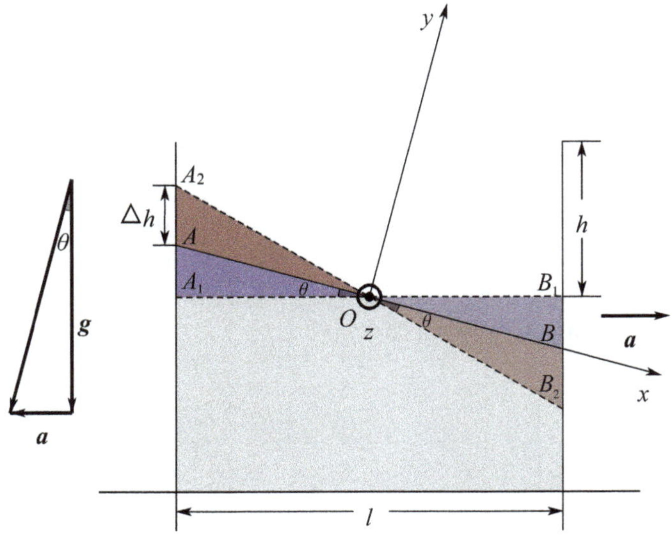

Figure 2

As shown in Fig. 2, the center of the free surface of water is denoted by O (according to the symmetry, Point O remains static in the water cup reference frame), the unit vector of the $-\boldsymbol{g}'$direction is denoted by $\boldsymbol{j}$, the unit vector perpendicular to the plane of the figure outward is denoted by $\boldsymbol{k}$. Establishing the Cartesian coordinate system with the frame $(O, \mathbf{i}, \boldsymbol{j}, \boldsymbol{k})$, where $\boldsymbol{i} = \boldsymbol{j} \times \boldsymbol{k}$, let x, y, z be the Cartesian coordinates.

The intersection line between the free surface of water and the O-xy plane is called the centerline. (a) When the normal direction of the free surface of the water is parallel to $\boldsymbol{j}$ (i.e., when the free surface of the water is in the O-xz plane), the two intersection points of the cup wall and the centerline is denoted by A and B, respectively [hereinafter referred to as Configuration AB]. (b) When the angle between the normal direction of the free surface of the water and the $\boldsymbol{j}$-direction is the maximum in the anticlockwise direction, the two intersection points of the cup wall and the centerline is denoted by A_1 and B_1, respectively [hereinafter referred to as Configuration A_1B_1]. (c) When the angle between the normal direction of the free surface of the water and the $\boldsymbol{j}$-direction is the maximum in the clockwise direction, the two intersection points of the cup wall and the centerline is denoted by A_2 and B_2, respectively [hereinafter referred to as Configuration A_2B_2].

Taking *the system composed of the water in the water cup and the equivalent gravity source* as the object of study. Studying *the process of the water's moving from Configuration A_1B_1 to Configuration A_2B_2* (or the inverse process). Considering ***law of conservation of mechanical energy*** [the work done by the external force acting on the system is zero (the normal contact force exerted on the water by the cup wall does none work) and the work done by the non-conservative internal force acting on the system is zero; therefore, the mechanical energy of the system is conserved]:

$$E_{p1} = E_{p2} \tag{3}$$

E_{p1} and E_{p2}—the equivalent gravitational potential energy of the system when the water in Configuration A_1B_1 and Configuration A_2B_2, respectively.

Defining ρ—the water density. (a) When the water is in Configuration A_1B_1, it can be regarded as the superposition of three parts, including (1) the water in Configuration AB, (2) the water of density ρ in prism with section ΔOBB_1, and (3) the water of density $-\rho$ in prism with section ΔOAA_1. (b) When the water is in Configuration A_2B_2, it can be regarded as the superposition three parts, including (1) the water is in Configuration AB, (2) the water of density ρ in prism with section ΔOAA_2, and (3) the water of density $-\rho$ in prism with section ΔOBB_2.

Considering ***theorem of gravitational potential energy***:

$$\begin{cases} E_{p1} = E_{p0} + (-\rho)V_{1L}g'y_{1L} + \rho V_{1R}g'y_{1R} & \text{(4a)} \\ E_{p2} = E_{p0} + \rho V_{2L}g'y_{2L} + (-\rho)V_{2R}g'y_{2R} & \text{(4b)} \end{cases}$$

E_{p0}—the equivalent gravitational potential energy of the system when the water in Configuration AB.

V_{1L} and V_{1R}—the volume of the water of density $-\rho$ in prism with section ΔOAA_1 and the water of density ρ in prism with section ΔOBB_1.

y_{1L} and y_{1R}—the y-coordinate of the water of density $-\rho$ in prism with section ΔOAA_1 and the water of density ρ in prism with section ΔOBB_1.

V_{2L} and V_{2R}—the volume of the water of density ρ in prism with section ΔOAA_2 and the water of density $-\rho$ in prism with section ΔOBB_2.

y_{2L} and y_{2R}—the y-coordinate of the water of density ρ in prism with section ΔOAA_2 and the water of density $-\rho$ in prism with section ΔOBB_2.

Considering the following geometric relations in Fig. 2:

$$V_{1L} = V_{1R} = \frac{1}{2}\frac{l}{2}\frac{l}{2}\tan\theta \cdot d \tag{5a}$$

$$y_{1L} = -y_{1R} = -\frac{1}{3}\frac{l}{2}\tan\theta \cdot \cos\theta \tag{5b}$$

$$V_{2L} = V_{2R} = \frac{1}{2}\Delta h \frac{l}{2} d \tag{6a}$$

$$y_{2L} = -y_{2R} = \frac{1}{3}\Delta h \cos\theta \tag{6b}$$

Δh—the distance between Point A and Point A_2.

d—the length of the edge of the water cup perpendicular to the plane of the figure.

$$\Rightarrow \quad \begin{cases} E_{p1} = E_{p0} + \dfrac{1}{24}\rho d l^3 \dfrac{\sin^2\theta}{\cos\theta} g' & \text{(7a)} \\ E_{p2} = E_{p0} + \dfrac{1}{6}\rho d l (\Delta h)^2 \cos\theta g' & \text{(7b)} \end{cases}$$

$$\Rightarrow \quad \Delta h = \frac{l}{2}\tan\theta = \frac{a}{2g} l \tag{8}$$

The condition that the water will not overflow the cup is that the height of Point A_2 is lower than the height of the cup edge:

$$\Delta h + \frac{l}{2}\tan\theta \le h \tag{9}$$

$$\Rightarrow \quad a \le \frac{h}{l} g \tag{10}$$

2.4.7 A Homogeneous Rope Sliding from a Horizontal Plane to a Slope (*d, B*)

As shown in Fig. 1, AB is a smooth horizontal plane, BC is a smooth inclined plane with the inclination angle of θ. Initially, a homogeneous rigid rope of length L is placed on AB plane and perpendicular to

B-edge, whose right end is aligned with B-edge. A small disturbance is given to the rope to make it slide down the inclined plane.

(1) Find the position of the rope when it separates from B-edge.
(2) Find the position of the rope when the tension in the rope at the bending point reaches the maximum value.

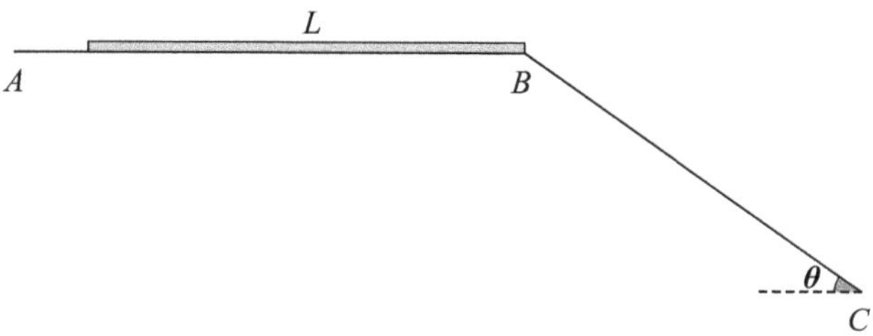

Figure 1

Solution [1] to Problem (1)

The time point when the rope starts to move is defined as Time zero.

As shown in Fig. 2, the unit vector in the horizontal right direction is denoted by $\boldsymbol{e}_1$, and the unit vector in the downward direction parallel to the inclined plane is denoted by $\boldsymbol{e}_2$. The cross section near B-edge (coplanar with the rope) is regarded as an arc of radius r and central angle θ, the center of which is denoted by O. Establishing the polar coordinate system, let r, φ be the polar coordinates.

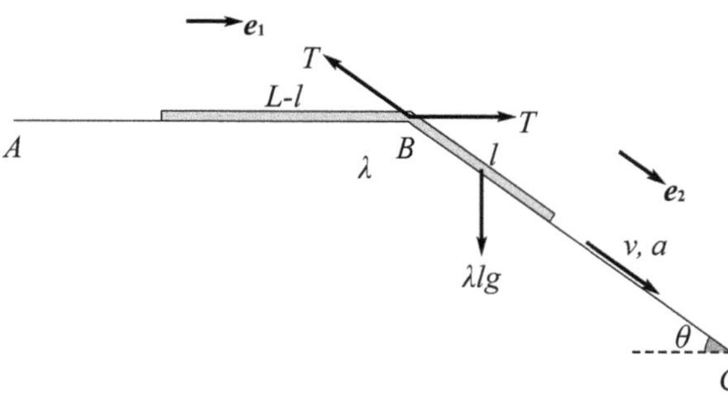

Figure 2

Considering the constraint conditions: the rope length remains unchanged, and the two parts of the rope keep in contact with the AB and BC planes, without rotation. It is easy to prove that at arbitrary

Time *t*, the velocity component and acceleration component of all the points on the rope in the direction perpendicular to the rope are all zero, the velocity component and acceleration component of all the points on the rope in the direction parallel to rope are, respectively, equal [proof omitted].

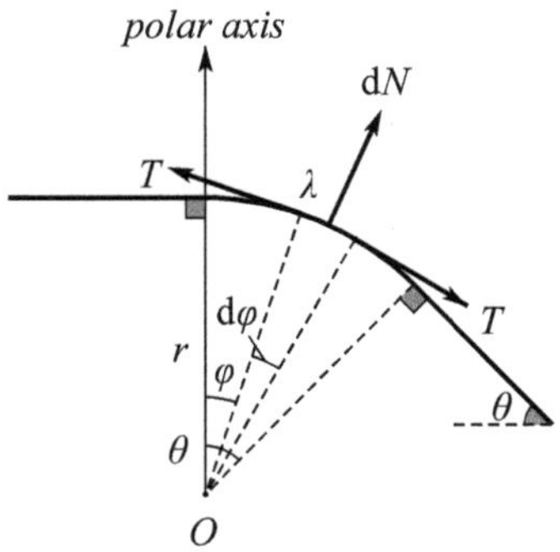

Figure 3

Taking *the system composed of the whole rope and the earth* as the object of study. Studying *the process from Time zero to arbitrary Time t*. Considering ***law of conservation of mechanical energy*** [the work done by the external force and the non-conservative internal force acting on the system is zero (the work done by the tension in rigid rope is zero); therefore, the mechanical energy of the system is conserved] (taking *AB* plane as the zero point of gravitational potential energy):

$$0=\int_{\Gamma}\frac{1}{2}(\lambda \mathrm{d}s)v^2-\lambda lg\cdot\frac{l}{2}\sin\theta=\frac{1}{2}\lambda Lv^2-\lambda lg\cdot\frac{l}{2}\sin\theta \tag{1}$$

λ—the line mass density of the rope.

l—the sliding distance of the rope at Time *t*.

v—the equal velocity component of all the points on the rope in the direction parallel to rope at Time *t*.

$$\Rightarrow \qquad v=l\sqrt{\frac{g\sin\theta}{L}} \tag{2}$$

Taking *the segment of the rope on AB plane and the segment of the rope on BC plane at arbitrary Time t (both can be regarded as a system of particles)* as the object of study, respectively. Studying *the state at Time t*. Considering ***theorem of motion of center of mass*** (the segment of the rope on AB plane: the $\boldsymbol{e}_1$-direction component equation, the segment of the rope on BC plane: the $\boldsymbol{e}_2$-direction component equation):

$$\begin{cases} T = \lambda(L-l)a & (3a) \\ \lambda lg\sin\theta - T = \lambda la & (3b) \end{cases}$$

T—the magnitude of the tension in rope at the bending point at Time t.

a—the equal acceleration component of the center of mass of the segment of the rope on AB plane and the segment of the rope on BC plane in the direction parallel to rope at Time t.

$$\Rightarrow \quad \begin{cases} T = \lambda \dfrac{l(L-l)}{L} g\sin\theta & (4a) \\ a = \dfrac{l}{L} g\sin\theta & (4b) \end{cases}$$

Taking *the rope infinitesimal element between φ and $\varphi + d\varphi$ at arbitrary Time t (can be regarded as a particle)* as the object of study (hereinafter referred to as element dφ). Studying *the state at Time t*. Considering ***Newton's second law*** [the gravity of element dφ is an infinitely small quantity, which can be ignored] (the $-\boldsymbol{r}^0$-direction component equation):

$$2T\sin\frac{\mathrm{d}\varphi}{2} - \mathrm{d}N = \lambda r\mathrm{d}\varphi \cdot \frac{v^2}{r} \tag{5}$$

dN—the $\boldsymbol{r}^0$-direction component of the normal contact force exerted on element dφ by B-edge at Time t.

dφ is a small quantity, taking the first-order approximation of $\sin\dfrac{\mathrm{d}\varphi}{\mathrm{d}t}$:

$$\sin\frac{\mathrm{d}\varphi}{2} \approx \frac{\mathrm{d}\varphi}{2} \tag{6}$$

$$\Rightarrow \quad \mathrm{d}N = \lambda \frac{l(L-2l)}{L} g\sin\theta\,\mathrm{d}\varphi \tag{7}$$

The condition for the rope to separate from B-edge is that the normal contact force exerted on the rope by B-edge is zero:

$$\mathrm{d}N = 0 \tag{8}$$

$$\Rightarrow \qquad l = \frac{L}{2} \tag{9}$$

Solution [2] to Problem (1)

Taking *the rope infinitesimal element moving from AB plane to BC plane in the instantaneous process (the small time interval dt) when the rope just separates from B-edge (can be regarded as a particle)* as the object of study. Studying *this instantaneous process*. Considering ***momentum theorem*** [in this process, the normal contact force exerted on the rope infinitesimal element by B-edge is zero, the gravity of the rope infinitesimal element is an infinitely small quantity, acting on an extremely short process, and its impulse is an infinitely small quantity, which can be ignored; therefore, only the impulse of the tension in the rope is considered] (the $\boldsymbol{e}_1$-direction component equation):

$$(T\cos\theta - T)\mathrm{d}t = \mathrm{d}Mv\cos\theta - \mathrm{d}Mv \tag{10}$$

$\mathrm{d}M$—the mass of the rope infinitesimal element.

Considering the definition of ***line mass density***:

$$\mathrm{d}M = \lambda v \mathrm{d}t \tag{11}$$

$$\Rightarrow \qquad T = \lambda\frac{l^2}{L}g\sin\theta \tag{12}$$

Combining (4a) and (12),

$$\Rightarrow \qquad l = \frac{L}{2} \tag{13}$$

Solution to Problem [2]

The sufficient and necessary conditions for T to take maximum value are as follows:

$$\begin{cases} \dfrac{\mathrm{d}T}{\mathrm{d}l} = 0 & \text{(14a)} \\[2ex] \dfrac{\mathrm{d}^2T}{\mathrm{d}l^2} < 0 & \text{(14b)} \end{cases}$$

$(4a) \rightarrow (14a)$,

$$\Rightarrow \qquad l = \frac{L}{2} \tag{15}$$

From (4a),

$$\Rightarrow \qquad \frac{\mathrm{d}^2T}{\mathrm{d}l^2} = -\frac{2\lambda}{L} g\sin\theta < 0 \tag{16}$$

$$\Rightarrow \qquad T_{\max}\Big|_{l=L/2} = \frac{1}{4}\lambda Lg\sin\theta \tag{17}$$

The Time when the magnitude of the tension in the rope at the bending point reaches the maximum value is the same as the Time when the rope just separates from the B-edge.

2.4.8 The Sliding or Rotating of a Special-Shaped Column (c, B)[5]

There is a homogeneous cylinder of radius R and height h resting on a rough horizontal ground, on the top of which a homogeneous hemisphere of radius R is fixed, the whole constitutes an operating body (as shown in Fig. 1). Initially, a homogeneous rigid rope of length L and the mass equal to the operating body is symmetrically placed on the operating body, $L = \pi R + h$. The rope starts to slide to the right along the operating body due to a small disturbance. Given the coefficient of friction between the cylinder and the ground is μ, and the friction between the rope and the operating body can be ignored. In the process before the right end of the rope falls to the ground.

(1) What conditions should be met for establishing the scenes (a, b) below, respectively?
   (a) The operating body will slide relative to the ground (assuming that the operating body will not rotate before this).
   (b) The operating body will rotate (assuming that the operating body will not slide relative to the ground before this).

(2) Find the condition for the operating body to slide relative to the ground before rotating.

Solution to Problem (1)

The study of Scene (a)

The time point when the rope starts to move is defined as Time zero.

[5]This problem is a test question of the 29th CphO (China Physics Olympiads) in 2012.

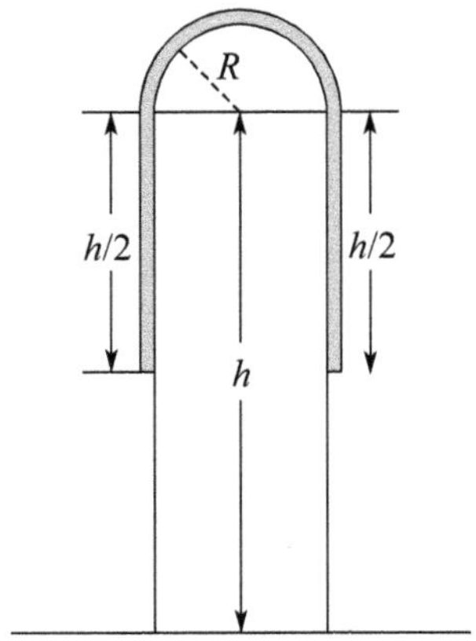

Figure 1

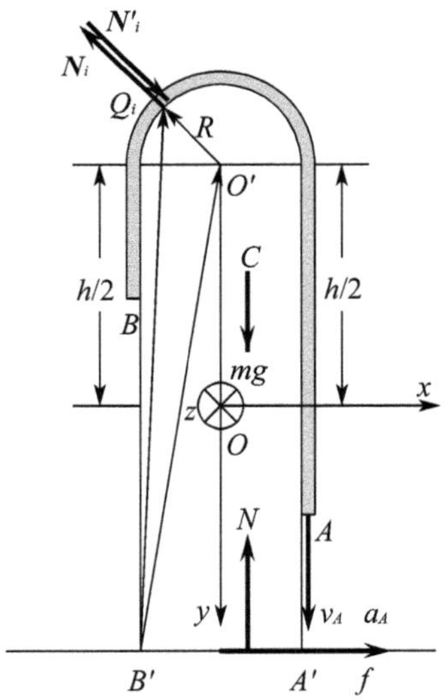

Figure 2

As shown in Fig. (2), the intersection of the central axis of the cylinder and the horizontal line linking the two ends of the rope initially is denoted by O, the unit vector in the horizontal right direction is denoted by $\boldsymbol{i}$, the unit vector in the vertical downward direction is denoted by $\boldsymbol{j}$, $\boldsymbol{i} \times \boldsymbol{j}$ is denoted by $\boldsymbol{k}$ (perpendicular to the plane of the figure inward). Establishing the Cartesian coordinate system with the frame $(O, \boldsymbol{i}, \boldsymbol{j}, \boldsymbol{k})$, let x, y, z be the Cartesian coordinates. The spherical center of the hemisphere is denoted by O', the center of mass of the rope is denoted by C, the two ends of the rope are denoted by A and B, respectively, the projection of Points A and B on the ground at Time zero are denoted by $A'$and B', respectively.

Defining m—the mass of the rope and the operating body, λ—the line mass density of the rope.

$$\lambda = \frac{m}{L} \tag{1}$$

Taking *the system composed of the rope, the operating body and the earth* as the object of study. Studying *the process from Time zero to arbitrary Time t before the operating body slides relative to the ground*. Considering ***law of conservation of mechanical energy*** [the work done by the external force and non-conservative internal force acting on the system is zero; therefore, the mechanical energy of the system is conserved] (taking the origin O as the zero point of gravitational potential energy) (the rope in the final state can be regarded as the superposition of the rope in the initial state and the rope of line mass density λ within the right descending area and the rope of line mass density $-\lambda$ within the left ascending area):

$$\frac{1}{2}mv_A^2 + E_{p0} + (\lambda y_A)g\left(-\frac{y_A}{2}\right) + (-\lambda y_A)g\left(\frac{y_A}{2}\right) = E_{p0} \tag{2}$$

v_A—the ***j***-direction component of the velocity of Point A at Time t.

E_{p0}—the system's gravitational potential energy at Time zero.

$$\Rightarrow \qquad v_A = \sqrt{\frac{2g}{L}}y_A \tag{3}$$

Considering the definition of ***acceleration***:

$$\Rightarrow \qquad a_A = \frac{\mathrm{d}v_A}{\mathrm{d}t} = \sqrt{\frac{2g}{L}}\frac{\mathrm{d}y_A}{\mathrm{d}t} = \sqrt{\frac{2g}{L}}v_A = \frac{2g}{L}y_A \tag{4}$$

a_A—the ***j***-direction component of the acceleration of Point A at arbitrary Time t.

Considering the definition of ***center of mass***:

$$\begin{cases} x_C = x_C\big|_{t=0} + \dfrac{(\lambda y_A)R + (-\lambda y_A)(-R)}{m} = x_C\big|_{t=0} + \dfrac{2R}{L}y_A & \text{(5a)} \\[2ex] y_C = y_C\big|_{t=0} + \dfrac{(\lambda y_A)\dfrac{y_A}{2} + (-\lambda y_A)\left(-\dfrac{y_A}{2}\right)}{m} = y_C\big|_{t=0} + \dfrac{y_A^2}{L} & \text{(5b)} \end{cases}$$

Considering the definition of ***velocity***:

$$\Rightarrow \begin{cases} v_{Cx} = \dfrac{\mathrm{d}x_C}{\mathrm{d}t} = \dfrac{2R}{L}\dfrac{\mathrm{d}y_A}{\mathrm{d}t} = \dfrac{2R}{L}v_A = \dfrac{2R}{L}\sqrt{\dfrac{2g}{L}}y_A & \text{(6a)} \\ v_{Cy} = \dfrac{\mathrm{d}y_C}{\mathrm{d}t} = 2\dfrac{y_A}{L}\dfrac{\mathrm{d}y_A}{\mathrm{d}t} = 2\dfrac{y_A}{L}v_A = \dfrac{2}{L}\sqrt{\dfrac{2g}{L}}y_A^2 & \text{(6b)} \end{cases}$$

v_{Cx} and v_{Cy}—the ***i***-direction and ***j***-direction component of the center-of-mass velocity of the rope at arbitrary Time *t*.

Considering the definition of ***acceleration***:

$$\Rightarrow \begin{cases} a_{Cx} = \dfrac{\mathrm{d}v_{Cx}}{\mathrm{d}t} = \dfrac{2R}{L}\sqrt{\dfrac{2g}{L}}\dfrac{\mathrm{d}y_A}{\mathrm{d}t} = \dfrac{2R}{L}\sqrt{\dfrac{2g}{L}}v_A = \dfrac{4Rg}{L^2}y_A & \text{(7a)} \\ a_{Cy} = \dfrac{\mathrm{d}v_{Cy}}{\mathrm{d}t} = \dfrac{4}{L}\sqrt{\dfrac{2g}{L}}y_A\dfrac{\mathrm{d}y_A}{\mathrm{d}t} = \dfrac{4}{L}\sqrt{\dfrac{2g}{L}}y_A v_A = \dfrac{8g}{L^2}y_A^2 & \text{(7b)} \end{cases}$$

a_{Cx} and a_{Cy}—the ***i***-direction and ***j***-direction component of the center-of-mass acceleration of the rope at arbitrary Time *t*.

The rope above the hemisphere is divided into a series of infinitesimal elements, which are sorted with positive integers clockwise. The position of the No. *i* rope infinitesimal element is defined by Q_i.

Taking *the rope (can be regarded as a system of particles)* as the object of study. Studying *the state at arbitrary Time t before the operating body slides relative to the ground.* Considering ***theorem of motion of center of mass*** (the ***i***-direction and ***j***-direction component equations):

$$\begin{cases} ma_{Cx} = \sum_i \boldsymbol{N}_i \cdot \boldsymbol{i} = \sum_i N_{ix} & \text{(8a)} \\ ma_{Cy} = mg + \sum_i \boldsymbol{N}_i \cdot \boldsymbol{j} = mg + \sum_i N_{iy} & \text{(8b)} \end{cases}$$

$\boldsymbol{N}_i$ (N_{ix}, N_{iy})—the normal contact force exerted on the No. *i* rope infinitesimal element by the hemisphere at Time *t*.

Taking *the operating body (can be regarded as a rigid body)* as the object of study. Studying *the state at arbitrary Time t before it slides relative to the ground.* Considering ***theorem of motion of center of mass*** (the ***i***-direction and ***j***-direction component equations):

$$\begin{cases} 0 = f + \sum_i \boldsymbol{N}_i' \cdot \boldsymbol{i} = f + \sum_i N_{ix}' & \text{(9a)} \\ 0 = mg - N + \sum_i \boldsymbol{N}_i' \cdot \boldsymbol{j} = mg - N + \sum_i N_{iy}' & \text{(9b)} \end{cases}$$

N—the $-\boldsymbol{j}$-direction component of the normal contact force exerted on the operating body by the ground at Time t.

f—the $\boldsymbol{i}$-direction component of the static friction force exerted on the operating body by the ground at Time t.

$\boldsymbol{N}'_i$ (N'_{ix}, N'_{iy})—the normal contact force exerted on the hemisphere by the No. i rope infinitesimal element at Time t.

Considering ***Newton's third law***:

$$\boldsymbol{N}_i = -\boldsymbol{N}_i' \Rightarrow \begin{cases} N_{ix} = -N_{ix}' & \text{(10a)} \\ N_{iy} = -N_{iy}' & \text{(10b)} \end{cases}$$

Combining (7) to (10),

$$\Rightarrow \begin{cases} f = \dfrac{4Ry_A}{L^2} \cdot mg & \text{(11a)} \\ N = \left(2 - \dfrac{8y_A^2}{L^2}\right) mg & \text{(11b)} \end{cases}$$

Considering ***friction law***:

$$f \le \mu N \tag{12}$$

(11) → (12),

$$\Rightarrow \quad 4y_A^2 + \frac{2R}{\mu} y_A - L^2 \le 0 \tag{13}$$

$$\Rightarrow \quad -\frac{R}{4\mu}\left(\sqrt{1 + 4\frac{\mu^2}{R^2}L^2} + 1\right) < y_A < \frac{R}{4\mu}\left(\sqrt{1 + 4\frac{\mu^2}{R^2}L^2} - 1\right) \tag{14}$$

Defining $y_{A\text{-sliding}}$—the y-coordinate of Point A when the operating body starts to slide relative to the ground.

$$\Rightarrow y_{A\text{-sliding}} = \frac{R}{4\mu}\left(\sqrt{1 + 4\frac{\mu^2}{R^2}L^2} - 1\right) = \frac{R}{4\mu}\left(\sqrt{1 + 4\mu^2\left(\pi + \frac{h}{R}\right)^2} - 1\right) \tag{15}$$

The condition for the operating body to slide relative to the ground before the right end of the rope falls to the ground is as follows:

$$y_{A\text{-sliding}} \leq \frac{h}{2} \tag{16}$$

$$\Rightarrow \qquad \mu \leq \frac{h}{\pi(\pi R + 2h)} \tag{17}$$

The study of Scene (b)

Defining $\boldsymbol{M}_{N}$—the $\boldsymbol{k}$-direction component of the torque of the normal contact force exerted on the operating body by the rope (taking Point B' as the reference point). Considering the definition of ***torque***:

$$\begin{aligned}\boldsymbol{M}_{N'} &= \sum_i \overrightarrow{B'Q_i} \times \boldsymbol{N}'_i = \sum_i \left(\overrightarrow{B'O'} + \overrightarrow{O'Q_i}\right) \times \boldsymbol{N}'_i \\ &= \sum_i [(R\boldsymbol{i} - h\boldsymbol{j}) + (R_{ix}\boldsymbol{i} + R_{iy}\boldsymbol{j})] \times (N'_{ix}\boldsymbol{i} + N'_{iy}\boldsymbol{j}) \\ &= \sum_i [(R + R_{ix})\boldsymbol{i} + (-h + R_{iy})\boldsymbol{j}] \times (N'_{ix}\boldsymbol{i} + N'_{iy}\boldsymbol{j}) \\ &= \sum_i [(R + R_{ix})N'_{iy}\boldsymbol{i} \times \boldsymbol{j} + (-h + R_{iy})N'_{ix}\boldsymbol{j} \times \boldsymbol{i}] \\ &= \sum_i \left[(R + R_{ix})N'_{iy} - (-h + R_{iy})N'_{ix}\right]\boldsymbol{k}\end{aligned} \tag{18}$$

(R_{ix}, R_{iy})—the $\boldsymbol{i}$-direction and $\boldsymbol{j}$-direction components of $\overrightarrow{O'Q_i}$.

Considering the geometric relation in Fig. (2):

$$\frac{N'_{ix}}{N'_{iy}} = \frac{R_{ix}}{R_{iy}} \tag{19}$$

$$\Rightarrow \qquad \boldsymbol{M}_{N'} = \sum_i (RN'_{iy} + hN'_{ix})\boldsymbol{k} = \left(R\sum_i N'_{iy} + h\sum_i N'_{ix}\right)\boldsymbol{k} \tag{20}$$

Defining $\boldsymbol{M}_g$—the $\boldsymbol{k}$-direction component of the torque of the gravity of the operating body (taking Point B' as the reference point). Considering the definition of ***torque***:

$$\boldsymbol{M}_g = mgR\boldsymbol{k} \tag{21}$$

The condition for the operating body not to rotate relative to Point B' is as follows:

$$(\boldsymbol{M}_{N'} + \boldsymbol{M}_g)\cdot \boldsymbol{k} \geq 0 \tag{22}$$

(20) and (21) $\rightarrow$ (22),

$$\Rightarrow \quad R\sum_i N'_{iy} + h\sum_i N'_{ix} + mgR \geq 0 \tag{23}$$

Combining (7) to (10), and (23),

$$\Rightarrow \quad 4y_A^2 + 2hy_A - L^2 \leq 0 \tag{24}$$

$$\Rightarrow \quad \frac{-h-\sqrt{h^2+4L^2}}{4} < y_A < \frac{-h+\sqrt{h^2+4L^2}}{4} \tag{25}$$

Defining $y_{A\text{-rotating}}$—the y-coordinate of Point A when the operating body starts to rotate.

$$\Rightarrow \quad y_{A\text{-rotating}} = \frac{-h+\sqrt{h^2+4L^2}}{4} = \frac{h}{4}\left[\sqrt{1+4\left(1+\pi\frac{R}{h}\right)^2} - 1\right] \tag{26}$$

The condition for the operating body to rotate before the right end of the rope falls to the ground is as follows:

$$y_{A\text{-rotating}} \leq \frac{h}{2} \tag{27}$$

$$\Rightarrow \quad \frac{\pi R}{h} \leq \sqrt{2} - 1 \tag{28}$$

Solution to Problem (2)

The condition for the operating body to slide relative to the ground before rotating is as follows:

$$y_{A\text{-sliding}} < y_{A\text{-rotating}} \tag{29}$$

(15) and (26) $\rightarrow$ (29),

$$\Rightarrow \quad \frac{R}{\mu} - h > \frac{R}{\mu}\sqrt{1+4\frac{\mu^2}{R^2}L^2} - \sqrt{h^2+4L^2} \tag{30}$$

Multiplying both sides of Eq. (30) by $\frac{R}{\mu}\sqrt{1+4\frac{\mu^2}{R^2}L^2} + \sqrt{h^2+4L^2}$:

$$\Rightarrow \quad \left[\frac{R}{\mu}\sqrt{1+4\frac{\mu^2}{R^2}L^2}+\sqrt{h^2+4L^2}\right]\left(\frac{R}{\mu}-h\right)>\left(\frac{R}{\mu}\right)^2-h^2$$

$$=\left(\frac{R}{\mu}+h\right)\left(\frac{R}{\mu}-h\right) \qquad (31)$$

Obviously:

$$\frac{R}{\mu}\sqrt{1+4\frac{\mu^2}{R^2}L^2}+\sqrt{h^2+4L^2}>\frac{R}{\mu}+h \qquad (32)$$

$$\Rightarrow \quad \frac{R}{\mu}-h>0$$

$$\Rightarrow \quad \mu<\frac{R}{h} \qquad (33)$$

2.4.9 A Light Rigid Rope with Heavy Objects Hanging at Its Both Ends Entangling on a Rough Cylinder (d, *C*)

Consider a light rigid rope tightly and densely twined on a cylinder of sectional radius R (perpendicular to the central axis of the cylinder), the two ends of which are, respectively, tied with Blocks 1 and 2 of mass m and M, $m < M$ (see Fig. 1), the surfaces of the cylinder and the rope are rough, the friction coefficient between them is μ. How many turns does the rope need to twin at least to keep the two blocks stationary?

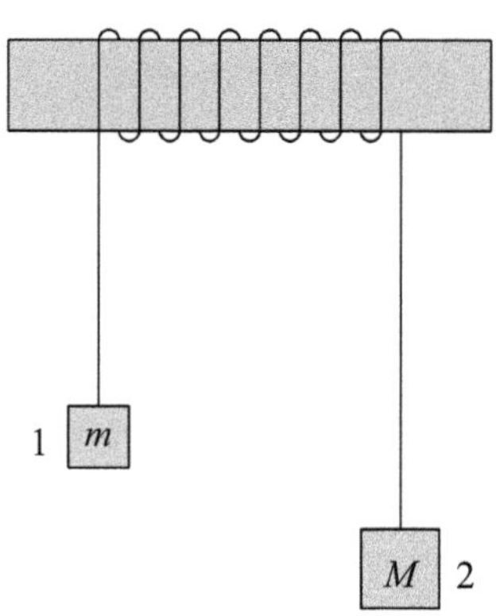

Figure 1

Solution

As shown in Fig. 2, the center of the circle surrounded by the rope on the cylinder is defined by O. Establishing the polar coordinate system, let r, θ be the polar coordinates.

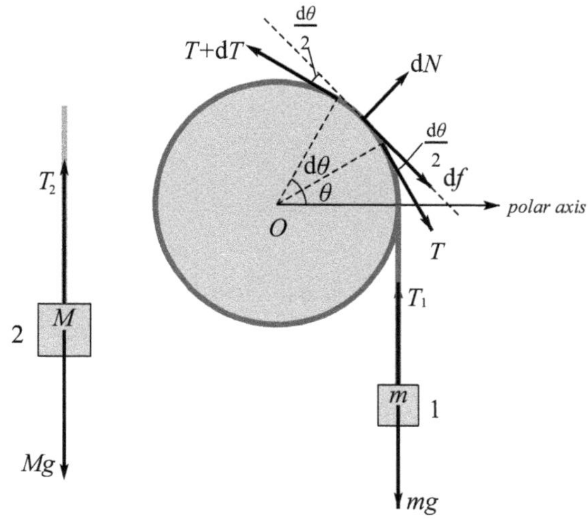

Figure 2

Taking *the rope infinitesimal element between θ and $\theta + d\theta$((hereinafter referred to as element $d\theta$) (can be regarded as a particle)* as the object of study. Studying *the state when the rope is about to slide relative to the cylinder.* (Since $m<M$, it can be inferred that in this state, the movement trend of Block 2 is downward, and the movement trend of Block 1 is upward) Considering ***Newton's second law*** [element $d\theta$ is in mechanical equilibrium] (the component equations in the ($\boldsymbol{r}^0$, $\boldsymbol{\theta}^0$)-directions):

$$\begin{cases} (T+dT)\sin\dfrac{d\theta}{2} + T\sin\dfrac{d\theta}{2} - dN = 0 & \text{(1a)} \\ (T+dT)\cos\dfrac{d\theta}{2} - T\cos\dfrac{d\theta}{2} - df = 0 & \text{(1b)} \end{cases}$$

T—the magnitude of the tension in the rope at θ.

dN—the $\boldsymbol{r}^0$-direction component of the normal contact force exerted on element $d\theta$ by the cylinder.

df—the $-\boldsymbol{\theta}^0$-direction component of the fiction exerted on element $d\theta$ by the cylinder.

Considering ***friction law***:

$$\mathrm{d}f = \mu \mathrm{d}N \tag{2}$$

$\mathrm{d}\theta$ is a small quantity, taking the following first-order approximations:

$$\begin{cases} \sin\dfrac{\mathrm{d}\theta}{2} \approx \dfrac{\mathrm{d}\theta}{2} & \text{(3a)} \\ \cos\dfrac{\mathrm{d}\theta}{2} \approx 1 & \text{(3b)} \end{cases}$$

(2) and (3) → (1),

$$\Rightarrow \quad \frac{\mathrm{d}T}{T} = \mu \mathrm{d}\theta \tag{4}$$

$$\Rightarrow \quad \int_{T_1}^{T_2} \frac{\mathrm{d}T}{T} = \mu \int_0^{\theta_{\max}} \mathrm{d}\theta$$

$$\Rightarrow \quad T_2 = T_1 e^{\mu\theta_{\max}} \tag{5}$$

T_1 and T_2—the magnitude of the tension in the rope acting on Block 1 and 2.

$\theta_{\max}$—the total twining angle of the rope on the cylinder.

Taking *Block 1 and 2 (both can be regarded as a particle)* as the object of study respectively. Studying *the state when the rope is about to slide relative to the cylinder*. Considering ***Newton's second law*** [the two blocks are both in mechanical equilibrium] (the component equation in the vertical up direction):

$$\begin{cases} T_1 - mg = 0 & \text{(6a)} \\ T_2 - Mg = 0 & \text{(6b)} \end{cases}$$

$$\Rightarrow \quad \begin{cases} T_1 = mg & \text{(7a)} \\ T_2 = Mg & \text{(7b)} \end{cases}$$

(7) → (5),

$$\Rightarrow \quad \theta_{\max} = \frac{1}{\mu} \ln \frac{M}{m} \tag{8}$$

Defining n—to keep the two blocks stationary, the number of turns the rope needs to twine at least.

$$\Rightarrow \quad n = \mathrm{Ceiling}\left(\frac{\theta_{\max}}{2\pi}\right) = \mathrm{Ceiling}\left(\frac{1}{2\pi\mu} \ln \frac{M}{m}\right) \tag{9}$$

where function Ceiling (x) represents the minimum integer not smaller than the variable x.

2.4.10 An Elastic Catenary (*b*, *A*)

An elastic rope of natural length L, line mass density λ (in natural length), and stiffness coefficient k (Young's modulus is uniformly distributed) is hung freely with its two ends A, B both fixed. When it reaches static equilibrium (hereinafter referred to as the final state), the anticlockwise direction angles from the vertical upward direction to the tangent lines at its two ends A, B are θ_1 and $\pi-\theta_2$, respectively.

(1) Find the length of the elastic rope at the final state.
(2) Find the shape of the elastic rope in the final state, then deduce its degeneration when $k \to \infty$ (it is considered that the elastic rope degenerates into a rigid rope).
(3) Find the elastic potential energy of the elastic rope in the final state (taking the shape of the elastic rope with no deformation as potential energy zero).
(4) Find the gravitational potential energy of the elastic rope in the final state (taking Point A as potential energy zero).

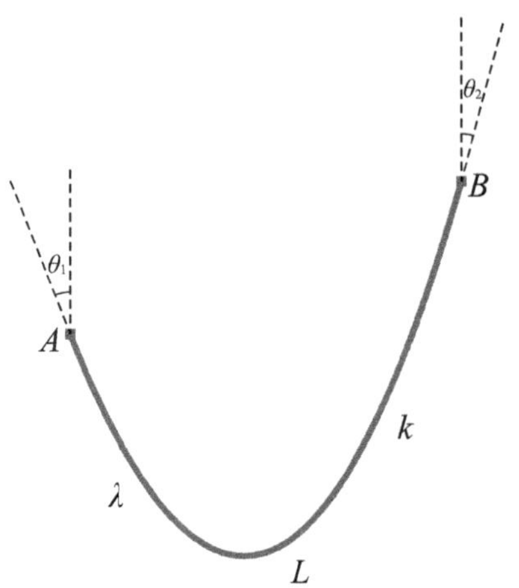

Figure 1

Solution to Problem (1)

Establishing the Cartesian coordinate system with the frame $(A, \boldsymbol{i}, \boldsymbol{j})$ (see Fig. 2), let x, y be the Cartesian coordinates.

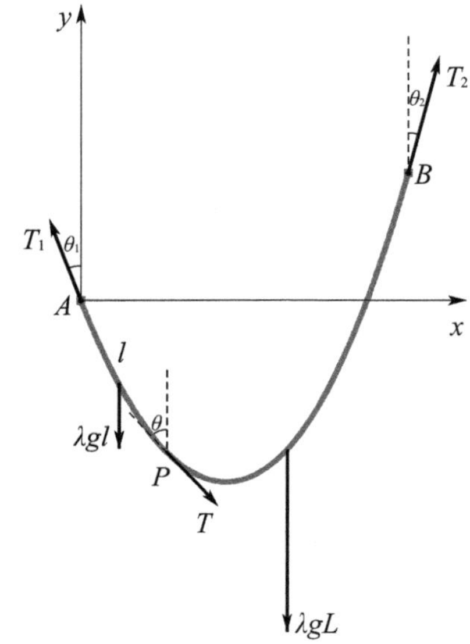

Figure 2

When the elastic rope has no deformation, the point on the elastic rope with the distance l from Point A is denoted by P.

Taking *the elastic rope (can be regarded as a system of particles)* as the object of study. Studying *the final state*. Considering ***Newton's second law*** [the elastic rope is in mechanical equilibrium] (the component equations in the $(\boldsymbol{i}, \boldsymbol{j})$-directions):

$$\begin{cases} T_1 \sin\theta_1 - T_2 \sin\theta_2 = 0 & \text{(1a)} \\ T_1 \cos\theta_1 + T_2 \cos\theta_2 - \lambda L g = 0 & \text{(1b)} \end{cases}$$

T_1 and T_2—the magnitude of the tension in the elastic rope at Point A and Point B in the final state.

$$\Rightarrow \quad \begin{cases} T_1 = \lambda L g \dfrac{\sin\theta_2}{\sin(\theta_1 + \theta_2)} & \text{(2a)} \\ T_2 = \lambda L g \dfrac{\sin\theta_1}{\sin(\theta_1 + \theta_2)} & \text{(2b)} \end{cases}$$

Defining the constant α:

$$\alpha=\frac{\sin\theta_1\sin\theta_2}{\sin(\theta_1+\theta_2)} \tag{3}$$

$$\Rightarrow \quad \begin{cases} T_1=\alpha\lambda Lg\csc\theta_1 & (4a) \\ T_2=\alpha\lambda Lg\csc\theta_2 & (4b) \end{cases}$$

Taking *the segment of the elastic rope between Point A and Point P (hereinafter referred to as AP segment) (can be regarded as a system of particles)* as the object of study. Studying *the final state*. Considering ***Newton's second law*** [*AP* segment is in mechanical equilibrium] (the component equations in the $(\boldsymbol{i},\boldsymbol{j})$-directions):

$$\begin{cases} T_1\sin\theta_1-T\sin\theta=0 & (5a) \\ T_1\cos\theta_1-T\cos\theta-\lambda lg=0 & (5b) \end{cases}$$

θ—the anticlockwise direction angle from the vertical upward direction to the tangent line at Point *P* in the final state.

$$\Rightarrow \quad \begin{cases} l=\alpha L(\cot\theta_1-\cot\theta) & (6a) \\ T=\alpha\lambda Lg\csc\theta & (6b) \end{cases}$$

Taking *the infinitesimal element with the natural length of dl at Point P on the elastic rope (hereinafter referred to as element dl)* as the object of study. Studying *the final state*. Considering **Hooke's law**:

$$T=k_{dl}ds \tag{7}$$

k_{dl}—the stiffness coefficient of element dl.

ds—the elongation of element dl.

Known that Young's modulus is uniformly distributed:

$$k_{dl}=\frac{L}{dl}k \tag{8}$$

$$\Rightarrow \quad T=\frac{L}{dl}kds \tag{9}$$

(6a) and (6b) $\rightarrow$ (9),

$$\Rightarrow \quad ds=\alpha^2\frac{\lambda Lg}{k}\csc^3\theta d\theta \tag{10}$$

$$\Rightarrow \; \Delta l = \int_0^{\Delta l} \mathrm{d}s$$

$$= \alpha^2 \frac{\lambda L g}{k} \int_{\theta_1}^{\theta} \csc^3\theta \,\mathrm{d}\theta$$

$$= \frac{1}{2}\alpha^2 \frac{\lambda L g}{k}\left[\ln\left(\frac{\csc\theta - \cot\theta}{\csc\theta_1 - \cot\theta_1}\right) - (\cos\theta\csc^2\theta - \cos\theta_1\csc^2\theta_1)\right] \tag{11}$$

Δl—the elongation of AP segment in the final state.

The operation above has taken into account the range of θ is $(0, \pi)$, thus simplifying the integral result, the same below.

$$\Rightarrow \; L' = L + \Delta l\big|_{\theta=\pi-\theta_2}$$

$$= L\left\{1 + \frac{1}{2}\alpha^2 \frac{\lambda g}{k}\left[\begin{matrix}\ln\left(\dfrac{\csc\theta_2 + \cot\theta_2}{\csc\theta_1 - \cot\theta_1}\right) + \\ (\cos\theta_1\csc^2\theta_1 + \cos\theta_2\csc^2\theta_2)\end{matrix}\right]\right\}$$

$$= L\left\{1 + \frac{1}{2}\alpha^2 \frac{\lambda g}{k}\left[\begin{matrix}\ln(\csc\theta_1 + \cot\theta_1)(\csc\theta_2 + \cot\theta_2) \\ +(\cos\theta_1\csc^2\theta_1 + \cos\theta_2\csc^2\theta_2)\end{matrix}\right]\right\} \tag{12}$$

L'—the length of the elastic rope in the final state.

Solution to Problem (2)

Defining l'—the length of AP segment in the final state.

$$\Rightarrow \; l' = l + \Delta l$$

$$= \alpha L(\cot\theta_1 - \cot\theta) +$$

$$\frac{1}{2}\alpha^2 \frac{\lambda L g}{k}\left[\ln\left(\frac{\csc\theta - \cot\theta}{\csc\theta_1 - \cot\theta_1}\right) - (\cos\theta\csc^2\theta - \cos\theta_1\csc^2\theta_1)\right] \tag{13}$$

$$\Rightarrow \qquad \mathrm{d}l' = \alpha L\left(\csc^2\theta + \alpha\frac{\lambda g}{k}\csc^3\theta\right)\mathrm{d}\theta \tag{14}$$

Considering the following geometric relations in Fig. 2:

$$\begin{cases}\mathrm{d}x = \mathrm{d}l'\sin\theta & (15a) \\ \mathrm{d}y = -\mathrm{d}l'\cos\theta & (15b)\end{cases}$$

$$\Rightarrow \quad \begin{cases} \mathrm{d}x = \alpha L\left(\csc\theta + \alpha\dfrac{\lambda g}{k}\csc^2\theta\right)\mathrm{d}\theta & (16a) \\ \mathrm{d}y = -\alpha L\left(\csc^2\theta\cos\theta + \alpha\dfrac{\lambda g}{k}\csc^3\theta\cos\theta\right)\mathrm{d}\theta & (16b) \end{cases}$$

$$\Rightarrow \quad \begin{cases} x = \displaystyle\int_0^x \mathrm{d}x \\ \quad = \alpha L\displaystyle\int_{\theta_1}^{\theta}\left(\csc\theta + \alpha\dfrac{\lambda g}{k}\csc^2\theta\right)d\theta \\ \quad = \alpha L\left[\ln\left(\dfrac{\csc\theta - \cot\theta}{\csc\theta_1 - \cot\theta_1}\right) - \alpha\dfrac{\lambda g}{k}(\cot\theta - \cot\theta_1)\right] & (17a) \\ y = \displaystyle\int_0^y \mathrm{d}y \\ \quad = -\alpha L\displaystyle\int_{\theta_1}^{\theta}\left(\csc^2\theta\cos\theta + \alpha\dfrac{\lambda g}{k}\csc^3\theta\cos\theta\right)d\theta \\ \quad = \alpha L(\csc\theta - \csc\theta_1)\left[1 + \alpha\dfrac{\lambda g}{2k}(\csc\theta + \csc\theta_1)\right] & (17b) \end{cases}$$

System of equations (17) are the curvilinear equations of the elastic rope in the final state (with θ as the parameter).

$$\Rightarrow \quad \begin{cases} x_{\text{rigid}} = x\big|_{k\to\infty} = \alpha L\ln\left(\dfrac{\csc\theta - \cot\theta}{\csc\theta_1 - \cot\theta_1}\right) \\ \qquad = \alpha L\left[\operatorname{arcsinh}(-\cot\theta) - \operatorname{arcsinh}(-\cot\theta_1)\right] & (18a) \\ y_{\text{rigid}} = y\big|_{k\to\infty} = \alpha L(\csc\theta - \csc\theta_1) & (18b) \end{cases}$$

System of equations (18) are the degeneration of system of equations (17) when $k \to \infty$ (that is, when the elastic rope degenerates into a rigid rope).

$$\Rightarrow \quad \operatorname{arcsinh}(-\cot\theta) = \frac{x_{\text{rigid}}}{\alpha L} + \operatorname{arcsinh}(-\cos\theta_1) \tag{19}$$

$$\Rightarrow \quad \cot\theta = -\sinh\left[\frac{x_{\text{rigid}}}{\alpha L} + \operatorname{arcsinh}(-\cot\theta_1)\right] \tag{20}$$

$$\Rightarrow \quad \csc\theta = \sqrt{\cot^2\theta + 1} = \sqrt{\sinh^2\left[\frac{x_{\text{rigid}}}{\alpha L} + \text{arsh}(-\cot\theta_1)\right] + 1}$$

$$= \cosh\left[\frac{x_{\text{rigid}}}{\alpha L} + \text{arcsinh}(-\cot\theta_1)\right]$$

$$= \cosh\left(\frac{x_{\text{rigid}}}{\alpha L} + \ln\tan\frac{\theta_1}{2}\right) \tag{21}$$

$$\Rightarrow \quad y_{\text{rigid}} = \alpha L\left[\cosh\left(\frac{x_{\text{rigid}}}{\alpha L} + \ln\tan\frac{\theta_1}{2}\right) - \csc\theta_1\right] \tag{22}$$

Solution to Problem (3)

Defining $\mathrm{d}E_{pk}$—the elastic potential energy of element $\mathrm{d}l$ in the final state, E_{pk}—the elastic potential energy of the elastic rope in the final state (taking the shape of the elastic rope with no deformation as potential energy zero). Considering ***theorem of elastic potential energy of spring***:

$$\Rightarrow \quad \mathrm{d}E_{pk} = \frac{1}{2}k_{\mathrm{d}l}(\mathrm{d}s)^2 = \frac{1}{2}\alpha^3\frac{\lambda^2L^2g^2}{k}\csc^4\theta\,\mathrm{d}\theta \tag{23}$$

$$\Rightarrow \quad E_{pk} = \int_0^{E_{pkL}}\mathrm{d}E_{pk}$$

$$= \frac{1}{2}\alpha^3\frac{\lambda^2L^2g^2}{k}\int_{\theta_1}^{\pi-\theta_2}\csc^4\theta\,\mathrm{d}\theta$$

$$= \frac{1}{6}\alpha^3\frac{\lambda^2L^2g^2}{k}[(\cot^3\theta_1 + 3\cot\theta_1) + (\cot^3\theta_2 + 3\cot\theta_2)] \tag{24}$$

Solution to Problem (4)

Defining $\mathrm{d}E_{pg}$—the gravitational potential energy of element $\mathrm{d}l$ in the final state, E_{pg}—the gravitational potential energy of the elastic rope in the final state (taking Point A as potential energy zero). Considering ***theorem of gravitational potential energy***:

$$\Rightarrow \quad \mathrm{d}E_{pg} = \lambda\mathrm{d}lg\cdot y$$

$$=\alpha^2\lambda L^2 g\begin{bmatrix}-\left(\csc\theta_1+\alpha\dfrac{\lambda g}{2k}\csc^2\theta_1\right)\csc^2\theta+\csc^3\theta+\\ \alpha\dfrac{\lambda g}{2k}\csc^4\theta\end{bmatrix}\mathrm{d}\theta \quad (25)$$

$$\Rightarrow\ E_{pg}=\int_0^{E_{pL}}\mathrm{d}E_{pg}$$

$$=\alpha^2\lambda L^2 g\int_{\theta_1}^{\pi-\theta_2}\begin{bmatrix}-\left(\csc\theta_1+\alpha\dfrac{\lambda g}{2k}\csc^2\theta_1\right)\csc^2\theta+\csc^3\theta+\\ \alpha\dfrac{\lambda g}{2k}\csc^4\theta\end{bmatrix}\mathrm{d}\theta$$

$$=\alpha^2\lambda L^2 g\left\{\begin{array}{l}-\left(\csc\theta_1+\alpha\dfrac{\lambda g}{2k}\csc^2\theta_1\right)(\cot\theta_1+\cot\theta_2)+\\ \dfrac{1}{2}\begin{bmatrix}\ln(\csc\theta_1+\cot\theta_1)(\csc\theta_2+\cot\theta_2)+\\ \left(\dfrac{\cos\theta_2}{\sin^2\theta_2}+\dfrac{\cos\theta_1}{\sin^2\theta_1}\right)\end{bmatrix}+\\ \dfrac{1}{3}\alpha\dfrac{\lambda g}{2k}\left[(\cot^3\theta_1+3\cot\theta_1)+(\cot^3\theta_2+3\cot\theta_2)\right]\end{array}\right\}$$

(26)

Another solution to T

As shown in Fig. 3, introducing the tangent direction $\boldsymbol{\tau}$ and normal direction $\boldsymbol{n}$ of the elastic rope at Point P in the final state.

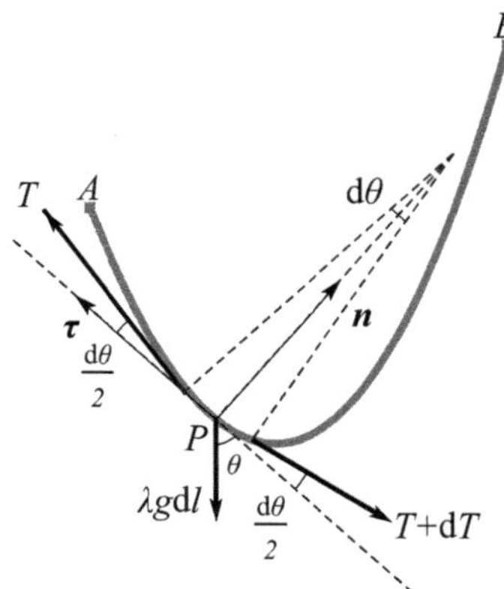

Figure 3

Taking *element dl (can be regarded as a particle)* as the object of study. Studying *the final state*. Considering ***Newton's second law*** [element dl is in mechanical equilibrium] (the component equations in the ($\boldsymbol{\tau}$, $\boldsymbol{n}$)-directions):

$$\begin{cases} T\cos\dfrac{\mathrm{d}\theta}{2}-(T+\mathrm{d}T)\cos\dfrac{\mathrm{d}\theta}{2}-\lambda \mathrm{d}lg\cos\theta=0 & (27a) \\ T\sin\dfrac{\mathrm{d}\theta}{2}+(T+\mathrm{d}T)\sin\dfrac{\mathrm{d}\theta}{2}-\lambda \mathrm{d}lg\sin\theta=0 & (27b) \end{cases}$$

d$\theta$is a small quantity, taking the following first-order approximations:

$$\begin{cases} \sin\dfrac{\mathrm{d}\theta}{2}\approx\dfrac{\mathrm{d}\theta}{2} & (28a) \\ \cos\dfrac{\mathrm{d}\theta}{2}\approx 1 & (28b) \end{cases}$$

$$\Rightarrow \quad \begin{cases} \mathrm{d}T+\lambda \mathrm{d}lg\cos\theta=0 & (29a) \\ T\mathrm{d}\theta-\lambda \mathrm{d}lg\sin\theta=0 & (29b) \end{cases}$$

$$\Rightarrow \quad \frac{\mathrm{d}T}{T}=-\cot\theta\mathrm{d}\theta \tag{26}$$

$$\Rightarrow \quad \int_{T_1}^{T}\frac{\mathrm{d}T}{T}=-\int_{\theta_1}^{\theta}\cot\theta\mathrm{d}\theta$$

$$\Rightarrow \quad T=\alpha\lambda Lg\csc\theta \tag{27}$$

$$\Rightarrow \quad \mathrm{d}l=\alpha L\csc^2\theta\mathrm{d}\theta \tag{28}$$

Further discussion

Due to the particularity of coordinate system selection, the form of Eq. (22) is different from that of traditional catenary equation. The following proves the equivalence of the two.

Let:

$$\theta_1=\theta_2=\theta_0 \tag{29}$$

As shown in Fig. 4, the point a below the lowest point of the rope is denoted by O, establishing the Cartesian coordinate system with the frame (O, $\boldsymbol{i}$, $\boldsymbol{j}$), let X, Y be the Cartesian coordinates.

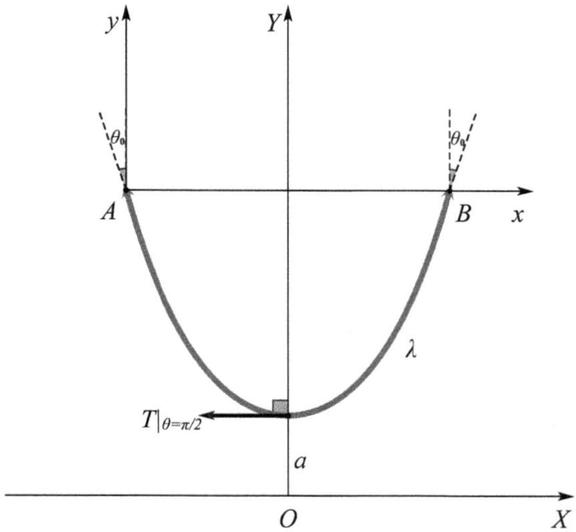

Figure 4

Considering the following geometric relations in Fig. 4:

$$\begin{cases} X = x_{\text{rigid}} - x_{\text{rigid}}\Big|_{\theta=\frac{\pi}{2}} & \text{(30a)} \\ Y = y_{\text{rigid}} + \left(-y_{\text{rigid}}\Big|_{\theta=\frac{\pi}{2}} + a \right) & \text{(30b)} \end{cases}$$

Defining a:

$$a = \frac{T\big|_{\theta=\frac{\pi}{2}}}{\lambda g} \tag{31}$$

(29) → (3),

⇒ $$\alpha\big|_{\theta_1=\theta_2=\theta_0} = \frac{1}{2}\tan\theta_0 \tag{32}$$

(32) → (6b),

⇒ $$T\big|_{\theta=\frac{\pi}{2}} = \alpha\big|_{\theta_1=\theta_2=\theta_0} \lambda Lg = \frac{1}{2}\tan\theta_0 \lambda Lg \tag{33}$$

⇒ $$a = \frac{1}{2}\tan\theta_0 L \tag{34}$$

(29), (32) → (18):

$$\Rightarrow \begin{cases} x_{\text{rigid}}\Big|_{\theta=\frac{\pi}{2}} = \dfrac{1}{2}\tan\theta_0 L \ln\left(\dfrac{1}{\csc\theta_0 - \cot\theta_0}\right) & (35a) \\ y_{\text{rigid}}\Big|_{\theta=\frac{\pi}{2}} = \dfrac{1}{2}\tan\theta_0 L(1-\csc\theta_0) & (35b) \end{cases}$$

(31), (35) → (30):

$$\Rightarrow \begin{cases} x_{\text{rigid}} = X + \dfrac{1}{2}\tan\theta_0 L \ln\left(\dfrac{1}{\csc\theta_0 - \cot\theta_0}\right) & (36a) \\ y_{\text{rigid}} = Y - \dfrac{1}{2}\tan\theta_0 L \csc\theta_0 & (36b) \end{cases}$$

(29), (32) → (22),

$$y_{\text{rigid}} = \frac{1}{2}\tan\theta_0 L\left[\cosh\left(\frac{x_{\text{rigid}}}{\frac{1}{2}\tan\theta_0 L} + \ln\tan\frac{\theta_0}{2}\right) - \csc\theta_0\right] \quad (37)$$

$$\Rightarrow Y = \frac{1}{2}\tan\theta_0 L\cosh\left(\frac{X}{\frac{1}{2}\tan\theta_0 L} + \ln\left(\frac{1}{\csc\theta_0 - \cot\theta_0}\right) + \ln\tan\frac{\theta_0}{2}\right)$$

$$= \frac{1}{2}\tan\theta_0 L\cosh\left(\frac{X}{\frac{1}{2}\tan\theta_0 L} + \ln\cot\frac{\theta_0}{2} + \ln\tan\frac{\theta_0}{2}\right)$$

$$= \frac{1}{2}\tan\theta_0 L\cosh\left(\frac{X}{\frac{1}{2}\tan\theta_0 L}\right) = a\cosh\left(\frac{X}{a}\right) \quad (38)$$

This equation is the traditional form of catenary equation.

Appendix

Defining the indefinite integral I_1:

$$I_1 = \int \csc^3\theta \, d\theta = \int \frac{d\theta}{\sin^3\theta} = \int \frac{\sin\theta}{\sin^4\theta} d\theta = -\int \frac{1}{(1-\cos^2\theta)^2} d(\cos\theta) \quad (1)$$

Defining the variable u:

$$u = \cos\theta \tag{2}$$

$$\Rightarrow \quad I_1 = -\int \frac{1}{(1-u^2)^2}\mathrm{d}u$$

$$= -\int\left[\frac{1}{(1-u)(1+u)}\right]^2 \mathrm{d}u = -\frac{1}{4}\int\left(\frac{1}{1-u}+\frac{1}{1+u}\right)^2 \mathrm{d}u$$

$$= -\frac{1}{4}\int\left[\frac{1}{(1-u)^2}+\frac{1}{(1+u)^2}+\frac{2}{1-u^2}\right]\mathrm{d}u$$

$$= -\frac{1}{4}\left[-\int\frac{1}{(1-u)^2}\mathrm{d}(1-u)+\int\frac{1}{(1+u)^2}\mathrm{d}(1+u)+2\int\frac{1}{1-u^2}\mathrm{d}u\right]$$

$$= -\frac{1}{4}\left(\frac{2u}{1-u^2}+\ln\left|\frac{1+u}{1-u}\right|\right) = -\frac{1}{4}\left(\frac{2\cos\theta}{1-\cos^2\theta}+\ln\left|\frac{1+\cos\theta}{1-\cos\theta}\right|\right)+C$$

$$= \frac{1}{2}\left(\ln\left|\csc\theta-\cot\theta\right|-\frac{\cos\theta}{\sin^2\theta}\right)+C \tag{3}$$

Defining the indefinite integral I_2:

$$I_2 = \int\csc^4\theta\,\mathrm{d}\theta = -\int\csc^2\theta\,\mathrm{d}(\cot\theta)$$

$$= -\int(\cot^2\theta+1)\mathrm{d}(\cot\theta) = -\frac{1}{3}(\cot^3\theta+3\cot\theta)+C \tag{4}$$

2.4.11 A Homogeneous Elastic Rope Suspended at One Side (*b*, *AC*)

An elastic rope of stiffness coefficient k (Young's modulus is uniformly distributed) and mass m (the mass is uniformly distributed) is suspended from the ceiling. Initially, the elastic rope remains static and no deformation under the action of external force. Removing the external force, after a sufficiently long time, the elastic rope reaches static equilibrium.

(1) Find the elongation of the elastic rope in the whole process.

(2) Find the elastic potential energy increment of the elastic rope in the whole process.

(3) Find the gravitational potential energy increment of the elastic rope in the whole process.

(4) Find the work done by non-conservative force in the whole process.

Solution [1] (type C)

Problem (1)

The state before the elastic rope is released is called the initial state, the static equilibrium state that the elastic rope reached after it is released is called the final state.

As shown in Fig. 1, the suspension end of the elastic rope is denoted by *O*, the unit vector in the vertical downward direction is denoted by ***i***, Establishing the linear coordinate system with Point *O* as the origin and the ***i***-direction as the positive direction of coordinate axis, let *x* be the coordinate.

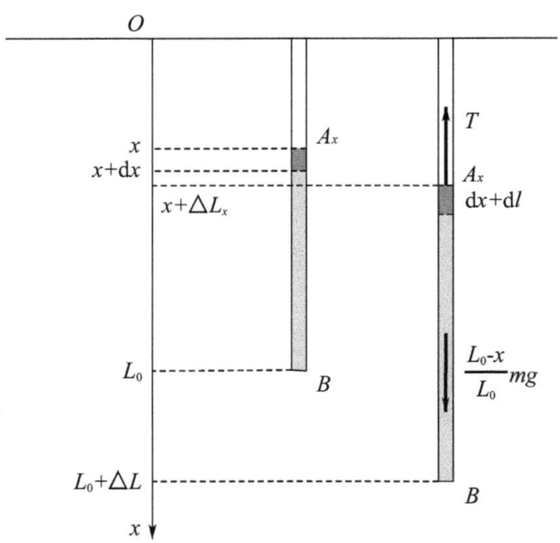

Figure 1

The point on the elastic rope at coordinate *x* in the initial state is denoted by A_x, the free end of the elastic rope is denoted by *B* (see Fig. 1).

Taking *the segment of the rope between Point* A_x *and Point B (can be regarded as a system of particles)* as the object of study. Studying *the final state*. Considering ***Newton's second law*** [the segment of the rope is in mechanical equilibrium] (the **−*i***-direction component equation):

$$T_x - \frac{L_0 - x}{L_0} mg = 0 \tag{1}$$

$$\Rightarrow \quad T_x = \frac{L_0 - x}{L_0} mg \tag{2}$$

L_0—the natural length of elastic rope.

T_x—the magnitude of the tension in the elastic rope at Point A_x in the final state.

Taking *the infinitesimal element on the elastic rope between x and x + dx in the initial state (hereinafter referred to as element dx)* as the object of study. Studying *the final state*. Considering ***Hooke's law***:

$$T_x = \frac{L_0}{\mathrm{d}x} k\mathrm{d}l \tag{3}$$

$\mathrm{d}l$—the elongation of element $\mathrm{d}x$ in the whole process.

$$\Rightarrow \quad \mathrm{d}l = \frac{mg}{kL_0^2}(L_0 - x)\mathrm{d}x \tag{4}$$

$$\Rightarrow \quad \Delta L = \int_0^{\Delta L} \mathrm{d}l = \frac{mg}{kL_0^2}\int_0^{L_0}(L_0 - x)\mathrm{d}x = \frac{mg}{2k} \tag{5}$$

ΔL—the elongation of the elastic rope in the whole process.

Problem (2)

Defining $\mathrm{d}E_{ps}$—the elastic potential energy increment of element $\mathrm{d}x$ in the whole process, ΔE_{ps}—the elastic potential energy increment of the elastic rope in the whole process. Considering ***theorem of elastic potential energy of spring***:

$$\mathrm{d}E_{ps} = \frac{1}{2}\frac{L_0}{dx} k(\mathrm{d}l)^2 \tag{6}$$

$$\Rightarrow \quad \mathrm{d}E_{ps} = \frac{m^2 g^2}{2kL_0^3}(L_0 - x)^2 \mathrm{d}x \tag{7}$$

$$\Rightarrow \quad \Delta E_{ps} = \int_0^{\Delta E_{ps}} \mathrm{d}E_{ps} = \frac{m^2 g^2}{2kL_0^3}\int_0^{L_0}(L_0 - x)^2 \mathrm{d}x = \frac{m^2 g^2}{6k} \tag{8}$$

Problem (3)

Defining ΔL_x—the falling distance of Point A_x (that is, the elongation of the segment of the elastic rope between Point O and Point A_x) in the whole process. From (4),

$$\Rightarrow \qquad \Delta L_x = \int_0^{\Delta L_x} \mathrm{d}l = \frac{mg}{kL_0^2}\int_0^x (L_0 - x)\mathrm{d}x = \frac{mg}{2kL_0^2}(2L_0 - x)x \tag{9}$$

Defining $\mathrm{d}E_{pg}$—the gravitational potential energy increment of element dx in the whole process, ΔE_{pg}—the gravitational potential energy increment of the elastic rope in the whole process. Considering ***theorem of gravitational potential energy***:

$$\mathrm{d}E_{pg} = -\frac{\mathrm{d}x}{L_0} mg\Delta L_x \tag{10}$$

$$\Rightarrow \qquad \mathrm{d}E_{pg} = -\frac{m^2g^2}{2kL_0^3}(2L_0 - x)x\mathrm{d}x \tag{11}$$

$$\Rightarrow \qquad \Delta E_{pg} = \int_0^{\Delta E_{pg}} \mathrm{d}E_{pg} = -\frac{m^2g^2}{2kL_0^3}\int_0^{L_0}(2L_0 - x)x\mathrm{d}x = -\frac{m^2g^2}{3k} \tag{12}$$

Problem (4)

Taking *the system composed of the elastic rope and the earth* as the object of study. Studying *the whole process*. Considering ***theorem of mechanical energy***:

$$W = \Delta E_{ps} + \Delta E_{pg} \tag{13}$$

W—the work done by non-conservative force in the whole process.

(8) and (12) $\rightarrow$ (13),

$$W = -\frac{m^2g^2}{6k} \tag{14}$$

Solution [2] (type A)

The elastic rope is divided into n segments of infinitesimal element with equal length (n is a large integer), which are sorted with positive integers downward from the origin O, the No. i infinitesimal element is denoted by element i.

Problem (1)

Taking *the system composed of element i to element n (can be regarded as a system of particles)* as the object of study. Studying *the final state.*

Considering ***Newton's second law*** [the system is in mechanical equilibrium] (the **−*i***-direction component equation):

$$T_i - \frac{n-(i-1)}{n}mg = 0 \tag{15}$$

$$\Rightarrow \qquad T_i = \frac{n-(i-1)}{n}mg \tag{16}$$

T_i—the magnitude of the tension in element i in the final state.

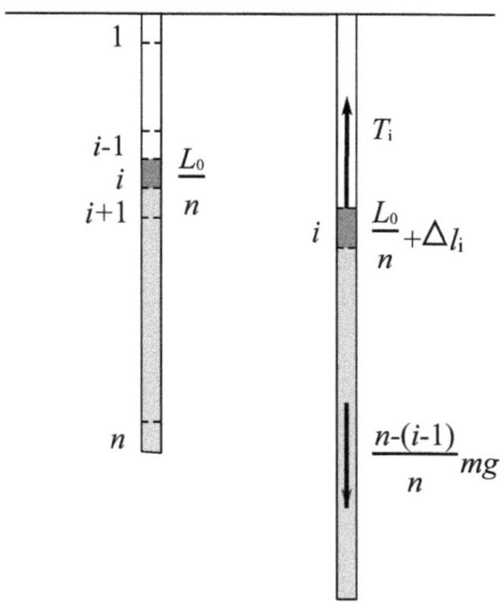

Figure 2

Taking *element i* as the object of study. Studying *the final state.* Considering ***Hooke's law***:

$$T_i = (nk)\Delta l_i \tag{17}$$

Δl_i—the elongation of element i in the final state.

$$\Rightarrow \qquad \Delta l_i = \frac{mg}{n^2 k}[(n+1)-i] \tag{18}$$

$$\Rightarrow \qquad \Delta L_{1\ to\ n} = \sum_{i=1}^{n} \Delta l_i = \frac{mg}{n^2 k}\sum_{i=1}^{n}[(n+1)-i] = \frac{mg}{2k}\frac{n+1}{n}$$

$$\Rightarrow \qquad \Delta L = \Delta L_{1\ to\ n}\Big|_{n\to\infty} = \frac{mg}{2k}\lim_{n\to\infty}\frac{n+1}{n} = \frac{mg}{2k}\cdot 1 = \frac{mg}{2k} \tag{19}$$

Problem (2)

Defining ΔE_{psi}—the elastic potential energy increment of element i in the whole process. Considering ***theorem of elastic potential energy of spring***:

$$\Delta E_{psi} = \frac{1}{2}(nk)\Delta l_i^2 \tag{20}$$

(18) → (20),

$$\Rightarrow \qquad \Delta E_{psi} = \frac{m^2 g^2}{2n^3 k}[(n+1)-i]^2 \tag{21}$$

$$\Rightarrow \qquad \Delta E_{ps(1\ to\ n)} = \sum_{i=1}^{n} \Delta E_{psi} = \frac{m^2 g^2}{2n^3 k}\sum_{i=1}^{n}[(n+1)-i]^2$$

$$= \frac{m^2 g^2}{2n^3 k}\sum_{i=1}^{n}[(n+1)^2 - 2(n+1)i + i^2]$$

$$= \frac{m^2 g^2}{12k}\frac{n(n+1)(2n+1)}{n^3}$$

$$\Rightarrow \qquad \Delta E_{ps} = \Delta E_{ps(1\ to\ n)}\Big|_{n\to\infty} = \frac{m^2 g^2}{12k}\lim_{n\to\infty}\frac{n(n+1)(2n+1)}{n^3}$$

$$= \frac{m^2 g^2}{12k}\cdot 2 = \frac{m^2 g^2}{6k} \tag{22}$$

Problem (3)

From (18),

$$\Rightarrow \quad \Delta L_j = \sum_{i=1}^{j}\Delta l_i = \frac{mg}{n^2 k}\sum_{i=1}^{j}[(n+1)-i] = \frac{mg}{2n^2 k}\left[(2n-1)j - j^2\right] \tag{23}$$

ΔL_j—the falling distance of element j in the whole process.

Defining ΔE_{pgj}—the gravitational potential energy increment of element j in the whole process. Considering ***theorem of gravitational potential energy***:

$$\Delta E_{pgj} = -\left(\frac{1}{n}mg\right)\Delta L_j \tag{24}$$

$$\Rightarrow \qquad \Delta E_{pgj} = -\frac{m^2 g^2}{2n^3 k}\left[(2n-1)j - j^2\right] \tag{25}$$

$$\Rightarrow \qquad \Delta E_{pg(1\ to\ n)} = \sum_{j=1}^{n} \Delta E_{pgj} = -\frac{m^2 g^2}{2n^3 k} \sum_{j=1}^{n} \left[(2n-1)j - j^2 \right]$$

$$= -\frac{m^2 g^2}{3k} \frac{(n-1)n(n+1)}{n^3}$$

$$\Rightarrow \qquad \Delta E_{pg} = \Delta E_{pg(1\ to\ n)} |_{n\to\infty} = -\frac{m^2 g^2}{3k} \lim_{n\to\infty} \frac{(n-1)n(n+1)}{n^3}$$

$$= -\frac{m^2 g^2}{3k} \cdot 1 = -\frac{m^2 g^2}{3k} \qquad (26)$$

2.5 Energy Conversion of Systems involving Electromagnetic Interaction

The magnetic force exerted on the free charges in a moving current-carrying conductor line element (generally part of closed circuit) in an external magnetic field can be divided into the component perpendicular to the conductor line element (determined by the current intensity, described by the Ampere's force) and the component parallel to the conductor line element (determined by the velocity of the conductor line element, described by the motional electromotive force). From the dynamic analysis in macro and micro perspectives, it can be found that this model has a feedback time-varying dynamic characteristic: the ampere force causes the change of conductor line element, resulting in the change of motional electromotive force, then the change of current intensity, finally the change of Ampere force ... This is an interesting and inspiring physical model. It is worth noting that the magnetic force acting on moving charge does no work; therefore, the energy conversion phenomenon in this model is also very special. In this section, we will study several specific problems of the model.

2.5.1 The Conductor Wire Frame's Motion Passing through a Magnetic Field (*b*, *AC*)

As shown in Fig. 1, apply a uniform magnetic field in space, the magnitude of whose magnetic induction intensity is B, and the direction is perpendicular to the plane of the figure inward. The upper boundary and the lower boundary of the magnetic field are two horizontal planes with distance of c. There is a rectangular conductor wire frame of mass m (the mass is uniformly distributed), resistance R, a length and b width, $b < c$. Place the wire frame in the plane of the figure, with its lower edge next to the upper boundary of the magnetic field area, and then release it freely. Given that the wire frame can leave the magnetic field area at a constant speed finally.

(1) Find the time taken for the wire frame to pass through the magnetic field area.

(2) Find the condition that the wire frame can leave the magnetic field area at a constant speed.

(3) Find the Joule heat generated by the wire frame in the whole process of its passing through the magnetic field area.

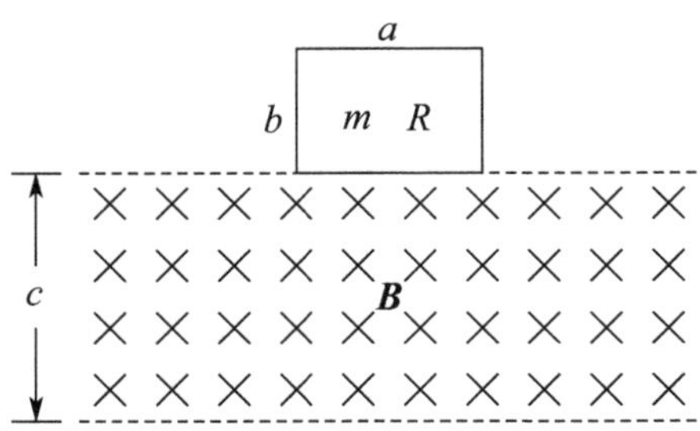

Figure 1

Solution to Problem (1) (type C)

The time point when the wire frame is released is defined as Time zero.

As shown in Fig. 2, taking an arbitrary point on the upper boundary of the magnetic field area and denoting it by O, and the unit vector in the vertical downward direction is denoted by $\boldsymbol{i}$. Establishing the linear coordinate system with Point O as the origin and the $\boldsymbol{i}$-direction as the positive direction of coordinate axis, let x be the coordinate.

The process of the wire frame entering the magnetic field area is denoted by Process 1. The process of the wire frame completely sitting in the magnetic field area is denoted by Process 2. The process of the wire frame leaving the magnetic field area at a constant speed is denoted by Process 3.

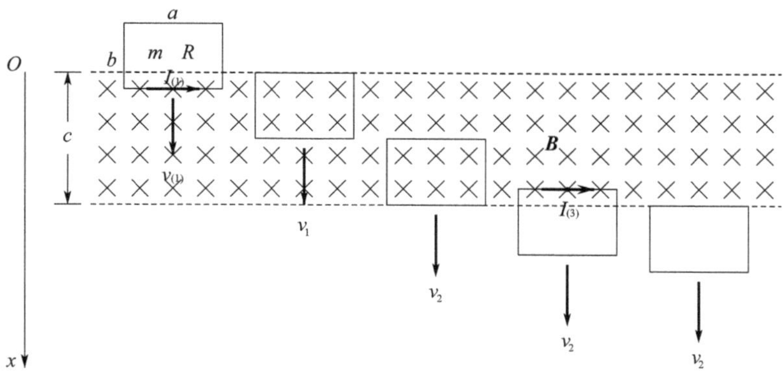

Figure 2

Taking *the wire frame [can be regarded as a rigid body and an electrocircuit]* as the object of study. Studying *the state at arbitrary Time t in Process 3*. Considering (a) ***theorem of motion of center of mass*** [the wire frame is in mechanical equilibrium] (***i***-direction component equation), (b) ***Kirchhoff's voltage law*** (clockwise):

$$mg - BI_{(3)}a = 0 \tag{1}$$

$$Bav_2 - I_{(3)}R = 0 \tag{2}$$

v_2—the ***i***-direction component of the center-of-mass velocity of the wire frame when it reaches the lower boundary of the magnetic field area.

$I_{(3)}$—the current intensity (clockwise) in the wire frame at Time t in Process 3.

$$\Rightarrow \quad \begin{cases} v_2 = \dfrac{mgR}{B^2a^2} & \text{(3a)} \\ I_{(3)} = \dfrac{mg}{Ba} & \text{(3b)} \end{cases}$$

Defining t_3—the time taken for Process 3. Considering ***the displacement-time relation of uniform linear motion***, then place (3a) in to obtain:

$$\Rightarrow \quad t_3 = \frac{b}{v_2} = \frac{B^2a^2b}{mgR} \tag{4}$$

When the wire frame entirely enters into the magnetic field area, no current is generated, and its motion is free falling [proof omitted]. Considering ***the displacement-velocity relation of uniformly variable motion***:

$$v_2^2 - v_1^2 = 2g(c-b) \tag{5}$$

v_1—the ***i***-direction component of the center-of-mass velocity of the wire frame when it just entirely enters into the magnetic field area.

$$\Rightarrow \qquad v_1 = \frac{mgR}{B^2a^2}\sqrt{1-2\frac{B^4a^4(c-b)}{m^2gR^2}} \tag{6}$$

Defining t_2—the time taken for Process 2. Considering ***the velocity-time relation of uniformly variable motion***, then place (3a) and (6) in to obtain:

$$\Rightarrow \qquad t_2 = \frac{v_2 - v_1}{g} = \frac{mR}{B^2a^2}\left[1-\sqrt{1-2\frac{B^4a^4(c-b)}{m^2gR^2}}\right] \tag{7}$$

Taking *the wire frame* as the object of study. Studying *the state at arbitrary Time t in Process 1*. Considering (a) ***theorem of motion of center of mass*** (the ***i***-direction component equation), (b) ***Kirchhoff's voltage law*** (anticlockwise):

$$mg - BI_{(1)}a = m\frac{\mathrm{d}v_{(1)}}{\mathrm{d}t} \tag{8}$$

$$Bav_{(1)} - I_{(1)}R = 0 \tag{9}$$

$v_{(1)}$—the ***i***-direction component of the center-of-mass velocity of the wire frame at Time t in Process 1.

$I_{(1)}$—the current intensity (anticlockwise) in the wire frame at Time t in Process 1.

$$\Rightarrow \qquad mg - \frac{B^2a^2v_{(1)}}{R} = m\frac{\mathrm{d}v_{(1)}}{\mathrm{d}t} \tag{10}$$

Considering the definition of ***velocity***:

$$\Rightarrow \qquad mg - \frac{B^2a^2}{R}\frac{\mathrm{d}x_{(1)}}{\mathrm{d}t} = m\frac{\mathrm{d}v_{(1)}}{\mathrm{d}t} \tag{11}$$

$$\Rightarrow \qquad mg\mathrm{d}t - \frac{B^2a^2}{R}\mathrm{d}x_{(1)} = m\mathrm{d}v_{(1)}$$

$$\Rightarrow \qquad mg\int_0^{t_1} \mathrm{d}t - \frac{B^2a^2}{R}\int_0^{b} \mathrm{d}x_{(1)} = m\int_0^{v_1} \mathrm{d}v_{(1)}$$

$$\Rightarrow \qquad mgt_1 - \frac{B^2a^2b}{R} = mv_1 \tag{12}$$

t_1—the time taken for Process 1.

(6) → (12),

$$\Rightarrow \qquad t_1 = \frac{mR}{B^2a^2}\sqrt{1-2\frac{B^4a^4(c-b)}{m^2gR^2}} + \frac{B^2a^2b}{mgR} \tag{13}$$

Defining t—the time taken for the wire frame to pass through the magnetic field area.

Combining (4), (7) and (13),

$$\Rightarrow \qquad t = t_1 + t_2 + t_3 = \frac{2B^2a^2b}{mgR} + \frac{mR}{B^2a^2} \tag{14}$$

Solution to Problem (2) (type A)

From (10),

$$\Rightarrow \qquad \frac{\mathrm{d}v_{(1)}}{v_{(1)} - \dfrac{mgR}{B^2a^2}} = -\frac{B^2a^2}{mR}\mathrm{d}t \tag{15}$$

$$\Rightarrow \qquad \int_0^{v_{(1)}} \frac{\mathrm{d}v_{(1)}}{v_{(1)} - \dfrac{mgR}{B^2a^2}} = -\frac{B^2a^2}{mR}\int_0^{t_{(1)}} \mathrm{d}t_{(1)}$$

$$\Rightarrow \qquad \ln\left(1 - \frac{B^2a^2}{mgR}v_{(1)}\right) = -\frac{B^2a^2}{mR}t_{(1)}$$

$$\Rightarrow \qquad v_{(1)} = \frac{mgR}{B^2a^2}\left[1 - \exp\left(-\frac{B^2a^2}{mR}t_{(1)}\right)\right] \tag{16}$$

(16) → (9),

$$\Rightarrow \qquad I_{(1)} = \frac{mg}{Ba}\left[1 - \exp\left(-\frac{B^2a^2}{mR}t_{(1)}\right)\right] \tag{17}$$

From (16),

$$\Rightarrow \qquad v_1 = v_{(1)}\Big|_{t_{(1)}=t_1} = \frac{mgR}{B^2a^2}\left[1-\exp\left(-\frac{B^2a^2}{mR}t_1\right)\right] \tag{18}$$

(6) and (13) $\rightarrow$ (18), we obtain the condition that the wire frame can leave the magnetic field area at a constant speed:

$$\Rightarrow \sqrt{1-2\frac{B^4a^4(c-b)}{m^2gR^2}} = 1-\exp\left(-\sqrt{1-2\frac{B^4a^4(c-b)}{m^2gR^2}}-\frac{B^4a^4b}{m^2gR^2}\right) \tag{19}$$

Defining the constant α:

$$\alpha = \frac{B^4}{m^2gR^2} \tag{20}$$

$$\Rightarrow \quad \sqrt{1-2\alpha a^4(c-b)} = 1-\exp\left[-\left(\sqrt{1-2\alpha a^4(c-b)}+\alpha a^4 b\right)\right] \tag{21}$$

Solution [1] to Problem (3) (type A)

Taking *the system composed of the wire frame and the earth* as the object of study. Studying *the whole process*. Considering ***law of conservation of energy*** [The energy contained in the system are the mechanical energy and the internal energy (the change part of which is Joule heat) of the wire frame. The work done by the magnetic force is zero (the work done by the magnetic force can be divided into two parts: the work done by the motional electromotive force and the work done by the Ampere force. The two parts have the equal magnitude and opposite signs); therefore, the Joule heat generated in the wire frame is equal to the loss of the mechanical energy of the system]:

$$Q = mg(b+c) - \frac{1}{2}mv_2^2 \tag{22}$$

Q—the Joule heat generated by the wire frame in the whole process of its passing through the magnetic field area.

(3a) $\rightarrow$ (22),

$$\Rightarrow \qquad Q = mg(b+c) - \frac{1}{2}\frac{m^3g^2R^2}{B^4a^4} \tag{23}$$

Solution [2] to Problem (3) (type A)

Defining $Q_{(3)}$—the Joule heat generated by the wire frame in Process 3. Considering ***Joule's law of heat***, then place (3b) and (4) in to obtain:

$$Q_{(3)} = I_{(3)}^2 R t_3 = mgb \tag{24}$$

Defining $Q_{(1)}$—the Joule heat generated by the wire frame in Process 1. Considering ***Joule's law of heat***, then place (17) in to obtain:

$$\Rightarrow \; Q_{(1)} = \int_0^{t_1} I_{(1)}^2 R\mathrm{d}t_{(1)} = \frac{m^2g^2R}{B^2a^2}\int_0^{t_1}\left[1-\exp\left(-\frac{B^2a^2}{mR}t_{(1)}\right)\right]^2 \mathrm{d}t_{(1)}$$

$$= \frac{m^2g^2R}{B^2a^2}\int_0^{t_1}\left[1-2\exp\left(-\frac{B^2a^2}{mR}t_{(1)}\right)+\exp\left(-2\frac{B^2a^2}{mR}t_{(1)}\right)\right]\mathrm{d}t_{(1)}$$

$$= \frac{m^2g^2R}{B^2a^2}\left\{\begin{array}{l} t_1+2\dfrac{mR}{B^2a^2}\left[\exp\left(-\dfrac{B^2a^2}{mR}t_1\right)-1\right]- \\ \dfrac{1}{2}\dfrac{mR}{B^2a^2}\left[\exp\left(-2\dfrac{B^2a^2}{mR}t_1\right)-1\right] \end{array}\right\} \tag{25}$$

$(6)\rightarrow(18)$,

$$\Rightarrow \qquad \exp\left(-\frac{B^2a^2}{mR}t_1\right) = 1-\sqrt{1-2\frac{B^4a^4(c-b)}{m^2gR^2}} \tag{26}$$

$$\Rightarrow \quad Q_{(1)} = \frac{m^2g^2R}{B^2a^2}\left[\begin{array}{l} \dfrac{mR}{B^2a^2}\sqrt{1-2\dfrac{B^4a^4(c-b)}{m^2gR^2}}+\dfrac{B^2a^2b}{mgR}- \\ 2\dfrac{mR}{B^2a^2}\sqrt{1-2\dfrac{B^4a^4(c-b)}{m^2gR^2}}+ \\ \dfrac{mR}{B^2a^2}\sqrt{1-2\dfrac{B^4a^4(c-b)}{m^2gR^2}}-\dfrac{1}{2}\dfrac{mR}{B^2a^2}+\dfrac{B^2a^2(c-b)}{mgR} \end{array}\right]$$

$$= \frac{m^2g^2R}{B^2a^2}\left(-\frac{1}{2}\frac{mR}{B^2a^2}+\frac{B^2a^2c}{mgR}\right) = -\frac{1}{2}\frac{m^3g^2R^2}{B^4a^4}+mgc \tag{27}$$

$$\Rightarrow \qquad Q = Q_{(1)}+Q_{(3)} = mg(b+c)-\frac{1}{2}\frac{m^3g^2R^2}{B^4a^4} \tag{28}$$

2.5.2 The Motion of a Pair of Conductor Rods on a Pair of Conductor Rails in a Magnetic Field (*b, AC*)

As shown in Fig. 1, there are two long, straight, and smooth conductor rails parallel to each other with the distance of l and fixed on a horizontal plane, whose resistances are ignored. There are two conductor rods 1 and 2 of mass m_1 and m_2, respectively, whose lengths are both slightly greater than l, they are statically and vertically placed on the two conductor rails initially, and the series resistance of the two conductor rods is R. Apply a uniform magnetic field in space, the magnitude of whose magnetic induction intensity is B, and the direction is vertical upward. The conductor rods 1 and 2 are subjected to the instantaneous impact at the same time and obtain the initial center-of-mass velocity $v_{10}\boldsymbol{i}$ and $v_{20}\boldsymbol{j}$, respectively.

(1) Find the relative displacement of the two conductor rods in the process from the initial time till arbitrary time.

(2) Find the Joule heat generated in the two conductor rods in the process from the initial time till arbitrary time.

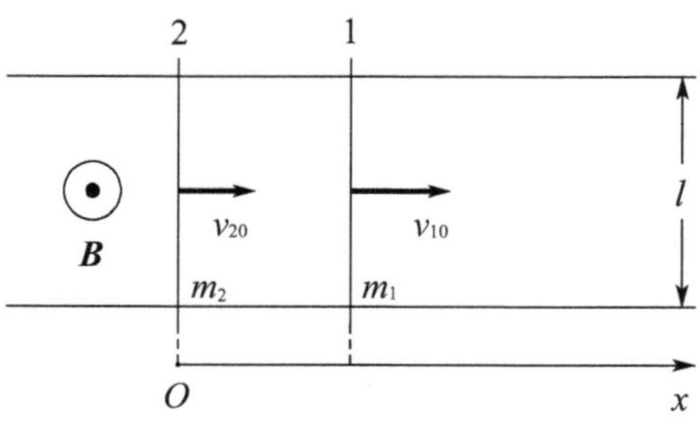

Figure 1

Solution to Problem (1) (type C)

The time point when the two conductor rods are impacted is defined as Time zero.

Taking *the circuit composed of the two conductor rails and the two conductor rods* as the object of study. Studying *the state at arbitrary Time t*. Considering ***Kirchhoff's voltage law*** (clockwise):

$$Blv_1 - Blv_2 - IR = 0 \tag{1}$$

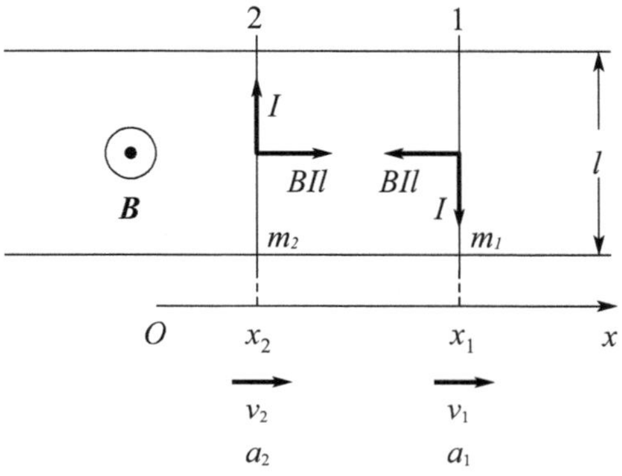

Figure 2

v_1 and v_2—the ***i***-direction component of the center-of-mass velocity of conductor rods 1 and 2 at Time *t*.

I—the current intensity in the circuit at Time *t* (clockwise).

$$\Rightarrow \qquad I = \frac{Bl(v_1 - v_2)}{R} \tag{2}$$

Taking *the conductor rods 1 and 2 [both can be regarded as a rigid body]* as the study objects, respectively. Studying *the state at arbitrary Time t*. Considering ***theorem of motion of center of mass*** (the ***i***-direction component equation):

$$\begin{cases} -BIl = m_1 a_1 & \text{(3a)} \\ BIl = m_2 a_2 & \text{(3b)} \end{cases}$$

a_1 and a_2—the ***i***-direction component of the center-of-mass acceleration of the conductor rods 1 and 2 at Time *t*.

$$\Rightarrow \qquad \begin{cases} -\dfrac{B^2 l^2 (v_1 - v_2)}{R} = m_1 \dfrac{\mathrm{d}v_1}{\mathrm{d}t} & \text{(4a)} \\ \dfrac{B^2 l^2 (v_1 - v_2)}{R} = m_2 \dfrac{\mathrm{d}v_2}{\mathrm{d}t} & \text{(4b)} \end{cases}$$

$$\Rightarrow \qquad \frac{\mathrm{d}}{\mathrm{d}t}(m_1 v_1 + m_2 v_2) = 0 \tag{5}$$

Equation (5) is the ***i***-direction component equation of momentum conservation of the system composed of two conductor rods.

$$\Rightarrow \qquad m_1 v_1 + m_2 v_2 = m_1 v_{10} + m_2 v_{20} \tag{6}$$

Defining m_μ—the reduced mass of the two conductor rods.

$$m_\mu = \frac{m_1 m_2}{m_1 + m_2} \tag{7}$$

Dividing both sides of Eq. (4a) by m_1, and dividing both sides of Eq. (4a) by m_2, and then subtracting, then place (7) in to obtain:

$$\Rightarrow \quad \frac{\mathrm{d}(v_1 - v_2)}{v_1 - v_2} = -\frac{B^2 l^2}{m_\mu R}\mathrm{d}t \tag{8}$$

$$\Rightarrow \quad \int_{v_{10}-v_{20}}^{v_1 - v_2} \frac{\mathrm{d}(v_1 - v_2)}{v_1 - v_2} = -\frac{B^2 l^2}{m_\mu R}\int_0^t \mathrm{d}t$$

$$\Rightarrow \quad \ln\frac{v_1 - v_2}{v_{10} - v_{20}} = -\frac{B^2 l^2}{m_\mu R}t$$

$$\Rightarrow \quad v_1 - v_2 = (v_{10} - v_{20})\exp\left(-\frac{B^2 l^2}{m_\mu R}t\right) \tag{9}$$

Considering ***theorem of the center-of-mass velocity***:

$$v_C = \frac{m_1 v_{10} + m_2 v_{20}}{m_1 + m_2} \tag{10}$$

v_C—the ***i***-direction component of the center-of-mass velocity of the two conductor rods at Time zero.

Combining (6), (9) and (10),

$$\Rightarrow \quad \begin{cases} v_1 = v_C + \dfrac{m_2}{m_1 + m_2}(v_{10} - v_{20})\exp\left(-\dfrac{B^2 l^2}{m_\mu R}t\right) & \text{(11a)} \\ v_2 = v_C - \dfrac{m_1}{m_1 + m_2}(v_{10} - v_{20})\exp\left(-\dfrac{B^2 l^2}{m_\mu R}t\right) & \text{(11b)} \end{cases}$$

Considering the definition of ***velocity***:

$$\Rightarrow \quad \begin{cases} \dfrac{\mathrm{d}x_1}{\mathrm{d}t} = v_C + \dfrac{m_2}{m_1 + m_2}(v_{10} - v_{20})\exp\left(-\dfrac{B^2 l^2}{m_\mu R}t\right) & \text{(12a)} \\ \dfrac{\mathrm{d}x_2}{\mathrm{d}t} = v_C - \dfrac{m_1}{m_1 + m_2}(v_{10} - v_{20})\exp\left(-\dfrac{B^2 l^2}{m_\mu R}t\right) & \text{(12b)} \end{cases}$$

Integrating the two equations above:

$$\Rightarrow \begin{cases} \Delta x_1 = v_C t + \dfrac{m_\mu R}{B^2 l^2} \dfrac{m_2}{m_1 + m_2} (v_{10} - v_{20}) \left[1 - \exp\left(-\dfrac{B^2 l^2}{m_\mu R} t \right) \right] & (13a) \\ \Delta x_2 = v_C t - \dfrac{m_\mu R}{B^2 l^2} \dfrac{m_1}{m_1 + m_2} (v_{10} - v_{20}) \left[1 - \exp\left(-\dfrac{B^2 l^2}{m_\mu R} t \right) \right] & (13b) \end{cases}$$

Δx_1 and Δx_2—the ***i***-direction component of the displacement of the center of mass of the conductor rods 1 and 2 in the process from Time zero to arbitrary Time *t*.

Defining Δx_{12}—the ***i***-direction component of the displacement of the center of mass of the conductor rod 1 relative to the conductor rod 2 in the process from Time zero to arbitrary Time *t*:

$$\Rightarrow \quad \Delta x_{12} = \Delta x_1 - \Delta x_2 = \frac{m_\mu R}{B^2 l^2} (v_{10} - v_{20}) \left[1 - \exp\left(-\frac{B^2 l^2}{m_\mu R} t \right) \right] \quad (14)$$

It can be seen that function $\Delta x_{12}(t)$ is a monotone increasing function.

$$\Rightarrow \quad \Delta x_{12\max}\Big|_{t \to \infty} = \frac{m_\mu R}{B^2 l^2} (v_{10} - v_{20}) \quad (15)$$

Solution [1] to Problem (2) (type C)

$(11) \to (1)$,

$$\Rightarrow \quad I = \frac{Bl}{R} (v_{10} - v_{20}) \exp\left(-\frac{B^2 l^2}{m_\mu R} t \right) \quad (16)$$

Considering ***Joule's law of heat***, then place (16) in to obtain:

$$\Rightarrow \quad Q = \int I^2 R \mathrm{d}t = \frac{B^2 l^2}{R} (v_{10} - v_{20})^2 \int_0^t \exp\left(-\frac{2B^2 l^2}{m_\mu R} t \right) \mathrm{d}t$$

$$= \frac{1}{2} m_\mu (v_{10} - v_{20})^2 \left[1 - \exp\left(-\frac{2B^2 l^2}{m_\mu R} t \right) \right] \quad (17)$$

Q—the Joule heat generated in the two conductor rods in the process from Time zero to arbitrary Time *t*.

Solution [2] to Problem (2) (type A)

Considering ***König theorem***:

$$\begin{cases} E_{k0} = E_{kC0} + E'_{k0} & (18a) \\ E_k = E_{kC} + E'_k & (18b) \end{cases}$$

E_{k0} and E_k—the total kinetic energy of the system composed of the two conductor rods in the ground reference frame at Time zero and arbitrary Time *t*.

E_{kC0} and E_{kC}—the translational kinetic energy of the system composed of the two conductor rods at Time zero and arbitrary Time *t*.

E'_{k0} and E'_k—the available energy of the system composed of the two conductor rods (the total kinetic energy of the system composed of the two conductor rods in the center-of-mass reference frame) at Time zero and arbitrary Time *t*.

Considering (a) the definition of ***translational kinetic energy***, (b) ***the form of available energy in the center-of-mass reference frame***:

$$\begin{cases} E_{kC0} = E_{kC} = \frac{1}{2}(m_1 + m_2)v_C^2 & (19a) \\ E'_{k0} = \frac{1}{2}m_\mu(v_{10} - v_{20})^2 & (19b) \\ E'_k = \frac{1}{2}m_\mu(v_1 - v_2)^2 & (19c) \end{cases}$$

Taking *the system composed of the two conductor rods and the circuit composed of the two conductor rails and the two conductor rods* as the object of study. Studying *the process from Time zero to arbitrary Time t*. Considering ***law of conservation of energy*** [The energy contained in the system are the mechanical energy and the internal energy (the change part of which is Joule heat) of the two conductor rods, The work done by the magnetic force is zero (the work done by the magnetic force can be divided into two parts: the work done by the motional electromotive force and the work done by the Ampere force, the two parts have the same magnitude and opposite signs); therefore, the Joule heat generated in the two conductor rods is equal to the loss of the mechanical energy of them], then place (18) and (19) in to obtain:

$$\Rightarrow \quad Q = E_{k0} - E_k = \frac{1}{2}m_\mu(v_{10} - v_{20})^2\left[1 - \exp\left(-\frac{2B^2l^2}{m_\mu R}t\right)\right] \quad (20)$$

2.5.3 The Motion of a Conductor Rod on a Pair of Inclined Smooth Conductor Rails Connected in Series by Specific Circuits in a Magnetic Field (d, *B*)

Two long, straight, and smooth conductor rails are fixed parallel to each other with the distance of L and with the included angle of θ with the horizontal plane (see Fig. 1). Branch 1 (a voltage source of electromotive force ε and resistance R in series) and Branch 2 (a capacitor of capacitance C) are connected between the two conductor rails in parallel. A conductor rod of mass m and length slightly longer than L is placed vertically on the two conductor rails. The resistance of the conductor rails and the conductor rod are ignored. Apply a uniform magnetic field in space, the magnitude of whose magnetic induction intensity is B, and the direction is vertical upward.

(1) Initially, the conductor rod is fixed. Switch on Branch 1, and release the conductor rod at the same time. Determine the maximum value of the center-of-mass speed of the conductor rod in the later motion process.

(2) When the center-of-mass speed of the conductor rod reaches the maximum value, Branch 2 is switched on. Find the electric quantity on the capacitor when the center of mass of the conductor rod further moves the distance s.

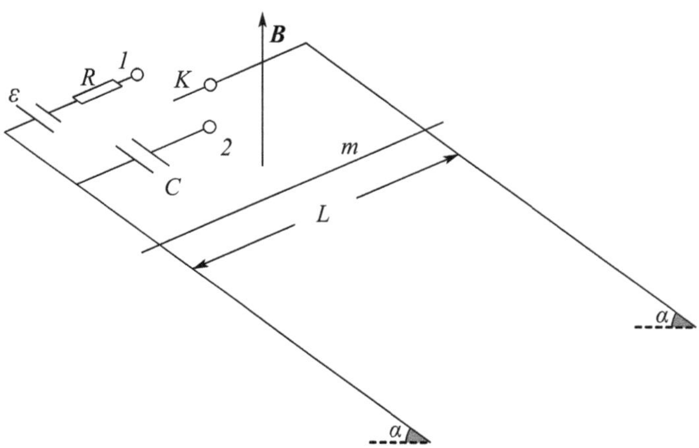

Figure 1

Solution to Problem (1)

When Branch 1 is switched on, the circuit composed of the conductor rod and Branch 1 is denoted by Circuit 1. When Branch 2 is switched on, the circuit composed of the conductor rod and Branch 2 is denoted by Circuit 2.

As shown in Fig. 2, at the initial time, taking an arbitrary point on the extension line of the conductor rod and denoting it by O, the unit vector parallel to the downward direction of the conductor rail is denoted by $\boldsymbol{i}$. Establishing the linear coordinate system with Point O as the origin and the $\boldsymbol{i}$-direction as the positive direction of coordinate axis, let x be the coordinate.

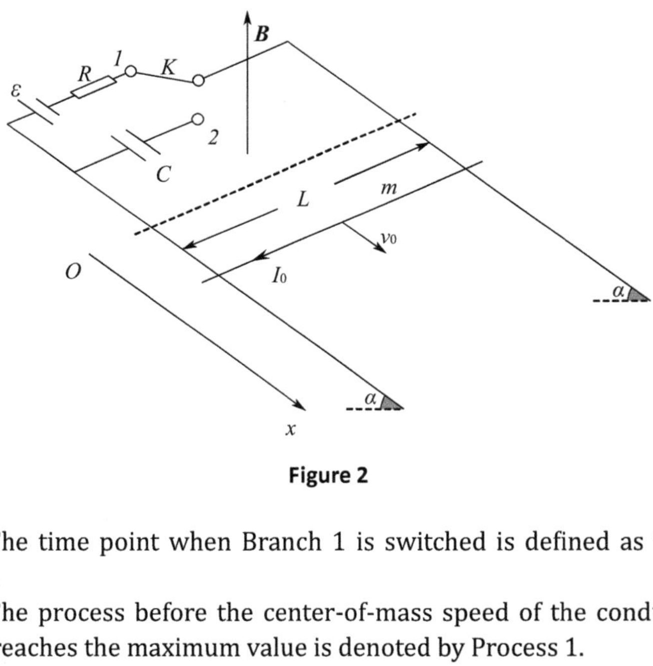

Figure 2

The time point when Branch 1 is switched is defined as Time zero.

The process before the center-of-mass speed of the conductor rod reaches the maximum value is denoted by Process 1.

The Time when the center-of-mass speed of the conductor rod reaches the maximum value (i.e., the Time when Branch 2 is switched on) is denoted by Time t_1.

After Time t_1, Circuit 2 will reach a steady state through a transient process, which is denoted by Process 2.

The Time when Circuit 2 is steady after the transient process is denoted by Time t_2.

After Time t_2, the process that the conductor rod further moves the distance s is denoted by Process 3.

The Time that the conductor rod further moves the distance s is denoted by Time t_3.

Taking *the conductor rod (can be regarded as a rigid body)* as the object of study. Studying *the state at Time t_1*. Considering ***theorem of motion of center of mass*** [the center-of-mass acceleration of the conductor rod is zero when its ***i***-direction component of the center-of-mass velocity reaches the maximum value] (the ***i***-direction component equation):

$$mg\sin\alpha - BI_0L\cos\alpha = 0 \tag{1}$$

I_0—the current intensity in Circuit 1 (the right-hand helix direction about the vertical downward direction) at Time t_1.

Taking *Circuit 1* as the object of study. Studying *the state at Time t_1*. Considering ***Kirchhoff's voltage law*** (the right-hand helix direction about the vertical upward direction):

$$\varepsilon - BLv_0\cos\alpha + I_0R = 0 \tag{2}$$

v_0—the ***i***-direction component of the center-of-mass velocity of the conductor rod at Time t_1.

$$\Rightarrow \qquad v_0 = \frac{mgR\sin\alpha + BL\varepsilon\cos\alpha}{B^2L^2\cos^2\alpha} \tag{3}$$

Solution to Problem (2)

Taking *Circuit 2* as the object of study. Studying *the state at Time t_2*. Considering ***Kirchhoff's voltage law*** (the right-hand helix direction about the vertical downward direction):

$$BLv_1\cos\alpha - \frac{Q_1}{C} = 0 \tag{4}$$

v_1—the ***i***-direction component of the center-of-mass velocity of the conductor rod at Time t_2.

Q_1—the electric quantity on the capacitor at Time t_2.

Taking *the conductor rod* as the object of study. Studying *Process 2*. Considering ***momentum theorem*** [In the transient process, an extremely large current will be generated in Circuit 2, the Ampere force acting on the conductor rod is an infinitely large quantity, acting on an extremely short process, its impulse is a finite quantity, which cannot be ignored. The gravity of the conductor rod is a finite quantity, acting on an extremely short process, its impulse is an infinitely small quantity, which can be ignored] (the ***i***-direction component equation):

$$
\begin{aligned}
mv_1 - mv_0 &= \int_{\Delta t} -BiL\cos\alpha \cdot \mathrm{d}t \\
&= -BL\cos\alpha \int_{\Delta t} i\mathrm{d}t \\
&= -BL\cos\alpha \int_0^{Q_1} \mathrm{d}Q \\
&= -BQ_1 L\cos\alpha \qquad (5)
\end{aligned}
$$

i—the current intensity of Circuit 2 (the right-hand helix direction about the vertical downward direction) at arbitrary Time t in Process 2.

$$
\Rightarrow \quad \begin{cases} v_1 = \dfrac{m}{m + CB^2L^2\cos^2\alpha} v_0 & (6a) \\ Q_1 = \dfrac{mv_0 CBL\cos\alpha}{m + CB^2L^2\cos^2\alpha} & (6b) \end{cases}
$$

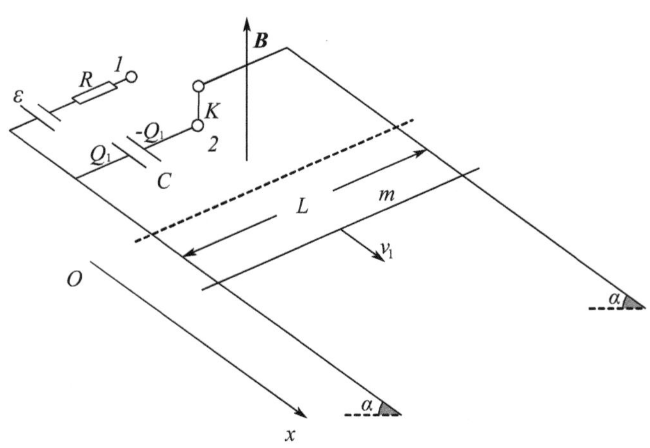

Figure 3

Taking *the system composed of the conductor rod and Circuit 2* as the object of study. Studying *Process 2*. Considering ***law of conservation of energy*** [(1) The energy contained in the system are the kinetic energy of the conductor rod, the electrostatic field energy of the charge on the capacitor, and the internal energy of the conductor rod (the change part of which is Joule heat—the current intensity in the transient process circuit is an infinitely large quantity, the resistance is an infinitely small quantity, the Joule heat is a finite quantity, which cannot be ignored). (2) The gravity of the conductor rod is a finite quantity, acting on an infinitely small displacement, its work is an infinitely small quantity, which can be ignored. (3) The work done by the magnetic force is zero (the work done by the magnetic force can be divided into two parts: the work done by the motional electromotive force and the work done by the Ampere force. The two parts have the same magnitude and opposite signs). Based on the above analyses, we can come to the conclusion that the energy of the system is conserved]:

$$\frac{1}{2}mv_0^2 = \left(\frac{1}{2}mv_1^2 + \frac{1}{2}\frac{Q_1^2}{C}\right) + \Delta E_{\text{Transient}} \tag{7}$$

$\Delta E_{\text{Transient}}$—the Joule heat generated by the conductor rod in Process 2.

$$\Rightarrow \qquad \Delta E_{\text{Transient}} = \frac{CB^2L^2\cos^2\alpha}{m + CB^2L^2\cos^2\alpha} \cdot \frac{1}{2}mv_0^2 \tag{8}$$

The result above retains v_0 determined by Eq. (3).

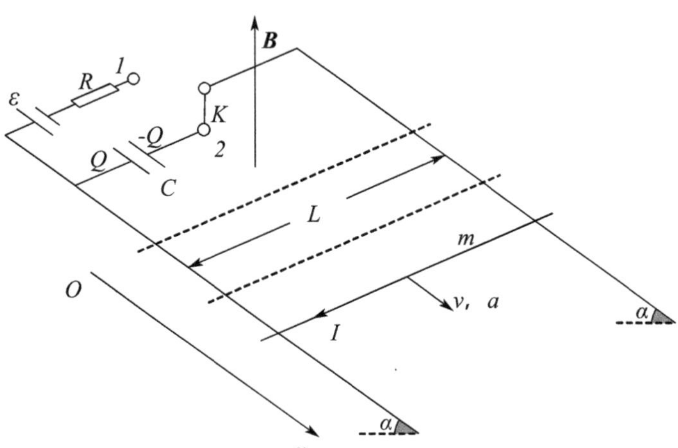

Figure 4

As shown in Fig. 4, taking *the conductor rod* as the object of study. Studying *the state at arbitrary Time t in Process 3*. Considering ***theorem of motion of center of mass*** (the ***i***-direction component equation):

$$mg\sin\alpha - BIL\cos\alpha = ma \tag{9}$$

a—the ***i***-direction component of the center-of-mass acceleration of the conductor rod at Time t in Process 3.

I—the current intensity of Circuit 2 (the right-hand helix direction about the vertical downward direction) at Time t in Process 3.

Taking *Circuit 2* as the object of study. Studying *the state at arbitrary Time t in Process 3.* Considering ***Kirchhoff's voltage law*** (the right-hand helix direction about the vertical downward direction):

$$BLv\cos\alpha - \frac{Q}{C} = 0 \tag{10}$$

v—the ***i***-direction component of the center-of-mass velocity of the conductor rod at Time t in Process 3.

Q—the charge on the capacitor at Time t in Process 3.

Considering the definition of ***current intensity*** and ***acceleration***:

$$\Rightarrow \qquad I = \frac{\mathrm{d}Q}{\mathrm{d}t} = CBL\cos\alpha\frac{\mathrm{d}v}{\mathrm{d}t} = CBL\cos\alpha\cdot a \tag{11}$$

Combining (9) and (11),

$$\Rightarrow \qquad \begin{cases} a = \dfrac{mg\sin\alpha}{m + CB^2L^2\cos^2\alpha} & \text{(12a)} \\ I = \dfrac{mgCBL\sin\alpha\cos\alpha}{m + CB^2L^2\cos^2\alpha} & \text{(12b)} \end{cases}$$

As shown in Fig. 5, considering ***the displacement-velocity relation of uniformly variable motion***:

$$v_2^2 - v_1^2 = 2as \tag{13}$$

v_2—the ***i***-direction component of the center-of-mass velocity of the conductor rod at Time t_3.

$$\Rightarrow \qquad v_2 = \frac{\sqrt{m^2v_0^2 + 2m(m + CB^2L^2\cos^2\alpha)gs\sin\alpha}}{m + CB^2L^2\cos^2\alpha} \tag{14}$$

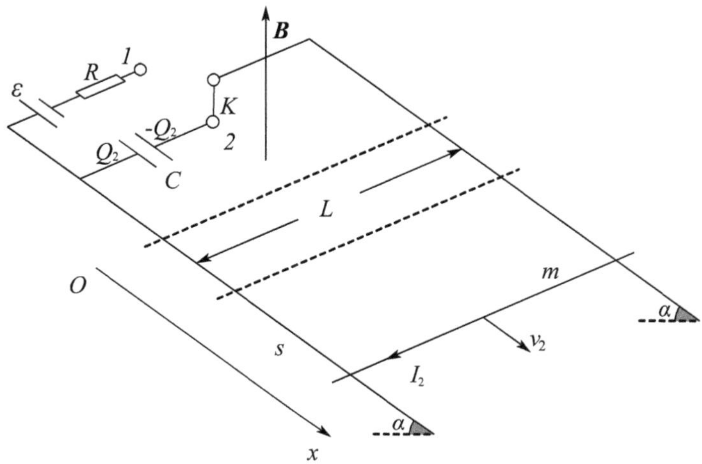

Figure 5

Taking *Circuit 2* as the object of study. Studying *the state at Time* t_3. Considering ***Kirchhoff's voltage law*** (the right-hand helix direction about the vertical downward direction):

$$BLv_2\cos\alpha - \frac{Q_2}{C} = 0 \tag{15}$$

Q_2—the charge on the capacitor at Time t_3.

$$\Rightarrow \quad Q_2 = CBL\cos\alpha\frac{\sqrt{m^2v_0^2 + 2m(m+CB^2L^2\cos^2\alpha)gs\sin\alpha}}{m+CB^2L^2\cos^2\alpha} \tag{16}$$

Another solution to Q_2

Taking *the system composed of the conductor rod and Circuit 2* as the object of study. Studying *Process 3*. Considering ***principle of energy-work*** [The energy contained in the system are the kinetic energy of the conductor rod, and the electrostatic field energy of the charge on the capacitor. The work done by the magnetic force is zero. Therefore, only the gravity of the rod changes the energy of the system]:

$$mgs\sin\alpha = \left(\frac{1}{2}mv_2^2 + \frac{1}{2}\frac{Q_2^2}{C}\right) - \left(\frac{1}{2}mv_1^2 + \frac{1}{2}\frac{Q_1^2}{C}\right) \tag{17}$$

Q_2 can be obtained by combining (15) and (17).

Another method to get Eq. (17)

Taking *the conductor rod* as the object of study. Studying *Process 3*. Considering ***theorem of kinetic energy:***

$$mgs\sin\alpha + W_{B1} = \frac{1}{2}mv_2^2 - \frac{1}{2}mv_1^2 \tag{18}$$

W_{B1}—the work done by Ampere force acting on the conductor rod in Process 3.

Taking *Circuit 2* as the object of study. Study *Process 3*. Considering ***principle of energy-work***:

$$W_{B2} = \frac{1}{2}\frac{Q_2^2}{C} - \frac{1}{2}\frac{Q_1^2}{C} \tag{19}$$

W_{B2}—the work done by the motional electromotive force in Circuit 2 in Process 3.

Considering that ***work done by magnetic force is zero***:

$$W_{B1} + W_{B2} = 0 \tag{20}$$

Equation (17) can be obtained by combining (18) to (20).

2.6 Mutual Transformation of Mechanical Energy and Internal Energy

According to König theorem, the mechanical energy of a system of particles is equal to the sum of its mechanical energy in its center-of-mass reference frame and its translational kinetic energy. When the system of particles is a classical molecular system (such as an ideal gas) in thermal equilibrium state, its mechanical energy in its center-of-mass reference frame is exactly the internal energy (take each atom that makes up a molecule as a particle), as the feature of small-scale disordered motion of the system of particles, it can be understood as the micro energy, and its translational kinetic energy as the feature of large-scale ordered motion of the system of particles can be understood as the macro energy. When the molecular system interacts with the outside, the energy change can be decomposed into the sum of macro energy and micro energy change. Based on this understanding, this section will study the energy conversion problems of some classical molecular systems with macroscopic motion.

2.6.1 The Proof of the Adiabatic Equation from the Micro Perspective (d, *C*)

Consider an air cylinder, one end of which is equipped with a sealed piston which can move along the air cylinder without friction. Given that the air cylinder and the piston are both adiabatic. A monatomic ideal gas is enclosed in the air cylinder. When the piston is pushed slowly deep into the air cylinder, the gas in the air cylinder will undergo a quasi-static process. From the microscopic perspective, the collision between an arbitrary gas molecule and the piston can be regarded as a completely elastic collision. The piston collides with a part of gas molecules as it moves deep into the cylinder, resulting in the increase of their kinetic energy. These gas molecules collide with other gas molecules, resulting in the increase of the average energy of all the gas molecules in the air cylinder. Please use this microscopic model to deduce the adiabatic equation of monatomic ideal gas.

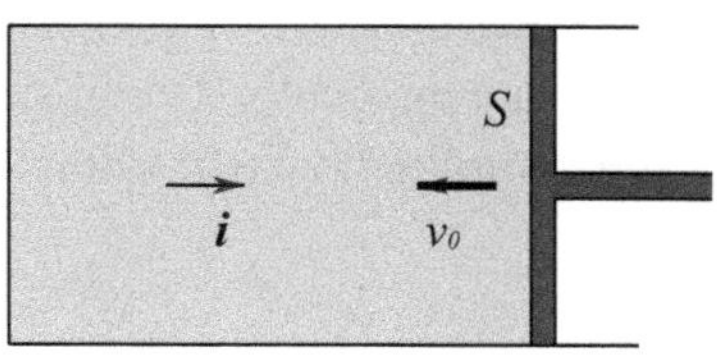

Figure 1

Solution

Time zero is arbitrarily set.

As shown in Fig. 1, the unit vector perpendicular to the piston from the inside of the air cylinder to the outside is denoted by ***i***. Considering ***Maxwell velocity distribution law*** (the distribution function of velocity component in Cartesian coordinate system):

$$f(v_x) = \left(\frac{m}{2\pi kT} \right)^{1/2} e^{-mv_x^2/2kT} \tag{1}$$

m—the molecular mass of the gas molecular.

T—the temperature of the gas at arbitrary Time t.

At arbitrary Time t, the gas molecules whose velocity component in the ***i***-direction is v_x and collides with the piston are denoted by molecules (v_x).

In the instantaneous process (the small time interval dt) at arbitrary Time t, the system composed of the gas molecules whose velocity components in the ***i***-direction are between v_x and $v_x + \mathrm{d}v_x$ and collides with the piston are denoted by molecules (dv_x).

Defining dN—the number of molecules (dv_x). Considering the definition of ***velocity distribution function***, then place (1) in to obtain:

$$\mathrm{d}N = n \cdot f(v_x)\mathrm{d}v_x \cdot S[v_x - (-v_0)]\mathrm{d}t = nS\mathrm{d}tf(v_x)(v_x + v_0)\mathrm{d}v_x \qquad (2)$$

n—the number density of molecules of the gas at arbitrary Time t.

v_0—the speed of the piston ($<<v_x$).

S—the area of the piston.

Taking *the system composed of one of the molecules (v_x) and the piston* as the object of study. Studying *the collision process between the two*. Considering ***Newton-collision-formula for completely elastic collision*** [for the gas molecule, its collision with the piston can be regarded as the completely elastic collision with the object with infinite mass]:

$$\frac{(-v_0) - v'_x}{v_x - (-v_0)} = 1 \qquad (3)$$

v'_x—the ***i***-direction component of the velocity of one of the molecules (v_x) at the time point after the collision.

$$\Rightarrow \qquad v'_x = -v_x - 2v_0 \qquad (4)$$

Defining dε_k—the kinetic energy increment that one of the molecules (v_x) obtained in the collision process.

$$\begin{aligned}\mathrm{d}\varepsilon_k &= \frac{1}{2}m\left(v_x'^2 + v_y^2 + v_z^2\right) - \frac{1}{2}m\left(v_x^2 + v_y^2 + v_z^2\right) \\ &= 2mv_xv_0 + 2mv_0^2 \approx 2mv_xv_0 \qquad (5)\end{aligned}$$

The operation above has considered $v_0<<v_x$ and taken the first-order approximation.

Defining dE_k—the kinetic energy increment that molecules (dv_x) obtained in the collision process.

From (2) and (5),

$$\Rightarrow \quad dE_k = dN \cdot d\varepsilon_k = 2mnSv_0 dt f(v_x)(v_x + v_0)v_x dv_x$$

$$\approx 2mnSv_0 dt f(v_x) v_x^2 dv_x \tag{6}$$

Defining V—the volume of the gas at arbitrary Time t.

$$dV = -Sv_0 dt \tag{7}$$

$$\Rightarrow \quad dE_k = -2mndVf(v_x)v_x^2 dv_x \tag{8}$$

Taking *the gas in the air cylinder* as the object of study. Studying *the instantaneous process (the small time interval dt) at arbitrary Time t*. Considering (a) ***law of conservation of energy*** [the air cylinder is adiabatic, the kinetic energy increment that the gas obtained in the collision process will be redistributed by the collisions process between all the gas molecules in the cylinder], (b) the definition of ***internal energy***, then place (1) and (8) in to obtain:

$$\Rightarrow \quad dU = \int_{v_x=0}^{v_x=\infty} dE_k = -2mndV \int_0^\infty f(v_x) v_x^2 dv_x$$

$$= -2mndV \int_0^\infty \left(\frac{m}{2\pi kT}\right)^{1/2} e^{-mv_x^2/2kT} v_x^2 dv_x$$

$$= -2mndV \cdot \frac{kT}{2m} = -nkTdV \tag{9}$$

U—the internal energy of the gas at Time t.

Considering ***equipartition theorem of energy***:

$$U = v\frac{i}{2}RT \tag{10}$$

i—the degree of freedom of the gas molecule (for monoatomic molecule, it is 3).

v—the number of moles of the gas.

Defining γ—the adiabatic coefficient of the gas. Considering ***the relation between adiabatic coefficient and degree of freedom of ideal gas molecule***:

$$\gamma = \frac{2}{i} + 1 \tag{11}$$

$$\Rightarrow \quad U = \frac{v}{\gamma - 1} RT \tag{12}$$

Combining (9) and (12),

$$\Rightarrow \qquad \frac{\nu}{\gamma-1}R\mathrm{d}T=-nkT\mathrm{d}V \qquad (13)$$

Considering the definition of ***number density of molecules***:

$$n=\frac{\nu N_A}{V} \qquad (14)$$

Considering the definition of ***Boltzmann's constant***:

$$k=\frac{R}{N_A} \qquad (15)$$

$$\Rightarrow \qquad \frac{\mathrm{d}T}{T}+(\gamma-1)\frac{\mathrm{d}V}{V}=0 \qquad (16)$$

$$\Rightarrow \qquad \int\frac{\mathrm{d}T}{T}+(\gamma-1)\int\frac{\mathrm{d}V}{V}=0$$

$$\Rightarrow \qquad TV^{\gamma-1}=\mathrm{const} \qquad (17)$$

QED.

2.6.2 The Quasi-Static Motion of a Sealed Divider Plate in a Gas Filled Adiabatic Cylinder Container (*a*, *A*)

As shown in Fig. 1, there is a cylindrical container of volume $2V_0$ and cross-sectional area *S*. A sealed thin divider plate of mass *M* is used to divide the container into two parts with the same volume up and down (remain stationary under external force). Given that the container and the divider plate are both adiabatic, and the divider plate can slide freely relative to the container wall without friction. The same ideal gas of number of moles *n* and adiabatic coefficient γ are injected into both sides of the divider plate. Initially, the gas pressure above and below the divider plate are both p_0, $Mg << p_0S$. The divider plate starts to move downward in quasi-static state under an external force until it reaches its position of mechanical equilibrium which no external force is needed, then fix the divider plate by a pin, and then use a built-in device to open a small hole in the divider plate to mix the gases on both sides. After a long enough time, the mixed gas reaches thermal equilibrium state.

(1) Find the pressure of the mixed gas in the final state.

(2) Find the work done by the external force that makes the divider plate perform the quasi-static motion.

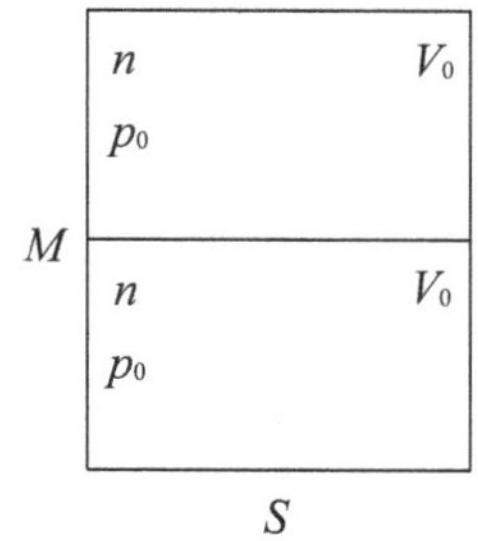

Figure 1

Solution to Problem (1)

The time point when the divider plate starts to move is defined as Time zero.

As shown in Fig. 2, taking an arbitrary point on the extension surface of the divider plate at Time zero and denoting it by *O*, the unit vector in vertical downward direction is denoted by ***i***. Establishing the liner coordinate system with Point *O* as the origin and the ***i***-direction as the positive direction of coordinate axis, let *x* be the coordinate.

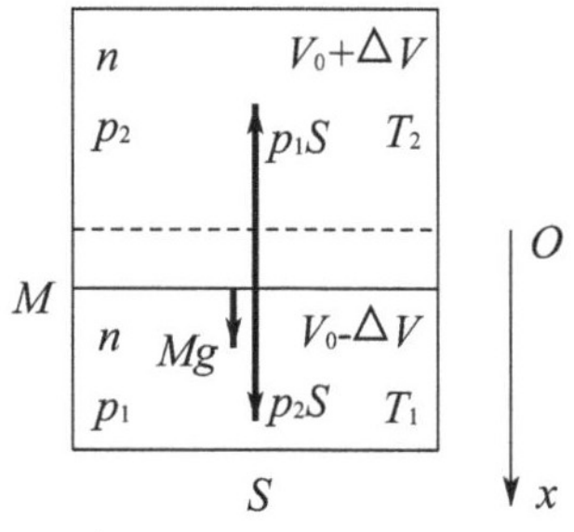

Figure 2

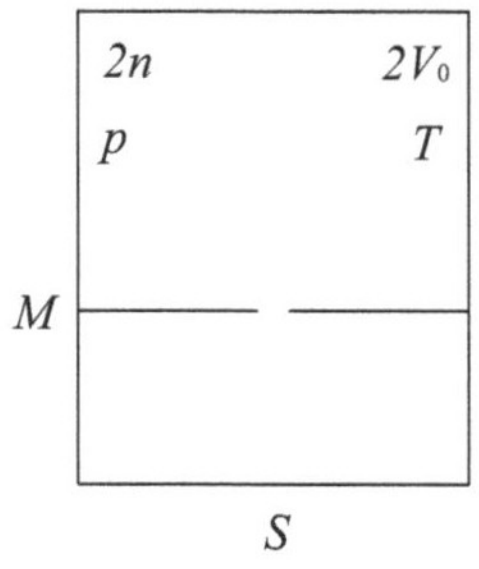

Figure 3

The whole motion process of the divider plate is denoted by Process 1.

The Time when the divider plate stops moving (the Time when the external force acting on the divider plate is removed) is denoted by Time t_1.

The process that the gases on both sides of the divider plate mix till the mixed gas reaches thermal equilibrium is denoted by Process 2.

The Time when the mixed gases reach thermal equilibrium is denoted by Time t_2.

Solution to Problem (1)

Taking *the gas on each side of the divider plate* as the object of study, respectively. Studying *Process 1*. Considering ***adiabatic equation of ideal gas***:

$$\begin{cases} p_0 V_0^{\gamma} = p_1 (V_0 - \Delta V)^{\gamma} & \text{(1a)} \\ p_0 V_0^{\gamma} = p_2 (V_0 + \Delta V)^{\gamma} & \text{(1b)} \end{cases}$$

p_1 and p_2—the pressure of the gas below and above at Time t_1.

ΔV—the decrease of the volume of the gas below (i.e., the increase of the volume of the gas above) in Process 1.

From $Mg \ll p_0 S$, it can be inferred that $\Delta V \ll V_0$.

$$\Rightarrow \begin{cases} p_1 = p_0 \left(\dfrac{V_0}{V_0 - \Delta V} \right)^{\gamma} \approx p_0 \left(1 - \dfrac{\Delta V}{V_0} \right)^{-\gamma} \\ \approx p_0 \left[1 + \gamma \dfrac{\Delta V}{V_0} + \dfrac{\gamma(\gamma+1)}{2} \left(\dfrac{\Delta V}{V_0} \right)^2 \right] & \text{(2a)} \\ p_2 = p_0 \left(\dfrac{V_0}{V_0 + \Delta V} \right)^{\gamma} \approx p_0 \left(1 + \dfrac{\Delta V}{V_0} \right)^{-\gamma} \\ \approx p_0 \left[1 - \gamma \dfrac{\Delta V}{V_0} + \dfrac{\gamma(\gamma+1)}{2} \left(\dfrac{\Delta V}{V_0} \right)^2 \right] & \text{(2b)} \end{cases}$$

The operations above have considered $\Delta V / V_0$ as a small quantity and taken the second-order approximations.

Taking *the divider plate* as the object of study. Studying *the state at Time t_1*. Considering ***theorem of motion of center of mass*** [the divider plate is in mechanical equilibrium without the action of external force] (the ***i***-direction component equation):

$$Mg + p_2 S - p_1 S = 0 \tag{3}$$

Combining (2) and (3),

$$\Rightarrow \begin{cases} p_1 = p_0 + \dfrac{1}{2}Mg/S + \dfrac{\gamma+1}{8\gamma}\dfrac{(Mg/S)^2}{p_0} & \text{(4a)} \\ p_2 = p_0 - \dfrac{1}{2}Mg/S + \dfrac{\gamma+1}{8\gamma}\dfrac{(Mg/S)^2}{p_0} & \text{(4b)} \\ \Delta V = \dfrac{1}{2\gamma}\dfrac{Mg/S}{p_0}V_0 & \text{(4c)} \end{cases}$$

It can be seen from (4c) that $\Delta V/V_0$ and $Mg/p_0 S$ are small quantities of the same order; therefore, the above assumption holds.

Taking *the gas on each side of the divider plate* as the object of study, respectively. Studying *the state at Time zero and Time t_1*, respectively. Considering ***state equation of ideal gas*** [all gases are in thermal equilibrium]:

$$\begin{cases} p_0 V_0 = nRT_0 & \text{(5a)} \\ \begin{cases} p_1(V_0 - \Delta V) = nRT_1 & \text{(5b)} \\ p_2(V_0 + \Delta V) = nRT_2 & \text{(5c)} \end{cases} \end{cases}$$

T_0—the temperature of the gases on both sides of the divider plate at Time zero.

T_1 and T_2—the temperature of the gases below and above at Time t_1.

Taking *the mixed gas in the container* as the object of study. Studying *the state at Time t_2*. Considering ***state equation of ideal gas***:

$$p \cdot 2V_0 = 2nRT \tag{6}$$

p—the pressure of the mixed gas at Time t_2.

T—the temperature of the mixed gas at Time t_2.

Taking *the system composed of gases on both sides of the divider plate* as the object of study. Study *Process 2*. Considering (a) ***law of conservation of energy*** [the work done by external force is zero, and the container is adiabatic; therefore, the energy of the system is conserved], (b) ***the internal energy of an ideal gas is a function only of its temperature*** [integrating into (a)]:

$$nC_V T_1 + nC_V T_2 = 2nC_V T \tag{7}$$

C_V—the molar heat capacity at constant volume of the gas.

Combining (4) to (7),

$$\Rightarrow \begin{cases} p = p_0 + \dfrac{\gamma-1}{8\gamma}\dfrac{(Mg/S)^2}{p_0} & (8a) \\ T = \dfrac{1}{nR}\left[p_0V_0 + \dfrac{\gamma-1}{8\gamma}\dfrac{(Mg/S)^2}{p_0}V_0\right] & (8b) \end{cases}$$

Solution [1] to Problem (2)

As shown in Fig. 4, taking *the gas on each side of the divider plate* as the object of study, respectively. Studying *the process from Time zero to arbitrary Time t in Process 1*. Considering ***adiabatic equation of ideal gas***:

$$\begin{cases} p_0V_0^\gamma = p_{1x}(V_0 - Sx)^\gamma & (9a) \\ p_0V_0^\gamma = p_{2x}(V_0 + Sx)^\gamma & (9b) \end{cases}$$

p_{1x} and p_{2x}—the pressure of the gases below and above at arbitrary Time t in Process 1.

$$\Rightarrow \begin{cases} p_{1x} = p_0\left(\dfrac{V_0}{V_0 - Sx}\right)^\gamma \approx p_0\left(1 - \dfrac{Sx}{V_0}\right)^{-\gamma} \\ \approx p_0\left[1 + \gamma\dfrac{Sx}{V_0} + \dfrac{\gamma(\gamma+1)}{2}\left(\dfrac{Sx}{V_0}\right)^2\right] & (10a) \\ p_{2x} = p_0\left(\dfrac{V_0}{V_0 + Sx}\right)^\gamma \approx p_0\left(1 + \dfrac{Sx}{V_0}\right)^{-\gamma} \\ \approx p_0\left[1 - \gamma\dfrac{Sx}{V_0} + \dfrac{\gamma(\gamma+1)}{2}\left(\dfrac{Sx}{V_0}\right)^2\right] & (10b) \end{cases}$$

The operations above have considered Sx/V_0 as a small quantity and taken the second-order approximations.

Taking *the divider plate* as the object of study. Studying *the state at arbitrary Time t in Process 1*. Considering ***theorem of motion of center of mass*** [the divider plate performs quasi-static motion, and at arbitrary Time t it is approximated in mechanical equilibrium] (the ***i***-direction component equation):

$$Mg + p_{2x}S - p_{1x}S - F_{-x} = 0 \qquad (11)$$

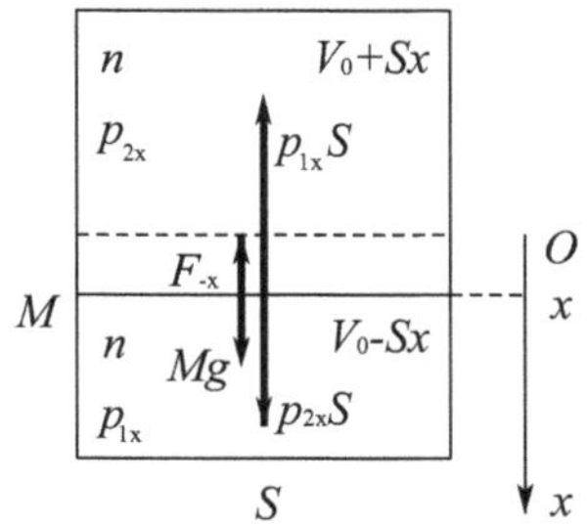

Figure 4

F_{-x}—the **−*i***-direction component of the external force acting on the divider plate at Time *t* in Process 1.

$$\Rightarrow \qquad F_{-x} = Mg - 2\gamma \frac{p_0 S^2}{V_0} x \tag{12}$$

Defining *W*—the work done by the external force acting on the divider plate in Process 1.

Considering the definition of ***work***:

$$\Rightarrow W = \int -F_{-x} \mathrm{d}x = -\int_0^{\frac{\Delta V}{S}} \left(Mg - 2\gamma \frac{p_0 S^2}{V_0} x \right) \mathrm{d}x = -\frac{1}{4\gamma} \frac{(Mg/S)^2}{p_0} V_0 \tag{13}$$

Solution [2] to Problem (2):

Taking *the system composed of the earth, the divider plate and the gases on both sides of the divider plate* as the object of study. Studying *the total process composed of Process 1 and Process 2*. Considering ***principle of energy-work*** [taking the position of the divider plate in the final state as the zero point of gravitational potential energy. the divider plate performs the quasi-static motion; therefore its kinetic energy can be ignored]:

$$W = 2nC_V T - \left(Mg \frac{\Delta V}{S} + 2nC_V T_0 \right) \tag{14}$$

(4c) and (8b) → (14),

$$\Rightarrow \qquad W = \frac{C_V}{R} \frac{\gamma - 1}{4\gamma} \frac{(Mg/S)^2}{p_0} V_0 - \frac{1}{2\gamma} \frac{(Mg/S)^2}{p_0} V_0 \tag{15}$$

Defining C_p—the molar heat capacity at constant pressure of the gas in the container. Considering ***the relation between molar heat***

capacity at constant volume and molar heat capacity at constant pressure:

$$\begin{cases} C_p - C_V = R & (16a) \\ \dfrac{C_p}{C_V} = \gamma & (16b) \end{cases}$$

$$\Rightarrow \quad W = \frac{C_V}{C_p - C_V} \frac{\gamma-1}{4\gamma} \frac{(Mg/S)^2}{p_0} V_0 - \frac{1}{2\gamma} \frac{(Mg/S)^2}{p_0} V_0$$

$$= \frac{1}{\gamma-1} \frac{\gamma-1}{4\gamma} \frac{(Mg/S)^2}{p_0} V_0 - \frac{1}{2\gamma} \frac{(Mg/S)^2}{p_0} V_0$$

$$= -\frac{1}{4\gamma} \frac{(Mg/S)^2}{p_0} V_0 \qquad (17)$$

2.6.3 The Repeated Collisions between an Adiabatic Container and Its Internal Ideal Gas (*a*, *A*)

A rectangular container of mass M is filled with a rigid molecule ideal gas of molecular mass m and molecular degree of freedom j, and the number of gas molecules is N. Given that the container is hermetically sealed and adiabatic, $M>>Nm$, it can be considered that in case there is a sudden change in the speed of the container due to the instantaneous impact from outside, with respect to the container, the gas can reach a thermal equilibrium state by an extremely short collision with the inner wall of the container.

Initially, the container is placed on a horizontal ground. The container is subjected to an instantaneous vertical upward impact, followed by an extremely short collision of the gas and the container, the gas reaches a thermal equilibrium state, at this time point the speed of the container is v_0 and the temperature of the gas is T_0.

(1) Find the speed of the container after an arbitrary collision with the gas.

(2) The container will eventually stop moving, find the total motion time of the container.

(3) Find the temperature of the gas at the time point after an arbitrary collision with the container.

Solution to Problem (1)

In this problem, it is assumed that an arbitrary collision process between the container and the ground and the subsequent collision process between the container and the gas occur successively in an extremely short period of time.

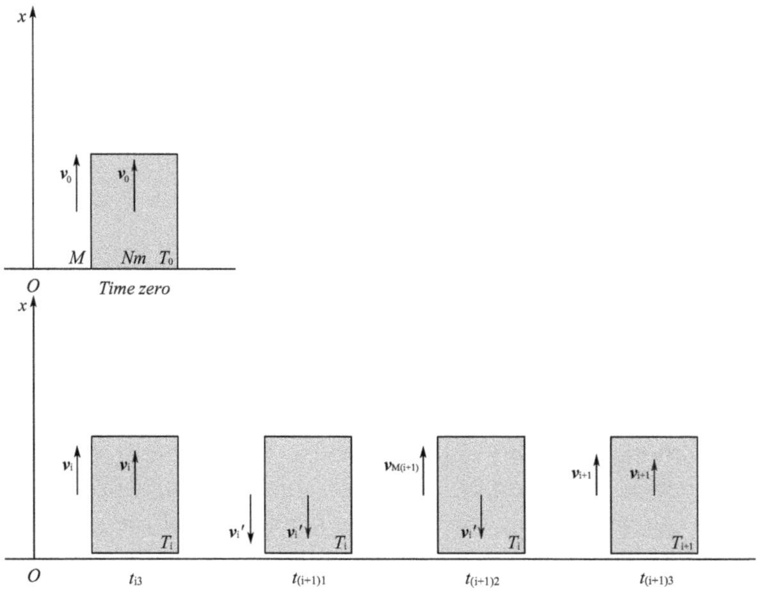

Figure 1

As shown in Fig. 1, Taking an arbitrary point on the ground and denoting it by *O*, and the unit vector in the vertical upward direction is denoted by ***i***. Establishing the linear coordinate system with Point *O* as the origin and the ***i***-direction as the positive direction of coordinate axis, let *x* be the coordinate.

The time point after the container is impacted is defined as Time zero.

The Time before the *i*-th collision between the container and the ground is denoted by Time t_{i1}.

The Time after the *i*-th collision between the container and the ground is denoted by Time t_{i2}.

The Time after the *i*-th collision between the container and the gas is denoted by Time t_{i3}.

According to the previous assumption, Time t_{i1}, t_{i2} and t_{i3} are three infinite approaching time points.

Considering ***the equation of motion of uniformly variable motion***:

$$v_i' = -v_i \tag{1}$$

v_i—the ***i***-direction component of the velocity of the container (i.e., the velocity of the center of mass of the gas) at Time t_{i3}.

v_i'—the ***i***-direction component of the velocity of the container (i.e., the velocity of the center of mass of the gas) at Time $t_{(i+1)1}$.

Considering that ***the collision between the container and the ground is completely elastic***:

$$v_{M(i+1)} = -v_i' \tag{2}$$

$v_{M(i+1)}$—the ***i***-direction component of the velocity of the container at Time $t_{(i+1)2}$.

Taking *the system composed of the container and the gas* as the object of study. Studying *the instantaneous process of the (i + 1)-th collision between the two*. Considering ***law of conservation of momentum*** [the gravity of the system is a finite quantity, acting on an extremely short process, its impulse is an infinitely small quantity, which can be ignored; therefore, the momentum of the system is approximately conserved]:

$$Mv_{M(i+1)} + Nmv_i' = (M + Nm)v_{i+1} \tag{3}$$

Combining (1) to (3),

$$\frac{v_{i+1}}{v_i} = \frac{M - Nm}{M + Nm} \tag{4}$$

Defining the constant α:

$$\alpha = \frac{M - Nm}{M + Nm} \tag{5}$$

$$\Rightarrow \qquad \frac{v_{i+1}}{v_i} = \alpha \tag{6}$$

Known $M \gg Nm$,

$$\Rightarrow \qquad 0 < \alpha < 1 \tag{7}$$

$$\Rightarrow \qquad \frac{v_n}{v_0} = \prod_{i=0}^{n-1} \frac{v_{i+1}}{v_i} = \prod_{i=0}^{n-1} \alpha = \alpha^n \tag{8}$$

$$\Rightarrow \qquad v_n = \alpha^n v_0 \tag{9}$$

Solution to Problem (2)

Considering the ***velocity-time relation of uniformly variable motion***:

$$t_{(i+1)3} - t_{i3} = \frac{2v_i}{g} \tag{10}$$

$$\Rightarrow \qquad t_{n3} = \sum_{i=0}^{n-1} (t_{(i+1)3} - t_{i3}) = \frac{2v_0}{g} \sum_{i=0}^{n-1} \alpha^i = \frac{2v_0}{g} \frac{1-\alpha^n}{1-\alpha} \tag{11}$$

$$\Rightarrow \qquad t_{n3}\Big|_{n\to\infty} = \frac{2v_0}{g} \frac{1}{1-\alpha} = \frac{2v_0}{g} \frac{M+Nm}{2Nm} \tag{12}$$

Solution to Problem (3)

Taking *the system composed of the container and the gas* as the object of study. Studying *the process from Time t_{i3} to Time $t_{(i+1)3}$*. Considering (a) ***law of conservation of energy*** [the work done by external force is zero, and the container is adiabatic; therefore, the energy of the system is conserved]. (b) ***König theorem*** [the gas is regarded as a system of particles, and the total kinetic energy of the gas in its center-of-mass reference frame (container reference frame) is the internal energy of the gas]:

$$\frac{1}{2}Mv_i^2 + \left(\frac{1}{2}Nmv_i^2 + N\overline{\varepsilon_i}\right) = \frac{1}{2}Mv_{i+1}^2 + \left(\frac{1}{2}Nmv_{i+1}^2 + N\overline{\varepsilon_{i+1}}\right) \tag{13}$$

$\overline{\varepsilon_i}$ and $\overline{\varepsilon_{i+1}}$ —the average energy of gas molecules at Time t_{i3} and Time $t_{(i+1)3}$ [in the center-of-mass reference frame when the gas is regarded as a system of particles].

Taking *the gas* as the object of study. Studying *the state at Time t_{i3} and Time $t_{(i+1)3}$*, respectively. Considering ***equipartition theorem of energy*** [the gas is in thermal equilibrium]:

$$\begin{cases} \overline{\varepsilon_i} = \dfrac{j}{2}kT_i & \text{(14a)} \\ \overline{\varepsilon_{i+1}} = \dfrac{j}{2}kT_{i+1} & \text{(14b)} \end{cases}$$

T_i and T_{i+1}—the temperature of the gas at Time t_{i3} and $t_{(i+1)3}$.

Defining the constant β:

$$\beta = \frac{1}{2}(M + Nm)v_0^2/N\frac{j}{2}k \tag{15}$$

$$\Rightarrow \quad T_{i+1} - T_i = \beta(1-\alpha^2)\alpha^{2i} \tag{16}$$

$$\Rightarrow \quad T_n - T_0 = \sum_{i=0}^{n-1}(T_{i+1} - T)_i = \beta(1-\alpha^2)\sum_{i=0}^{n-1}\alpha^{2i}$$

$$= \beta(1-\alpha^2)\frac{1-\alpha^{2n}}{1-\alpha^2} = \beta(1-\alpha^{2n}) \tag{17}$$

$$\Rightarrow \quad T_n = T_0 + \beta(1-\alpha^{2n}) \tag{18}$$

$$\Rightarrow \quad T_n\big|_{n\to\infty} = T_0 + \beta \tag{19}$$

2.6.4 The Ideal Gas in a Rotating Cylindrical Adiabatic Container (*a*, *A*)

Consider a cylindrical container of radius R placed horizontally, which is filled with a rigid diatomic molecule ideal gas of molecular mass m_0. The container is adiabatic. Initially, the container and the gas inside it are at rest, and the temperature of the gas is T_1. Use external force to make the container rotate around its central axis at the constant angular velocity of the magnitude ω and the direction vertical upward (see Fig. 1). After a sufficiently long time, the gas in the container will reach thermal equilibrium, determine the equation that its temperature should satisfy, and find the approximate solution when ω is small enough.

Solution

The container reference frame is denoted by Reference frame S'.

As shown in Fig. 2, the bottom center of the container is denoted by O. Establishing the cylindrical coordinate system fixed to Reference frame S', let z, r, θ be the cylindrical coordinates.

Defining dm_1—the mass of the gas infinitesimal element located in the interval r to $r + dr$, θ to $\theta + d\theta$, z to $z + dz$ in the initial state (hereinafter referred to as the initial element).

Defining n—the molecular number density of the gas in the initial state [in the initial state, the gas is uniform distributed]. Considering the definition of ***molecular number density***:

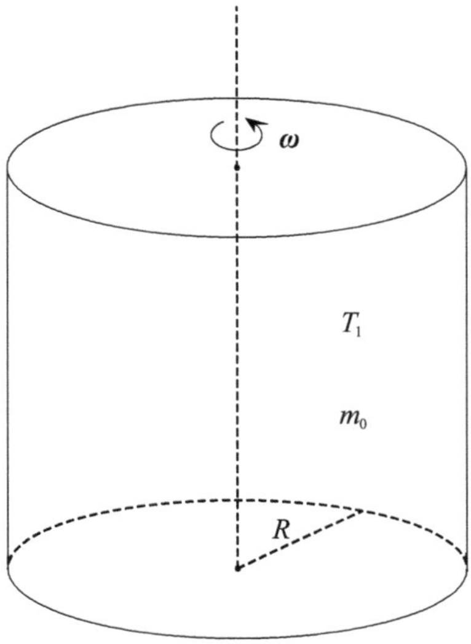

Figure 1

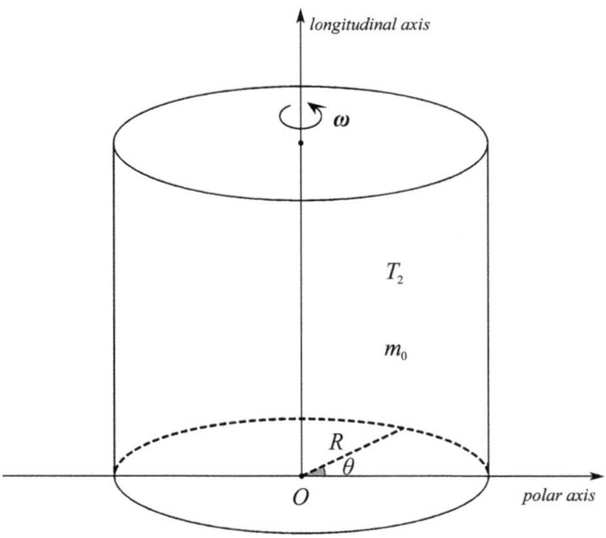

Figure 2

$$dm_1 = m_0 n r dr d\theta dz \tag{1}$$

Defining dE_{k1}, dE_{p1} and dE_1—the kinetic energy, inertial centrifugal force potential energy (taking $r = 0$ as potential energy zero) and the mechanical energy of the initial element in Reference frame S'. Considering (a) the definition of ***particle kinetic energy***, (b) ***theorem of inertial centrifugal potential energy***:

$$\begin{cases} \mathrm{d}E_{k1} = \dfrac{1}{2}\mathrm{d}m_1\omega^2 r^2 & \text{(2a)} \\ \mathrm{d}E_{p1} = -\dfrac{1}{2}\mathrm{d}m_1\omega^2 r^2 & \text{(2b)} \end{cases}$$

$$\Rightarrow \qquad \mathrm{d}E_1 = \mathrm{d}E_{k1} + \mathrm{d}E_{p1} = 0 \qquad (3)$$

Defining E_1—the total mechanical energy of the gas in the container in the initial state.

$$\Rightarrow \qquad E_1 = \iiint \mathrm{d}E_1 = 0 \qquad (4)$$

Defining U_1—the internal energy of the gas in the initial state. Considering ***equipartition theorem of energy*** [the degree of freedom of rigid diatomic molecules is 5]:

$$U_1 = \frac{5}{2}kT_1 \cdot n\pi R^2 H \qquad (5)$$

H—the height of the container.

Defining E'_1—the total energy of the gas in the initial state.

$$\Rightarrow \qquad E'_1 = E_1 + U_1 = \frac{5}{2}kT_1 \cdot n\pi R^2 H \qquad (6)$$

Taking *the gas* as the object of study. Studying *the final state*. Considering ***the distribution of molecular number density of isothermal gases as a function on the potential energy in conservative force field*** [according to spatial symmetry, the molecular number density is independent of the polar angle θ and coordinate z, and is the function of the polar radius r]:

$$n_r = n_0 \exp\left(\frac{1}{2}m_0\omega^2 r^2 / kT_2\right) \qquad (7)$$

T_2—the gas temperature in the final state.

n_r—the molecular number density at the polar radius r in the final state.

Taking *the gas* as the object of study. Studying *the whole process*. Considering ***particle conservation***:

$$n\pi R^2 H = \iiint n_r r\mathrm{d}r\mathrm{d}\theta\mathrm{d}z \tag{8}$$

$$\Rightarrow \quad n\pi R^2 H = n_0 \int_0^R \exp\left(\frac{1}{2}m_0\omega^2 r^2/kT_2\right) r\mathrm{d}r \int_0^{2\pi} \mathrm{d}\theta \int_0^H \mathrm{d}z$$

$$= n_0 \cdot 2\pi H \frac{kT_2}{m_0\omega^2}\left[\exp\left(\frac{1}{2}m_0\omega^2 R^2/kT_2\right) - 1\right] \tag{9}$$

$$\Rightarrow \quad n_0 = n\frac{m_0\omega^2 R^2}{2kT_2}\frac{1}{\exp\left(\frac{1}{2}m_0\omega^2 R^2/kT_2\right) - 1} \tag{10}$$

(10) → (7),

$$\Rightarrow \quad n_r = n\frac{m_0\omega^2 R^2}{2kT_2}\frac{\exp\left(\frac{1}{2}m_0\omega^2 R^2/kT_2\right)}{\exp\left(\frac{1}{2}m_0\omega^2 R^2/kT_2\right) - 1} \tag{11}$$

Defining $\mathrm{d}m_1$—the mass of the gas infinitesimal element located in the interval r to $r + \mathrm{d}r$, θ to $\theta + \mathrm{d}\theta$, z to $z + \mathrm{d}z$ in the Reference frame S' in the final state (hereinafter referred to as the final element).

Considering the definition of ***molecular number density***:

$$\mathrm{d}m_2 = m_0 n_r r\mathrm{d}r\mathrm{d}\theta\mathrm{d}z \tag{12}$$

Defining $\mathrm{d}E_{k2}$, $\mathrm{d}E_{p1}$ and $\mathrm{d}E_2$—the kinetic energy, inertial centrifugal force potential energy and mechanical energy of the final element in Reference frame S'. Considering (a) the definition of ***particle kinetic energy***, (b) ***theorem of inertial centrifugal potential energy*** (taking $r = 0$ as potential energy zero):

$$\begin{cases} \mathrm{d}E_{k2} = 0 & \text{(13a)} \\ \mathrm{d}E_{p2} = -\frac{1}{2}\mathrm{d}m_2\omega^2 r^2 = -\frac{1}{2}m_0 n_r \omega^2 r^3 \mathrm{d}r\mathrm{d}\theta\mathrm{d}z & \text{(13b)} \end{cases}$$

$$\Rightarrow \quad \mathrm{d}E_2 = \mathrm{d}E_{k2} + \mathrm{d}E_{p2} = -\frac{1}{2}m_0 n_r \omega^2 r^3 \mathrm{d}r\mathrm{d}\theta\mathrm{d}z \tag{14}$$

Defining E_2—the total mechanical energy of the gas in the final state, then place (11) in to obtain:

$$\Rightarrow \quad E_2 = \iiint \mathrm{d}E_2 = -\frac{1}{2}m_0\omega^2 \iiint n_r r^3 \mathrm{d}r\mathrm{d}\theta\mathrm{d}z$$

$$=-\frac{1}{2}m_0\omega^2 n\frac{m_0\omega^2R^2}{2kT_2}\frac{\int_0^R \exp\left(\frac{1}{2}m_0\omega^2r^2/kT_2\right)r^3\mathrm{d}r\int_0^{2\pi}\mathrm{d}\theta\int_0^H\mathrm{d}z}{\exp\left(\frac{1}{2}m_0\omega^2R^2/kT_2\right)-1} \tag{15}$$

Defining the definite integral I:

$$I=\int_0^R \exp\left(\frac{1}{2}m_0\omega^2r^2/kT_2\right)r^3\mathrm{d}r \tag{16}$$

Defining the constant α:

$$\alpha=\frac{1}{2}m_0\omega^2R^2 \tag{17}$$

$$\begin{aligned}
\Rightarrow\quad I&=\int_0^R \exp(\alpha r^2)r^3\mathrm{d}r\\
&=\frac{1}{2\alpha}\int_0^R r^2\exp(\alpha r^2)\mathrm{d}(\alpha r^2)\\
&=\frac{1}{2\alpha}\int_0^R r^2\mathrm{d}\left[\exp(\alpha r^2)\right]\\
&=\frac{1}{2\alpha}\left\{r^2\exp(\alpha r^2)|_0^R-\int_0^R\left[\exp(\alpha r^2)\right]\mathrm{d}(r^2)\right\}\\
&=\frac{1}{2\alpha}\left\{r^2\exp(\alpha r^2)|_0^R-\frac{1}{\alpha}\int_0^R\left[\exp(\alpha r^2)\right]\mathrm{d}(\alpha r^2)\right\}\\
&=\frac{1}{2\alpha}\left\{r^2\exp(\alpha r^2)|_0^R-\frac{1}{\alpha}\exp(\alpha r^2)|_0^R\right\}\\
&=\frac{1}{2\alpha}\left\{R^2\exp(\alpha R^2)-\frac{1}{\alpha}\left[\exp(\alpha R^2)-1\right]\right\}\\
&=\frac{kT_2}{m_0\omega^2}\left\{\begin{array}{l}R^2\exp\left(\frac{1}{2}m_0\omega^2R^2/kT_2\right)\\ -\frac{2kT_2}{m_0\omega^2}\left[\exp\left(\frac{1}{2}m_0\omega^2R^2/kT_2\right)-1\right]\end{array}\right\}\\
&=\frac{kT_2}{m_0\omega^2}\left[\left(R^2-\frac{2kT_2}{m_0\omega^2}\right)\exp\left(\frac{1}{2}m_0\omega^2R^2/kT_2\right)+\frac{2kT_2}{m_0\omega^2}\right]
\end{aligned} \tag{18}$$

$$\Rightarrow E_2 = -\frac{1}{2}m_0\omega^2 \frac{\left\{\left(R^2 - \frac{2kT_2}{m_0\omega^2}\right)\exp\left(\frac{1}{2}m_0\omega^2R^2/kT_2\right) + \frac{2kT_2}{m_0\omega^2}\right\}}{\exp\left(\frac{1}{2}m_0\omega^2R^2/kT_2\right) - 1} n\pi R^2 H$$

$$= \left[-\frac{1}{2}m_0\omega^2R^2 \frac{\exp\left(\frac{1}{2}m_0\omega^2R^2/kT_2\right)}{\exp\left(\frac{1}{2}m_0\omega^2R^2/kT_2\right) - 1} + kT_2\right] n\pi R^2 H \quad (19)$$

Defining U_2—the internal energy of the gas in the final state. Considering ***equipartition theorem of energy*** [the degree of freedom of rigid diatomic molecules is 5]:

$$U_2 = \frac{5}{2}kT_2 \cdot n\pi R^2 H \quad (20)$$

Defining E'_2—the total energy of the gas in the final state.

$$\Rightarrow E'_2 = E_2 + U_2 = \left[-\frac{1}{2}m_0\omega^2R^2 \frac{\exp\left(\frac{1}{2}m_0\omega^2R^2/kT_2\right)}{\exp\left(\frac{1}{2}m_0\omega^2R^2/kT_2\right) - 1} + \frac{7}{2}kT_2\right] n\pi R^2 H \quad (21)$$

Taking *the gas* as the object of study. Studying *the whole process.* Considering ***law of conservation of energy*** [the work done by external force is zero, and the container is adiabatic; therefore, the energy of the system is conserved]:

$$E'_1 = E'_2 \quad (22)$$

(6) and (21) → (22), we obtain the equation that T_2 should satisfy:

$$\Rightarrow \quad \frac{5}{2}kT_1 = -\frac{1}{2}m_0\omega^2R^2 \frac{\exp\left(\frac{1}{2}m_0\omega^2R^2/kT_2\right)}{\exp\left(\frac{1}{2}m_0\omega^2R^2/kT_2\right) - 1} + \frac{7}{2}kT_2$$

$$\approx -\frac{1}{2}m_0\omega^2R^2 \frac{1 + \frac{1}{2}m_0\omega^2R^2/kT_2}{\left(1 + \frac{1}{2}m_0\omega^2R^2/kT_2\right) - 1} + \frac{7}{2}kT_2$$

$$= -\frac{1}{2}m_0\omega^2R^2 + \frac{5}{2}kT_2 \tag{23}$$

The operation above has taken the low order approximation of ω.

$$\Rightarrow \qquad T_2 \approx T_1 + \frac{m_0\omega^2R^2}{5k} \tag{24}$$

2.6.5 Gas Passing through an Electric Resistance Wire Mesh in a Pipeline (*c*, *B*)[6]

Consider an adiabatic cylindrical pipeline of cross-sectional area S fixed in the horizontal direction. There is a rigid diatomic molecule ideal gas moving in the pipeline at the velocity of the magnitude v_0 and the direction parallel to the central axis of the pipeline towards the right, given the molecular mass of the gas is m, the molecular number density is n_0, the temperature is T_0. An electric resistance wire mesh is equipped at a cross section of the pipeline, the resisting effect of the mesh to the gas can be ignored. The resistance wire mesh is energized and its Joule heat power is P, supposing that all heat released by the resistance wire mesh is absorbed by the gas, the heat transfer between the gas can be ignored, and the gas passing though the resistance wire mesh can reach thermal equilibrium immediately. Find the velocity of the gas on the right side of the resistance wire mesh.

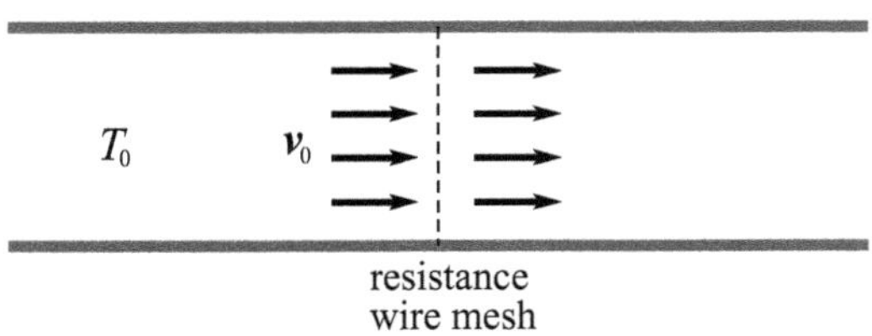

Figure 1

Solution

The time point when the resistance wire mesh is energized is defined as Time zero.

As shown in Fig. 2, the unit vector parallel to the pipeline central axis towards the right is denoted by $\boldsymbol{i}$.

---

[6]Yunhao Qin (2004). *General Physics Course, Heat, 2nd ed. Guidance of Exercise and Thinking Problem* (Higher Education Press, China).

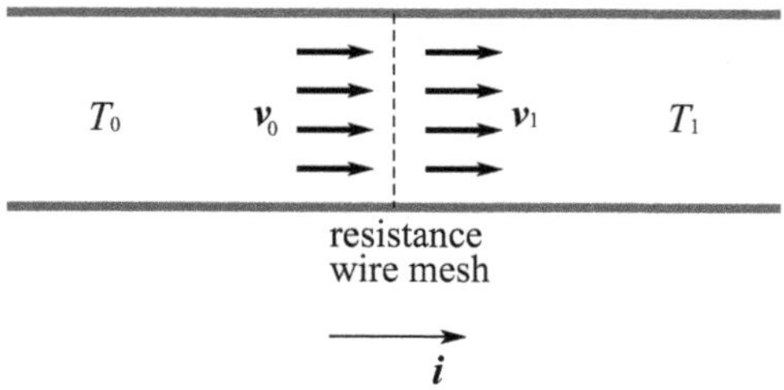

Figure 2

Taking *the gas on each side of the resistance wire mesh at arbitrary Time t* as the object of study, respectively. Studying *the state at Time t*. Considering ***state equation of ideal gas***:

$$\begin{cases} p_0 = n_0 k T_0 & \text{(1a)} \\ p_1 = n_1 k T_1 & \text{(1b)} \end{cases}$$

p_0 and p_1—the pressure of the gases on the left and right at Time t.

n_1—the molecular number density of the gas on the right at Time t.

T_1—the temperature of the gas on the right at Time t.

Defining dN—the number of molecules of the gas infinitesimal element passing through the resistance wire mesh in the instantaneous process (the small time interval dt) at arbitrary Time t. Considering ***particle conservation***:

$$\mathrm{d}N = n_0 \cdot S v_0 \mathrm{d}t = n_1 \cdot S v_1 \mathrm{d}t \quad (2)$$

v_1—the ***i***-direction component of the velocity of the gas on the right side of the resistance wire mesh at Time t.

Taking *the gas infinitesimal element passing through the resistance wire mesh in the instantaneous process (the small time interval dt) at arbitrary Time t* as the object of study. Studying *that instantaneous process*. Considering (a) ***momentum theorem*** (the ***i***-direction component equation), (b) ***principle of energy-work*** [the energy of the gas includes the kinetic energy of macroscopic motion and the kinetic energy of microscopic motion (internal energy) [note: see further important discussion at the end of this problem], (c) ***König theorem*** [the gas is regarded as a system of particles, and the total kinetic energy of the gas in its center-of-mass reference frame (i.e., the translational reference frame defined by its macroscopic motion velocity) is just its internal energy] [integrating into (a)]:

$$\begin{cases}(p_0S-p_1S)\mathrm{d}t=\mathrm{d}N\cdot mv_1-\mathrm{d}N\cdot mv_0 & \text{(3a)}\\ (p_0Sv_0\mathrm{d}t-p_1Sv_1\mathrm{d}t)+P\mathrm{d}t & \\ =\left[\frac{1}{2}(\mathrm{d}Nm)v_1^2+\mathrm{d}N\cdot\overline{\varepsilon_1}\right]-\left[\frac{1}{2}(\mathrm{d}Nm)v_0^2+\mathrm{d}N\cdot\overline{\varepsilon_0}\right] & \text{(3b)}\end{cases}$$

$\overline{\varepsilon_0}$ and $\overline{\varepsilon_1}$ —the average kinetic energy of the gas molecules on the left and right at Time t [in the center-of-mass reference frame when the gas is regarded as a system of particles].

Taking *the gas on each side of the resistance wire mesh* as the object of study, respectively. Studying *the state at arbitrary Time t*. Considering ***equipartition theorem of energy*** [the degree of freedom of diatomic molecules is 5]:

$$\begin{cases}\overline{\varepsilon_0}=\frac{5}{2}kT_0 & \text{(4a)}\\ \overline{\varepsilon_1}=\frac{5}{2}kT_1 & \text{(4b)}\end{cases}$$

Combining (1) to (4),

$$\begin{cases}kT_0-\frac{v_0}{v_1}kT_1=mv_0(v_1-v_0) & \text{(5a)}\\ \frac{P}{n_0v_0S}=\frac{1}{2}mv_1^2-\frac{1}{2}mv_0^2+\frac{7}{2}kT_1-\frac{7}{2}kT_0 & \text{(5b)}\end{cases}$$

$$\Rightarrow \qquad T_1=T_0+\frac{1}{7k}\left[\frac{2P}{n_0v_0S}-m(v_1^2-v_0^2)\right] \tag{6}$$

(6) → (5b)

$$\Rightarrow \qquad 6mv_0v_1^2-7(kT_0+mv_0^2)v_1+mv_0^3+7kT_0v_0+\frac{2P}{n_0S}=0 \tag{7}$$

$$\Rightarrow \quad v_{1(\pm)}=\frac{7(kT_0+mv_0^2)\pm\sqrt{\begin{array}{l}49(kT_0+mv_0^2)^2-\\ 24mv_0(mv_0^3+7kT_0v_0+2P/Sn_0)\end{array}}}{12mv_0}$$

$$=\frac{7(kT_0+mv_0^2)\pm\sqrt{49k^2T_0^2-70kT_0mv_0^2+25m^2v_0^4-48mv_0P/Sn_0}}{12mv_0}$$

$$= \frac{7(kT_0 + mv_0^2) \pm \sqrt{(7kT_0 - 5mv_0^2)^2 - 48mv_0P/Sn_0}}{12mv_0} \tag{8}$$

The following judges which solution is reasonable:

When $P = 0$, v_1 should be equal to v_0:

$$v_0 = v_1\big|_{P=0} \tag{9}$$

From (8),

$$\Rightarrow \qquad v_{1(\pm)}\Big|_{P=0} = \frac{7(kT_0 + mv_0^2) \pm \left|7kT_0 - 5mv_0^2\right|}{12mv_0} \tag{10}$$

From (4a),

$$\Rightarrow \qquad \frac{1}{2}m\bar{u}_0^2 = \frac{5}{2}kT_0 \tag{11}$$

$\sqrt{\bar{u}_0^2}$ —the root-mean-square speed of the thermal motion of the gas molecules on the left.

Considering that ***the root-mean-square speed of the thermal motion of gas molecules should be much greater than the macroscopic speed of the gas***:

$$\bar{u}_0 \gg v_0 \tag{12}$$

$$\Rightarrow \qquad 7kT_0 > 5mv_0^2 \tag{13}$$

$$\Rightarrow \begin{cases} v_{1(+)}\Big|_{P=0} = \dfrac{7(kT_0 + mv_0^2) + (7kT_0 - 5mv_0^2)}{12mv_0} = \dfrac{7kT_0 + mv_0^2}{6mv_0} & \text{(14a)} \\[2ex] v_{1(-)}\Big|_{P=0} = \dfrac{7(kT_0 + mv_0^2) - (7kT_0 - 5mv_0^2)}{12mv_0} = v_0 & \text{(14b)} \end{cases}$$

Therefore, the solution $v_{1(-)}$ is reasonable.

***Note*: Further discussion:**

The popular solution to this problem considers that the pressure of the gas in the pipeline is equal everywhere, and the momentum and energy changes of the gas passing through the resistance wire mesh only originate from the heating of the resistance wire mesh. But the author of this book thinks that the gas undergoes a extremely short non-equilibrium process when it passes through the resistance wire mesh, transitioning from the original equilibrium state to the new equilibrium state, and the pressures of these two states should be unequal.

2.7 Variable Mass System

As a special continuum (system of particles with continuous mass distribution) of time-varying boundary, variable mass system can be widely used to describe many practical phenomena, such as a rocket in jetting fuel, a falling raindrop, etc. Its dynamic information is described by Mischelski equation, as the degradation of the basic dynamic equation, this description scheme is relatively simple. This section will introduce the general theory of variable quality system, and study some classic variable quality system problems.

2.7.1 Dynamics of Variable Mass System (d, *C*)

A system of particles (which can be regarded as a particle in the scale of research problem) whose mass changes with time continuously due to the continuous incorporation of external particles into it on its boundary or the continuous separation of the particles on its boundary form it. This kind of system of particles with time-varying mass is called the main body, and the system of particles that are continuously incorporating into or separating from the main body are called minor body. This model is called variable mass system model.

(1) Find the dynamical equation of variable mass system.
(2) Find the energy equation of variable mass system.

Solution to Problem (1)

Time zero is arbitrarily set.

Taking *the system composed of the main body at arbitrary Time t and the minor body incorporating into or separating from the main body in the instantaneous process (the small time interval dt) at Time t* as the object of study. Studying *that instantaneous process.* Considering ***momentum theorem***:

$$\boldsymbol{F}\mathrm{d}t = (M+m)(\boldsymbol{v}+\mathrm{d}\boldsymbol{v}) - (M\boldsymbol{v}+m\boldsymbol{u}) \tag{1}$$

M—the mass of the main body at Time t.

m—the mass of the minor body incorporating into main body in the instantaneous process at Time t, let: $m>0$ for incorporation, $m<0$ for separation.

$\boldsymbol{F}$—the resultant external force acting on the system at Time t.

$\boldsymbol{v}$—the velocity of the main body at Time t.

$\boldsymbol{u}$—the velocity of the minor body incorporating into or separating from the main body in the instantaneous process at Time t.

Known that the change of the mass of the main body with time is continuous, Considering ***law of conservation of mass***:

$$m = \mathrm{d}M \tag{2}$$

$$\Rightarrow \quad \boldsymbol{F}\mathrm{d}t = (M+\mathrm{d}M)(\boldsymbol{v}+\mathrm{d}\boldsymbol{v}) - (M\boldsymbol{v}+\mathrm{d}M\boldsymbol{u}) \approx M\mathrm{d}\boldsymbol{v} + \mathrm{d}M(\boldsymbol{v}-\boldsymbol{u}) \tag{3}$$

We obtain the dynamical equation of variable mass system as follows (it's called ***Mischelski equation***):

$$\Rightarrow \quad \boldsymbol{F} = M\frac{\mathrm{d}\boldsymbol{v}}{\mathrm{d}t} + \frac{\mathrm{d}M}{\mathrm{d}t}(\boldsymbol{v}-\boldsymbol{u}) \tag{4}$$

Defining $\boldsymbol{F}_{\mathrm{recoil}}$—the recoil force of the minor body acting on the main body.

$$\boldsymbol{F}_{\mathrm{recoil}} = \frac{\mathrm{d}M}{\mathrm{d}t}(\boldsymbol{u}-\boldsymbol{v}) \tag{5}$$

We obtain the equivalent form of Eq. (4), which can be regarded as the dynamical equation of the main body:

$$\Rightarrow \quad \boldsymbol{F}+\boldsymbol{F}_{\mathrm{recoil}} = M\frac{\mathrm{d}\boldsymbol{v}}{\mathrm{d}t} \tag{6}$$

Solution to Problem (2)

Taking *the system composed of the main body at arbitrary Time t and the minor body incorporating into or separating from the main body in the instantaneous process (the small time interval dt) at Time t* as the object of study. Studying *that instantaneous process*. Considering ***principle of energy-work***:

$$\boldsymbol{F}\cdot\boldsymbol{v}\mathrm{d}t = \left[\frac{1}{2}(M+\mathrm{d}M)(v+\mathrm{d}v)^2 - \left(\frac{1}{2}Mv^2 + \frac{1}{2}\mathrm{d}Mu^2\right)\right] + \mathrm{d}E$$

$$\approx \frac{1}{2}\mathrm{d}Mv^2 + Mv\mathrm{d}v - \frac{1}{2}\mathrm{d}Mu^2 + \mathrm{d}E \tag{7}$$

$\mathrm{d}E$—the increment of the internal energy of the system in the instantaneous process at Time t.

$$\Rightarrow \qquad \boldsymbol{F}\cdot\boldsymbol{v} = \frac{1}{2}\frac{\mathrm{d}M}{\mathrm{d}t}v^2 + Mv\frac{\mathrm{d}v}{\mathrm{d}t} - \frac{1}{2}\frac{\mathrm{d}M}{\mathrm{d}t}u^2 + \frac{\mathrm{d}E}{\mathrm{d}t}$$

$$= \frac{1}{2}\frac{\mathrm{d}M}{\mathrm{d}t}v^2 + \frac{1}{2}M\frac{\mathrm{d}v^2}{\mathrm{d}t} - \frac{1}{2}\frac{\mathrm{d}M}{\mathrm{d}t}u^2 + \frac{\mathrm{d}E}{\mathrm{d}t}$$

$$= \frac{\mathrm{d}}{\mathrm{d}t}\left(\frac{1}{2}Mv^2\right) - \frac{1}{2}\frac{\mathrm{d}M}{\mathrm{d}t}u^2 + \frac{\mathrm{d}E}{\mathrm{d}t} \tag{8}$$

$(4) \rightarrow (8)$, we obtain the energy equation of the variable mass system:

$$\Rightarrow \qquad \frac{\mathrm{d}E}{\mathrm{d}t} = M\frac{\mathrm{d}\boldsymbol{v}}{\mathrm{d}t}\cdot\boldsymbol{v} + \frac{\mathrm{d}M}{\mathrm{d}t}(\boldsymbol{v}\cdot\boldsymbol{v} - \boldsymbol{u}\cdot\boldsymbol{v}) - \frac{\mathrm{d}}{\mathrm{d}t}\left(\frac{1}{2}Mv^2\right) + \frac{1}{2}\frac{\mathrm{d}M}{\mathrm{d}t}u^2$$

$$= \frac{1}{2}\frac{\mathrm{d}M}{\mathrm{d}t}v^2 - \frac{\mathrm{d}M}{\mathrm{d}t}\boldsymbol{u}\cdot\boldsymbol{v} + \frac{1}{2}\frac{\mathrm{d}M}{\mathrm{d}t}u^2$$

$$= \frac{1}{2}\frac{\mathrm{d}M}{\mathrm{d}t}(v^2 - 2\boldsymbol{u}\cdot\boldsymbol{v} + u^2)$$

$$= \frac{1}{2}\frac{\mathrm{d}M}{\mathrm{d}t}(\boldsymbol{v} - \boldsymbol{u})^2 \tag{9}$$

$$\Rightarrow \qquad \frac{\mathrm{d}E}{\mathrm{d}t}\begin{cases} >0, & \dfrac{\mathrm{d}M}{\mathrm{d}t} > 0 \\ <0, & \dfrac{\mathrm{d}M}{\mathrm{d}t} < 0 \end{cases} \tag{10}$$

It can be seen that the internal energy of the system will increase when the minor body incorporates into the main body (from the perspective in the macro scale, the process of the minor body incorporating into the main body is similar to a completely inelastic collision, in which the interaction between the main body and the minor body will do negative work on the macro system and convert the mechanical energy into internal energy); the internal energy of the system will decrease when the minor body separates from the main body (the process of the minor body separating from the main body is similar to explosion, which can be regarded as the inverse process of a completely inelastic collision, in which the interaction between the main body and the minor body will do positive work on the macro system and convert the internal energy into mechanical energy).

2.7.2 The Motion of Jet Rocket in Gravity Field (d, *C*)

A rocket of mass M_s with the fuel of mass M_f is launched vertically from the ground in a static state. The rocket jets the fuel vertically downward at the relative speed v', and the mass of the fuel jetted in unit time is a, $av' > (M_s + M_f)g$. It is assumed that the gravity field is uniform in the range of the rocket flight. Find the displacement and velocity of the rocket at arbitrary time before the fuel is exhausted.

Solution

The time point when the rocket is launched is defined as Time zero.

It can be considered that the fuel distribution is approximately continuous in space; therefore, the system composed of rocket and the remaining fuel in the rocket at arbitrary Time t can be regarded as a variable mass system.

Taking an arbitrary point on the ground and denoting it by O, the unit vector in the vertical upward direction is denoted by $\boldsymbol{i}$. Establishing the linear coordinate system with Point O as the origin and the $\boldsymbol{i}$-direction as the positive direction of coordinate axis, let x be the coordinate.

Defining M_0—the total mass of the rocket and the fuel before the launch:

$$M_0 = M_s + M_f \tag{1}$$

Taking *the system composed of the rocket and the remaining fuel in the rocket at arbitrary Time t* as the object of study. Studying *the state at Time t*. Considering ***Mischelski equation*** (the $\boldsymbol{i}$-direction component equation):

$$-Mg = M\frac{\mathrm{d}v}{\mathrm{d}t} + \frac{\mathrm{d}M}{\mathrm{d}t}v' \tag{2}$$

M—the total mass of the rocket and the remaining fuel in the rocket at Time t.

v—the $\boldsymbol{i}$-direction component of the velocity of the rocket at Time t.

Known:

$$M = M_0 - at \tag{3}$$

(3) → (2),

$$\Rightarrow \qquad \mathrm{d}v = \left(\frac{av'}{M_0 - at} - g \right) \mathrm{d}t$$

$$\Rightarrow \qquad v = \int_0^v \mathrm{d}v = \int_0^t \left(\frac{av'}{M_0 - at} - g \right) \mathrm{d}t$$

$$= v' \ln \frac{M_0}{M_0 - at} - gt = -v' \ln \left(1 - \frac{a}{M_0} t \right) - gt \tag{4}$$

Considering the definition of ***velocity***:

$$v = \frac{\mathrm{d}x}{\mathrm{d}t} \tag{5}$$

$$\Rightarrow \qquad \mathrm{d}x = -\left[v' \ln \left(1 - \frac{a}{M_0} t \right) + gt \right] \mathrm{d}t$$

$$\Rightarrow \qquad x = \int_0^x \mathrm{d}x = -\int_0^t \left[v' \ln \left(1 - \frac{a}{M_0} t \right) + gt \right] \mathrm{d}t$$

$$= v't + v' \left(\frac{M_0}{a} - t \right) \ln \left(1 - \frac{a}{M_0} t \right) - \frac{1}{2} g t^2 \tag{6}$$

Defining X—the x-coordinate of the rocket when the fuel is exhausted. V—the ***i***-direction component of the velocity of the rocket when the fuel is exhausted.

$$\Rightarrow \qquad \begin{cases} X = x|_{t=M_f/a} = v' \left(\dfrac{M_f}{a} + \dfrac{M_s}{a} \ln \dfrac{M_s}{M_0} \right) - \dfrac{1}{2} g \left(\dfrac{M_f}{a} \right)^2 & \text{(7a)} \\ V = v|_{t=M_f/a} = v' \ln \dfrac{M_0}{M_s} - g \dfrac{M_f}{a} & \text{(7b)} \end{cases}$$

2.7.3 Pushing Sand Off a Flatcar (*a*, *A*)

A railroad flatcar of mass M_0 can roll without friction along a straight horizontal track, on the flatcar there is a pile of sand of mass m_0. Initially, the flatcar and all the sand are at rest. A machine with negligible mass fixed on the flatcar continuously pushes the sand off the flatcar in the same horizontal direction (parallel to the track) at a relative speed of u'. Find the velocity of the flatcar when all the sand is pushed off.

Solution [1]

The time point when the machine starts to push the sand is defined as Time zero.

The unit vector of the velocity direction of the sand pushed off the flatcar relative to the flatcar is denoted by ***i***.

It can be considered that the sand distribution is approximately continuous in space; therefore, the system composed of flatcar and the remaining sand on the flatcar at arbitrary Time t can be regarded as a variable mass system.

Taking *the system composed of the flatcar and the remaining sand on the flatcar at arbitrary Time t* as the object of study. Studying *the state at Time t*. Considering ***Mischelski equation*** (the ***i***-direction component equation):

$$0 = M\frac{\mathrm{d}v}{\mathrm{d}t} + \frac{\mathrm{d}M}{\mathrm{d}t}(v - u) \tag{1}$$

M—the total mass of the flatcar and the remaining sand on the flatcar at Time t.

v—the ***i***-direction component of the velocity of the flatcar at Time t.

u—the ***i***-direction component of the velocity of the sand infinitesimal element pushed off the flatcar in the instantaneous process at Time t.

Taking *the sand infinitesimal element pushed off the flatcar in the instantaneous process at arbitrary Time t* as the object of study. Studying *the state at Time t*. Considering ***theorem for composition of velocities of a particle*** [taking the ground as stationary reference frame, and the flatcar as the moving reference frame] (the ***i***-direction component equation):

$$u = v + u' \tag{2}$$

$$\Rightarrow \qquad M\frac{\mathrm{d}v}{\mathrm{d}t} = u'\frac{\mathrm{d}M}{\mathrm{d}t} \tag{3}$$

$$\Rightarrow \qquad \mathrm{d}v = u'\frac{\mathrm{d}M}{M} \tag{4}$$

$$\Rightarrow \qquad v_{\text{final}} = \int_0^{v_{\text{final}}} \mathrm{d}v = u' \int_{M_0+m_0}^{M_0} \frac{\mathrm{d}M}{M} = -u' \ln \frac{M_0+m_0}{M_0} \tag{5}$$

v_{final}—the $\boldsymbol{i}$-direction component of the velocity of the flatcar when all the sand is pushed out.

Solution [2]

The following studies a situation that the machine pushes the sand off the flatcar discontinuously: it is assumed that the sand is divided into N equal parts (N is a positive integer), these piles of sand are pushed off the flatcar one by one by the machine, it's instantaneously done for pushing the sand every time.

Taking *the system composed of the NO. (i + 1) pile of sand pushed off the flatcar, the flatcar and the remaining sand on the flatcar* as the object of study. Studying *the process of pushing the NO. (i + 1) pile of sand*. Considering ***law of conservation of momentum*** [the $\boldsymbol{i}$-direction component of the resultant external force acting on the system is zero; therefore, the $\boldsymbol{i}$-direction component of the momentum of the system is conserved] (the $\boldsymbol{i}$-direction component equation):

$$\left[M_0+(N-i)\delta\right]v_i = \left[M_0+(N-i-1)\delta\right]v_{i+1} + \delta u_{i+1} \tag{6}$$

δ—the mass of each pile of sand.

v_i—the $\boldsymbol{i}$-direction component of the velocity of the flatcar at the time point after the NO. i pile of sand is pushed off the flatcar.

u_{i+1}—the $\boldsymbol{i}$-direction component of the velocity of the NO. (i + 1) pile of sand at the time point after it is pushed off the flatcar.

Taking *the NO. (i + 1) pile of sand* as the object of study. Studying *the state at the time point after it is pushed off the flatcar*. Considering ***theorem for composition of velocities of a particle*** [taking the ground as stationary reference frame, and the flatcar as the moving reference frame] (the $\boldsymbol{i}$-direction component equation):

$$u_{i+1} = v_{i+1} + u' \tag{7}$$

$$\Rightarrow \qquad v_i - v_{i+1} = \frac{\delta}{M_0+(N-i)\delta} u' \tag{8}$$

$$\Rightarrow \qquad v_0 - v_N = \sum_{i=0}^{N-1}(v_i - v_{i+1}) = \left[\sum_{i=0}^{N-1} \frac{\delta}{M_0+(N-i)\delta}\right] u' \tag{9}$$

Known:

$$v_0 = 0 \tag{10}$$

$$\Rightarrow \qquad v_{\text{final}} = v_N = -\left[\sum_{i=0}^{N-1} \frac{\delta}{M_0 + (N-i)\delta}\right] u' \tag{11}$$

Known:

$$N\delta = m_0 \tag{12}$$

Defining m—the total mass of the front i piles of sand that pushed off the flatcar:

$$m = i\delta \tag{13}$$

When N approaches infinity, m can be regarded as a continuous variable, the relation between m and i can be regarded as a continuous function, and δ can be regarded as the differential of m:

$$\Rightarrow \qquad \mathrm{d}m = \delta \tag{14}$$

Based on this understanding, the summation operation in Eq. (11) can be regarded as the definite integral with m as the variable:

$$\Rightarrow \qquad v_{\text{final}} = v_N = -u' \int_0^{m_0} \frac{\mathrm{d}m}{M_0 + (m_0 - m)} = -u' \ln \frac{M_0 + m_0}{M_0} \tag{15}$$

2.7.4 A Homogeneous Rope Whose Two Ends are Hung on the Ceiling, One End of Which is Released and Falls (d, *AC*)

There is a homogeneous rigid rope of mass m and length l, the two ends of which are hung on a small hook on a ceiling. Release one end of the rope from the hook. Find the force exerted on the fixed end of the rope by the hook when the free end falls to an arbitrary position.

Solution [1] (type C)

Let the right end of the rope be released (see Fig. 1), and the time point when it is released is defined as Time zero.

As shown in Fig. 1, taking an arbitrary point on the ceiling and denoting it by O, and the unit vector in the vertical downward direction is denoted by $\boldsymbol{i}$. Establishing the linear coordinate system with Point O as the origin and the $\boldsymbol{i}$-direction as the positive direction of coordinate axis, let x be the coordinate.

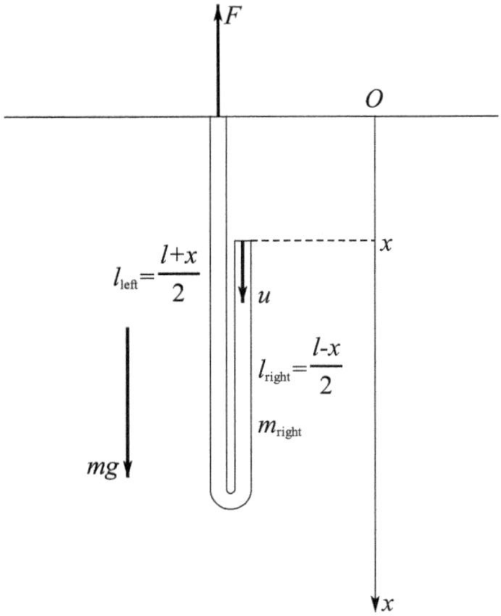

Figure 1

Taking *the whole rope* as the object of study. Studying *the instantaneous process (the small time interval dt) at arbitrary Time t.* Considering ***theorem of momentum*** (the $\boldsymbol{i}$-direction component equation):

$$mg - F = \frac{\mathrm{d}}{\mathrm{d}t}(m_{\text{right}}\, u) \tag{1}$$

m_{right}—the mass of the right part of the rope at Time t.

u—the $\boldsymbol{i}$-direction component of the velocity of the free end of the rope at Time t.

F—the $-\boldsymbol{i}$-direction component of the force exerted on the fixed end of the rope by the hook at Time t.

Defining λ—the line mass density of the rope.

$$\lambda = \frac{m}{l} \tag{2}$$

Defining l_{right} and l_{left}—the length of the right part and left part of the rope at Time t.

From the geometric relation, we can know that:

$$\begin{cases} l_{\text{right}} = \dfrac{l-x}{2} & \text{(3a)} \\ l_{\text{left}} = \dfrac{l+x}{2} & \text{(3b)} \end{cases}$$

$$\Rightarrow \qquad m_{\text{right}} = \lambda l_{\text{right}} = \lambda \frac{l-x}{2} \tag{4}$$

$$\Rightarrow \qquad F = \lambda \left[lg + \frac{1}{2}\frac{\mathrm{d}x}{\mathrm{d}t}u - \frac{1}{2}(l-x)\frac{\mathrm{d}u}{\mathrm{d}t} \right] \tag{5}$$

The free end of the rope can be regarded as a free-falling body. Considering ***the velocity-displacement relation of uniformly variable motion***:

$$u = \sqrt{2gx} \tag{6}$$

Considering the definition of ***velocity*** and ***acceleration***:

$$u = \frac{\mathrm{d}x}{\mathrm{d}t} \tag{7}$$

$$g = \frac{\mathrm{d}u}{\mathrm{d}t} \tag{8}$$

Combining all the above equations,

$$\Rightarrow \qquad F = \frac{1}{2}\lambda(l+3x)g = \frac{l+3x}{2l}mg \tag{9}$$

Solution [2] (type C)

Considering the rope infinitesimal element that moving from the right to the left in the instantaneous process (the small time interval dt) at arbitrary Time t, the two ends of which are denoted by A and B, respectively. The rope infinitesimal element is denoted by element AB.

Taking *element AB* as the object of study. Studying *the instantaneous process of its moving from the right to the left*. Considering ***theorem of momentum*** [compared with the tension at Point A acting on it, its gravity is an infinitely small quantity, which can be ignored] (the $-\boldsymbol{i}$-direction component equation):

$$T\mathrm{d}t = 0 - (-m_{AB}u) \tag{10}$$

m_{AB}—the mass of element *AB*.

T—the $-\boldsymbol{i}$-direction component of the tension at the *A* end of element *AB*.

Considering ***law of conservation of mass***:

$$m_{AB} = \mathrm{d}m_{\text{left}} = \lambda \mathrm{d}l_{\text{left}} \tag{11}$$

m_{left}—the mass of the left part of the rope at Time *t*.

From (3b),

$$\mathrm{d}l_{\text{left}} = \frac{1}{2}\mathrm{d}x \tag{12}$$

Combining (6), (7), (10) to (12),

$$\Rightarrow \qquad T = \lambda x g \tag{13}$$

Taking *the left part of the rope at arbitrary Time t* as the object of study. Studying *the state at Time t*. Considering ***Newton's second law*** [the left part of the rope is in mechanical equilibrium] (the $-\boldsymbol{i}$-direction component equation):

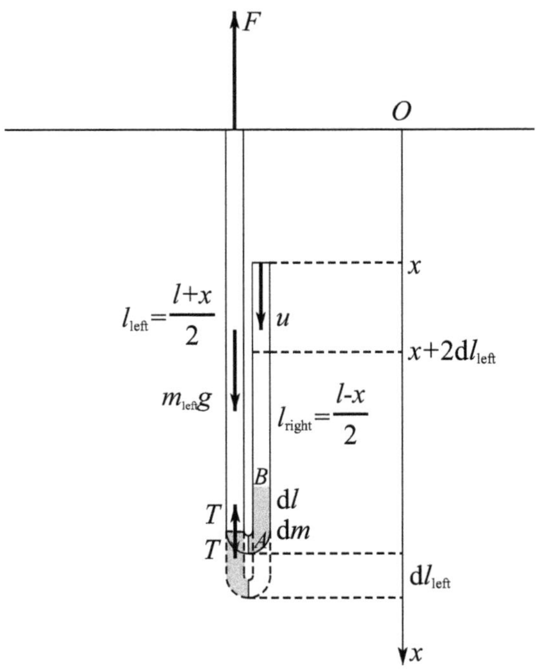

Figure 2

$$F - m_{\text{left}}g - T = 0 \tag{14}$$

From (3b),

$$\Rightarrow \qquad m_{\text{left}} = \lambda l_{\text{left}} = \lambda \frac{l + x}{2} \tag{15}$$

(13) and (15) → (14), then place (2) in to obtain:

$$\Rightarrow \qquad F = \lambda \frac{l + 3x}{2} g = \frac{l + 3x}{2l} mg \tag{16}$$

Solution [3] (type A)

The left part of the rope at arbitrary Time t can be regarded as a variable mass system.

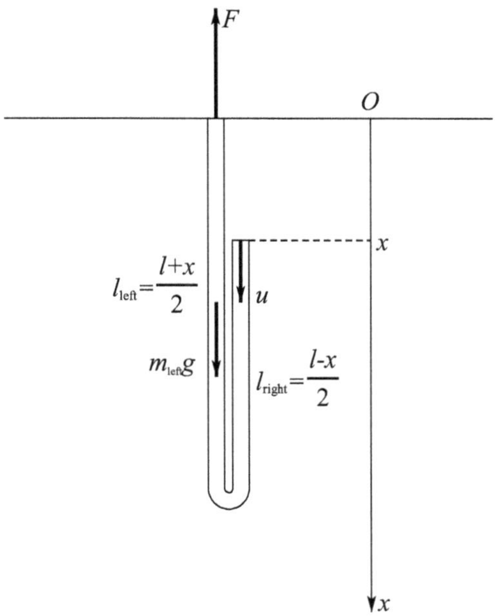

Figure 3

Taking *the left part of the rope at arbitrary Time t* as the object of study. Studying *the state at Time t*. Considering ***Mischelski equation*** (the −***i***-direction component equation):

$$F - m_{\text{left}}g = m_{\text{left}} \frac{\mathrm{d}v}{\mathrm{d}t} + \frac{\mathrm{d}m_{\text{left}}}{\mathrm{d}t}(v - u') \tag{17}$$

v—the $-\boldsymbol{i}$-direction component of the velocity of the center of mass of the left part of the rope at Time t.

u'—the $-\boldsymbol{i}$-direction component of the velocity of the rope infinitesimal element incorporated from the right into the left in the instantaneous process at Time t.

From (15),

$$\Rightarrow \qquad \frac{\mathrm{d}m_{\mathrm{left}}}{\mathrm{d}t}=\frac{1}{2}\lambda\frac{\mathrm{d}x}{\mathrm{d}t} \tag{18}$$

Known:

$$v=0 \tag{19}$$

$$u'=-u \tag{20}$$

(6), (7), (18), (19) and (20) → (17), then place (2) in to obtain

$$\Rightarrow \qquad F=\lambda\frac{l+3x}{2}g=\frac{l+3x}{2l}mg \tag{21}$$

2.7.5 A Homogeneous Rope Falling through a Hole (*b*, *A*)

There is a small hole in a horizontal table top, and a rigid rope of length L and line mass density ρ is densely piled on the table, tightly next to the hole. Let a small segment of one side of the rope pass through the hole, then release the rope freely, the rope will slide down through the hole. Given that the height of the table is greater than the length of the rope, and the friction between the rope and the hole can be ignored.

(1) Find the position and velocity of the lower end of the rope at arbitrary time before the rope passes through the hole completely.

(2) Find the internal energy generated in the whole process of the rope passing through the hole.

Solution to Problem (1)

The time point when the rope is released is defined as Time zero.

The position of the hole is denoted by O, the unit vector in the vertical downward direction is denoted by $\boldsymbol{i}$. Establishing the linear coordinate system with Point O as the origin and the $\boldsymbol{i}$-direction as the positive direction of coordinate axis, let x be the coordinate.

The segment of the rope under the table at arbitrary Time t can be regarded as a variable mass system.

Taking *the segment of the rope under the table at arbitrary Time t* as the object of study. Studying *the state at Time t*. Considering ***Mischelski equation*** (the ***i***-direction component equation):

$$Mg = M\frac{\mathrm{d}v}{\mathrm{d}t} + \frac{\mathrm{d}M}{\mathrm{d}t}(v - u) \tag{1}$$

M—the mass of the segment of the rope under the table at Time t.

v—the ***i***-direction component of the velocity of the lower end of the rope at Time t.

u—the ***i***-direction component of the velocity of the rope infinitesimal element passing through the hole (incorporating into the segment of the rope under the table) in the instantaneous process at Time t.

Considering the definition of ***line mass density***:

$$M = \rho x \tag{2}$$

x—the coordinate of the lower end of the rope at arbitrary Time t.

Known that the segment of the rope on the table is static:

$$u = 0 \tag{3}$$

(2) and (3) $\rightarrow$ (1),

$$\Rightarrow \qquad \frac{\mathrm{d}v}{\mathrm{d}t} + \frac{1}{x}\frac{\mathrm{d}x}{\mathrm{d}t}v = g \tag{4}$$

Considering the definition of ***velocity***:

$$v = \frac{\mathrm{d}x}{\mathrm{d}t} \tag{5}$$

(5) $\rightarrow$ (4),

$$\Rightarrow \qquad \frac{\mathrm{d}^2x}{\mathrm{d}t^2} + \frac{1}{x}\left(\frac{\mathrm{d}x}{\mathrm{d}t}\right)^2 = g \tag{6}$$

Let the trial solution of Eq. (6) be:

$$x = At^2 + Bt + C \tag{7}$$

Considering the initial conditions:

$$\begin{cases} x\big|_{t=0} = 0 & \text{(8a)} \\ \left.\dfrac{\mathrm{d}x}{\mathrm{d}t}\right|_{t=0} = 0 & \text{(8b)} \end{cases}$$

$(7) \rightarrow (8)$,

$$\Rightarrow \qquad \begin{cases} C = 0 & (9a) \\ B = 0 & (9b) \end{cases}$$

$$\Rightarrow \qquad x = At^2 \qquad (10)$$

$(10) \rightarrow (6)$:

$$\Rightarrow \qquad A = \frac{1}{6}g \qquad (11)$$

$$\Rightarrow \qquad x = \frac{1}{6}gt^2 \qquad (12)$$

$$\Rightarrow \qquad v = \frac{1}{3}gt \qquad (13)$$

Solution [1] to Problem (2)

Defining T—the time taken for the rope to pass through the hole completely.

From (12),

$$\Rightarrow \qquad T = t\big|_{x=L} = \sqrt{\frac{6L}{g}} \qquad (14)$$

Defining V—the ***i***-component of the velocity of the lower end of the rope when the rope passes through the hole completely.
From (13) and (14),

$$\Rightarrow \qquad V = v\big|_{t=T} = \sqrt{\frac{2gL}{3}} \qquad (15)$$

Taking *the system composed of the rope and the earth* as the object of study. Studying *the whole process of the rope's passing through the hole*. Considering ***law of conservation of energy*** (taking the center of mass of the rope when it passes through the hole completely as the zero point of gravitational potential energy):

$$\rho Lg \cdot \frac{1}{2}L = \frac{1}{2}\rho LV^2 + \Delta E \qquad (16)$$

ΔE—the internal energy generated in the whole process of the rope passing through the hole.

(15) → (16),

$$\Rightarrow \qquad \Delta E = \frac{1}{6}\rho L^2 g \qquad (17)$$

Solution [2] to Problem (2)

Taking *the system composed of the segment of the rope under the table at arbitrary Time t and the rope infinitesimal element incorporating into it in the instantaneous process (the small time interval dt) at Time t* as the object of study. Studying *the state at Time t*. Considering ***theorem of mechanical energy loss rate of variable mass system***:

$$\frac{\mathrm{d}E}{\mathrm{d}t} = Mgv - \frac{\mathrm{d}}{\mathrm{d}t}\left(\frac{1}{2}Mv^2\right) + \frac{1}{2}\frac{\mathrm{d}M}{\mathrm{d}t}u^2 \qquad (18)$$

$\mathrm{d}E$—the increment of the internal energy of the system in the instantaneous process at Time *t*.

(2), (3), (12) and (13) → (18),

$$\Rightarrow \qquad \frac{\mathrm{d}E}{\mathrm{d}t} = \rho xgv - \frac{\mathrm{d}}{\mathrm{d}t}\left(\frac{1}{2}\rho xv^2\right)$$

$$= \rho gxv - \frac{1}{2}\rho v\left(v^2 + 2x\frac{\mathrm{d}v}{\mathrm{d}t}\right)$$

$$= \frac{1}{54}\rho g^3 t^3$$

$$\Rightarrow \qquad \mathrm{d}E = \frac{1}{54}\rho g^3 t^3 \mathrm{d}t \qquad (19)$$

From (14) and (19),

$$\Rightarrow \qquad \Delta E = \int_0^{\Delta E}\mathrm{d}E = \frac{1}{54}\rho g^3\int_0^T t^3\mathrm{d}t = \frac{1}{6}\rho L^2 g \qquad (20)$$

2.7.6 The Falling of Raindrop (d, *C*)

In a very dense fog, a very small "raindrop" is condensed from some tiny water drops and starts to fall from rest. Given that it absorbs all the tiny water drops in its moving path to make its volume continuously increase, and remains spherical shape throughout the falling. The viscous force of the air can be ignored. The initial mass and initial radius of the raindrop are denoted by M_0 and r_0, respectively, which are regarded as small quantities. Please prove that after a long enough time, the acceleration of the raindrop will tend to be stable, and calculate this stable value.

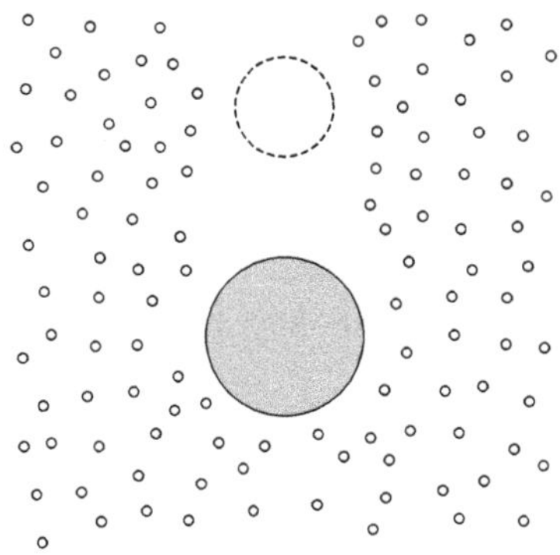

Figure 1

Solution [1]

The time point when the raindrop starts to fall is defined as Time zero.

The position where the raindrop starts to fall is defined as *O*, the unit vector in the vertical downward direction is denoted by ***i***. Establishing the linear coordinate system with Point *O* as the origin and the ***i***-direction as the positive direction of coordinate axis, let *x* be the coordinate.

It can be considered that the fog distribution is approximately continuous in space; therefore, the raindrop can be regarded as a variable mass system.

Taking *the raindrop at arbitrary Time t* as the object of study. Studying *the instantaneous process (the small time interval dt) at Time t*. Considering ***momentum theorem*** (the ***i***-direction component equation):

$$Mg\mathrm{d}t = \mathrm{d}(Mv) \tag{1}$$

M—the mass of the raindrop at Time *t*.

v—the ***i***-direction component of the velocity of the raindrop at Time *t*.

Considering the definition of ***mass density***:

$$M = \rho_{\text{water}} \cdot \frac{4}{3}\pi r^3 \tag{2}$$

ρ_{water}—the density of the raindrop.

r—the radius of the raindrop at arbitrary Time t.

Considering (a) ***law of conservation of mass*** (the increment of the mass of the raindrop in the instantaneous process at arbitrary Time t is equal to the mass of the small drops of water adsorbed by the raindrop in its path), (b) the definition of ***mass density***:

$$\mathrm{d}M = \rho_{\text{fog}} \cdot \pi r^2 v \cdot \mathrm{d}t \tag{3}$$

ρ_{fog}—the water density of the fog.

Combining (1) and (3),

$$\Rightarrow \quad Mg = \frac{\mathrm{d}(Mv)}{\mathrm{d}t} = \frac{\mathrm{d}(Mv)}{\mathrm{d}M}\frac{\mathrm{d}M}{\mathrm{d}t} = \frac{\mathrm{d}(Mv)}{\mathrm{d}M}\rho_{\text{fog}} \cdot \pi r^2 v \tag{4}$$

Combining (2) and (4),

$$\Rightarrow \quad (Mv)\mathrm{d}(Mv) = \frac{g}{\pi\rho_{\text{fog}}}\sqrt[3]{\frac{16}{9}\pi^2\rho_{\text{water}}^2}\, M^{4/3}\mathrm{d}M \tag{5}$$

$$\Rightarrow \quad \int_0^{Mv}(Mv)\mathrm{d}(Mv) = \frac{g}{\pi\rho_{\text{fog}}}\sqrt[3]{\frac{16}{9}\pi^2\rho_{\text{water}}^2}\int_{M_0}^{M} M^{4/3}\mathrm{d}M$$

$$\Rightarrow \quad \frac{1}{2}(Mv)^2 = \frac{3}{7}\frac{g}{\pi\rho_{\text{fog}}}\sqrt[3]{\frac{16}{9}\pi^2\rho_{\text{water}}^2}\,(M^{7/3} - M_0^{7/3})$$

$$\Rightarrow \quad v^2 = \frac{6}{7}\frac{g}{\pi\rho_{\text{fog}}}\sqrt[3]{\frac{16}{9}\pi^2\rho_{\text{water}}^2}\,\frac{M^{7/3} - M_0^{7/3}}{M^2}$$

$$\underset{M \gg M_0}{\approx} \frac{6}{7}\frac{g}{\pi\rho_{\text{fog}}}\sqrt[3]{\frac{16}{9}\pi^2\rho_{\text{water}}^2}\, M^{1/3} \tag{6}$$

(2) $\rightarrow$ (6),

$$\Rightarrow \quad v = \sqrt{\frac{8g}{7}\frac{\rho_{\text{water}}}{\rho_{\text{fog}}}r} \tag{7}$$

(2) $\rightarrow$ (3),

$$\Rightarrow \quad v = \frac{4\rho_{\text{water}}}{\rho_{\text{fog}}}\frac{\mathrm{d}r}{\mathrm{d}t} \tag{8}$$

$$\Rightarrow \qquad \frac{\mathrm{d}r}{\sqrt{r}} = \sqrt{\frac{g}{14}\frac{\rho_{\text{fog}}}{\rho_{\text{water}}}}\mathrm{d}t \tag{9}$$

$$\Rightarrow \qquad \int_{r_0}^{r}\frac{\mathrm{d}r}{\sqrt{r}} = \sqrt{\frac{g}{14}\frac{\rho_{\text{fog}}}{\rho_{\text{water}}}}\int_0^t \mathrm{d}t$$

$$\Rightarrow \qquad 2\left(\sqrt{r}-\sqrt{r_0}\right) = \sqrt{\frac{g}{14}\frac{\rho_{\text{fog}}}{\rho_{\text{water}}}}\,t$$

$$\Rightarrow \qquad r \underset{r \gg r_0}{\approx} \frac{\rho_{\text{fog}}}{56\rho_{\text{water}}} g t^2 \tag{10}$$

$(10) \to (7)$,

$$\Rightarrow \qquad v = \frac{g}{7}t \tag{11}$$

Defining a—the $\boldsymbol{i}$-direction component of the acceleration of the raindrop at Time t. Considering the definition of ***acceleration***, then place (11) in to obtain:

$$a = \frac{\mathrm{d}v}{\mathrm{d}t} = \frac{g}{7} \tag{12}$$

That is, the acceleration of raindrop will tend to this stable value after a long enough time.

Solution [2]

Taking *the raindrop at arbitrary Time t* as the object of study. Studying *the state at Time t*. Considering ***Mischelski equation*** (the $\boldsymbol{i}$-direction component equation):

$$Mg = M\frac{\mathrm{d}v}{\mathrm{d}t} + \frac{\mathrm{d}M}{\mathrm{d}t}(v-u) \tag{13}$$

u—the $\boldsymbol{i}$-direction component of the velocity of the small drops of water adsorbed by the raindrop in the instantaneous process at arbitrary Time t.

Known:

$$u = 0 \tag{14}$$

(2) and (14) → (13):

$$\Rightarrow \qquad \frac{dv}{dt}+\frac{3}{r}v\frac{dr}{dt}=g \tag{15}$$

(8) → (15):

$$\Rightarrow \qquad \frac{d^2r}{dt^2}+\frac{3}{r}\left(\frac{dr}{dt}\right)^2=\frac{\rho_{\text{fog}}}{4\rho_{\text{warter}}}g \tag{16}$$

Let the trial solution of Eq. (16) be:

$$r=At^2+Bt+C \tag{17}$$

Considering the initial conditions:

$$\begin{cases} r|_{t=0}=r_0 & (18a) \\ v|_{t=0}=0 & (18b) \end{cases}$$

(8) → (18b),

$$\Rightarrow \qquad \left.\frac{dr}{dt}\right|_{t=0}=0 \tag{19}$$

(17) → (18a) and (19),

$$\Rightarrow \qquad \begin{cases} C=r_0 \\ B=0 \end{cases} \tag{20}$$

$$\Rightarrow \qquad r=At^2+r_0\approx At^2 \tag{21}$$

(21) → (16),

$$\Rightarrow \qquad A=\frac{\rho_{\text{fog}}}{56\rho_{\text{water}}}g \tag{22}$$

$$\Rightarrow \qquad r=\frac{\rho_{\text{fog}}}{56\rho_{\text{water}}}gt^2 \tag{23}$$

This result is the same as Eq. (10).

Chapter 3

Rigid Body

3.1 The Fixed-Axis Rotation of Rigid Body

The motion of an arbitrary point on a rigid body can be described jointly by two parts: (1) the translation of an arbitrarily selected base point on the rigid body; and (2) the rotation of the studied point relative to the base point. For the simple principle of mathematics, the center of mass or the instantaneous center (if it exists) of the rigid body is the preferred object of the base point. The dynamic information of the translation and the rotation of a rigid body can be described by theorem of motion of center of mass and Euler dynamic equation, respectively.

The rotation of rigid body about a base point takes the fixed-point rotation as the general form, the fixed-axis rotation is the most typical degradation of the fixed-point rotation, its Euler dynamics equation degenerates into a linear form (generally known as Newton's second law for rotation).

This section will focus on some simple rigid bodies (such as homogeneous rod, homogeneous cylinder, homogeneous sphere, etc.), aiming at the dynamic problems involving the fixed-axis rotation of rigid body.

Solving Physics Problems: Exploring New Thinking Paradigms
Yipeng Gu

ISBN 978-981-4877-41-1 (Hardcover), 978-1-003-16254-4 (eBook)
www.jennystanford.com

3.1.1 The Rotating of a Homogeneous Rod around the Rough Edge of a Table (d, *B*)

As shown in Fig. 1, place the left segment of a homogeneous rod of length L on a horizontal table, and hold the right end of the rod so that it is perpendicular to the table edge. The center of mass of the rod is denoted by C, the contact point between the rod and the table edge is denoted by O. Given $OC = r$, the coefficient of static friction between the rod and the table edge is μ. Release the right end of the rod. Find the position that the rod begins to slide relative to the table edge.

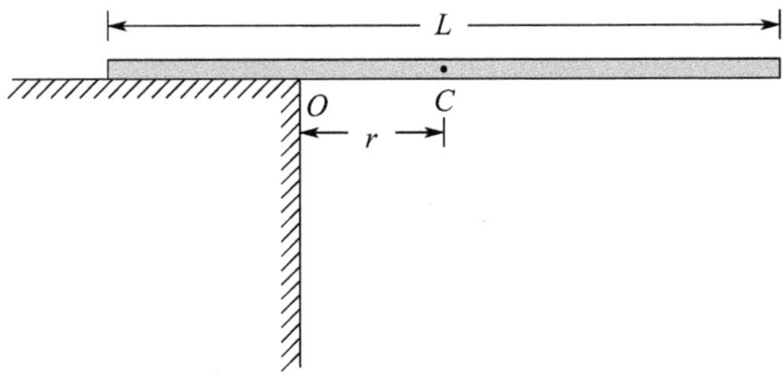

Figure 1

Solution

The time point when the right end of the rod is released is defined as Time zero.

As shown in Fig. 2, the unit vector perpendicular to the plane of the figure inward is denoted by $\boldsymbol{k}$. Taking Point O as the pole, the horizontal right direction as the positive direction of polar axis, the right-hand helix direction with regard to the $\boldsymbol{k}$-direction as the positive direction of polar angle, the $\boldsymbol{k}$-direction as the positive direction of the longitudinal axis, establishing the cylindrical coordinate system, let r, θ, z be the cylindrical coordinates.

Taking *the rod* as the object of study. Studying *the process from Time zero to arbitrary Time t*. Considering (a) ***theorem of kinetic energy***, (b) ***the form of kinetic energy of a rigid body performing fixed-axis rotation***, (c) ***the form of moment of inertia of a***

homogeneous rod about one of its central axes, (d) ***parallel-axis theorem***:

$$mgr\sin\theta = E_k \tag{1}$$

$$E_k = \frac{1}{2}I_O\omega^2 \tag{2}$$

$$\begin{cases} I_C = \dfrac{1}{12}mL^2 & (3a) \\ I_O = I_C + mr^2 & (3b) \end{cases}$$

m—the mass of the rod.

I_O—the moment of inertia of the rod about the z-axis.

I_C—the moment of inertia of the rod about its central axis parallel to the z-axis.

ω—the $\boldsymbol{k}$-direction component of the rotating angular velocity of the rod at Time t.

E_k—the kinetic energy of the rod at Time t.

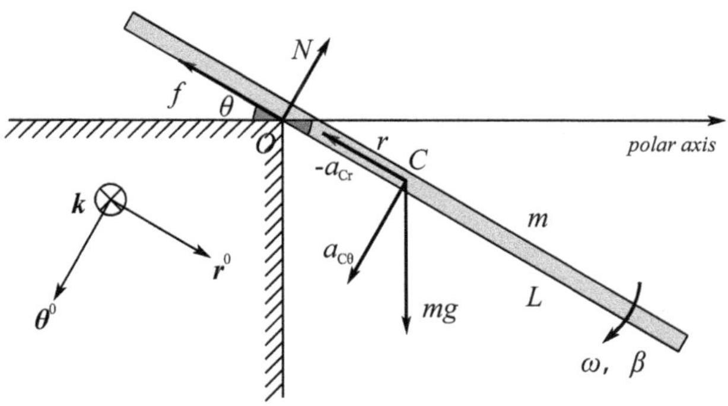

Figure 2

Combining (1), (2) and (3),

$$\Rightarrow \quad mgr\sin\theta = \frac{1}{2}\left(\frac{1}{12}mL^2 + mr^2\right)\omega^2$$

$$\Rightarrow \quad \omega = \sqrt{\frac{24r}{L^2+12r^2}g\sin\theta} \tag{4}$$

Taking *the rod* as the object of study. Studying *the state at arbitrary Time t*. Considering (a) ***theorem of motion of center of mass*** (the $-\boldsymbol{r}^0$-direction and $\boldsymbol{\theta}^0$-direction component equations), (b) ***law of rotation*** [taking the z-axis as the rotation axis] (the $\boldsymbol{k}$-direction component equation):

$$\begin{cases}\begin{cases} f - mg\sin\theta = m(-a_{Cr}) & (5a) \\ mg\cos\theta - N = ma_{C\theta} & (5b) \end{cases} \\ mgr\cos\theta = I_O\beta & (5c) \end{cases}$$

N and f—the $-\boldsymbol{\theta}^0$-direction component of the normal contact force exerted on the rod by the table edge and the $-\boldsymbol{r}^0$-direction component of the static friction exerted on the rod by the table edge at Time t.

a_{Cr} and $a_{C\theta}$—the $\boldsymbol{r}^0$-direction and $\boldsymbol{\theta}^0$-direction component of the center-of-mass acceleration of the rod at Time t.

β—the $\boldsymbol{k}$-direction component of the rotating angular acceleration of the rod at Time t.

Considering ***the components form of the acceleration of a particle in polar coordinate system*** [the constraint condition: the length of the rod remains unchanged, and there is no relative sliding between the rod and the table edge]:

$$\begin{cases} a_{Cr} = -\omega^2 r & (6a) \\ a_{C\theta} = \beta r & (6b) \end{cases}$$

Combining (4), (5) and (6),

$$\Rightarrow \begin{cases} N = \dfrac{L^2}{L^2 + 12r^2} mg\cos\theta & (7a) \\ f = \dfrac{L^2 + 36r^2}{L^2 + 12r^2} mg\sin\theta & (7b) \end{cases}$$

Considering ***friction law***:

$$f \le \mu N \quad (8)$$

(7) → (8),

$$\Rightarrow \quad \tan\theta \le \frac{\mu L^2}{L^2 + 36r^2} \quad (9)$$

$$\Rightarrow \qquad \theta_{\max} = \arctan\left(\frac{\mu L^2}{L^2 + 36r^2}\right) \tag{10}$$

$\theta_{\max}$—the polar angle of the rod when it begins to slide relative to the table edge.

Another solution to E_k

Taking *the rod* as the object of study. Studying *the state at arbitrary Time t*. Considering (a) ***König theorem***, (b) the definition of ***translational kinetic energy***, (c) ***the form of kinetic energy of a rigid body performing fixed-axis rotation***:

$$E_k = E_{kC} + E_k' \tag{11}$$

$$\begin{cases} E_{Ck} = \dfrac{1}{2}m(v_{Cr}^2 + v_{C\theta}^2) & (12a) \\ E_k' = \dfrac{1}{2}I_C\omega^2 & (12b) \end{cases}$$

E_{kC} and E'_k—the translational kinetic energy of the rod and the kinetic energy of the rod in its center-of-mass reference frame at Time *t*.

v_{Cr} and $v_{C\theta}$—the $\boldsymbol{r}^0$-direction and $\boldsymbol{\theta}^0$-direction component of the center-of-mass velocity of the rod at Time *t*.

Considering ***the components form of the velocity of a particle in polar coordinate system*** [the constraint condition: the length of the rod remains unchanged, and there is no relative sliding between the rod and the table edge]:

$$\begin{cases} v_{Cr} = 0 & (13a) \\ v_{C\theta} = \omega r & (13b) \end{cases}$$

E_k can obtained by combining (11) to (13).

Another representation of law of rotation

Taking *the rod* as the object of study. Studying *the state at arbitrary Time t*. Considering (a) ***law of rotation*** [in its center-of-mass reference frame, taking its central axis parallel to the *z*-axis as the rotation axis] (the $\boldsymbol{k}$-direction component equation), (b) ***in the center-of-mass reference frame, the resultant torque of the inertia force acting on system of particles is zero***:

$$Nr = I_C\beta \tag{14}$$

3.1.2 The Crawling of a Beetle Along a Rotating Homogeneous Rod (d, *AC*)

As shown in Fig. 1, a homogeneous rod of length l can rotate freely around its central axis fixed in the horizontal direction (perpendicular to the plane of the figure). A beetle with the same mass as the rod falls on the rod with the distance of 1/4 of the rod length to its right end, with the impact velocity of the magnitude v_0 and the direction vertical downward. The beetle immediately crawls toward the right end of the rod after falling on it. In order to keep the rotating angular velocity of the rod constant, how should the beetle crawl? (Determine the kinematics equation and trajectory equation of the beetle.)

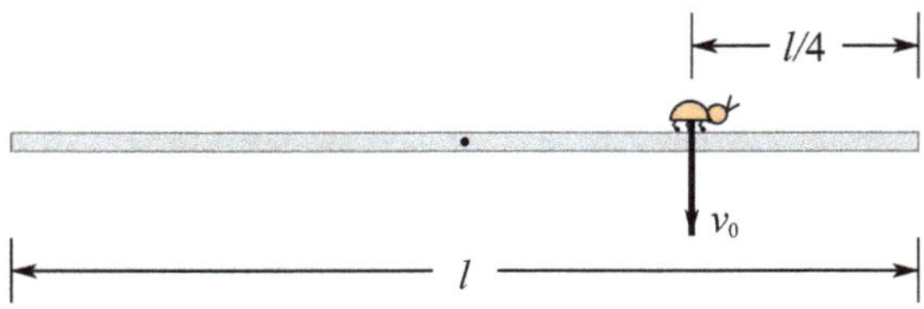

Figure 1

Solution [1] (type C)

The time point after the beetle collides with the rod is defined as Time zero.

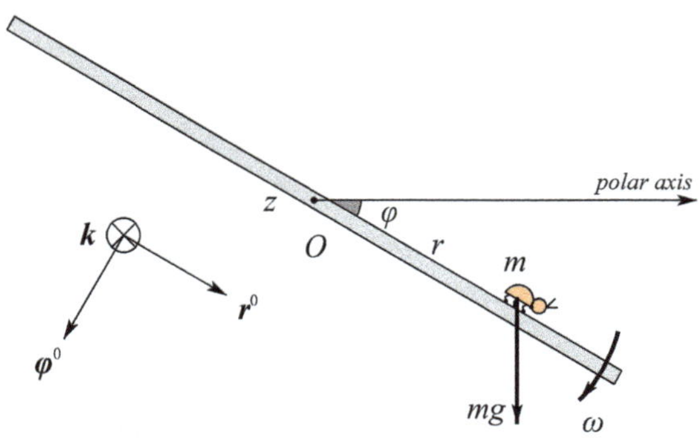

Figure 2

As shown in Fig. 2, the center of the rod is denoted by O, the unit vector perpendicular to the plane of the figure inward is denoted by $\boldsymbol{k}$. Taking Point O as the pole, the horizontal right direction as the positive direction of polar axis, the right-hand helix direction with regard to the $\boldsymbol{k}$-direction as the positive direction of polar angle, the $\boldsymbol{k}$-direction as the positive direction of the longitudinal axis, establishing the cylindrical coordinate system, let r, φ, z be the cylindrical coordinates.

Taking *the system composed of the beetle and the rod* as the object of study. Studying *the instantaneous process of the collision between the two*. Considering (a) ***law of conservation of angular momentum*** [taking the z-axis as the axis of rotation, only the beetle's gravity provides the moment of impulse, the beetle's gravity is a finite quantity, acting on an extremely short process, its moment of impulse is an infinitely small quantity, which can be ignored; therefore the angular momentum of the system is approximately conserved] (the $\boldsymbol{k}$-direction component equation), (b) ***the form of angular momentum of a rigid body performing fixed-axis rotation*** [integrating into (a)], (c) ***the transverse component form of velocity*** [integrating into (a)], (d) ***the form of moment of inertia of a homogeneous rod about one of its central axes***:

$$mv_0\frac{l}{4}=I_0\omega+m\left(\omega\frac{l}{4}\right)\frac{l}{4} \tag{1}$$

$$I_0=\frac{1}{12}ml^2 \tag{2}$$

m—the mass of the beetle and the rod.

ω—the $\boldsymbol{k}$-direction component of the rotating angular velocity of the rod at Time zero.

I_0—the moment of inertia of the rod about the z-axis.

$$\Rightarrow \qquad \omega=\frac{12v_0}{7l} \tag{3}$$

Known that the rotating angular velocity of the rod remains constant, considering ***the angular displacement-time relation of uniform rotation***:

$$\varphi=\omega t \tag{4}$$

Taking *the system composed of the beetle and the rod* as the object of study. Studying *the instantaneous process (the small time interval dt) at arbitrary Time t.* Considering (a) ***theorem of angular momentum*** [taking the z-axis as the axis of rotation] (the $\boldsymbol{k}$-direction component equation), (b) ***the form of angular momentum of a rigid body performing fixed-axis rotation*** [integrating into (a)], (c) ***the transverse component form of velocity*** [integrating into (a)]:

$$mgr\cos\varphi\cdot \mathrm{d}t = \mathrm{d}[I_0\omega + m(\omega r)r] \tag{5}$$

(2) → (5):

$$\Rightarrow \qquad \frac{\mathrm{d}r}{\mathrm{d}t} = \frac{g}{2\omega}\cos\varphi \tag{6}$$

(4) → (6),

$$\Rightarrow \qquad \frac{\mathrm{d}r}{\mathrm{d}t} = \frac{g}{2\omega}\cos(\omega t)$$

$$\Rightarrow \qquad \mathrm{d}r = \frac{g}{2\omega}\cos(\omega t)\mathrm{d}t$$

$$\Rightarrow \qquad \int_{\frac{l}{4}}^{r}\mathrm{d}r = \frac{g}{2\omega}\int_0^t\cos(\omega t)\mathrm{d}t$$

$$\Rightarrow \qquad r = \frac{l}{4} + \frac{g}{2\omega^2}\sin\omega t \tag{7}$$

Equations (4) and (7) are the kinematics equations of the beetle.

Combining (4) and (7), we obtain the trajectory equation of the beetle:

$$\Rightarrow \qquad r = \frac{l}{4} + \frac{g}{2\omega^2}\sin\varphi \tag{8}$$

Solution [2] (type A)

Taking *the beetle* as the object of study (see Fig. 3). Studying *the state at arbitrary Time t.* Considering ***Newton's second law*** (the $\boldsymbol{\varphi}^0$-direction component equation):

$$mg\cos\varphi - F_{21\varphi} = ma_\varphi \tag{9}$$

$F_{21\varphi}$—the $-\boldsymbol{\varphi}^0$-direction component of the force exerted on the beetle by the rod at Time t.

a_φ—the $\boldsymbol{\varphi}^0$-direction component of the acceleration of the beetle at Time t.

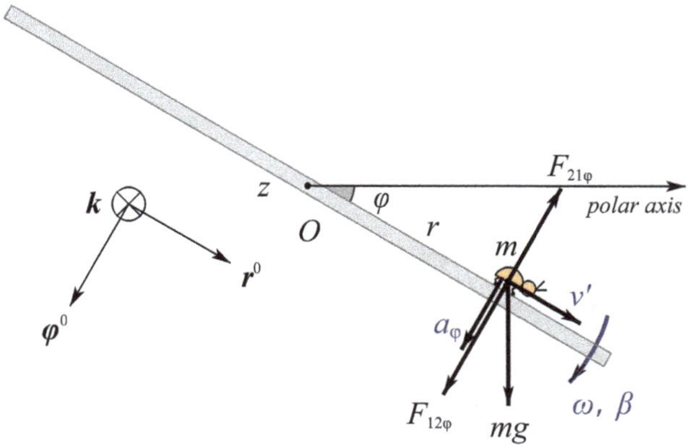

Figure 3

Taking *the rod* as the object of study. Studying *the state at arbitrary Time t*. Considering ***law of rotation*** [taking the z-axis as the rotation axis] (the $\boldsymbol{k}$-direction component equation):

$$F_{12\varphi}r = I_0\beta \tag{10}$$

$F_{12\varphi}$—the $\boldsymbol{\varphi}^0$-direction component of the force exerted on the rod by the beetle at Time t.

β—the $\boldsymbol{k}$-direction component of the angular acceleration of the rod at Time t.

Considering ***Newton's third law***:

$$F_{21\varphi} = F_{12\varphi} \tag{11}$$

The condition for the rotating angular velocity of the rod to be constant is as follows:

$$\beta = 0 \tag{12}$$

Combining (10) to (12),

$$\Rightarrow \quad F_{21\varphi} = 0 \tag{13}$$

Considering ***the transverse component form of acceleration*** and ***the radial component form of velocity***:

$$a_\varphi = 2v_r\omega = 2\frac{\mathrm{d}r}{\mathrm{d}t}\omega \tag{14}$$

v_r—the $\boldsymbol{r}^0$-direction component of the velocity of the beetle at arbitrary Time t.

(13) and (14) $\rightarrow$ (9),

$$\Rightarrow \qquad \frac{\mathrm{d}r}{\mathrm{d}t}=\frac{g}{2\omega}\cos\varphi \tag{15}$$

This result is the same as Eq. (6).

3.1.3 The Most Easily Broken Point of a Rotating Homogeneous Rod (d, *C*)

As shown in Fig. 1, a homogeneous rod can rotate freely around a fixed rotating shaft (perpendicular to the plane of the figure) on a ground. Initially, the rod is at rest in the vertical direction. Give the rod a weak disturbance to make it fall to the right.

(1) Find the angular velocity and angular acceleration of the rod when it rotates to an arbitrary position.
(2) Find the elastic force in the rod at an arbitrary point when it is rotates to an arbitrary position.
(3) Find the most easily broken point of the rod in its falling process.

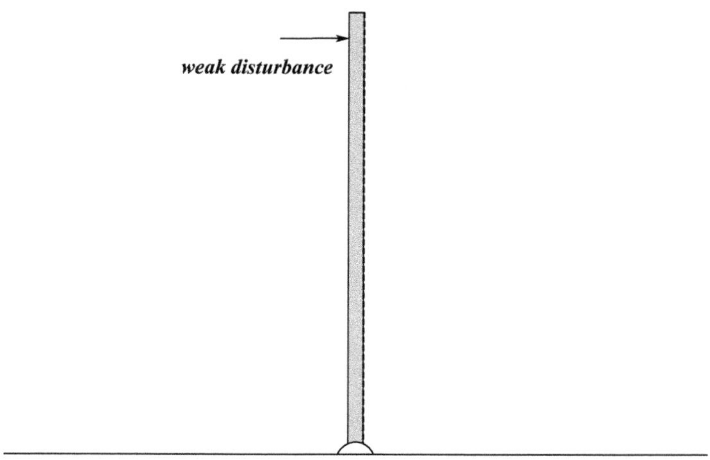

Figure 1

Solution to Problem (1)

The time point when the rod starts to rotate is defined as Time zero.

As shown in Fig. 2, the position of the fixed rotating shaft is denoted by O, the unit vector perpendicular to the plane of the figure inward is denoted by $\boldsymbol{k}$. Taking Point O as the pole, the vertical upward direction as the positive direction of polar axis, the right-hand helix direction with regard to the $\boldsymbol{k}$-direction as the positive direction of polar angle, the $\boldsymbol{k}$-direction as the positive direction of the longitudinal axis, establishing the cylindrical coordinate system, let r, θ, z be the cylindrical coordinates.

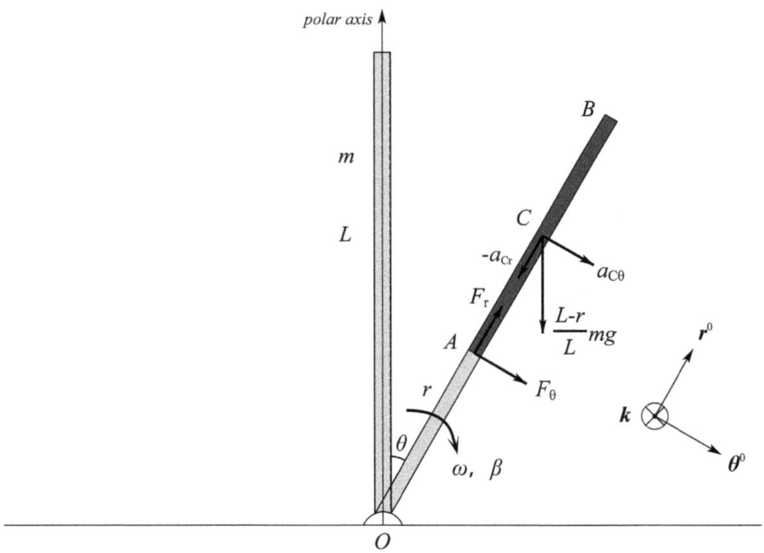

Figure 2

Taking *the system composed of the rod and the earth* as the object of study. Studying *the process from Time zero to arbitrary Time t*. Considering (a) ***law of conservation of mechanical energy*** [the work done by the external force and non-conservative internal force acting on the system is zero; therefore the mechanical energy of the system is conserved] (taking the ground as the zero point of gravitational potential energy), (b) ***the form of kinetic energy of a rigid body performing fixed-axis rotation*** [integrating into (a)], (c) ***the form of moment of inertia of a homogeneous rod about one of its perpendicular axes passing through its endpoint***:

$$mg\frac{L}{2}=\frac{1}{2}I_O\omega^2+mg\frac{L}{2}\cos\theta \tag{1}$$

$$I_O = \frac{1}{3}mL^2 \tag{2}$$

m and L—the mass and length of the rod.

I_O—the moment of inertia of the rod about the z-axis.

ω—the $\boldsymbol{k}$-direction component of the rotating angular velocity of the rod at Time t.

$$\Rightarrow \qquad mg\frac{L}{2} = \frac{1}{2}\cdot\frac{1}{3}mL^2\cdot\omega^2 + mg\frac{L}{2}\cos\theta$$

$$\Rightarrow \qquad \omega = \sqrt{3\frac{g}{L}(1-\cos\theta)} \tag{3}$$

Taking *the rod* as the object of study. Studying *the state at arbitrary Time t*. Considering ***law of rotation*** [taking the z-axis as the rotation axis] (the $\boldsymbol{k}$-direction component equation):

$$mg\frac{L}{2}\sin\theta = I_O\beta \tag{4}$$

β—the $\boldsymbol{k}$-direction component of the rotating angular acceleration of the rod at Time t.

(2) → (4),

$$\Rightarrow \qquad \beta = \frac{3}{2}\frac{g}{L}\sin\theta \tag{5}$$

Solution to Problem (2)

Taking an arbitrary point on the rod and denoting it by A, the free end of the rod is denoted by B, the segment of the rod from Point A to Point B is denoted by segment AB, whose center of mass is denoted by C.

Taking *segment AB* as the object of study. Studying *the state at arbitrary Time t*. Considering ***theorem of motion of center of mass*** (the $-\boldsymbol{r}^0$-direction and $\boldsymbol{\theta}^0$-direction component equations):

$$\begin{cases} \dfrac{L-r}{L}mg\cos\theta - F_r = \dfrac{L-r}{L}m(-a_{Cr}) & \text{(6a)} \\[2ex] \dfrac{L-r}{L}mg\sin\theta + F_\theta = \dfrac{L-r}{L}ma_{C\theta} & \text{(6b)} \end{cases}$$

F_r and F_θ—the $\boldsymbol{r}^0$-direction and $\boldsymbol{\theta}^0$-direction components of the elastic force at Point A acting on segment AB at Time t.

a_{Cr} and $a_{C\theta}$—the $\boldsymbol{r}^0$-direction and $\boldsymbol{\theta}^0$-direction component of the center-of-mass acceleration of segment AB at Time t.

Considering ***the components form of the acceleration of a particle in polar coordinate system*** (the constraint condition: the rod length remains unchanged):

$$\begin{cases} a_{Cr} = -\omega^2 \dfrac{L+r}{2} & (7\text{a}) \\ a_{C\theta} = \beta \dfrac{L+r}{2} & (7\text{b}) \end{cases}$$

[(3) → (7a), (5) → (7b)] → (6):

$$\Rightarrow \quad \begin{cases} F_r = \dfrac{L-r}{L} mg\left[\cos\theta - \dfrac{3}{2}\dfrac{L+r}{L}(1-\cos\theta)\right] & (8\text{a}) \\ F_\theta = \dfrac{1}{4}\dfrac{(L-r)(3r-L)}{L^2} mg\sin\theta & (8\text{b}) \end{cases}$$

$$\Rightarrow \quad F_\theta \begin{cases} >0, & \text{when } r > \dfrac{L}{3} \\ =0, & \text{when } r = \dfrac{L}{3} \\ <0, & \text{when } r < \dfrac{L}{3} \end{cases} \qquad (9)$$

Solution to Problem (3)

Taking *segment AB* as the object of study. Studying *the state at arbitrary Time t*. Considering (a) ***law of rotation*** [in its center-of-mass reference frame, taking its central axis parallel to the z-axis as the rotation axis] (the $\boldsymbol{k}$-direction component equation), (b) ***in the center-of-mass reference frame, the resultant torque of the inertia force acting on system of particles is zero***, (c) ***the form of moment of inertia of a homogeneous rod about one of its central axes***:

$$-F_\theta \cdot \frac{1}{2}(L-r) + M = I_C\beta \qquad (10)$$

$$I_C = \frac{1}{12} \cdot \frac{L-r}{L} m \cdot (L-r)^2 \tag{11}$$

M—at arbitrary Time t, the $\boldsymbol{k}$-direction component of the torque acting on segment AB due to the non-uniform distribution of F_r [taking its central axis parallel to the z-axis as the rotation axis].

I_C—the moment of inertia of segment AB about its central axis parallel to the z-axis.

(5), (8b), (11) $\rightarrow$ (10),

$$\Rightarrow \qquad M = \frac{1}{4} \frac{r(L-r)^2}{L^2} mg \sin\theta \tag{12}$$

The point where M takes the maximum value is the point where the rod is most easily broken.

$$\Rightarrow \qquad \frac{\mathrm{d}M}{\mathrm{d}r} = \frac{1}{4} \frac{3r^2 - 4Lr + L^2}{L^2} mg \sin\theta \tag{13}$$

$$\Rightarrow \qquad \frac{\mathrm{d}^2 M}{\mathrm{d}r^2} = \frac{1}{4} \frac{6r - 4L}{L^2} mg \sin\theta \tag{14}$$

The necessary condition for M to take the maximum value is as follows:

$$\frac{\mathrm{d}M}{\mathrm{d}r} = 0 \tag{15}$$

$$\Rightarrow \qquad r = \frac{L}{3} \tag{16}$$

$$\Rightarrow \qquad \left. \frac{\mathrm{d}^2 M}{\mathrm{d}r^2} \right|_{r=L/3} = -\frac{1}{2L} mg \sin\theta < 0 \tag{17}$$

$$\Rightarrow \qquad M_{\max} = M\big|_{r=\frac{L}{3}} = \frac{1}{27} mgL \sin\theta \tag{18}$$

The existence of torque M originates from the non-uniform distribution of F_r on the cross section of the rod. It can be seen from Eq. (12) that M is always positive. According to this conclusion, we can draw the distribution schematic diagram of F_r on an arbitrary cross section (see Fig. 3):

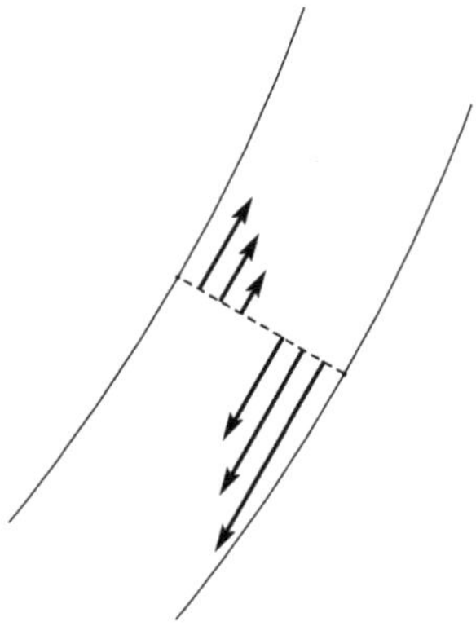

Figure 3

3.1.4 The Translation and Rotation of a Homogeneous Rod in the Process of Toppling toward the Ground (*b*, *AC*)

A homogeneous rod of mass m and length l is statically placed on a smooth horizontal ground in a vertical direction (see Fig. 1), apply a force of the magnitude F and the direction horizontally to the right on its lower end.

(1) Find the normal contact force exerted on the lower end of the rod by the ground at arbitrary time before the rod separates from the ground (the angle between the rod and the ground can be used as the parameter variable).

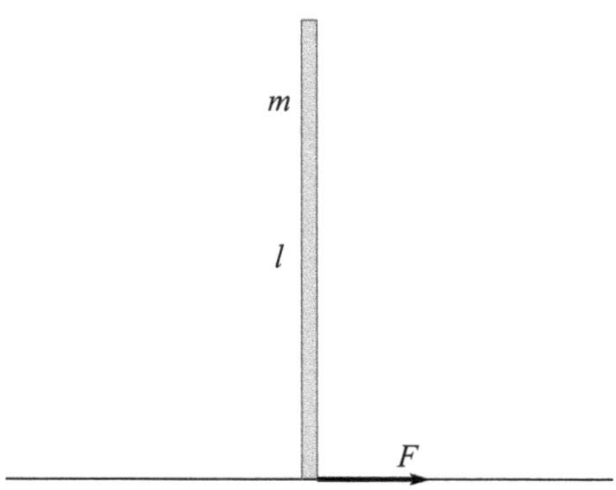

Figure 1

(2) Find the differential equation of the angle between the rod and the ground with time.

Solution to Problem (1) (type C)

As shown in Fig. 2, the center of mass of the rod is denoted by C, the lower end of the rod is denoted by P, the point on the ground that coincides with the lower end of the rod at Time zero is denoted by O. Establishing the Cartesian coordinate system with the frame (O, $\boldsymbol{i}$, $\boldsymbol{j}$, $\boldsymbol{k}$), let x, y, z be the Cartesian coordinates.

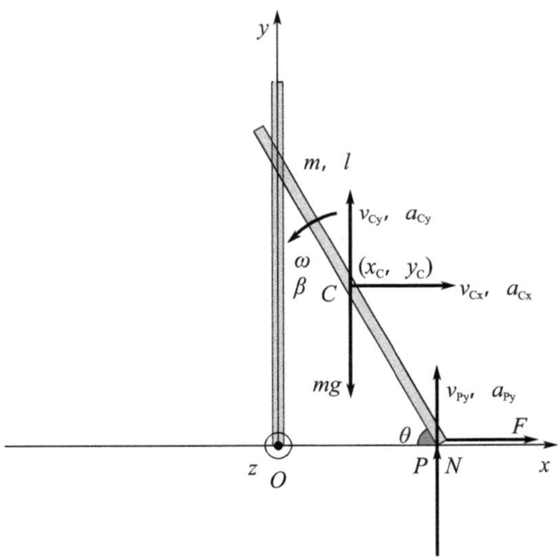

Figure 2

Taking *the rod* as the object of study. Studying *the process from Time zero to arbitrary Time t before it separates from the ground.* Considering (a) ***theorem of kinetic energy***, (b) ***König theorem*** [integrating into (a)], (c) the definition of ***translational kinetic energy***, (d) ***the form of kinetic energy of a rigid body performing fixed-axis rotation***, (e) ***the form of moment of inertia of a homogeneous rod about one of its central axes***:

$$mg \cdot \frac{1}{2} l(1-\sin\theta) + Fx_P = E_{kC} + E_k' \tag{1}$$

$$\begin{cases} E_{kC} = \dfrac{1}{2} m(v_{Cx}^2 + v_{Cy}^2) & \text{(2a)} \\ E_k' = \dfrac{1}{2} I\omega^2 & \text{(2b)} \end{cases}$$

$$I = \frac{1}{12} ml^2 \tag{3}$$

θ—the angle between the rod and the ground at Time t.

E_{kC} and E'_k—the translational kinetic energy of the rod and the kinetic energy of the rod in its center-of-mass reference frame at Time t.

v_{Cx} and v_{Cy}—the ***i***-direction and ***j***-direction velocity component of the center of mass C at Time t.

ω—the ***k***-direction component of the angular velocity of the rod at Time t.

I—the moment of inertia of the rod about its central axis parallel to the z-axis.

Combining (1), (2) and (3),

$$\Rightarrow \quad mg \cdot \frac{1}{2} l(1 - \sin\theta) + Fx_P = \frac{1}{2} m(v_{Cx}^2 + v_{Cy}^2) + \frac{1}{2} \cdot \frac{1}{12} ml^2 \omega^2 \tag{4}$$

Taking *the rod* as the object of study. Studying *the state at arbitrary Time t*. Considering (a) ***theorem of motion of center of mass*** (the component equations in the (***i***, ***j***)-directions), (b) ***law of rotation*** [in its center-of-mass reference frame, taking its central axis parallel to the z-axis as the rotation axis] (the ***k***-direction component equation), (c) ***in the center-of-mass reference frame, the resultant torque of the inertia force acting on system of particles is zero***:

$$\begin{cases} \begin{cases} F = ma_{Cx} & \text{(5a)} \\ N - mg = ma_{Cy} & \text{(5b)} \end{cases} \\ N \cdot \frac{l}{2} \cos\theta + F \cdot \frac{l}{2} \sin\theta = I\beta & \text{(6)} \end{cases}$$

N—the ***j***-direction component of the normal contact force exerted on the lower end of the rod by the ground at Time t.

a_{Cx} and a_{Cy}—the ***i***-direction and ***j***-direction acceleration component of the center of mass C at Time t.

β—the ***k***-direction component of the angular acceleration of the rod at Time t.

Taking *the center of mass C* as the object of study. Studying *the state at arbitrary Time t*. Considering ***theorem for composition***

of velocities of a particle and theorem for the composition of accelerations of a particle [taking the ground as the stationary reference frame, and the translational reference frame that is at rest with respect to Point P as the moving reference frame] (the ***j***-direction component equation):

$$\begin{cases} v_{Cy} = v_{Py} + v_{CPy} & (7a) \\ a_{Cy} = a_{Py} + a_{CPy} & (7b) \end{cases}$$

v_{Cy} and a_{Cy}—the ***j***-direction component of the velocity and acceleration of the center of mass C at Time t.

v_{Py} and a_{Py}—the ***j***-direction component of the velocity and acceleration of Point P at Time t.

v_{CPy} and a_{CPy}—the ***j***-direction component of the velocity and acceleration of the center of mass C relative to Point P at Time t.

Considering the constraint condition: the lower end of the rod is in contact with the ground all the time before it separates from the ground:

$$\begin{cases} v_{Py} = 0 & (8a) \\ a_{Py} = 0 & (8b) \end{cases}$$

Considering ***the components form of the velocity and acceleration of a particle in polar coordinate system*** [the constraint condition: the rod length remains unchanged]:

$$\begin{cases} v_{CPy} = -\omega \dfrac{l}{2} \cos\theta & (9a) \\ a_{CPy} = -\left(\omega^2 \dfrac{l}{2} \sin\theta + \beta \dfrac{l}{2} \cos\theta \right) & (9b) \end{cases}$$

(8) and (9) $\rightarrow$ (7),

$$\Rightarrow \begin{cases} v_{Cy} = -\omega \dfrac{l}{2} \cos\theta & (10a) \\ a_{Cy} = -\left(\omega^2 \dfrac{l}{2} \sin\theta + \beta \dfrac{l}{2} \cos\theta \right) & (10b) \end{cases}$$

Considering the following geometric relations in Fig. 2:

$$\begin{cases} x_C = x_P - \dfrac{l}{2}\cos\theta & \text{(11a)} \\ y_C = \dfrac{l}{2}\sin\theta & \text{(11b)} \end{cases}$$

Considering ***the velocity-displacement relation of uniformly variable motion***:

$$v_{Cx}^2 = 2a_{Cx}x_C \tag{12}$$

Combining (3) to (6), (10), (11a) and (12),

$$\Rightarrow \quad N = \frac{mg[(1+3\cos^2\theta) - 6(1-\sin\theta)\sin\theta] - 9F(1+\cos^2\theta)\sin\theta\cos\theta}{(1+3\cos^2\theta)^2} \tag{13}$$

Solution to Problem (2) (type A)

Considering the definition of ***acceleration*** and ***angular acceleration***:

$$\begin{cases} \begin{cases} a_{Cx} = \dfrac{\mathrm{d}^2 x_C}{\mathrm{d}t^2} & \text{(14a)} \\ a_{Cy} = \dfrac{\mathrm{d}^2 y_C}{\mathrm{d}t^2} & \text{(14b)} \end{cases} \\ \beta = -\dfrac{\mathrm{d}^2\theta}{\mathrm{d}t^2} & \text{(15)} \end{cases}$$

(14) $\rightarrow$ (5), (15) $\rightarrow$ (6),

$$\Rightarrow \quad \begin{cases} \begin{cases} F = m\dfrac{\mathrm{d}^2 x_C}{\mathrm{d}t^2} & \text{(16a)} \\ N - mg = m\dfrac{\mathrm{d}^2 y_C}{\mathrm{d}t^2} & \text{(16b)} \end{cases} \\ N\cdot\dfrac{l}{2}\cos\theta + F\cdot\dfrac{l}{2}\sin\theta = -I\dfrac{\mathrm{d}^2\theta}{\mathrm{d}t^2} & \text{(17)} \end{cases}$$

Combining (3), (11), (16) and (17), we obtain the differential equation of the angle between the rod and the ground with time:

$$\Rightarrow \quad (1+3\cos^2\theta)\frac{\mathrm{d}^2\theta}{\mathrm{d}t^2} - 3\sin\theta\cos\theta\left(\frac{\mathrm{d}\theta}{\mathrm{d}t}\right)^2 + 6\left(\frac{g}{l}\cos\theta + \frac{F}{ml}\sin\theta\right) = 0 \tag{18}$$

3.1.5 The Motion of the System Composed of a Homogeneous Rod and a Charged Small Ring Sleeved on It in a Magnetic Field (d, *C*)

As shown in Fig. 1, a homogeneous insulating rod of mass M and length l can rotate freely around its central axis perpendicular to the plane of the figure. A charged small ring of mass m and charge q ($q > 0$) is sleeved on the rod, the ring can slide along the rod without friction. Apply a uniform magnetic field in space, the magnitude of whose magnetic induction intensity is B, and the direction is perpendicular to the plane of the figure inward. Initially, the ring is at rest at the center of the rod, and the rod performs the uniform rotation with the angular velocity of the magnitude ω_0 and the direction perpendicular to the plane of the figure outward. Give the ring a small disturbance along the rod.

(1) Find the condition that the ring can move outward along the rod.

(2) When the condition of Problem (1) holds, determine the condition that the ring will not separate from the rod in the later motion process.

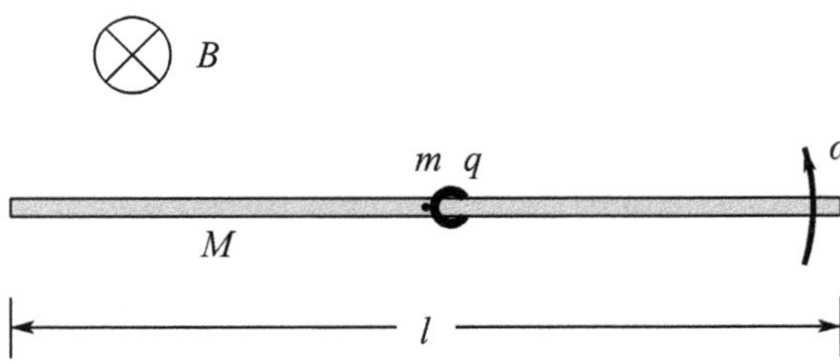

Figure 1

Solution to Problem (1)

The time point after the disturbance is defined as Time zero.

As shown in Fig. 2, the center of mass of the rod is denoted by O, the unit vector perpendicular to the plane of the figure outward is denoted by $\boldsymbol{k}$. Taking Point O as the pole, the horizontal right direction as the positive direction of polar axis, the right-hand helix direction with regard to the $\boldsymbol{k}$-direction as the positive direction of polar angle, the $\boldsymbol{k}$-direction as the positive direction of the longitudinal axis, establishing the cylindrical coordinate system, let r, θ, z be the cylindrical coordinates.

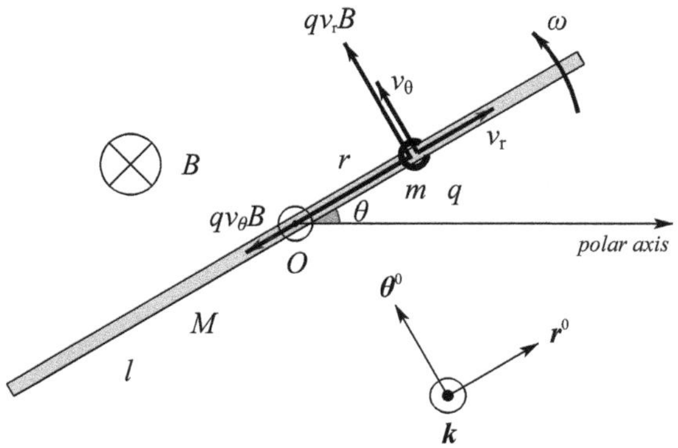

Figure 2

Taking *the ring* as the object of study. Studying *the state at Time zero*. Considering ***Newton's second law*** [in the rod reference frame, the disturbance is quite small; therefore, the rotating angular velocities of the rod at the time point before and after the disturbance are approximately the same] (the $\boldsymbol{r}^0$-direction component equation):

$$m\omega_0^2\Delta r_0 - qv_{\theta 0}B = ma'_{r0} \tag{1}$$

Δr_0—the small polar radius of the ring at Time zero.

$v_{\theta 0}$—the $\boldsymbol{\theta}^0$-direction component of the velocity of the ring at Time zero.

a'_{r0}—the $\boldsymbol{r}^0$-direction component of the acceleration of the ring in the rod reference frame at Time zero.

Considering ***the transverse component form of velocity***:

$$v_{\theta 0} = \omega_0 \Delta r_0 \tag{2}$$

The condition that the ring can move outward along the rod is as follows:

$$a'_{r0} > 0 \tag{3}$$

(2) and (3) → (1):

$$\Rightarrow \qquad \omega_0 > \frac{qB}{m} \tag{4}$$

Solution to Problem (2)

It is assumed that the ring will not separate from the rod in the later motion process, that is, the small ring will move to a point farthest from Point O on the rod.

Taking *the system composed of the ring and the rod* as the object of study. Studying *the process from Time zero to the Time when the distance between the ring and the pole O is the farthest.* Considering (a) ***law of conservation of mechanical energy*** [the work done by the external force and non-conservative internal force acting on the system is zero; therefore, the mechanical energy of the system is conserved], (b) ***theorem of angular momentum*** [taking the z-axis as the axis of rotation] (the $\boldsymbol{k}$-direction component equation), (c) ***the form of kinetic energy of a rigid body performing fixed-axis rotation*** [integrating into (a)], (d) ***the form of angular momentum of a rigid body performing fixed-axis rotation*** [integrating into (b)]:

$$\begin{cases} \frac{1}{2}I\omega_0^2 = \left.\left[\frac{1}{2}I\omega^2 + \frac{1}{2}m(v_r^2 + v_\theta^2)\right]\right|_{\text{last}} & \text{(5a)} \\ \left.(I\omega + mv_\theta r)\right|_{\text{last}} - I\omega_0 = \int qv_r B \cdot r\text{d}t & \text{(5b)} \end{cases}$$

I—the moment of inertia of the rod about the z-axis.

v_r and v_θ—the $\boldsymbol{r}^0$-direction and $\boldsymbol{\theta}^0$-direction component of the velocity of the ring at Time t.

ω—the $\boldsymbol{k}$-direction component of the rotating angular velocity of the rod at Time t.

Considering ***the form of moment of inertia of a homogeneous rod about one of its central axes***:

$$I = \frac{1}{12}Ml^2 \tag{6}$$

Considering ***the components form of the velocity of a particle in polar coordinate system***:

$$\begin{cases} v_r = \frac{\text{d}r}{\text{d}t} & \text{(7a)} \\ v_\theta = \omega r & \text{(7b)} \end{cases}$$

(6) and (7) → (5),

$$\Rightarrow \begin{cases} \frac{1}{2}\cdot\frac{1}{12}Ml^2\omega_0^2 = \frac{1}{2}\cdot\frac{1}{12}Ml^2\omega_{\text{last}}^2 + \frac{1}{2}m\left[\left(\frac{\text{d}r}{\text{d}t}\right)\Big|_{\text{last}}^2 + (\omega_{\text{last}}r_{\text{last}})^2\right] & \text{(8a)} \\ \left(\frac{1}{12}Ml^2\omega_{\text{last}} + m\omega_{\text{last}}r_{\text{last}}^2\right) - \frac{1}{12}Ml^2\omega_0 \\ = \int qv_rB\cdot r\text{d}t = \int q\frac{\text{d}r}{\text{d}t}B\cdot r\text{d}t = \int_0^{r_{\text{last}}} qBr\text{d}r = \frac{1}{2}qBr_{\text{last}}^2 & \text{(8b)} \end{cases}$$

The necessary condition for the distance between the ring and the pole O to take the maximum value is as follows:

$$\left(\frac{\text{d}r}{\text{d}t}\right)\Big|_{\text{last}} = 0 \tag{9}$$

Combining (8) and (9),

$$\Rightarrow \begin{cases} r_{\text{last1}} = 0, \omega_{\text{last1}} = \omega_0 & \text{(10a)} \\ r_{\text{last2}} = \sqrt{\frac{1}{3}\frac{M\omega_0}{qB}\left(\frac{m\omega_0}{qB} - 1\right)}l, \omega_{\text{last2}} = \frac{1}{2\frac{m\omega_0}{qB} - 1}\omega_0 & \text{(10b)} \end{cases}$$

$$\Rightarrow \qquad r_{\max} = r_{\text{last2}} = \sqrt{\frac{1}{3}\frac{M\omega_0}{qB}\left(\frac{m\omega_0}{qB} - 1\right)}l \tag{11}$$

The condition that the ring will not separate from the rod in the later motion process is as follows:

$$r_{\max} < \frac{l}{2} \tag{12}$$

$$\Rightarrow \qquad \frac{M\omega_0}{qB}\left(\frac{m\omega_0}{qB} - 1\right) < \frac{3}{4} \tag{13}$$

3.1.6 Two Fingers Moving Alternately from Both Ends of a Rough Homogeneous Rod to the Middle of the Rod (d, *C*)

There is a homogeneous rod of mass m and length l, both ends of which are supported horizontally by one's two fingers. With the mechanical equilibrium of the rod kept all the time, moves the two fingers alternately slowly toward the center of mass of the rod (every time moving a finger, be sure to move to the furthest distance), and finally let the two fingers meet at the center of mass of the rod. The coefficient of static friction and coefficient of sliding friction between the fingers and the rod are μ_0 and μ, respectively. Find the work done by the friction between the fingers and the rod in the whole process.

Solution

The following studies the situation of firstly moving the right finger.

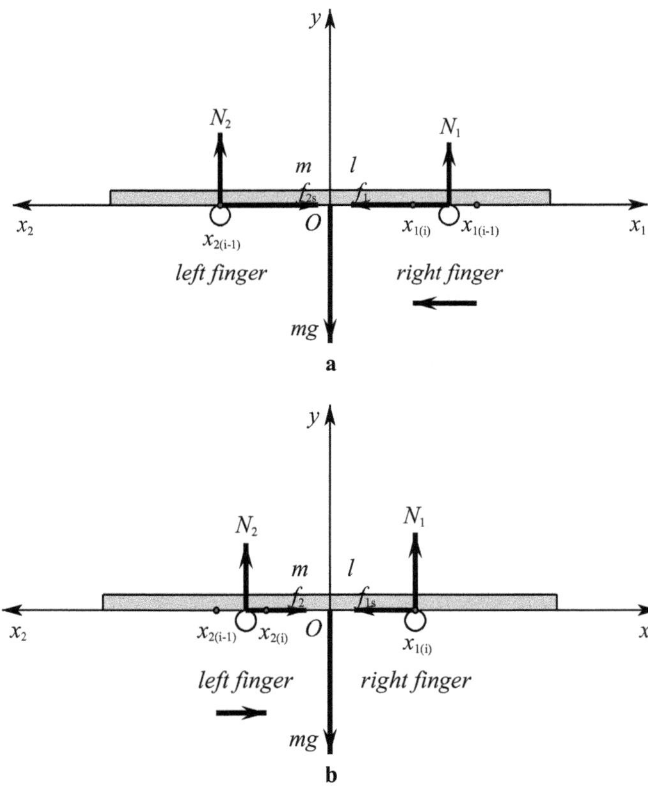

Figure 1a, b

As shown in Fig. 1, the center of mass of the rod is denoted by O, the unit vector parallel the rod to horizontal right is denoted by $\boldsymbol{i}_1$, the unit vector parallel the rod to horizontal left is denoted by

$\boldsymbol{i}_2$, the unit vector in the vertical upward direction is denoted by $\boldsymbol{j}$. Establishing the Cartesian coordinate systems 1 with the frame $(O, \boldsymbol{i}_1, \boldsymbol{j})$, let x_1, y be the Cartesian coordinates, and establishing the Cartesian coordinate systems 2 with the frame $(O, \boldsymbol{i}_2, \boldsymbol{j})$, let x_2, y be the Cartesian coordinates.

Taking *the rod* as the object of study. Studying *the state at arbitrary Time t*. Considering (a) ***theorem of motion of center of mass*** [the rod is in mechanical equilibrium] (the $\boldsymbol{j}$-direction component equation), (b) ***law of rotation*** [the rod is in mechanical equilibrium] [taking its central axis perpendicular to the plane of the figure as the rotation axis] (the anticlockwise direction component equation):

$$\begin{cases} N_1 + N_2 - mg = 0 & \text{(1a)} \\ N_1 x_1 - N_2 x_2 = 0 & \text{(1b)} \end{cases}$$

N_1 and N_2—the $\boldsymbol{j}$-direction component of the normal contact force exerted on the rod by the right finger and the left finger at Time t.

$$\Rightarrow \begin{cases} N_1 = \dfrac{x_2}{x_1 + x_2} mg & \text{(2a)} \\ N_2 = \dfrac{x_1}{x_1 + x_2} mg & \text{(2b)} \end{cases}$$

Defining f_1—the $-\boldsymbol{i}_1$-direction component of the sliding friction force exerted on the rod by the right finger at arbitrary Time t during the arbitrary motion process of the right finger. f_2—the $-\boldsymbol{i}_2$-direction component of the sliding friction force exerted on the rod by the left finger at arbitrary Time t during the arbitrary motion process of the left finger. Considering ***friction law***:

$$\begin{cases} f_1 = \mu N_1 & \text{(3a)} \\ f_2 = \mu N_2 & \text{(3b)} \end{cases}$$

$$\Rightarrow \begin{cases} f_1 = \mu m g \dfrac{x_2}{x_1 + x_2} & \text{(4a)} \\ f_2 = \mu m g \dfrac{x_1}{x_1 + x_2} & \text{(4b)} \end{cases}$$

Defining f_{1s}—the $-\boldsymbol{i}_1$-direction component of the static friction force exerted on the rod by the right finger at arbitrary Time t during the arbitrary motion process of the left finger. f_{2s}—the

$-\boldsymbol{i}_2$-direction component of the static friction force exerted on the rod by the left finger at arbitrary Time t during the arbitrary motion process of the right finger. Considering ***friction law***:

$$\begin{cases} f_{1s} \le \mu_0 N_1 & (5a) \\ f_{2s} \le \mu_0 N_2 & (5b) \end{cases}$$

$$\Rightarrow \begin{cases} f_{1s} \le \mu_0 mg \dfrac{x_2}{x_1 + x_2} & (6a) \\ f_{2s} \le \mu_0 mg \dfrac{x_1}{x_1 + x_2} & (6b) \end{cases}$$

Taking *the rod* as the object of study. Studying (A) *the state at arbitrary Time t in the process of the i-th motion of the right finger*. Considering **theorem of motion of center of mass** [the rod is in mechanical equilibrium] (the $\boldsymbol{i}_1$-direction component equation). (B) *the state at arbitrary Time t in the process of the i-th motion of the left finger*. Considering ***theorem of motion of center of mass*** [the rod is in mechanical equilibrium] (the $\boldsymbol{i}_2$-direction component equation):

$$\begin{cases} f_{2s}\big|_{x_2 = x_{2(i-1)}} - f_1\big|_{x_2 = x_{2(i-1)}} = 0 & (7a) \\ f_{1s}\big|_{x_1 = x_{1i}} - f_2\big|_{x_1 = x_{1i}} = 0 & (7b) \end{cases}$$

$x_{1(i)}$—the coordinate of the right finger on the x_1-axis at the time point when it just completes its i-th motion.

$x_{2(i)}$—the coordinate of the left finger on the x_2-axis at the time point when it just completes its i-th motion.

(4) and (6) → (7), we obtain the range of motion of one finger when another finger is stationary:

$$\Rightarrow \begin{cases} \dfrac{x_1\big|_{x_2 = x_{2(i-1)}}}{x_{2(i-1)}} \ge \dfrac{\mu}{\mu_0} = k & (8a) \\ \dfrac{x_2\big|_{x_1 = x_{1(i)}}}{x_{1(i)}} \ge \dfrac{\mu}{\mu_0} = k & (8b) \end{cases}$$

$$\Rightarrow \begin{cases} \dfrac{x_{1(i)}}{x_{2(i-1)}} = \dfrac{\left[x_1\big|_{x_2=x_{2(i-1)}}\right]_{\min}}{x_{2(i-1)}} = k & \text{(9a)} \\ \dfrac{x_{2(i)}}{x_{1(i)}} = \dfrac{\left[x_2\big|_{x_1=x_{1(i)}}\right]_{\min}}{x_{1(i)}} = k & \text{(9b)} \end{cases}$$

$$\Rightarrow \qquad \frac{x_{2(i)}}{x_{2(i-1)}} = k^2 \tag{10}$$

Equation (10) is the recursion satisfied by $x_{2(i)}$.

$$\Rightarrow \qquad \frac{x_{2(n)}}{x_{2(0)}} = \prod_{i=1}^{n} \frac{x_{2(i)}}{x_{2(i-1)}} = k^{2n} \qquad n = 0,1,2,\cdots \tag{11}$$

Known:

$$x_{2(0)} = \frac{l}{2} \tag{12}$$

$$\Rightarrow \qquad x_{2(n)} = \frac{l}{2}k^{2n} \qquad n = 0,1,2,\cdots \tag{13}$$

(13) → (9b),

$$\Rightarrow \qquad x_{1(n)} = \begin{cases} \dfrac{l}{2}, & \text{when } n = 0 & \text{(14a)} \\ \dfrac{l}{2}k^{2n-1}, & \text{when } n \geq 1 & \text{(14b)} \end{cases}$$

Considering the definition of ***work***, then place (4) and (14) in to obtain:

$$\Rightarrow \qquad W = -\int_{x_{1(0)}}^{x_{1(1)}} f_1\big|_{x_2=x_{2(0)}} \mathrm{d}x_1 - \sum_{n=2}^{\infty} \int_{x_{1(n-1)}}^{x_{1(n)}} f_1\big|_{x_2=x_{2(n-1)}} \mathrm{d}x_1$$

$$- \sum_{n=1}^{\infty} \int_{x_{2(n-1)}}^{x_{2(n)}} f_2\big|_{x_1=x_{1(n)}} \mathrm{d}x_2$$

$$= -\mu mg \frac{l}{2}\left[\ln\frac{1+k}{2} + \ln k\left(\sum_{n=2}^{\infty} k^{2n-2} + \sum_{n=1}^{\infty} k^{2n-1}\right)\right]$$

$$= -\mu mg \frac{l}{2}\left(\ln\frac{1+k}{2} + \ln k \sum_{m=1}^{\infty} k^m\right)$$

$$= -\mu mg\frac{l}{2}\left(\ln\frac{1+k}{2} + \frac{k}{1-k}\ln k\right) \tag{15}$$

3.1.7 A Ladder Placed in a Special Way between a Vertical Wall and a Horizontal Ground (*a*, *A*)

A long and narrow ladder (which can be regarded as a homogeneous rod) is obliquely placed with its endpoint A against a vertical wall and its endpoint B on a horizontal ground (see Fig. 1). The friction coefficient between the ladder and the wall is μ_1, and the friction coefficient between the ladder and the ground is μ_2. The projection of Point A on the ground is denoted by O, the projection of Point B on the wall is denoted by O', the angle between BO and BO' is denoted by α, the angle between the ladder and the ground is denoted by θ. A man of mass m (which can be regarded as a particle) stands on the ladder. To ensure that the ladder will not slide when the person stands at an arbitrary position of the ladder, determine the minimum value of θ (with α unchanged).

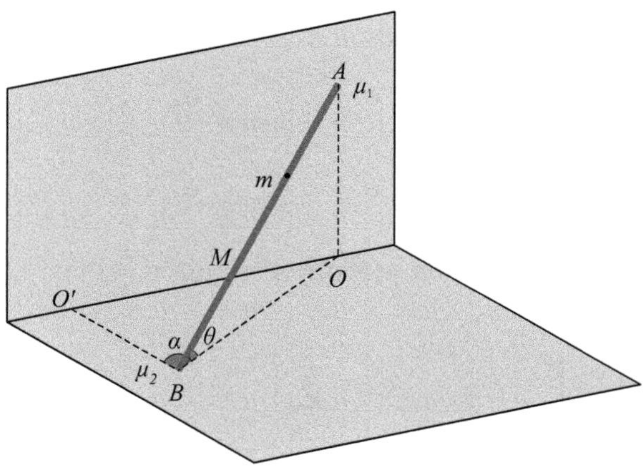

Figure 1

Solution

The following is a study of the minimum value of θ in an extreme case—the person stands at the top of the ladder, in which case the minimum value of θ is less than all other possible cases (proof omitted).

As shown in Fig. 2, the center of mass of the ladder is denoted by C, and the center of mass of the system composed of the man and the ladder is denoted by D.

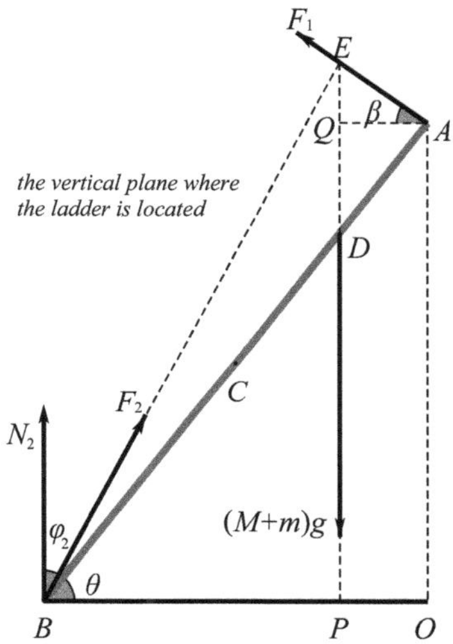

Figure 2

Defining the following interaction forces:

$\boldsymbol{G}$—the resultant force of the gravity of the man and the ladder.

$\boldsymbol{F}_1$ and $\boldsymbol{N}_1$—the total antiforce and normal contact force exerted on the ladder at its endpoint A by the wall.

$\boldsymbol{F}_2$ and $\boldsymbol{N}_2$—the total antiforce and normal contact force exerted on the ladder at its endpoint B by the ground.

Taking *the system composed of the man and the ladder* as the object of study. Studying *the state when the system is in mechanical equilibrium*. Considering ***three force balance intersection theorem*** [the three force acting on the system are taken as $\boldsymbol{G}$, $\boldsymbol{F}_1$, $\boldsymbol{F}_2$] (see Fig. 2).

The intersection point of the three force is denoted by E, the projection of Point E on the ground is denoted by P, and the projection of Point A on EP is denoted by Q.

As shown in Fig. 3, considering ***three cosine theorem***:

$$\cos\beta = \frac{\cos\varphi_1}{\cos\alpha} = \frac{1}{\cos\alpha}\frac{1}{\sqrt{1+\tan^2\varphi_1}} \tag{1}$$

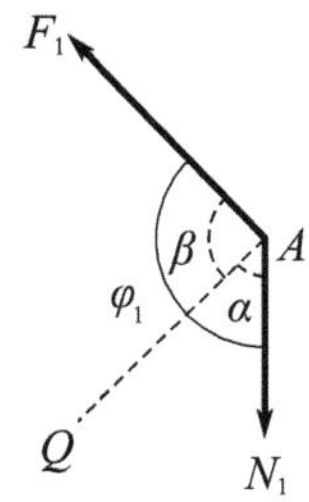

Figure 3

φ_1—the angle between $\boldsymbol{F}_1$ and $\boldsymbol{N}_1$.

β—the angle between $\boldsymbol{F}_1$ and AQ.

$$\Rightarrow \qquad \tan\beta = \sqrt{\frac{1}{\cos^2\beta} - 1} = \sqrt{(1+\tan^2\varphi_1)\cos^2\alpha - 1} \tag{2}$$

Considering the following geometric relations in Fig. 2:

$$\tan\theta = \frac{\overline{AO}}{\overline{BO}} = \frac{\overline{EP} - \overline{EQ}}{\overline{BO}} = \frac{\overline{BP}}{\overline{BO}}\frac{\overline{EP}}{\overline{BP}} - \frac{\overline{PO}}{\overline{BO}}\frac{\overline{EQ}}{\overline{PO}} = \frac{\overline{BD}}{\overline{AB}}\frac{\overline{EP}}{\overline{BP}} - \frac{\overline{AD}}{\overline{AB}}\frac{\overline{EQ}}{\overline{PO}} \tag{3}$$

Considering the definition of ***center of mass***:

$$\frac{\overline{AD}}{\overline{CD}} = \frac{M}{m} \tag{4}$$

Considering the following geometric relations in Fig. 2:

$$\overline{BD} - \overline{CD} = \overline{AD} + \overline{CD} = \frac{1}{2}AB \tag{5}$$

Combining (4) and (5),

$$\Rightarrow \qquad \begin{cases} \dfrac{\overline{BD}}{\overline{AB}} = \dfrac{M+2m}{2(M+m)} & \text{(6a)} \\[2ex] \dfrac{\overline{AD}}{\overline{AB}} = \dfrac{M}{2(M+m)} & \text{(6b)} \end{cases}$$

Considering the following geometric relations in Fig. 2:

$$\begin{cases} \tan\varphi_2 = \dfrac{\overline{BP}}{\overline{EP}} & \text{(7a)} \\[2ex] \tan\beta = \dfrac{\overline{EQ}}{\overline{PO}} & \text{(7b)} \end{cases}$$

(6) and (7) $\to$ (3),

$$\Rightarrow \quad \tan\theta = \frac{M+2m}{2(M+m)}\frac{1}{\tan\varphi_2} - \frac{M}{2(M+m)}\tan\beta$$

$$= \frac{M+2m}{2(M+m)}\frac{1}{\tan\varphi_2} - \frac{M}{2(M+m)}\sqrt{(1+\tan^2\varphi_1)\cos^2\alpha - 1} \quad (8)$$

Considering ***theorem of friction angle***:

$$\begin{cases} \tan\varphi_1 \le \mu_1 & (9a) \\ \tan\varphi_2 \le \mu_2 & (9b) \end{cases}$$

$$\Rightarrow \quad \tan\theta \ge \frac{M+2m}{2(M+m)}\frac{1}{\mu_2} - \frac{M}{2(M+m)}\sqrt{(1+\mu_1^2)\cos^2\alpha - 1} \quad (10)$$

$$\Rightarrow \theta_{\min} = \arctan\left[\frac{M+2m}{2(M+m)}\frac{1}{\mu_2} - \frac{M}{2(M+m)}\sqrt{(1+\mu_1^2)\cos^2\alpha - 1}\right] \quad (11)$$

3.1.8 The Stability of Equilibrium of a Homogeneous Square Column in Liquid (*c*, *AC*)[1]

Consider a homogeneous square column of density ρ placed in a homogeneous liquid of density ρ_0, $\rho < \rho_0/2$. The column reaches mechanical equilibrium in the configuration shown in Fig. 1. Please discuss the stability of equilibrium of the column under the tiny rotating disturbance.

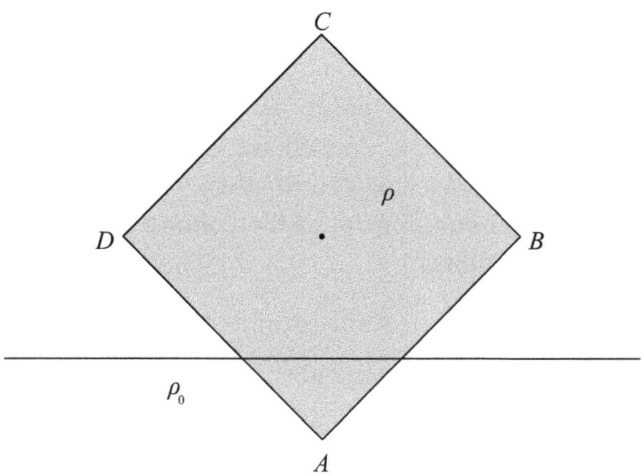

Figure 1

[1]This problem is originally proposed by Changchun Shi, a member of the Chinese team who participated in the 23th IPhO (International Physics Olympiad) in 1992.

Solution [1] (type C)

The following studies the situation of the column is disturbed in the clockwise direction.

As shown in Fig. 2, the projection of the center of mass of the column on the bottom *ABCD* is denoted by P_1, the projection of the geometric center (i.e., the center of buoyancy) of the immersed part of the column on the bottom *ABCD* is denoted by P_2, the dynamic intersection of side *AB* and the liquid level is denoted by *E*, and the dynamic intersection of side *AD* and the liquid level is denoted by *F*, the projection of Point P_1 on side *AB* is denoted by *G*, the projection of Point P_2 on P_1G is denoted by *Q*.

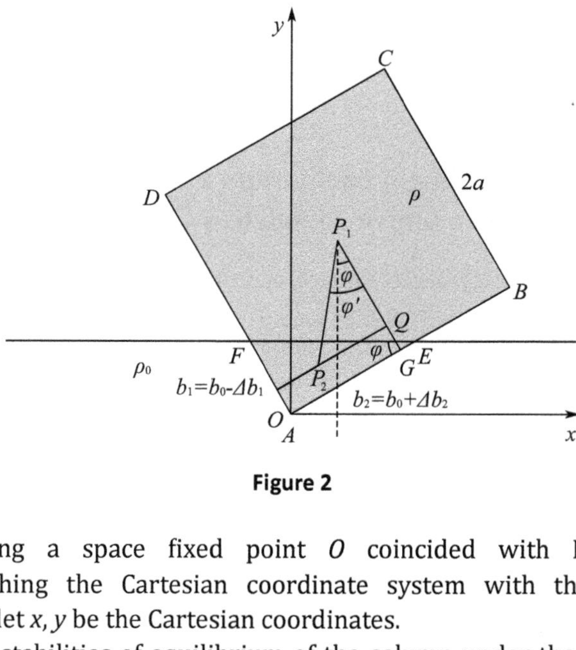

Figure 2

Taking a space fixed point *O* coincided with Point *A*. Establishing the Cartesian coordinate system with the frame $(O, \boldsymbol{i}, \boldsymbol{j})$, let x, y be the Cartesian coordinates.

The stabilities of equilibrium of the column under the rotating disturbance are as follows:

$$x_{P_1}\begin{cases} < x_{P_2} & \text{stable equilibrium} \quad (1a) \\ = x_{P_2} & \text{neutral equilibrium} \quad (1b) \\ > x_{P_2} & \text{unstable equilibrium} \quad (1c) \end{cases}$$

Taking *the column* as the object of study. Studying *the state before and after the disturbance, respectively*. Considering ***theorem of motion***

of center of mass [the rotating disturbance maintains the center-of-mass acceleration of the column to be zero] (the ***j***-direction component equation):

$$\begin{cases} \rho_0 \cdot \frac{1}{2} b_0^2 Lg - \rho \cdot 4a^2 Lg = 0 & \text{(2a)} \\ \rho_0 \cdot \frac{1}{2} b_1 b_2 Lg - \rho \cdot 4a^2 Lg = 0 & \text{(2b)} \end{cases}$$

L—the height of the column.

$2a$—the side length of the bottom of the column.

b_0—the length of AF and AE before the disturbance.

b_1 and b_2—the length of AF and AE after the disturbance.

Defining the variables Δb_1 and Δb_2:

$$\begin{cases} b_1 = b_0 - \Delta b_1 & \text{(3a)} \\ b_2 = b_0 + \Delta b_2 & \text{(3b)} \end{cases}$$

(3) → (2),

$$\Rightarrow \qquad \Delta b_2 - \Delta b_1 = \frac{\Delta b_1 \Delta b_2}{b_0} \tag{4}$$

$$\Rightarrow \qquad \Delta b_2 > \Delta b_1 \tag{5}$$

Considering the following geometric relations in Fig. 3:

$$\begin{cases} \tan\varphi = \frac{\overline{AF}}{\overline{AE}} = \frac{b_1}{b_2} = \frac{b_0 - \Delta b_1}{b_0 + \Delta b_2} & \text{(6a)} \\ \tan\varphi' = \frac{\overline{P_2Q}}{\overline{P_1Q}} = \frac{a - b_2/3}{a - b_1/3} = \frac{a - (b_0 + \Delta b_2)/3}{a - (b_0 - \Delta b_1)/3} & \text{(6b)} \end{cases}$$

φ—the angle between P_1G and the y-axis.

φ'—the angle between P_1G and P_1P_2.

From the geometric relation in Fig. 2 and Eq. (1), we can know that:

$$\Rightarrow \qquad \varphi' \begin{cases} < \varphi & \text{stable equilibrium} & \text{(7a)} \\ = \varphi & \text{neutral equilibrium} & \text{(7b)} \\ > \varphi & \text{unstable equilibrium} & \text{(7c)} \end{cases}$$

Combining (6) and (7),

$$\left(a-\frac{b_0-\Delta b_1}{3}\right)(b_0-\Delta b_1)$$

$$\Rightarrow \begin{cases} >\left(a-\dfrac{b_0+\Delta b_2}{3}\right)(b_0+\Delta b_2) & \text{stable equilibrium} \quad (8a) \\ =\left(a-\dfrac{b_0+\Delta b_2}{3}\right)(b_0+\Delta b_2) & \text{neutral equilibrium} \quad (8b) \\ <\left(a-\dfrac{b_0+\Delta b_2}{3}\right)(b_0+\Delta b_2) & \text{unstable equilibrium} \quad (8c) \end{cases}$$

Defining the function $f(u)$:

$$f(u)=\left(a-\frac{u}{3}\right)u \tag{9}$$

The terms on both sides of Inequality (8) are, respectively, equal to the values of the function $f(u)$ when $u = b_0 - \Delta b_1$ and $u = b_0 + \Delta b_2$.

The monotonicity of $f(u)$ are as follows:

$$\begin{cases} \left.\dfrac{\mathrm{d}f(u)}{\mathrm{d}x}\right|_{u>3a/2} <0 & f(u) \text{ is monotonically decreasing} \quad (10a) \\ \left.\dfrac{\mathrm{d}f(u)}{\mathrm{d}x}\right|_{u=3a/2} =0 & f(u) \text{ take maximum value} \quad (10b) \\ \left.\dfrac{\mathrm{d}f(u)}{\mathrm{d}x}\right|_{u<3a/2} >0 & f(u) \text{ is monotonically increasing} \quad (10c) \end{cases}$$

Combining (8), (9) and (10):

$$\Rightarrow \quad b_0 \begin{cases} >\dfrac{3}{2}a & \text{stable equilibrium} \quad (11a) \\ \le\dfrac{3}{2}a & \text{unstable equilibrium} \quad (11b) \end{cases}$$

$$\Rightarrow \quad \frac{\rho_0}{\rho} \begin{cases} <\dfrac{32}{9} & \text{stable equilibrium} \quad (12a) \\ \ge\dfrac{32}{9} & \text{unstable equilibrium} \quad (12b) \end{cases}$$

Since $\Delta b_2 > \Delta b_1$, the neutral equilibrium is excluded.

Solution [2] (type A)

Considering the following geometric relation in Fig. 3:

$$\tan\left(\frac{\pi}{4}-\theta\right)=\frac{b_1}{b_2} \tag{13}$$

θ—the angle between AP_1 and the y-axis.

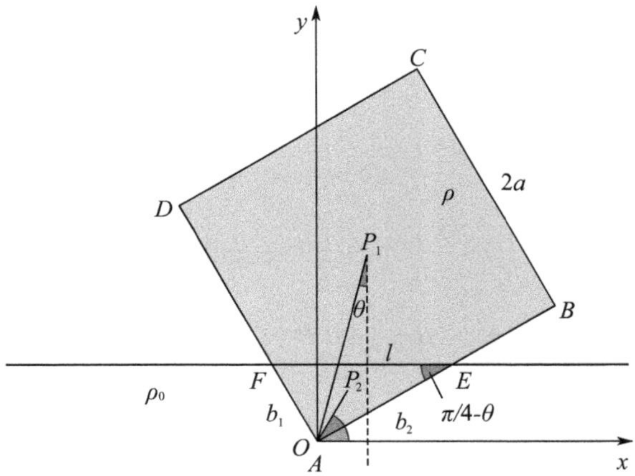

Figure 3

$$\Rightarrow \qquad \frac{b_1}{b_2}=\frac{1-\tan\theta}{1+\tan\theta}\approx\frac{1-\theta}{1+\theta}\approx(1-\theta)^2\approx 1-2\theta \tag{14}$$

$(14)\to(2)$,

$$\Rightarrow \qquad \begin{cases} b_1\approx b_0(1-\theta) & (15\text{a})\\ b_2\approx b_0(1+\theta) & (15\text{b})\end{cases}$$

Defining l—the length of EF after the disturbance. Considering the following geometric relation in Fig. 3:

$$\Rightarrow \qquad l=\sqrt{b_1^2+b_2^2}=b_0^2\sqrt{(1-\theta)^2+(1+\theta)^2}$$
$$\approx b_0^2\sqrt{(1-2\theta)+(1+2\theta)}=\sqrt{2}b_0 \tag{16}$$

Considering the following geometric relations in Fig. 3:

$$\begin{cases} x_{P_1}=\sqrt{2}a\sin\theta\approx\sqrt{2}a\theta & (17\text{a})\\ x_{P_2}=\frac{2}{3}\cdot\frac{1}{2}l\cos\left[2\left(\frac{\pi}{4}-\theta\right)\right]=\frac{\sqrt{2}}{3}b_0\sin(2\theta)\approx\frac{2\sqrt{2}}{3}b_0\theta & (17\text{b})\end{cases}$$

$(17) \rightarrow (1)$,

$$\Rightarrow \qquad b_0 \begin{cases} > \dfrac{3}{2}a & \text{stable equilibrium} & (18a) \\ = \dfrac{3}{2}a & \text{neutral equilibrium} & (18b) \\ < \dfrac{3}{2}a & \text{unstable equilibrium} & (18c) \end{cases}$$

$$\Rightarrow \qquad \frac{\rho_0}{\rho} \begin{cases} < \dfrac{32}{9} & \text{stable equilibrium} & (19a) \\ = \dfrac{32}{9} & \text{neutral equilibrium} & (19b) \\ > \dfrac{32}{9} & \text{unstable equilibrium} & (19c) \end{cases}$$

The operations (14), (16) and (17) have taken the first-order approximation of θ.

The reason why **Solution 2** obtains the situation of neutral equilibrium that is not obtained by **Solution 1** is that **Solution 2** performs a strict first-order approximation.

3.1.9 The Motion of a Homogeneous Sphere on a Rough Inclined Plane (d, *B*)

A homogeneous sphere of mass m and radius r is located on a long enough inclined plane of inclination angle θ (see Fig. 1). Initially, the center of mass of the sphere is at rest, rotating with the initial angular velocity of the magnitude ω_0 and the direction clockwise. The coefficient of friction between the sphere and the inclined plane is μ, $\mu > \tan\theta$, given that μ is large enough that the sphere can change

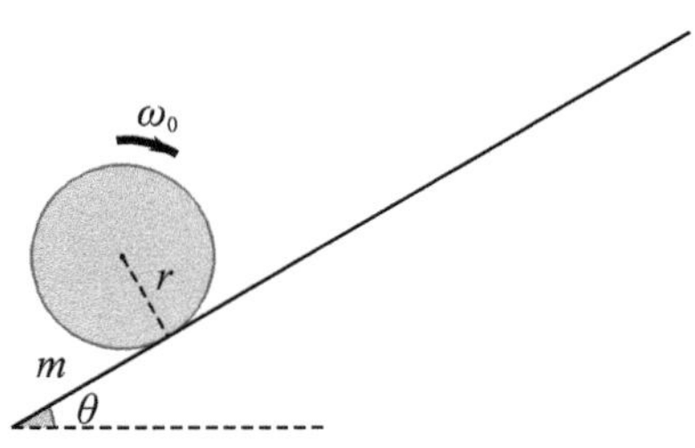

Figure 1

from sliding rolling to pure rolling relative to the inclined plane before it reaches the highest point. Find the maximum displacement of the center of mass of the sphere in the upward direction along the inclined plane.

Solution [1]

The initial time is defined as Time zero.

As shown in Fig. 2, the center of mass of the sphere is denoted by C, and the space point that coincides with the center of mass C at Time zero is denoted by O. Establishing the Cartesian coordinate system with the frame (O, $\boldsymbol{i}$, $\boldsymbol{j}$, $\boldsymbol{k}$), let x, y, z be the Cartesian coordinates.

Firstly, the sphere performs sliding rolling relative to the inclined plane, the process of which is denoted by Process 1.

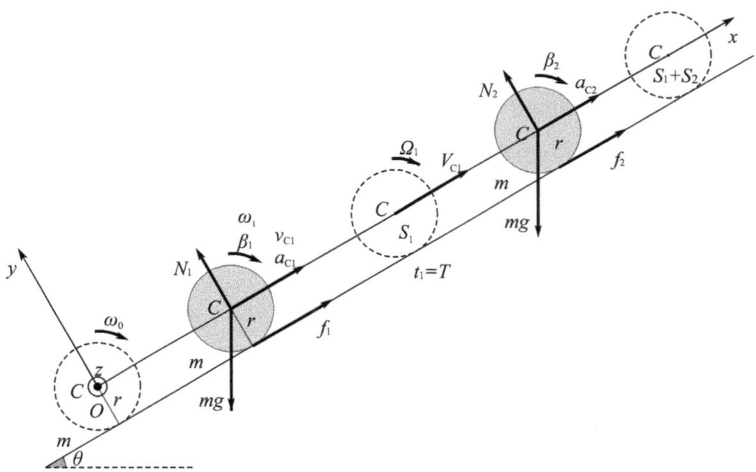

Figure 2

Taking *the sphere* as the object of study. Studying *the state at arbitrary Time t in Process 1*. Considering (a) ***theorem of motion of center of mass*** (the $\boldsymbol{i}$-direction and $\boldsymbol{j}$-direction component equations), (b) ***law of rotation*** [in its center-of-mass reference frame, taking its central axis parallel to the z-axis as the rotation axis] (the $-\boldsymbol{k}$-direction component equation), (c) ***in the center-of-mass reference frame, the resultant torque of the inertia force acting on system of particles is zero***, (d) ***friction law***, (e) ***the form of moment of inertia of a homogeneous sphere about one of its central axes***:

$$\begin{cases} \begin{cases} f_1 - mg\sin\theta = ma_{C1} & \text{(1a)} \\ N_1 - mg\cos\theta = 0 & \text{(1b)} \end{cases} \\ -f_1 r = I\beta_1 & \text{(1c)} \end{cases}$$

$$f_1 = \mu N_1 \tag{2}$$

$$I = \frac{2}{5}mr^2 \tag{3}$$

N_1—the ***j***-direction component of the normal contact force exerted on the sphere by the inclined plane at Time t in Process 1.

f_1—the ***i***-direction component of the friction exerted on the sphere by the inclined plane at Time t in Process 1.

a_{C1}—the ***i***-direction component of the center-of-mass acceleration of the sphere at Time t in Process 1.

β_1—the −***k***-direction component of the rotating angular acceleration of the sphere at Time t in Process 1.

I—the moment of inertia of the sphere about its central axis parallel to the z-axis.

Combining (1), (2) and (3),

$$\Rightarrow \quad \begin{cases} a_{C1} = (\mu\cos\theta - \sin\theta)g & \text{(4a)} \\ \beta_1 = -\dfrac{5\mu\cos\theta}{2}\dfrac{g}{r} & \text{(4b)} \end{cases}$$

Considering ***the kinematics equation of uniformly variable motion and uniformly variable rotation***:

$$\begin{cases} v_{C1} = a_{C1}t_1 & \text{(5a)} \\ \omega_1 = \omega_0 + \beta_1 t_1 & \text{(5b)} \end{cases}$$

$$s_1 = \frac{1}{2}a_{C1}t_1^2 \tag{5c}$$

v_{C1}—the ***i***-direction component of the center-of-mass velocity of the sphere at Time t_1 in Process 1.

ω_1—the −***k***-direction component of the rotating angular velocity of the sphere at Time t_1 in Process 1.

s_1—the ***i***-direction component of the center-of-mass displacement of the sphere in the process from Time zero to Time t_1 in Process 1.

The condition for the sphere to perform pure rolling relative to the inclined plane is as follows:

$$(v_{C1} - \omega_1 r)\big|_{t_1=T} = 0 \tag{6}$$

T—the Time when the sphere reaches pure rolling relative to the inclined plane.

The process that the sphere performs pure rolling relative to the inclined plane is denoted by Process 2.

(5a) and (5b) $\rightarrow$ (6),

$$\Rightarrow \qquad T = \frac{2}{7\mu\cos\theta - 2\sin\theta}\frac{\omega_0 r}{g} \tag{7}$$

(7) $\rightarrow$ (5),

$$\Rightarrow \begin{cases} V_{C1} = v_{C1}\big|_{t_1=T} = \dfrac{2(\mu\cos\theta - \sin\theta)}{7\mu\cos\theta - 2\sin\theta}\omega_0 r & \text{(8a)} \\ \Omega_1 = \omega_1\big|_{t_1=T} = \dfrac{2(\mu\cos\theta - \sin\theta)}{7\mu\cos\theta - 2\sin\theta}\omega_0 & \text{(8b)} \\ S_1 = s_1\big|_{t_1=T} = \dfrac{2(\mu\cos\theta - \sin\theta)}{(7\mu\cos\theta - 2\sin\theta)^2}\dfrac{\omega_0^2 r^2}{g} & \text{(8c)} \end{cases}$$

V_{C1}—the ***i***-direction component of the center-of-mass velocity of the sphere when it reaches pure rolling relative to the inclined plane.

Ω_1—the −***k***-direction component of the rotating angular velocity of the sphere when it reaches pure rolling relative to the inclined plane.

S_1—the ***i***-direction component of the center-of-mass displacement of the sphere in the whole sliding rolling process.

Taking *the sphere* as the object of study. Studying *the state at arbitrary Time t in Process 2*. Considering (a) ***theorem of motion of center of mass*** (the ***i***-direction and ***j***-direction component equations), (b) ***law of rotation*** [in its center-of-mass reference frame, taking its central axis parallel to the *z*-axis as the rotation axis] (the −***k***-direction component equation), (c) ***in the center-of-mass reference frame, the resultant torque of the inertia force acting on system of particles is zero***:

$$\begin{cases} f_2 - mg\sin\theta = ma_{C2} & \text{(9a)} \\ -f_2 r = I\beta_2 & \text{(9b)} \end{cases}$$

f_2—the ***i***-direction component of the friction exerted on the sphere by the inclined plane at Time t in Process 2.

a_{C2}—the ***i***-direction component of the center-of-mass acceleration of the sphere at Time t in Process 2.

β_2—the $-\boldsymbol{k}$-direction component of the rotating angular acceleration of the sphere at Time t in Process 2.

The condition for the sphere to perform pure rolling relative to the inclined plane all the time is as follows:

$$a_{C2} - \beta_2 r = 0 \tag{10}$$

Combining (9) and (10),

$$\Rightarrow \quad a_{C2} = -\frac{5\sin\theta}{7}g \tag{11}$$

Considering ***the displacement-velocity relation of uniformly variable motion***:

$$0 - V_{C1}^2 = 2a_{C2}S_2 \tag{12}$$

S_2—the $\boldsymbol{i}$-direction component of the center-of-mass displacement of the sphere in the whole pure rolling process.

(8a) and (11) $\rightarrow$ (12),

$$\Rightarrow \quad S_2 = \frac{14(\mu\cos\theta - \sin\theta)^2}{5\sin\theta(7\mu\cos\theta - 2\sin\theta)^2}\frac{\omega_0^2 r^2}{g} \tag{13}$$

Defining S—the $\boldsymbol{i}$-direction component of the center-of-mass displacement of the sphere in the whole process.

Combining (8c) and (13),

$$\Rightarrow \quad S = S_1 + S_2 = \frac{2(\mu\cos\theta - \sin\theta)}{5\sin\theta(7\mu\cos\theta - 2\sin\theta)}\frac{\omega_0^2 r^2}{g} \tag{14}$$

Solution [2]

Taking *the system composed of the sphere and the earth* as the object of study. Studying *the whole pure rolling process*. Considering ***law of conservation of mechanical energy*** [the work done by the external force and non-conservative internal force acting on the system is zero (no work is done by the static friction force exerted on the sphere by the inclined plane); therefore, the mechanical energy of the system is conserved] (taking the position of center of mass C at Time T as the zero point of gravitational potential energy), (b) ***König theorem*** [integrating into (a)], (c) ***the form of kinetic energy of a rigid body performing fixed-axis rotation*** [integrating into (b)]:

$$\frac{1}{2}mV_{C1}^2 + \frac{1}{2}I\Omega_1^2 = mgS_2\sin\theta \tag{15}$$

(8a) and (8b) $\rightarrow$ (15),

$$\Rightarrow \qquad S_2 = \frac{14(\mu\cos\theta - \sin\theta)^2}{5\sin\theta(7\mu\cos\theta - 2\sin\theta)^2}\frac{\omega_0^2 r^2}{g} \qquad (16)$$

3.1.10 The Motion of a Homogeneous Sphere on a Rough Spherical Surface (d, *C*)

A homogeneous sphere is statically placed at the top of a fixed spherical shell (see Fig. 1). Due to a small disturbance, the sphere rolls down from the top of the spherical shell. The friction coefficient between the sphere and the spherical shell is μ. Determine the position of the sphere when it separates from the spherical shell.

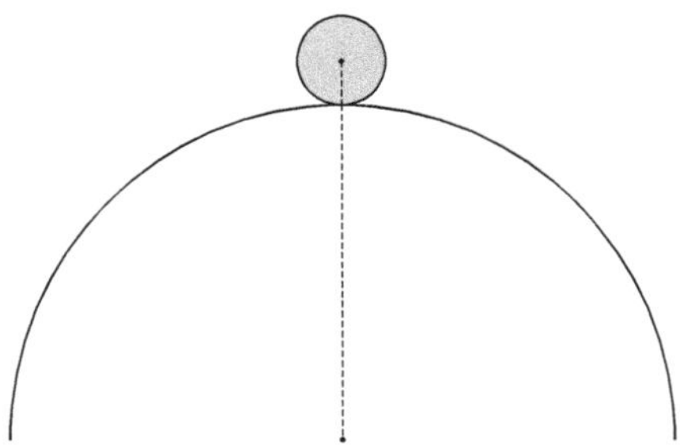

Figure 1

Solution

Let the sphere be disturbed to the right in the figure.

The time point when the sphere starts to move is defined as Time zero.

As shown in Fig. 2, the center of the spherical shell is denoted by O, the center of mass of the sphere is denoted by C, the unit vector perpendicular to the plane of the figure inward is denoted by $\boldsymbol{k}$. Taking Point O as the pole, the vertical upward direction as the positive direction of the polar axis, the right-hand helix direction with regard to $\boldsymbol{k}$ as the positive direction of polar angle, the $\boldsymbol{k}$-direction as the positive direction of the longitudinal axis, establishing the cylindrical coordinate system, let r, φ, z be the cylindrical coordinates.

Assumption I: firstly, the sphere performs pure rolling relative to the spherical shell, the process of which is denoted by Process 1.

The point on the spherical shell in contact with the sphere at arbitrary Time t is denoted by P.

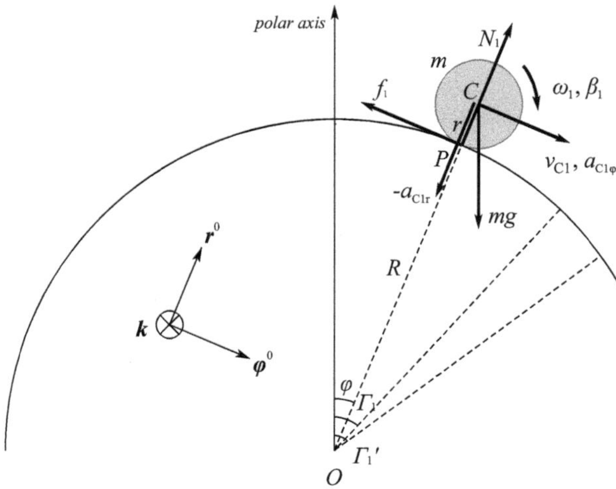

Figure 2

Taking *the system composed of the sphere and the earth* as the object of study. Studying *the process from Time zero to arbitrary Time t in Process 1*. Considering (a) ***law of conservation of mechanical energy*** [the work done by the external force and non-conservative internal force acting on the system is zero (the work done by the static friction force exerted on the sphere by the spherical shell is zero, and its effect is to convert the translational kinetic energy of the sphere into the rotational kinetic energy); therefore, the mechanical energy of the system is conserved] (taking the initial position of the center of mass C as the zero of gravitational potential energy), (b) ***the form of kinetic energy of a rigid body performing fixed-axis rotation***, (c) ***parallel-axis theorem***, (d) ***the form of moment of inertia of a homogeneous sphere about one of its central axes***:

$$E_k - mg(R+r)(1-\cos\varphi) = 0 \tag{1}$$

$$E_k = \frac{1}{2} I_P \omega_1^2 \tag{2}$$

$$\begin{cases} I_P = I_C + mr^2 & \text{(3a)} \\ I_C = \dfrac{2}{5} mr^2 & \text{(3b)} \end{cases}$$

m—the mass of the sphere.

I_P—the moment of inertia of the sphere about the tangent passing through Point P and parallel to the z-axis.

I_C—the moment of inertia of the sphere about its central axis parallel to the z-axis.

ω—the $\boldsymbol{k}$-direction component of the angular velocity of the sphere at Time t in Process 1.

E_k—the kinetic energy of the sphere at Time t in Process 1.

Combining (1), (2) and (3),

$$\Rightarrow \qquad \frac{1}{2}\left(\frac{2}{5}mr^2+mr^2\right)\omega_1^2-mg(R+r)(1-\cos\varphi)=0$$

$$\Rightarrow \qquad \omega_1=\sqrt{\frac{10(R+r)g(1-\cos\varphi)}{7r^2}} \tag{4}$$

Taking *the sphere* as the object of study. Studying *the state at arbitrary Time t*. Considering (a) ***theorem of motion of center of mass*** (the $-\boldsymbol{r}^0$-direction and $\boldsymbol{\varphi}^0$-direction component equations), (b) ***law of rotation*** [taking the line passing through Point P and parallel to the z-axis as the rotation axis] (the $\boldsymbol{k}$-direction component equation):

$$\begin{cases}\begin{cases} mg\cos\varphi-N_1=m(-a_{C1r}) & \text{(5a)}\\ mg\sin\varphi-f_1=ma_{C1\varphi} & \text{(5b)}\end{cases}\\ mgr\sin\varphi=I_P\beta_1 & \text{(5c)}\end{cases}$$

N_1 and f_1—the $\boldsymbol{r}^0$-direction component of the normal contact force exerted on the sphere by the spherical shell, and the $-\boldsymbol{\varphi}^0$-direction component of the friction exerted on the sphere by the spherical shell at Time t in Process 1.

a_{C1r} and $a_{C1\varphi}$—the $\boldsymbol{r}^0$-direction and $\boldsymbol{\varphi}^0$-direction component of the center-of-mass acceleration of the sphere at Time t in Process 1.

β_1—the $\boldsymbol{k}$-direction component of the angular acceleration of the sphere at Time t in Process 1.

Considering ***the radial component form of acceleration*** (the constraint condition: both the sphere and the spherical shell are rigid bodies; therefore, the distance between the center of mass C and Point O remains unchanged):

$$a_{C1r}=-\frac{v_{C1}^2}{R+r} \tag{6}$$

v_{C1}—the $\boldsymbol{\varphi}^0$-direction component of the center-of-mass velocity of the sphere at Time *t* in Process 1.

The conditions for the sphere to perform pure rolling relative to the spherical shell all the time in Process 1 are as follows:

$$\begin{cases} v_{C1} - \omega_1 r = 0 & \text{(7a)} \\ a_{C1\varphi} - \beta_1 r = 0 & \text{(7b)} \end{cases}$$

Combining (3) to (7),

$$\Rightarrow \quad \begin{cases} N_1 = \dfrac{mg(17\cos\varphi - 10)}{7} & \text{(8a)} \\ f_1 = \dfrac{2mg\sin\varphi}{7} & \text{(8b)} \end{cases}$$

Considering ***friction law***:

$$f_1 \le \mu N_1 \tag{9}$$

$$\Rightarrow \quad (289\mu^2 + 4)\cos^2\varphi - 340\mu^2\cos\varphi + 100\mu^2 - 4 \ge 0$$

$$\Rightarrow \quad \cos\varphi \ge \frac{170\mu^2 + 2\sqrt{189\mu^2 + 4}}{289\mu^2 + 4} \tag{10}$$

Define Γ_1—the maximum polar angle of the center of mass of the sphere in Process 1.

$$^*\Gamma_1 = \arccos\frac{170\mu^2 + 2\sqrt{189\mu^2 + 4}}{289\mu^2 + 4} \tag{11}$$

Assumption II: the sphere will separate from the spherical shell before the sphere enters sliding rolling relative to the spherical shell.

The condition for the sphere to separate from the spherical shell in Process 1 is that the normal contact force exerted on the sphere by the spherical shell is zero:

$$N_1\big|_{\varphi=\Gamma_1'} = 0 \tag{12}$$

Γ'_1—in the situation of assumption II, the polar angle of the center of mass of the sphere when it separates from the spherical shell.

$$\Rightarrow \quad \cos\Gamma_1' = \frac{10}{17}$$

$$\Rightarrow \quad \Gamma_1' = \arccos\frac{10}{17} \tag{13}$$

Comparing Γ_1 and Γ'_1:

$$\frac{170\mu^2+2\sqrt{189\mu^2+4}}{289\mu^2+4}-\frac{10}{17}=\frac{34\sqrt{189\mu^2+4}-40}{17(289\mu^2+4)}>0 \quad (14)$$

$$\Rightarrow \qquad \Gamma_1<\Gamma_1' \quad (15)$$

Therefore, Assumption II doesn't hold, when the center of mass C reaches the polar angle Γ_1, the sphere will enter sliding rolling relative to the spherical shell, the process of which is denoted by Process 2.

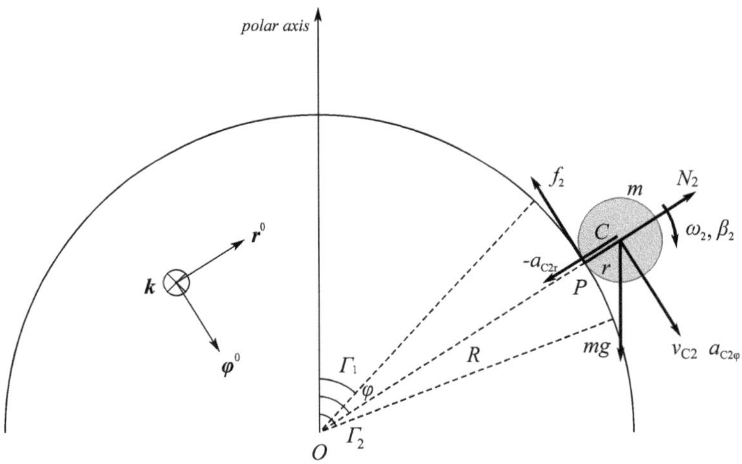

Figure 3

Taking *the sphere* as the object of study (see Fig. 3). Studying *the state at arbitrary Time t in Process 2*. Considering (a) ***theorem of motion of center of mass*** (the $-\boldsymbol{r}^0$-direction and $\boldsymbol{\varphi}^0$-direction component equations), (b) ***friction law***:

$$\begin{cases} mg\cos\varphi-N_2=m(-a_{C2r}) & (16a)\\ mg\sin\varphi-f_2=ma_{C2\varphi} & (16b)\end{cases}$$

$$f_2=\mu N_2 \quad (16c)$$

N_2 and f_2—the $\boldsymbol{r}^0$-direction component of the normal contact force exerted on the sphere by the spherical shell, and the $-\boldsymbol{\varphi}^0$-direction component of the friction exerted on the sphere by the spherical shell at Time t in Process 2.

a_{C2r} and $a_{C2\varphi}$—the $\boldsymbol{r}^0$-direction and $\boldsymbol{\varphi}^0$-direction component of the center-of-mass acceleration of the sphere at Time t in Process 2.

Considering ***the radial component form of acceleration*** (the constrained condition: both the sphere and the spherical shell are rigid bodies; therefore, the distance between the center of mass C and Point O remains unchanged):

$$a_{C2r} = -\frac{v_{C2}^2}{R+r} \tag{17}$$

v_{C2}—the $\boldsymbol{\varphi}^0$-direction component of the center-of-mass velocity of the sphere at arbitrary Time t in Process 2.

Considering ***the transverse component form of acceleration*** (the distance between the center of mass C and Point O remains unchanged):

$$a_{C2\varphi} = \frac{\mathrm{d}v_{C2}}{\mathrm{d}t} = \frac{\mathrm{d}v_{C2}}{\mathrm{d}\varphi}\frac{\mathrm{d}\varphi}{\mathrm{d}t} \tag{18}$$

Considering ***the transverse component form of velocity***:

$$v_{C2} = \frac{\mathrm{d}\varphi}{\mathrm{d}t}(R+r) \tag{19}$$

$$\Rightarrow \qquad a_{C2\varphi} = \frac{\mathrm{d}v_{C2}}{\mathrm{d}\varphi}\frac{v_{C2}}{R+r} = \frac{1}{2(R+r)}\frac{\mathrm{d}v_{C2}^2}{\mathrm{d}\varphi} \tag{20}$$

(17) and (20) → (16),

$$\Rightarrow \qquad \frac{\mathrm{d}v_{C2}^2}{\mathrm{d}\varphi} - 2\mu v_{C2}^2 + 2(R+r)g(\mu\cos\varphi - \sin\varphi) = 0 \tag{21}$$

The homogeneous equation of Eq. (21) is as follows:

$$\frac{\mathrm{d}v_{C2}^2}{\mathrm{d}\varphi} - 2\mu v_{C2}^2 = 0 \tag{22}$$

Let the solution of Eq. (22) be:

$$\Rightarrow \qquad v_{C2(1)}^2 = Ae^{2\mu\varphi} \tag{23}$$

Let the particular solution of Eq. (21) be:

$$v_{C2(2)}^2 = B\cos\varphi + C\sin\varphi \tag{24}$$

(24) → (21),

$$\Rightarrow \qquad -B\sin\varphi + C\cos\varphi - 2\mu(B\cos\varphi + C\sin\varphi) + 2(R+r)g(\mu\cos\varphi - \sin\varphi) = 0$$

$$\Rightarrow \quad [B+2\mu C+2(R+r)g]\sin\varphi-[C-2\mu B+2\mu(R+r)g]\cos\varphi=0 \qquad (25)$$

$$\Rightarrow \qquad \begin{cases} B+2\mu C+2(R+r)\,g=0 \\ C-2\mu B+2\mu(R+r)\,g=0 \end{cases}$$

$$\Rightarrow \qquad \begin{cases} B=\dfrac{2(2\mu^2-1)(R+r)\,g}{4\mu^2+1} & (26a) \\ C=-\dfrac{6\mu(R+r)\,g}{4\mu^2+1} & (26b) \end{cases}$$

$$\Rightarrow \qquad v_{C2(2)}^2=\frac{(R+r)g}{4\mu^2+1}[2(2\mu^2-1)\cos\varphi-6\mu\sin\varphi] \qquad (27)$$

The general solution of Eq. (21) can be obtained from (23) and (27):

$$\Rightarrow \quad v_{C2}^2=v_{C2(1)}^2+v_{C2(2)}^2=Ae^{2\mu\varphi}+\frac{(R+r)g}{4\mu^2+1}[2(2\mu^2-1)\cos\varphi-6\mu\sin\varphi] \qquad (28)$$

Known:

$$v_{C1}\big|_{\varphi=\Gamma_1}=v_{C2}\big|_{\varphi=\Gamma_1} \qquad (29)$$

From (4), (7a), (11), (28) and (29),

$$\Rightarrow \quad A=\frac{(R+r)g}{4\mu^2+1}\left[\begin{array}{l}\dfrac{10(4\mu^2+1)(1-\cos\Gamma_1)}{7}-2(2\mu^2-1)\cos\Gamma_1+ \\ 6\mu\sin\Gamma_1\end{array}\right]e^{-2\mu\Gamma_1} \qquad (30)$$

The condition for the sphere to separate from the spherical shell in Process 2 is that the normal contact force exerted on the sphere by the spherical shell is zero:

$$N_2\big|_{\varphi=\Gamma_2}=0 \qquad (31)$$

Γ_2—the polar angle of the center of mass of the sphere when it separates from the spherical shell.

From (16a), (17) and (31),

$$\Rightarrow \qquad mg\cos\Gamma_2=m\frac{v_{C2}^2\big|_{\varphi=\Gamma_2}}{R+r} \qquad (32)$$

$(28) \to (32)$,

$$\Rightarrow \quad \cos\Gamma_2 = \frac{A}{(R+r)g} e^{2\mu\Gamma_2} + \frac{1}{4\mu^2+1}[2(2\mu^2-1)\cos\Gamma_2 - 6\mu\sin\Gamma_2] \tag{33}$$

$(30) \to (33)$, we obtain the transcendental equation satisfied by Γ_2:

$$\cos\Gamma_2 =$$

$$\Rightarrow \quad \frac{1}{4\mu^2+1}\left\{\begin{array}{l}\left[\begin{array}{l}\dfrac{10(1+4\mu^2)(1-\cos\Gamma_1)}{7} - 2(2\mu^2-1)\cos\Gamma_1 \\ +6\mu\sin\Gamma_1\end{array}\right] e^{2\mu(\Gamma_2-\Gamma_1)} \\ +\left[2(2\mu^2-1)\cos\Gamma_2 - 6\mu\sin\Gamma_2\right]\end{array}\right\} \tag{34}$$

3.1.11 The Collision between Two Homogeneous Spheres (*c*, *B*)[2]

On a horizontal ground there are two identical homogeneous spheres 1 and 2. Initially, Sphere 2 is at rest, Sphere 1 performs the pure rolling with the center-of-mass velocity of the magnitude v and the direction pointing to the center of Sphere 2. There is a completely elastic collision between the two spheres. Given that the coefficient of friction between the two spheres and the ground is μ, the friction between the two spheres can be ignored. Find the center-of-mass velocity of the two spheres when they, respectively, reach the pure rolling after the collision.

Solution [1]

As shown in Fig. 1 taking an arbitrary point on the ground and denoting it by O, establishing the Cartesian coordinate system with the frame $(O, \boldsymbol{i}, \boldsymbol{j}, \boldsymbol{k})$, let x, y, z be the Cartesian coordinates.

The condition for Sphere 1 to perform the pure rolling relative to the ground before the collision is as follows:

$$v - \omega R = 0 \tag{1}$$

R—the radius of Sphere 1 and Sphere 2.

[2]This problem is a test question of CUSPEA (China-U.S. Physics Examination and Application) in 1984.

ω—the $\boldsymbol{j}$-direction component of the rotating angular velocity of Sphere 1 at arbitrary time before the collision.

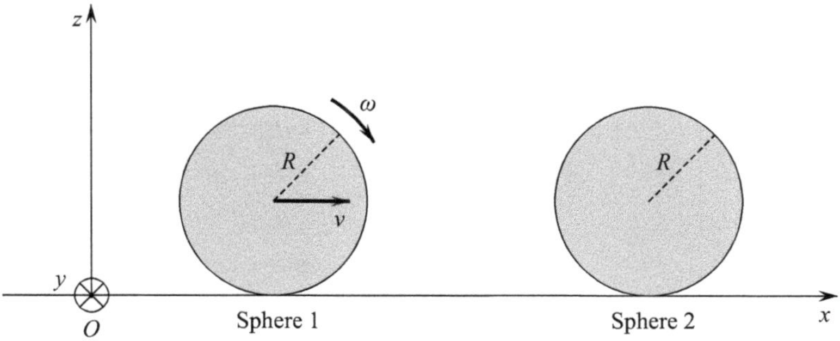

Figure 1

$$\Rightarrow \qquad \omega = \frac{v}{R} \tag{2}$$

Taking *the system composed of the two spheres (see Fig. 2)* as the object of study. Studying *the instantaneous process of the collision*. Considering (a) ***law of conservation of momentum*** [the friction exerted on each sphere by the ground is a finite quantity, acting in an extremely short process, its impulse is an infinitely small quantity, which can be ignored; therefore, the momentum of the system is approximately conserved] (the $\boldsymbol{i}$-direction component equation), (b) ***law of conservation of mechanical energy*** [the collision between the two spheres is completely elastic, the friction exerted on each sphere by the ground is a finite quantity, acting on small displacement, its work is an infinitely small quantity, which can be ignored; therefore, the mechanical energy of the system is approximately conserved], (c) ***the total momentum of the system of particles is equal to the center-of-mass momentum*** [the two spheres themselves are both system of particles] [integrating into (a)], (d) ***König theorem*** [integrating into (b)], (e) ***the form of kinetic energy of a rigid body performing fixed-axis rotation*** [integrating into (d)]:

$$\begin{cases} mv = mv_{10} + mv_{20} & \text{(3a)} \\ \dfrac{1}{2}mv^2 + I_1\omega^2 = \left(\dfrac{1}{2}mv_{10}^2 + \dfrac{1}{2}I\omega_{10}^2\right) + \left(\dfrac{1}{2}mv_{20}^2 + \dfrac{1}{2}I\omega_{20}^2\right) & \text{(3b)} \end{cases}$$

m—the mass of Sphere 1 and Sphere 2.

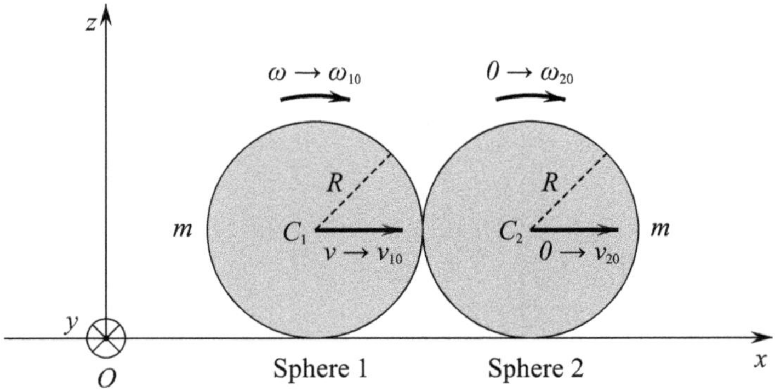

Figure 2

v_{10} and v_{20}—the ***i***-direction component of the center-of-mass velocity of Sphere 1 and Sphere 2 at the time point after the collision.

ω_{10} and ω_{20}—the ***j***-direction component of the rotating angular velocity of Sphere 1 and Sphere 2 at the time point after the collision.

I_1 and I_2—the moment of inertia of Sphere 1 and Sphere 2 about one of their central axes.

Taking *Sphere 1 and Sphere 2* as the object of study, respectively. Studying *the instantaneous process of the collision*. Considering ***law of conservation of angular momentum*** [in their respective center-of-mass reference frames, taking each central axis parallel to the y-axis as the corresponding rotation axis, the friction exerted on each sphere by the ground is a finite quantity, acting in an extremely short process, its impulse moment is an infinitely small quantity, which can be ignored; therefore, the angular momentum of each sphere is approximately conserved] (the ***j***-direction component equation):

$$\begin{cases} I_1\omega = I_1\omega_{10} & (4a) \\ 0 = I_2\omega_{20} & (4b) \end{cases}$$

Considering ***the form of moment of inertia of a homogeneous sphere about one of its central axes***:

$$I_1 = I_2 = \frac{2}{5}mR^2 \tag{5}$$

Combining (2) to (5),

$$\Rightarrow \quad \begin{cases} v_{10} = 0 & \text{(6a)} \\ \omega_{10} = v/R & \text{(6b)} \\ v_{20} = v & \text{(6c)} \\ \omega_{20} = 0 & \text{(6d)} \end{cases}$$

Assumption: Sphere 1 and Sphere 2 will not collide again.

The time point after the collision between the two spheres is defined as Time zero.

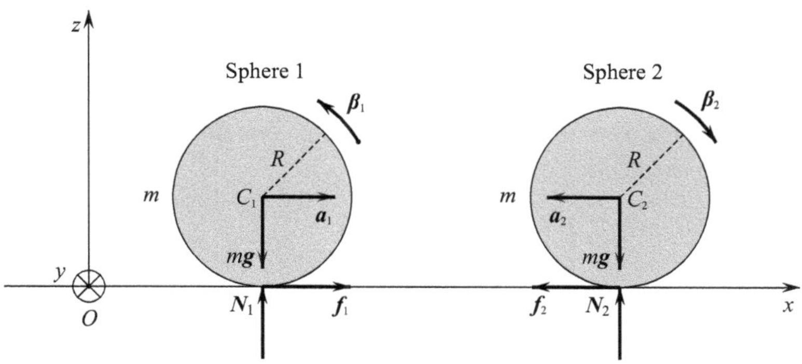

Figure 3

Taking *Sphere 1* as the object of study (see Fig. 3). Studying *the state at arbitrary Time t*. Considering (a) ***theorem of motion of center of mass*** (the ***i***-direction and ***k***-direction component equations), (b) ***law of rotation*** [in its center-of-mass reference frame, taking its central axis parallel to the *y*-axis as the rotation axis] (the ***j***-direction component equation), (c) ***in the center-of-mass reference frame, the resultant torque of the inertia force acting on system of particles is zero***:

$$\begin{cases} f_1 = ma_1 & \text{(7a)} \\ N_1 - mg = 0 & \text{(7b)} \end{cases}$$

$$-f_1 R = I_1 \beta_1 \qquad \text{(7c)}$$

N_1—the ***k***-direction component of the normal contact force exerted on Sphere 1 by the ground at Time *t*.

f_1—the ***i***-direction component of the friction exerted on Sphere 1 by the ground at Time *t*.

a_1—the ***i***-direction component of the center-of-mass acceleration of Sphere 1 at Time *t*.

β_1—the ***j***-direction component of the rotating angular acceleration of Sphere 1 at Time *t*.

Considering ***friction law***:

$$f_1 = \mu N_1 \tag{8}$$

Combining (5), (7) and (8),

$$\Rightarrow \quad \begin{cases} a_1 = \mu g & (9a) \\ \beta_1 = -\dfrac{5\mu g}{2R} & (9b) \end{cases}$$

Considering ***the velocity-time relation of uniformly variable motion and the angular velocity-time relation of uniformly variable rotation***:

$$\Rightarrow \quad \begin{cases} v_1 = v_{10} + a_1 t = \mu g t & (10a) \\ \omega_1 = \omega_{10} + \beta_1 t = \dfrac{v}{R} - \dfrac{5\mu g}{2R} t & (10b) \end{cases}$$

v_1—the ***i***-direction component of the center-of-mass velocity of Sphere 1 at arbitrary Time *t*.

ω_1—the ***j***-direction component of the rotating angular velocity of Sphere 1 at arbitrary Time *t*.

The condition for Sphere 1 to perform the pure rolling relative to the ground is as follows:

$$v_1\big|_{t=t_1} - \omega_1\big|_{t=t_1} R = 0 \tag{11}$$

t_1—the Time when Sphere 1 reaches the pure rolling relative to the ground.

$$\Rightarrow \quad t_1 = \frac{2v}{7\mu g} \tag{12}$$

$$\Rightarrow \quad v_1\big|_{t=t_1} = \frac{2}{7}v \tag{13}$$

Taking *Sphere 2* as the object of study. Studying *the state at arbitrary Time t*. Considering (a) ***theorem of motion of center of mass*** (the ***i***-direction and ***k***-direction component equations), (b) ***law of rotation*** [in its center-of-mass reference frame, taking its central axis parallel to the y-axis as the rotation axis] (the $\boldsymbol{j}$-direction component equation), (c) ***in the center-of-mass reference frame, the resultant torque of the inertia force acting on system of particles is zero***:

$$\begin{cases} -f_2 = ma_2 & (14a) \\ N_2 - mg = 0 & (14b) \end{cases}$$

$$f_2 R = I\beta_2 \quad (14c)$$

N_2—the $\boldsymbol{k}$-direction component of the normal contact force exerted on Sphere 2 by the ground at Time t.

f_2—the $-\boldsymbol{i}$-direction component of the friction exerted on Sphere 2 by the ground at Time t.

a_2—the $\boldsymbol{i}$-direction component of the center-of-mass acceleration of Sphere 2 at Time t.

β_2—the $\boldsymbol{j}$-direction component of the rotating angular acceleration of Sphere 2 at Time t.

Considering ***friction law***:

$$f_2 = \mu N_2 \quad (15)$$

Combining (5), (14) and (15),

$$\Rightarrow \begin{cases} a_2 = -\mu g & (16a) \\ \beta_2 = \dfrac{5\mu g}{2R} & (16b) \end{cases}$$

Considering ***the velocity-time relation of uniformly variable motion and the angular velocity-time relation of uniformly variable rotation***:

$$\Rightarrow \begin{cases} v_2 = v_{20} + a_2 t = v - \mu g t & (17a) \\ \omega_2 = \omega_{20} + \beta_2 t = \dfrac{5\mu g}{2R} t & (17b) \end{cases}$$

v_2—the $\boldsymbol{i}$-direction component of the center-of-mass velocity of the Sphere 2 at arbitrary Time t.

ω_2—the $\boldsymbol{j}$-direction component of the rotating angular velocity of Sphere 2 at arbitrary Time t.

The condition for Sphere 2 to perform the pure rolling relative to the ground is as follows:

$$v_2\big|_{t=t_2} - \omega_2\big|_{t=t_2} R = 0 \quad (18)$$

t_2—the Time when Sphere 2 reaches pure rolling relative to the ground.

$$\Rightarrow \qquad t_2 = \frac{2v}{7\mu g} \tag{19}$$

$$\Rightarrow \qquad v_2\big|_{t=t_2} = \frac{5}{7}v \tag{20}$$

Combining (10a) and (17a), and combining (13) and (20),

$$\Rightarrow \qquad v_2 - v_1 = v - 2\mu g t \begin{cases} > 0 & t < \dfrac{2v}{7\mu g} & (21a) \\ = \dfrac{3}{7}v & t \geq \dfrac{2v}{7\mu g} & (21b) \end{cases}$$

That is, after the first collision between the two spheres, the ***i***-component of the center-of-mass velocity of Sphere 2 relative to Sphere 1 is greater than zero all the time; therefore, the second collision between the two spheres will not occur, the assumption above holds.

Solution [2]

Taking *Sphere 1* as the object of study (see Fig. 4). Studying *the process from Time zero to the time point when Sphere 1 reaches pure rolling relative to the ground*. Considering (a) ***law of conservation of angular momentum*** [taking an arbitrary straight line on the ground parallel to the y-axis as the rotation axis, the ***j***-direction component of the resultant external torque acting on Sphere 1 is zero; therefore, the ***j***-direction component of its angular momentum is conserved] (the ***j***-direction component equation), (b) ***theorem for the composition of angular momentums based on center-of-mass reference frame*** [integrating into (a)], (c) ***the form of angular momentum of a rigid body performing fixed-axis rotation*** [integrating into (b)]:

$$mv_{10}R + I_1\omega_{10} = mV_1R + I_1\Omega_1 \tag{23}$$

V_1—the ***i***-direction component of the center-of-mass velocity of Sphere 1 at the time point when it reaches pure rolling relative to the ground.

Ω_1—the ***j***-direction component of the rotating angular velocity of Sphere 1 at the time point when it reaches pure rolling relative to the ground.

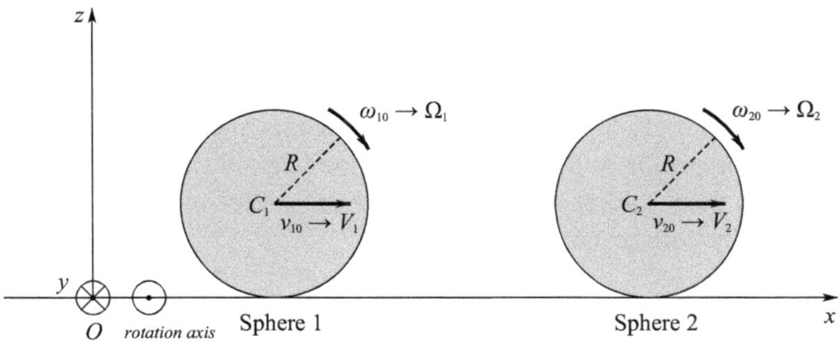

Figure 4

The condition for Sphere 1 to perform the pure rolling relative to the ground is as follows:

$$V_1 - \Omega_1 R = 0 \tag{24}$$

Combining (5), (6a), (6b), (23) and (24),

$$\Rightarrow \qquad V_1 = \frac{2}{7}v \tag{25}$$

Employing a similar analysis for Ball 2, we obtain:

$$mv_{20}R + I_2\omega_{20} = mV_2R + I_2\Omega_2 \tag{26}$$

$$V_2 - \Omega_2 R = 0 \tag{27}$$

V_2—the ***i***-direction component of the center-of-mass velocity of Sphere 2 at the time point when it reaches pure rolling relative to the ground.

Ω_2—the ***j***-direction component of the rotating angular velocity of Sphere 2 at the time point when it reaches pure rolling relative to the ground.

Combining (5), (6c), (6d), (26) and (27),

$$\Rightarrow \qquad V_2 = \frac{5}{7}v \tag{28}$$

3.1.12 The Motion of Two Homogeneous Wheels Connected with Each Other on a Rough Ground (d, *C*)

There are two identical homogeneous cylindrical wheels 1 and 2, the central axes of which are parallel to each other, connected by a light rigid rod (see Fig. 1). Let the two wheels landing gently at the same time with the initial center-of-mass velocity of the magnitude v_0 and the direction horizontal right, without rotation. Given that the friction coefficient between the two wheels and the ground is μ_1 and μ_2, respectively, $\mu_1 > \mu_2$. Find the time taken for the two wheels to reach pure rolling, respectively, and the center-of-mass velocity when they both reach pure rolling.

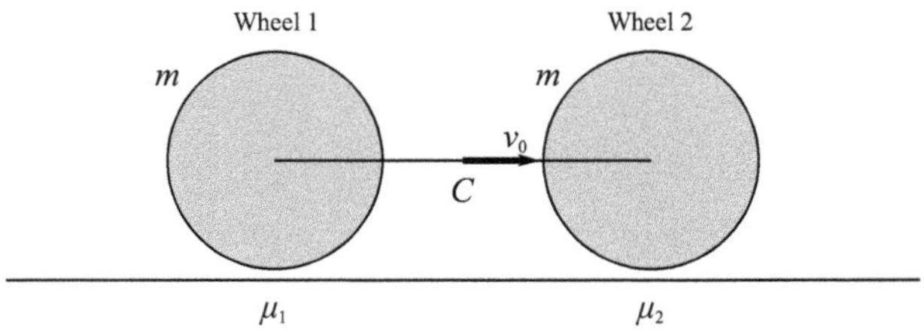

Figure 1

Solution [1]

The time point of the two wheels landing is defined as Time zero.

As shown in Fig. 2, taking an arbitrary point on the ground and denoting it by O, establishing the Cartesian coordinate system with the frame $(O, \boldsymbol{i}, \boldsymbol{j}, \boldsymbol{k})$, let x, y, z be the Cartesian coordinates.

The process before one of the two wheels reaches pure rolling firstly is denoted by Process 1.

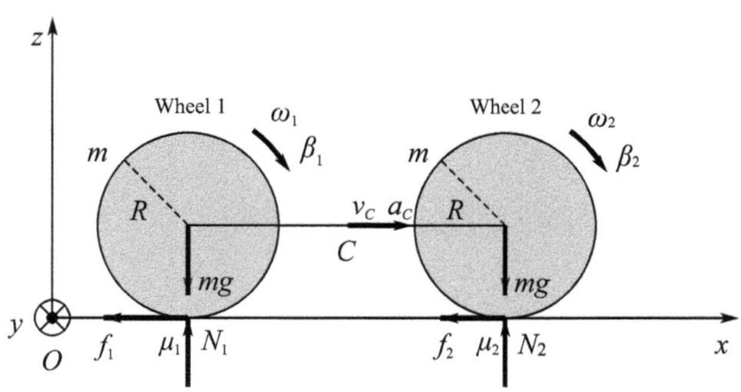

Figure 2

Taking *the system composed of the two wheels and the light rod* as the object of study. Studying *the state at arbitrary Time t in Process*

1. Considering ***theorem of motion of center of mass*** (the ***i***-direction component equation):

$$-f_1 - f_2 = 2ma_C \tag{1}$$

f_1 and f_2—the −***i***-direction component of the sliding friction force exerted on Wheel 1 and Wheel 2 by the ground at Time *t* in Process 1.

a_C—the ***i***-direction component of the center-of-mass acceleration of the system at Time *t* in Process 1.

Taking *Wheel 1 and Wheel 2* as the object of study, respectively. Studying *the state at arbitrary Time t in Process 1*. Considering (a) ***theorem of motion of center of mass*** (the ***k***-direction component equation), (b) ***law of rotation*** [in their respective center-of-mass reference frames, taking each central axis as the corresponding rotation axis] (the ***j***-direction component equation), (c) ***in the center-of-mass reference frame, the resultant torque of the inertia force acting on system of particles is zero***, (d) ***the form of moment of inertia of a homogeneous cylinder about its central axis*** [integrating into (b)]:

$$\begin{cases} N_1 - mg = 0 & \text{(2a)} \\ f_1 R = \dfrac{1}{2} mR^2 \beta_1 & \text{(2b)} \end{cases}$$

$$\begin{cases} N_2 - mg = 0 & \text{(3a)} \\ f_2 R = \dfrac{1}{2} mR^2 \beta_2 & \text{(3b)} \end{cases}$$

N_1 and N_2—the ***k***-direction component of the normal contact force exerted on Wheel 1 and Wheel 2 by the ground at Time *t* in Process 1.

β_1 and β_2—the ***j***-direction component of the angular acceleration of Wheel 1 and Wheel 2 at Time *t* in Process 1.

Considering ***friction law***:

$$\begin{cases} f_1 = \mu_1 N_1 & \text{(4a)} \\ f_2 = \mu_2 N_2 & \text{(4b)} \end{cases}$$

Combining (1) to (4),

$$\Rightarrow \quad \begin{cases} a_C = -\dfrac{\mu_1+\mu_2}{2}g & \text{(5a)} \\ \beta_1 = \dfrac{2\mu_1 g}{R} & \text{(5b)} \\ \beta_2 = \dfrac{2\mu_2 g}{R} & \text{(5c)} \end{cases}$$

Considering (a) ***the velocity-time relation of uniformly variable motion***. (b) ***the angular velocity-time relation of uniformly variable rotation***:

$$\Rightarrow \quad \begin{cases} v_C = v_0 + a_C t = v_0 - \dfrac{\mu_1+\mu_2}{2}gt & \text{(6a)} \\ \omega_1 = \beta_1 t = \dfrac{2\mu_1 g}{R}t & \text{(6b)} \\ \omega_2 = \beta_2 t = \dfrac{2\mu_2 g}{R}t & \text{(6c)} \end{cases}$$

v_C—the ***i***-direction component of the center-of-mass velocity of the system at Time t in Process 1.

ω_1 and ω_2—the ***j***-direction component of the angular velocity of Wheel 1 and Wheel 2 at Time t in Process 1.

Known:

$$\mu_1 > \mu_2 \tag{7}$$

$$\Rightarrow \quad \omega_1 > \omega_2 \tag{8}$$

It can be seen that Wheel 1 reaches pure rolling firstly.

The condition for Wheel 1 to reach pure rolling relative to the ground is as follows:

$$v_C\big|_{t=t_1} - \omega_1\big|_{t=t_1} R = 0 \tag{9}$$

t_1—the Time when Wheel 1 reaches pure rolling relative to the ground.

(6a) and (6b) $\rightarrow$ (9),

$$\Rightarrow \quad t_1 = \frac{2v_0}{(5\mu_1+\mu_2)g} \tag{10}$$

(10) $\rightarrow$ (6a),

$$\Rightarrow \qquad v_C\Big|_{t=t_1} = \frac{4\mu_1}{5\mu_1 + \mu_2} v_0 \qquad (11)$$

The process from Wheel 1 reaching pure rolling to Wheel 2 reaching pure rolling is denoted by Process 2.

It is assumed that in Process 2, Wheel 1 performs the pure rolling relative to the ground all the time, and the rotation law of Wheel 2 is the same as that in Process 1.

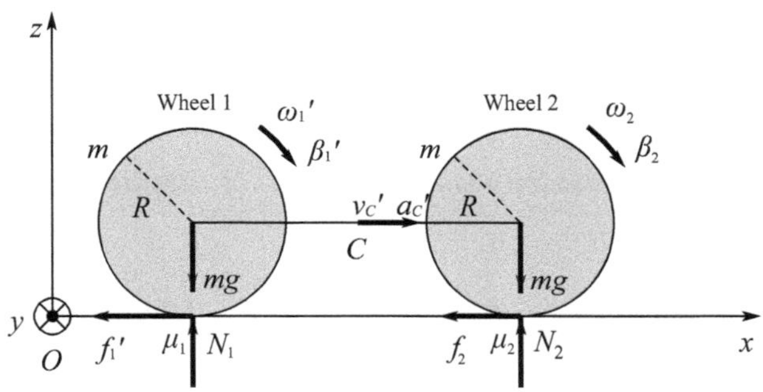

Figure 3

Taking *the system composed of the two wheels and the light rod* as the object of study (see Fig. 3). Studying *the state at arbitrary Time t′ in Process 2*. Considering ***the theorem of motion of center of mass*** (the ***i***-direction component equation):

$$-f_1' - f_2 = 2ma_C' \qquad (12)$$

f'_1—the −***i***-direction component of the sliding friction force exerted on Wheel 1 by the ground at Time t' in Process 2.

a'_C—the ***i***-direction component of the center-of-mass acceleration of the system at Time t' in Process 2.

Taking *Wheel 1* as the object of study. Studying *the state at arbitrary Time t′ in Process 2*. Considering (a) ***law of rotation*** [in its center-of-mass reference frame, taking its central axis as the rotation axis] (the ***j***-direction component equation), (b) ***in the center-of-mass reference frame, the resultant torque of the inertia force acting on system of particles is zero***, (c) ***the form of moment of inertia of a homogeneous cylinder about its central axis*** [integrating into (a)].

$$f_1{}'R = \frac{1}{2}mR^2\beta_1{}' \tag{13}$$

β'_1—the ***j***-direction component of the angular acceleration of Wheel 1 at Time t' in Process 2.

The condition for Wheel 1 to perform pure rolling relative to the ground all the time in Process 2 is as follows:

$$a'_C - \beta_1{}'R = 0 \tag{14}$$

Combining (3), (4b), (12) to (14),

$$\Rightarrow \quad a'_C = -\frac{2}{5}\mu_2 g \tag{15}$$

Considering ***the velocity-time relation of uniformly variable motion***:

$$\Rightarrow \quad v'_C = v_C\big|_{t=t_1} + a'_C t' = \frac{4\mu_1}{5\mu_1+\mu_2}v_0 - \frac{2}{5}\mu_2 g t' \tag{16}$$

v'_C—the ***i***-direction component of the center-of-mass velocity of the system at arbitrary Time t' in Process 2.

The condition for Wheel 2 to reach pure rolling relative to the ground is as follows:

$$v'_C\big|_{t'=t_2} - \omega_2\big|_{t=t_1+t_2}R = 0 \tag{17}$$

t_2—the Time when Wheel 2 reaches pure rolling relative to the ground.

(6c) and (16) $\rightarrow$ (17),

$$\Rightarrow \quad \left(\frac{4\mu_1}{5\mu_1+\mu_2}v_0 - \frac{2}{5}\mu_2 g t_2\right) - \frac{2\mu_2 g}{R}(t_1+t_2)\cdot R = 0 \tag{18}$$

(10) $\rightarrow$ (18),

$$\Rightarrow \quad t_2 = \frac{5(\mu_1-\mu_2)v_0}{3(5\mu_1+\mu_2)\mu_2 g} \tag{19}$$

(19) $\rightarrow$ (16),

$$\Rightarrow \quad v'_C\big|_{t'=t_2} = \frac{2}{3}v_0 \tag{20}$$

Solution [2]

Taking *the system composed of the two wheels and the light rod* as the object of study (see Fig. 4). Studying *the total process consisting of*

Process 1 and Process 2. Considering ***law of conservation of angular momentum*** [taking an arbitrary line parallel to the y-axis on the ground as the rotation axis, the resultant external torque acting on the system is zero; therefore, the angular momentum of the system is conserved] (the $\boldsymbol{j}$-direction component equation):

$$2mv_0R = 2\left(mV_CR + \frac{1}{2}mR^2\Omega\right) \tag{21}$$

V_C—the $\boldsymbol{i}$-direction component of the center-of-mass velocity of the system when both wheels reach pure rolling relative to the ground.

Ω—the $\boldsymbol{j}$-direction component of the angular velocity of two wheels when both wheels reach pure rolling relative to the ground.

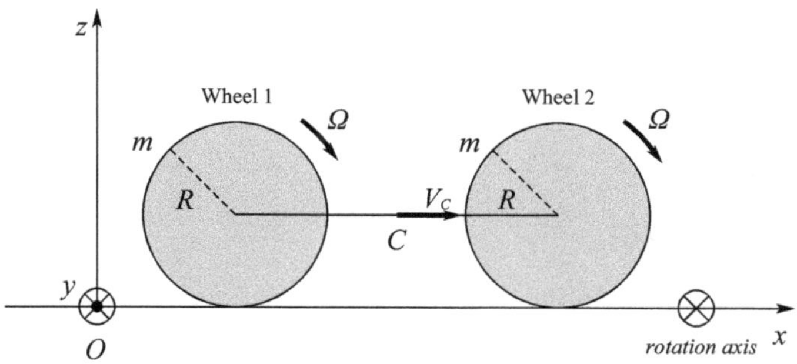

Figure 4

The condition for both wheels to reach pure rolling relative to the ground is as follows:

$$V_C - \Omega R = 0 \tag{22}$$

$$\Rightarrow \quad V_C = \frac{2}{3}v_0 \tag{23}$$

3.1.13 The Constrained Motions of a Homogeneous Hollow Cylinder and a Small Ball (d, *AC*)

As shown in Fig. 1, there is a homogeneous hollow cylinder of mass M, sectional radius R and height h, with an equidistant spiral groove on its inner surface, whose pitch is just h, that is, the line segment between the highest point P_1 and the lowest point P_2 of the groove is in the vertical direction, the helix direction is as shown in the figure. The cylinder is statically placed on a smooth horizontal ground. Release a small ball of mass m from the highest point of the groove, the ball can slide along the groove without friction. Find the rotation angle of the plane between the ball and the central axis of the cylinder when the ball lands.

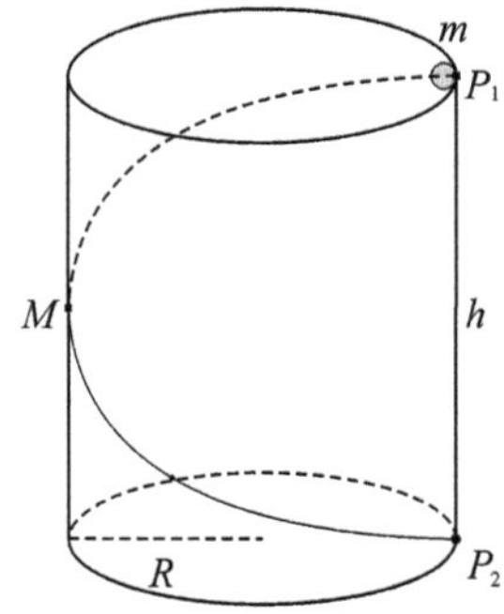

Figure 1

Solution (type C)

The time point when the ball is released is defined as Time zero.

As shown in Fig. 2, the center of mass of the system composed of the cylinder and the ball is denoted by C, the projection of which on the ground is denoted by O.

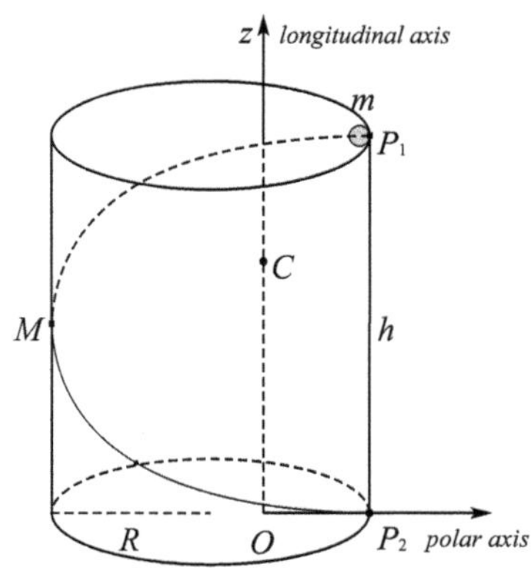

Figure 2

Taking *the system composed of the cylinder and the ball* as the object of study. Studying *the process from Time zero to arbitrary Time*

t. Considering (a) ***law of conservation of momentum*** [the horizontal component of the resultant external force acting on the system is zero; therefore, the horizontal component of the momentum of the system is conserved] (the horizontal projection equation), (b) ***the total momentum of the system of particles is equal to the center-of-mass momentum*** [integrating into (a)]:

$$(M + m)\, \boldsymbol{v}_C = 0 \tag{1}$$

$\boldsymbol{v}_C$—the horizontal projection of the center-of-mass velocity of the system at Time t.

$$\Rightarrow \qquad \boldsymbol{v}_C = 0 \tag{2}$$

Therefore, the projection of the center of mass C on the ground, i.e., Point O remains stationary.

At arbitrary Time t, the position of the ball is denoted by A, the point on the groove in contact with the ball is denoted by P, the projection of the ball on the central axis of the cylinder is denoted by A', the projection of Point O on AA' is denoted by O', the unit vector in the vertical upward direction is denoted by $\boldsymbol{k}$. Taking Point O as the pole, the direction of $\overrightarrow{OP_2}$ at Time zero as the positive direction of the polar axis, the right-hand helix direction with regard to $\boldsymbol{k}$ as the positive direction of polar angle, the $\boldsymbol{k}$-direction as the positive direction of the longitudinal axis, establishing cylindrical coordinate system, let r, θ, z be the cylindrical coordinates.

Taking *the system composed of the cylinder and the ball* as the object of study. Studying *the process from Time zero to arbitrary Time t*. Considering ***law of conservation of angular momentum*** [taking the z-axis as the rotation axis, the $\boldsymbol{k}$-direction component of the resultant external torque acting on the system relative to the z-axis is zero; therefore, the $\boldsymbol{k}$-direction component of the angular momentum of the system relative to the z-axis is conserved] (the $\boldsymbol{k}$-direction component equation):

$$L_M + L_m = 0 \tag{3}$$

L_M and L_m—the $\boldsymbol{k}$-direction component of the angular momentum of the cylinder and the ball relative to the z-axis at Time t.

Considering ***theorem for the composition of angular momentums based on center-of-mass reference frame***:

$$L_M = L_{MC_M} + L'_M \tag{4}$$

L_{MC_M}—the $\boldsymbol{k}$-direction component of the orbital angular momentum of the cylinder at arbitrary Time t.

L'_M—the $\boldsymbol{k}$-direction component of the spin angular momentum of the cylinder at arbitrary Time t.

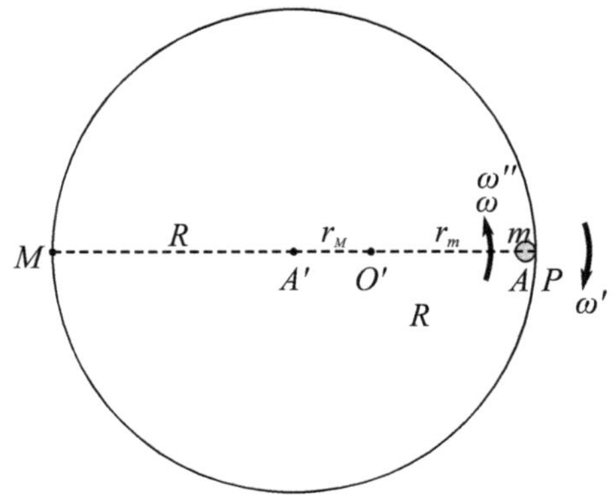

Figure 3

As shown in Fig. 3, considering (a) the definition of ***angular momentum of particle***, (b) the definition of ***orbital angular momentum***, (c) ***the form of angular momentum of a rigid body performing fixed-axis rotation***.

$$L_m = m\omega r_m \cdot r_m \tag{5}$$

$$L_{MC_M} = M\omega r_M \cdot r_M \tag{6}$$

$$L'_M = -I\omega' \tag{7}$$

ω—the $\boldsymbol{k}$-direction component of the angular velocity of AA' at arbitrary Time t.

ω'—the $-\boldsymbol{k}$-direction component of the angular velocity of PA' at arbitrary Time t.

r_m and r_M—the length of AO' and $A'O'$.

I—the moment of inertia of the cylinder about its central axis.

Considering ***the form of moment of inertia of a homogeneous hollow cylinder about its central axis***:

$$I = MR^2 \tag{8}$$

Combining (3) to (8),

$$\Rightarrow \qquad (M\omega r_M \cdot r_M - MR^2\omega') + m\omega r_m \cdot r_m = 0 \tag{9}$$

Considering the definition of ***center of mass***:

$$\begin{cases} r_M = \dfrac{m}{M+m}R & \text{(10a)} \\ r_m = \dfrac{M}{M+m}R & \text{(10b)} \end{cases}$$

$$\Rightarrow \qquad \omega' = \frac{m}{M+m}\omega \tag{11}$$

Taking *the line segment AA′* as the object of study. Studying *the state at arbitrary Time t*. Considering ***theorem for the composition of angular velocities*** [taking the ground as the stationary reference frame, and the cylinder reference frame as the moving reference frame] (the ***k***-direction component equation):

$$\omega = -\omega' + \omega'' \tag{12}$$

ω''—the ***k***-direction component of the angular velocity of AA' relative to PA' at Time t.

$$\Rightarrow \qquad \omega = \frac{M+m}{M+2m}\omega'' \tag{13}$$

Considering the definition of ***angular velocity***:

$$\begin{cases} \omega = \dfrac{\mathrm{d}\theta}{\mathrm{d}t} & \text{(14a)} \\ \omega'' = \dfrac{\mathrm{d}\theta''}{\mathrm{d}t} & \text{(14b)} \end{cases}$$

$\mathrm{d}\theta$—the ***k***-direction component of the infinitesimal angular displacement of AA' in the instantaneous process (the small time interval $\mathrm{d}t$) at arbitrary Time t.

$\mathrm{d}\theta''$—the ***k***-direction component of the infinitesimal angular displacement of AA' relative to PA' in the instantaneous process (the small time interval $\mathrm{d}t$) at arbitrary Time t.

$$\Rightarrow \qquad \mathrm{d}\theta = \frac{M+m}{M+2m}\mathrm{d}\theta'' \tag{15}$$

$$\Rightarrow \qquad \theta_{\text{landing}} = \int_0^{\theta_{\text{landing}}} \mathrm{d}\theta = \frac{M+m}{M+2m}\int_0^{2\pi} \mathrm{d}\theta'' = \frac{M+m}{M+2m}\cdot 2\pi \tag{16}$$

Another method to get Eq. (12) (type A)

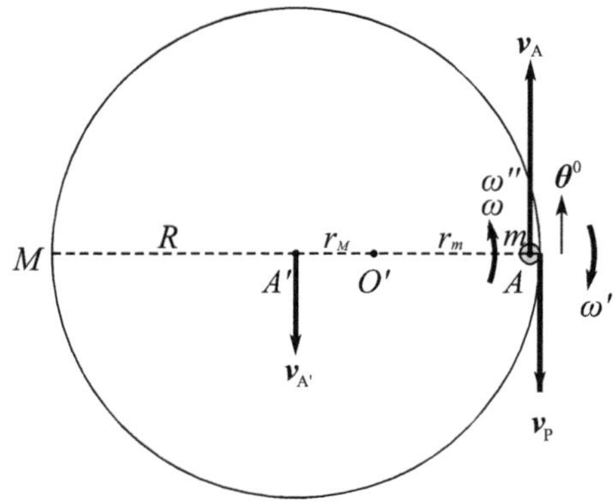

Figure 4

Taking *Point A* as the object of study. Studying *the state at arbitrary Time t.* Considering ***theorem for composition of velocities of a particle*** (see Fig. 4) [taking the ground as the stationary reference frame, and the translational reference frame that is at rest with respect to Point *P* as the moving reference frame] (the $\boldsymbol{\theta}^0$-direction component equation):

$$v_A = v_P + v_{AP} \tag{17}$$

v_A, v_P and v_{AP}—the $\boldsymbol{\theta}^0$-direction component of the velocity of Point *A*, Point *P*, and Point *A* relative to Point *P* at Time *t*.

Taking *Point P* as the object of study. Studying *the state at arbitrary Time t.* Considering ***theorem for composition of velocities of a particle*** [taking the ground as the stationary reference frame, and the translational reference frame that is at rest with respect to Point *A′* as the moving reference frame] (the $\boldsymbol{\theta}^0$-direction component equation):

$$v_P = v_{A'} + v_{PA'} \tag{18}$$

$v_{A'}$ and $v_{PA'}$—the $\boldsymbol{\theta}^0$-direction component of the velocity of Point *A′*, and Point *P* relative to Point *A′* at Time *t*.

Considering ***the transverse component form of velocity***:

$$\begin{cases} v_A = \omega r_m & (19a) \\ v_{A'} = -\omega r_M & (19b) \\ v_{PA'} = -\omega' R & (19c) \\ v_{AP} = \omega'' R & (19d) \end{cases}$$

Combining (17) to (19):

$$\Rightarrow \qquad \omega r_m = (-\omega r_M - \omega' R) + \omega'' R \qquad (20)$$

(10) $\rightarrow$ (20),

$$\Rightarrow \qquad \omega = \omega'' - \omega' \qquad (21)$$

3.1.14 One Cylinder Rolling over Another Cylinder without Relative Sliding (*a, A*)

Consider a homogeneous cylinder A of sectional radius R performing pure rolling on a horizontal ground, with the center-of-mass velocity of the magnitude v_0 and the direction horizontally to the right (see Fig. 1). In front of A, there is a long cylinder B of sectional radius r fixed on the ground, parallel to A. There will be a completely inelastic collision between them, and the friction coefficient between them is large enough to prevent the relative sliding between them.

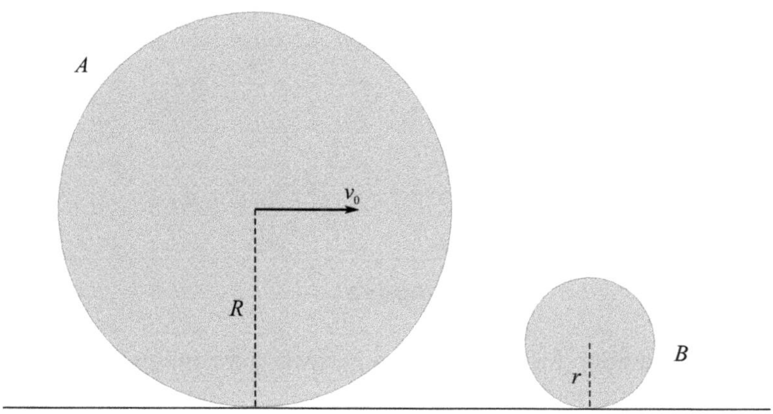

Figure 1

(1) Suppose Cylinder A can roll over Cylinder B in a way to be in contact with Cylinder B all the time. When Cylinder A moves to an arbitrary position, find: (i) the normal contact force exerted on Cylinder A by Cylinder B, (ii) the rotating angular velocity and rotating angular acceleration of Cylinder A.

(2) Find the conditions for Cylinder A to be able to roll over Cylinder B in constant contact with Cylinder B.

Solution to Problem (1)

As shown in Fig. 2, the center of mass of Cylinder A is denoted by O_A, the projection of O_A on the central axis of Cylinder B is denoted by O_B, when Cylinder A and B collide, the intersection of their contact line and O_AO_B is denoted by C. The unit vector perpendicular to the plane of the figure inward is denoted by $\boldsymbol{k}$. Taking Point O_B as the pole, the horizontal left direction as the positive direction of the polar axis, the right-hand helix direction with regard to the $\boldsymbol{k}$-direction as the positive direction of polar angle, the $\boldsymbol{k}$-direction as the positive direction of the longitudinal axis, establishing the cylindrical coordinate system, let r, θ, z be the cylindrical coordinates.

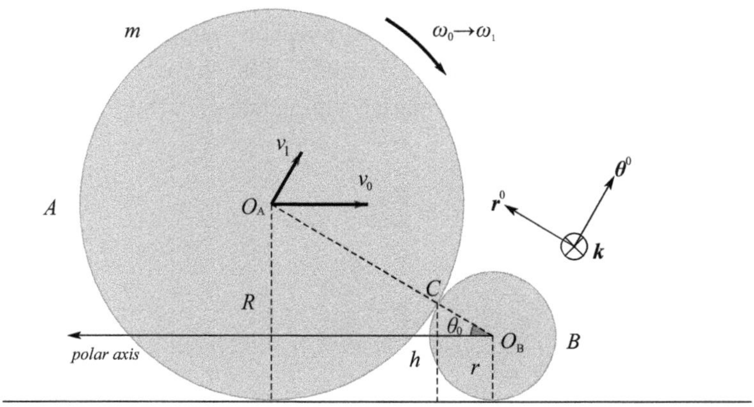

Figure 2

The condition for Cylinder A to perform pure rolling relative to the ground is as follows:

$$v_0 - \omega_0 R = 0 \tag{1}$$

ω_0—the $\boldsymbol{k}$-direction component of the rotating angular velocity of Cylinder A at arbitrary time before the collision.

Taking *Cylinder A* as the object of study. Studying *the instantaneous process of the collision between Cylinders A and B*. Considering (a) ***law of conservation of angular momentum*** [taking the contact line of Cylinder A and B as the rotation axis, only the gravity of Cylinder A provides the moment of impulse, which is a finite quantity, acting on an extremely short process, its moment of impulse is an infinitely small quantity, which can be ignored; therefore, the $\boldsymbol{k}$-direction component of the angular momentum of Cylinder A is approximately conserved] (the $\boldsymbol{k}$-direction component equation), (b) ***theorem for the composition of angular momentums based on center-of-mass reference frame***:

$$L_{C0} = L_{C1} \tag{2}$$

$$\begin{cases} L_{C0} = L_{O_A-C0} + L'_{O_A 0} & \text{(3a)} \\ L_{C1} = L_{O_A-C1} + L'_{O_A 1} & \text{(3b)} \end{cases}$$

L_{C0} and L_{C1}—the $\boldsymbol{k}$-direction component of the angular momentum of Cylinder A at the time point before and after the collision [taking the contact line of Cylinder A and B as the rotation axis].

L_{O_A-C0} and L_{O_A-C1}—the $\boldsymbol{k}$-direction component of the orbital angular momentum of Cylinder A at the time point before and after the collision.

$L'_{O_A 0}$ and $L'_{O_A 1}$—the $\boldsymbol{k}$-direction component of the spin angular momentum of Cylinder A at the time point before and after the collision.

Considering (a) the definition of ***orbital angular momentum***, (b) ***the form of angular momentum of a rigid body performing fixed-axis rotation***:

$$\begin{cases} L_{O_A-C0} = mv_0(R-h) & \text{(4a)} \\ L_{O_A-C1} = mv_1 R & \text{(4b)} \end{cases}$$

$$\begin{cases} L'_{O_A 0} = I_{O_A}\omega_0 & \text{(5a)} \\ L'_{O_A 1} = I_{O_A}\omega_1 & \text{(5b)} \end{cases}$$

m—the mass of Cylinder A.

I_{O_A}—the moment of inertia of Cylinder A about its central axis.

ω_1—the $\boldsymbol{k}$-direction component of the rotating angular velocity of Cylinder A at the time point after the collision.

v_1—the $\boldsymbol{\theta}^0$-direction component of the center-of-mass velocity of Cylinder A at the time point after the collision.

h—the distance from Point C to the ground.

Considering ***the form of moment of inertia of a homogeneous cylinder about its central axis***:

$$I_{O_A} = \frac{1}{2}mR^2 \tag{6}$$

The condition for Cylinder A to perform pure rolling relative to Cylinder B is as follows:

$$v_1 - \omega_1 R = 0 \tag{7}$$

Considering the following geometric relation in Fig. 2:

$$h = r + r\sin\theta_0 = r + r\frac{R-r}{R+r} = r\left(1+\frac{R-r}{R+r}\right) \tag{8}$$

Combining (1) to (8),

$$\Rightarrow \qquad \omega_1 = \left[1 - \frac{4r}{3(R+r)}\right]\frac{v_0}{R} \tag{9}$$

The time point after the collision between Cylinder A and B is defined as Time zero.

As shown in Fig. 3, the intersection of the contact line of the two cylinders and O_AO_B at arbitrary Time t is denoted by C'.

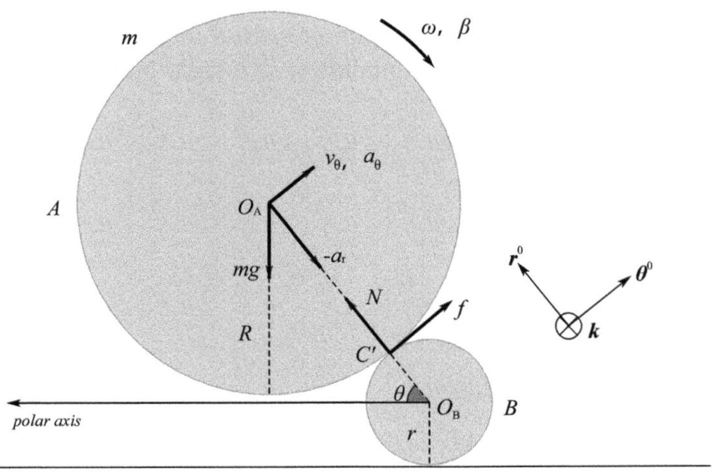

Figure 3

Taking *the system composed of Cylinder A and the earth* as the object of study. Studying *the process from Time zero to arbitrary Time t*. Considering ***law of conservation of mechanical energy*** [the work done by the external force and non-conservative internal force acting on the system is zero (no work is done by the static friction force exerted on Cylinder A by Cylinder B); therefore, the mechanical energy of the system is conserved] (taking the ground as the zero point of gravitational potential energy):

$$E_{k1}+mgR=E_k+mg\left[r+(R+r)\sin\theta\right] \tag{10}$$

E_{k1} and E_k—the kinetic energy of Cylinder A at Time zero and Time t.

Considering (a) ***the form of kinetic energy of a rigid body performing fixed-axis rotation***, (b) ***parallel-axis theorem***:

$$\begin{cases} E_{k1}=\dfrac{1}{2}I_C\omega_1^2 & \text{(11a)} \\ E_k=\dfrac{1}{2}I_{C'}\omega^2 & \text{(11b)} \end{cases}$$

$$I_C=I_{C'}=I_{O_A}+mR^2 \tag{12}$$

ω—the $\boldsymbol{k}$-direction component of the rotating angular velocity of Cylinder A at arbitrary Time t.

I_C—the moment of inertia of Cylinder A about the contact line of Cylinder A and B at Time zero.

$I_{C'}$—the moment of inertia of Cylinder A about the contact line of Cylinder A and B at arbitrary Time t.

Combining (6), (9) to (12),

$$\Rightarrow \qquad \omega=\sqrt{\frac{4}{3}\frac{g\left[R-r-(R+r)\sin\theta\right]}{R^2}+\left[1-\frac{4r}{3(R+r)}\right]^2\frac{v_0^2}{R^2}} \tag{13}$$

Taking *Cylinder A* as the object of study. Studying *the state at arbitrary Time t*. Considering (a) ***theorem of motion of center of mass*** (the $-\boldsymbol{r}^0$-direction and $\boldsymbol{\theta}^0$-direction component equations), (b) ***law of rotation*** [in its center-of-mass reference frame, taking its central axis as the rotation axis] (the $\boldsymbol{k}$-direction component equation), (c) ***in the center-of-mass reference frame, the resultant torque of the inertia force acting on system of particles is zero***:

$$\begin{cases}\begin{cases} mg\sin\theta - N = m(-a_r) & (14a)\\ f - mg\cos\theta = ma_\theta & (14b)\end{cases}\\ -fR = I_{O_A}\beta & (14c)\end{cases}$$

N—the $\boldsymbol{r}^0$-direction component of the normal contact force exerted on Cylinder A by Cylinder B at Time t.

f—the $\boldsymbol{\theta}^0$-direction component of the fiction exerted on Cylinder A by Cylinder B at Time t.

a_r and a_θ—the $\boldsymbol{r}^0$-direction and $\boldsymbol{\theta}^0$-direction component of the center-of-mass acceleration of Cylinder A at Time t.

β—the $\boldsymbol{k}$-direction component of the rotating angular acceleration of Cylinder A at Time t.

Considering ***the radial component form of acceleration*** [the constraint condition: the distance between the center of mass O_A and Point O_B remains unchanged]:

$$a_r = -\frac{v_\theta^2}{R+r} \tag{15}$$

v_θ—the $\boldsymbol{\theta}^0$-direction component of the velocity of center of mass O_A at arbitrary Time t.

The condition for Cylinder A to perform pure rolling relative to Cylinder B all the time is as follows:

$$\begin{cases} v_\theta - \omega R = 0 & (16a)\\ a_\theta - \beta R = 0 & (16b)\end{cases}$$

Combining (6), (13) to (16),

$$\Rightarrow \begin{cases}\begin{cases} N = mg\sin\theta - \\ \quad m\frac{R^2}{R+r}\left\{\frac{4}{3}\frac{g[R-r-(R+r)\sin\theta]}{R^2} + \left[1-\frac{4r}{3(R+r)}\right]^2\frac{v_0^2}{R}\right\} & (17a)\\ f = \frac{1}{3}mg\cos\theta & (17b)\end{cases}\\ \beta = -\frac{2}{3}\frac{g}{R}\cos\theta & (18)\end{cases}$$

Solution to Problem (2)

The condition that Cylinder A can roll over Cylinder B is that at arbitrary Time t before Cylinder A rolls to the top of Cylinder B, the $\boldsymbol{k}$-direction component of the rotating angular velocity of Cylinder A is greater than zero, that is:

$$\omega_{\min} > 0 \tag{19}$$

From (13),

$$\Rightarrow \qquad \omega_{\min}^2 = \omega^2\Big|_{\theta=\frac{\pi}{2}} = -\frac{8}{3}\frac{gr}{R^2} + \left[1 - \frac{4r}{3(R+r)}\right]^2 \frac{v_0^2}{R^2} \tag{20}$$

$$\Rightarrow \qquad v_0 > \frac{3(R+r)}{3R-r}\sqrt{\frac{8}{3}gr} \tag{21}$$

The condition that Cylinder A to be in contact with Cylinder B all the time is that at arbitrary Time t before Cylinder A rolls to the top of Cylinder B, the $\boldsymbol{r}^0$-direction component of normal contact force exerted on Cylinder A by Cylinder B is greater than zero, that is:

$$N_{\min} > 0 \tag{22}$$

From (17a),

$$\Rightarrow \qquad N_{\min} = N\Big|_{\theta=\theta_0} = mg\frac{R-r}{R+r} - m\frac{v_0^2}{R+r}\left[1 - \frac{4r}{3(R+r)}\right]^2 \tag{23}$$

$$\Rightarrow \qquad v_0 < \frac{3(R+r)}{3R-r}\sqrt{g(R-r)} \tag{24}$$

Combining (21) and (24),

$$\Rightarrow \qquad \frac{3(R+r)}{3R-r}\sqrt{\frac{8}{3}gr} < v_0 < \frac{3(R+r)}{3R-r}\sqrt{g(R-r)}$$

$$\Rightarrow \qquad \sqrt{\frac{8}{3}gr} < \sqrt{g(R-r)}$$

$$\Rightarrow \qquad R > \frac{11}{3}r \tag{25}$$

The conditions for Cylinder A to be able to roll over Cylinder B in constant contact with Cylinder B are inequalities (21), (24) and (25).

3.1.15 A Plank's Motion on Rollers Arranged on an Inclined Plane (d, *AC*)

As shown in Fig. 1, on an inclined plane of inclination angle α, there are many rollers of mass m and sectional radius r, whose central axes are all horizontal and parallel to each other, and arranged with spacing d, the surfaces of the rollers are all covered with the rough rubber. A plank of mass M and length far greater than d is released from the top of the inclined plane, and slides down along the inclined plane. After a sufficiently long time, the velocity of the plank will be approximately stable, in this stage, at the time point after the plank drives each roller newly encountered, its velocity is the same, which is called the steady velocity of the plank. Find the steady velocity.

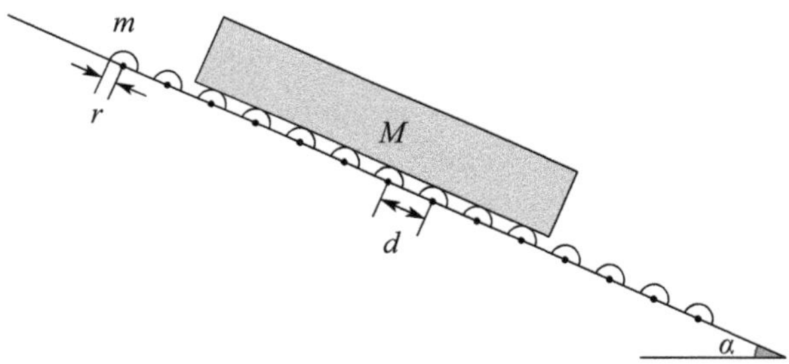

Figure 1

Solution [1] (type C)

As shown in Fig. 2, establishing the Cartesian coordinate system with the frame $(O, \mathbf{i}, \mathbf{j}, \mathbf{k})$, let x, y, z be the Cartesian coordinates.

When the plank just contacts an arbitrary roller in front, an instantaneous relative sliding process will first occur between them.

Sorting the rollers with positive integers from the highest one downward along the x-axis. Suppose that the velocity of the plank reaches the steady velocity after its interacting with No. $(n-1)$ roller. When the instantaneous relative sliding between the plank and No. $(n-1)$ roller completes, the plank will continue to move and then contacts No. n roller, the process of which is denoted by Process *A*. Then there is the instantaneous relative sliding process between the plate and No. n roller, which is denoted by Process *B*.

Taking *No. n roller* as the object of study (see Fig. 3). Studying *the state at arbitrary Time t in Process B*. Considering (a) ***law of***

rotation [taking its central axis as the rotation axis] (the ***j***-direction component equation), (b) ***the form of moment of inertia of a homogeneous cylinder about its central axis***:

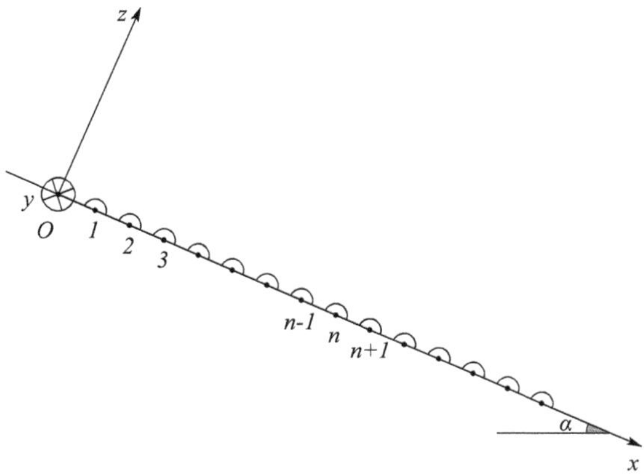

Figure 2

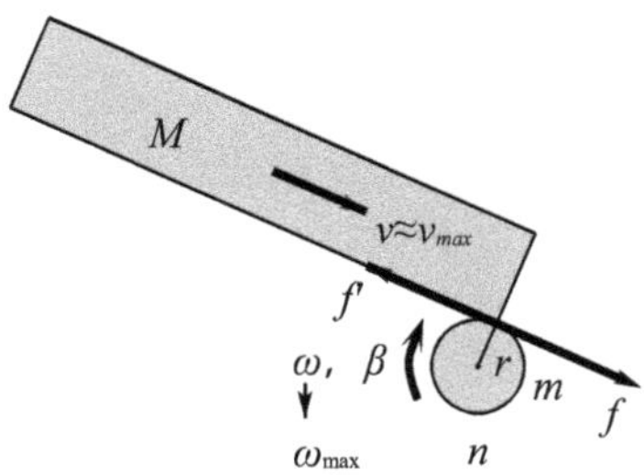

Figure 3

$$fr = I\beta \tag{1}$$

$$I = \frac{1}{2}mr^2 \tag{2}$$

f—the ***i***-direction component of the friction exerted on No. *n* roller by the plank at Time *t* in Process *B*.

β—the ***j***-direction component of the rotating angular acceleration of No. *n* roller at Time *t* in Process *B*.

I—the moment of inertia of No. *n* roller about its central axis.

$$\Rightarrow \qquad f = \frac{1}{2}mr\beta \tag{3}$$

Taking *the system composed of the plank and No. n roller* as the object of study. Studying *the whole process consisting of Process A and Process B*. Considering ***theorem of kinetic energy*** [the velocities of the plank are equal at the beginning and end of this process]:

$$Mgd\sin\theta - \int_0^{\Delta t_B} f'(v-\omega r)\,\mathrm{d}t = \frac{1}{2}I\omega_{\max}^2 \tag{4}$$

f'—the $-\boldsymbol{i}$-direction component of the friction exerted on the plank by No. n roller at arbitrary Time t in Process B.

Δt_B—the time taken of Process B.

v—the $\boldsymbol{i}$-direction component of the velocity of the plank at Time t in Process B.

ω—the $\boldsymbol{j}$-direction component of the rotating angular velocity of No. n roller at Time t in Process B.

$\omega_{\max}$—the $\boldsymbol{j}$-direction component of the rotating angular velocity of No. n roller at the end of Process B.

Considering ***Newton's third law***:

$$f = f' \tag{5}$$

Defining $v_{\max}$—the steady velocity of the plank.

At arbitrary Time t in Process B, the velocity of the plank is approximately equal to the steady velocity:

$$v \approx v_{\max} \tag{6}$$

(3), (5) and (6) $\rightarrow$ (4):

$$\frac{1}{2}I\omega_{\max}^2 = Mgd\sin\theta - \frac{1}{2}mr\int_0^{\Delta t_B}(v_{\max}-\omega r)\beta\,\mathrm{d}t \tag{7}$$

Considering the definition of ***angular acceleration***:

$$\beta = \frac{\mathrm{d}\omega}{\mathrm{d}t} \tag{8}$$

$$\Rightarrow \quad \frac{1}{2}I\omega_{\max}^2 = Mgd\sin\theta - \frac{1}{2}mr\int_0^{\omega_{\max}}(v_{\max}-\omega r)\,\mathrm{d}\omega$$

$$= Mgd\sin\theta - \frac{1}{2}mr\left(v_{\max}\omega_{\max} - \frac{1}{2}r\omega_{\max}^2\right) \tag{9}$$

At the end of Process *B*, No. *n* roller reaches pure rolling relative to the plank, and the condition is as follow:

$$v_{\max} - \omega_{\max} r = 0 \tag{10}$$

Combining (9) and (10):

$$\Rightarrow \qquad v_{\max} = \sqrt{\frac{M}{m} \cdot 2gd\sin\theta} \tag{11}$$

Solution [2] (type A)

Defining Δv_A and Δv_B—the ***i***-direction component of the velocity increment of the plank in Process *A* and Process *B*.

When the velocity of the plank is stable, Δv_A and Δv_B should be opposite to each other:

$$\Delta v_A = -\Delta v_B \tag{12}$$

Δv_A is a small quantity, considering ***the displacement-time relation and the velocity-time relation of uniformly variable motion***:

$$\begin{cases} d \approx v_{\max}\Delta t_A + \dfrac{1}{2} g\sin\theta \Delta t_A^2 \approx v_{\max}\Delta t_A & \text{(13a)} \\ \Delta v_A = g\sin\theta \Delta t_A & \text{(13b)} \end{cases}$$

Δt_A—the time taken of Process *A*.

Taking *the plank* as the object of study. Studying *Process B*. Considering ***momentum theorem*** [the gravity of the plank is a finite quantity, acting on an extremely short process, its impulse is an infinitely small quantity, which can be ignored] (the ***i***-direction component equation):

$$-\int_0^{\Delta t_B} f'\mathrm{d}t = M\Delta v_B \tag{14}$$

Taking *No. n roller* as the object of study. Studying *Process B*. Considering ***theorem of angular momentum*** [taking its central axis as the rotation axis] (the ***j***-direction component equation):

$$\int_0^{\Delta t_B} fr\mathrm{d}t = I\omega_{\max} \tag{15}$$

Equation (11) can obtain by combining Eqs. (2), (5), (10), (12) to (15).

3.1.16 The Infinite Collisions between a Rotating Homogeneous Sphere and a Rough Ground (*b, A*)[3]

A homogeneous sphere of radius of R is located at h above a horizontal ground initially, with its center-of-mass static and the rotating angular velocity of the magnitude ω_0 and the direction clockwise (see Fig. 1). The sphere will collide with the ground, given that the coefficient of friction between them is μ and the collision coefficient of restitution between them is e. Find the center-of-mass displacement of the sphere in the horizontal right direction after infinite collisions.

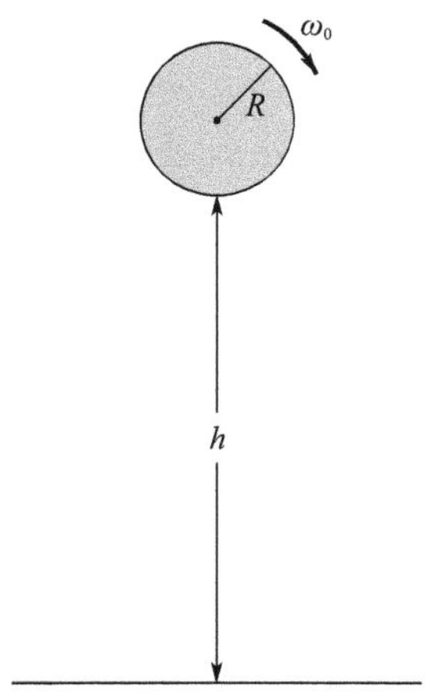

Figure 1

Solution

The initial time is defined as Time zero.

As shown in Fig. 2, the center of mass of the sphere is denoted by C, the projection of which on the ground at Time zero is denoted by O, establishing the Cartesian coordinate system with the frame (O, $\boldsymbol{i}$, $\boldsymbol{j}$, $\boldsymbol{k}$), let x, y, z be the Cartesian coordinates.

The collision process in which there is relative sliding all the time between the sphere and the ground is called sliding collision process. The collision process in which the sphere changes from relative

[3]This problem is adapted from a test question of the 22th IPhO (International Physics Olympiad) in 1991 by the author.

sliding to pure rolling relative to the ground is called transitional collision process (including turning into pure rolling at the end of collision). The collision process in which the sphere is pure rolling all the time relative to the ground is called rolling collision process.

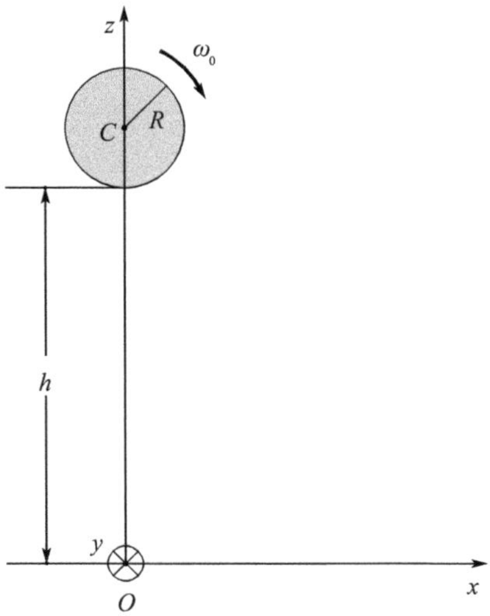

Figure 2

According to the above classification, the first collision process can only be the sliding collision process or the transitional collision process.

Considering ***the velocity-time relation and displacement-time relation of uniformly variable motion***:

$$v'_{z(i)} = v_{z(i)} - gt_i \tag{1}$$

$$\begin{cases} s_i = v_{x(i)}t_i & \text{(2a)} \\ 0 = v_{z(i)}t_i - \dfrac{1}{2}gt_i^2 & \text{(2b)} \end{cases}$$

$v_{z(i)}$ and $v'_{z(i)}$—the ***k***-direction component of the center-of-mass velocity of the sphere at the time point after the *i*-th collision and before the (*i*+1)-th collision.

$v_{x(i)}$—the ***i***-direction component of the center-of-mass velocity of the sphere at the time point after the *i*-th collision.

t_i—the time interval between the *i*-th collision and the (*i*+1)-th collision.

s_i—the ***i***-direction component of the center-of-mass displacement of the sphere in the process from the i-th collision to the $(i+1)$-th collision.

$$\Rightarrow \quad \begin{cases} t_i = \dfrac{2v_{z(i)}}{g} & \text{(3a)} \\ v'_{z(i)} = -v_{z(i)} & \text{(3b)} \\ s_i = \dfrac{2v_{x(i)}v_{z(i)}}{g} & \text{(3c)} \end{cases}$$

Considering ***Newton-collision-formula***:

$$e = \frac{0 - v_{z(i+1)}}{v'_{z(i)} - 0} \tag{4}$$

$$\Rightarrow \quad e = \frac{v_{z(i+1)}}{v_{z(i)}} \tag{5}$$

$$\Rightarrow \quad \frac{v_{z(j)}}{v_{z(0)}} = \prod_{i=0}^{j-1} \frac{v_{z(i+1)}}{v_{z(i)}} = \prod_{i=0}^{j-1} e = e^j \tag{6}$$

$$\Rightarrow \quad v_{z(j)} = e^j v_{z(0)} \tag{7}$$

Considering ***the displacement-velocity relation of uniformly variable motion***:

$$v_{z(0)} = \sqrt{2gh} \tag{8}$$

$$\Rightarrow \quad v_{z(i)} = e^i \sqrt{2gh} \tag{9}$$

Taking *the sphere* as the object of study (see Fig. 3). Studying *the i-th sliding collision process*. Considering (a) ***momentum theorem*** (the ***i***-direction and ***k***-direction component equations), (b) ***theorem of angular momentum*** [in its center-of-mass reference frame, taking its center of mass *C* as the reference point] (the ***j***-direction component equation), (c) ***in the center-of-mass reference frame, the resultant torque of the inertia force acting on system of particles is zero***, (d) ***the form of angular momentum of a rigid body performing fixed-axis rotation*** (integrating into (b)), (e) ***the form of moment***

of inertia of a homogeneous sphere about one of its central axes (integrating into (d)):

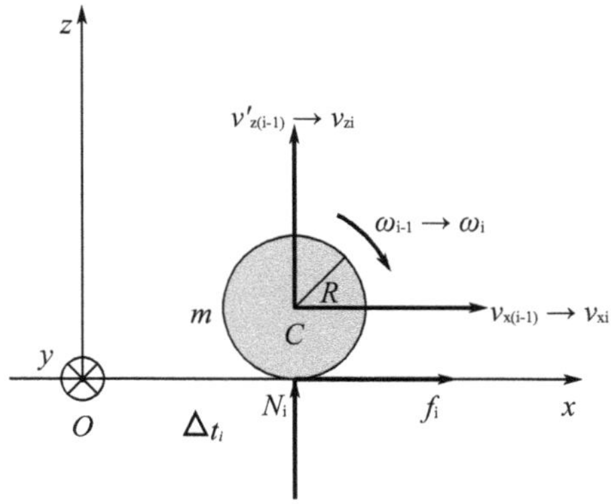

Figure 3

$$\begin{cases} \begin{cases} \int_0^{\Delta t_i} f_i \mathrm{d}t_i = mv_{x(i)} - mv_{x(i-1)} & (10a) \\ \int_0^{\Delta t_i} N_i \mathrm{d}t_i = mv_{z(i)} - mv'_{z(i-1)} & (10b) \end{cases} \\ \int_0^{\Delta t_i} -f_i R \mathrm{d}t_i = \frac{2}{5} mR^2 \omega_i - \frac{2}{5} mR^2 \omega_{i-1} & (10c) \end{cases}$$

Δt_i—the time taken of the i-th sliding collision process.

f_i and N_i—the ***i***-direction component of the friction and the ***k***-direction component of the normal contact force exerted on the sphere by the ground at arbitrary Time t in the i-th sliding collision process.

ω_i—the ***j***-direction component of the angular velocity of the sphere at the time point after the i-th sliding collision.

Considering ***friction law***:

$$f_i = \mu N_i \tag{11}$$

Combining (10) and (11), we obtain the recursion satisfied by $v_{x(i)}$ and ω_i in the process of sliding collision:

$$\Rightarrow \begin{cases} v_{x(i)} - v_{x(i-1)} = \mu(e^i + e^{i-1}) v_{z(0)} & (12a) \\ \omega_i - \omega_{i-1} = -\frac{5}{2} \mu(e^i + e^{i-1}) \frac{v_{z(0)}}{R} & (12b) \end{cases}$$

$$\Rightarrow \begin{cases} v_{x(j)} - v_{x(0)} = \sum_{i=1}^{j}(v_{x(i)} - v_{x(i-1)}) = \mu v_{z(0)}\left(\sum_{i=1}^{j} e^{i} + \sum_{i=1}^{j} e^{i-1}\right) \\ \quad = \mu\frac{1+e}{1-e}(1-e^{j})v_{z(0)} & (13a) \\ \omega_j - \omega_0 = \sum_{i=1}^{j}(\omega_i - \omega_{i-1}) = -\frac{5}{2}\mu\frac{v_{z(0)}}{R}\left(\sum_{i=1}^{j} e^{i} + \sum_{i=1}^{j} e^{i-1}\right) \\ \quad = -\frac{5}{2}\mu\frac{1+e}{1-e}(1-e^{j})\frac{v_{z(0)}}{R} & (13b) \end{cases}$$

$$\Rightarrow \begin{cases} v_{x(j)} = v_{x(0)} + \mu\frac{1+e}{1-e}(1-e^{j})v_{z(0)} = \mu\frac{1+e}{1-e}(1-e^{j})v_{z(0)} & (14a) \\ \omega_j = \omega_0 - \frac{5}{2}\mu\frac{1+e}{1-e}(1-e^{j})\frac{v_{z(0)}}{R} & (14b) \end{cases}$$

The point on the sphere in contact with the ground at the end of the i-th collision is denoted by P_i.

Taking *Point* P_j as the object of study. Studying *the state at the end of the j-th collision*. Considering ***theorem for composition of velocities of a particle*** [taking the ground as the stationary reference frame, and the center-of-mass reference frame of the sphere as the moving reference frame] (the ***i***-direction component equation):

$$v_{P_j x(j)} = v_{x(j)} - \omega_j R \tag{15}$$

$v_{P_j x(j)}$ —the ***i***-direction component of the velocity of Point P_j at the time point after the j-th collision.

For an arbitrary sliding collision process, the following relation holds:

$$v_{P_j x(j)} < 0 \tag{16}$$

$$\Rightarrow \quad v_{x(j)} - \omega_j R < 0 \tag{17}$$

$$\Rightarrow \quad e^{j} > 1 - \frac{2}{7}\frac{1-e}{1+e}\frac{\omega_0 R}{\mu v_{z(0)}} \tag{18}$$

Situation 1, under the following condition:

$$1 - \frac{2}{7}\frac{1-e}{1+e}\frac{\omega_0 R}{\mu v_{z(0)}} < 0 \tag{19}$$

Under this condition, inequality (18) always holds, that is, all the collision processes that can occur are sliding collision processes (i.e., the transitional collision processes and rolling collision processes will not occur).

(8), (9) and (14a) $\rightarrow$ (3a):

$$\Rightarrow \qquad s_i = 4\mu\frac{1+e}{1-e}(e^i - e^{2i})h \qquad (20)$$

Defining s—the ***i***-direction component of the center-of-mass displacement of the sphere after infinite collisions.

$$\Rightarrow \quad s = \sum_{i=0}^{\infty} s_i = \frac{2}{g}\sum_{i=0}^{\infty}\left[\frac{1+e}{1-e}(e^i - e^{2i})\right]h$$

$$= 4\mu\frac{1+e}{1-e}\left(\frac{1}{1-e} - \frac{1}{1-e^2}\right)h = \frac{4\mu eh}{(1-e)^2} \qquad (21)$$

Situation 2, under the following condition:

$$1 - \frac{2}{7}\frac{1-e}{1+e}\frac{\omega_0 R}{\mu v_{z(0)}} \geq 0 \qquad (22)$$

Under this condition, inequality (18) has nontrivial solution, that is, the three kinds of processes will occur in the sequential order.

From (18),

$$\Rightarrow \qquad j\lg e > \lg\left(1 - \frac{2}{7}\frac{1-e}{1+e}\frac{\omega_0 R}{\mu v_{z(0)}}\right) \qquad (23)$$

$$\Rightarrow \quad j < \frac{\lg\left(1 - \frac{2}{7}\frac{1-e}{1+e}\frac{\omega_0 R}{\mu v_{z(0)}}\right)}{\lg e} = \log_e\left(1 - \frac{2}{7}\frac{1-e}{1+e}\frac{\omega_0 R}{\mu v_{z(0)}}\right)$$

$$= \log_e\left(1 - \frac{2}{7}\frac{1-e}{1+e}\frac{\omega_0 R}{\mu\sqrt{2gh}}\right) \equiv Z \qquad (24)$$

Defining N—the total number of the sliding collision processes. N is a non-negative integer (when N is zero, the first collision is the transitional collision process).

$$\Rightarrow \qquad N=\begin{cases}\text{Floor}[Z] & \text{when } Z\notin \mathbf{N}\\ Z-1, & \text{when } Z\in \mathbf{N}\end{cases} \tag{25}$$

where function Floor (x) represents the maximum integer not greater than the variable x.

For the transitional collision process and an arbitrary rolling collision process, the following relation holds:

$$v_{Px(j)}=0 \quad \text{when } j>N \tag{26}$$

(15) → (26):

$$\Rightarrow \qquad v_{x(j)}-\omega_j R=0 \quad \text{when } j>N \tag{27}$$

Taking *the sphere* as the object of study. Studying *the (N+1)-th collision process (i.e., the transitional collision process)*. Considering (a) ***momentum theorem*** (the ***i***-direction component equation), (b) ***theorem of angular momentum*** [in its center-of-mass reference frame, taking its center of mass C as the reference point] (the ***j***-direction component equation), (c) ***in the center-of-mass reference frame, the resultant torque of the inertia force acting on system of particles is zero***, (d) ***the form of angular momentum of a rigid body performing fixed-axis rotation*** (integrating into (b)), (e) ***the form of moment of inertia of a homogeneous sphere about one of its central axes*** (integrating into (d)):

$$\begin{cases}\displaystyle\int_0^{\Delta t_{N+1}} f_{N+1}\mathrm{d}t_i = mv_{x(N+1)}-mv_{x(N)} & \text{(28a)}\\ \displaystyle\int_0^{\Delta t_{N+1}} -f_{N+1}R\mathrm{d}t=\frac{2}{5}mR^2\omega_{N+1}-\frac{2}{5}mR^2\omega_N & \text{(28b)}\end{cases}$$

Δt_{N+1}—the duration of the relative sliding in the $(N+1)$-th collision process.

From (27),

$$v_{x(N+1)}-\omega_{N+1}R=0 \tag{29}$$

$$\Rightarrow \qquad \begin{cases}\displaystyle v_{x(N+1)}=\frac{5}{7}v_{x(N)}+\frac{1}{7}R\omega_N & \text{(30a)}\\ \displaystyle \omega_{N+1}=\frac{5}{7}\frac{v_{x(N)}}{R}+\frac{1}{7}\omega_N & \text{(30b)}\end{cases}$$

$(14) \to (30)$:

$$\Rightarrow \begin{cases} v_{x(N+1)} = -\dfrac{5}{14}\mu\dfrac{1+e}{1-e}(1-e^N)v_{z(0)} + \dfrac{1}{7}\omega_0 R & (31a) \\ \omega_{N+1} = -\dfrac{5}{14}\mu\dfrac{1+e}{1-e}(1-e^N)\dfrac{v_{z(0)}}{R} + \dfrac{1}{7}\omega_0 & (31b) \end{cases}$$

For an arbitrary rolling collision process, the following relations hold:

$$\begin{cases} v_{x(j)} = v_{x(N+1)} & (32a) \\ \omega_j = \omega_{N+1} & (32b) \end{cases} \quad \text{when } j > N+1$$

(8), (14) and $(32) \to (3c)$:

$$\Rightarrow s_i = \begin{cases} 4\mu\dfrac{1+e}{1-e}(e^i - e^{2i})h & \text{when } i \le N \quad (33a) \\ \left[-\dfrac{10}{7}\mu\dfrac{1+e}{1-e}(1-e^N)h + \dfrac{2}{7}\omega_0 R\sqrt{\dfrac{2h}{g}}\right]e^i & \text{when } i > N \quad (33b) \end{cases}$$

$$\Rightarrow \quad s = \sum_{i=0}^{\infty} s_i$$

$$= 4\mu\frac{1+e}{1-e}h\sum_{i=0}^{N}(e^i - e^{2i}) +$$

$$\left[-\frac{10}{7}\mu\frac{1+e}{1-e}\left(1-e^N\right)h + \frac{2}{7}\omega_0 R\sqrt{\frac{2h}{g}}\right]\sum_{i=N+1}^{\infty} e^i$$

$$= 4\mu\frac{1+e}{1-e}\left(\frac{1-e^{N+1}}{1-e} - \frac{1-e^{2(N+1)}}{1-e^2}\right)h +$$

$$\frac{e^{N+1}}{1-e}\left[-\frac{10}{7}\mu\frac{1+e}{1-e}(1-e^N)h + \frac{2}{7}\omega_0 R\sqrt{\frac{2h}{g}}\right]$$

$$= \mu\frac{1+e}{1-e}\left[4\left(\frac{1-e^{N+1}}{1-e} - \frac{1-e^{2(N+1)}}{1-e^2}\right) - \frac{10}{7}\frac{e^{N+1}}{1-e}(1-e^N)\right]h +$$

$$\frac{2}{7}\frac{e^{N+1}}{1-e}\omega_0 R\sqrt{\frac{2h}{g}} \tag{34}$$

3.1.17 The Rolling of a Homogeneous Rigid Cylinder on a Soft Plane (c, C)[4]

Ideally, a homogeneous rigid cylinder can roll on a rigid horizontal plane without energy dissipation. However, in reality, the cylinder will eventually stop moving, which indicates that there are dissipation effects related to rolling in reality. There are many reasons for this phenomenon, the obvious reasons are the adhesion and the electrostatic force between the horizontal plane and the cylinder, and there are some other reasons unrelated to these two effects, the most important one is that the actual cylinder and the horizontal plane are not ideal rigid bodies. When the cylinder rolls on the horizontal plane, both of them will inevitably deform, which will produce a torque to hinder the motion of the cylinder.

In this problem, given that the hardness of the cylinder is much greater than that of the horizontal plane, and the cylinder can be regarded as an ideal rigid body. Figure. 1 is a schematic diagram of a hard hollow cylinder of mass M and radius R performing pure rolling on a soft horizontal plane. The Cartesian coordinate system with the frame $(O, \boldsymbol{i}, \boldsymbol{j}, \boldsymbol{k})$ is established, at arbitrary Time t, the x-coordinate of the center of mass C is denoted by x_C, and the $\boldsymbol{i}$-direction component of the velocity of the center of mass C is denoted by v_C, the coordinates of the frontmost contact point and the rearmost contact point between the cylinder and the horizontal plane are denoted by x_1 and x_2, respectively, the y-coordinate of an arbitrary contact point (denoted by P) between the cylinder and the horizontal plane is denoted by $y\ (x, t)$, the descent distance of the central axis of the cylinder due to the plane deformation is denoted by h. In this problem, it is considered that the motion of the cylinder is quite slow, and the deformation of the plane is quite small.

As shown in Fig. 2, the deformed part of the plane can be regarded as a series of vertical spring oscillators (each composed of a light spring and a particle on the top) arranged densely, ignoring the lateral coupling of the spring oscillators, the particle moves only in the vertical direction. Given the line mass density of the particles in the $\boldsymbol{i}$-direction is λ, and the total coefficient of stiffness of the springs

---

[4]This problem is a test question of the Taiwan China team selection exam for the 11th AphO (Asian Physics Olympiads) and the 22th IPhO (International Physics Olympiad)

within the ***i***-direction unit length is k. When a particle moves in the vertical direction, it will be subject to the resistance proportional to its speed, the ***j***-component of the resistance acting on the particles within the ***i***-direction unit length is $-b\dot{y}\ (x,\ t)$, where $b>0$, is the damping coefficient of the springs in the unit length. The gravity of the particles can be ignored.

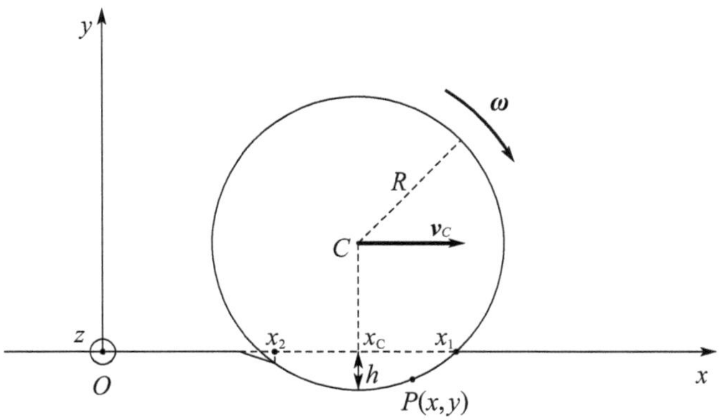

Figure 1

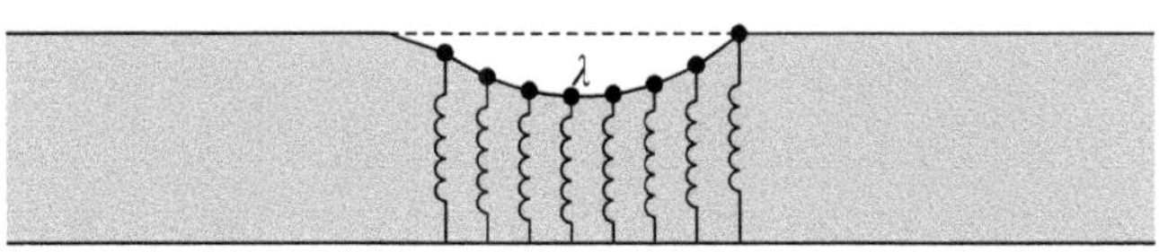

k (unit length)

Figure 2

(1) Prove: $y(x,t) \approx \dfrac{(x-x_C)^2}{2R} - h$.

(2) Find the normal contact force exerted on the particles within the ***i***-direction unit length by the cylinder at an arbitrary contact point other than the frontmost contact point at arbitrary Time t, then find the expressions of x_1 and x_2 on this basis.

(3) The particles in front of the cylinder are impacted by the cylinder in a static state in succession and reach the bottom of the cylinder, find the impact force exerted on the particles within the ***i***-direction unit length by the cylinder at the frontmost contact point at arbitrary Time t.

(4) Find the expression of h.
(5) Find the velocity of center of mass C as a function of time.
(6) Defining the rolling friction coefficient μ_r at arbitrary Time t:

$\tau_{\mu_r}\omega = \dfrac{\mathrm{d}E}{\mathrm{d}t}$, where $\tau_{\mu_r} = -\mu_r Mg$, which can be regarded as the $-\boldsymbol{k}$-direction component of the equivalent torque of the force exerted on the cylinder by the horizontal plane at arbitrary Time t, ω is the $\boldsymbol{k}$-direction component of the rotating angular velocity of the cylinder at arbitrary Time t, E is the kinetic energy of the cylinder at arbitrary Time t. Find the expression of μ_r.

Solution to Problem (1)

Considering the following geometric relation in Fig. 3:

$$y(x,t) = -\left[\sqrt{R^2-(x-x_C)^2}-(R-h)\right]$$

$$= -\left[R\sqrt{1-\frac{(x-x_C)^2}{R^2}}-(R-h)\right]$$

$$\approx -\left\{R\left[1-\frac{(x-x_C)^2}{2R^2}\right]-(R-h)\right\} = \frac{(x-x_C)^2}{2R}-h \qquad (1)$$

This operation has considered that $|x - x_C| << R$ and taken the low order approximation.

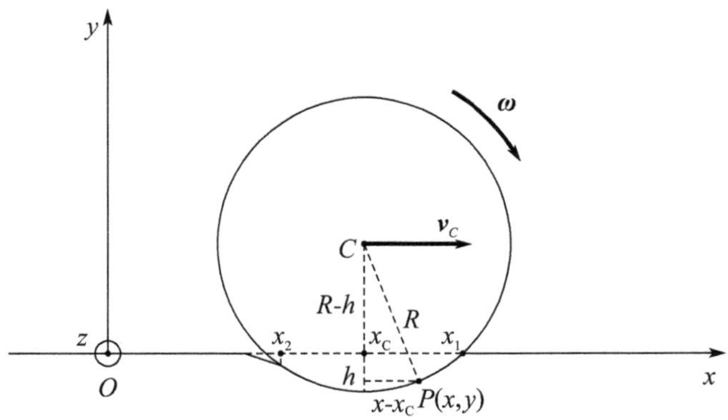

Figure 3

Solution to Problem (2)

The mass infinitesimal element within the ***i***-direction unit length at an arbitrary contact point with coordinate *x* other than the frontmost contact point is called the pressed element *x* for short.

Taking *the pressed element x* as the object of study. Studying *the state at arbitrary Time t*. Considering ***Newton's second law*** (the ***j***-direction component equation):

$$f(x,t)-b\dot{y}(x,t)-ky(x,t)=\lambda\ddot{y}(x,t) \tag{2}$$

f (*x*, *t*)—the ***j***-direction component of the normal contact force exerted on the pressed element *x* by the cylinder at Time *t*.

From (1),

$$\Rightarrow \qquad \dot{y}(x,t)=-\frac{x-x_C}{R}\dot{x}_C=-\frac{x-x_C}{R}v_C \tag{3}$$

$$\Rightarrow \qquad \ddot{y}(x,t)=\frac{1}{R}\dot{x}_C v_C=\frac{v_C^2}{R} \tag{4}$$

$$\Rightarrow \qquad f(x,t)=\frac{\lambda v_C^2}{R}-\frac{b(x-x_C)v_C}{R}+\frac{k(x-x_C)^2}{2R}-kh \tag{5}$$

Considering the boundary condition:

$$y(x_1,t)=0 \tag{6}$$

(1) → (6),

$$\Rightarrow \qquad \frac{(x_1-x_C)^2}{2R}-h=0$$

$$\Rightarrow \qquad x_1=x_C+\sqrt{2Rh} \tag{7}$$

The condition for the cylinder to separate from the horizontal plane at the rearmost contact point is that the mutual normal contact forces of them at that point are zero:

$$f(x_2,t)=0 \tag{8}$$

(5) → (8),

$$\Rightarrow \qquad \frac{\lambda v_C^2}{R}-\frac{b(x_2-x_C)v_C}{R}+\frac{k(x_2-x_C)^2}{2R}-kh=0$$

$$\Rightarrow \qquad x_2=x_C+\frac{bv_C}{k}-\sqrt{2hR+\left(\frac{b^2}{k^2}-\frac{2\lambda}{k}\right)v_C^2} \tag{9}$$

Solution to Problem (3)

The mass infinitesimal element of the soft horizontal plane that reach the bottom of the cylinder after being impacted by the cylinder in the instantaneous process (the small time interval dt) at arbitrary Time t is called the impacted element dt.

Taking *the impacted element dt* as the object of study. Studying *the instantaneous process of its being impacted by the cylinder.* Considering ***momentum theorem*** (the ***j***-direction component equation):

$$F(x_1)\mathrm{d}t = \mathrm{d}m \cdot \dot{y}(x_1, t) \tag{10}$$

dm—the mass of the impacted element dt.

$F(x_1)$—the ***j***-component of the impact force exerted on the impacted element dt by the cylinder.

Considering the definition of ***line mass density***:

$$\mathrm{d}m = \lambda v_C \mathrm{d}t \tag{11}$$

(3) and (11) → (10),

$$\Rightarrow \qquad F(x_1) = -\lambda \frac{(x_1 - x_C)v_C^2}{R} \tag{12}$$

Solution to Problem (4)

Taking *the cylinder* as the object of study. Studying *the state at arbitrary Time t*. Considering ***theorem of motion of center of mass*** (the ***j***-direction component equation):

$$\int_{x_2}^{x_1} -f(x,t)\mathrm{d}x - F(x_1) - Mg = 0 \tag{13}$$

Defining the variable x':

$$x' = x - x_C \tag{14}$$

Combining (5) and (14),

$$\Rightarrow \qquad \int_{x_2}^{x_1} f(x,t)\mathrm{d}x = \int_{x_2 - x_C}^{x_1 - x_C} \left(\frac{\lambda v_C^2}{R} - \frac{b v_C}{R} x' + \frac{k}{2R} x'^2 - kh \right) \mathrm{d}x'$$

$$\approx \int_{-\sqrt{2hR}}^{\sqrt{2hR}} \left(\frac{k}{2R} x'^2 - kh \right) \mathrm{d}x'$$

$$= \frac{k}{3R}\left(\sqrt{2Rh}\right)^3 - 2kh\sqrt{2Rh}$$

$$= -\frac{4}{3}kh\sqrt{2Rh} \tag{15}$$

This operation has considered that v_C is a small quantity, and ignored its infinitesimal of higher order.

Taking the low order approximation of $F(x_1)$:

$$F(x_1) \approx 0 \tag{16}$$

(15) and (16) $\rightarrow$ (13),

$$\Rightarrow \qquad \frac{4}{3}kh\sqrt{2Rh} \approx Mg$$

$$\Rightarrow \qquad h \approx \left(\frac{3Mg}{4k\sqrt{2R}}\right)^{2/3} \tag{17}$$

Solution to Problem (5)

Taking *the impacted element dt* as the object of study. Studying *the instantaneous process of its being impacted by the cylinder.* Considering ***theorem of kinetic energy***:

$$\text{đ}W_1 = \frac{1}{2}\mathrm{d}m\dot{y}^2(x_1,t) \tag{18}$$

$\text{đ}W_1$—the work done by the cylinder on the impacted element dt in the impact process.

Defining $\text{đ}W_2$—the work done by the cylinder on all particles beneath it in the instantaneous process at arbitrary Time t.

$$\text{đ}W_2 = \int_{x_2}^{x_1}\left[f(x,t)\mathrm{d}y(x,t)\right]\mathrm{d}x \tag{19}$$

Taking *the cylinder* as the object of study. Studying *the instantaneous process at arbitrary Time t*. Considering ***theorem of kinetic energy***:

$$-\text{đ}W_1 - \text{đ}W_2 = \mathrm{d}E \tag{20}$$

Combining (3), (5), (11), (18) to (20):

$$\Rightarrow \qquad \frac{\mathrm{d}E}{\mathrm{d}t} = -\frac{1}{2}\lambda v_C\dot{y}^2(x_1,t) - \int_{x_2}^{x_1}\left[f(x,t)\dot{y}(x,t)\right]\mathrm{d}x$$

$$=-\frac{1}{2}\lambda\left(\frac{x_1-x_C}{R}\right)^2 v_C^3+$$

$$\frac{1}{R}\int_{x_2-x_C}^{x_1-x_C}\left[\left(\frac{\lambda v_C^3}{R}-khv_C\right)x'-\frac{bv_C^2}{R}x'^2+\frac{kv_C}{2R}x'^3\right]\mathrm{d}x'$$

$$\approx\frac{1}{R}\int_{-\sqrt{2hR}}^{\sqrt{2hR}}\left(-khv_Cx'-\frac{bv_C^2}{R}x'^2+\frac{kv_C}{2R}x'^3\right)\mathrm{d}x$$

$$=-\frac{2bv_C^2}{3R^2}\left(\sqrt{2hR}\right)^3 \tag{21}$$

This operation has ignored the infinitesimal of higher order of (v_C^2).

(17) → (21),

$$\Rightarrow \qquad \frac{\mathrm{d}E}{\mathrm{d}t}=-\frac{bv_C^2}{Rk}Mg \tag{22}$$

Considering (a) ***König theorem***, (b) ***the form of kinetic energy of a rigid body performing fixed-axis rotation*** [integrating into (a)]:

$$E=\frac{1}{2}Mv_C^2+\frac{1}{2}I\omega^2 \tag{23}$$

I—the moment of inertia of the cylinder about its central axis.

Considering ***the form of moment of inertia of a homogeneous hollow cylinder about its central axis***:

$$I = MR^2 \tag{24}$$

The condition for the cylinder to perform pure rolling relative to the horizontal plane is as follows:

$$v_C - \omega R = 0 \tag{25}$$

$$\Rightarrow \qquad E = Mv_C^2 \tag{26}$$

(26) → (22),

$$\Rightarrow \qquad M\frac{\mathrm{d}(v_C^2)}{\mathrm{d}t}=-\frac{bv_C^2}{Rk}Mg \tag{27}$$

$$\Rightarrow \qquad \frac{\mathrm{d}(v_C^2)}{v_C^2}=-\frac{bg}{Rk}\mathrm{d}t$$

$$\Rightarrow \qquad \int_{v_{C0}}^{v_C}\frac{\mathrm{d}(v_C^2)}{v_C^2}=-\frac{bg}{Rk}\int_0^t\mathrm{d}t$$

$$\Rightarrow \qquad v_C = v_{C0} e^{-\frac{bg}{2Rk}t} \tag{28}$$

Solution to Problem (6)

Considering the definition of ***rolling friction coefficient***:

$$\tau_{\mu_r}\omega = \frac{\mathrm{d}E}{\mathrm{d}t} \tag{29}$$

$$\tau_{\mu_r} = -\mu_r Mg \tag{30}$$

$$\Rightarrow \qquad \mu_r = -\frac{1}{Mg\omega}\frac{\mathrm{d}E}{\mathrm{d}t} \tag{31}$$

(22) and (25) $\rightarrow$ (31)

$$\Rightarrow \qquad \mu_r = \frac{bv_C}{k} \tag{32}$$

3.1.18 The Rolling of a Uniformly Charged Ring in a Uniform Magnetic Field (*c*, *B*)[5]

A uniform magnetic field is applied in space, the magnitude of whose magnetic induction intensity is B, and the direction is perpendicular to the plane of the figure inward (see Fig. 1). Consider a homogeneous ring of mass m, which is uniformly charged with electric quantity Q, $Q > 0$. Initially, the ring is resting on a horizontal ground (in the vertical plane of the figure), it obtains a center-of-mass velocity of the magnitude v_0 and the direction horizontally to the right due to an instant impact. Given the friction coefficient between the ring and the ground is μ, $v_0 = mg/2QB$. Find the time taken for the ring to reach pure rolling.

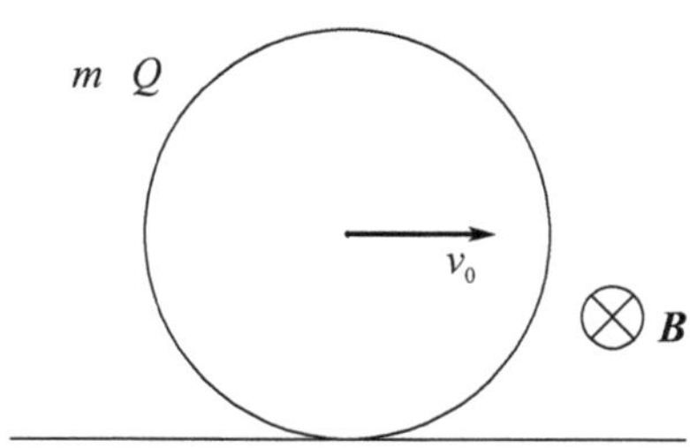

Figure 1

[5]This problem is a test question of the Physics experiment class of Peking University.

Solution

The time point when the ring starts to move is defined as Time zero.

As shown in Fig. 2, the center of the ring is denoted by *O*, establishing the Cartesian coordinate system with the frame (*O*, ***i***, ***j***, ***k***), let *x*, *y*, *z* be the Cartesian coordinates, and the polar coordinate system with the center *O* as the pole and the positive *z*-axis as the polar axis, let *r*, θ be the polar coordinates.

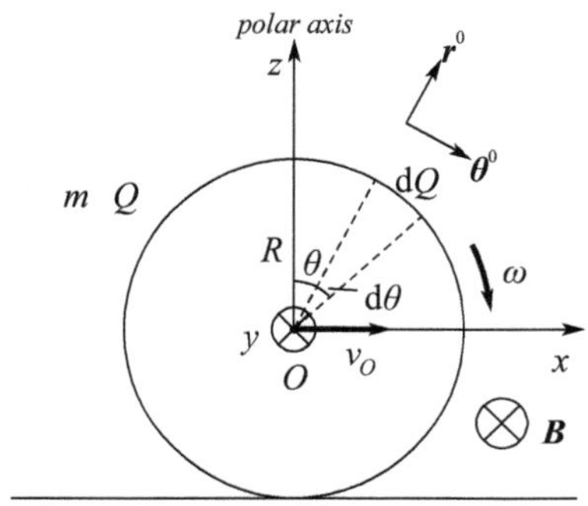

Figure 2

Taking *the mass infinitesimal element between* θ *and* $\theta + d\theta$ *on the ring at arbitrary Time t (hereinafter referred to as element d*θ*)* as the object of study. Studying *the state at Time t*. Considering (a) ***theorem for composition of velocities of a particle*** [taking the ground as the stationary reference frame, and the center-of-mass reference frame of the ring as the moving reference frame], (b) ***the components form of the velocity of a particle in polar coordinate system*** [integrating into (a)]:

$$\boldsymbol{v}_\theta = v_O \boldsymbol{i} + \omega R \boldsymbol{\theta}^0 \tag{1}$$

$\boldsymbol{v}_\theta$—the velocity of element dθ at Time *t*.

v_O—the ***i***-direction component of the center-of-mass velocity of the ring at Time *t*.

ω—the ***j***-direction component of the angular velocity of the ring at Time *t*.

R—the radius of the ring.

Defining η—the linear charge density of the ring:

$$\eta = \frac{Q}{2\pi R} \tag{2}$$

Considering ***Lorentz's formula of force***:

$$F_B = \oint_L (\mathrm{d}Q_\theta)\boldsymbol{v}_\theta \times \boldsymbol{B} = \oint_L (\eta R\mathrm{d}\theta)\boldsymbol{v}_\theta \times (B\boldsymbol{j}) \tag{3}$$

$\mathrm{d}Q_\theta$—the electric quantity of element $\mathrm{d}\theta$.

$\boldsymbol{F}_B$—the resultant force of the magnetic field force acting on the charges on the ring at Time t.

(1) → (3),

$$\Rightarrow \quad \boldsymbol{F}_B = \oint_L (\eta R\mathrm{d}\theta)(v_O\boldsymbol{i})\times(B\boldsymbol{j}) + \oint_L (\eta R\mathrm{d}\theta)(\omega R\boldsymbol{\theta}^0)\times(B\boldsymbol{j}) \tag{4}$$

The first and second terms in the equation above are defined as $\boldsymbol{F}_{B-V_O}$ and $\boldsymbol{F}_{B\text{-}\omega}$, respectively:

$$\begin{cases} \boldsymbol{F}_{B-V_O} = \oint_L (\eta R\mathrm{d}\theta)(v_O\boldsymbol{i})\times(B\boldsymbol{j}) & \text{(5a)} \\ \boldsymbol{F}_{B-\omega} = \oint_L (\eta R\mathrm{d}\theta)(\omega R\boldsymbol{\theta}^0)\times(B\boldsymbol{j}) & \text{(5b)} \end{cases}$$

Taking the integral with vector operation rules:

$$\boldsymbol{F}_{B-V_O} = \eta R v_O B(\boldsymbol{i}\times\boldsymbol{j})\oint_L \mathrm{d}\theta = \eta\cdot 2\pi R v_O B\boldsymbol{k} = Q v_O B\boldsymbol{k} \tag{6}$$

$$\boldsymbol{F}_{B-\omega} = \eta R^2\omega B\oint_L (\boldsymbol{\theta}^0\times\boldsymbol{j})\mathrm{d}\theta = \eta R^2\omega B\oint_L \boldsymbol{r}^0\mathrm{d}\theta = 0 \tag{7}$$

$$\Rightarrow \quad \boldsymbol{F}_B = Q v_O B\boldsymbol{k} \tag{8}$$

That is, at arbitrary Time t, the resultant force of the magnetic field force acting on the charges on the ring is equivalent to the magnetic force acting on them when they concentrate in the center O.

Considering the definition of ***torque***, then place (3) in to obtain:

$$\Rightarrow \quad \boldsymbol{M}_B = \oint_L \boldsymbol{R}\times\mathrm{d}\boldsymbol{F}_B = \oint_L \boldsymbol{R}\times\left[(\mathrm{d}Q)\boldsymbol{v}_\theta\times\boldsymbol{B}\right]$$

$$= \oint_L R\boldsymbol{r}^0\times\left[(\eta R\mathrm{d}\theta)\boldsymbol{v}_\theta\times(B\boldsymbol{j})\right] \tag{9}$$

$\boldsymbol{M}_B$—the resultant torque of the magnetic field force acting on the charges on the ring [taking the center O as the reference point] at arbitrary Time t.

(1) → (9),

$$\Rightarrow \quad \boldsymbol{M}_B = \oint_L R\boldsymbol{r}^0\times\left[(\eta R\mathrm{d}\theta)(v_O\boldsymbol{i})\times(B\boldsymbol{j})\right] +$$

$$\oint_L R\boldsymbol{r}^0\times\left[(\eta R\mathrm{d}\theta)(\omega R\boldsymbol{\theta}^0)\times(B\boldsymbol{j})\right] \tag{10}$$

The first and second terms in the equation above are defined as $\boldsymbol{M}_{B-V_O}$ and $\boldsymbol{M}_{B-\omega}$, respectively:

$$\begin{cases} \boldsymbol{M}_{B-V_O} = \oint_L R\boldsymbol{r}^0 \times \left[(\eta R \mathrm{d}\theta)(v_O \boldsymbol{i}) \times (B\boldsymbol{j}) \right] & (11a) \\ \boldsymbol{M}_{B-\omega} = \oint_L R\boldsymbol{r}^0 \times \left[(\eta R \mathrm{d}\theta)(\omega R \boldsymbol{\theta}^0) \times (B\boldsymbol{j}) \right] & (11b) \end{cases}$$

Taking the integral with vector operation rules:

$$\boldsymbol{M}_{B-V_O} = \eta R^2 v_O B \oint_L \boldsymbol{r}^0 \times (\boldsymbol{i} \times \boldsymbol{j})\, \mathrm{d}\theta = \eta R^2 v_O B \oint_L \boldsymbol{r}^0 \times \boldsymbol{k} \mathrm{d}\theta$$

$$= -\eta R^2 v_O B \boldsymbol{j} \oint_L \sin\theta \mathrm{d}\theta = 0 \quad (12)$$

$$\boldsymbol{M}_{B-\omega} = \eta \omega R^3 B \oint_L \boldsymbol{r}^0 \times (\boldsymbol{\theta}^0 \times \boldsymbol{j}) \mathrm{d}\theta = \eta \omega R^3 B \oint_L \boldsymbol{r}^0 \times \boldsymbol{r}^0 \mathrm{d}\theta = 0 \quad (13)$$

$$\Rightarrow \quad \boldsymbol{M}_B = 0 \quad (14)$$

That is, the resultant torque of the magnetic field force acting on the charges on the ring [taking the center O as the reference point] at arbitrary Time t is zero.

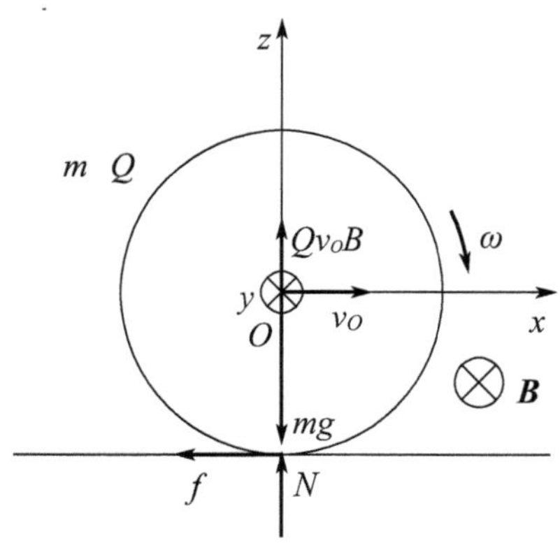

Figure 3

Taking *the ring* as the object of study (see Fig. 3). Studying *the state at arbitrary Time t before it reaches pure rolling.* Considering ***theorem of motion of center of mass*** (the ***i***-direction and ***j***-direction component equations):

$$\begin{cases} -f = m\dfrac{\mathrm{d}v_O}{\mathrm{d}t} & (15a) \\ N + Qv_O B - mg = 0 & (15b) \end{cases}$$

N—the $\boldsymbol{k}$-direction component of the normal contact force exerted on the ring by the ground at Time t.

f—the $-\boldsymbol{i}$-direction component of the friction force exerted on the ring by the ground at Time t.

Considering ***friction law***:

$$f = \mu N \tag{16}$$

$$\Rightarrow \qquad \frac{\mathrm{d}v_O}{v_O - \dfrac{mg}{QB}} = \frac{\mu QB}{m}\mathrm{d}t \tag{17}$$

$$\Rightarrow \qquad \int_{v_0}^{v_O} \frac{\mathrm{d}v_O}{v_O - \dfrac{mg}{QB}} = \frac{\mu QB}{m}\int_0^t \mathrm{d}t$$

$$\Rightarrow \qquad \ln\frac{v_O - \dfrac{mg}{QB}}{v_0 - \dfrac{mg}{QB}} = \frac{\mu QB}{m}t$$

$$\Rightarrow \qquad v_O = \frac{mg}{QB} - \left(\frac{mg}{QB} - v_0\right)e^{\frac{\mu QB}{m}t} \tag{18}$$

Taking *the ring* as the object of study. Studying *the state at arbitrary Time t before it reaches pure rolling*. Considering (a) ***law of rotation*** [in its center-of-mass reference frame, taking the y-axis as the rotation axis] (the $\boldsymbol{j}$-direction component equation), (b) ***in the center-of-mass reference frame, the resultant torque of the inertia force acting on system of particles is zero***, (c) ***the form of moment of inertia of a homogeneous ring about its central axis perpendicular to the plane of the ring***:

$$fR = I\frac{\mathrm{d}\omega}{\mathrm{d}t} \tag{19}$$

$$I = mR^2 \tag{20}$$

I—the moment of inertia of the ring about the y-axis.

Combining (15a), (19) and (20),

$$\Rightarrow \qquad \frac{\mathrm{d}v_O}{\mathrm{d}t} = -R\frac{\mathrm{d}\omega}{\mathrm{d}t} \tag{21}$$

$$\Rightarrow \qquad \mathrm{d}v_O = -R\mathrm{d}\omega$$

$$\Rightarrow \qquad \int_{v_0}^{v_O} \mathrm{d}v_O = -R\int_0^{\omega} \mathrm{d}\omega$$

$$\Rightarrow \qquad v_O = v_0 - \omega R \tag{22}$$

The condition for the ring to perform pure rolling relative to the ground is as follows:

$$v_O\big|_{t=T} - \omega\big|_{t=T} R = 0 \tag{23}$$

T—the Time when the ring reaches pure rolling.

Combining (22) and (23),

$$\Rightarrow \qquad v_O\big|_{t=T} = \frac{v_0}{2} \tag{24}$$

(18) → (24),

$$\frac{v_0}{2} = \frac{mg}{QB} - \left(\frac{mg}{QB} - v_0\right)e^{\frac{\mu QB}{m}T} \tag{25}$$

Known:

$$v_0 = \frac{mg}{2QB} \tag{26}$$

$$\Rightarrow \qquad T = \frac{m}{\mu QB}\ln\frac{3}{2} \tag{27}$$

Another method to get Eq. (7)

$$\boldsymbol{\theta}^0\Big|_{\theta} = -\boldsymbol{\theta}^0\Big|_{\theta+\pi} \tag{28}$$

$$\Rightarrow \qquad \oint_L (\boldsymbol{\theta}^0 \times \boldsymbol{j})\,\mathrm{d}\theta = 0 \times \boldsymbol{j} = 0 \tag{29}$$

$$\Rightarrow \qquad \boldsymbol{F}_{B-\omega} = 0 \tag{30}$$

3.1.19 The Stability Condition of the Mechanical Energy of Planet-Moon System (*d*, *C*)

Consider a planet performing self-rotation, and its only moon performing revolution around it (circular motion). Given that the above two rotations have the same angular velocity direction and different angular velocity magnitudes, the moon's mass is much less than the planet. Consider the tidal friction caused by the moon on the planet, the angular velocity magnitudes of the above two rotations will change with time. Under what circumstances can the mechanical energy of this planet-moon system reach stable?

Solution

Time zero is arbitrarily set.

The center of the planet is denoted by O, the unit vector in the positive direction of the angular velocity of the planet's self-rotation and the moon's revolution is denoted by $\boldsymbol{k}$. Taking Point O as the pole, an arbitrary direction in the orbital plane of the moon's revolution as the positive direction of the polar axis, the right-hand helix direction with regard to the $\boldsymbol{k}$-direction as the positive direction of polar angle, the $\boldsymbol{k}$-direction as the positive direction of the longitudinal axis, establishing the cylindrical coordinate system, let r, φ, z be the cylindrical coordinates.

Taking *the moon* as the object of study. Studying *the state at arbitrary Time t*. Considering ***Newton's second law*** (the $-\boldsymbol{r}^0$-direction component equation):

$$G\frac{Mm}{r^2} = m\omega^2 r \tag{1}$$

M and m—the mass of the planet and the moon.

ω—the $\boldsymbol{k}$-direction component of the moon's revolution angular velocity at Time t.

$$\Rightarrow \qquad -3G\frac{M}{r^4}\mathrm{d}r = 2\omega \mathrm{d}\omega \tag{2}$$

$$\Rightarrow \qquad \frac{\mathrm{d}r}{\mathrm{d}\omega} = -\frac{2\omega}{3G\frac{M}{r^4}} = -\frac{2}{3}\frac{r}{\omega} \tag{3}$$

Taking *the planet-moon system* as the object of study. Studying *the instantaneous process (the small time interval dt) at arbitrary Time t*. Considering ***law of conservation of angular momentum*** (differential form) [taking Point O as the reference point, the resultant external torque acting on the system is zero; therefore, the angular momentum of the system is conserved] (the $\boldsymbol{k}$-direction component equation):

$$\frac{\mathrm{d}L}{\mathrm{d}t} = 0 \tag{4}$$

L—the $\boldsymbol{k}$-direction component of the angular momentum of the planet-moon system at Time t [taking Point O as the reference point].

Considering ***the form of angular momentum of a rigid body performing fixed-axis rotation***:

$$L = I\Omega + mr^2\omega \tag{5}$$

I—the moment of inertia of the planet about the z-axis.

Ω—the $\boldsymbol{k}$-direction component of the planet's self-rotation angular velocity at Time t.

$$\Rightarrow \qquad I\mathrm{d}\Omega + 2mr\omega\mathrm{d}r + mr^2\mathrm{d}\omega = 0 \tag{6}$$

(3) $\rightarrow$ (6),

$$\Rightarrow \qquad I\mathrm{d}\Omega - \frac{1}{3}mr^2\mathrm{d}\omega = 0 \tag{7}$$

$$\Rightarrow \qquad \frac{\mathrm{d}\Omega}{\mathrm{d}\omega} = \frac{mr^2}{3I} \tag{8}$$

Taking *the planet-moon system* as the object of study. Studying *the instantaneous process at arbitrary Time t*. Considering ***theorem of mechanical energy***:

$$đW = -\mathrm{d}E \tag{9}$$

$đW$ —the work done by the tidal force in the instantaneous process at Time t.

E—the mechanical energy of the planet-moon system at Time t.
Considering (a) the definition of ***mechanical energy***, (b) ***the form of kinetic energy of a rigid body performing fixed-axis rotation***, (c) ***the form of the universal gravitational potential energy of a homogeneous sphere at an arbitrary point outside it*** (taking the configuration in which the planet and the moon are infinitely apart as potential energy zero):

$$E = \frac{1}{2}I\Omega^2 + \frac{1}{2}m(\omega r)^2 - G\frac{Mm}{r} \tag{10}$$

$$\Rightarrow \qquad \mathrm{d}E = I\Omega\mathrm{d}\Omega + m\omega^2 r\mathrm{d}r + mr^2\omega\mathrm{d}\omega + G\frac{Mm}{r^2}\mathrm{d}r \tag{11}$$

(1) $\rightarrow$ (11),

$$\Rightarrow \qquad \mathrm{d}E = I\Omega\mathrm{d}\Omega + 2m\omega^2 r\mathrm{d}r + mr^2\omega\mathrm{d}\omega$$

$$= \left(I\Omega \frac{\mathrm{d}\Omega}{\mathrm{d}\omega} + 2m\omega^2 r \frac{\mathrm{d}r}{\mathrm{d}\omega} + m\omega r^2 \right) \mathrm{d}\omega \tag{12}$$

(3) and (8) $\rightarrow$ (12),

$$\Rightarrow \qquad \mathrm{d}E = \frac{1}{3} mr^2 (\Omega - \omega)\, \mathrm{d}\omega \tag{13}$$

When:

$$\mathrm{d}E = 0 \tag{14}$$

$$\Rightarrow \qquad \Omega = \omega \tag{15}$$

Therefore, when the planet's self-rotation angular velocity and the moon's revolution angular velocity are equal, the mechanical energy of the system can reach stable.

By analyzing from another perspective, we can get a more intuitive understanding of the above result: that the angular velocities of the above two rotations reach the same is the condition for the tide to disappear and stop doing negative work on the system.

3.2 The Fixed-Point Rotation of Rigid Body

The fixed-point rotation of a rigid body can show rich forms based on different external conditions. As one of the most classical rigid body models, gyroscope is the main representative of the fixed-point rotation model of rigid body. Some simple models involving the general motion of a rigid body, such as the general rolling of a homogeneous sphere on a specific surface (it can be regarded as performing the fixed-point rotation in its center-of-mass system), are also of great research value. In this section, some simple but representative problems involving the fixed-point rotation of rigid body will be studied.

3.2.1 The Fixed-Point Rotation of a Rotational Symmetric Rigid Body without External Torque (d, *C*)

Consider a rotationally symmetric rigid body rotating around a fixed point of its axis of symmetry without external torque. Determine the directional relations of the angular velocity, angular momentum, and axis of symmetry of the rigid body.

Solution

Time zero is arbitrarily set.

The fixed point of the rigid body is denoted by *O*, taking the unit vector of one of the directions of the axis of symmetry of the rigid body and denoting it by ***k***, and in the rigid body reference frame, taking two unit vectors perpendicular to ***k*** and mutually perpendicular to each other, and denoting them by ***i*** and ***j***, respectively, ***i*** × ***j*** = ***k***. Establishing the body coordinate system with the frame (*O*, ***i***, ***j***, ***k***), let *x*, *y*, *z* be the Cartesian coordinates, it is the coordinate system of principal axis of the rigid body which can be seen form the symmetry of the rigid body.

Taking *the rigid body* as the object of study. Studying the state at arbitrary Time *t*. Considering ***Euler dynamical equations*** [taking Point *O* as the reference point] (the component equations in the (***i***, ***j***, ***k***)-directions):

$$\begin{cases} 0 = I_1\dot{\omega}_1 - (I_2 - I_3)\omega_2\omega_3 & \text{(1a)} \\ 0 = I_2\dot{\omega}_2 - (I_3 - I_1)\omega_3\omega_1 & \text{(1b)} \\ 0 = I_3\dot{\omega}_3 - (I_1 - I_2)\omega_1\omega_2 & \text{(1c)} \end{cases}$$

(I_1, I_2, I_3)—the principal moments of inertia of the rigid body about the axes *x*, y and *z* of the body coordinate system.

$(\omega_1, \omega_2, \omega_3)$—the components of the angular velocity of the rigid body in the (***i***, ***j***, ***k***)-directions at Time *t*.

It can be seen form the symmetry that:

$$I_1 = I_2 \tag{2}$$

$$\Rightarrow \quad \begin{cases} I_1\dot{\omega}_1 - (I_1 - I_3)\omega_2\omega_3 = 0 & \text{(3a)} \\ I_1\dot{\omega}_2 - (I_3 - I_1)\omega_3\omega_1 = 0 & \text{(3b)} \\ I_3\dot{\omega}_3 = 0 & \text{(3c)} \end{cases}$$

From (3c),

$$\Rightarrow \quad \omega_3 = \text{const} \tag{4}$$

Defining the constant Ω:

$$\Omega = \frac{I_3 - I_1}{I_1}\omega_3 \tag{5}$$

$$\Rightarrow \qquad \Rightarrow \begin{cases} \ddot{\omega}_1 + \Omega^2 \omega_1 = 0 & \text{(6a)} \\ \ddot{\omega}_2 + \Omega^2 \omega_2 = 0 & \text{(6b)} \end{cases}$$

The general solutions of Eqs. (6a) and (6b) are as follows:

$$\Rightarrow \begin{cases} \omega_1 = A_1 \cos(\Omega t + \alpha_1) & \text{(7a)} \\ \omega_2 = A_2 \cos(\Omega t + \alpha_2) & \text{(7b)} \end{cases}$$

(A_1, α_1) and (A_2, α_2)—the integral constants of Eqs. (6a) and (6b).

(7) $\rightarrow$ (3),

$$\Rightarrow \qquad A_1^2 \sin[2(\Omega t + \alpha_1) = -A_2^2 \sin[2(\Omega t + \alpha_2) \tag{8}$$

$$\Rightarrow \begin{cases} A_1 = A_2 & \text{(9a)} \\ \alpha_1 = \alpha_2 + \dfrac{\pi}{2} & \text{(9b)} \end{cases}$$

Defining the constants A and α:

$$A_1 = A_2 = A \tag{10a}$$

$$\alpha_1 = \alpha \tag{10b}$$

$$\Rightarrow \begin{cases} \omega_1 = A\cos(\Omega t + \alpha) & \text{(11a)} \\ \omega_2 = A\sin(\Omega t + \alpha) & \text{(11b)} \\ \omega_3 = \text{const} & \text{(11c)} \end{cases}$$

Defining ω_{12}—the component of the angular velocity of the rigid body in the O-xy plane at arbitrary Time t.

$$\Rightarrow \qquad \omega_{12} = \sqrt{\omega_1^2 + \omega_2^2} = A = \text{const} \tag{12}$$

Defining $\theta_{\omega-z}$—the angle between the angular velocity of the rigid body and $\boldsymbol{k}$ at arbitrary Time t.

$$\Rightarrow \qquad \tan\theta_{\omega-z} = \frac{\omega_{12}}{\omega_3} = \frac{A}{\omega_3} = \text{const} \tag{13}$$

It can be seen from (11) to (13) that the angular velocity magnitude of the rigid body is constant, and the angular velocity vector rotates at a uniform angular velocity around the axis of symmetry, depicting a cone in space. This cone is called the body cone. The corresponding schematic diagram is as follows (see Fig. 1):

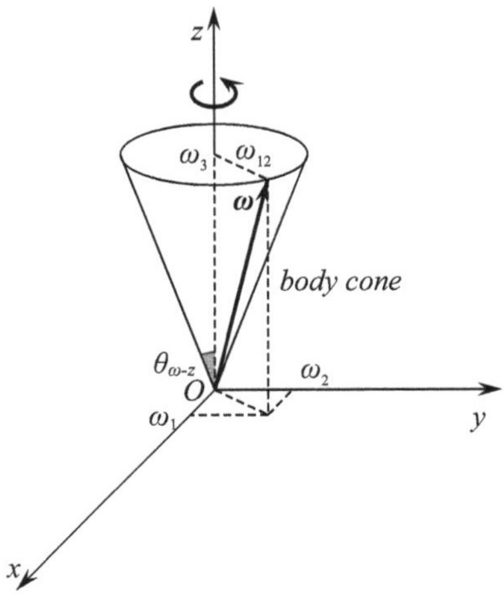

Figure 1

Defining (L_1, L_2, L_3)—the components of the angular momentum of the rigid body in the (***i***, ***j***, ***k***)-directions at arbitrary Time *t*. Considering ***the form of the angular momentum of a rigid body performing fixed-point rotation based on the principal moments of inertia***:

$$\begin{cases} L_1 = I_1 \omega_1 & (14a) \\ L_2 = I_2 \omega_2 & (14b) \\ L_3 = I_3 \omega_3 & (14c) \end{cases}$$

(11) → (14),

$$\Rightarrow \quad \begin{cases} L_1 = I_1 A \cos(\Omega t + \alpha) & (15a) \\ L_2 = I_1 A \sin(\Omega t + \alpha) & (15b) \\ L_3 = I_3 \omega_3 = \text{const} & (15c) \end{cases}$$

Defining L_{12}—the component of the angular momentum of the rigid body in the *O-xy* plane at arbitrary Time *t*.

$$\Rightarrow \quad L_{12} = \sqrt{L_1^2 + L_2^2} = I_1 A = \text{const} \tag{16}$$

Defining $\theta_{L\text{-}z}$—the angle between the angular momentum of the rigid body and ***k*** at arbitrary Time t.

$$\Rightarrow \quad \tan\theta_{L-z} = \frac{L_{12}}{L_3} = \frac{I_1 A}{I_3 \omega_3} = \text{const} \tag{17}$$

It can be seen from (15) to (17) that the angular momentum magnitude of the rigid body is constant, and the angular momentum vector rotates at a uniform angular velocity around the axis of symmetry, depicting a cone in space. This law is similar to the law satisfied by the angular velocity vector of the rigid body. The corresponding schematic diagram is as follows (see Fig. 2):

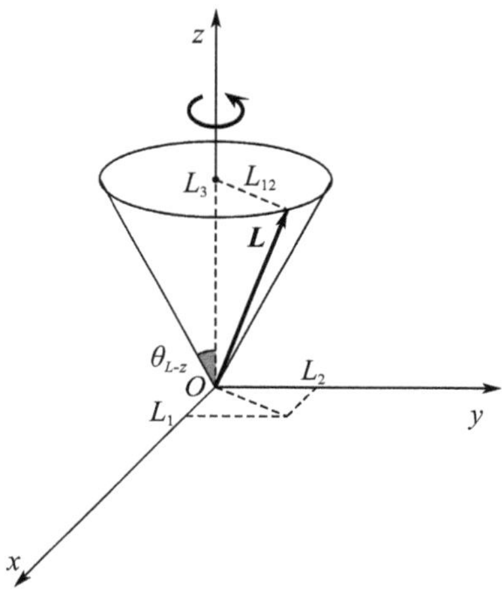

Figure 2

Taking *the rigid body* as the object of study. Studying *an arbitrary process*. Considering (a) ***law of conservation of mechanical energy*** [the work done by the external force acting on the rigid body is zero (the external force acting on the rigid body at Point *O* does none work); therefore, its mechanical energy is conserved] (b) ***law of conservation of angular momentum*** [taking Point *O* as the reference point, the resultant external torque acting on the rigid body is zero; therefore, its angular momentum is conserved]:

$$\begin{cases} E_k = \text{const} & (18a) \\ \boldsymbol{L} = \text{const} & (18b) \end{cases}$$

E_k and $\boldsymbol{L}$—the kinetic energy and the angular momentum [taking Point *O* as the reference point] of the rigid body at Time *t*.

Considering ***the relation of the kinetic energy and the angular momentum of a rigid body performing fixed-point rotation***:

$$E_k = \frac{1}{2}\boldsymbol{\omega}\cdot\boldsymbol{L} = \frac{1}{2}\omega L\cos\theta_{\omega-L} \tag{19}$$

$\theta_{\omega\text{-}L}$—the angle between the angular velocity and the angular momentum of the rigid body at arbitrary Time t.

$$\Rightarrow \qquad \cos\theta_{\omega-L} = \text{const} \tag{20}$$

That is, the angle between the angular velocity and the angular momentum of the rigid body remains constant.

The unit vector in the direction of the angular momentum is denoted by $\boldsymbol{k}'$, taking the two unit vectors perpendicular to $\boldsymbol{k}'$ and mutually perpendicular to each other, and denoting them by $\boldsymbol{i}'$ and $\boldsymbol{j}'$, respectively, $\boldsymbol{i}' \times \boldsymbol{j}' = \boldsymbol{k}'$. Establishing the space coordinate system with the frame $(O, \boldsymbol{i}', \boldsymbol{j}', \boldsymbol{k}')$, let X, Y, Z be the Cartesian coordinates.

Considering ***the formula of the rate of change of Euler angle with respect to time***:

$$\begin{cases} \dot{\theta} = \omega_1 \cos\psi - \omega_2 \sin\psi & (21a) \\ \dot{\varphi} = \omega_1 \csc\theta \sin\psi + \omega_2 \csc\theta \cos\psi & (21b) \\ \dot{\psi} = -\omega_1 \cot\theta \sin\psi - \omega_2 \cot\theta \cos\psi + \omega_3 & (21c) \end{cases}$$

$(11) \to (21)$,

$$\Rightarrow \qquad \begin{cases} \dot{\theta} = A\cos(\Omega t + \psi + \alpha) & (22a) \\ \dot{\varphi} = \csc\theta A\sin(\Omega t + \psi + \alpha) & (22b) \\ \dot{\psi} = \omega_3 - \cot\theta A\sin(\Omega t + \psi + \alpha) & (22c) \end{cases}$$

Considering the definition of ***Euler angle***:

$$\theta = \theta_{L-z} = \text{const} \tag{23}$$

$$\Rightarrow \qquad \dot{\theta} = 0 \tag{24}$$

$(24) \to (22a)$,

$$\Rightarrow \qquad \begin{cases} \psi = -\Omega t & (25a) \\ \alpha = \dfrac{\pi}{2} & (25b) \end{cases}$$

$$\Rightarrow \qquad \dot{\psi} = -\Omega \tag{26}$$

$(25) \to (22b)$,

$$\Rightarrow \qquad \dot{\varphi} = A\csc\theta \tag{27}$$

Combining (17), (23) and (27),

$$\Rightarrow \qquad \dot{\varphi} = \frac{I_3}{I_1 \cos\theta}\omega_3 \tag{28}$$

It can be seen from (24), (27) and (28) that the nutation angle of the rigid body remains unchanged, and its rotation and precession are both uniform. This form of motion is called regular precession.

Considering ***the expression of the angular velocity of a rigid body performing fixed-point rotation based on Euler angle***, then place (24) in to obtain:

$$\boldsymbol{\omega} = \dot{\psi}\boldsymbol{k} + \dot{\theta}\boldsymbol{n} + \dot{\varphi}\boldsymbol{k}' = \dot{\psi}\,\boldsymbol{k} + \dot{\varphi}\boldsymbol{k}' \tag{29}$$

$$\Rightarrow \qquad \omega = \sqrt{\dot{\psi}^2 + 2\dot{\psi}\dot{\varphi}\cos\theta + \dot{\varphi}^2} = \text{const} \tag{30}$$

It can be seen from (20) and (30) that the angular velocity magnitude of the rigid body is constant, and the angular velocity vector rotates at a uniform angular velocity around the Z-axis, depicting a cone in space. This cone is called the space cone. The corresponding schematic diagram is as follows (see Fig. 3):

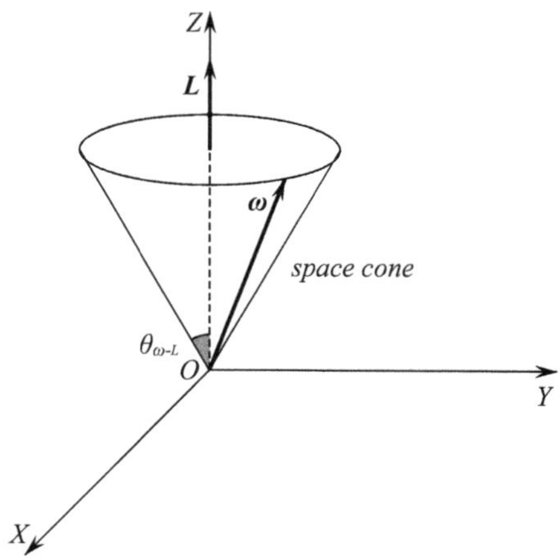

Figure 3

Considering the following vector operation:

$$\begin{aligned} \boldsymbol{L}\cdot(\boldsymbol{\omega}\times\boldsymbol{k}) &= (I_1\omega_1\boldsymbol{i} + I_2\omega_2\boldsymbol{j} + I_3\omega_3\boldsymbol{k})\cdot\left[(\omega_1\boldsymbol{i} + \omega_2\boldsymbol{j} + \omega_3\boldsymbol{k})\times\boldsymbol{k}\right] \\ &= \left[I_1(\omega_1\boldsymbol{i} + \omega_2\boldsymbol{j}) + I_3\omega_3\boldsymbol{k}\right]\cdot(-\omega_1\boldsymbol{j} + \omega_2\boldsymbol{i}) \\ &= 0 \end{aligned} \tag{31}$$

That is, the angular momentum, angular velocity, and axis of symmetry of the rigid body are coplanar all the time.

Combining (13), (17) and (23),

$$\Rightarrow \qquad \tan\theta_{\omega-z} = \frac{I_3}{I_1}\tan\theta_{L-z} = \frac{I_3}{I_1}\tan\theta \tag{32}$$

$$\Rightarrow \qquad \theta_{\omega-z}\begin{cases} < \theta_{L-z} = \theta, & \text{when } I_3 < I_1 \\ > \theta_{L-z} = \theta, & \text{when } I_3 > I_1 \end{cases} \tag{33}$$

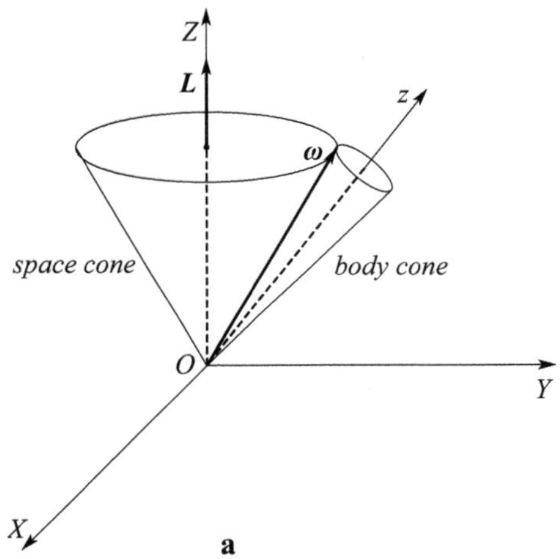

a

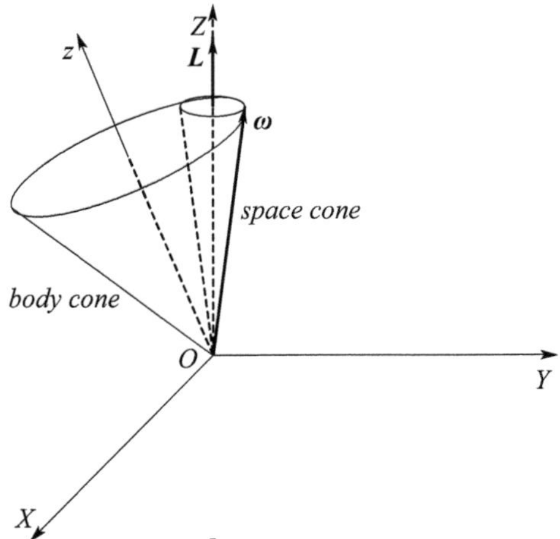

b

Figure 4a, b

That is, when $I_3 < I_1$, $\boldsymbol{\omega}$ is between $\boldsymbol{L}$ and the axis of symmetry, and the body cone and the space cone are externally tangent; when $I_3 > I_1$, $\boldsymbol{L}$ is between $\boldsymbol{\omega}$ and the axis of symmetry, and the body cone is internally tangent to the space cone. In both situations, there is no relative sliding between the body cone and the space cone. The corresponding schematic diagrams are as follows (see Fig. 4a, b).

3.2.2 The Motion of a High-Speed Self-Rotation Gyro (*c*, *B*)[6]

Consider a rotationally symmetric gyro of mass m, the distance from its center of mass C to the lower end O of its rotation axis of symmetry is l_C (see Fig. 1). Let the gyro perform high-speed self-rotation on the ground around the rotation axis of symmetry at the angular velocity of the magnitude ω, the angle between its direction and the vertical direction is θ.

(1) Find the precession angular velocity of the gyro.
(2) If the friction between the lower end of the rotation axis and the ground cannot be ignored, please prove there will be an additional motion of the gyro towards the vertical direction, and calculate the time taken for the motion of gyro's rotation axis from the angle θ_0 with the vertical direction to the vertical direction.

Figure 1

Solution [1] to Problem (1)

Time zero is arbitrarily set.

[6]This problem is a test question from University of California, Berkeley, in 1976. It is also included in the following book: Yongde Zhang, et al. (1988). *American Physics Test Questions and Solutions* (University of Science and Technology of China Press).

As shown in Fig. 2, establishing the body coordinate system with the frame $(O, \boldsymbol{i}, \boldsymbol{j}, \boldsymbol{k})$, let x, y, z be the Cartesian coordinates (the coordinate system of principal axis of the gyro), and establishing the space coordinate system with the frame $(O, \boldsymbol{i}', \boldsymbol{j}', \boldsymbol{k}')$, let X, Y, Z be the Cartesian coordinates.

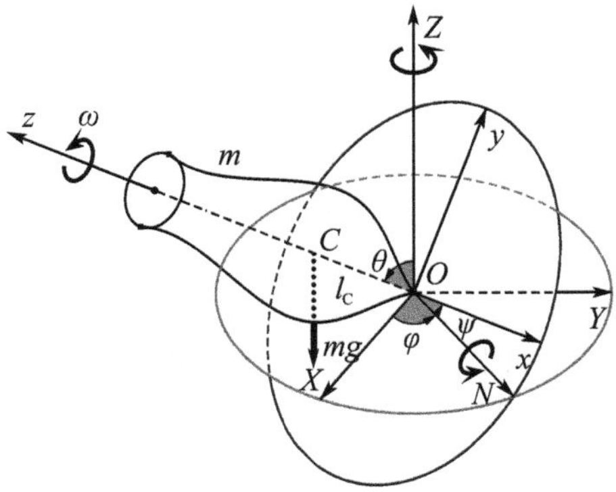

Figure 2

Taking *the gyro* as the object of study. Studying *the state at arbitrary Time t*. Considering ***Euler dynamical equations*** [taking Point O as the basic point] (the component equations in the $(O, \boldsymbol{i}, \boldsymbol{j}, \boldsymbol{k})$-directions):

$$\begin{cases} mgl_C \sin\theta\cos\psi = I_1\dot{\omega}_x - (I_2 - I_3)\omega_y\omega_z & \text{(1a)} \\ -mgl_C \sin\theta\sin\psi = I_2\dot{\omega}_y - (I_3 - I_1)\omega_z\omega_x & \text{(1b)} \\ 0 = I_3\dot{\omega}_z - (I_1 - I_2)\omega_x\omega_y & \text{(1c)} \end{cases}$$

(I_1, I_2, I_3)—the principal moments of inertia of the rigid body about the axes x, y and z of the body coordinate system.

$(\omega_x, \omega_y, \omega_z)$—the components of the angular velocity of the rigid body in the $(O, \boldsymbol{i}, \boldsymbol{j}, \boldsymbol{k})$-directions at Time t.

It can be seen from the symmetry that:

$$I_1 = I_2 \tag{2}$$

$$\Rightarrow \quad \begin{cases} mgl_C \sin\theta\cos\psi = I_1\dot{\omega}_x - (I_1 - I_3)\omega_y\omega_z & \text{(3a)} \\ -mgl_C \sin\theta\sin\psi = I_1\dot{\omega}_y - (I_3 - I_1)\omega_z\omega_x & \text{(3b)} \\ \dot{\omega}_z = 0 & \text{(3c)} \end{cases}$$

Considering ***Euler kinematical equations***:

$$\begin{cases} \omega_x = \dot{\varphi}\sin\theta\sin\psi + \dot{\theta}\cos\psi & (4a) \\ \omega_y = \dot{\varphi}\sin\theta\cos\psi - \dot{\theta}\sin\psi & (4b) \\ \omega_z = \dot{\psi} + \dot{\varphi}\cos\theta & (4c) \end{cases}$$

Known:

$$\begin{cases} \dot{\theta} = 0 & (5a) \\ \dot{\psi} = \omega & (5b) \end{cases}$$

$$\Rightarrow \quad \begin{cases} \omega_x = \dot{\varphi}\sin\theta\sin\psi & (6a) \\ \omega_y = \dot{\varphi}\sin\theta\cos\psi & (6b) \\ \omega_z = \omega + \dot{\varphi}\cos\theta & (6c) \end{cases}$$

Combining (3), (5) and (6):

$$\Rightarrow \quad \begin{cases} mgl_C\cos\psi = I_1(\ddot{\varphi}\sin\psi + \omega\dot{\varphi}\cos\psi) - \\ \qquad (I_1 - I_3)\dot{\varphi}\cos\psi(\omega + \dot{\varphi}\cos\theta) & (7a) \\ -mgl_C\sin\psi = I_1(\ddot{\varphi}\cos\psi - \omega\dot{\varphi}\sin\psi) - \\ \qquad (I_3 - I_1)\dot{\varphi}\sin\psi(\omega + \dot{\varphi}\cos\theta) & (7b) \end{cases}$$

Multiplying both sides of Eq. (7a) by $\cos\psi$, and multiplying both sides of Eq. (7b) by $\sin\psi$, and then subtracting:

$$\Rightarrow \quad (I_1 - I_3)\cos\theta\dot{\varphi}^2 - I_3\omega\dot{\varphi} + mgl_C = 0 \tag{8}$$

$$\Rightarrow \quad \dot{\varphi} = \frac{I_3\omega - \sqrt{I_3^2\omega^2 - 4(I_1 - I_3)mgl_C\cos\theta}}{2(I_1 - I_3)\cos\theta}$$

$$= \frac{I_3\omega}{2(I_1 - I_3)\cos\theta}\left[1 - \sqrt{1 - \frac{4(I_1 - I_3)mgl_C\cos\theta}{I_3^2\omega^2}}\right]$$

$$\approx \frac{I_3\omega}{2(I_1 - I_3)\cos\theta}\left[1 - \left(1 - \frac{2(I_1 - I_3)mgl_C\cos\theta}{I_3^2\omega^2}\right)\right]$$

$$= \frac{mgl_C}{I_3\omega} \tag{9}$$

This operation has considered $1/\omega$ as a small quantity and taken the first-order approximation.

Solution [2] to Problem (1)

Taking *the gyro* as the object of study. Studying *the instantaneous process (the small time interval dt) at arbitrary Time t*. Considering ***theorem of angular momentum*** [taking Point O as the reference point]:

$$\boldsymbol{M}=\frac{\mathrm{d}\boldsymbol{L}}{\mathrm{d}t} \tag{10}$$

$\boldsymbol{M}$—the resultant external torque acting on the gyro at Time t [taking Point O as the reference point].

$\boldsymbol{L}$—the angular momentum of the gyro at Time t [taking Point O as the reference point].

Considering the definition of ***torque***:

$$\boldsymbol{M}=l_C\boldsymbol{k}\times(-mg\boldsymbol{k}')=mgl_C\boldsymbol{k}'\times\boldsymbol{k} \tag{11}$$

Considering ***the form of the angular momentum of a rigid body performing fixed-point rotation based on the principal moments of inertia***:

$$\begin{aligned}\Rightarrow\quad \boldsymbol{L}&=I_1\omega_x\boldsymbol{i}+I_2\omega_y\boldsymbol{j}+I_3\omega_z\boldsymbol{k}\\&=I_1\dot{\varphi}\sin\theta\sin\psi\boldsymbol{i}+I_2\dot{\varphi}\sin\theta\cos\psi\boldsymbol{j}+I_3(\omega+\dot{\varphi}\cos\theta)\boldsymbol{k}\\&\approx I_3\omega\boldsymbol{k}\end{aligned} \tag{12}$$

This operation has considered $\omega >> \dot{\varphi}$, the angular momentum components in the $\boldsymbol{i}$-direction and $\boldsymbol{j}$-direction are ignored.

Considering Poisson formula:

$$\Rightarrow\quad \frac{\mathrm{d}\boldsymbol{L}}{\mathrm{d}t}=\dot{\varphi}\boldsymbol{k}'\times\boldsymbol{L}=I_3\omega\dot{\varphi}\boldsymbol{k}'\times\boldsymbol{k} \tag{13}$$

(11) and (13) $\rightarrow$ (10):

$$\Rightarrow\quad \dot{\varphi}=\frac{mgl}{I_3\omega} \tag{14}$$

Solution to Problem (2)

The time point when the angle between the rotation axis of the gyro and the vertical direction is θ_0 is defined as Time zero.

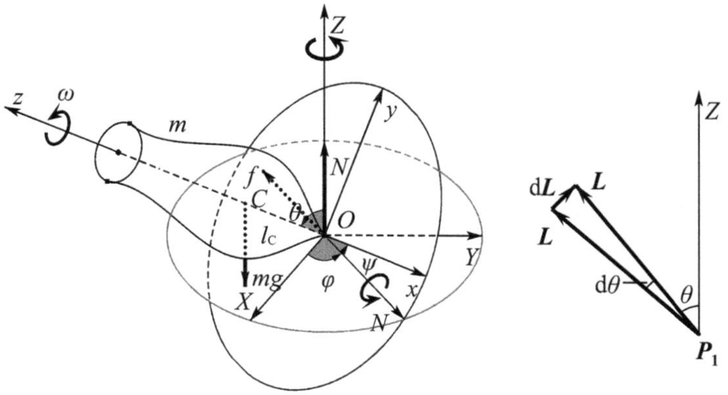

Figure 3

Taking *the gyro* as the object of study (see Fig. 3). Studying *the state at arbitrary Time t.* Considering ***theorem of motion of center of mass*** (the ***k***′-direction component equation):

$$N - mg = ma_{CZ} \tag{15}$$

N—the ***k***′-direction component of the normal contact force exerted on the gyro by the ground at Time *t*.

a_{CZ}—the ***k***′-direction component of the center-of-mass acceleration of the gyro at Time *t*.

Considering ***the components form of the acceleration of a particle in polar coordinate system***:

$$a_{CZ} = -\ddot{\theta} l_C \sin\theta - \dot{\theta}^2 l_C \cos\theta \approx 0 \tag{16}$$

This operation has considered $\dot{\theta}$ and $\ddot{\theta}$ as small quantities and taken the low order approximation.

$$\Rightarrow \qquad N \approx mg \tag{17}$$

Considering ***friction law***:

$$\boldsymbol{f} = -\mu N \boldsymbol{n} \tag{18}$$

f—the fiction exerted on the gyro by the ground at arbitrary Time *t*.

$$\Rightarrow \qquad \boldsymbol{f} \approx -\mu mg \boldsymbol{n} \tag{19}$$

Taking *the gyro* as the object of study. Studying *the instantaneous process (the small time interval dt) at arbitrary Time t.* Considering (a) ***theorem of angular momentum*** [in its center-of-mass reference frame, taking the center of mass C as the reference point] [note: see further important discussion at the end of this problem], (b) ***in the center-of-mass reference frame, the resultant torque of the inertia force acting on system of particles is zero***:

$$-l_C\boldsymbol{k}\times(\boldsymbol{f}+N\boldsymbol{k}')\mathrm{d}t=\mathrm{d}\boldsymbol{L} \tag{20}$$

Taking the projection of the equation above in the plane determined by $\boldsymbol{k}$ and $\boldsymbol{k}'$:

$$\Rightarrow \quad (-l_C\boldsymbol{k})\times(-\mu mg\boldsymbol{n})\mathrm{d}t\approx L(-\mathrm{d}\theta)(\boldsymbol{k}\times\boldsymbol{n}) \tag{21}$$

$$\Rightarrow \quad \mathrm{d}\theta=-\mu\frac{mgl_C}{I_3\omega}\mathrm{d}t \tag{22}$$

$$\Rightarrow \quad \int_{\theta_0}^{0}\mathrm{d}\theta=-\mu\frac{mgl_C}{I_3\omega}\int_0^T\mathrm{d}t$$

$$\Rightarrow \quad T=\frac{I_3\omega\theta_0}{\mu mgl_C} \tag{23}$$

T—the time taken for the motion of gyro's rotation axis from the angle θ_0 with the vertical direction to the vertical direction.

***Note*: Further discussion:**

The popular solution to this problem only considers the torque of the friction exerted on the gyro by the ground, but does not consider the torque of the normal contact force exerted on the gyro by the ground. Therefore, the "theorem of angular momentum" expressed in the popular solution is only the projection form of the complete "theorem of angular momentum" in the plane determined by $\boldsymbol{k}$ and $\boldsymbol{k}'$.

3.2.3 The Regular Precession of a Homogeneous Thin Circular Disk (d, *B*)

As shown in Fig. 1, there is a homogeneous thin circular disk of radius r performing the pure rolling on a horizontal ground. The angle between the disk and the vertical direction is θ all the time, which is a small quantity. The herpolhode of the disk on the ground is a circle of radius R. The centers of circle of the disk and its herpolhode are denoted by O_1 and O_2, respectively. Find the precession angular velocity of the disk.

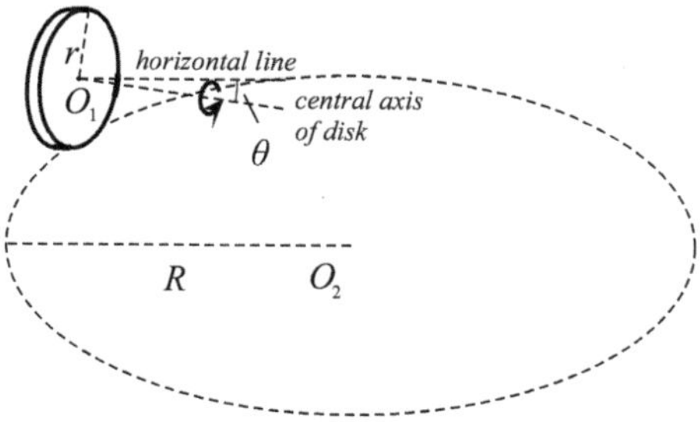

Figure 1

Solution

Time zero is arbitrarily set.

As shown in Fig. 2, the center of the circular trajectory of Point O_1 is denoted by O_3, the intersection point of the central axis of the disk and the central axis of the herpolhode on the ground is denoted by O. Establishing the body coordinate system with the frame (O, $\boldsymbol{i}$, $\boldsymbol{j}$, $\boldsymbol{k}$), let x, y, z be the Cartesian coordinates (the coordinate system of principal axis of the disk).

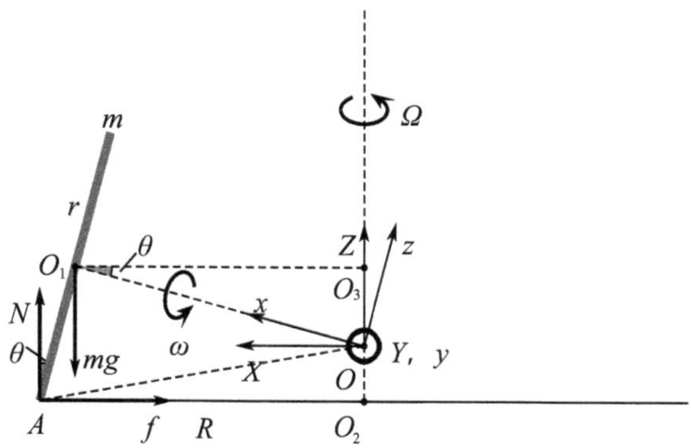

Figure 2

Once the y-axis is in horizontal direction is denoted by Time T. Establishing the space coordinate system with the frame $(O, \boldsymbol{i}', \boldsymbol{j}', \boldsymbol{k}')$, let X, X, Z be the Cartesian coordinates, where $\boldsymbol{j}$ coincides with $\boldsymbol{j}'$ at Time T. The point on the disk in contact with the ground at Time T is denoted by A.

θ is a small quantity, considering the first-order approximation of the following trigonometric functions:

$$\begin{cases} \sin\theta \approx \theta & (1a) \\ \cos\theta \approx 1 & (1b) \\ \tan\theta \approx \theta & (1c) \end{cases}$$

Considering ***Euler kinematical equations***:

$$\begin{cases} \omega_x = -\omega + \Omega\sin\theta \approx -\omega + \Omega\theta & (2a) \\ \omega_y = 0 & (2b) \\ \omega_z = \Omega\cos\theta \approx \Omega & (2c) \end{cases}$$

$(\omega_x, \omega_y, \omega_z)$—the components of the angular velocity of the disk in the $(\boldsymbol{i}, \boldsymbol{j}, \boldsymbol{k})$-directions at Time T.

ω—the $-\boldsymbol{i}$-direction component of the self-rotation angular velocity of the disk at Time T.

Taking *Point A* as the object of study. Studying *the state at Time T*. Considering ***theorem for composition of velocities of a particle*** [taking the ground as the stationary reference frame, and the translational reference frame that is at rest with respect to Point O_1 as the moving reference frame]:

$$\boldsymbol{v}_A = \boldsymbol{v}_{O_1} + \boldsymbol{v}_{AO_1} \quad (3)$$

$\boldsymbol{v}_A$—the velocity of Point A at Time T.

$\boldsymbol{v}_{O_1}$ —the velocity of Point O_1 at Time T.

$\boldsymbol{v}_{AO_1}$ —the velocity of Point A relative to Point O_1 at Time T.

The condition for the disk to be in contact with the ground all the time without relative slipping is as follows:

$$\boldsymbol{v}_A = 0 \quad (4)$$

Considering the definition of ***angular velocity*** and Poisson formula, then place (1a) in to obtain:

$$\Rightarrow \begin{cases} \boldsymbol{v}_{O_1} = \Omega\boldsymbol{k}' \times \overrightarrow{O_3O_1} = \Omega\boldsymbol{k}' \times (R - r\sin\theta)\boldsymbol{i}' \approx \Omega(R - r\theta)\boldsymbol{j}' & (5a) \\ \boldsymbol{v}_{AO_1} = (\omega_x\boldsymbol{i} + \omega_y\boldsymbol{j} + \omega_z\boldsymbol{k}) \times r(-\boldsymbol{k}) = (-\omega r + \Omega r\theta)\boldsymbol{j} & \\ \quad = (-\omega r + \Omega r\theta)\boldsymbol{j}' & (5b) \end{cases}$$

(4) and (5) → (3),

$$\Rightarrow \qquad \omega = \frac{R}{r}\Omega \qquad (6)$$

$$\Rightarrow \qquad \omega_x = \left(\theta - \frac{R}{r}\right)\Omega \qquad (7)$$

Taking *the disk* as the object of study. Studying *the state at Time T*. Considering (a) ***theorem of motion of center of mass*** (the $\boldsymbol{i}'$-direction and $\boldsymbol{k}'$-direction component equations), (b) ***the radial component form of acceleration*** [integrating into (a)]:

$$\begin{cases} \boldsymbol{f} = -m\Omega^2 \overline{O_1O_3}\boldsymbol{i}' = -m\Omega^2 (R - r\sin\theta)\boldsymbol{i}' & \text{(8a)} \\ \boldsymbol{N} + mg(-\boldsymbol{k}') = 0 & \text{(8b)} \end{cases}$$

m—the mass of the disk.

$\boldsymbol{N}$—the normal contact force exerted on the disk by the ground at Time T.

$\boldsymbol{f}$—the friction exerted on the disk by the ground at Time T.

$$\Rightarrow \qquad \boldsymbol{N} = mg\boldsymbol{k}' \qquad (9)$$

Taking *the disk* as the object of study. Studying *the instantaneous process (the small time interval dt) at Time T.* Considering ***theorem of angular momentum*** [taking Point O as the reference point]:

$$\boldsymbol{M} = \frac{\mathrm{d}\boldsymbol{L}}{\mathrm{d}t} = \left(\frac{\mathrm{d}}{\mathrm{d}t}\right)' \boldsymbol{L} + \Omega\boldsymbol{k}' \times \boldsymbol{L} = \Omega\boldsymbol{k}' \times \boldsymbol{L} \qquad (10)$$

$\boldsymbol{M}$—the resultant external torque acting on the disk [taking Point O as the reference point] at Time T.

$\boldsymbol{L}$—the angular momentum of the disk [taking Point O as the reference point] at Time T.

Considering the definition of ***torque***:

$$\Rightarrow \quad \boldsymbol{M} = \overrightarrow{OO_1} \times mg(-\boldsymbol{k}') + \overrightarrow{OA} \times \boldsymbol{N} + \overrightarrow{OA} \times \boldsymbol{f}$$

$$= \frac{R - r\sin\theta}{\cos\theta}\boldsymbol{i} \times mg(-\boldsymbol{k}') + \left[\frac{R - r\sin\theta}{\cos\theta}\boldsymbol{i} + r(-\boldsymbol{k})\right] \times mg\boldsymbol{k}' +$$

$$\left[\frac{R - r\sin\theta}{\cos\theta}\boldsymbol{i} + r(-\boldsymbol{k})\right] \times m\Omega^2 (R - r\sin\theta)(-\boldsymbol{i}')$$

$$= -mgr\sin\theta\boldsymbol{j} - m\Omega^2(R - r\sin\theta)^2\tan\theta\boldsymbol{j} +$$
$$m\Omega^2(R - r\sin\theta)r\cos\theta\boldsymbol{j}$$
$$\approx -mgr\theta\boldsymbol{j} + m\Omega^2[Rr - (R^2 + r^2)\theta]\boldsymbol{j} \tag{11}$$

This operation has taken the first-order approximation of θ.

Considering ***the form of the angular momentum of a rigid body performing fixed-point rotation based on the principal moments of inertia.***

$$\boldsymbol{L} = I_1\omega_x\boldsymbol{i} + \boldsymbol{I}_2\omega_y\boldsymbol{j} + \boldsymbol{I}_3\omega_z\boldsymbol{k} \tag{12}$$

(I_1, I_2, I_3)—the principal moments of inertia of the disk about the axes x, y and z of the body coordinate system.

Considering (a) ***the form of moment of inertia of a homogeneous thin circular disk about its central axis perpendicular to the plane of the disk***, (b) ***the form of moment of inertia of a homogeneous thin circular disk about one of its central axes in the plane of the disk***, (c) ***parallel-axis theorem*** [integrating into (b)]:

$$\begin{cases} I_1 = \dfrac{1}{2}mr^2 & \text{(13a)} \\ I_2 = I_3 = \dfrac{1}{4}mr^2 + m\overline{OO_1}^2 = \dfrac{1}{4}mr^2 + m\left(\dfrac{R - r\sin\theta}{\cos\theta}\right)^2 \\ \qquad \approx \dfrac{1}{4}mr^2 + m(R - r\theta)^2 \approx \dfrac{1}{4}mr^2 + m(R^2 - 2Rr\theta) & \text{(13b)} \end{cases}$$

(2) and (13) $\rightarrow$ (12):

$$\Rightarrow \quad \boldsymbol{L} = \frac{1}{2}m(r^2\theta - Rr)\Omega\boldsymbol{i} + \left[\frac{1}{4}mr^2 + m(R^2 - 2Rr\theta)\right]\Omega\boldsymbol{k} \tag{14}$$

(11) and (14) $\rightarrow$ (10):

$$\Rightarrow \quad -mgr\theta + m\Omega^2[Rr - (R^2 + r^2)\theta]$$
$$= \frac{1}{2}m(r^2\theta - Rr)\Omega^2\cos\theta - \left[\frac{1}{4}mr^2 + m(R^2 - 2Rr\theta)\right]\sin\theta\Omega^2$$
$$\approx m\Omega^2\left[-\frac{1}{2}Rr + \left(\frac{1}{4}r^2 - R^2\right)\theta\right] \tag{15}$$

$$\Rightarrow \qquad \Omega = \sqrt{\frac{g\theta}{\frac{3}{2}R - \frac{5}{4}r\theta}} = \sqrt{\frac{2g\theta}{3R}\left(1 + \frac{5r\theta}{6R}\right)} \approx \sqrt{\frac{2g\theta}{3R}} \qquad (16)$$

This operation has taken the first-order approximation of θ.

3.2.4 The Curvilinear Motion of a Billiard (*b*, *B*)[7]

Consider a billiard ball (can be regarded as a homogeneous sphere of radius R) moving on a billiard table, whose initial center-of-mass velocity is $\boldsymbol{v}_{C0}$ and initial rotating angular velocity is $\boldsymbol{\omega}_0$ (see Fig. 1), $\boldsymbol{v}_{C0}$ and $\boldsymbol{\omega}_0$ are both in the horizontal plane, the included angle between them is θ. The coefficient of static friction between the billiard ball and the billiard table is μ. Analyze the motion of the billiard ball.

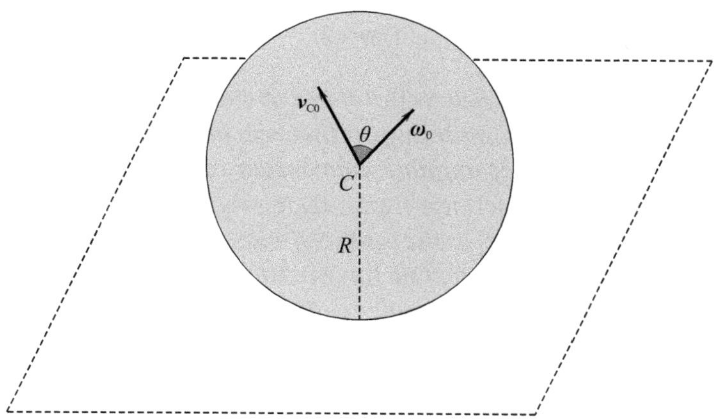

Figure 1

Solution

The initial time is defined as Time zero of the process of the billiard ball's sliding rolling relative to the billiard table.

As shown in Fig. 2, the point of the billiard ball in contact with the billiard table at arbitrary Time t is denoted by P_t, the projection of Point P_t on the billiard table is denoted by Q_t (the component of the acceleration of Point P_t in vertical direction is generally not zero), the unit vector in the vertical downward direction is denoted by $\boldsymbol{e}_g$,

[7]Gnadig, Peter (2016). *200 More Puzzling Physics Problems* (Cambridge University Press).

the unit vector in the velocity direction of Point Q_t at arbitrary Time t is denoted by $\boldsymbol{e}_{Q_t}$.

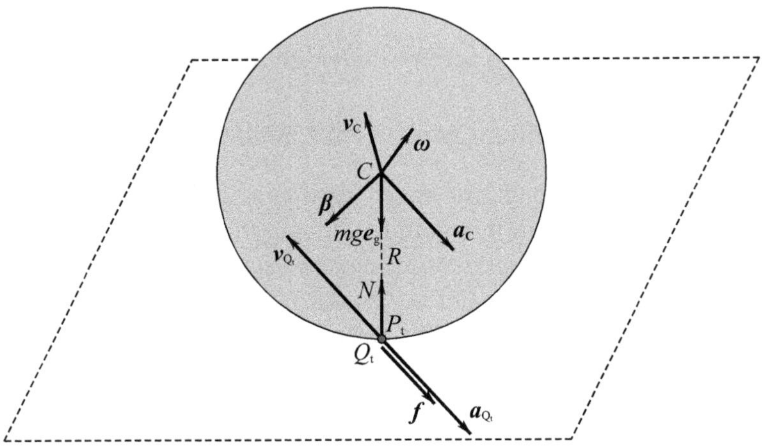

Figure 2

Taking *the billiard ball* as the object of study. Studying *the state at arbitrary Time t*. Considering (a) ***theorem of motion of center of mass***, (b) ***theorem of angular momentum*** (differential form) [in its center-of-mass reference frame, taking the center of mass C as the reference point], (c) ***in the center-of-mass reference frame, the resultant torque of the inertia force acting on system of particles is zero***, (d) ***the form of moment of inertia of a homogeneous sphere about one of its central axes***, (e) ***friction law***:

$$\begin{cases} \begin{cases} \boldsymbol{f} = m\boldsymbol{a}_C & \text{(1a)} \\ \boldsymbol{N} - mg\boldsymbol{e}_g = 0 & \text{(1b)} \end{cases} \\ R\boldsymbol{e}_g \times \boldsymbol{f} = I\boldsymbol{\beta} & \text{(1c)} \end{cases}$$

$$I = \frac{2}{5}mR^2 \quad (2)$$

$$\boldsymbol{f} = -\mu N\boldsymbol{e}_{Q_t} \quad (3)$$

$\boldsymbol{f}$ and $\boldsymbol{N}$—the sliding friction force and the normal contact force exerted on the billiard ball by the billiard table at Time t.

$\boldsymbol{a}_C$ and $\boldsymbol{\beta}$—the center-of-mass acceleration and the rotating angular acceleration of the billiard ball at Time t.

I—the moment of inertia of the billiard ball about one of its central axes.

Combining (1), (2) and (3),

$$\Rightarrow \quad \begin{cases} \begin{cases} \boldsymbol{N} = mg\boldsymbol{e}_g & \text{(4a)} \\ \boldsymbol{f} = -\mu mg\boldsymbol{e}_{Q_t} & \text{(4b)} \end{cases} \\ \begin{cases} \boldsymbol{a}_C = -\mu g\boldsymbol{e}_{Q_t} & \text{(4c)} \\ \boldsymbol{\beta} = -\dfrac{5\mu g}{2R}(\boldsymbol{e}_g \times \boldsymbol{e}_{Q_t}) & \text{(4d)} \end{cases} \end{cases}$$

Taking *Point* P_t as the object of study. Studying *the state at Time t*. Considering ***theorem for the composition of accelerations of a particle*** [taking the billiard table as the stationary reference frame, and the center-of-mass reference frame of the billiard ball as the moving reference frame]:

$$\boldsymbol{a}_{P_t} = \boldsymbol{a}_C + \boldsymbol{\beta} \times R\boldsymbol{e}_g + \boldsymbol{\omega} \times \boldsymbol{v}'_{P_t} \tag{5}$$

$\boldsymbol{a}_{P_t}$ —the acceleration of Point P_t.

$\boldsymbol{v}'_{P_t}$ —the velocity of Point P_t relative the center-of-mass C.

$\boldsymbol{\omega}$—the rotating angular velocity of the billiard ball at Time t.

Considering the definition of ***angular acceleration***:

$$\boldsymbol{\beta} = \frac{\mathrm{d}\boldsymbol{\omega}}{\mathrm{d}t} \tag{6}$$

It can be seen from Eq. (4d) that $\boldsymbol{\beta}$ is in the horizontal plane all the time, and known that $\boldsymbol{\omega}_0$ is in the horizontal plane; therefore, $\boldsymbol{\omega}$ is in the horizontal plane all the time. It can be further seen that the term $\boldsymbol{\omega} \times \boldsymbol{v}'_{P_t}$ is in the vertical direction all the time.

$$\Rightarrow \quad \boldsymbol{a}_{Q_t} = (\boldsymbol{a}_{P_t})_{\text{horizontal}} = \boldsymbol{a}_C + \boldsymbol{\beta} \times R\boldsymbol{e}_g \tag{7}$$

$\boldsymbol{a}_{Q_t}$ —the acceleration of Point Q_t.

(4c) and (4d) → (7),

$$\Rightarrow \quad \boldsymbol{a}_{Q_t} = -\mu g\boldsymbol{e}_{Q_t} - \frac{5\mu g}{2}(\boldsymbol{e}_g \times \boldsymbol{e}_{Q_t}) \times \boldsymbol{e}_g \tag{8}$$

Considering the following vector operations:

$$(\boldsymbol{e}_g \times \boldsymbol{e}_{Q_t}) \times \boldsymbol{e}_g = \boldsymbol{e}_{Q_t}(\boldsymbol{e}_g \cdot \boldsymbol{e}_g) - \boldsymbol{e}_g(\boldsymbol{e}_g \cdot \boldsymbol{e}_{Q_t}) = \boldsymbol{e}_{Q_t} \tag{9}$$

$$\Rightarrow \quad \boldsymbol{a}_{Q_t} = -\frac{7\mu g}{2}\boldsymbol{e}_{Q_t} \tag{10}$$

Considering the definition of ***acceleration***:

$$\boldsymbol{a}_{Q_t} = \frac{\mathrm{d}\boldsymbol{v}_{Q_t}}{\mathrm{d}t} = \frac{\mathrm{d}v_{Q_t}}{\mathrm{d}t}\boldsymbol{e}_{Q_t} + v_{Q_t}\frac{\mathrm{d}\boldsymbol{e}_{Q_t}}{\mathrm{d}t} \tag{11}$$

$\boldsymbol{v}_{Q_t}$ —the velocity of Point Q_t.

Combining (10) and (11),

$$\Rightarrow \qquad \frac{\mathrm{d}\boldsymbol{e}_{Q_t}}{\mathrm{d}t} = -\frac{1}{v_{Q_t}}\left(\frac{\mathrm{d}v_{Q_t}}{\mathrm{d}t} + \frac{7\mu g}{2}\right)\boldsymbol{e}_{Q_t} \tag{12}$$

Considering the definition of ***derivative***:

$$\frac{\mathrm{d}\boldsymbol{e}_{Q_t}}{\mathrm{d}t} = \frac{\left|\boldsymbol{e}_{Q_t}\right|\mathrm{d}\varphi}{\mathrm{d}t}\frac{\mathrm{d}\boldsymbol{e}_{Q_t}}{\left|\mathrm{d}\boldsymbol{e}_{Q_t}\right|} = \frac{\mathrm{d}\varphi}{\mathrm{d}t}\frac{\mathrm{d}\boldsymbol{e}_{Q_t}}{\left|\mathrm{d}\boldsymbol{e}_{Q_t}\right|} \tag{13}$$

$\mathrm{d}\varphi$—the infinitesimal angular displacement of $\boldsymbol{e}_{Q_t}$ in time interval $\mathrm{d}t$ in the $\mathrm{d}\boldsymbol{e}_{Q_t}$-direction.

Combining (12) and (13),

$$\Rightarrow \qquad \frac{\mathrm{d}\varphi}{\mathrm{d}t}\frac{\mathrm{d}\boldsymbol{e}_{Q_t}}{\left|\mathrm{d}\boldsymbol{e}_{Q_t}\right|} = -\frac{1}{v_{Q_t}}\left(\frac{\mathrm{d}v_{Q_t}}{\mathrm{d}t} + \frac{7\mu g}{2}\right)\boldsymbol{e}_{Q_t} \tag{14}$$

The condition for the equation above to hold is as follows:

$$\Rightarrow \qquad \frac{\mathrm{d}\varphi}{\mathrm{d}t} = \frac{\mathrm{d}v_{Q_t}}{\mathrm{d}t} + \frac{7\mu g}{2} = 0 \tag{15}$$

(15) $\rightarrow$ (12),

$$\Rightarrow \qquad \frac{\mathrm{d}\boldsymbol{e}_{Q_t}}{\mathrm{d}t} = 0 \tag{16}$$

$$\Rightarrow \qquad \boldsymbol{e}_{Q_t} = \boldsymbol{e}_{Q_0} \tag{17}$$

That is, the velocity direction of the point of the billiard ball in contact with the billiard table at arbitrary Time t remains unchanged [note: see further important discussion at the end of this problem].

(17) $\rightarrow$ (4c), (4d) and (10),

$$\Rightarrow \qquad \begin{cases} \boldsymbol{a}_C = -\mu g\boldsymbol{e}_{Q_0} & (18\text{a}) \\ \boldsymbol{\beta} = -\dfrac{5\mu g}{2R}(\boldsymbol{e}_g \times \boldsymbol{e}_{Q_0}) & (18\text{b}) \\ \boldsymbol{a}_{Q_t} = -\dfrac{7\mu g}{2}\boldsymbol{e}_{Q_0} & (18\text{c}) \end{cases}$$

That is, $\boldsymbol{a}_{Q_t}$, $\boldsymbol{a}_C$ and $\boldsymbol{\beta}$ are all constant in the process of the billiard ball's sliding rolling relative to the billiard table. The center-of-mass of the billiard ball performs the uniformly variable motion, and the billiard ball performs the uniformly variable rotation.

Considering the definition of ***acceleration***:

$$\Rightarrow \qquad \boldsymbol{v}_{Q_t} = \boldsymbol{v}_{Q_0} + \int_0^t \boldsymbol{a}_{Q_t}\,\mathrm{d}t = \boldsymbol{v}_{Q_0} - \frac{7\mu g}{2}\boldsymbol{e}_{Q_0} t \tag{19}$$

Taking *Point* P_0 as the object of study. Studying *the state at Time zero*. Considering ***theorem for composition of velocities of a particle*** [taking the billiard table as the stationary reference frame, and the center of mass reference frame of the billiard ball as the moving reference frame]:

$$\boldsymbol{v}_{Q_0} = \boldsymbol{v}_{P_0} = \boldsymbol{v}_{C0} + \boldsymbol{\omega}_0 \times R\boldsymbol{e}_g \tag{20}$$

Considering the following geometric relation in Fig. 3:

$$\Rightarrow \qquad \boldsymbol{v}_{Q_0} = \left[v_{C0}^2 + \omega_0^2 R^2 - 2v_{C0}\omega_0 R\cos\left(\frac{\pi}{2}+\theta\right)\right]\boldsymbol{e}_{Q_0}$$

$$= (v_{C0}^2 + \omega_0^2 R^2 + 2v_{C0}\omega_0 R\sin\theta)\boldsymbol{e}_{Q_0} \tag{21}$$

(21) $\rightarrow$ (19),

$$\Rightarrow \qquad \boldsymbol{v}_{Q_t} = \left[(v_{C0}^2 + \omega_0^2 R^2 + 2v_{C0}\omega_0 R\sin\theta) - \frac{7\mu g}{2}t \right]\boldsymbol{e}_{Q_0} \tag{22}$$

It can be seen that $\boldsymbol{v}_{Q_t}$ will be reduced to zero when t is large enough, and then the billiard ball will start pure rolling relative to the billiard table.

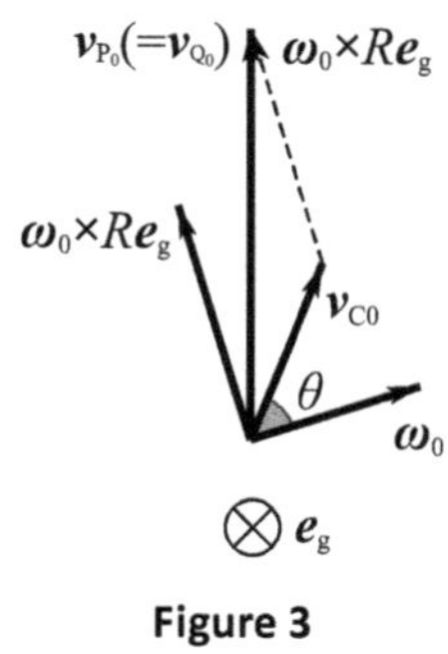

Figure 3

Defining T—the Time when the billiard ball reaches pure rolling relative to the billiard table.

$$\Rightarrow \qquad T = t\big|_{v_{Q_t}=0} = \frac{2}{7\mu g}(v_{C0}^2 + \omega_0^2 R^2 + 2v_{C0}\omega_0 R\sin\theta) \tag{23}$$

Considering ***the velocity-time relation of uniformly variable motion and the angular velocity-time relation of the uniformly variable rotation***, then place (18a) and (18b) in to obtain:

$$\Rightarrow \quad \begin{cases} \boldsymbol{v}_C = \boldsymbol{v}_{C0} + \boldsymbol{a}_C t = \boldsymbol{v}_{C0} - \mu g \boldsymbol{e}_{Q_0} t & (24a) \\ \boldsymbol{\omega} = \boldsymbol{\omega}_0 + \boldsymbol{\beta} t = \boldsymbol{\omega}_0 - \dfrac{5\mu g}{2R}(\boldsymbol{e}_g \times \boldsymbol{e}_{Q_0})t & (24b) \end{cases}$$

$\boldsymbol{v}_C$ and $\boldsymbol{\omega}$—the center-of-mass velocity and rotating angular velocity of the billiard ball at arbitrary Time t.

$$\Rightarrow \quad \begin{cases} \boldsymbol{V}_C = \boldsymbol{v}_C\big|_{t=T} = \boldsymbol{v}_{C0} - \dfrac{2}{7}(v_{C0}^2 + \omega_0^2 R^2 + 2v_{C0}\omega_0 R\sin\theta)\boldsymbol{e}_{Q_0} & (25a) \\ \boldsymbol{\Omega} = \boldsymbol{\omega}\big|_{t=T} = \boldsymbol{\omega}_0 - \dfrac{5}{7R}(v_{C0}^2 + \omega_0^2 R^2 + 2v_{C0}\omega_0 R\sin\theta)(\boldsymbol{e}_g \times \boldsymbol{e}_{Q_0}) & (25b) \end{cases}$$

$\boldsymbol{V}_C$ and $\boldsymbol{\Omega}$—the center-of-mass velocity and rotating angular velocity of the billiard ball when it reaches pure rolling relative to the billiard table.

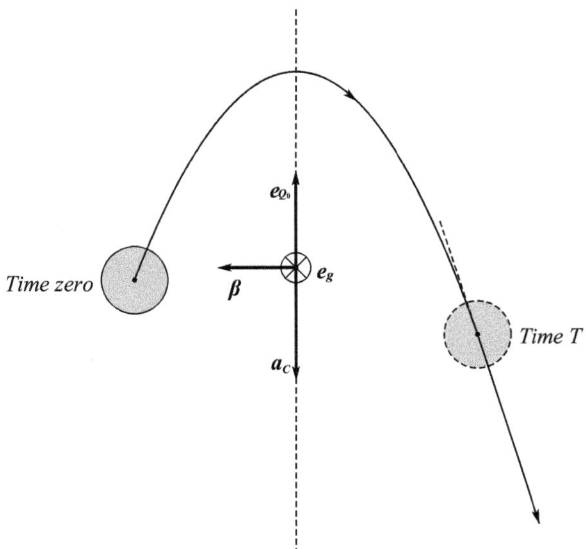

Figure 4

The schematic diagram (top view) of the billiard ball's motion is as follows (see Fig. 4):

***Note*: Further discussion:**

The interesting conclusion of this problem comes from the setting that the direction of the initial angular velocity of the billiard ball is in the horizontal plane. According to the solution provided in this book, the acceleration of the lowest point of the billiard ball relative to the center of mass of the billiard ball is divided into two terms (the terms "$\boldsymbol{\beta} \times R\boldsymbol{e}_g$" and "$\boldsymbol{\omega} \times \boldsymbol{v}'_{P_t}$" in Eq. (5)), The term "$\boldsymbol{\omega} \times \boldsymbol{v}'_{P_t}$" generally has both horizontal and vertical components, only when the angular velocity of the billiard ball is just horizontal, the horizontal component of this term could degenerate to zero, only in this case can we further infer that the projection of the acceleration at the lowest point of the billiard ball in the horizontal direction is opposite to its velocity direction all the time, and the angular velocity of the billiard ball is horizontal all the time.

The prototype of this problem sets the initial angular velocity of the billiard ball to be in an arbitrary direction. The existing solution by other authors to this prototype problem lost the term "$\boldsymbol{\omega} \times \boldsymbol{v}'_{P_t}$" when representing the acceleration of the lowest point of the ball. So, this solution still can deduce the conclusion "the projection of the acceleration at the lowest point of the billiard ball in the horizontal direction is opposite to its velocity direction all the time ". Certainly, this is an obvious mistake.

3.2.5 The Pure Rolling of a Homogeneous Ball on a Turntable (d, *C*)

Consider a disk located in a horizontal ground, it is rotating around its central axis of the vertical direction at a constant angular velocity of the magnitude Ω and the vertically downward direction (see Fig. 1). Place a homogeneous sphere gently on the disk, the friction coefficient between the sphere and the disk is large enough that there is no relative sliding between them. Prove that the center of mass of the sphere will perform a uniform circular motion and find its angular velocity.

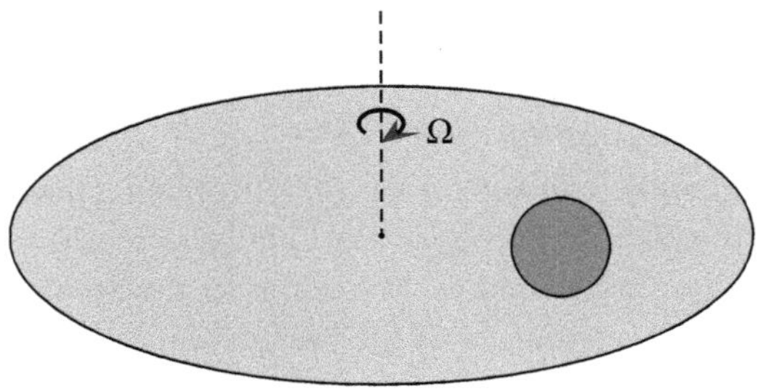

Figure 1

Solution

The time point when the sphere is placed on the disk is defined as Time zero.

As shown in Fig. 2, the centers of the disk and the sphere are denoted by O and C, respectively, the projection of Point C on the central axis of the disk is denoted by O', the point of the sphere in contact with the disk at arbitrary Time t is denoted by P_t, and the point of the disk in contact with the sphere at arbitrary Time t is denoted by S_t. The unit vector in the vertical downward direction is denoted by $\boldsymbol{e}_g$.

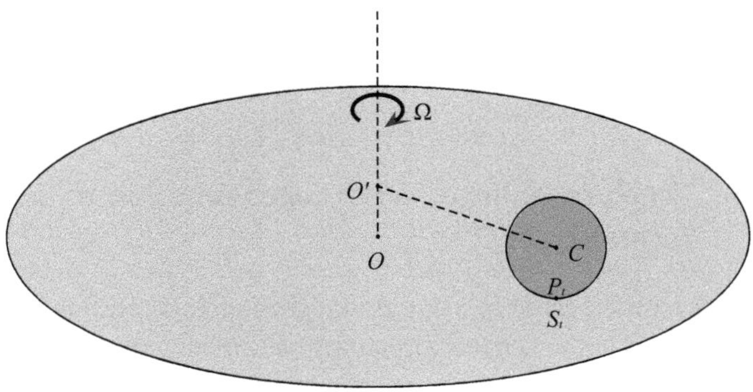

Figure 2

Taking *the sphere* as the object of study. Studying *the state at arbitrary Time t*. Considering (a) ***theorem of motion of center of mass***, (b) ***theorem of angular momentum*** (differential form) [in its center-of-mass reference frame, taking the center of mass C as the reference point], (c) ***in the center-of-mass reference frame, the***

resultant torque of the inertia force acting on system of particles is zero:

$$\begin{cases} mg\boldsymbol{e}_g - N\boldsymbol{e}_g + \boldsymbol{F} = m\ddot{\boldsymbol{r}} & \text{(1a)} \\ R\boldsymbol{e}_g \times \boldsymbol{F} = I\dot{\boldsymbol{\omega}} & \text{(1b)} \end{cases}$$

m—the mass of the sphere.

R—the radius of the sphere.

I—the moment of inertia of the sphere about one of its central axes.

$\boldsymbol{r}$—the position vector of the center of mass of the sphere relative to Point O' at Time t.

$\boldsymbol{\omega}$—the rotating angular velocity of the sphere in its center-of-mass reference frame at Time t.

N—the $-\boldsymbol{e}_g$-direction component of the normal contact force exerted on the sphere by the disk at Time t.

$\boldsymbol{F}$—the static friction force exerted on the sphere by the disk at Time t.

Considering the constraint conditions: the sphere is a rigid body and in contact with the disk all the time; therefore $\dot{\boldsymbol{r}}$ and $\ddot{\boldsymbol{r}}$, are perpendicular to $\boldsymbol{e}_g$ all the time:

$$\Rightarrow \quad \begin{cases} \dot{\boldsymbol{r}} \cdot \boldsymbol{e}_g = 0 & \text{(2a)} \\ \ddot{\boldsymbol{r}} \cdot \boldsymbol{e}_g = 0 & \text{(2b)} \end{cases}$$

$$\Rightarrow \quad \boldsymbol{F} = m\ddot{\boldsymbol{r}} \tag{3}$$

Combining (1b) and (3),

$$\Rightarrow \quad \dot{\boldsymbol{\omega}} = \frac{m}{I}(R\boldsymbol{e}_g \times \ddot{\boldsymbol{r}}) \tag{4}$$

Considering the constraint condition: there is no relative sliding between the sphere and the disk:

$$\boldsymbol{v}_{P_t} = \boldsymbol{v}_{S_t} \tag{5}$$

$\boldsymbol{v}_{P_t}$ and $\boldsymbol{v}_{S_t}$ —the velocity of Point P_t and Point S_t.

Considering the definition of ***velocity*** and Poisson formula:

$$\boldsymbol{v}_{S_t} = \Omega\boldsymbol{e}_g \times \boldsymbol{r} \tag{6}$$

Taking *Point P_t* as the object of study. Studying *the state at Time t*. Considering ***theorem for composition of velocities of a particle*** [taking the ground as the stationary reference frame, and the center-of-mass reference frame of the sphere as the moving reference frame]:

$$\boldsymbol{v}_{P_t} = \dot{\boldsymbol{r}} + \boldsymbol{\omega} \times (R\boldsymbol{e}_g) \tag{7}$$

$$\Rightarrow \quad \dot{\boldsymbol{r}} + \boldsymbol{\omega} \times (R\boldsymbol{e}_g) = \Omega \boldsymbol{e}_g \times \boldsymbol{r} \tag{8}$$

$$\Rightarrow \quad \ddot{\boldsymbol{r}} + \dot{\boldsymbol{\omega}} \times (R\boldsymbol{e}_g) = \Omega \boldsymbol{e}_g \times \dot{\boldsymbol{r}} \tag{9}$$

$$\Rightarrow \quad \ddot{\boldsymbol{r}} + \frac{mR^2}{I}(\boldsymbol{e}_g \times \ddot{\boldsymbol{r}}) \times \boldsymbol{e}_g = \Omega \boldsymbol{e}_g \times \dot{\boldsymbol{r}} \tag{10}$$

Considering the following vector operations:

$$(\boldsymbol{e}_g \times \ddot{\boldsymbol{r}}) \times \boldsymbol{e}_g = \ddot{\boldsymbol{r}}(\boldsymbol{e}_g \cdot \boldsymbol{e}_g) - \boldsymbol{e}_g(\boldsymbol{e}_g \cdot \ddot{\boldsymbol{r}}) = \ddot{\boldsymbol{r}} - \boldsymbol{e}_g(\boldsymbol{e}_g \cdot \ddot{\boldsymbol{r}}) \tag{11}$$

(2b) $\rightarrow$ (11),

$$\Rightarrow \quad (\boldsymbol{e}_g \times \ddot{\boldsymbol{r}}) \times \boldsymbol{e}_g = \ddot{\boldsymbol{r}} \tag{12}$$

(12) $\rightarrow$ (10),

$$\Rightarrow \quad \left(1 + \frac{mR^2}{I}\right)\ddot{\boldsymbol{r}} = \Omega \boldsymbol{e}_g \times \dot{\boldsymbol{r}} \tag{13}$$

$$\Rightarrow \quad \left(1 + \frac{mR^2}{I}\right)\dot{\boldsymbol{r}} \cdot \ddot{\boldsymbol{r}} = \dot{\boldsymbol{r}} \cdot (\Omega \boldsymbol{e}_g \times \dot{\boldsymbol{r}}) = 0 \tag{14}$$

$$\Rightarrow \quad \dot{\boldsymbol{r}} \cdot \ddot{\boldsymbol{r}} = 0 \tag{15}$$

$$\Rightarrow \quad \dot{\boldsymbol{r}} \cdot \frac{\mathrm{d}\dot{\boldsymbol{r}}}{\mathrm{d}t} = \frac{1}{2}\frac{\mathrm{d}(\dot{\boldsymbol{r}} \cdot \dot{\boldsymbol{r}})}{\mathrm{d}t} = \frac{\mathrm{d}|\dot{\boldsymbol{r}}|^2}{\mathrm{d}t} = 0 \tag{16}$$

$$\Rightarrow \quad |\dot{\boldsymbol{r}}| = \text{const} \tag{17}$$

That is, the center of mass of the sphere moves at a uniform speed.

From (13),

$$\Rightarrow \quad \left(1 + \frac{mR^2}{I}\right)|\ddot{\boldsymbol{r}}| = \Omega |\boldsymbol{e}_g \times \dot{\boldsymbol{r}}| = \Omega |\dot{\boldsymbol{r}}| \tag{18}$$

$$\Rightarrow \quad |\ddot{\boldsymbol{r}}| = \text{const} \tag{19}$$

It can be seen from (17) and (19) that the center of mass of the sphere performs uniform circular motion.

The center of the circular trajectory of the center of mass of the sphere is denoted by O''.

Taking *the center of mass C* as the object of study. Studying *the state at arbitrary Time t*. Considering ***theorem for composition of velocities of a particle*** [taking the ground as stationary reference frame, and the translational reference frame that is at rest with respect to the center O'' as the moving reference frame]:

$$\dot{\boldsymbol{r}} = \dot{\boldsymbol{r}}_{O''} + \dot{\boldsymbol{\rho}} \tag{20}$$

$\boldsymbol{r}_{O''}$—the position vector of the center O'' relative to the center O at Time t.

$\boldsymbol{\rho}$—the position vector of the center of mass of the sphere relative to the center O'' at Time t.

The center O'' is static:

$$\dot{\boldsymbol{r}}_{O''} = 0 \tag{21}$$

Considering the definition of ***velocity*** and Poisson formula:

$$\dot{\boldsymbol{\rho}} = \omega_C \boldsymbol{e}_g \times \boldsymbol{\rho} \tag{22}$$

ω_C—the $\boldsymbol{e}_g$-direction component of the angular velocity of the uniform circular motion of the center of mass of the sphere.

$$\Rightarrow \quad \dot{\boldsymbol{r}} = \omega_C \boldsymbol{e}_g \times \boldsymbol{\rho} \tag{23}$$

Considering Poisson formula:

$$\Rightarrow \quad \ddot{\boldsymbol{r}} = \frac{\mathrm{d}}{\mathrm{d}t}(\omega_C \boldsymbol{e}_g \times \boldsymbol{\rho}) = \omega_C \boldsymbol{e}_g \times (\omega_C \boldsymbol{e}_g \times \boldsymbol{\rho})$$
$$= \omega_C^2 \boldsymbol{e}_g \times (\boldsymbol{e}_g \times \boldsymbol{\rho}) = \omega_C^2 [\boldsymbol{e}_g(\boldsymbol{e}_g \cdot \boldsymbol{\rho}) - \boldsymbol{\rho}(\boldsymbol{e}_g \cdot \boldsymbol{e}_g)] \tag{24}$$

$\boldsymbol{\rho}$ and $\boldsymbol{e}_g$ are perpendicular to each other:

$$\boldsymbol{\rho} \cdot \boldsymbol{e}_g = 0 \tag{25}$$

Defining $\boldsymbol{e}_\rho$ and $\boldsymbol{e}_\theta$ (the angular unit vector and the radial unit vector of the center of mass of the sphere relative to the center O''):

$$\begin{cases} \boldsymbol{e}_\rho = \dfrac{\boldsymbol{\rho}}{\rho} & \text{(26a)} \\ \boldsymbol{e}_\theta = \boldsymbol{e}_g \times \boldsymbol{e}_\rho & \text{(26b)} \end{cases}$$

$$\Rightarrow \quad \begin{cases} \dot{\boldsymbol{r}} = \omega_C \rho \boldsymbol{e}_\theta & \text{(27a)} \\ \ddot{\boldsymbol{r}} = -\omega_C^2 \rho \boldsymbol{e}_\rho & \text{(27b)} \end{cases}$$

(27) → (18),

$$\Rightarrow \qquad \left(1+\frac{mR^2}{I}\right)\omega_C^2\,\rho=\Omega\omega_C\,\rho \tag{28}$$

$$\Rightarrow \qquad \omega_C=\frac{\Omega}{1+\dfrac{mR^2}{I}} \tag{29}$$

Considering ***the form of moment of inertia of a homogeneous sphere about one of its central axes***:

$$I=\frac{2}{5}mR^2 \tag{29}$$

$$\Rightarrow \qquad \omega_C=\frac{5}{7}\Omega \tag{30}$$

3.2.6 The Pure Rolling of a Homogeneous Sphere on the Rough Inner Wall of a Cylinder (d, *C*)

Consider a homogeneous sphere of radius a and mass m moving on the inner wall of a vertically fixed hollow cylinder of sectional radius b. The inner wall of the cylinder is rough enough that the sphere can slide on it without friction. Prove that the motion of the center of mass of the sphere in the vertical direction is a simple harmonic vibration.

Solution

Time zero is arbitrarily set.

As shown in Fig. 1, the center of the bottom of cylinder is denoted by O, the center of mass of the sphere is denoted by C, the projection of the center of mass C on the central axis of the cylinder at arbitrary Time t is denoted by C'. the unit vector in the vertical upward direction is denoted by $\boldsymbol{k}$. Taking Point O as the pole, an arbitrary direction in the horizontal plane as the positive direction of the polar axis, the right-hand helix direction with regard to the $\boldsymbol{k}$-direction as the positive direction of polar angle, the $\boldsymbol{k}$-direction as the positive direction of the longitudinal axis, establishing the cylindrical coordinate system, let r, θ, Z be the cylindrical coordinates.

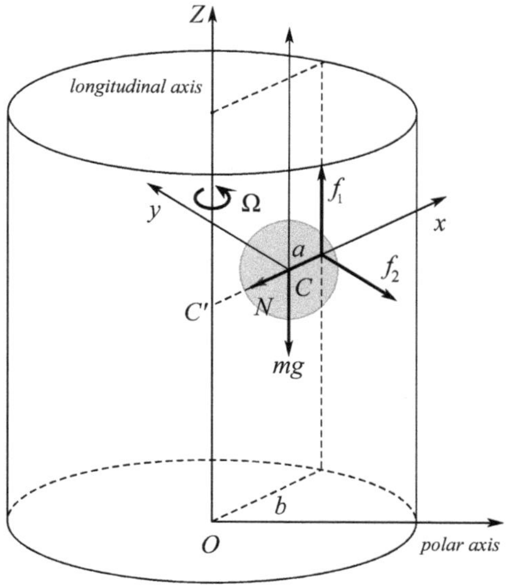

Figure 1

Taking *the sphere* as the object of study. Studying *the state at arbitrary Time t*. Considering ***theorem of motion of center of mass*** (the component equations in the ($\boldsymbol{r}^0$, $\boldsymbol{\theta}^0$, $\boldsymbol{k}$)-directions):

$$\begin{cases} -N = ma_r & \text{(1a)} \\ -f_2 = ma_\theta & \text{(1b)} \\ f_1 - mg = ma_Z & \text{(1c)} \end{cases}$$

N—the $-\boldsymbol{r}^0$-direction component of the normal contact force exerted on the sphere by the inner wall of the cylinder at Time t.

f_1 and f_2—the $\boldsymbol{k}$-direction and $-\boldsymbol{\theta}^0$-direction component of the friction exerted on the sphere by the inner wall of the cylinder at Time t.

(a_r, a_θ, a_Z)—the components of the acceleration of the center of mass C in the ($\boldsymbol{r}^0$, $\boldsymbol{\theta}^0$, $\boldsymbol{k}$)-directions at Time t.

Considering ***the components form of the acceleration of a particle in cylindrical coordinate system*** (the constraint condition: both the sphere and the cylinder are rigid bodies, and they are in contact with each other all the time; therefore, the distance between the center of mass C and the z-axis remains constant):

$$\begin{cases} a_r = -\Omega^2(b-a) & \text{(2a)} \\ a_\theta = \dot{\Omega}(b-a) & \text{(2b)} \\ a_Z = \ddot{Z} & \text{(2c)} \end{cases}$$

Ω—the $\boldsymbol{k}$-direction component of the rotating angular velocity of CC' at arbitrary Time t.

$$\Rightarrow \quad \begin{cases} N = m\Omega^2(b-a) & \text{(3a)} \\ -f_2 = m\dot{\Omega}(b-a) & \text{(3b)} \\ f_1 - mg = m\ddot{Z} & \text{(3c)} \end{cases}$$

Establishing the Cartesian coordinate system with the frame (C, $\boldsymbol{i}$, $\boldsymbol{j}$, $\boldsymbol{k}$), where $\boldsymbol{i} \equiv \boldsymbol{r}^0$, $\boldsymbol{j} \equiv \boldsymbol{\theta}^0$. let x, y, z be the Cartesian coordinates.

Taking *the sphere* as the object of study. Studying *the state at arbitrary Time t*. Considering (a) ***theorem of angular momentum*** [in its center-of-mass reference frame, taking its center of mass C as the reference point], (b) ***in the center-of-mass reference frame, the resultant torque of the inertia force acting on system of particles is zero***:

$$\boldsymbol{M} = \frac{\mathrm{d}\boldsymbol{L}}{\mathrm{d}t} = \left(\frac{\mathrm{d}}{\mathrm{d}t}\right)' \boldsymbol{L} + \Omega \boldsymbol{k} \times \boldsymbol{L} \tag{4}$$

$\boldsymbol{M}$—the resultant external torque acting on the sphere at Time t [in its center-of-mass reference frame, taking the center of mass C as the reference point].

$\boldsymbol{L}$—the angular momentum of the sphere at Time t [taking the center of mass C as the reference point].

Considering the definition of ***torque***:

$$\boldsymbol{M} = a\boldsymbol{i} \times (f_1\boldsymbol{k} - f_2\boldsymbol{j}) = -af_1\boldsymbol{j} - af_2\boldsymbol{k} \tag{5}$$

Considering ***the form of the angular momentum of rigid body performing fixed-point rotation based on the principal moments of inertia*** (Using the coordinate system of principal axis of the sphere which coincides with the Cartesian coordinate system with the frame ($\boldsymbol{C}$, $\boldsymbol{i}$, $\boldsymbol{j}$, $\boldsymbol{k}$) instantaneously):

$$\boldsymbol{L} = I\omega_x\boldsymbol{i} + I\omega_y\boldsymbol{j} + I\omega_z\boldsymbol{k} = I(\omega_x\boldsymbol{i} + \omega_y\boldsymbol{j} + \omega_z\boldsymbol{k}) \tag{6}$$

I—the principal moment of inertia of the sphere.

$(\omega_x, \omega_y, \omega_z)$—the components of the angular velocity of the sphere in the $(\boldsymbol{i}, \boldsymbol{j}, \boldsymbol{k})$-directions at *arbitrary Time t.*

$$\Rightarrow -af_1\boldsymbol{j} - af_2\boldsymbol{k} = I(\dot{\omega}_x\boldsymbol{i} + \dot{\omega}_y\boldsymbol{j} + \dot{\omega}_z\boldsymbol{k}) + \Omega\boldsymbol{k} \times I(\omega_x\boldsymbol{i} + \omega_y\boldsymbol{j} + \omega_z\boldsymbol{k})$$

$$= (I\dot{\omega}_x - I\Omega\omega_y)\boldsymbol{i} + (I\dot{\omega}_y + I\Omega\omega_x)\boldsymbol{j} + I\dot{\omega}_z\boldsymbol{k} \tag{7}$$

$$\Rightarrow \begin{cases} I\dot{\omega}_x - I\Omega\omega_y = 0 & (8a) \\ -af_1 = I\dot{\omega}_y + I\Omega\omega_x & (8b) \\ -af_2 = I\dot{\omega}_z & (8c) \end{cases}$$

The condition for the sphere to perform pure rolling relative to the inner wall of the cylinder is as follows:

$$\boldsymbol{v}_C + (\omega_x\boldsymbol{i} + \omega_y\boldsymbol{j} + \omega_z\boldsymbol{k}) \times a\boldsymbol{i} = 0 \tag{9}$$

Considering ***the components form of the velocity of a particle in cylindrical coordinate system*** (the constraint condition: both of the sphere and the cylinder are rigid bodies, and they are in contact with each other all the time; therefore, the distance between the center of mass C and the central axis of the hollow cylinder remains constant):

$$\Rightarrow \quad \boldsymbol{v}_C = \Omega\boldsymbol{k} \times (b-a)\boldsymbol{r}^0 + \dot{Z}\boldsymbol{k} = \Omega\boldsymbol{k} \times (b-a)\boldsymbol{i} + \dot{Z}\boldsymbol{k} \tag{10}$$

Combining (9) and (10),

$$\Rightarrow \quad [\Omega(b-a) + \omega_z a]\boldsymbol{j} + (\dot{Z} - \omega_y a)\boldsymbol{k} = 0 \tag{11}$$

$$\Rightarrow \begin{cases} \Omega(b-a) + \omega_z a = 0 & (12a) \\ \dot{Z} - \omega_y a = 0 & (12b) \end{cases}$$

Combining (3b) and (8c),

$$\Rightarrow \quad m\dot{\Omega}a(b-a) - I\dot{\omega}_z = 0 \tag{13}$$

From (12a),

$$\Rightarrow \quad \dot{\Omega}(b-a) + \dot{\omega}_z a = 0 \tag{14}$$

$$\Rightarrow \quad -(ma^2 + I)\dot{\omega}_z = 0 \tag{15}$$

$$\Rightarrow \quad \dot{\omega}_z = 0 \tag{16}$$

$$\Rightarrow \quad \dot{\Omega} = 0 \tag{17}$$

It can be seen that CC' performs the uniform rotation.

Combining (3c) and (8b),

$$\Rightarrow \qquad I\dot{\omega}_y + I\Omega\omega_x + ma(g + \ddot{Z}) = 0 \tag{18}$$

Combining (8a) and (12b),

$$\Rightarrow \qquad \dot{\omega}_x = \frac{\Omega}{a}\dot{Z} \tag{19}$$

$$\Rightarrow \qquad \omega_x = \frac{\Omega}{a}Z + \text{const} \tag{20}$$

(12b) and (20) $\rightarrow$ (18):

$$\Rightarrow \qquad (I + ma^2)\ddot{Z} + I\Omega^2 Z = \text{const}' \tag{21}$$

Considering ***the form of moment of inertia of a homogeneous sphere about one of its central axes***:

$$I = \frac{2}{5}ma^2 \tag{22}$$

$$\Rightarrow \qquad \ddot{Z} + \frac{2}{7}\Omega^2 Z = \text{const}'' \tag{23}$$

Equation (23) is the differential equation of the center-of-mass of the sphere performing simple harmonic vibration in the vertical direction, and its angular frequency ω_0 is:

$$\Rightarrow \qquad \omega_0 = \sqrt{\frac{2}{7}}\Omega \tag{24}$$

Chapter 4

Vibration

4.1 Single-Degree of Freedom Vibration

As a kind of special physical model in which the kinematics quantities changes repeatedly, mechanical vibration models can broadly describe many practical phenomena from macro to micro, such as the ups and downs of the waves, earthquakes, the swing of the clock, the beating of the heart, thermal motion of microparticles, etc. The response of any vibration system under specific excitation can be described by the dynamic equation (differential form or integral form) which contains the characteristics of the system itself and the excitation conditions. According to different excitation conditions and response characteristics, the mechanical vibration model can be divided into many subcategories (for example, according to different excitation methods, it can be divided into free vibration, forced vibration and self-excited vibration; according to different response characteristics, it can be divided into periodic vibration and random vibration; according to the mathematical characteristics of the dynamic equation, it can be divided into linear vibration and nonlinear vibration, etc.). The natural frequency determined by the characteristics of the system itself is the research focus of all vibration problems.

Solving Physics Problems: Exploring New Thinking Paradigms
Yipeng Gu

ISBN 978-981-4877-41-1 (Hardcover), 978-1-003-16254-4 (eBook)
www.jennystanford.com

The division of vibration types according to the degrees of freedom of the vibration system is conducive to the expanding of the theory, and the study of single-degree of freedom vibration has a basic position. As the most typical subcategory of a single degree of freedom vibration, periodic vibration can be expanded into a series of simple harmonic vibrations whose periods are commensurable by the spectrum method, so simple harmonic vibration has a very important research value. Spring vibrator and small-angle pendulum are the most classical simple harmonic vibration models.

From the mathematical point of view, the difficulty of solving linear vibration is obviously lower than that of non-linear vibration. It is worth noting that the vibration of most vibration systems will degenerate into linear vibration under the tiny vibration approximation, so the tiny vibration approximation becomes an ideal method to deal with tiny vibration problems approximately.

In this section, some representative single-degree of freedom vibration problems will be studied.

4.1.1 Simple Pendulum (d, *C*)

Find the motion period of the simple pendulum with pendulum length l and maximum pendulum angle θ_0, and find its approximation under the condition of small angle amplitude (θ_0 is a small quantity).

Solution [1]

As shown in Fig. 1, the hanging point of the simple pendulum is denoted by O. Taking Point O as the pole, the vertical downward direction as the positive direction of the polar axis, the anticlockwise direction as the positive direction of polar angle, establishing the polar coordinate system, let r, θ be the polar coordinates.

The time point when the pendulum ball is at the polar angle θ_0 is defined as Time zero.

Taking *the system composed of the pendulum ball and the earth* as the object of study. Studying *the process from Time zero to arbitrary Time t*. Considering (a) ***law of conservation of mechanical energy*** [the work done by the external force and non-conservative internal force acting on the system is zero (the tension in the string does none work); therefore, the mechanical energy of the system is conserved] (taking Point (l, 0) as the zero point of gravitational potential energy), (b) ***the components form of the velocity of a particle in polar coordinate system*** [the constraint condition: the length of the string remains constant] [integrating into (a)]:

$$\frac{1}{2}m(l\dot{\theta})^2 + mgl(1-\cos\theta) = mgl(1-\cos\theta_0) \tag{1}$$

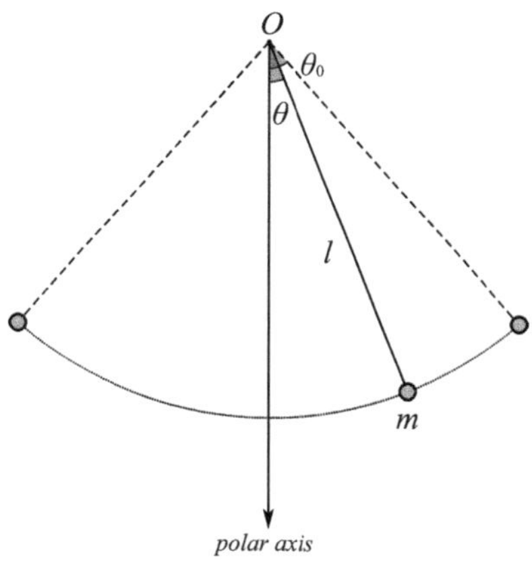

Figure 1

m—the mass of the pendulum ball.

$$\Rightarrow \qquad \mathrm{d}t = \sqrt{\frac{l}{g}}\frac{\mathrm{d}\theta}{\sqrt{2(\cos\theta-\cos\theta_0)}} \tag{2}$$

$$\Rightarrow \qquad \int_0^{T/4}\mathrm{d}t = \sqrt{\frac{l}{g}}\int_0^{\theta_0}\frac{\mathrm{d}\theta}{\sqrt{2(\cos\theta-\cos\theta_0)}} \tag{3}$$

T—the motion period of the simple pendulum.

$$\Rightarrow \qquad T = 2\sqrt{\frac{l}{g}}\int_0^{\theta_0}\frac{\mathrm{d}\theta}{\sqrt{\sin^2\frac{\theta_0}{2}-\sin^2\frac{\theta}{2}}} \tag{4}$$

Defining the definite integral I:

$$I = \int_0^{\theta_0}\frac{\mathrm{d}\theta}{\sqrt{\sin^2\frac{\theta_0}{2}-\sin^2\frac{\theta}{2}}} \tag{5}$$

Defining the variable φ:

$$\sin\frac{\theta}{2}=\sin\frac{\theta_0}{2}\sin\varphi \tag{6}$$

$$\Rightarrow \qquad \mathrm{d}\theta=2\cos\varphi\mathrm{d}\varphi\frac{\sin\frac{\theta_0}{2}}{\sqrt{1-\sin^2\frac{\theta_0}{2}\sin^2\varphi}} \tag{7}$$

$$\Rightarrow \qquad I=2\int_0^{\frac{\pi}{2}}\frac{\mathrm{d}\varphi}{\sqrt{1-\sin^2\frac{\theta_0}{2}\sin^2\varphi}} \tag{8}$$

This is a complete elliptic integral of the first kind.

Considering the result of incomplete elliptic integral of the first kind:

$$\int_0^{\varphi}\frac{\mathrm{d}\varphi}{\sqrt{1-k^2\sin^2\varphi}}=\sum_{n=0}^{\infty}\frac{(2n)!}{(2^n\cdot n!)^2}k^{2n}\int_0^{\varphi}\sin^{2n}\varphi\mathrm{d}\varphi \tag{9}$$

$$\Rightarrow \qquad I=2\sum_{n=0}^{\infty}\frac{(2n)!}{(2^n\cdot n!)^2}\sin^{2n}\frac{\theta_0}{2}\int_0^{\frac{\pi}{2}}\sin^{2n}\varphi\mathrm{d}\varphi \tag{10}$$

$$\Rightarrow \qquad T=2\pi\sqrt{\frac{l}{g}}\cdot\frac{2}{\pi}\sum_{n=0}^{\infty}\frac{(2n)!}{(2^n\cdot n!)^2}\sin^{2n}\frac{\theta_0}{2}\int_0^{\frac{\pi}{2}}\sin^{2n}\varphi\mathrm{d}\varphi \tag{11}$$

This is the general form of T.

We can further obtain the second-order approximation and zero-order approximation of T when θ_0 is a small quantity:

$$\Rightarrow \qquad T\Big|_{\theta_0\ll 1}\approx 2\pi\sqrt{\frac{l}{g}}\left(1+\frac{\theta_0^2}{16}\right)\approx 2\pi\sqrt{\frac{l}{g}} \tag{12}$$

Solution [2] (For small angle swing)

Defining E—the mechanical energy of the system composed of the pendulum ball and the earth (taking Point $(l, 0)$ as the zero point of gravitational potential energy) at arbitrary Time t:

$$\Rightarrow \qquad E=\frac{1}{2}m(l\dot{\theta})^2+mgl(1-\cos\theta)$$

$$\approx \frac{1}{2}m(l\dot{\theta})^2 + mgl\left[1-\left(1-\frac{1}{2}\theta^2\right)\right]$$

$$= \frac{1}{2}m(l\dot{\theta})^2 + \frac{1}{2}mgl\theta^2 \tag{13}$$

This operation has taken the second-order approximation of θ.

Taking *the system composed of the pendulum ball and the earth* as the object of study. Studying *the instantaneous process (the small time interval dt) at arbitrary Time t.* Considering ***law of conservation of mechanical energy***:

$$\frac{dE}{dt} = 0 \tag{14}$$

$$\Rightarrow \qquad ml\dot{\theta}(l\ddot{\theta} + g\theta) = 0$$

$$\Rightarrow \qquad \ddot{\theta} + \frac{g}{l}\theta = 0 \tag{15}$$

Equation (15) is the differential equation of the string performing the tiny rotating simple harmonic vibration in the vertical plane near the polar axis, and its angular frequency ω is:

$$\Rightarrow \qquad \omega = \sqrt{\frac{g}{l}} \tag{16}$$

$$\Rightarrow \qquad T = \frac{2\pi}{\omega} = 2\pi\sqrt{\frac{l}{g}} \tag{17}$$

Solution [3] (For small angle swing)

As shown in Fig. 3, Point $(l, 0)$ is denoted by O', the unit vector in the horizontal right direction is denoted by $\boldsymbol{i}$. Establishing the linear coordinate system with Point O' as the origin and the $\boldsymbol{i}$-direction as the positive direction of coordinate axis, let x be the coordinate.

The time point when the pendulum ball is at the origin O' is defined as Time zero.

Taking *the pendulum ball* as the object of study. Studying *the state at arbitrary Time t.* Considering (a) ***Newton's second law*** (the $\boldsymbol{i}$-direction and $-\boldsymbol{r}^0$-direction component equations), (b) ***the radial component form of acceleration*** [the constraint condition: the string length remains constant] [integrating into (a)]:

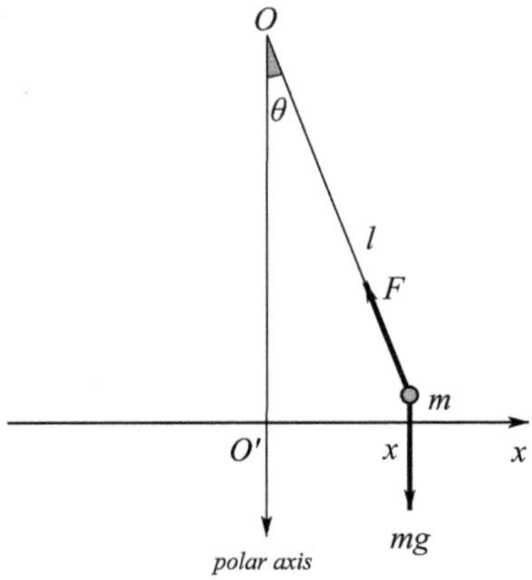

Figure 2

$$\begin{cases} -F\sin\theta = m\ddot{x} & \text{(18a)} \\ F - mg\cos\theta = ml\dot{\theta}^2 & \text{(18b)} \end{cases}$$

F—the magnitude of the tension in the string at Time t.

From (18b),

$\Rightarrow$
$$F \approx mg \tag{19}$$

(19) $\rightarrow$ (18a),

$\Rightarrow$
$$\ddot{x} \approx -g\theta \tag{20}$$

The two operations above have taken the first-order approximations of θ.

Considering the following geometric relation in Fig. 2:

$$\theta \approx \frac{x}{l} \tag{21}$$

$\Rightarrow$
$$\ddot{x} + \frac{g}{l}x = 0 \tag{22}$$

Equation (22) is the differential equation of the pendulum ball performing the simple harmonic vibration in the horizontal direction near the origin O', and its angular frequency ω is:

$\Rightarrow$
$$\omega = \sqrt{\frac{g}{l}} \tag{23}$$

4.1.2 The Motion of a Simple Pendulum Suspended on a Free Sliding Block (d, *C*)

A sliding block of mass M can slide freely on a smooth and straight guide rail in the horizontal direction. There is a light rigid string of length l, one end of which is tied with the sliding block and the other end is tied with a small ball of mass m. The system can be regarded as a simple pendulum with the sliding block as its suspension point. Initially, the simple pendulum is at rest in the vertical direction, give it a small disturbance, find the angular frequency of its tiny simple harmonic vibration.

Solution [1]

The time point when the simple pendulum is disturbed is defined as Time zero.

As shown in Fig. 1, the mechanical equilibrium position of the sliding block is denoted by O, the position of the sliding block at arbitrary Time t is denoted by O', the unit vector in the horizontal right direction is denoted by $\boldsymbol{i}$. Establishing three coordinate systems: (i) the linear coordinate system with Point O as the origin and the $\boldsymbol{i}$-direction as the positive direction of coordinate axis (let x be the coordinate) in the ground reference frame, (ii) the linear coordinate system (fixed to the sliding block) with Point O' as the origin and the $\boldsymbol{i}$-direction as the positive direction of coordinate axis (let x' be the coordinate), (iii) the polar coordinate system (fixed to the sliding block) with Point O' as the pole and vertical downward as the positive direction of polar axis (let r, θ be the polar coordinates).

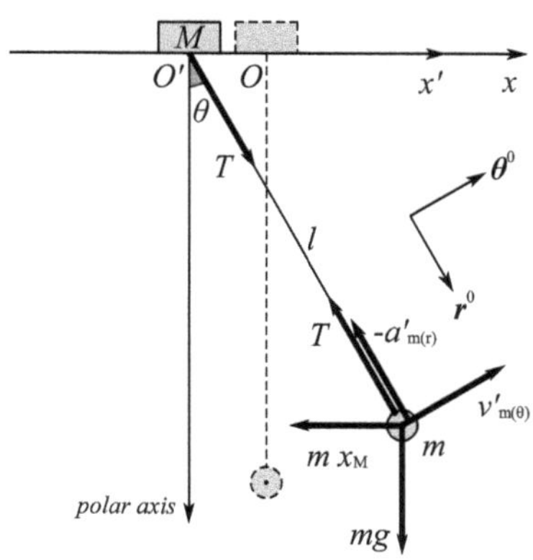

Figure 1

Taking *the sliding block and the ball* as the object of study respectively. Studying *the state at arbitrary Time t*. Considering ***Newton's second law*** [for the sliding block: in the ground reference frame, for the ball: in the sliding block reference frame] (for the sliding block: the ***i***-direction component equation, for the ball: the ***i***-direction and $-\boldsymbol{r}^0$-direction component equations):

$$T\sin\theta = M\ddot{x}_M \tag{1a}$$

$$\begin{cases} -T\sin\theta - m\ddot{x}_M = m\ddot{x}'_m & \text{(1b)} \\ T + m\ddot{x}_M\sin\theta - mg\cos\theta = m(-a'_{m(r)}) & \text{(1c)} \end{cases}$$

T—the magnitude of the tension in the string at Time t.

$a'_{m(r)}$—the $\boldsymbol{r}^0$-direction component of the acceleration of the ball in the sliding block reference frame at Time t.

Considering ***the radial component form of acceleration*** [the constraint condition: the length of the string remains unchanged]:

$$a'_{m(r)} = -l\dot{\theta}^2 \tag{2}$$

θ is a small quantity, taking the following approximations:

$$\begin{cases} \sin\theta \approx \theta & \text{(3a)} \\ \cos\theta \approx 1 - \dfrac{1}{2}\theta^2 \approx 1 & \text{(3b)} \\ \dot{\theta}^2 \approx 0 & \text{(3c)} \end{cases}$$

Combining (1) to (3):

$$\Rightarrow \qquad \ddot{x}'_m \approx -\frac{M+m}{M}g\theta \tag{4}$$

Considering the following geometric relation in Fig. 2:

$$\theta \approx \frac{x'}{l} \tag{5}$$

$$\Rightarrow \qquad \ddot{x}'_m + \frac{M+m}{M}\frac{g}{l}x'_m = 0 \tag{6}$$

Equation (6) is the differential equation of the ball performing the tiny simple harmonic vibration in the horizontal direction near Point (l, 0) in the sliding block reference frame, and its angular frequency ω is:

$$\Rightarrow \qquad \omega = \sqrt{\frac{M+m}{M}\frac{g}{l}} \tag{7}$$

Solution [2]

For the system composed of the sliding block, the ball and the earth at arbitrary Time t, defining E_k—the kinetic energy of the system in the center-of-mass reference frame of the sliding block and the ball, E_p—the gravitational potential energy [taking Point $(l, 0)$ as the zero point of potential energy], and E—the mechanical energy of the system in the center-of-mass reference frame of the sliding block and the ball. Considering (a) ***the form of available energy in the center-of-mass reference frame***, (b) ***theorem of gravitational potential energy***:

$$\begin{cases} E_k = \dfrac{1}{2}\dfrac{Mm}{M+m}v_m'^2 & \text{(9a)} \\ E_p = Mgl + mgl(1-\cos\theta) \approx Mgl + \dfrac{1}{2}mgl\theta^2 & \text{(9b)} \end{cases}$$

v'_m—the relative speed of the ball and the sliding block (i.e., the speed of the ball in the sliding block reference frame) at Time t.

$$\Rightarrow \qquad E = E_k + E_p = \frac{1}{2}\frac{Mm}{M+m}v_m'^2 + Mgl + \frac{1}{2}mgl\theta^2 \tag{10}$$

Taking *the system composed of the sliding block, the ball and the earth* as the object of study. Studying *the instantaneous process (the small time interval dt) at arbitrary Time t*. Considering ***law of conservation of mechanical energy*** [in the center-of-mass reference frame, the work done by the external force and non-conservative internal force acting on the system is zero; therefore, the mechanical energy of the system is conserved]:

$$\frac{\mathrm{d}E}{\mathrm{d}t} = 0 \tag{11}$$

Considering ***the components form of velocity in polar coordinate system*** [the constraint condition: the string length remains unchanged; therefore, the radial component of the velocity of the ball relative the sliding block is zero]:

$$v'_m = \left|v'_{m(\theta)}\right| = \left|l\dot{\theta}\right| \tag{12}$$

$$\Rightarrow \qquad \frac{\mathrm{d}}{\mathrm{d}t}\left[\frac{1}{2}\frac{Mm}{M+m}(l\dot{\theta})^2 + Mgl + \frac{1}{2}mgl\theta^2\right] = 0 \qquad (13)$$

$$\Rightarrow \qquad \ddot{\theta} + \frac{M+m}{M}\frac{g}{l}\theta = 0 \qquad (14)$$

Equation (14) is the differential equation of the string performing the tiny rotational simple harmonic vibration in the vertical plane near the polar axis, and its angular frequency ω is:

$$\Rightarrow \qquad \omega = \sqrt{\frac{M+m}{M}\frac{g}{l}} \qquad (15)$$

4.1.3 Cycloid Pendulum (d, *AC*)

As shown in Fig. 1, a light rigid string of length $4a$, one end of which is fixed at the origin O of the Cartesian coordinate system provided in the figure, and the other end is tied with a pendulum ball, forming a single pendulum. Its motion range is limited by the two baffles whose shape are described by the cycloid equations $\begin{cases} x = a(\varphi - \sin\varphi) \\ y = a(1-\cos\varphi) \end{cases}$ and $\begin{matrix} x = -a(\varphi - \sin\varphi) \\ y = a(1-\cos\varphi) \end{matrix}$, respectively, where $0 \le \varphi \le 2\pi$. Given that the speed of the pendulum ball is just zero each time it reaches the right baffle or left baffle.

(1) Prove that the trajectory of the pendulum is a cycloid.

(2) Find the angular frequency of the swing of the simple pendulum.

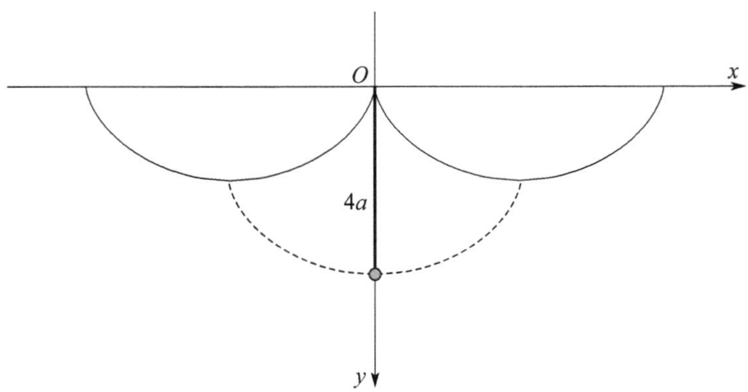

Figure 1

Solution to Problem (1) (type C)

Once the pendulum ball reaches the right baffle is defined as Time zero.

As shown in Fig. 2, at arbitrary Time t before the pendulum ball reaches its lowest point for the first time, the position of the pendulum ball is denoted by A, the tangent point of the string and the right baffle is denoted by P, drawing the circle that depicts the cycloid in the purely rolling manner, whose angle of rotation is denoted by φ.

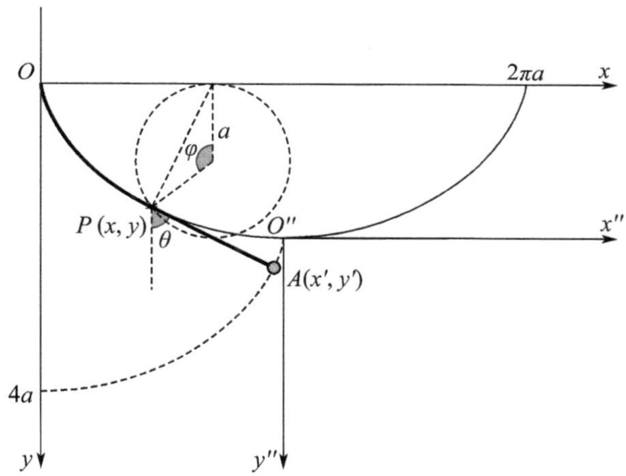

Figure 2

Considering ***the geometric meaning of derivative***:

$$\tan\theta = \frac{\mathrm{d}x}{\mathrm{d}y} = \frac{\mathrm{d}x/\mathrm{d}\varphi}{\mathrm{d}y/\mathrm{d}\varphi} \tag{1}$$

θ—the angle between $\overrightarrow{PA}$ and $\boldsymbol{j}$.

Known:

$$\begin{cases} x = a(\varphi - \sin\varphi) & (2a) \\ y = a(1 - \cos\varphi) & (2b) \end{cases}$$

$$\Rightarrow \qquad \tan\theta = \frac{a(1-\cos\varphi)}{a\sin\varphi} = \frac{2\sin^2\dfrac{\varphi}{2}}{2\sin\dfrac{\varphi}{2}\cos\dfrac{\varphi}{2}} = \tan\frac{\varphi}{2} \tag{3}$$

$$\Rightarrow \qquad \theta = \frac{\varphi}{2} \tag{4}$$

Considering the following geometric relations in Fig. 2:

$$\begin{cases} x' = x + \overline{PA}\sin\theta & (5a) \\ y' = y + \overline{PA}\cos\theta & (5b) \end{cases}$$

Considering the constraint condition: the string length remains unchanged:

$$\overline{PA} = 4a - \int_{OP} \mathrm{d}s \tag{6}$$

Considering ***the length formula of the curved element***, then place (2) in to obtain:

$$\int_{OP} \mathrm{d}s = \int_0^{\varphi} \sqrt{\left(\frac{\mathrm{d}x}{\mathrm{d}\varphi}\right)^2 + \left(\frac{\mathrm{d}y}{\mathrm{d}\varphi}\right)^2}\,\mathrm{d}\varphi = a\int_0^{\varphi} \sqrt{(1-\cos\varphi)^2 + \sin^2\varphi}\,\mathrm{d}\varphi$$

$$= a\int_0^{\varphi} \sqrt{2(1-\cos\varphi)}\,\mathrm{d}\varphi = 2a\int_0^{\varphi} \sin\frac{\varphi}{2}\,\mathrm{d}\varphi = 4a\left(1-\cos\frac{\varphi}{2}\right) \tag{7}$$

$$\Rightarrow \qquad \overline{PA} = 4a\cos\frac{\varphi}{2} \tag{8}$$

$$\Rightarrow \qquad \begin{cases} x' = a(\varphi + \sin\varphi) & (9a) \\ y' = a(3 + \cos\varphi) & (9b) \end{cases}$$

Defining the variables φ', x'', y'':

$$\begin{cases} \varphi' = \varphi - \pi & (10a) \\ x'' = x' - a\pi & (10b) \\ y'' = y' - 2a & (10c) \end{cases}$$

$$\Rightarrow \qquad \begin{cases} x'' = a(\varphi' - \sin\varphi') & (11a) \\ y'' = a(1 - \cos\varphi') & (11b) \end{cases}$$

It will be seen from Eqs. (11a) and (11b) that the trajectory of the pendulum ball (described by the new variables φ', x'', y'') is a cycloid.

Solution to Problem (2) (type C)

Defining E—the mechanical energy of the system composed of the pendulum ball and the earth at arbitrary Time t [taking the origin O

as the zero point of gravitational potential energy]:

$$E=\frac{1}{2}mv^2-mgy' \tag{12}$$

v—the speed of the pendulum ball at Time t.

Considering the definition of ***speed***:

$$v=\frac{|\mathrm{d}s|}{\mathrm{d}t} \tag{13}$$

Considering ***the length formula of the curved element***, then place (9) in to obtain:

$$\Rightarrow \quad |\mathrm{d}s|=\sqrt{\left(\frac{\mathrm{d}x'}{\mathrm{d}\varphi}\right)^2+\left(\frac{\mathrm{d}y'}{\mathrm{d}\varphi}\right)^2}\,|\mathrm{d}\varphi|=a\sqrt{(1+\cos\varphi)^2+\sin^2\varphi}\,|\mathrm{d}\varphi|$$

$$=a\sqrt{2(1+\cos\varphi)}\,|\mathrm{d}\varphi|=a\sqrt{4\cos^2\frac{\varphi}{2}}\,|\mathrm{d}\varphi|=2a\cos\frac{\varphi}{2}|\mathrm{d}\varphi|$$

$$=4a\cos\theta|\mathrm{d}\theta| \tag{14}$$

$$\Rightarrow \qquad v=4a\cos\theta\left|\frac{\mathrm{d}\theta}{\mathrm{d}t}\right| \tag{15}$$

$$\Rightarrow \qquad E=8ma^2\cos^2\theta\left(\frac{\mathrm{d}\theta}{\mathrm{d}t}\right)^2-mga(3+\cos2\theta)$$

$$=8ma^2\left[\frac{\mathrm{d}(\sin\theta)}{\mathrm{d}t}\right]^2-2mga(2-\sin^2\theta) \tag{16}$$

Taking *the system composed of the pendulum ball and the earth* as the object of study. Studying *the instantaneous process (the small time interval dt) at arbitrary Time t.* Considering ***law of conservation of mechanical energy*** [the work done by the external force and non-conservative internal force acting on the system is zero (the tension in the string does none work); therefore, the mechanical energy of the system is conserved]:

$$\frac{\mathrm{d}E}{\mathrm{d}t}=0 \tag{17}$$

$$\Rightarrow \qquad \frac{\mathrm{d}(\sin\theta)}{\mathrm{d}t}\left[\frac{\mathrm{d}^2(\sin\theta)}{\mathrm{d}t^2}+\frac{g}{4a}\sin\theta\right]=0 \tag{18}$$

$$\Rightarrow \qquad \frac{d^2(\sin\theta)}{dt^2} + \frac{g}{4a}\sin\theta = 0 \qquad (19)$$

It can be seen that the motion described by $\sin\theta$ is a simple harmonic vibration, and its angular frequency ω is:

$$\Rightarrow \qquad \omega = \sqrt{\frac{g}{4a}} \qquad (20)$$

Another method to get *v* (type A)

(4) → (8),

$$\Rightarrow \qquad \overline{PA} = 4a\cos\theta \qquad (21)$$

Considering ***the components form of velocity in polar coordinate system*** (the constraint condition: the string length remains unchanged; therefore, the radial component of the pendulum ball's velocity is zero):

$$v = |v_\theta| = \overline{PA}\left|\frac{d\theta}{dt}\right| \qquad (22)$$

$$\Rightarrow \qquad v = 4a\cos\theta\left|\frac{d\theta}{dt}\right| \qquad (23)$$

4.1.4 The Motion of a Spring Vibrator under Friction Resistance (*c*, *B*)[1]

As shown in Fig. 1, consider a light elastic rope of natural length approximately zero and stiffness coefficient k, one end of which is fixed at a point (denoted by O) on a horizontal ground, and the other end is tied with a small ball of mass m. The friction coefficient between the ball and the ground is μ. Make the ball rest at Point A on the right side of Point O with the distance l, $l > 2\ \mu mg/k$, then release the ball.

(1) Draw the phase trajectory of the ball's motion.
(2) Determine the total number of times of the ball's one-way motions (right and left).

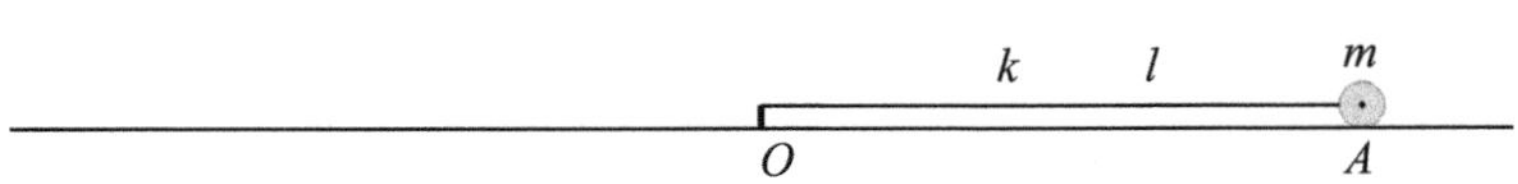

Figure 1

[1]This problem is a test question of the 7th AphO (Asian Physics Olympiads) in 2006.

Solution

The time point when the ball is released is defined as Time zero.

As shown in Fig. 2, the unit vector in the horizontal right direction is denoted by ***i***, the unit vector in the vertical upward direction is denoted by ***j***. Establishing the Cartesian coordinate system with the frame $(O, \boldsymbol{i}, \boldsymbol{j})$, let x, y be the Cartesian coordinates.

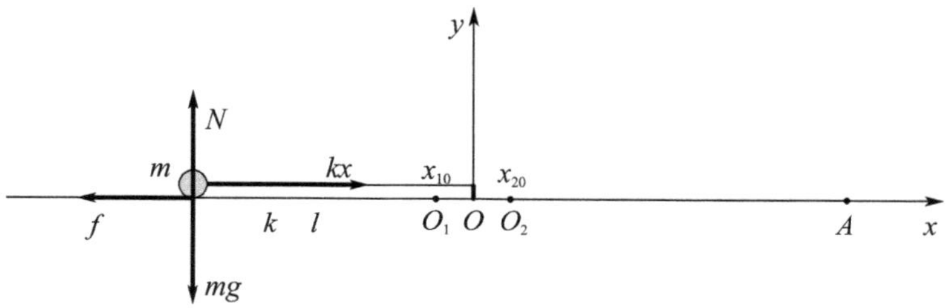

Figure 2

Taking *the ball* as the object of study. Studying *the state at arbitrary Time t during its moving*. Considering ***Newton's second law*** (the component equations in the $(\boldsymbol{i}, \boldsymbol{j})$-directions):

$$\begin{cases} -kx - |\boldsymbol{f}| = m\ddot{x} & \text{when } \dot{x} > 0 \quad (1a) \\ -kx + |\boldsymbol{f}| = m\ddot{x} & \text{when } \dot{x} < 0 \quad (1b) \end{cases}$$

$$N - mg = 0 \quad (2)$$

N—the ***j***-direction component of the normal contact force exerted on the ball by the ground at Time t.

f—the sliding friction force exerted on the ball by the ground at Time t.

Considering ***friction law***:

$$\boldsymbol{f} = -\mu N \boldsymbol{e}_v \quad (3)$$

$\boldsymbol{e}_v$—the unit vector in the velocity direction of the ball at arbitrary Time t.

Combining (1), (2) and (3):

$$\Rightarrow \quad \begin{cases} \ddot{x} + \dfrac{k}{m}x = -\mu g & \text{when } \dot{x} > 0 \quad (4a) \\ \ddot{x} + \dfrac{k}{m}x = \mu g & \text{when } \dot{x} < 0 \quad (4b) \end{cases}$$

Defining the constants x_{10} and x_{20}:

$$\begin{cases} x_{10} = -\dfrac{\mu mg}{k} & \text{(5a)} \\ x_{20} = \dfrac{\mu mg}{k} & \text{(5b)} \end{cases}$$

Point $(x_{10}, 0)$ and Point $(x_{20}, 0)$ are denoted by O_1 and O_2, respectively. Defining the variables x_1 and x_2 (the x-coordinate of the ball in the Cartesian coordinate systems with the frame $(O_1, \boldsymbol{i}, \boldsymbol{j})$ and $(O_2, \boldsymbol{i}, \boldsymbol{j})$):

$$\begin{cases} x_1 = x - x_{10} & \text{(6a)} \\ x_2 = x - x_{20} & \text{(6b)} \end{cases}$$

$$\Rightarrow \quad \begin{cases} \ddot{x}_1 + \dfrac{k}{m} x_1 = 0 \quad \text{when } \dot{x} > 0 & \text{(7a)} \\ \ddot{x}_2 + \dfrac{k}{m} x_2 = 0 \quad \text{when } \dot{x} < 0 & \text{(7b)} \end{cases}$$

Equations (7a) and (7b) are the differential equations of the ball performing simple harmonic vibration along the x-axis with Point O_1 and Point O_2 as the equilibrium position, respectively, and the equal angular frequency ω_0 is:

$$\omega_0 = \sqrt{\frac{k}{m}} \tag{8}$$

Studying the simple harmonic vibration model of a particle of mass m which described by the following differential equation:

$$\ddot{x} + \frac{k}{m} x = 0 \tag{9}$$

where k is a positive real number.

By integrating the equation above, we can obtain:

$$\frac{1}{2} m (\dot{x})^2 + \frac{1}{2} k x^2 = C \tag{10}$$

C—the integral constant.

where the term $\frac{1}{2} kx^2$ does not necessarily mean the potential energy of the physical entity.

Defining p—the $\boldsymbol{i}$-direction component of momentum of the particle at arbitrary Time t. Considering the definition of ***momentum***:

$$p = m\dot{x} \tag{11}$$

$$\Rightarrow \quad \frac{x^2}{2C/k} + \frac{p^2}{2mC} = 1 \tag{12}$$

That is, the phase trajectory of the particle is an ellipse, as shown in Fig. 3.

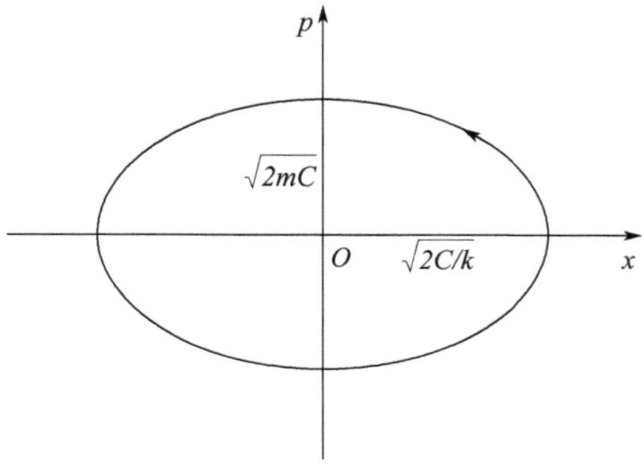

Figure 3

It can be seen that the phase trajectory of the ball in this problem is composed of a series of semi-ellipse phase trajectories corresponding to each one-way motion, as shown in Fig. 4.

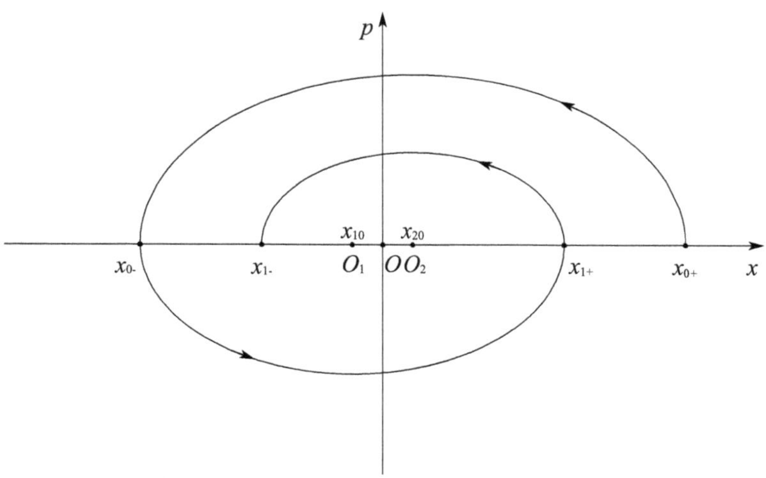

Figure 4

Defining x_{i+}—the x-coordinate of the ball when it stops at the positive x-axis for the i_+-th time. Considering the following geometric relation in Fig. 4:

$$x_{(i-1)+} - x_{i+} = 2(x_{(i-1)+} - x_{20}) - 2(x_{i+} - x_{10}) \tag{13}$$

(5) → (13):

$$\Rightarrow \quad x_{(i-1)+} - x_{i+} = 2(x_{20} - x_{10}) = 4\frac{\mu mg}{k} \tag{14}$$

$$\Rightarrow \quad x_{0+} - x_{n+} = \sum_{i=1}^{n_+}(x_{(i-1)+} - x_{i+}) = 4n_+\frac{\mu mg}{k} \tag{15}$$

$$\Rightarrow \quad x_{n+} = x_{0+} - 4n_+\frac{\mu mg}{k} \tag{16}$$

This is the general form of x_{n+}.

Defining x_{0-}—the x-coordinate of the ball when it stops at the negative x-axis for the first time. Considering the following geometric relation in Fig. 4, then place (5) in to obtain:

$$\Rightarrow \quad x_{0-} = x_{0+} - 2x_{0+} - x_{20}(x_{0+} - x_{20}) = 2\frac{\mu mg}{k} - x_{0+} \tag{17}$$

Defining x_{i-}—the x-coordinate of the ball when it stops at the negative x-axis for the i_--th time (taking Point (x_{0-}, 0) as the initial position). Considering the following geometric relation in Fig. 4:

$$x_{i-} - x_{(i-1)-} = 2(x_{10} - x_{(i-1)-}) - 2(x_{20} - x_{i-}) \tag{18}$$

(5) → (18):

$$\Rightarrow \quad x_{i-} - x_{(i-1)-} = 2(x_{20} - x_{10}) = 4\frac{\mu mg}{k} \tag{19}$$

$$\Rightarrow \quad x_{n-} - x_{0-} = \sum_{i=1}^{n-}(x_{i-} - x_{(i-1)-}) = 4n_-\frac{\mu mg}{k} \tag{20}$$

$$\Rightarrow \quad x_{n-} = -x_{0+} + (4n_- + 2)\frac{\mu mg}{k} \tag{21}$$

This is the general form of x_{n-}.

The condition for the ball to stop reciprocating motion is that the ball once enters the closed interval $[x_{10}, x_{20}]$ and stops there. That is, the following relations need to be satisfied before the reciprocating motion stops:

$$\begin{cases} x_{n+} > x_{20} & (22a) \\ x_{n-} < x_{10} & (22b) \end{cases}$$

(16) → (22a), (21) → (22b):

$$\Rightarrow \begin{cases} n_+ < \dfrac{x_{0+}}{4\dfrac{\mu mg}{k}} - \dfrac{1}{4} & (23a) \\ n_- < \dfrac{x_{0+}}{4\dfrac{\mu mg}{k}} - \dfrac{3}{4} & (23b) \end{cases}$$

Defining the constant X:

$$X = -x_{10} = x_{20} = \frac{\mu mg}{k} \tag{24}$$

Defining N_+—the number of times of the ball's one-way motions till it stops at the positive x-axis for the n_+-th time, and N_-—the number of times of the ball's one-way motions till it stops at the negative x-axis for the n_--th time (number n_- does not include the first time).

$$\Rightarrow \begin{cases} N_+ = 2n_+ < \dfrac{1}{2}\left(\dfrac{x_{0+}}{X} - 1\right) & (25a) \\ N_- = 2n_- + 1 < \dfrac{1}{2}\left(\dfrac{x_{0+}}{X} - 1\right) & (25b) \end{cases}$$

Defining N—the total number of times of the ball's one-way motions, from the above equations, we can obtain:

$$\Rightarrow \qquad N = \text{Ceiling}\left[\frac{1}{2}\left(\frac{x_{0+}}{X} - 1\right)\right] = \text{Ceiling}\left[\frac{1}{2}\left(\frac{l}{X} - 1\right)\right] \tag{26}$$

where function Ceiling (x) represents the minimum integer not smaller than the variable x.

4.1.5 The Tiny Expansion-Contraction Vibration of a Soap Bubble (*a*, *A*)

Consider a spherical soap bubble composed of the soapy water of mass m and surface tension coefficient σ, which is filled with air of negligible mass, and the outside is vacuum. It is approximately assumed that the temperature of the whole system remains constant. Find the angular frequency of tiny vibration (expanding and contracting repeatedly) of the soap bubble.

Solution

As shown in Fig. 1, the center of the soap bubble is denoted by O, taking an arbitrary point on the soap bubble and denoting it by A. The radial outward direction at Point A is denoted by $\boldsymbol{k}$, establishing the Cartesian coordinate system with the frame $(O, \boldsymbol{i}, \boldsymbol{j}, \boldsymbol{k})$, let x, y, z be the Cartesian coordinates. Taking the origin O as the pole, the $\boldsymbol{k}$-direction as the zenith direction, the $\boldsymbol{i}$-direction as the reference direction on the reference plane, establishing the spherical coordinate system, let r, φ, θ be the spherical coordinates.

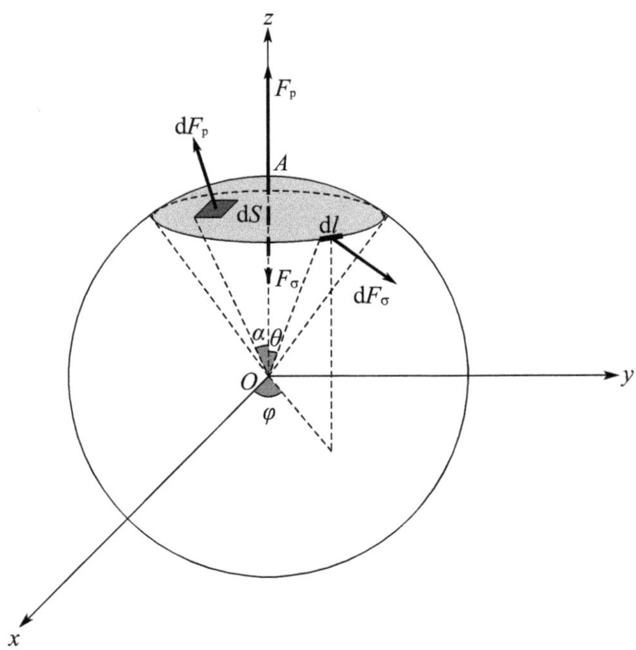

Figure 1

Taking *the soap bubble's spherical crown infinitesimal element with Point A as the center and with a small half top angle θ (hereinafter referred to as the spherical crown element)* as the object of study.

Studying *the state once it is in mechanical equilibrium*. Considering ***Newton's second law*** (the ***k***-direction component equation):

$$F_p\Big|_{r=r_0} - F_\sigma\Big|_{r=r_0} = 0 \quad (1)$$

r—the radius of the soap bubble at arbitrary Time t.

r_0—the radius of the soap bubble when the spherical crown element is in mechanical equilibrium.

F_p—the ***k***-direction component of the pressure exerted on the spherical crown element by the air inside of the soap bubble at arbitrary Time t.

F_σ—the −***k***-direction component of the surface tension exerted on the spherical crown element by the soap water around it at arbitrary Time t.

As shown in Fig. 1, taking the surface element of area dS at the polar angle α (hereinafter referred to as element dS), considering the definition of ***pressure***:

$$\Rightarrow \quad F_p = \int_\Sigma p\mathrm{d}S\cos\alpha = p\int_\Sigma \mathrm{d}S\cos\alpha = p\cdot\pi(r\sin\theta)^2 \approx \pi p r^2\theta^2 \quad (2)$$

p—the pressure of the air inside of the soap bubble at arbitrary Time t.

As shown in Fig. 1, taking *the soap water on both sides of the line element of length* dl *at azimuthal angle* φ *on the boundary of the spherical crown element (hereinafter referred to as element* dl*)* as the object of study. Studying *the state at arbitrary Time t*. Considering ***surface tension theorem*** (Need to consider both the inner and outer surfaces):

$$\mathrm{d}F_\sigma = 2\sigma\cdot\mathrm{d}l \quad (3)$$

dF_σ—the $\boldsymbol{\theta}^0$-direction component of the surface tension exerted on the soap water on the inside of spherical crown element at element dl by the soap water on the outside at arbitrary Time t.

$$\Rightarrow F_\sigma = \oint \mathrm{d}F_\sigma\sin\theta = 2\sigma\sin\theta\oint\mathrm{d}l = 2\sigma\sin\theta\cdot 2\pi r\sin\theta \approx 4\pi\sigma r\theta^2 \quad (4)$$

(2) and (4) $\rightarrow$ (1),

$$\Rightarrow \quad p\Big|_{r=r_0} = \frac{4\sigma}{r_0} \quad (5)$$

Taking *the air inside the soap bubble* as the object of study. Studying *the process of the soap bubble's radius changing from r_0 to $r_0 + \Delta r$*. Considering ***Boyle-Mariotte's law***:

$$p\big|_{r=r_0} \cdot \frac{4}{3}\pi r_0^3 = p\big|_{r=r_0+\Delta r} \cdot \frac{4}{3}\pi (r_0+\Delta r)^3 \tag{6}$$

$$\Rightarrow \qquad p\big|_{r=r_0+\Delta r} = \left(\frac{r_0}{r_0+\Delta r}\right)^3 p\big|_{r=r_0} = \left(1+\frac{\Delta r}{r_0}\right)^{-3} p\big|_{r=r_0}$$

$$\approx \left(1-\frac{3\Delta r}{r_0}\right) p\big|_{r=r_0} = \frac{4\sigma}{r_0}\left(1-\frac{3\Delta r}{r_0}\right) \tag{7}$$

$(7) \to (2)$:

$$\Rightarrow \; F_p\big|_{r=r_0+\Delta r} = \pi\, p\big|_{r=r_0+\Delta r} (r_0+\Delta r)^2 \theta^2 = 4\pi\sigma r_0 \theta^2 \left(1-\frac{3\Delta r}{r_0}\right)\left(1+\frac{\Delta r}{r_0}\right)^2$$

$$\approx 4\pi\sigma r_0\theta^2\left(1-\frac{3\Delta r}{r_0}\right)\left(1+\frac{2\Delta r}{r_0}\right) \approx 4\pi\sigma r_0\theta^2\left(1-\frac{\Delta r}{r_0}\right) \tag{8}$$

Taking *the spherical crown element* as the object of study. Studying *the state at arbitrary Time t*. Considering ***Newton's second law*** (the ***k***-direction component equation):

$$F_P\big|_{r=r_0+\Delta r} - F_\sigma\big|_{r=r_0+\Delta r} = \Delta m \Delta\ddot{r} \tag{9}$$

Δm—the mass of the spherical crown element.

Considering that the mass of the soap bubbles is uniformly distributed:

$$\Delta m \approx \frac{\pi (r\sin\theta)^2}{4\pi r^2} m \approx \frac{\pi r^2\theta^2}{4\pi r^2} m = \frac{\theta^2}{4} m \tag{10}$$

(4), (8) and $(10) \to (9)$,

$$\Rightarrow \qquad \Delta\ddot{r} + \frac{32\pi\sigma}{m}\Delta r = 0 \tag{11}$$

Equation (11) is the differential equation of the spherical crown element performing the simple harmonic vibration in the radial direction near Point $(r_0, 0, 0)$, and its angular frequency [i.e., the angular frequency of the tiny vibration (expanding and contracting repeatedly) of the soap bubble] ω is:

$$\Rightarrow \qquad \omega = 4\sqrt{\frac{2\pi\sigma}{m}} \tag{12}$$

Operations (2), (4) and (10) have taken the second-order approximation of θ. Operations (7), (8) have taken the first-order approximation of Δr.

4.1.6 The Tiny Vibration of Liquid in a Cuboid Container (d, *C*)

A cuboid container of cross-sectional side length L is fixed on a horizontal table, in which there is a homogeneous liquid of height h (see Fig. 1). Let the liquid perform the tiny oscillation in the direction parallel to one side of the container, find the angular frequency.

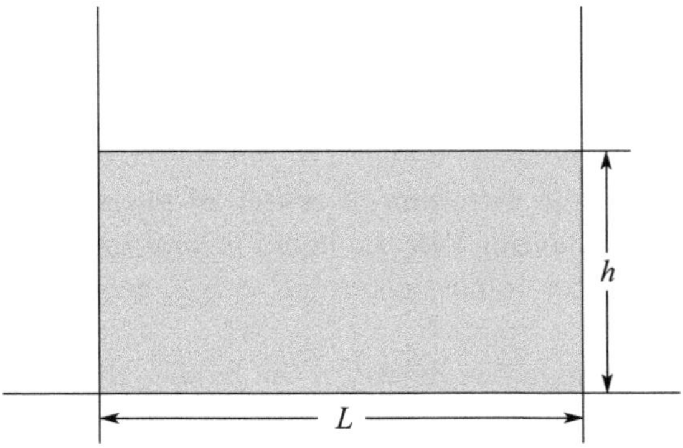

Figure 1

Solution

In the process of the liquid's oscillation, once the free surface of the liquid is in the horizontal plane is defined as Time zero.

As shown in Fig. 2, the position of the geometric center of the liquid at Time zero is denoted by O, the unit vector in the direction in which the liquid is disturbed is denoted by $\boldsymbol{i}$, the unit vector in the vertical upward direction is denoted by $\boldsymbol{j}$. Establishing the Cartesian coordinate system with the frame $(O, \boldsymbol{i}, \boldsymbol{j})$, let x, y be the Cartesian coordinates. The section of the O-xy plane on the liquid at Time zero is denoted by ABB_1A_1. At arbitrary Time t, the section of the O-xy plane on the liquid is denoted by ABB_2A_2, the center of mass of the liquid is denoted by C, the center of the free surface of the liquid is denoted by O_1.

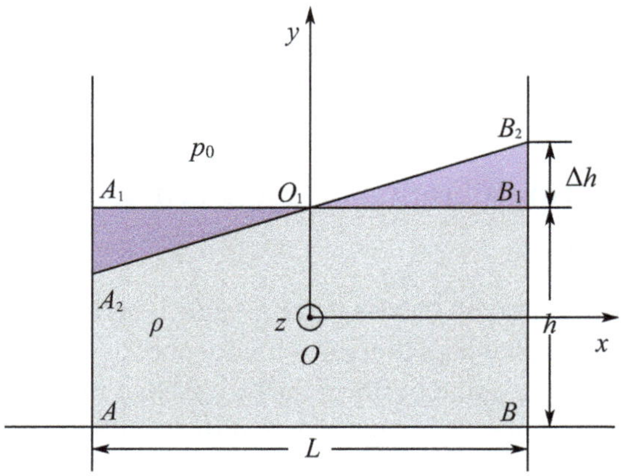

Figure 2

Considering the conservation of the liquid volume and spatial symmetry, it can be seen that Point O_1 remains static.

Considering the definition of ***center of mass*** and ***geometric center***, and considering that the liquid is homogeneous, it can be seen that its center of mass coincides with its geometric center all the time.

Considering spatial symmetry, it can be seen that the geometric center of the liquid coincides with the geometric center of Section ABB_2A_2 all the time. Section ABB_2A_2 can be regarded as the superposition of three parts, including (1) Section $\square ABB_1A_1$ with positive area, (2) Section $\Delta O_1B_1B_2$ with positive area, (3) Section $\Delta O_1A_1A_2$ with negative area:

$$\begin{cases} x_C = \dfrac{S_0 \cdot x_0 + S_1 \cdot x_1 + S_2 \cdot x_2}{S_0} & \text{(1a)} \\ y_C = \dfrac{S_0 \cdot y_0 + S_1 \cdot y_1 + S_2 \cdot y_2}{S_0} & \text{(1b)} \\ z_C = 0 & \text{(1c)} \end{cases}$$

S_0, S_1, S_2—the area of $\square ABB_1A_1$(taking positive value), $\Delta O_1B_1B_2$ (taking positive value) and $\Delta O_1A_1A_2$ (taking negative value) at arbitrary Time t.

(x_0, y_0), (x_1, y_1), (x_2, y_2)—the coordinates of the geometric center of $\square ABB_1A_1$, $\Delta O_1B_1B_2$ and $\Delta O_1A_1A_2$ at arbitrary Time t.

Considering the following geometric relations in Fig. 2:

$$\begin{cases} S_0 = Lh & \text{(2a)} \\ S_1 = \dfrac{1}{2}\cdot\dfrac{1}{2}L\Delta h & \text{(2b)} \\ S_2 = -\dfrac{1}{2}\cdot\dfrac{1}{2}L\Delta h & \text{(2c)} \end{cases}$$

$$\begin{cases} (x_0, y_0) = (0,0) & \text{(3a)} \\ (x_1, y_1) = \left(\dfrac{1}{3}L, \dfrac{h}{2} + \dfrac{1}{3}\Delta h\right) & \text{(3b)} \\ (x_2, y_2) = \left(-\dfrac{1}{3}L, \dfrac{h}{2} - \dfrac{1}{3}\Delta h\right) & \text{(3c)} \end{cases}$$

Δh—the projected length of $\overrightarrow{B_1B_2}$ on the z-axis ($\Delta h > 0$, indicating B_2 is above B_1; $\Delta h < 0$, indicating B_2 is below B_1).

$$\Rightarrow \quad \begin{cases} x_C = \dfrac{L}{6h}\Delta h & \text{(4a)} \\ y_C = \dfrac{1}{6h}\Delta h^2 & \text{(4b)} \end{cases}$$

Considering Δh is a small quantity:

$$\Rightarrow \quad \frac{y_C}{x_C} = \frac{\Delta h}{L} \ll 1 \tag{5}$$

It can be seen that the center-of-mass displacement of the liquid in the $\boldsymbol{j}$-direction is far less than that in the $\boldsymbol{i}$-direction; therefore, its motion in the $\boldsymbol{j}$-direction can be ignored.

Taking *the liquid* as the object of study. Studying *the state at arbitrary Time t*. Considering ***theorem of motion of center of mass*** (the $\boldsymbol{i}$-direction component equation):

$$F_L + F_R + F_U = \rho L^2 h \ddot{x}_C \tag{6}$$

F_L—the $\boldsymbol{i}$-direction component of the pressure exerted on the liquid by the left inner wall of the container at Time t.

F_R—the $\boldsymbol{i}$-direction component of the pressure exerted on the liquid by the right inner wall of the container at Time t.

F_U—the ***i***-direction component of the pressure exerted on the liquid by the atmosphere at Time t.

ρ—the density of the liquid.

Considering (a) the definition of ***pressure***, (b) ***basic equation of hydrostatic pressure*** (the motion of the liquid's center of mass in the ***j***-direction can be ignored; therefore, the liquid pressure can approximately take the hydrostatic pressure):

$$F_L = \int_{\Sigma_L} (p_0 + \rho g y)\mathrm{d}S_L = \int_0^{h-\Delta h} (p_0 + \rho g y)L\mathrm{d}y$$

$$= \left[p_0 + \frac{1}{2}\rho g(h-\Delta h)\right](h-\Delta h)L$$

$$\approx \left(p_0 h + \frac{1}{2}\rho g h^2\right)L - (p_0 + \rho g h)\Delta h L \tag{7}$$

$$F_R = -\int_{\Sigma_R} (p_0 + \rho g y)\mathrm{d}S_R = -\int_0^{h+\Delta h} (p_0 + \rho g y)L\mathrm{d}y$$

$$= -\left[p_0 + \frac{1}{2}\rho g(h+\Delta h)\right](h+\Delta h)L$$

$$\approx -\left(p_0 h + \frac{1}{2}\rho g h^2\right)L + (p_0 + \rho g h)\Delta h L \tag{8}$$

$$F_U = \int_{\Sigma_U} p_0 \mathrm{d}S_U \cos\angle O_1 A_2 A_1 = p_0 \int_{\Sigma_U} \mathrm{d}S_U \cos\angle O_1 A_2 A_1$$

$$= p_0 \cdot 2\Delta h L = 2 p_0 \Delta h L \tag{9}$$

p_0—the atmospheric pressure.

$$\Rightarrow \qquad \ddot{x}_C = -\frac{2g}{L}\Delta h = -\frac{12gh}{L^2}x_C \tag{10}$$

$$\Rightarrow \qquad \ddot{x}_C + \frac{12gh}{L^2}x_C = 0 \tag{11}$$

Equation (11) is the differential equation of the liquid's center of mass performing tiny simple harmonic vibration parallel to the x-axis near the origin O, and its angular frequency ω is:

$$\Rightarrow \qquad \omega = \frac{2\sqrt{3gh}}{L} \tag{12}$$

4.1.7 The Tiny Vibration of Liquid in a Cylindrical Cup (*a*, *A*)

A cylindrical water cup of sectional radius R is fixed on a horizontal table, in which there is a homogeneous liquid of height H. Find the angular frequency of the tiny oscillation of the liquid in the water cup.

Solution

In the process of the liquid's oscillation, once the free surface of the liquid is in the horizontal plane is defined as time zero.

As shown in Fig. 1, the center of the bottom of the cup is denoted by O, the unit vector in the vertical upward direction is denoted by $\boldsymbol{k}$. Taking the origin O as the pole, the $\boldsymbol{k}$-direction as the positive direction of the longitudinal axis, the direction in which the liquid is disturbed (horizontally to the right in the figure) as the positive direction of the polar axis, the right-hand helix direction with regard to the $\boldsymbol{k}$-direction as the positive direction of polar angle, establishing the cylindrical coordinate system, let r, φ, z be the cylindrical coordinates (at arbitrary Time t, taking the point on the boundary of the liquid's free surface with polar angle φ, let Z_φ be its z-coordinate.)

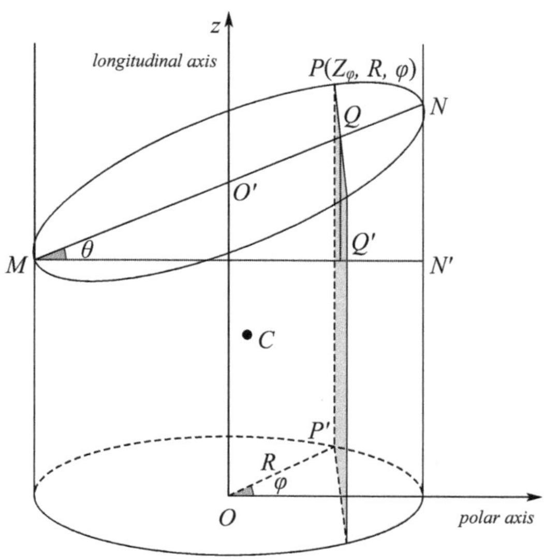

Figure 1

At arbitrary Time t, the angle between the free surface of the liquid and the horizontal plane is denoted by θ (θ is a small quantity, which is positive in the configuration of the figure), the center of the free surface of the liquid is denoted by O', the leftmost and rightmost points of the free surface of the liquid are, respectively, denoted by M and N, the symmetry point of Point M relative to z-axis is denoted by N', taking an arbitrary point on the boundary of the free surface of the liquid and denoting it by P, the projection of Point P on the bottom of the cup is denoted by P', and the projection of Point P on the plane $\varphi = 0$ (the plane determined by the polar axis and longitudinal axis) is denoted by Q, the projection of Point Q on MN' is denoted by Q', the center of mass of the liquid is denoted by C.

Considering the conservation of the liquid volume and spatial symmetry, it can be seen that Point O' remains static.

Considering $\Delta MNN' \sim \Delta MQQ'$:

$$\frac{\overline{NN'}}{\overline{QQ'}} = \frac{\overline{MN'}}{\overline{MQ'}} \tag{1}$$

$$\Rightarrow \quad \frac{2R\sin\theta}{Z_\varphi - (H - R\sin\theta)} = \frac{2R}{R + R\cos\varphi} \tag{2}$$

$$\Rightarrow \quad Z_\varphi = H + R\sin\theta\cos\varphi \approx H + R\theta\cos\varphi \tag{3}$$

This operation has taken the first-order approximation of θ.

Defining $\mathrm{d}V_\varphi$—the volume of the thin layer of liquid between two cross sections that are perpendicular to the polar axis and, respectively, passing through Point (Z_φ, R, φ) and Point $(Z_{\varphi+\mathrm{d}\varphi}, R, \varphi + \mathrm{d}\varphi)$ at arbitrary Time t.

Considering the following geometric relation in Fig 1:

$$\mathrm{d}V_\varphi = 2R\sin\varphi \cdot R\mathrm{d}\varphi\sin\varphi \cdot Z_\varphi$$

$$= 2R^2(H\sin^2\varphi + R\theta\sin^2\varphi\cos\varphi)\mathrm{d}\varphi \tag{4}$$

Considering the definition of ***center of mass*** and ***geometric center***, and considering that the liquid is homogeneous, it can be seen that its center of mass coincides with its geometric center all the time:

$$\Rightarrow \quad x_C = \frac{1}{\pi R^2 H}\int \mathrm{d}V_\varphi \cdot R\cos\varphi$$

$$=\frac{2R}{\pi H}\int_0^{\pi}(H\sin^2\varphi\cos\varphi+R\theta\sin^2\varphi\cos^2\varphi)\mathrm{d}\varphi$$

$$=\frac{R^2}{4H}\theta \tag{5}$$

Considering the definition of ***pressure***:

$$F_x=-\int_0^{2\pi}\int_0^{Z_\varphi}p(R,\varphi,z)\cdot R\mathrm{d}\varphi\cdot\mathrm{d}z\cdot\cos\varphi \tag{6}$$

F_x—the ***i***-direction component of the pressure exerted on the liquid by the inner wall of the cup at arbitrary Time *t*.

$p\ (R,\varphi,z)$—the liquid pressure at an arbitrary point on the inner wall of the cup at arbitrary Time *t*.

Taking *the liquid in the cup* as the object of study. Studying *the state at arbitrary Time t*. Considering ***basic equation of hydrostatic pressure*** (use the same analysis as the previous problem "***the tiny vibration of liquid in a cuboid container***", it can be seen that the ***k***-direction component of the acceleration of an arbitrary liquid infinitesimal element is a high-order small quantity at arbitrary Time *t*; therefore, the liquid pressure can approximately take the hydrostatic pressure):

$$p(R,\varphi,z)=\rho g(Z_\varphi-z) \tag{7}$$

$$\Rightarrow \qquad F_x=-\rho gR\int_0^{2\pi}\cos\varphi\mathrm{d}\varphi\int_0^{Z_\varphi}(Z_\varphi-z)\mathrm{d}z$$

$$=-\frac{1}{2}\rho gR\int_0^{2\pi}Z_\varphi^2\cos\varphi\mathrm{d}\varphi$$

$$=-\frac{1}{2}\rho gR\int_0^{2\pi}(H+R\theta\cos\varphi)^2\cos\varphi\mathrm{d}\varphi$$

$$\approx-\frac{1}{2}\rho gR\left(H^2\int_0^{2\pi}\cos\varphi\mathrm{d}\varphi+2HR\theta\int_0^{2\pi}\cos^2\varphi\mathrm{d}\varphi\right)$$

$$=-\rho g\cdot\pi R^2H\theta \tag{8}$$

Taking *the liquid in the cup* as the object of study. Studying *the state at arbitrary Time t*. Considering ***theorem of motion of center of mass*** (the ***i***-direction component equation):

$$F_x=\rho\cdot\pi R^2H\cdot\ddot{x}_C \tag{9}$$

Combining (5), (8) and (9),

$$\Rightarrow \qquad \ddot{x}_C + \frac{4H}{R^2} g x_C = 0 \qquad (10)$$

Equation (10) is the differential equation of the liquid's center of mass performing tiny simple harmonic vibration parallel to the polar axis, and its angular frequency ω is:

$$\Rightarrow \qquad \omega = \sqrt{\frac{4H}{R^2} g} \qquad (11)$$

4.1.8 The Motion of a conductor Ring over a Cylindrical Magnet (d, *B*)

A cylindrical magnet is fixed in the vertical direction (see Fig. 1), taking a point over the center of the top of the magnet (close enough to the magnet) and denoting it by *O*, the unit vector in the vertical upward direction is denoted by $\boldsymbol{k}$, establishing the right-handed cylindrical coordinate system, let *r*, θ, *z* be the cylindrical coordinates. Given the magnetic induction intensity of the magnetic field excited by the magnet is $\boldsymbol{B} = B_0\beta r\boldsymbol{r}^0 + B_0(1 - \alpha z)\boldsymbol{k}$, where α and

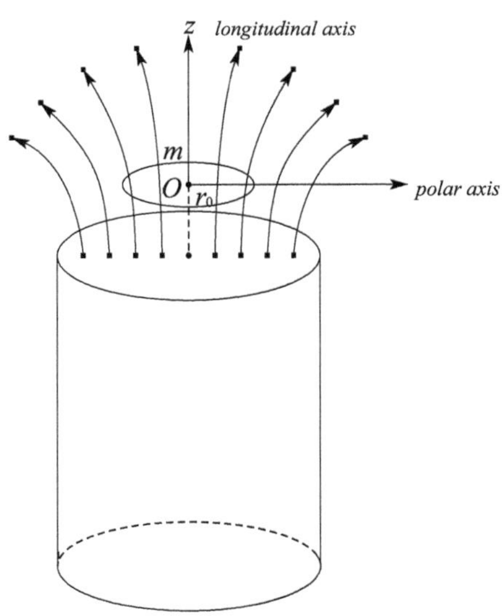

Figure 1

β are constants. There is a homogeneous conductor ring of mass m, radius R, self-inductance coefficient L and negligible resistance. Place the ring statically and horizontally with its center O' coincides with the Pole O, and then release it. Given that R is far smaller than the sectional radius of the magnet. Determine the position of the center O' and the current intensity in the ring at arbitrary time.

Solution [1]:

The time point when the conductor ring is released is defined as Time zero.

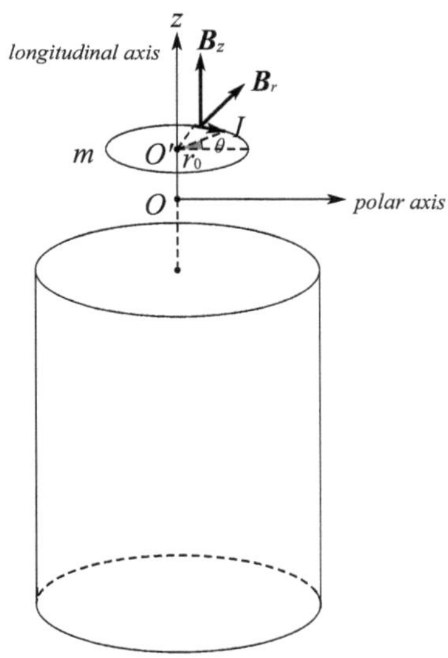

Figure 2

Taking *the conductor ring* as the object of study (see Fig. 2). Studying *the state at arbitrary Time t*. Considering (a) ***Kirchhoff's voltage law*** (opposite to the angular direction), (b) ***theorem of motional electromotive force*** [integrating into (a)], (c) ***theorem of self-induction electromotive force*** [integrating into (a)]:

$$\oint_L \boldsymbol{v}\times\boldsymbol{B}\cdot \mathrm{d}\boldsymbol{l} - L\dot{I} = 0 \tag{1}$$

$$\Rightarrow \quad 0 = \oint_L \dot{z}\boldsymbol{k}\times\left(B_r\big|_{r_0,z}\boldsymbol{r}^0 + B_z\big|_{r_0,z}\boldsymbol{k}\right)\cdot(-\mathrm{d}l\boldsymbol{\theta}^0) - L\dot{I}$$

$$= -\oint_L B_r\big|_{r_0,z}\dot{z}\mathrm{d}l - L\dot{I} = -B_r\big|_{r_0,z}\dot{z}\oint_L \mathrm{d}l - L\dot{I}$$

$$= -B_r\big|_{r_0,z}\,\dot{z}\cdot 2\pi r_0 - L\dot{I} \tag{2}$$

I—the current intensity in the ring at Time t (opposite to the angular direction).

Taking *the conductor ring* as the object of study. Studying *the state at arbitrary Time t*. Considering ***theorem of motion of center of mass*** (the $\boldsymbol{k}$-direction component equation), (b) ***Ampere force law*** [integrating into (a)]:

$$\oint_L (I\mathrm{d}\boldsymbol{l}\times\boldsymbol{B})\cdot\boldsymbol{k} - mg = m\ddot{z} \tag{3}$$

$$\Rightarrow \qquad m\ddot{z} = \oint_L\left[I\left(-\mathrm{d}l\boldsymbol{\theta}^0\right)\times\left(B_r\big|_{r_0,z}\boldsymbol{r}^0 + B_z\big|_{r_0,z}\boldsymbol{k}\right)\right]\cdot\boldsymbol{k} - mg$$

$$= \oint_L B_r\big|_{r_0,z} I\mathrm{d}l - mg = B_r\big|_{r_0,z} I\oint_L \mathrm{d}l - mg$$

$$= B_r\big|_{r_0,z} I\cdot 2\pi r_0 - mg \tag{4}$$

Known:

$$B_r\big|_{r_0,z} = B_0\beta r_0 \tag{5}$$

Combining (1) to (5):

$$\Rightarrow \qquad \ddot{I} + \frac{4\pi^2\beta^2 B_0^2 r_0^4}{mL} I = \frac{2\pi\beta B_0 r_0^2 g}{L} \tag{6}$$

Defining the variable I^*:

$$I^* = I - \frac{mg}{2\pi\beta B_0 r_0^2} \tag{7}$$

$$\Rightarrow \qquad \ddot{I}^* + \frac{4\pi^2\beta^2 B_0^2 r_0^4}{mL} I^* = 0 \tag{8}$$

That is, I^* varies harmonically with time, the angular frequency ω is:

$$\omega = \frac{2\pi\beta B_0 r_0^2}{\sqrt{mL}} \tag{9}$$

The general solution of Eq. (8) is as follows:

$$\Rightarrow \qquad I^* = I_0\cos(\omega t + \varphi) \tag{10}$$

I_0 and φ—the peak value and the initial phase of I^*.

Considering the initial conditions:

$$\begin{cases} \dot{z}\big|_{t=0} = 0 & \text{(11a)} \\ I\big|_{t=0} = 0 & \text{(11b)} \end{cases}$$

(11a) → (2),

⇒ $$\dot{I}\big|_{t=0} = 0 \qquad (12)$$

⇒ $$\begin{cases} I^*\big|_{t=0} = -\dfrac{mg}{2\pi\beta B_0 r_0^2} & \text{(13a)} \\ \dot{I}^*\big|_{t=0} = 0 & \text{(13b)} \end{cases}$$

(10) → (13),

⇒ $$\begin{cases} I_0 = \dfrac{mg}{2\pi\beta B_0 r_0^2} & \text{(14a)} \\ \varphi = \pi & \text{(14b)} \end{cases}$$

⇒ $$I^* = \frac{mg}{2\pi\beta B_0 r_0^2}\cos(\omega t + \pi) = -\frac{mg}{2\pi\beta B_0 r_0^2}\cos\omega t \qquad (15)$$

⇒ $$I = \frac{mg}{2\pi\beta B_0 r_0^2}(1 - \cos\omega t) \qquad (16)$$

(16) → (2):

⇒ $$\mathrm{d}z = -\frac{L}{2\pi\beta B_0 r_0^2}\mathrm{d}I \qquad (17)$$

⇒ $$z = \int_0^z \mathrm{d}z = -\frac{L}{2\pi\beta B_0 r_0^2}\int_0^I \mathrm{d}I = -\frac{L}{2\pi\beta B_0 r_0^2}I \qquad (18)$$

⇒ $$z = \frac{g}{\omega^2}(\cos\omega t - 1) \qquad (19)$$

That is, Point O' performs the simple harmonic vibration in the vertical direction.

Solution [2]:

Taking *the conductor ring* as the object of study. Studying *the state at arbitrary Time t*. Considering (a) ***Kirchhoff's voltage law*** (opposite to the angular direction), (b) ***Faraday's law of electromagnetic***

induction [taking the ***k***-direction as the positive direction of the surface integral] [according to the fact that r is a small quantity, it can be seen that B_z on the circular plane surrounded by the conductor ring is approximately equal everywhere, the value of B_z at an arbitrary point on the circular plane can approximately take the value of B_z at the center] [integrating into (a)], (c) ***theorem of self-induction electromotive force*** [integrating into (a)]:

$$0=-\left(-\frac{\mathrm{d}}{\mathrm{d}t}\iint_{\Sigma}\boldsymbol{B}\cdot\mathrm{d}\boldsymbol{S}\right)-L\dot{I}=\frac{\mathrm{d}}{\mathrm{d}t}\iint_{\Sigma}\left(B_r\big|_{0,z}\boldsymbol{r}^0+B_z\big|_{0,z}\boldsymbol{k}\right)\cdot(\mathrm{d}S)\boldsymbol{k}-L\dot{I}$$

$$=\frac{\mathrm{d}\left(B_z\big|_{0,z}\right)}{\mathrm{d}t}\cdot\pi r_0^2-L\dot{I} \tag{20}$$

Known:

$$B_z\big|_{0,z}=B_0(1-\alpha z) \tag{21}$$

Combining (4), (5), (20) and (21):

$$\Rightarrow \qquad \ddot{I}+\frac{2\pi^2\alpha\beta B_0^2r_0^4}{mL}I=\frac{\pi\alpha B_0r_0^2g}{L} \tag{22}$$

Taking *the magnetic field* as the object of study. Studying *the state at arbitrary Time t*. Considering ***Gauss's law for magnetism*** [the closed integral surface is taken the cylindrical surface with the z-axis as the central axis, r as the radius (r is a small quantity), and whose bottom and top are located at z and z + dz, respectively] [it can be seen from the symmetry that B_r at the side of the cylinder is equal everywhere, according to the fact that r is a small quantity, it can be seen that B_z at the bottom and top is approximately equal everywhere, respectively]:

$$\Rightarrow \qquad (B_z+\mathrm{d}B_z)\cdot\pi r^2-B_z\cdot\pi r^2+B_r\cdot 2\pi r\mathrm{d}z=0 \tag{23}$$

$$\Rightarrow \qquad B_r=-\frac{1}{2}\frac{\mathrm{d}B_z}{\mathrm{d}z}r \tag{24}$$

$$\Rightarrow \qquad \alpha=2\beta \tag{25}$$

That is, α and β are not independent.

$$\Rightarrow \qquad \ddot{I}+\frac{4\pi^2\beta^2B_0^2r_0^4}{mL}I=\frac{2\pi\beta B_0r_0^2g}{L} \tag{26}$$

This result is the same as Eq. (6).

4.1.9 A Rotating Spring Vibrator (d, *C*)

A light spring of stiffness coefficient k and natural length l_0 is placed on a smooth horizontal ground, one end of which is fixed, and the other end is tied with a small ball of mass m. Initially, the spring is the natural length and the ball is at rest, give the ball a weak instantaneous impact to make it obtain an initial velocity of the magnitude v_0 and the direction perpendicular to the spring (the top view is shown in Fig. 1). Determine the kinematics equation of the ball.

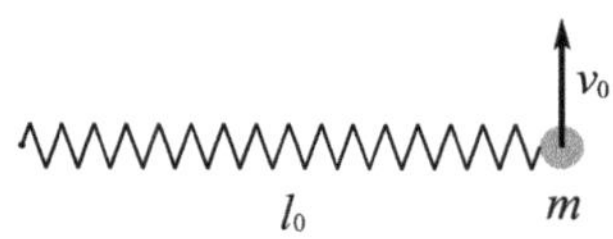

Figure 1

Solution

The time point when the ball is impacted is defined as Time zero.

As shown in Fig. 2, the fixed end of the spring is denoted by *O*, the unit vector in the vertical upward direction is denoted by $\boldsymbol{k}$. Taking Point *O* as the pole, the direction of the pole *O* pointing to the initial position of the ball as the positive direction of the polar axis, the right-hand helix direction with regard to the $\boldsymbol{k}$-direction as the positive direction of polar angle, establishing the polar coordinate system, let r, θ be the polar coordinates.

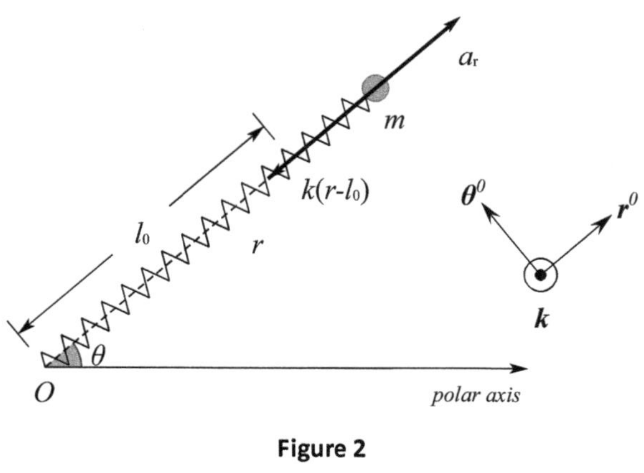

Figure 2

Taking *the ball* as the object of study. Studying *the state at arbitrary Time t*. Considering ***Newton's second law*** (the $\mathbf{r}^0$-direction component equation):

$$-k(r-l_0)=ma_r \tag{1}$$

a_r—the $\mathbf{r}^0$-direction component of the acceleration of the ball at Time t.

Considering ***the radial component form of acceleration***:

$$a_r=\ddot{r}-r\dot{\theta}^2 \tag{2}$$

Defining the variable x:

$$x=r-l_0 \tag{3}$$

Combining (1) to (3),

$$\Rightarrow \qquad \ddot{x}-(l_0+x)\dot{\theta}^2+\frac{k}{m}x=0 \tag{4}$$

Taking *the ball* as the object of study. Studying *the process from Time zero to arbitrary Time t*. Considering ***law of conservation of angular momentum*** [taking the pole O as the reference point, the resultant external torque acting on the ball is zero; therefore, the angular momentum of the ball is conserved] (the $\mathbf{k}$-direction component equation):

$$mv_0l_0=mr^2\dot{\theta} \tag{5}$$

(3) $\rightarrow$ (5),

$$\Rightarrow \qquad \dot{\theta}=\frac{v_0l_0}{(l_0+x)^2} \tag{6}$$

(6) $\rightarrow$ (4),

$$\Rightarrow \qquad \ddot{x}-\frac{v_0^2l_0^2}{(l_0+x)^3}+\frac{k}{m}x=0 \tag{7}$$

x is a small quantity, taking the following approximation:

$$\frac{1}{(l_0+x)^3}=\frac{1}{l_0^3}\left(1+\frac{x}{l_0}\right)^{-3}\approx\frac{1}{l_0^3}\left(1-\frac{3x}{l_0}\right) \tag{8}$$

$$\Rightarrow \qquad \ddot{x}+\left(\frac{k}{m}+3\frac{v_0^2}{l_0^2}\right)x=\frac{v_0^2}{l_0} \tag{9}$$

Defining the variable X:

$$X = x - \frac{\frac{v_0^2}{l_0}}{\frac{k}{m} + 3\frac{v_0^2}{l_0^2}} \tag{10}$$

$$\Rightarrow \qquad \ddot{X} + \left(\frac{k}{m} + 3\frac{v_0^2}{l_0^2}\right)X = 0 \tag{11}$$

Equation (11) is the differential equation of the ball performing the tiny simple harmonic vibration in the radial direction, and its angular frequency ω is:

$$\omega = \sqrt{\frac{k}{m} + 3\frac{v_0^2}{l_0^2}} = \sqrt{\frac{k}{m}\left(1 + 3\frac{mv_0^2}{kl_0^2}\right)} \approx \sqrt{\frac{k}{m}}\left(1 + \frac{3}{2}\frac{mv_0^2}{kl_0^2}\right) \tag{12}$$

This operation has considered $\frac{mv_0^2}{kl_0^2}$ as a small quantity and taken the first-order approximation.

The solution of Eq. (11) is as follows:

$$\Rightarrow \qquad X = A\cos(\omega t + \phi) \tag{13}$$

A and φ—the amplitude and the initial phase of the simple harmonic vibration of the ball.

Considering the initial conditions:

$$\begin{cases} r\big|_{t=0} = l_0 & (14a) \\ \dot{r}\big|_{t=0} = 0 & (14b) \end{cases}$$

$$\Rightarrow \qquad \begin{cases} x\big|_{t=0} = 0 & (15a) \\ \dot{x}\big|_{t=0} = 0 & (15b) \end{cases}$$

$$\Rightarrow \qquad \begin{cases} X\big|_{t=0} = -\frac{v_0^2}{\omega^2 l_0} & (16a) \\ \dot{X}\big|_{t=0} = 0 & (16b) \end{cases}$$

(13) → (16),

$$\Rightarrow \quad \begin{cases} A = \dfrac{v_0^2}{\omega^2 l_0} & (17a) \\ \varphi = \pi & (17b) \end{cases}$$

$$\Rightarrow \quad X = \frac{v_0^2}{\omega^2 l_0}\cos(\omega t + \pi) = -\frac{v_0^2}{\omega^2 l_0}\cos\omega t \quad (18)$$

(18) → (10),

$$\Rightarrow \quad x = \frac{v_0^2}{\omega^2 l_0}(1 - \cos\omega t) \quad (19)$$

(19) → (3),

$$\Rightarrow \quad r = l_0 + \frac{v_0^2}{\omega^2 l_0}(1 - \cos\omega t) \quad (20)$$

Taking the first-order approximation of Eq. (6):

$$\Rightarrow \quad \frac{\mathrm{d}\theta}{\mathrm{d}t} = \frac{v_0}{l_0}\left(1 + \frac{x}{l_0}\right)^{-2} \approx \frac{v_0}{l_0}\left(1 - \frac{2x}{l_0}\right) \quad (21)$$

(19) → (21),

$$\Rightarrow \quad \frac{\mathrm{d}\theta}{\mathrm{d}t} = \frac{v_0}{l_0}\left[1 - \frac{2v_0^2}{\omega^2 l_0^2}(1 - \cos\omega t)\right] \quad (22)$$

$$\Rightarrow \quad \mathrm{d}\theta = \frac{v_0}{l_0}\left[1 - \frac{2v_0^2}{\omega^2 l_0^2}(1 - \cos\omega t)\right]\mathrm{d}t$$

$$\Rightarrow \quad \theta = \int_0^\theta \mathrm{d}\theta = \frac{v_0}{l_0}\int_0^t \left[1 - \frac{2v_0^2}{\omega^2 l_0^2}(1 - \cos\omega t)\right]\mathrm{d}t$$

$$= \frac{v_0}{l_0}\left(1 - \frac{2v_0^2}{\omega^2 l_0^2}\right)t + \frac{2v_0^3}{\omega^3 l_0^3}\sin\omega t \quad (23)$$

Equations (20) and (23) are the kinematics equations of the ball.

4.1.10 A Rotating Simple Pendulum (d, *C*)

Consider a disk located on a horizontal ground rotating around its central axis at the rotating angular velocity of the magnitude Ω and the direction vertical upward. There are two vertical supports of equal height fixed on the disk, which support a lightweight T-shaped rod structure, the length of the rotatable rod is l, a small ball of mass m is tied at its lower end, forming a simple pendulum, the suspension point O' is just located on the central axis of the disk.

(1) Find the mechanical equilibrium position of the ball.
(2) Find the angular frequency of the tiny vibration of the ball.

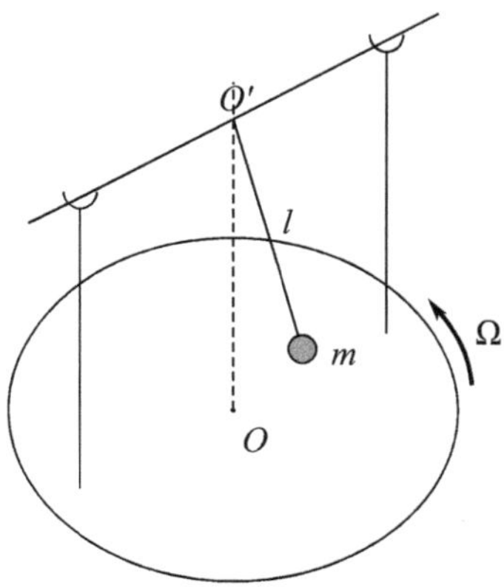

Figure 1

Solution to Problem (1)

As shown in Fig. 2, establishing the polar coordinate system fixed to the disk reference frame, let r, θ be the polar coordinates.

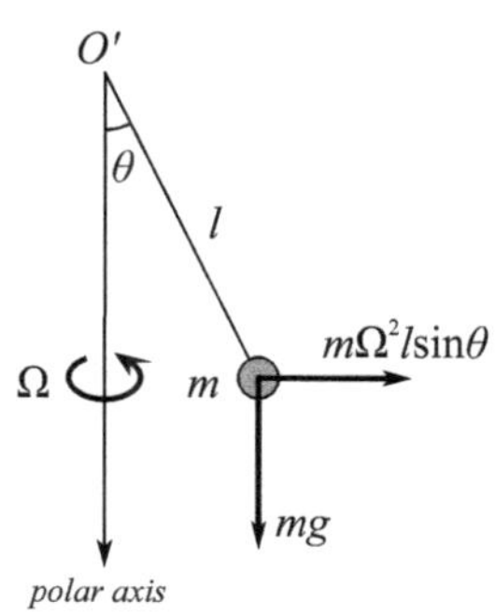

Figure 2

For the system composed of the ball and the earth at arbitrary Time t, defining E_g—the gravitational potential energy [taking Point O' as the potential energy zero], E_Ω—the centrifugal potential in the disk reference frame [taking the central axis as the potential energy zero], and E_p—the total potential energy in the disk reference frame.

$$E_p = E_g + E_\Omega \tag{1}$$

Considering ***theorem of gravitational potential energy***:

$$E_g = -mgl\cos\theta \tag{2}$$

Considering (a) the definition of ***potential energy***, (b) ***theorem of inertial centrifugal force*** [integrating into (a)]:

$$E_\Omega = \int_{l\sin\theta}^{0} m\Omega^2 r dr = -\frac{1}{2}m\Omega^2 l^2 \sin^2\theta \tag{3}$$

$$\Rightarrow \quad E_p = -mgl\cos\theta - \frac{1}{2}m\Omega^2 l^2 \sin^2\theta \tag{4}$$

Considering ***principle of minimum potential energy***:

$$\begin{cases} \left.\dfrac{\mathrm{d}E_p}{\mathrm{d}\theta}\right|_{\theta_0} = 0 & \text{(5a)} \\ \left.\dfrac{\mathrm{d}^2E_p}{\mathrm{d}\theta^2}\right|_{\theta_0} > 0 & \text{(5b)} \end{cases}$$

θ_0—the polar angle of the ball when it is in stable equilibrium.
From (4),

$$\Rightarrow \quad \frac{\mathrm{d}E_p}{\mathrm{d}\theta} = mgl\sin\theta - m\Omega^2 l^2 \sin\theta\cos\theta = ml^2\sin\theta\left(\frac{g}{l} - \Omega^2 l\cos\theta\right) \tag{6}$$

$$\Rightarrow \quad \frac{\mathrm{d}^2E_p}{\mathrm{d}\theta^2} = ml^2\left[\frac{g}{l}\cos\theta + \Omega^2(\sin^2\theta - \cos^2\theta)\right] \tag{7}$$

(6) → (5a),

$$\Rightarrow \quad \theta_0 = 0,\ \pi,\ \arccos\frac{g}{\Omega^2 l} \tag{8}$$

$$\Rightarrow \begin{cases} \left.\dfrac{d^2E}{d\theta^2}\right|_{\theta_0=0} = ml^2\left(\dfrac{g}{l}-\Omega^2\right)\begin{cases} >0 & \Omega<\sqrt{\dfrac{g}{l}} & (9a) \\ <0 & \Omega>\sqrt{\dfrac{g}{l}} & (9b) \end{cases} \\ \left.\dfrac{d^2E}{d\theta^2}\right|_{\theta_0=\pi} = -ml^2\left(\dfrac{g}{l}+\Omega^2\right)<0 & (10) \\ \left.\dfrac{d^2E}{d\theta^2}\right|_{\theta_0=\arccos\frac{g}{\Omega^2 l}} = ml^2\Omega^2\sin^2\theta_0>0 \quad \Omega>\sqrt{\dfrac{g}{l}} & (11) \end{cases}$$

That is, when $\Omega< \sqrt{\dfrac{g}{l}}$, the stable equilibrium position of the ball is $\theta = 0$; when $\Omega > \sqrt{\dfrac{g}{l}}$, the stable equilibrium position of the ball is $\theta = \arccos \dfrac{g}{\Omega^2 l}$. In any case, $\theta = \pi$ is the unstable equilibrium position of the ball.

Solution to Problem (2)

Defining E_k—the kinetic energy of the ball in the disk reference frame at arbitrary Time *t*.

$$E_k = \frac{1}{2}m(l\dot{\theta})^2 \tag{12}$$

Defining *E*—the mechanical energy of the system composed of the ball and the earth in the disk reference frame at arbitrary Time *t*.

$$\Rightarrow \quad E = E_k + E_p = \frac{1}{2}ml^2\dot{\theta}^2 - mgl\cos\theta - \frac{1}{2}m\Omega^2 l^2\sin^2\theta \tag{13}$$

Taking *the system composed of the ball and the earth in the disk reference frame* as the object of study. Studying *the instantaneous process (the small time interval dt) at arbitrary Time t.* Considering ***law of conservation of mechanical energy*** [in the disk reference frame, the work done by the external force and non-conservative internal force is zero; therefore, the mechanical energy of the system is conserved]:

$$\frac{\mathrm{d}E}{\mathrm{d}t}=0 \tag{14}$$

(I) **Studying the situation of $\Omega < \sqrt{\frac{g}{l}}$ (the stable equilibrium position of the ball is $\theta = 0$)**

θ is a small quantity, taking the following low order approximations:

$$\sin\theta \approx \theta, \cos\theta \approx 1-\frac{1}{2}\theta^2 \tag{15a, b}$$

(15) $\rightarrow$ (13),

$$\Rightarrow \qquad E \approx \frac{1}{2}ml^2\left[\dot{\theta}^2+\left(\frac{g}{l}-\Omega^2\right)\theta^2\right] \tag{16}$$

(16) $\rightarrow$ (14):

$$\Rightarrow \qquad \ddot{\theta}+\left(\frac{g}{l}-\Omega^2\right)\theta=0 \tag{17}$$

Equation (17) is the differential equation of the ball performing the tiny simple harmonic vibration near $\theta = 0$ in the disk reference frame, and its angular frequency ω_1 is:

$$\Rightarrow \qquad \omega_1=\Omega\sqrt{\frac{g}{\Omega^2 l}-1} \tag{18}$$

(II) **Studying the situation of $\Omega > \sqrt{\frac{g}{l}}$ (the stable equilibrium position of the ball is $\theta = \arccos\frac{g}{\Omega^2 l}$)**

Defining θ'—the small angular displacement of the ball near the stable equilibrium position:

$$\theta=\theta_0+\theta' \tag{19}$$

(19) $\rightarrow$ (13),

$$\Rightarrow \qquad E=ml^2\left[\frac{1}{2}\dot{\theta}'^2-\frac{g}{l}\cos(\theta_0+\theta')-\frac{1}{2}\Omega^2\sin^2(\theta_0+\theta')\right] \tag{20}$$

θ' is a small quantity, taking the following low order approximations:

$$\sin\theta' \approx \theta', \cos\theta' \approx 1 - \frac{1}{2}\theta'^2 \tag{21a, b}$$

$$\Rightarrow \quad E = ml^2\left[\frac{1}{2}\dot{\theta}'^2 - \frac{g}{l}\cos(\theta_0 + \theta') - \frac{1}{4}\Omega^2\left[1 - \cos(2\theta_0 + 2\theta')\right]\right]$$

$$= ml^2\left\{\begin{array}{l}\frac{1}{2}\dot{\theta}'^2 - \frac{g}{l}(\cos\theta_0\cos\theta' - \sin\theta_0\sin\theta') \\ -\frac{1}{4}\Omega^2\left[1 - (\cos 2\theta_0 \cos 2\theta' - \sin 2\theta_0 \sin 2\theta')\right]\end{array}\right\}$$

$$\approx ml^2\left\{\begin{array}{l}\frac{1}{2}\dot{\theta}'^2 - \frac{g}{l}\left[\left(1 - \frac{1}{2}\theta'^2\right)\cos\theta_0 - \theta'\sin\theta_0\right] \\ -\frac{1}{4}\Omega^2\left[1 - (1 - 2\theta'^2)\cos 2\theta_0 + 2\theta'\sin 2\theta_0\right]\end{array}\right\}$$

$$= ml^2\left\{\begin{array}{l}\frac{1}{2}\dot{\theta}'^2 + \frac{1}{2}\theta'^2\left[\frac{g}{l}\cos\theta_0 - \Omega^2(2\cos^2\theta_0 - 1)\right] \\ +\theta'\sin\theta_0\left(\frac{g}{l} - \Omega^2\cos\theta_0\right) - \frac{g}{l}\cos\theta_0 - \frac{1}{4}\Omega^2(1 - \cos 2\theta_0)\end{array}\right\} \tag{22}$$

This operation has taken the second-order approximation of θ'.

(22) → (14),

$$\Rightarrow \quad \dot{\theta}'\left\{\ddot{\theta}' + \theta'\left[\frac{g}{l}\cos\theta_0 - \Omega^2(2\cos^2\theta_0 - 1)\right]\right\} = 0$$

$$\Rightarrow \quad \ddot{\theta}' + \Omega^2\left(1 - \frac{g^2}{\Omega^4 l^2}\right)\theta' = 0 \tag{23}$$

Equation (23) is the differential equation of the ball performing the tiny simple harmonic vibration near $\theta = \arccos\dfrac{g}{\Omega^2 l}$ in the disk reference frame, and its angular frequency ω_2 is:

$$\Rightarrow \quad \omega_2 = \Omega\sqrt{1 - \frac{g^2}{\Omega^4 l^2}} \tag{24}$$

4.1.11 The Constrained Vibration of Two Particles Connected by a Light Rigid String (*c*, *C*)[2]

There is a horizontal table top with a small hole on it (see Fig. 1), a light rigid string passes through the hole, a small ball of mass m on the table and a block of mass M under the table are, respectively, tied to the two ends of the string. Initially, the ball performs the uniform circular motion on the table with the hole as the center at the angular velocity $\boldsymbol{\omega}_0$, and the block is at rest. The friction between the string and the hole can be ignored. If the ball is slightly disturbed in the radial direction, analyze the motion of the ball and the block.

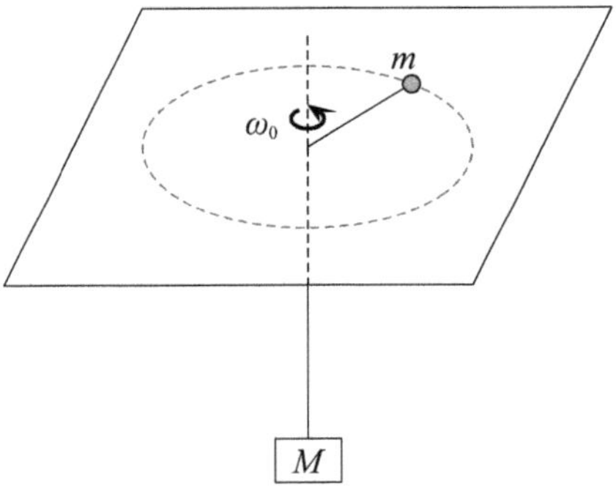

Figure 1

Solution

The time point before the ball is disturbed is defined as Time zero.

As shown in Fig. 2, the position of the hole is denoted by O, the unit vector in the vertical upward direction is denoted by $\boldsymbol{k}$. Taking Point O as the pole, the direction of the pole O pointing to the ball at Time zero as the positive direction of the polar axis, the right-hand helix direction with regard to the $\boldsymbol{k}$-direction as the positive direction of polar angle, the $\boldsymbol{k}$-direction as the positive direction of the longitudinal axis, establishing the cylindrical coordinate system, let r, θ, z be the cylindrical coordinates.

Taking *the block and the ball* as the object of study, respectively. Studying *the state at Time zero*. Considering ***Newton's second law*** [for the block: in the ground reference frame, for the ball: in the rotating reference frame with the hole as the center and with

[2]This problem is a test question from Princeton University in 1977. It is also included in the following book: Yongde Zhang, et al. (1988). *American Physics Test Questions and Solutions* (University of Science and Technology of China Press).

the same rotating angular velocity as the string] (for the block: the $\boldsymbol{k}$-direction component equation, for the ball: the $\boldsymbol{r}^0$-direction component equation):

$$\begin{cases} T_0 - Mg = 0 & (1a) \\ m\omega_0^2 r_0 - T_0 = 0 & (1b) \end{cases}$$

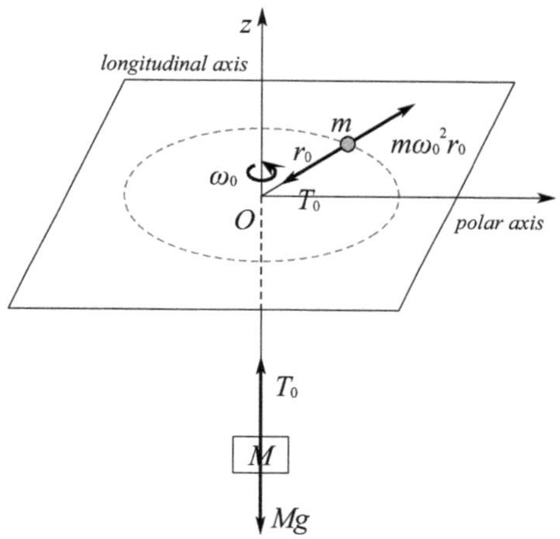

Figure 2

T_0—the magnitude of the tension in the string at Time zero.

r_0—the radius of the circular trajectory of the ball before it is disturbed.

$$\Rightarrow \qquad r_0 = \frac{Mg}{m\omega_0^2} \qquad (2)$$

Taking *the ball* as the object of study. Studying *the process from Time zero to arbitrary Time t*. Considering ***law of conservation of angular momentum*** [taking the pole *O* as the reference point, the resultant torque acting on the ball is zero (the ball is disturbed in the radial direction); therefore, its angular momentum is conserved] (the $\boldsymbol{k}$-direction component equation):

$$m\omega_0 r_0^2 = m\omega r^2 \qquad (3)$$

Defining Δr — the $\boldsymbol{r}^0$-direction component of the displacement of the ball at Time *t*.

$$\Rightarrow \qquad \omega = \frac{M^2 g^2}{m^2 \omega_0^3 r^2} = \frac{M^2 g^2}{m^2 \omega_0^3 (r_0 + \Delta r)^2} \qquad (4)$$

Taking *the block and the ball* as the object of study, respectively (see Fig. 3). Studying *the state at arbitrary Time t*. Considering ***Newton's second law*** [for the block: in the ground reference frame, for the ball: in the rotating reference frame with the hole as the center and with the same rotating angular velocity as the string] (for the block: the ***k***-direction component equation, for the ball: the $\boldsymbol{r}^0$-direction component equation):

$$\begin{cases} T - Mg = M\ddot{z}_M & \text{(5a)} \\ m\omega^2 r - T = m\ddot{r} & \text{(5b)} \end{cases}$$

T—the magnitude of the tension in the string at Time *t*.

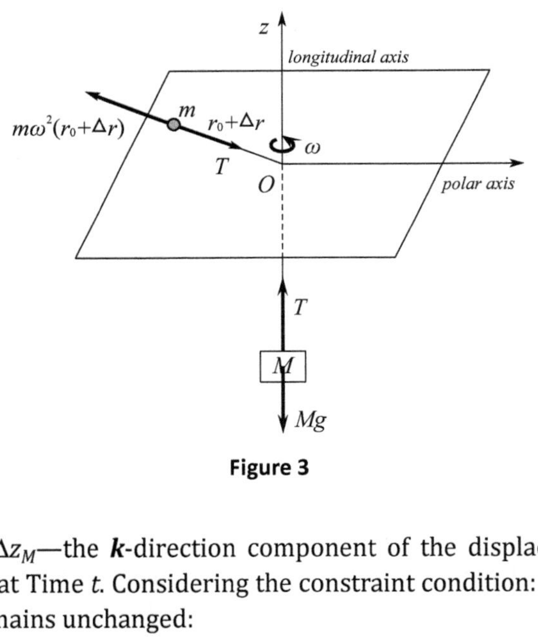

Figure 3

Defining Δz_M—the ***k***-direction component of the displacement of the block at Time *t*. Considering the constraint condition: the string length remains unchanged:

$$\Delta z_M = \Delta r \tag{6}$$

$$\Rightarrow \qquad \ddot{z}_M = \ddot{r} \tag{7}$$

Combining (2) to (7),

$$\Rightarrow \qquad \ddot{r} = \frac{M^4 g^4}{(M+m)m^3\omega_0^6 r^3} - \frac{M}{M+m}g$$

$$=\frac{M^4g^4}{(M+m)m^3\omega_0^6(r_0+\Delta r)^3}-\frac{M}{M+m}g$$

$$=\frac{M^4g^4}{(M+m)m^3\omega_0^6r_0^3\left(1+\dfrac{\Delta r}{r_0}\right)^3}-\frac{M}{M+m}g$$

$$\approx\frac{M^4g^4}{(M+m)m^3\omega_0^6r_0^3}\left(1-3\frac{\Delta r}{r_0}\right)-\frac{M}{M+m}g$$

$$=\frac{M}{M+m}g\left(1-3\frac{\Delta r}{r_0}\right)-\frac{M}{M+m}g$$

$$=-\frac{3M}{M+m}\frac{g}{r_0}\Delta r=-\frac{3m}{M+m}\omega_0^2\Delta r \tag{8}$$

This operation has taken the first-order approximation of Δr.

$$\Rightarrow \qquad \ddot{r}+\frac{3m}{M+m}\omega_0^2\Delta r=0 \tag{9}$$

$$\Rightarrow \qquad \ddot{z}+\frac{3m}{M+m}\omega_0^2\Delta z=0 \tag{10}$$

Equation (9) is the differential equation of the ball performing the tiny simple harmonic vibration in the radial direction, and Eq. (10) is the differential equation of the block performing the tiny simple harmonic vibration in the vertical direction, their equal angular frequency ω' is:

$$\Rightarrow \qquad \omega'=\sqrt{\frac{3m}{M+m}}\omega_0 \tag{11}$$

4.1.12 The Periodic Motion of a Small Ball on the Inner Wall of a Cone (*a*, *A*)

Consider an inverted hollow cone of vertex angle 2θ fixed in space, whose inner wall is smooth (see Fig. 1). A small ball of mass m performs the uniform circular motion on the inner wall of the cone, whose plane of motion is z_0 above the vertex of the cone. The ball is subjected to an instantaneous impact in the same direction as its velocity, causing its speed $\sqrt{1+\alpha}$ times of the original, $\alpha>0$.

(1) Find the velocity and acceleration of the ball when it moves to an arbitrary position.
(2) Find the normal contact force exerted on the ball by the inner wall of the cone when it moves to an arbitrary position.
(3) Find the highest and lowest position of the ball in its motion process after the impact.
(4) If α is a small quantity, please find the condition for the ball to perform periodic motion.

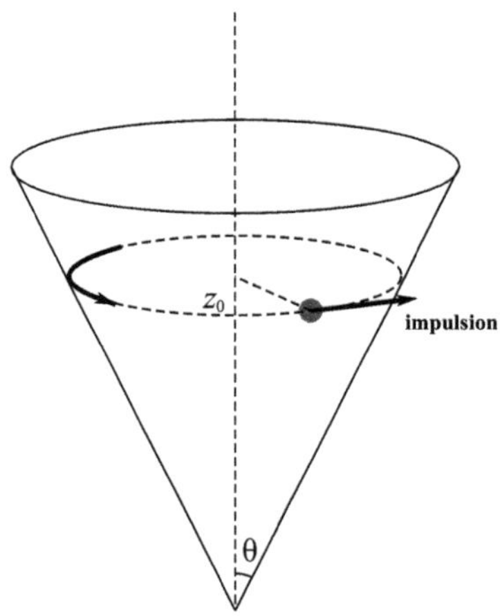

Figure 1

Solution to Problem (1)

As shown in Fig. 2, before the ball is impacted, the center of its circular trajectory is denoted by O, the position where the ball is impacted is denoted by A, the vertex of the cone is denoted by O', the projection of Point A on the horizontal plane where Point O' is located is denoted by A'.

The unit vector in vertical upward direction is denoted by $\boldsymbol{k}$. Taking Point O' as the pole, the direction of $\overrightarrow{O'A'}$ as the positive direction of the polar axis, the right-hand helix direction with regard to $\boldsymbol{k}$ as the positive direction of polar angle, the $\boldsymbol{k}$-direction as the positive direction of the longitudinal axis, establishing the cylindrical coordinate system, let r, φ, z be the cylindrical coordinates.

The time point before the ball is impacted is defined as Time zero.

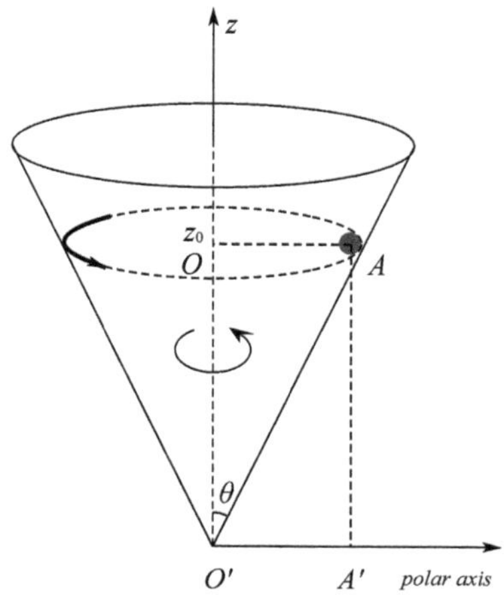

Figure 2

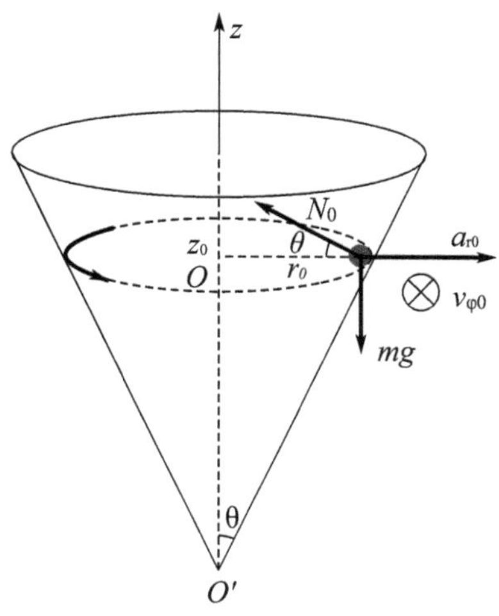

Figure 3

Taking *the ball* as the object of study (see Fig. 3). Studying *the state at Time zero*. Considering ***Newton's second law*** (the $\boldsymbol{r}^0$-direction and $\boldsymbol{k}$-direction component equations):

$$\begin{cases} -N_0\cos\theta = ma_{r0} & \text{(1a)} \\ N_0\sin\theta - mg = 0 & \text{(1b)} \end{cases}$$

N_0—the normal contact force exerted on the ball by the inner wall of the cone at Time zero.

a_{r0}—the $\boldsymbol{r}^0$-direction component of the ball's acceleration at Time zero.

Considering ***the radial component form of acceleration***:

$$a_{r0} = -\frac{v_{\varphi 0}^2}{r_0} \tag{2}$$

$v_{\varphi 0}$—the $\boldsymbol{\varphi}^0$-direction component of the ball's velocity at Time zero.

r_0—the radius of the circular trajectory of the ball before the impact.

Considering the following geometric relation in Fig. 3:

$$r_0 = z_0 \tan\theta \tag{3}$$

Combining (1), (2) and (3),

$$\Rightarrow \qquad v_{\varphi 0} = \sqrt{g z_0} \tag{4}$$

Taking *the ball* as the object of study (see Fig. 4). Studying *the process from the time point after the impact to arbitrary Time t*. Considering ***law of conservation of angular momentum*** [taking the z-axis as the rotation axis, the $\boldsymbol{k}$-direction component of the resultant torque acting on the ball is zero; therefore, the $\boldsymbol{k}$-direction component of its angular momentum is conserved] (the $\boldsymbol{k}$-direction component equation). Taking *the system composed of the ball and earth* as the object of study. Studying *the process from the time point after the impact to arbitrary Time t*. Considering ***law of conservation of mechanical energy*** [the work done by the external force and non-conservative internal force is zero (the normal contact force exerted on the ball by the inner wall of the cone does none work), therefore, the mechanical energy of the system is conserved] (taking Point O' as the zero point of gravitational potential energy):

$$\begin{cases} m\sqrt{1+\alpha}\, v_{\varphi 0} r_0 = m v_\varphi r & \text{(5a)} \\ \frac{1}{2} m(1+\alpha) v_{\varphi 0}^2 + mgz_0 = \frac{1}{2} m(v_r^2 + v_\varphi^2 + v_z^2) + mgz & \text{(5b)} \end{cases}$$

(v_r, v_φ, v_z)—the components of the ball's velocity in the $(\boldsymbol{r}^0, \boldsymbol{\varphi}^0, \boldsymbol{k})$-directions at Time t.

Considering the constraint condition: the ball is in contact with the inner wall of the cone all the time, it can be seen that the component of the ball's velocity in the normal direction of the inner wall of the cone is zero at arbitrarily Time t.

$$v_z \sin\theta - v_r \cos\theta = 0 \tag{6}$$

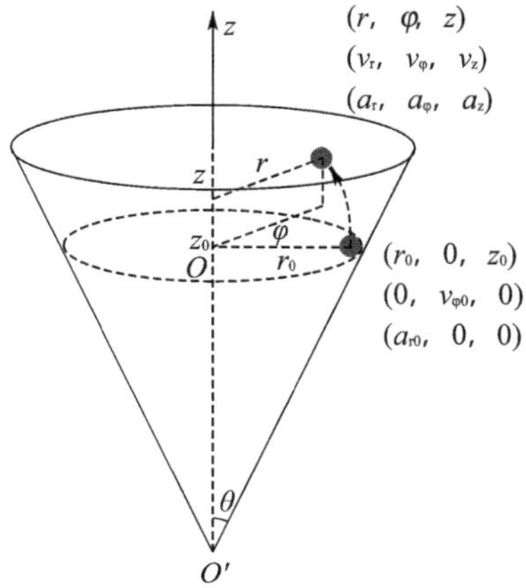

Figure 4

Considering the following geometric relation in Fig. 4:

$$r = z\tan\theta \tag{7}$$

Combining (3) to (7), we obtain the ball's velocity when it moves to an arbitrary position:

$$\Rightarrow \quad \begin{cases} v_z = \sqrt{gz_0\left[(1+\alpha)\left(1-\dfrac{z_0^2}{z^2}\right)+2\left(1-\dfrac{z}{z_0}\right)\right]}\cos\theta & \text{(8a)} \\ v_r = \sqrt{gz_0\left[(1+\alpha)\left(1-\dfrac{z_0^2}{z^2}\right)+2\left(1-\dfrac{z}{z_0}\right)\right]}\sin\theta & \text{(8b)} \\ v_\varphi = \sqrt{gz_0(1+\alpha)}\dfrac{z_0}{z} & \text{(8c)} \end{cases}$$

Considering ***the components form of the velocity and the acceleration of a particle in cylindrical coordinate system***:

$$\begin{cases} v_z = \dot{z} & \text{(9a)} \\ v_r = \dot{r} & \text{(9b)} \\ v_\varphi = r\dot{\varphi} & \text{(9c)} \end{cases}$$

$$\begin{cases} a_z = \ddot{z} & \text{(10a)} \\ a_r = \ddot{r} - r\dot{\varphi}^2 & \text{(10b)} \\ a_\varphi = r\ddot{\varphi} + 2\dot{r}\dot{\varphi} & \text{(10c)} \end{cases}$$

$(8) \to (9)$:

$$\Rightarrow \begin{cases} \dot{z} = \sqrt{gz_0\left[(1+\alpha)\left(1-\dfrac{z_0^2}{z^2}\right)+2\left(1-\dfrac{z}{z_0}\right)\right]}\cos\theta & (11a) \\ \dot{r} = \sqrt{gz_0\left[(1+\alpha)\left(1-\dfrac{z_0^2}{z^2}\right)+2\left(1-\dfrac{z}{z_0}\right)\right]}\sin\theta & (11b) \\ \dot{\varphi} = \sqrt{gz_0(1+\alpha)}\dfrac{z_0}{z^2\tan\theta} & (11c) \end{cases}$$

$$\Rightarrow \begin{cases} \ddot{z} = g\left[(1+\alpha)\dfrac{z_0^3}{z^3}-1\right]\cos^2\theta & (12a) \\ \ddot{r} = g\left[(1+\alpha)\dfrac{z_0^3}{z^3}-1\right]\sin\theta\cos\theta & (12b) \\ \ddot{\varphi} = -2g\sqrt{1+\alpha}\dfrac{z_0^2}{z^3}\sqrt{(1+\alpha)\left(1-\dfrac{z_0^2}{z^2}\right)+2\left(1-\dfrac{z}{z_0}\right)}\cos\theta\cot\theta & (12c) \end{cases}$$

(7), (11) and (12) $\to$ (10), we obtain the ball's acceleration when it moves to an arbitrary position:

$$\Rightarrow \begin{cases} a_z = g\left[(1+\alpha)\dfrac{z_0^3}{z^3}-1\right]\cos^2\theta & (13a) \\ a_r = -g\left[(1+\alpha)\dfrac{z_0^3}{z^3}\cos^2\theta+\sin^2\theta\right]\cot\theta & (13b) \\ a_\varphi = 0 & (13c) \end{cases}$$

Solution to Problem (2)

Taking *the ball* as the object of study (see Fig. 5). Studying *the state at arbitrarily Time t after the impact.* Considering ***Newton's second law*** (the component equations in the ($\boldsymbol{r}^0$, $\boldsymbol{\varphi}^0$, $\boldsymbol{k}$)-directions):

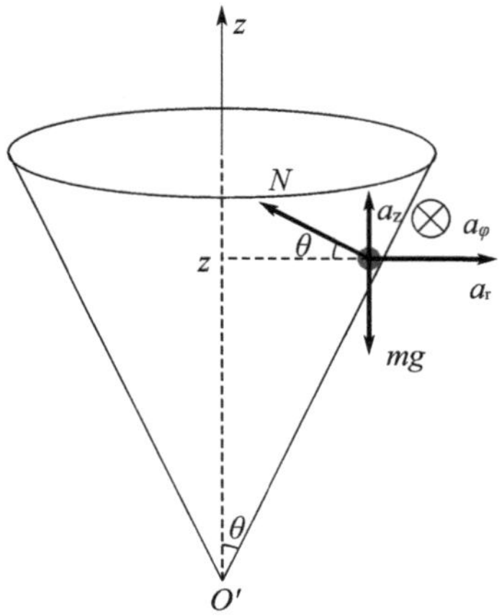

Figure 5

$$\begin{cases} N\sin\theta - mg = ma_z & (14a) \\ -N\cos\theta = ma_r & (14b) \\ 0 = ma_\varphi & (14c) \end{cases}$$

(13a) → (14a),

$$\Rightarrow \qquad N = \left[(1+\alpha)\frac{z_0^3}{z^3}\cos^2\theta + \sin^2\theta\right]\frac{mg}{\sin\theta} \tag{15}$$

Solution to Problem (3)

The necessary condition for the ball to move to the highest and lowest points is as follows:

$$\dot{z}\big|_{z=Z} = 0 \tag{16}$$

Z—the z-coordinate of the ball when it moves to the highest or lowest point.

(9a) → (16),

$$\Rightarrow \qquad v_z\big|_{z=Z} = 0 \tag{17}$$

(8a) → (17),

$$\Rightarrow \qquad (z_0 - Z)[2Z^2 - (1+\alpha)z_0 Z - (1+\alpha)z_0^2] = 0 \tag{18}$$

$$\Rightarrow \begin{cases} Z_1 = z_0 & (19a) \\ Z_2 = \frac{1}{4}[(1+\alpha) - \sqrt{(1+\alpha)^2 + 8(1+\alpha)}]z_0 & (19b) \\ Z_3 = \frac{1}{4}[(1+\alpha) + \sqrt{(1+\alpha)^2 + 8(1+\alpha)}]z_0 & (19c) \end{cases}$$

Known:

$$\alpha > 1 \quad (20)$$

$$\Rightarrow \begin{cases} Z_2 < 0 & (21) \\ Z_3 > Z_1 & (22) \end{cases}$$

Z_2 is an invalid solution.

$$\Rightarrow \begin{cases} Z_{\text{min}} = Z_1 = z_0 & (23a) \\ Z_{\text{max}} = Z_3 = \frac{1}{4}\left[(1+\alpha) + \sqrt{(1+\alpha)^2 + 8(1+\alpha)}\right]z_0 & (23b) \end{cases}$$

Solution to Problem (4)

Defining the variable Δz—the $\boldsymbol{k}$-direction component of the ball's displacement at arbitrary Time t after the impact, which is a small quantity.

From (13a),

$$a_z = g\left[(1+\alpha)\frac{z_0^3}{(z_0 + \Delta z)^3} - 1\right]\cos^2\theta$$

$$= g\left[(1+\alpha)\left(1 + \frac{\Delta z}{z_0}\right)^{-3} - 1\right]\cos^2\theta$$

$$\approx g\left[(1+\alpha)\left(1 - \frac{3\Delta z}{z_0}\right) - 1\right]\cos^2\theta$$

$$\approx g\left[\left(1 - \frac{3\Delta z}{z_0} + \alpha\right) - 1\right]\cos^2\theta$$

$$= g\left(-\frac{3\Delta z}{z_0} + \alpha\right)\cos^2\theta$$

$$\Rightarrow \qquad a_z + \frac{3g}{z_0}\Delta z = \alpha g \cos^2\theta \qquad (24)$$

This operation has considered Δz and α as small quantities and taken the first-order approximations.

Defining the variable $\Delta z'$:

$$\Delta z' = \Delta z - \frac{\alpha}{3} z_0 \cos^2\theta \qquad (25)$$

$$\Rightarrow \qquad a_z' + \frac{3g}{z_0}\Delta z' = 0 \qquad (26)$$

Equation (26) is the differential equation of the ball performing tiny simple harmonic vibration in the vertical direction, and its angular frequency ω^* is:

$$\omega^* = \sqrt{\frac{3g}{z_0}} \qquad (27)$$

The solution of Eq. (26) is as follows:

$$\Rightarrow \qquad \Delta z' = A\cos(\omega^* t + \delta) \qquad (28)$$

A and δ—the amplitude and initial phase of the simple harmonic vibration of the ball.

Considering the initial conditions:

$$\begin{cases} \Delta z\big|_{t=0} = 0 & (29a) \\ v_z\big|_{t=0} = \dot{\Delta z}\Big|_{t=0} = 0 & (29b) \end{cases}$$

(25) $\rightarrow$ (29),

$$\Rightarrow \qquad \begin{cases} \Delta z'\big|_{t=0} = -\frac{\alpha}{3} z_0 \cos^2\theta & (30a) \\ v_z'\big|_{t=0} = \dot{\Delta z'}\Big|_{t=0} = 0 & (30b) \end{cases}$$

$$\Rightarrow \qquad \begin{cases} A = \frac{\alpha}{3} z_0 \cos^2\theta & (31a) \\ \delta = \pi & (31b) \end{cases}$$

$$\Rightarrow \qquad \Delta z' = \frac{\alpha}{3} z_0 \cos^2\theta \cos(\omega^* t + \pi) = -\frac{\alpha}{3} z_0 \cos^2\theta \cos\omega^* t \qquad (32)$$

(32) → (25),

$$\Rightarrow \qquad \Delta z = \frac{\alpha}{3} z_0 \cos^2\theta(1-\cos\omega^* t) \tag{33}$$

Considering the following geometric relation in Fig. 4, then place (33) in to obtain:

$$\Rightarrow \quad r = (z_0 + \Delta z)\tan\theta = \left[1 + \frac{\alpha}{3}\cos^2\theta(1-\cos\omega^* t)\right] z_0 \tan\theta \tag{34}$$

It can be seen that the projection motion of the ball in the radial direction is also a simple harmonic vibration with the angular frequency ω^*.

Defining ω_0 and ω —the ***k***-direction component of the angular velocity of the ball rotating around the z axis at Time zero and arbitrary Time t. Considering ***the transverse component form of velocity*** (the equivalent form):

$$\begin{cases} \omega_0 = \dfrac{v_{\varphi 0}}{r_0} & \text{(35a)} \\ \omega = \dfrac{v_\varphi}{r} & \text{(35b)} \end{cases}$$

Combining (3), (5), (7) and (35),

$$\begin{aligned} \Rightarrow \qquad \omega &= \omega_0 \left(1 + \frac{\Delta z}{z_0}\right)^{-2} \sqrt{1+\alpha} \\ &\approx \omega_0 \left(1 - 2\frac{\Delta z}{z_0}\right)\left(1 + \frac{1}{2}\alpha\right) \\ &\approx \omega_0 \left(1 - 2\frac{\Delta z}{z_0} + \frac{1}{2}\alpha\right) \\ &= \omega_0 \left[1 + \left(\frac{1}{2} - \frac{2}{3}\cos^2\theta\right)\alpha + \frac{2}{3}\alpha\cos^2\theta\cos\omega^* t\right] \end{aligned} \tag{36}$$

This operation has considered Δz and α as small quantities and taken the first-order approximations.

Considering the definition of ***angular velocity***:

$$\omega = \frac{\mathrm{d}\varphi}{\mathrm{d}t} \tag{37}$$

$$\Rightarrow \quad \mathrm{d}\varphi = \omega_0 \left[1 + \left(\frac{1}{2} - \frac{2}{3}\cos^2\theta\right)\alpha + \frac{2}{3}\alpha\cos^2\theta\cos\omega^* t\right]\mathrm{d}t$$

$$\Rightarrow \quad \varphi = \int_0^{\varphi} \mathrm{d}\varphi = \omega_0 \int_0^t \left[1 + \left(\frac{1}{2} - \frac{2}{3}\cos^2\theta \right)\alpha + \frac{2}{3}\alpha\cos^2\theta\cos\omega^* t \right] \mathrm{d}t$$

$$= \omega_0 \left\{ \left[1 + \left(\frac{1}{2} - \frac{2}{3}\cos^2\theta \right)\alpha \right] t + \frac{2}{3\omega^*}\alpha\cos^2\theta\sin\omega^* t \right\} \tag{38}$$

Defining T—the motion period of the ball.

Supposing that the ball has the motion pattern that satisfies the following conditions:

$$\begin{cases} \Delta z_{t+T} - \Delta z_T = 0 & (39a) \\ \varphi_{t+T} - \varphi_T = N_\varphi \cdot 2\pi \qquad N_\varphi \in \mathbf{N}_+ & (39b) \end{cases}$$

(33) → (39a),

$$\Rightarrow \qquad T = N_{\Delta z} \cdot \frac{2\pi}{\omega^*} \qquad N_{\Delta z} \in \mathbf{N}_+ \tag{40}$$

That is, T is a positive integer times of the motion period of the ball's simple harmonic vibration in the vertical direction.

(38) → (39b),

$$\Rightarrow \omega_0 \left\{ \left[1 + \left(\frac{1}{2} - \frac{2}{3}\cos^2\theta \right)\alpha \right] T + \frac{2}{3\omega^*}\alpha\cos^2\theta \left[\sin\omega^*(t+T) - \sin\omega^* t \right] \right\}$$

$$= N_\varphi \cdot 2\pi \tag{41}$$

Equation (41) holds for arbitrarily Time t, which can be understood as an identical equation for t, and the necessary conditions for it to hold are as follows:

$$\Rightarrow \qquad \begin{cases} T = \dfrac{N_\varphi}{1 + \left(\dfrac{1}{2} - \dfrac{2}{3}\cos^2\theta \right)\alpha} \dfrac{2\pi}{\omega_0} & (42a) \\ T = n \cdot \dfrac{2\pi}{\omega^*} \qquad n \in \mathbf{N}_+ & (42b) \end{cases}$$

Combining (40) and (42b),

$$\Rightarrow \qquad n = N_{\Delta z} \tag{43}$$

Combining (42) and (43),

$$\Rightarrow \qquad \frac{N_\varphi}{N_{\Delta z}} = \frac{\omega_0}{\omega^*} \left[1 + \left(\frac{1}{2} - \frac{2}{3}\cos^2\theta \right)\alpha \right] \tag{44}$$

Combining (3), (4), (27), (35a) and (44),

$$\Rightarrow \qquad \frac{N_\varphi}{N_{\Delta z}} = \frac{1+\left(\frac{1}{2}-\frac{2}{3}\cos^2\theta\right)\alpha}{\sqrt{3}\tan\theta} \qquad (45)$$

When θ takes the values determined by all positive integer arrays ($N_{\Delta z}$, N_φ) according to the equation above, the ball can perform periodic motion after being impacted.

Figure 6 is the schematic diagram of the ball's trajectory in the special case of $\{N_{\Delta z} = 2, N_\varphi = 3, \theta \approx 21.05°\}$ [the z'-coordinate of Point 1 is zero. The path of the ball's motion is $1 \to 2 \to 3 \to 4 \to 5 \to 6 \to 7 \to 8 \to 1$].

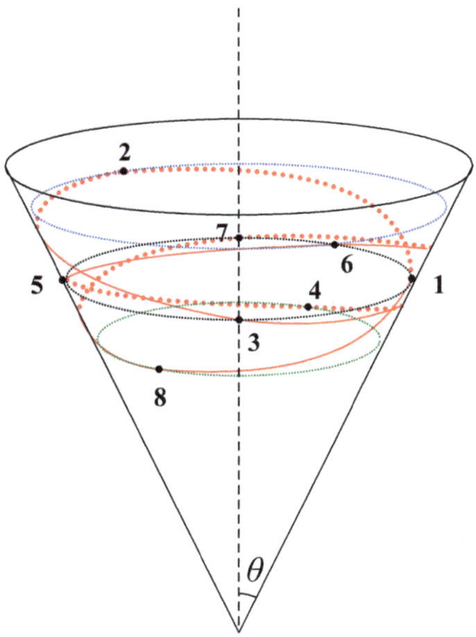

Figure 6

4.1.13 The Tiny Rotational Vibration of a Homogeneous Rotational Ellipsoid on a Horizontal Plane (*a*, *A*)

A homogeneous rotational ellipsoid is statically placed on a horizontal ground (see Fig. 1). In the Cartesian coordinate system with the frame (O, $\boldsymbol{i}$, $\boldsymbol{j}$, $\boldsymbol{k}$) provided in the figure, the surface equation of the ellipsoid is: $\frac{x^2}{a^2}+\frac{y^2}{b^2}+\frac{z^2}{b^2}=1$, $a > b$. The friction coefficient between the ellipsoid and the ground is large enough. Find the angular frequency of the tiny rotational vibration of the ellipsoid in the O-xy plane.

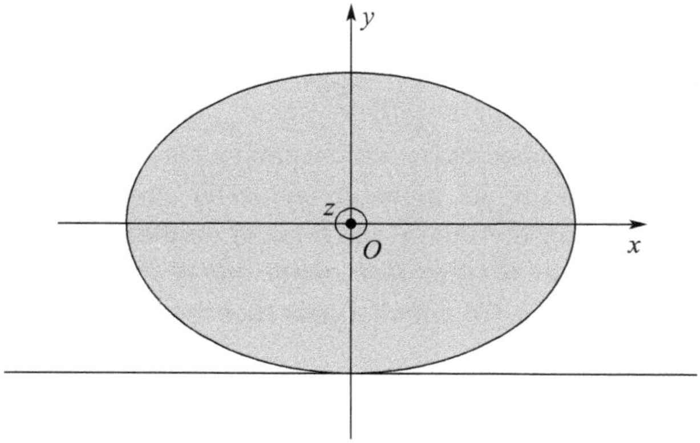

Figure 1

Solution [1]

In the process of the ellipsoid's rotating, once its major axis is in the horizontal direction is defined as Time zero.

As shown in Fig. 2, the center of mass of the ellipsoid is denoted by C, the points on the ellipsoid in contact with the ground at Time zero and arbitrary Time t are denoted by P_0 and P, respectively.

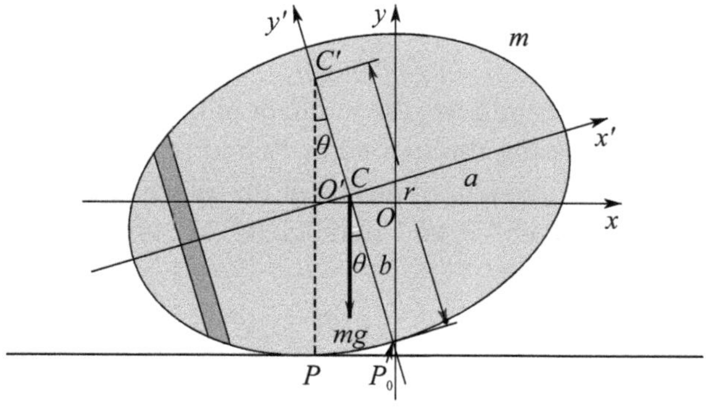

Figure 2

Establishing the body coordinate system of the ellipsoid with the frame $(O, \boldsymbol{i}', \boldsymbol{j}', \boldsymbol{k}')$, let x', y', z' be the Cartesian coordinates, which coincides with the space coordinate system with the frame $(O, \boldsymbol{i}, \boldsymbol{j}, \boldsymbol{k})$ at Time zero.

Defining m—the mass of the ellipsoid, ρ—the mass density of the ellipsoid.

$$\rho = \frac{m}{\frac{4}{3}\pi ab^2} \tag{1}$$

Taking the disk-shaped mass infinitesimal element of the ellipsoid between x and $x+dx$, its moment of inertia about the z'-axis is defined as dI_O. Considering (a) ***the form of moment of inertia of a homogeneous thin circular disk about one of its central axes in the plane of the disk***, (b) ***parallel-axis theorem***:

$$\mathrm{d}I_O = \frac{1}{4}\rho\cdot\pi y'^2\mathrm{d}x'\cdot y'^2 + \rho\cdot\pi y'^2\mathrm{d}x'\cdot x'^2 \tag{2}$$

Defining I_O—the moment of inertia of the ellipsoid about the z'-axis.

$$\Rightarrow \qquad I_O = \int \mathrm{d}I_O = \int \frac{1}{4}\rho\cdot\pi y'^2\mathrm{d}x'\cdot y'^2 + \rho\cdot\pi y'^2\mathrm{d}x'\cdot x'^2 \tag{3}$$

Known:

$$\frac{x'^2}{a^2} + \frac{y'^2}{b^2} = 1 \tag{4}$$

$$\Rightarrow \qquad I_O = \frac{1}{4}\pi\rho\int_{-a}^{a}\left[\left(b^2 - \frac{b^2}{a^2}x'^2\right)^2 + 4\left(b^2 - \frac{b^2}{a^2}x'^2\right)x'^2\right]\mathrm{d}x'$$

$$= \frac{4}{15}\pi\rho(ab^4 + a^3b^2) = \frac{1}{5}m(a^2 + b^2) \tag{5}$$

Defining I_{P_0} — at Time zero, the moment of inertia of the ellipsoid about the axis passing through Point P_0 and parallel to the z-axis, I_P—at Time t, the moment of inertia of the ellipsoid about the axis passing through Point P and parallel to the z-axis. Considering ***parallel-axis theorem***:

$$\Rightarrow \qquad I_P \approx I_{P_0} = I_O + mb^2 = \frac{1}{5}m(a^2 + 6b^2) \tag{6}$$

Defining r—the radius of curvature of the ellipse $\frac{x'^2}{a^2} + \frac{y'^2}{b^2} = 1$ at the vertex of its minor axis. The corresponding center of curvature is denoted by C'. Considering the form of the radius of curvature of an ellipse at its vertex:

$$r = \frac{a^2}{b} \tag{7}$$

Defining M_P—the $\boldsymbol{k}$-direction component of the resultant external torque acting on the ellipsoid at arbitrary Time t [taking the axis passing through Point P and parallel to the z-axis as the rotation axis]. Considering the definition of ***torque***:

$$\Rightarrow \quad M_P = -mg(P_0C' - P_0C)\sin\theta = -mg(r-b)\sin\theta$$

$$= -mg\frac{c^2}{b}\sin\theta \approx -mg\frac{c^2}{b}\theta \tag{8}$$

Taking *the ellipsoid* as the object of study. Studying *the state at arbitrary Time t*. Considering ***law of rotation*** [taking the axis passing through Point P and parallel to the z-axis as the rotation axis] (the $\boldsymbol{k}$-direction component equation):

$$M_P = I_P\ddot{\theta} \tag{9}$$

$$\Rightarrow \quad \ddot{\theta} + \frac{mg}{I_P}\frac{c^2}{b}\theta = 0 \tag{10}$$

Equation (10) is the differential equation of the ellipsoid performing the tiny rotational simple harmonic vibration in the O-xy plane, and its angular frequency ω is as follows, then place (6) in to obtain:

$$\Rightarrow \quad \omega = \sqrt{\frac{mg}{I_P}\frac{c^2}{b}} = \sqrt{\frac{5gc^2}{(a^2+6b^2)b}} \tag{11}$$

Solution [2]

Considering the following geometric relations in Fig. 3:

$$\begin{cases} x_C = -[PP_0(1-\cos\theta) + P_0C\sin\theta] \\ \quad \approx -[r\theta(1-\cos\theta) + b\sin\theta] \approx -b\theta & \text{(12a)} \\ y_C = PP_0\sin\theta + P_0C\cos\theta - b \approx r\theta\sin\theta + b\cos\theta - b \approx 0 & \text{(12b)} \end{cases}$$

$$\Rightarrow \quad \begin{cases} \ddot{x}_C = -\ddot{\theta}b & \text{(13a)} \\ \ddot{y}_C = 0 & \text{(13b)} \end{cases}$$

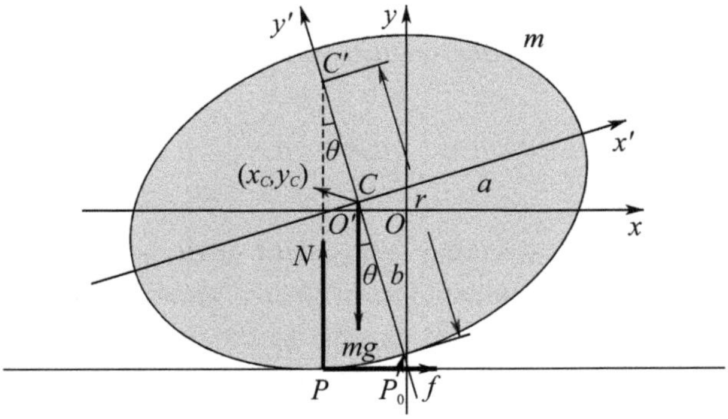

Figure 3

Taking *the ellipsoid* as the object of study. Studying *the state at arbitrary Time t*. Considering ***theorem of motion of center of mass*** (the ***i***-direction and ***j***-direction component equations):

$$\begin{cases} f = m\ddot{x}_C & (14a) \\ N - mg = m\ddot{y}_C & (14b) \end{cases}$$

f—the ***i***-direction component of the static friction force exerted on the ellipsoid by the ground at Time *t*.

N—the ***j***-direction component of the normal contact force exerted on the ellipsoid by the ground at Time *t*.

$$\Rightarrow \quad \begin{cases} N = mg & (15a) \\ f = -m\ddot{\theta}b & (15b) \end{cases}$$

Defining M_C—the ***k***-direction component of the resultant external torque acting on the ellipsoid at arbitrary Time *t* [taking the z'-axis as the rotation axis]. Considering the definition of ***torque***:

$$\begin{aligned} M_C &= -N(P_0C' - P_0C)\sin\theta + f(PP_0\sin\theta + P_0C\cos\theta) \\ &\approx -N(r-b)\sin\theta + f(r\theta\sin\theta + b\cos\theta) \\ &= -mg(r-b)\sin\theta - m\ddot{\theta}b(r\theta\sin\theta + b\cos\theta) \\ &\approx -mg\frac{c^2}{b}\theta - mb^2\ddot{\theta} \qquad (16) \end{aligned}$$

Taking *the ellipsoid* as the object of study. Studying *the state at arbitrary Time t*. Considering (a) ***law of rotation*** [in its center-of-mass reference frame, taking the z'-axis as the rotation axis] (the **k**-direction component equation), (b) ***in the center-of-mass reference frame, the resultant torque of the inertia force acting on system of particles is zero***:

$$M_C = I_O \ddot{\theta} \tag{17}$$

$$\Rightarrow \quad \ddot{\theta} + \frac{mg}{I_O + mb^2} \frac{c^2}{b} \theta = 0 \tag{18}$$

Equation (18) is the differential equation of the ellipsoid performing the tiny rotational simple harmonic vibration in the O-xy plane, and its angular frequency ω is as follows, then place (5) in to obtain:

$$\Rightarrow \quad \omega = \sqrt{\frac{mg}{I_O + mb^2} \frac{c^2}{b}} = \sqrt{\frac{5g}{(a^2 + 6b^2)} \frac{c^2}{b}} \tag{19}$$

Operations (8), (12) and (16) have taken the first-order approximation of θ.

4.1.14 The Tiny Rotational Vibration of a Dumbbell-Shaped Artificial Satellite (d, *C*)

Consider a dumbbell-shaped artificial satellite, which can be regarded as composed of two homogeneous spheres of equal mass m and equal radius a which connected by a light rigid rod of length l. The satellite moves in a circular orbit of radius R, and it can freely rotate around its center of mass in the orbit plane. The mass of the earth is denoted by M. Find the angular frequency of the satellite performing tiny rotational vibration.

Solution

The rotating reference frame with the geocenter as the center and with the satellite's revolution angular velocity as the rotating angular velocity is denoted by S'-system. As shown in Fig. 1, the geocenter is denoted by O, and the center of mass of the satellite is denoted by C. Establishing the polar coordinate system 1 with the geocenter O as the pole and with an arbitrary radial direction of the satellite's orbit plane as the positive direction of the polar axis (let r, θ be the polar coordinates), and the polar coordinate system 2 fixed to S'-system with the center of mass C as the pole and with the direction of $\overrightarrow{OC}$ as the positive direction of the polar axis (let ρ, φ be the polar coordinates).

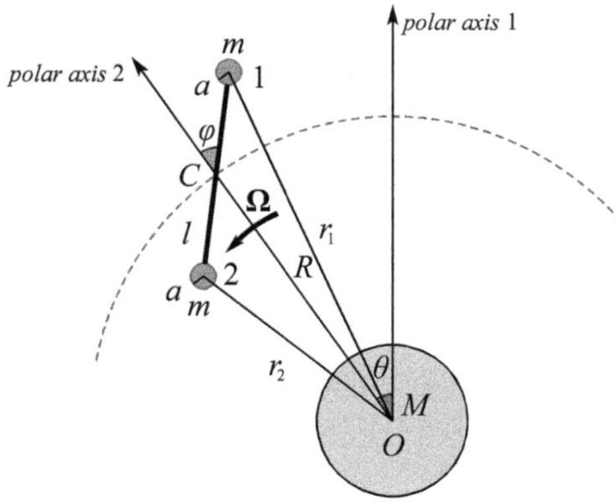

Figure 1

Taking *the satellite* as the object of study. Studying *the state at arbitrary Time t*. Considering ***theorem of motion of center of mass*** (the $-\boldsymbol{r}^0$-direction component equation):

$$G\frac{M\cdot 2m}{R^2}=2m\Omega^2 R \tag{1}$$

Ω—the angular velocity magnitude of the center of mass C performing the uniform circular motion around the earth.

$$\Rightarrow \qquad \Omega=\sqrt{\frac{GM}{R^3}} \tag{2}$$

For the system composed of the earth and the satellite at arbitrary Time t, defining E_G—the universal gravitational potential energy [taking the configuration in which the earth and the satellite are infinitely apart as potential energy zero], E_Ω—the centrifugal potential in S'-system [taking the geocenter as potential energy zero], and E_p—the total potential energy in S'-system:

$$E_p=E_G+E_\Omega \tag{3}$$

Considering ***theorem of universal gravitational potential energy***:

$$E_G=-G\frac{Mm}{r_1}-G\frac{Mm}{r_2} \tag{4}$$

r_1 and r_2—the polar radius of Ball 1 and Ball 2 in the polar coordinate system 1 at arbitrary Time t.

Considering (a) the definition of ***potential energy***, (b) ***theorem of inertial centrifugal force*** [integrating into (a)]:

$$\Rightarrow \qquad E_\Omega = \int_{r_1}^{0} m\Omega^2 r dr + \int_{r_2}^{0} m\Omega^2 r dr = -\frac{1}{2} m\Omega^2 (r_1^2 + r_2^2)$$

$$= -\frac{GMm}{2R^3}(r_1^2 + r_2^2) \tag{5}$$

Considering the following geometric relation in Fig. 1:

$$r_1 = \sqrt{R^2 + Rl\cos\varphi + \left(\frac{l}{2}\right)^2} = R\sqrt{1 + \frac{l}{R}\cos\varphi + \frac{l^2}{4R^2}}$$

$$\approx R\left[1 + \frac{1}{2}\left(\frac{l}{R}\cos\varphi + \frac{l^2}{4R^2}\right) + \frac{\frac{1}{2}\left(-\frac{1}{2}\right)}{2!}\left(\frac{l}{R}\cos\varphi + \frac{l^2}{4R^2}\right)^2\right]$$

$$\approx R\left(1 + \frac{l}{2R}\cos\varphi + \frac{l^2}{8R^2} - \frac{l^2}{8R^2}\cos^2\varphi\right)$$

$$= R\left(1 + \frac{l}{2R}\cos\varphi + \frac{l^2}{8R^2}\sin^2\varphi\right) \tag{6a}$$

In the same way we obtain:

$$\Rightarrow \qquad r_2 \approx R\left(1 - \frac{l}{2R}\cos\varphi + \frac{l^2}{8R^2}\sin^2\varphi\right) \tag{6b}$$

(6) → (4),

$$\Rightarrow \qquad E_G = -G\frac{Mm}{R}\left[\begin{array}{l}\left(1 + \frac{l}{2R}\cos\varphi + \frac{l^2}{8R^2}\sin^2\varphi\right)^{-1} \\ + \left(1 - \frac{l}{2R}\cos\varphi + \frac{l^2}{8R^2}\sin^2\varphi\right)^{-1}\end{array}\right]$$

$$\approx -G\frac{Mm}{R}\left(\begin{array}{l}1 - \frac{l}{2R}\cos\varphi - \frac{l^2}{8R^2}\sin^2\varphi + \frac{l^2}{4R^2}\cos^2\varphi \\ +1 + \frac{l}{2R}\cos\varphi - \frac{l^2}{8R^2}\sin^2\varphi + \frac{l^2}{4R^2}\cos^2\varphi\end{array}\right)$$

$$= -G\frac{Mm}{R}\left(2 - \frac{l^2}{4R^2}\sin^2\varphi + \frac{2l^2}{4R^2}\cos^2\varphi\right)$$

$$= -G\frac{Mm}{R}\left(2-\frac{l^2}{4R^2}+\frac{3l^2}{4R^2}\cos^2\varphi\right) \tag{7}$$

(6) → (5),

$$\Rightarrow \qquad E_\Omega = -G\frac{Mm}{R}\left(1+\frac{l^2}{4R^2}\right) \tag{8}$$

$$\Rightarrow \qquad E_p = -G\frac{3Mm}{R}\left(1+\frac{l^2}{4R^2}\cos^2\varphi\right) \tag{9}$$

Considering ***principle of minimum potential energy***:

$$\begin{cases} \left.\dfrac{\mathrm{d}E_p}{\mathrm{d}\varphi}\right|_{\varphi_0} = 0 & (10\text{a}) \\ \left.\dfrac{\mathrm{d}^2E_p}{\mathrm{d}\varphi^2}\right|_{\varphi_0} > 0 & (10\text{b}) \end{cases}$$

φ_0—the polar angle of Ball 1 in the polar coordinate system 2 when the satellite is in stable equilibrium in S'-system.

(9) → (10a),

$$\Rightarrow \qquad \varphi_0 = 0, \frac{\pi}{2}, \pi, \frac{3\pi}{2} \tag{11}$$

$$\Rightarrow \qquad \frac{\mathrm{d}^2E_p}{\mathrm{d}\varphi^2} = \begin{cases} >0 & \varphi_0 = 0, \pi & (12\text{a}) \\ <0 & \varphi_0 = \dfrac{\pi}{2}, \dfrac{3\pi}{2} & (12\text{b}) \end{cases}$$

It can be seen that the stable equilibrium position of the satellite is $\varphi_0 = 0, \pi$.

Defining E_k—the kinetic energy of the satellite in S'-system at arbitrary Time t. Considering ***the form of kinetic energy of a rigid body performing fixed-axis rotation***:

$$E_k = \frac{1}{2}I\dot{\varphi}^2 \tag{13}$$

I—the moment of inertia of the satellite about its central axis perpendicular to the plane of the figure.

Considering (a) ***parallel-axis theorem***, (b) ***the form of moment of inertia of a homogeneous sphere about one of its central axes*** [integrating into (a)]:

$$I = 2\left[\frac{2}{5}ma^2 + m\left(\frac{l}{2}\right)^2\right] \tag{14}$$

$$\Rightarrow \quad E_k = m\left(\frac{2}{5}a^2 + \frac{1}{4}l^2\right)\dot{\varphi}^2 \tag{15}$$

Defining E—the kinetic energy of the system composed of the satellite and the earth in S'-system at arbitrary Time t.

$$\Rightarrow \quad E = E_k + E_p = m\left(\frac{2}{5}a^2 + \frac{1}{4}l^2\right)\dot{\varphi}^2 - G\frac{3Mm}{R}\left(1 + \frac{l^2}{4R^2}\cos^2\varphi\right) \tag{16}$$

Taking *the system composed of the earth and the satellite in S′-system* as the object of study. Studying *an arbitrary instantaneous process.* Considering ***law of conservation of mechanical energy*** [the work done by the external force and non-conservative internal force acting on the system is zero; therefore, the mechanical energy of the system is conserved]:

$$\frac{\mathrm{d}E}{\mathrm{d}\varphi} = 0 \tag{17}$$

$$\Rightarrow \quad \dot{\varphi}\left[\left(\frac{4}{5}a^2 + \frac{1}{2}l^2\right)\ddot{\varphi} + G\frac{3Ml^2}{4R^3}\sin 2\varphi\right] = 0 \tag{18}$$

$$\Rightarrow \quad \left(\frac{4}{5}a^2 + \frac{1}{2}l^2\right)\ddot{\varphi} + G\frac{3Ml^2}{4R^3}\sin 2\varphi = 0 \tag{19}$$

φ is a small quantity, taking the first-order approximation of $\sin 2\varphi$:

$$\sin 2\varphi \approx 2\varphi \tag{20}$$

$$\Rightarrow \quad \left(\frac{4}{5}a^2 + \frac{1}{2}l^2\right)\ddot{\varphi} + G\frac{3Ml^2}{2R^3}\varphi = 0 \tag{21}$$

Equation (21) is the differential equation of the satellite performing the tiny rotational vibration near the mechanical equilibrium position in S'-system, and its angular frequency ω is:

$$\Rightarrow \quad \omega = \sqrt{\frac{G\dfrac{3Ml^2}{R^3}}{\dfrac{8}{5}a^2 + l^2}} \tag{22}$$

Combining (2) and (22) we can further obtain the relation between ω and Ω:

$$\Rightarrow \qquad \omega = \Omega\sqrt{\frac{3l^2}{\frac{8}{5}a^2 + l^2}} \approx \sqrt{3}\Omega \qquad (23)$$

Operations (6a) and (7) has considered $l << R$ and taken the second-order approximations.

4.1.15 The Pure Rolling Vibration of a Homogeneous Cylinder on the Inner Wall of a Freely Rotating Homogeneous Cylindrical shell (*c*, *C*)[3]

A homogeneous cylindrical shell of mass M and sectional radius R can rotate freely around its central axis fixed in the horizontal direction (see Fig. 1). Inside, there is a homogeneous cylinder of mass m and sectional radius r ($<R$) placed in parallel with its central axis. The friction coefficient between the cylindrical shell and the cylinder is large enough to prevent the relative sliding. Initially, the cylinder is located at the bottom of the cylindrical shell, both are static. Find the angular frequency of the tiny vibration of the system under a small disturbance.

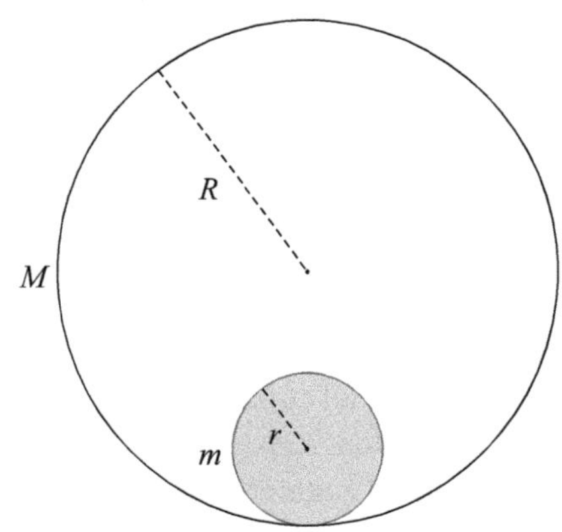

Figure 1

Solution

The time point when the system is disturbed is defined as Time zero.

[3]This problem is a test question of the 10th AphO (Asian Physics Olympiads) in 2009.

As shown in Fig. 2, the center of mass of the cylinder is denoted by O_1; the projection of Point O_1 on the central axis of the cylindrical shell is denoted by O_2; taking the line on the cylinder in contact with the inner wall of the cylindrical shell at Time zero, denoting its intersection point with O_1O_2 as P_{01}; taking the line on the inner wall of the cylindrical shell in contact with the cylinder at Time zero, denoting its intersection point with O_1O_2 as P_{02}; taking the line on the inner wall of the cylindrical shell in contact with the cylinder at arbitrary Time t, denoting its intersection point with O_1O_2 as P.

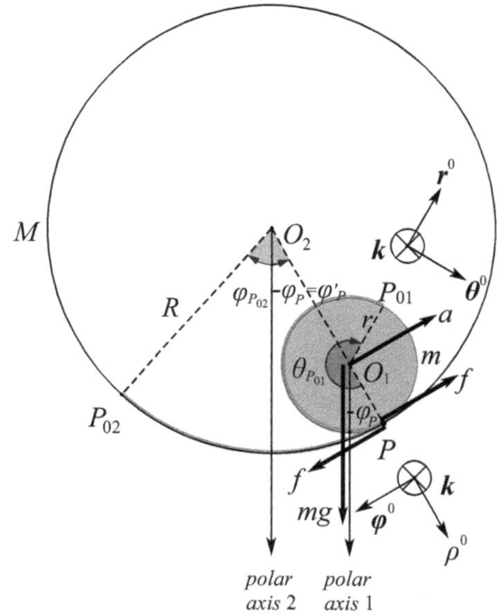

Figure 2

The unit vector perpendicular to the plane of the figure inward is denoted by $\boldsymbol{k}$. Establishing the cylindrical coordinate system 1 fixed to the cylinder's center-of-mass reference frame with Point O_1 as the pole, with the vertical downward direction as the positive direction of the polar axis and with the $\boldsymbol{k}$-direction as the positive direction of the longitudinal axis (let r, θ, z be the cylindrical coordinates), and the cylindrical coordinate system 2 with Point O_2 as the pole, with the vertical downward direction as the positive direction of the polar axis and with the $\boldsymbol{k}$-direction as the positive direction of the longitudinal axis (let ρ, φ, z be the cylindrical coordinates).

Considering the constraint condition: there is no relative sliding between the cylinder and cylindrical shell:

$$\widehat{P_{01}P} = \widehat{P_{02}P} \tag{1}$$

$$\Rightarrow \qquad [\theta_{P_{01}} + (-\varphi_P)]r = [(-\varphi_P) + \varphi_{P_{02}}]R \tag{2}$$

Defining φ'_P:

$$\Rightarrow \qquad \varphi'_P = -\varphi_P \tag{3}$$

$$\Rightarrow \qquad \theta_{P_{01}} = \frac{R-r}{r}\varphi'_P + \frac{R}{r}\varphi_{P_{02}} \tag{4}$$

Taking *the cylindrical shell* as the object of study. Studying *the state at arbitrary Time t.* Considering (a) ***law of rotation*** [taking its central axis as the rotation axis] (the ***k***-direction component equation), (b) ***the form of moment of inertia of a homogeneous cylindrical shell about its central axis***:

$$fR = I_M \ddot{\varphi}_{P_{02}} \tag{5}$$

$$I_M = MR^2 \tag{6}$$

f—the $\boldsymbol{\varphi}^0$-direction component of the friction exerted on the cylindrical shell by the cylinder at Time t.

I_M—the moment of inertia of the cylindrical shell about its central axis.

Taking *the cylinder* as the object of study. Studying *the state at arbitrary Time t.* Considering (a) ***theorem of motion of center of mass*** (the $-\boldsymbol{\varphi}^0$-direction component equation) (b) ***law of rotation*** [in its center-of-mass reference frame, taking its central axis as the rotation axis] (the ***k***-direction component equation), (c) ***in the center-of-mass reference frame, the resultant torque of the inertia force acting on system of particles is zero***, (d) ***the form of moment of inertia of a homogeneous cylinder about its central axis***:

$$f - mg\sin\varphi'_P = ma \tag{7}$$

$$-fr = I_m \ddot{\theta}_{P_{01}} \tag{8}$$

$$I_m = \frac{1}{2}mr^2 \tag{9}$$

a—the $-\boldsymbol{\varphi}^0$-direction component of the center-of-mass acceleration of the cylinder at Time t.

I_m—the moment of inertia of the cylinder about its central axis.

Considering ***the transverse component form of acceleration***:

$$a = \ddot{\varphi}'_P (R - r) \tag{10}$$

From (4),

$$\Rightarrow \qquad \ddot{\theta}_{P_{01}} = \frac{R-r}{r} \ddot{\varphi}'_P + \frac{R}{r} \ddot{\varphi}_{P_{02}} \tag{11}$$

Combining (5), (6), (8), (9) and (11),

$$\Rightarrow \qquad \begin{cases} f = -\dfrac{Mm}{2M+m}(R-r)\ddot{\varphi}'_P & (12) \\ \ddot{\beta} = -\dfrac{m}{2M+m}\dfrac{R-r}{R}\ddot{\varphi}'_P & (13) \end{cases}$$

(10) and (12) $\rightarrow$ (7),

$$\Rightarrow \qquad \ddot{\varphi}'_P + \frac{2M+m}{3M+m}\frac{g}{R-r}\sin\varphi'_P = 0 \tag{14}$$

φ'_P is a small quantity, taking the first-order approximation of $\sin\varphi'_P$:

$$\sin\varphi'_P \approx \varphi'_P \tag{15}$$

$$\Rightarrow \qquad \ddot{\varphi}'_P + \frac{2M+m}{3M+m}\frac{g}{R-r}\varphi'_P = 0 \tag{16}$$

Equation (16) is the differential equation of the system performing the tiny rotational simple harmonic vibration, and its angular frequency ω is:

$$\Rightarrow \qquad \omega = \sqrt{\frac{2M+m}{3M+m}\frac{g}{R-r}} \tag{17}$$

4.1.16 The Tiny Translational and Rotational Vibrations of an Electrified Ring above an Infinite Superconducting Plane (*c*, *B*)[4]

An infinite superconducting plane lies in a horizontal plane (see Fig. 1), a homogeneous conductor ring of mass m and radius r can reach mechanical equilibrium in the plane h above the

[4]This problem is a test question from Princeton University in 1977. It is also included in the following book: Yongde Zhang, et al. (1988). *American Physics Test Questions and Solutions* (University of Science and Technology of China Press).

superconducting plane, $h << r$, when a constant current is electrified into it.

(1) Find the current intensity in the ring.

(2) Find the angular frequency of the tiny vibration of the ring in the vertical direction near its mechanical equilibrium position.

(3) Find the angular frequency of the tiny rotational vibration of the ring around one of its center axes in the horizontal direction.

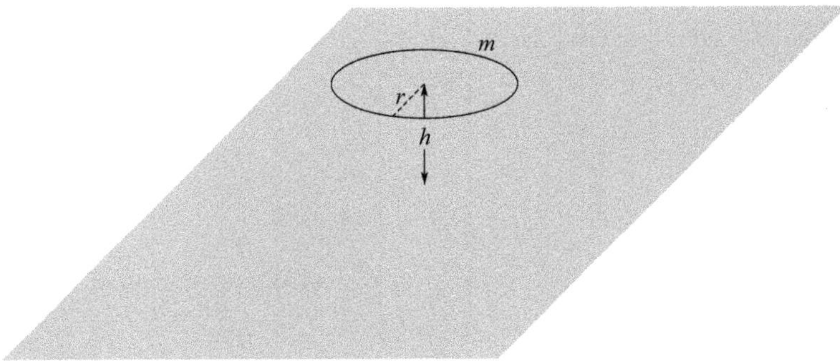

Figure 1

Solution to Problem (1):

As shown in Fig. 2, the center of the ring is denoted by O, the projection of Point O on the superconducting plane is denoted by O', establishing the Cartesian coordinate system with the frame (O', $\boldsymbol{i}$, $\boldsymbol{j}$, $\boldsymbol{k}$), let x, y, z be the Cartesian coordinates, and establishing the polar coordinate system fixed to the ring with the center O as the pole, with the $\boldsymbol{i}$-direction as the positive direction of the polar axis, and with the right-hand helix direction with regard to the $\boldsymbol{k}$-direction as the positive direction of polar angle, let r, φ be the polar coordinates.

The following studies the situation that the current direction in the ring is the right-hand helix direction with regard to the $\boldsymbol{k}$-direction.

Taking *the current in the ring, the induced current in the superconducting plane, and the magnetic field in space* as the object of study. Studying *the state when the system reaches magnetostatic equilibrium*. Considering (a) ***Biot–Savart law***, (b) ***boundary***

condition of magnetic induction intensity. It can be seen that the magnetic field generated by the induced current in the superconducting plane in the $z > 0$ region can be equivalent to the magnetic field generated by the image current of the current in the ring relative to the superconducting plane (i.e., the current in the image ring of the real ring, whose current intensity is equal to that in the real ring, and the direction is opposite).

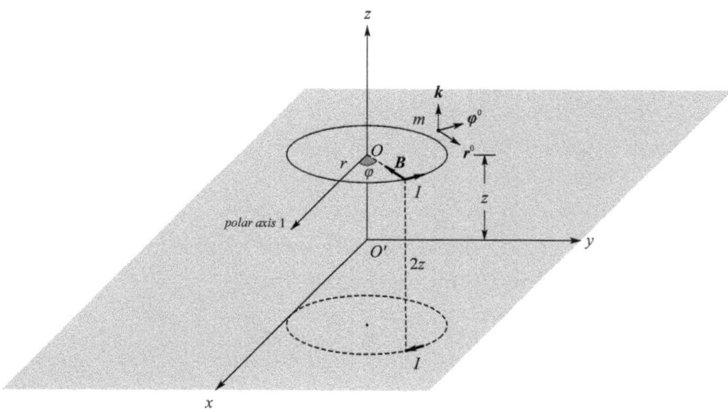

Figure 2

Because $h << r$, for an arbitrary point on the real ring, the image current can be regarded as distributed in the infinite straight conductor wire.

Taking *the image current and its magnetic field at an arbitrary point on the real ring* as the object of study. Studying *the state when the system reaches magnetostatic equilibrium (the ring is very close to the superconducting plane)*. Considering ***the form of magnetic the induction intensity excited by the current in the infinite straight conductor wire at an arbitrary point in space***:

$$\boldsymbol{B} = -\frac{\mu_0 I}{2\pi \cdot 2z}\boldsymbol{r}^0 \tag{1}$$

B—the magnetic induction intensity generated by the image current at an arbitrary point on the real ring when the system reaches magnetostatic equilibrium.

Taking *the real ring* as the object of study. Studying *the state when it is in mechanical equilibrium*. Considering ***theorem of motion of center of mass*** (the ***k***-direction component equation):

$$\oint_L (Id\boldsymbol{l} \times \boldsymbol{B}|_{z=h}) \cdot \boldsymbol{k} - mg = 0 \tag{2}$$

$$\Rightarrow 0 = \int_L (-B_r)|_{z=h} Idl - mg = (-B_r)|_{z=h} I \cdot 2\pi r - mg = \frac{\mu_0 I^2 r}{2h} - mg \tag{3}$$

I—the current intensity in the ring when it is in mechanical equilibrium.

$$\Rightarrow \qquad I = \sqrt{\frac{2mgh}{\mu_0 r}} \ll \sqrt{\frac{mg}{\mu_0}} \tag{4}$$

Solution to Problem (2)

The time point when the ring is disturbed in the vertical direction is defined as Time (1) zero.

Taking *the real ring* as the object of study. Studying *the state at arbitrary Time (1) t*. Considering ***theorem of motion of center of mass*** (the $\boldsymbol{k}$-direction component equation):

$$\oint_L (Id\boldsymbol{l} \times \boldsymbol{B}) \cdot \boldsymbol{k} - mg = m\ddot{z} \tag{5}$$

$$\Rightarrow \qquad \ddot{z} = \frac{1}{m}\oint_L (-B_r)|_{z=h+z'} Idl - g = \frac{1}{m}(-B_r)|_{z=h+z'} I \cdot 2\pi r - g$$

$$= \frac{1}{m}\frac{\mu_0 I^2 r}{2(h+z')} - g = \frac{1}{m}\frac{\mu_0 I^2 r}{2h}\left(1+\frac{z'}{h}\right)^{-1} - g$$

$$\approx \frac{1}{m}\frac{\mu_0 I^2 r}{2h}\left(1-\frac{z'}{h}\right) - g = -\frac{g}{h}z' \tag{6}$$

z'—the $\boldsymbol{k}$-direction component of the small displacement of the center O at Time (1) t.

$$\Rightarrow \qquad \ddot{z} + \frac{g}{h}z' = 0 \tag{7}$$

$$\Rightarrow \qquad \ddot{z}' + \frac{g}{h}z' = 0 \tag{8}$$

Equation (8) is the differential equation of Point O performing the tiny simple harmonic vibration in the vertical direction near the mechanical equilibrium position of the ring, and its angular frequency ω_1 is:

$$\Rightarrow \qquad \omega_1 = \sqrt{\frac{g}{h}} \tag{9}$$

Solution to Problem (3)

The time point when the ring is disturbed by rotation is defined as Time (2) zero.

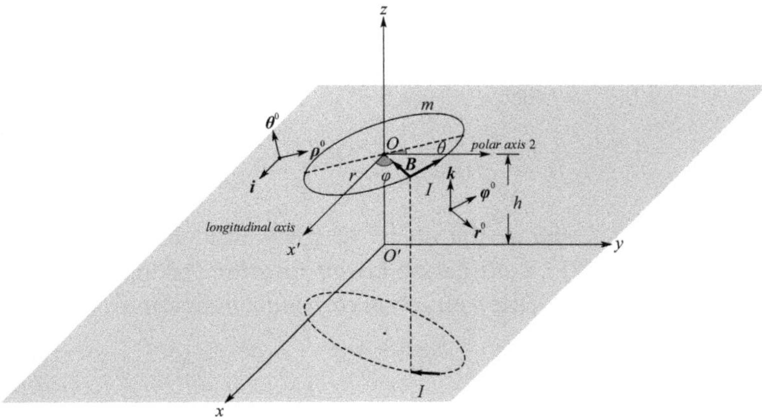

Figure 3

As shown in Fig. 3, establishing the cylindrical coordinate system with the center O as the pole, with the ***j***-direction as the positive direction of the polar axis, with the right-hand helix direction with regard to the ***i***-direction as the positive direction of polar angle, and the ***i***-direction as the positive direction of the longitudinal axis, let ρ, θ, x' be the cylindrical coordinates. Let the rotation disturbance be around the x'-axis.

Defining $\mathrm{d}F_z$—the ***k***-component of the Ampere force exerted on the current infinitesimal element at an arbitrary point in the real ring by the magnetic field generated by the image current. Considering ***Ampere force law***:

$$\Rightarrow \qquad \mathrm{d}F_z = (I\mathrm{d}\boldsymbol{l}\times\boldsymbol{B})\cdot\boldsymbol{k} = (-B_r)\big|_{z=h+r\sin\varphi\sin\theta} Ir\mathrm{d}\varphi$$

$$= \frac{\mu_0 I^2 r}{4\pi h\left(1+\frac{r}{h}\sin\varphi\sin\theta\right)}\mathrm{d}\varphi \approx \frac{\mu_0 I^2 r}{4\pi h\left(1+\frac{r}{h}\theta\sin\varphi\right)}\mathrm{d}\varphi$$

$$\approx \frac{\mu_0 I^2 r}{4\pi h}\left(1-\frac{r}{h}\theta\sin\varphi\right)\mathrm{d}\varphi \qquad (10)$$

Defining M_x—the $\boldsymbol{i}$-component of the resultant torque exerted on the current in the real ring by the Ampere force (taking the x'-axis as the rotation axis). Considering the definition of ***torque***:

$$\Rightarrow \qquad M_x = \oint_L (\boldsymbol{r}\times \mathrm{d}\boldsymbol{F})\cdot \boldsymbol{i} = \oint_L \mathrm{d}F_z \cdot r\sin\varphi \cdot \cos\theta$$

$$= \frac{\mu_0 I^2 r^2}{4\pi h}\cos\theta \int_0^{2\pi} \sin\varphi\left(1-\frac{r}{h}\theta\sin\varphi\right)d\varphi$$

$$= -\frac{\mu_0 I^2 r^3}{4h^2}\theta\cos\theta \approx -\frac{\mu_0 I^2 r^3}{4h^2}\theta = -\frac{mgr^2}{2h}\theta \qquad (11)$$

Taking *the real ring* as the object of study. Studying *the state at arbitrary Time (2) t*. Considering ***law of rotation*** (taking the x'-axis as the rotation axis) (the $\boldsymbol{i}$-direction component equation):

$$M_x = J\ddot{\theta} \qquad (12)$$

J—the moment of inertia of the real ring about the x'-axis.

Considering ***the form of moment of inertia of a homogeneous ring about one of its central axes in the plane of the ring***:

$$J = \frac{1}{2}mr^2 \qquad (13)$$

Combining (11) to (13),

$$\Rightarrow \qquad \ddot{\theta} + \frac{g}{h}\theta = 0 \qquad (14)$$

Equation (14) is the differential equation of the ring performing the tiny rotational simple harmonic vibration around the x'-axis, and its angular frequency ω_2 is:

$$\Rightarrow \qquad \omega_2 = \sqrt{\frac{g}{h}} \qquad (15)$$

4.1.17 A Squirrel Running in a Cage (*c*, *C*)[5]

A cylindrical shell squirrel cage is horizontally fixed on the table, inside it a small squirrel is running within a vertical plane (see Fig. 1). Given the sectional radius of the cage is R, the moment of inertia of the cage about its central axis is I, the mass of the squirrel is m, the

[5]This problem is a test question from University of Wisconsin in 1973. It is also included in the following book: Yongde Zhang, et al. (1988). *American Physics Test Questions and Solutions* (University of Science and Technology of China Press).

cage is subjected to a resistance torque proportional to its angular speed when rotating, and the proportion coefficient is k ($k > 0$), the speed of the squirrel relative to point on the cage in contact with it is v_0 all the time. Initially, the cage is at rest, the squirrel starts to run to the right from the bottom of the cage. Please analyze the motion of the squirrel under the condition of weak damping.

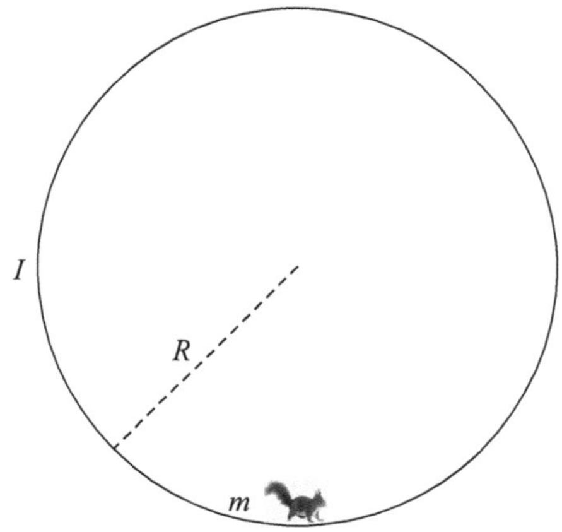

Figure 1

Solution

The time point when the squirrel starts to run is defined as Time zero.

As shown in Fig. 2, the center of the circular trajectory of the squirrel is denoted by O, the points on the cage in contact with the squirrel at Time zero and arbitrary Time t are denoted by P_0 and P, respectively, the unit vector perpendicular to the plane of the figure outward is denoted by $\boldsymbol{k}$. Taking Point O as the pole, the vertical downward direction as the positive direction of the polar axis, the right-hand helix direction with regard to the $\boldsymbol{k}$-direction as the positive direction of polar angle, the $\boldsymbol{k}$-direction as the positive direction of the longitudinal axis, establishing the cylindrical coordinate system, let r, θ, z be the cylindrical coordinates.

Taking *the squirrel and the cage* as the object of study, respectively. Studying *the state at arbitrary Time t*. Considering (a) ***Newton's second law*** (for the squirrel) (the $\boldsymbol{\theta}^0$-direction component equation), (b) ***law of rotation*** (for the cage) [taking the z-axis as the rotation axis] (the $\boldsymbol{k}$-direction component equation):

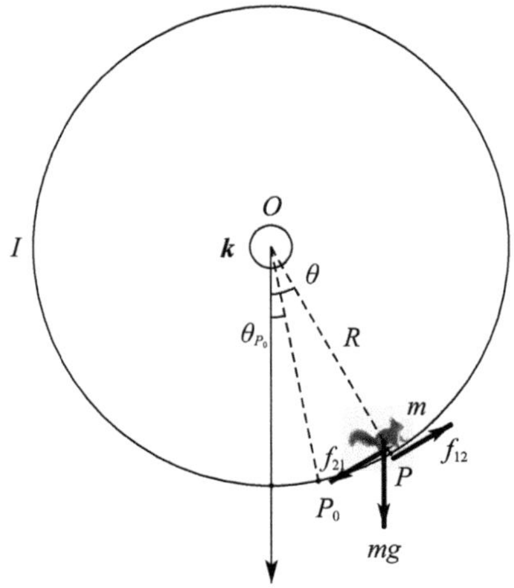

Figure 2

$$-f_{21} - mg\sin\theta = mR\ddot{\theta} \tag{1}$$

$$f_{12}R - k\dot{\theta}_{P_0} = I\ddot{\theta}_{P_0} \tag{2}$$

f_{21}—the $-\boldsymbol{\theta}^0$-component of the friction exerted on the squirrel by the cage at Time t.

f_{12}—the $\boldsymbol{\theta}^0$-component of the friction exerted on the cage by the squirrel at Time t.

Considering ***Newton's third law***:

$$f_{12} = f_{21} \tag{3}$$

Known that the speed of the squirrel relative to the point on the cage in contact with it is v_0 all the time:

$$v_0 = R\dot{\theta} - R\dot{\theta}_P = R\dot{\theta} - R\dot{\theta}_{P_0} \tag{4}$$

$$\Rightarrow \quad \dot{\theta}_{P_0} = \dot{\theta} - \frac{v_0}{R} \tag{5}$$

$$\Rightarrow \quad \ddot{\theta}_{P_0} = \ddot{\theta} \tag{6}$$

Combining (1), (2), (3), (5) and (6):

$$\Rightarrow \quad (I + mR^2)\ddot{\theta} + k\dot{\theta} + mgR\sin\theta - \frac{kv_0}{R} = 0 \tag{7}$$

θ is a small quantity, taking the first-order approximation of $\sin\theta$:

$$\sin\theta \approx \theta \tag{8}$$

$$\Rightarrow \quad (I+mR^2)\ddot{\theta}+k\dot{\theta}+mgR\theta-\frac{kv_0}{R}=0 \tag{9}$$

Defining the variable θ':

$$\theta' = \theta - \frac{kv_0}{mgR^2} \tag{10}$$

$$\Rightarrow \quad \ddot{\theta}'+\frac{k}{I+mR^2}\dot{\theta}'+\frac{mgR}{I+mR^2}\theta'=0 \tag{11}$$

The motion described by θ' is a damped vibration.

Defining ω_0 — the intrinsic angular frequency, β — the equivalent damping coefficient:

$$\begin{cases} \omega_0 = \sqrt{\dfrac{mgR}{I+mR^2}} & (12) \\ \beta = \dfrac{k}{2(I+mR^2)} & (13) \end{cases}$$

The condition of weak damping is as follows:

$$\beta < \omega_0 \tag{14}$$

Defining ω—the angular frequency of the weak damping vibration:

$$\omega = \sqrt{\omega_0^2-\beta^2} = \sqrt{\frac{mgR}{I+mR^2}\left(1-\frac{k}{2mgR}\right)} \tag{15}$$

The general solution of Eq. (11) is as follows:

$$\Rightarrow \quad \theta' = e^{-\beta t}(A\sin\omega t + B\cos\omega t) \tag{16}$$

A and B — the integral constants.

(16) $\rightarrow$ (10),

$$\Rightarrow \quad \theta = \frac{kv_0}{mgR^2}+e^{-\beta t}(A\sin\omega t + B\cos\omega t) \tag{17}$$

Considering the initial conditions, then place (5) in to obtain:

$$\begin{cases} \theta\big|_{t=0} = 0 & (18a) \\ \dot{\theta}_{P_0}\big|_{t=0} = 0 \;\Rightarrow\; \dot{\theta}\big|_{t=0} = \dfrac{v_0}{R} & (18b) \end{cases}$$

(17) → (18),

$$\Rightarrow \quad \begin{cases} A = \dfrac{kv_0}{mgR^2}\left(\dfrac{mgR}{k\omega} - \dfrac{\beta}{\omega}\right) & \text{(19a)} \\ B = -\dfrac{kv_0}{mgR^2} & \text{(19b)} \end{cases}$$

$$\Rightarrow \quad \theta = \frac{kv_0}{mgR^2}\left\{1 - e^{-\beta t}\left[\cos\omega t + \left(\frac{\beta}{\omega} - \frac{mgR}{k\omega}\right)\sin\omega t\right]\right\} \tag{20}$$

This is the kinematics equation of the squirrel. If there is no damping ($k = 0$), it will degenerate to:

$$\Rightarrow \quad \theta\big|_{k\to 0} = \frac{v_0}{\omega_0 R}\sin\omega_0 t \tag{21}$$

4.1.18 The Driving of a Car on a Hillside (*c*, *C*)[6]

Consider a car going over a hillside along its top line, whose curve equation in the Cartesian coordinate system provided is: $y_0 = H\left[1 - \cos\left(\dfrac{\pi x}{l}\right)\right]$ (see Fig. 1). The car can be approximately regarded as a vertical spring oscillator composed of a block of mass m fixed on a light spring of stiffness coefficient k and natural length l_0. Given that the ***i***-direction component of the velocity of the car is v all the time when going over the hillside. Find the kinematics equation of the car in the ***j***-direction when going over the hillside.

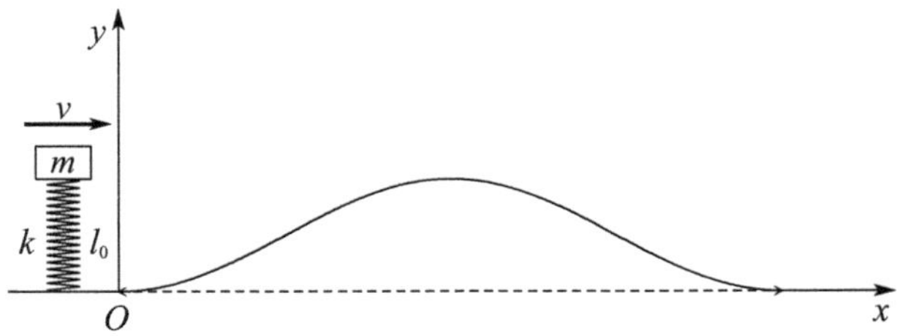

Figure 1

Solution

The time point when the car starts to climb the hillside is defined as Time zero.

[6]This problem is a test question from MIT in 1979. It is also included in the following book: Yongde Zhang, et al. (1988). *American Physics Test Questions and Solutions*, (University of Science and Technology of China Press).

The lower end of the spring oscillator is located on the hillside all the time, its trajectory equation is consistent with the curve equation of the top line of the hillside:

$$y_0 = H\left[1-\cos\left(\frac{\pi x}{l}\right)\right] \tag{1}$$

Considering ***the displacement-time relation of uniform linear motion***:

$$x = vt \tag{2}$$

$$\Rightarrow \quad y_0 = H\left[1-\cos\left(\frac{\pi v}{l}t\right)\right] \tag{3}$$

Defining the constant ω:

$$\omega = \frac{\pi v}{l} \tag{4}$$

$$\Rightarrow \quad y_0 = H(1-\cos\omega t) \tag{5}$$

Taking *the block* as the object of study. Studying *the state at arbitrary Time t*. Considering ***Newton's second law*** (the ***j***-direction component equation):

$$k\left[l_0-(y-y_0)\right]-mg = m\ddot{y} \tag{6}$$

$$\Rightarrow \quad \ddot{y}+\frac{k}{m}\left(y-H-l_0+\frac{mg}{k}\right) = -\frac{k}{m}H\cos\omega t \tag{7}$$

Defining the variable Y:

$$Y = y-H-l_0+\frac{mg}{k} \tag{8}$$

$$\Rightarrow \quad \ddot{Y}+\frac{k}{m}Y = -\frac{k}{m}H\cos\omega t \tag{9}$$

Equation (9) is a differential equation of forced vibration, that is, the motion of the block described by Y is a forced vibration, and its angular frequency ω_0 is:

$$\omega_0 = \sqrt{\frac{k}{m}} \tag{10}$$

Defining the constant f_0:

$$f_0 = -\frac{k}{m}H \tag{11}$$

$$\Rightarrow \qquad \ddot{Y} + \omega_0^2 Y = f_0 \cos\omega t \tag{12}$$

The particular solution of Eq. (12) is as follows:

$$Y^* = c\cos(\omega t - \varphi) \tag{13}$$

(13) → (12),

$$\Rightarrow \qquad c = \frac{f_0}{\sqrt{(\omega_0^2 - \omega^2)^2}} = \frac{kl^2}{\left|\pi^2 mv^2 - kl^2\right|}H$$

$$= \begin{cases} \dfrac{kl^2}{\pi^2 mv^2 - kl^2}H & \pi^2 mv^2 \geq kl^2 \\ -\dfrac{kl^2}{\pi^2 mv^2 - kl^2}H & \pi^2 mv^2 < kl^2 \end{cases} \tag{14}$$

$$\begin{cases} \tan\varphi = 0 & (15a) \\ \cos\varphi = \dfrac{\omega_0^2 - \omega^2}{\sqrt{(\omega_0^2 - \omega^2)^2}} = \dfrac{\pi^2 mv^2 - kl^2}{\left|\pi^2 mv^2 - kl^2\right|} = \begin{cases} 1 & \pi^2 mv^2 \geq kl^2 \\ -1 & \pi^2 mv^2 < kl^2 \end{cases} & (15b) \end{cases}$$

$$\Rightarrow \qquad \varphi = \begin{cases} 0 & \pi^2 mv^2 \geq kl^2 \\ \pi & \pi^2 mv^2 < kl^2 \end{cases} \tag{16}$$

$$\Rightarrow \qquad Y^* = \frac{kl^2}{\pi^2 mv^2 - kl^2}H\cos\omega t \tag{17}$$

The general solution of Eq. (12) is as follows:

$$Y = a\cos\omega_0 t + b\sin\omega_0 t + Y^* \tag{18}$$

a and b—the integration constants.

$$\Rightarrow \quad y = a\cos\omega_0 t + b\sin\omega_0 t + H\left(\frac{kl^2}{\pi^2 mv^2 - kl^2}\cos\omega t + 1\right) + l_0 - \frac{mg}{k} \tag{19}$$

Considering the initial conditions:

$$\begin{cases} \dot{y}\big|_{t=0} = 0 & (20a) \\ \ddot{y}\big|_{t=0} = 0 & (20b) \end{cases}$$

(20b) → (7),

$$\Rightarrow \qquad y\big|_{t=0} = l_0 - \frac{mg}{k} \qquad (21)$$

(19) → (20a) and (21),

$$\Rightarrow \qquad \begin{cases} a = \dfrac{\pi^2 mv^2}{kl^2 - \pi^2 mv^2} H & (22a) \\ b = 0 & (22b) \end{cases}$$

$$\Rightarrow \quad y = H\left[(1-\cos\omega t) + \frac{\pi^2 mv^2}{kl^2 - \pi^2 mv^2}(\cos\omega_0 t - \cos\omega t)\right] + l_0 - \frac{mg}{k} \qquad (23)$$

4.1.19 A Spring Oscillator in a Liquid Storage Container Driven by an External Force (d, *C*)

A rectangular container filled with some liquid is placed on the ground, and in which there is a spring oscillator fixed on its right inner wall (see Fig. 1). The spring oscillator is composed of a block of mass m and a spring of stiffness coefficient k and natural length x_0, the damping coefficient of the liquid to the block is γ (weak damping). The container is driven by an external force to vibrate in the horizontal direction, the kinematics equation is $x = F\cos\omega t$, $F > 0$ (in the linear coordinate system shown in the figure). Find the kinematics equation of the block and find its degenerate form when ω is large enough.

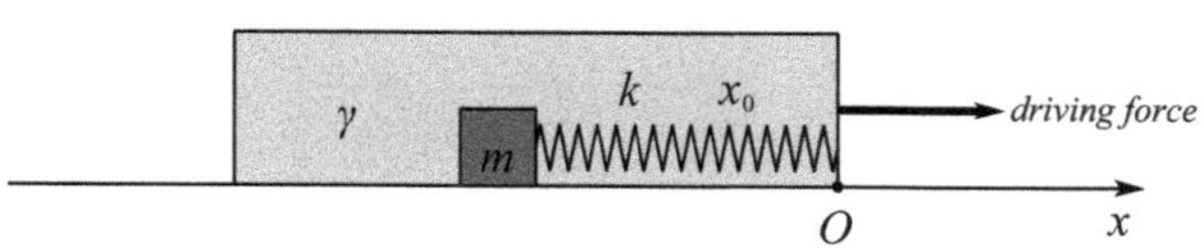

Figure 1

Solution

As shown in Fig. 2, the positions of the fixed end of the spring and the block at arbitrary Time t are denoted by A and B, respectively.

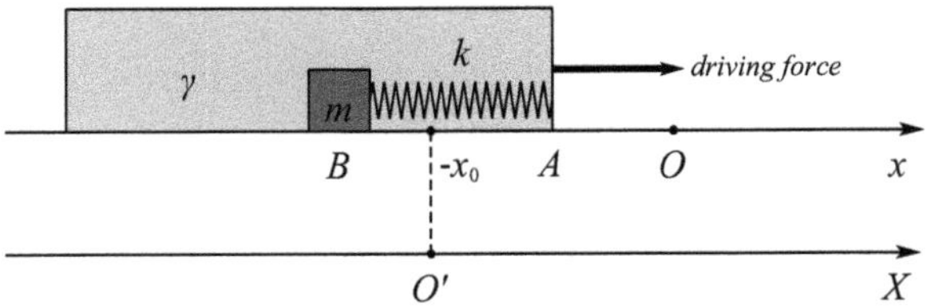

Figure 2

Taking *the block* as the object of study. Studying *the state at arbitrary Time t*. Considering ***Newton's second law*** (the ***i***-direction component equation):

$$k(x_A - x_B - x_0) - \gamma(\dot{x}_B - \dot{x}_A) = m\ddot{x}_B \tag{1}$$

Defining the variable X_B:

$$X_B = x_B + x_0 \tag{2}$$

X_B can be understood as the coordinate of the block in the new linear coordinate system with Point O' (the point obtained by translating the origin O a distance x_0 to the left) as the origin and the ***i***-direction as the positive direction of coordinate axis.

$$\Rightarrow \qquad k(x_A - X_B) - \gamma(\dot{X}_B - \dot{x}_A) = m\ddot{X}_B \tag{3}$$

$$\Rightarrow \qquad \ddot{X}_B + \frac{\gamma}{m}\dot{X}_B + \frac{k}{m}X_B = \frac{\gamma}{m}\dot{x}_A + \frac{k}{m}x_A \tag{4}$$

Known:

$$x_A = F\cos\omega t \tag{5}$$

Defining ω_0—the intrinsic angular frequency of the spring oscillator, β—the equivalent damping coefficient:

$$\omega_0 = \sqrt{\frac{k}{m}} \tag{6}$$

$$2\beta = \frac{\gamma}{m} \tag{7}$$

$$\Rightarrow \qquad \ddot{X}_B + 2\beta\dot{X}_B + \omega_0^2 X_B = \omega_0^2 F\cos\omega t - 2\beta\omega F\sin\omega t \tag{8}$$

Defining the constants $f\,(>0)$ and φ_1:

$$\begin{cases} \omega_0^2 F = f\cos\varphi_1 & \text{(9a)} \\ 2\beta\omega F = f\sin\varphi_1 & \text{(9b)} \end{cases}$$

$$\Rightarrow \quad \begin{cases} f = \sqrt{\omega_0^4 + 4\beta^2\omega^2}\,F & \text{(10a)} \\ \tan\varphi_1 = \dfrac{2\beta\omega}{\omega_0^2} & \text{(10b)} \end{cases}$$

$$\Rightarrow \quad \ddot{X}_B + 2\beta\dot{X}_B + \omega_0^2 X_B = f\cos\varphi_1\cos\omega t - f\sin\varphi_1\sin\omega t$$

$$= f\cos(\omega t + \varphi_1) \qquad (11)$$

The motion described by X_B is the weak damping vibration driven by a harmonic force.

The condition of weak damping is as follows:

$$\beta < \omega_0 \qquad (12)$$

Defining ω_γ—the angular frequency of the weak damping vibration:

$$\omega_\gamma = \sqrt{\omega_0^2 - \beta^2} \qquad (13)$$

The particular solution of Eq. (11) is:

$$X_B^* = B\cos(\omega t + \varphi_1 + \varphi_2) \qquad (14)$$

$(14) \rightarrow (11)$:

$$\Rightarrow \quad \begin{cases} B = \dfrac{f}{\sqrt{(\omega_0^2 - \omega^2)^2 + 4\beta^2\omega^2}} = \dfrac{\sqrt{\omega_0^4 + 4\beta^2\omega^2}}{\sqrt{(\omega_0^2 - \omega^2)^2 + 4\beta^2\omega^2}}F & \text{(15a)} \\ \begin{cases} \tan\varphi_2 = -\dfrac{2\beta\omega}{\omega_0^2 - \omega^2} & \text{(15b)} \\ \cos\varphi_2 = \dfrac{\omega_0^2 - \omega^2}{\sqrt{(\omega_0^2 - \omega^2)^2 + 4\beta^2\omega^2}} & \text{(15c)} \end{cases} \end{cases}$$

The general solution of Eq. (11) is as follows:

$$X_B = Ae^{-\beta t}\cos(\omega_\gamma t + \varphi_\gamma) + X_B^*$$

$$= Ae^{-\beta t}\cos(\omega_\gamma t + \varphi_\gamma) + B\cos(\omega t + \varphi_1 + \varphi_2) \qquad (16)$$

A and φ_γ—the integral constants.

This equation retains B and φ_2 determined by Eq. (15).

The following study the case that ω is large enough:

$$\omega \gg \omega_0 \qquad (17)$$

Combining (12) and (17):

$$\Rightarrow \qquad \beta \ll \omega \tag{18}$$

(17) $\rightarrow$ (10b):

$$\Rightarrow \qquad \tan\varphi_1 \rightarrow \infty$$

$$\Rightarrow \qquad \varphi_1 \approx \frac{\pi}{2} \tag{19}$$

From (15), (17) and (18)

$$\Rightarrow \begin{cases} B \approx \dfrac{2\beta\omega}{\sqrt{\omega^4 + 4\beta^2\omega^2}}F = \dfrac{2\beta}{\omega\sqrt{1+\dfrac{4\beta^2}{\omega^2}}}F \approx \dfrac{2\beta}{\omega}\left(1 - \dfrac{2\beta^2}{\omega^2}\right)F \\ \approx \dfrac{2\beta}{\omega}F & (20a) \\ \begin{cases} \tan\varphi_2 \approx \dfrac{2\beta}{\omega} \approx 0 \\ \cos\varphi_2 \approx -\dfrac{\omega}{\sqrt{\omega^2 + 4\beta^2}} \approx -1 \end{cases} \Rightarrow \varphi_2 \approx -\pi & (20b) \end{cases}$$

(19) and (20) $\rightarrow$ (16):

$$\Rightarrow \qquad X_B = Ae^{-\beta t}\cos(\omega_\gamma t + \varphi_\gamma) - \frac{2\beta}{\omega}F\sin\omega t \tag{21}$$

4.2 Multi-Degree of Freedom Vibration

The multi-degree of freedom vibration model is used to describe the vibration of discrete systems (with discontinuous distribution of mass and elastic properties) in the real world. As an important subcategory, linear multi-degree of freedom vibration has the property of "excellence"—it's any possible vibration state can be regarded as the linear superposition of main vibrations at each natural frequency (the number of which is equal to the degrees of freedom of the vibration system), In view of this property, linear algebra theory has been deeply applied in this problem.

The molecular model is a most typical discrete system, and its vibration problem has been deeply studied based on the multi-degree of freedom vibration theory.

In this section, some representative linear multi-degree of freedom vibration problems will be studied.

4.2.1 The Vibration of a Pair of conductor Rods on a Pair of Conductor Rails in a Magnetic Field (*c*, *C*)[7]

Two long, straight and smooth conductor rails with the distance of l are fixed on a horizontal plane, parallel to each other (see Fig. 1). Two conductor rods 1 and 2 of equal mass m and equal length slightly greater than l are placed vertically on the two conductor rails. An inductor of self-inductance coefficient L and neglected mass is connected in series into the circuit composed of the two conductor rails and the two conductor rods. The resistance of the conductor rails and the conductor rods and the inductor can be ignored. Apply a uniform magnetic field in space, the magnitude of whose magnetic induction intensity is B, and the direction is vertical downward. Initially, the conductor rods 1 and 2 are static at $x = x_0$ and $x = 0$, respectively (in the linear coordinate system shown in the figure). The conductor rod 1 is subjected to an instantaneous impact and obtains the initial center-of-mass velocity $v_0\boldsymbol{i}$. Find the position of each conductor rod and the current intensity in the circuit at arbitrary time.

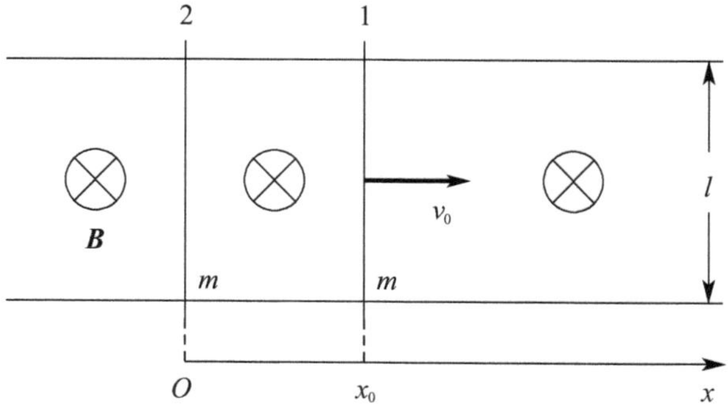

Figure 1

Solution

The time point when the conductor rod 1 is impacted is defined as Time zero.

[7]This problem is a test question of the 21th CphO (China Physics Olympiads) in 2004.

Taking *the circuit composed of the two conductor rails, the two conductor rods and the inductor* as the object of study (see Fig. 2). Studying *the state at arbitrary Time t.* Considering (a) ***Kirchhoff's voltage law*** (anticlockwise), (b) ***theorem of motional electromotive force*** [integrating into (a)], (c) ***theorem of self-induction electromotive force*** [integrating into (a)]:

$$Bl\dot{x}_1 - Bl\dot{x}_2 - L\dot{I} = 0 \tag{1}$$

I—the current intensity in the circuit at Time t (anticlockwise).

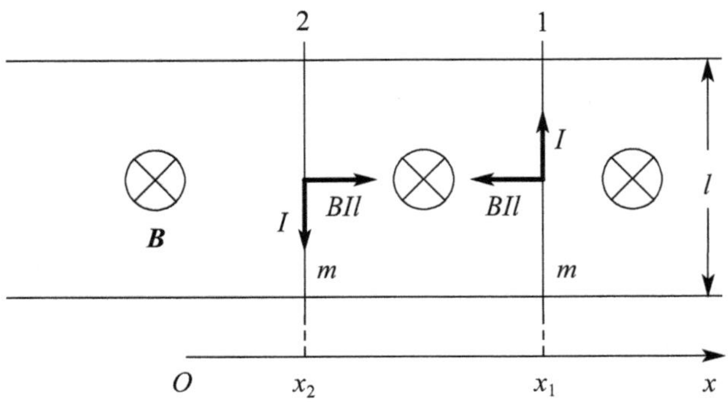

Figure 2

Defining the variable X':

$$x_1 - x_2 = x_0 + X' \tag{2}$$

$$\Rightarrow \quad Bl\dot{X}' = L\dot{I} \tag{3}$$

$$\Rightarrow \quad Bl\mathrm{d}X' = L\mathrm{d}I$$

$$\Rightarrow \quad BlX' = Bl\int_0^{X'} \mathrm{d}X' = L\int_0^{I} \mathrm{d}I = LI$$

$$\Rightarrow \quad I = \frac{Bl}{L}X' \tag{4}$$

Taking *the conductor rods 1 and 2* as the object of study, respectively. Studying *the state at arbitrary Time t.* Considering ***theorem of motion of center of mass*** (the ***i***-direction component equation):

$$\begin{cases} -BIl = m\ddot{x}_1 & \text{(5a)} \\ BIl = m\ddot{x}_2 & \text{(5b)} \end{cases}$$

Combining (4) and (5):

$$\Rightarrow \qquad \ddot{X}' + \frac{2B^2l^2}{mL}X' = 0 \tag{6}$$

It can be seen that the relative motion of the two conductor rods is a simple harmonic vibration, and its angular frequency ω is:

$$\Rightarrow \qquad \omega = \sqrt{\frac{2B^2l^2}{mL}} \tag{7}$$

The general solution of Eq. (6) is as follows:

$$\Rightarrow \qquad X' = A\cos(\omega t + \varphi) \tag{8}$$

A and φ—the amplitude and initial phase of the simple harmonic vibration described by X'.

Considering the initial conditions:

$$\begin{cases} (x_1 - x_2)\big|_{t=0} = x_0 & \text{(9a)} \\ (\dot{x}_1 - \dot{x}_2)\big|_{t=0} = v_0 & \text{(9b)} \end{cases}$$

$$\Rightarrow \qquad \begin{cases} X'\big|_{t=0} = 0 & \text{(10a)} \\ \dot{X}'\big|_{t=0} = v_0 & \text{(10b)} \end{cases}$$

(8) → (10),

$$\Rightarrow \qquad \begin{cases} A = \dfrac{v_0}{Bl}\sqrt{\dfrac{mL}{2}} & \text{(11a)} \\ \varphi = \dfrac{3\pi}{2} & \text{(11b)} \end{cases}$$

$$\Rightarrow \qquad X' = \frac{v_0}{Bl}\sqrt{\frac{mL}{2}}\sin\left(\sqrt{\frac{2B^2l^2}{mL}}t\right) \tag{12}$$

$$\Rightarrow \qquad x_1 - x_2 = x_0 + \frac{v_0}{Bl}\sqrt{\frac{mL}{2}}\sin\left(\sqrt{\frac{2B^2l^2}{mL}}t\right) \tag{13}$$

Equation (13) is the kinematics equation of the relative motion of the two conductor rods.

Taking *the system composed of the two conductor rods* as the study object. Studying *the state at arbitrary Time t*. Considering

theorem of motion of center of mass (the ***i***-direction component equation):

$$-BIl + BIl = 2m\ddot{x}_C \tag{14}$$

$$\Rightarrow \quad \ddot{x}_C = 0 \tag{15}$$

$$\Rightarrow \quad \dot{x}_C = \dot{x}_C\big|_{t=0} \tag{16}$$

$$\Rightarrow \quad x_C = x_C\big|_{t=0} + \dot{x}_C\big|_{t=0} t \tag{17}$$

Considering (a) the definition of the ***center of mass***, (b) ***theorem of center-of-mass velocity***:

$$\begin{cases} x_C = \dfrac{mx_1 + mx_2}{2m} = \dfrac{1}{2}(x_1 + x_2) & (18) \\ x_C\big|_{t=0} = \dfrac{mx_0 + m\cdot 0}{2m} = \dfrac{1}{2}x_0 & (19) \\ \dot{x}_C\big|_{t=0} = \dfrac{mv_0 + m\cdot 0}{2m} = \dfrac{1}{2}v_0 & (20) \end{cases}$$

$$\Rightarrow \quad \frac{1}{2}(x_1 + x_2) = \frac{1}{2}x_0 + \frac{1}{2}v_0 t \tag{21}$$

Combining (13) and (21):

$$\Rightarrow \begin{cases} x_1 = x_0 + \dfrac{1}{2}v_0 t + \dfrac{v_0}{2Bl}\sqrt{\dfrac{mL}{2}}\sin\left(\sqrt{\dfrac{2B^2l^2}{mL}}t\right) & (22a) \\ x_2 = \dfrac{1}{2}v_0 t - \dfrac{v_0}{2Bl}\sqrt{\dfrac{mL}{2}}\sin\left(\sqrt{\dfrac{2B^2l^2}{mL}}t\right) & (22b) \end{cases}$$

It can be seen that the motion of each conductor rod is the superposition of a uniform linear motion and a simple harmonic vibration.

Combining (2), (4) and (22):

$$\Rightarrow \quad I = v_0\sqrt{\frac{m}{2L}}\sin\left(\sqrt{\frac{2B^2l^2}{mL}}t\right) \tag{23}$$

4.2.2 Coupled Spring Oscillator System 1 (d, *C*)

As shown in Fig. 1, there are two blocks 1 and 2 of mass m_1 and m_2, which are, respectively, tied to the two ends of a light spring of stiffness coefficient k and natural length l_0, placed on a smooth horizontal ground as a whole. Initially, Block 1 and Block 2 are at rest, the spring is the natural length. On the left side of Block 1, there is a block of mass m (called the impacting block), which collides with Block 1 completely elastically with the initial velocity of magnitude v_0 and pointing to Block 1. Find the kinematics equation of Block 1 and Block 2 after the collision.

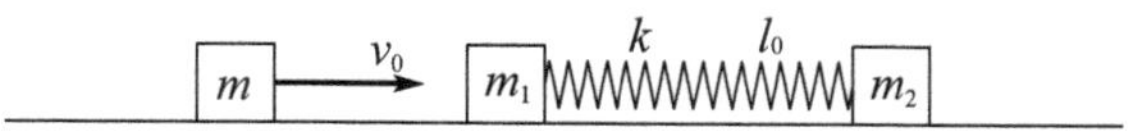

Figure 1

Solution

The time point after the impacting block collides with Block 1 is defined as Time zero.

As shown in Fig. 2, the center of mass of the system composed of Block 1 and Block 2 is denoted by C, the space point that coincides with the center of mass C at Time zero is denoted by O, the unit vector of Block 1 pointing to Block 2 is denoted by $\boldsymbol{i}$. Establishing the linear coordinate system with Point O as the origin and the $\boldsymbol{i}$-direction as the positive direction of coordinate axis, let x be the coordinate.

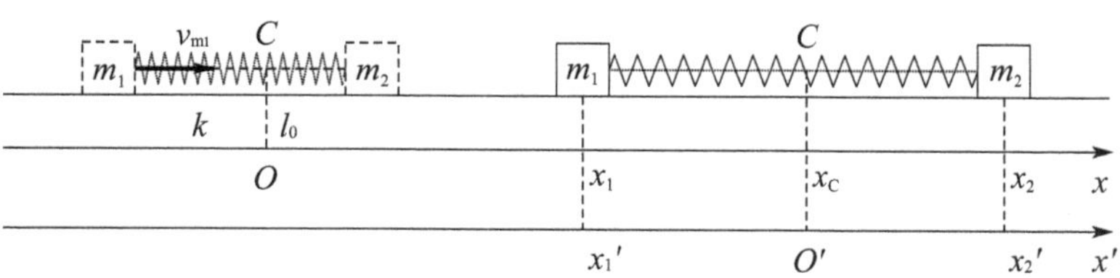

Figure 2

Taking *the system composed of the impacting block and Block 1* as the object of study. Studying *the instantaneous process of the collision between them*. Considering ***velocity theorem of completely elastic collision between two particles*** (the $\boldsymbol{i}$-direction component equation):

$$\begin{cases} v_m = \dfrac{m-m_1}{m+m_1} v_0 & \text{(1a)} \\ v_{m_1} = \dfrac{2m}{m+m_1} v_0 & \text{(1b)} \end{cases}$$

v_m and v_{m_1}—the ***i***-direction component of the velocity of the impacting block and Block 1 at the time point after the collision.

Taking *the system composed of Block 1 and Block 2 and the spring* as the object of study. Studying *the process from Time zero to arbitrary Time t*. Considering (a) ***law of conservation of momentum*** [the resultant external force acting on the system is zero; therefore, the momentum of the system is conserved] (the ***i***-direction component equation), (b) ***the total momentum of system of particles is equal to the center-of-mass momentum*** [integrating into (a)]:

$$m_1 v_{m_1} = (m_1 + m_2) v_C \tag{2}$$

v_C—the ***i***-direction component of the center-of-mass velocity of the system at Time t.

$$\Rightarrow \qquad v_C = \frac{2mm_1}{(m+m_1)(m_1+m_2)} v_0 \tag{3}$$

Defining the constant α, and m_μ—the reduced mass of Block 1 and Block 2:

$$\alpha = \frac{2m}{m+m_1} \tag{4}$$

$$m_\mu = \frac{m_1 m_2}{m_1 + m_2} \tag{5}$$

$$\Rightarrow \qquad v_C = \alpha \frac{m_\mu}{m_2} v_0 \tag{6}$$

As shown in Fig. 2, the space point that coincides with the center of mass C at arbitrary Time t is denoted by O'. Establishing the linear coordinate system with Point O' as the origin and the ***i***-direction as the positive direction of coordinate axis, let x' be the coordinate.

Taking *Block 1 and Block 2* as the object of study, respectively. Studying *the state at arbitrary Time t*. Considering ***Galileo transformation*** [taking the ground as the stationary reference frame, and the center-of-mass reference frame of the system composed of Block 1 and Block 2 as the moving reference frame]:

$$\begin{cases} x_1 = x_C + x_1' & \text{(7a)} \\ x_2 = x_C + x_2' & \text{(7b)} \end{cases}$$

Considering ***the displacement-time relation of uniform linear motion***:

$$x_C = v_C t \tag{8}$$

$$\Rightarrow \qquad x_C = \alpha \frac{m_\mu}{m_2} v_0 t \tag{9}$$

Taking *Block 1 and Block 2* as the object of study, respectively. Studying *the state at arbitrary Time t*. Considering ***Newton's second law*** (the ***i***-direction component equation), then place (5) in to obtain:

$$\begin{cases} k\left[(x_2' - x_1') - l_0\right] = m_1 \ddot{x}_1' & (10a) \\ -k\left[(x_2' - x_1') - l_0\right] = m_2 \ddot{x}_2' & (10b) \end{cases}$$

$$\Rightarrow \qquad \begin{cases} \ddot{x}_1' + \dfrac{k}{m_\mu} x_1' = -\dfrac{k l_0}{m_1} & (11a) \\ \ddot{x}_2' + \dfrac{k}{m_\mu} x_2' = \dfrac{k l_0}{m_2} & (11b) \end{cases}$$

Defining the variables X'_1 and X'_2:

$$\begin{cases} X_1' = x_1' - \left(-\dfrac{m_\mu}{m_1} l_0\right) & (12a) \\ X_2' = x_2' - \dfrac{m_\mu}{m_2} l_0 & (12b) \end{cases}$$

$$\Rightarrow \qquad \begin{cases} \ddot{X}_1' + \dfrac{k}{m_\mu} X_1' = 0 & (13a) \\ \ddot{X}_2' + \dfrac{k}{m_\mu} X_2' = 0 & (13b) \end{cases}$$

It can be seen that the motions described by X'_1 and X'_2, respectively, are both simple harmonic vibrations, and their equal angular frequency ω is:

$$\Rightarrow \qquad \omega = \sqrt{\frac{k}{m_\mu}} \tag{14}$$

The general solutions of Eqs. (13a) and (13b) are as follows:

$$\Rightarrow \qquad \begin{cases} X_1' = A_1 \cos(\omega t + \varphi_1) & (15a) \\ X_2' = A_2 \cos(\omega t + \varphi_2) & (15b) \end{cases}$$

A_1 and φ_1—the amplitude and the initial phase of the simple harmonic vibration described by X'_1.

A_2 and φ_2—the amplitude and the initial phase of the simple harmonic vibration described by X'_2.

Considering the definition of ***center of mass***, then place (5) in to obtain:

$$\begin{cases} x_1'\big|_{t=0} = -\dfrac{m_2}{m_1+m_2} l_0 = -\dfrac{m_\mu}{m_1} l_0 & \text{(16a)} \\[2ex] x_2'\big|_{t=0} = \dfrac{m_1}{m_1+m_2} l_0 = \dfrac{m_\mu}{m_2} l_0 & \text{(16b)} \end{cases}$$

Taking *Block 1 and Block 2* as the object of study, respectively. Studying *the state at arbitrary Time t*. Considering ***theorem for composition of velocities of a particle*** [taking the ground as the stationary reference frame, and the center-of-mass reference frame as the moving reference frame] (the ***i***-direction component equation), then place (1b) and (3) in to obtain:

$$\begin{cases} v_{m_1} = v_C + \dot{x}_1'\big|_{t=0} & \text{(17a)} \\ 0 = v_C + \dot{x}_2'\big|_{t=0} & \text{(17b)} \end{cases}$$

$$\Rightarrow \begin{cases} \dot{x}_1'\big|_{t=0} = \alpha \dfrac{m_\mu}{m_1} v_0 & \text{(18a)} \\[2ex] \dot{x}_2'\big|_{t=0} = -\alpha \dfrac{m_\mu}{m_2} v_0 & \text{(18b)} \end{cases}$$

$$\Rightarrow \begin{cases} \begin{cases} X_1\big|_{t=0} = 0 & \text{(19a)} \\ \dot{X}_1'\big|_{t=0} = \alpha \dfrac{m_\mu}{m_1} v_0 & \text{(19b)} \end{cases} \\ \begin{cases} X_2'\big|_{t=0} = 0 & \text{(20a)} \\ \dot{X}_2'\big|_{t=0} = -\alpha \dfrac{m_\mu}{m_2} v_0 & \text{(20b)} \end{cases} \end{cases}$$

(15a) → (19), (15b) → (20),

$$\Rightarrow \begin{cases} \begin{cases} A_1 = \sqrt{\dfrac{m_\mu}{k}} \alpha \dfrac{m_\mu}{m_1} v_0 & \text{(21a)} \\ \varphi_1 = \dfrac{3\pi}{2} & \text{(21b)} \end{cases} \\ \begin{cases} A_2 = \sqrt{\dfrac{m_\mu}{k}} \alpha \dfrac{m_\mu}{m_2} v_0 & \text{(22a)} \\ \varphi_2 = \dfrac{\pi}{2} & \text{(22b)} \end{cases} \end{cases}$$

$$\Rightarrow \begin{cases} X_1' = \alpha \dfrac{m_\mu}{m_1} \sqrt{\dfrac{m_\mu}{k}} v_0 \sin\left(\sqrt{\dfrac{k}{m_\mu}} t\right) & \text{(23a)} \\ X_2' = -\alpha \dfrac{m_\mu}{m_2} \sqrt{\dfrac{m_\mu}{k}} v_0 \sin\left(\sqrt{\dfrac{k}{m_\mu}} t\right) & \text{(23b)} \end{cases}$$

(23) → (12), we obtain the kinematics equations of Block 1 and Block 2 in the center-of-mass reference frame:

$$\Rightarrow \begin{cases} x_1' = \alpha \dfrac{m_\mu}{m_1} \sqrt{\dfrac{m_\mu}{k}} v_0 \sin\left(\sqrt{\dfrac{k}{m_\mu}} t\right) - \dfrac{m_\mu}{m_1} l_0 & \text{(24a)} \\ x_2' = -\alpha \dfrac{m_\mu}{m_2} \sqrt{\dfrac{m_\mu}{k}} v_0 \sin\left(\sqrt{\dfrac{k}{m_\mu}} t\right) + \dfrac{m_\mu}{m_2} l_0 & \text{(24b)} \end{cases}$$

(9) and (24) → (7), we obtain the kinematics equations of Block 1 and Block 2 in the ground reference frame:

$$\Rightarrow \begin{cases} x_1 = \alpha \dfrac{m_\mu}{m_2} v_0 t + \alpha \dfrac{m_\mu}{m_1} \sqrt{\dfrac{m_\mu}{k}} v_0 \sin\left(\sqrt{\dfrac{k}{m_\mu}} t\right) - \dfrac{m_\mu}{m_1} l_0 & \text{(25a)} \\ x_2 = \alpha \dfrac{m_\mu}{m_2} v_0 t - \alpha \dfrac{m_\mu}{m_2} \sqrt{\dfrac{m_\mu}{k}} v_0 \sin\left(\sqrt{\dfrac{k}{m_\mu}} t\right) + \dfrac{m_\mu}{m_2} l_0 & \text{(25b)} \end{cases}$$

It can be seen that the motion of each block in the ground reference frame is the superposition of a uniform linear motion and a simple harmonic vibration.

4.2.3 Coupled Spring Oscillator System 2 (d, *C*)

Linear triatomic molecules can be simulated by the following model: a small ball of mass M and two small balls of equal mass m are connected by two light springs of equal stiffness coefficient k in collinear, placed on a smooth horizontal plane as a whole (see Fig. 1). Find the characteristic frequencies and vibration modes of the tiny vibration of the system.

Figure 1

Solution

As shown in Fig. 2, sorting the three balls from left to right with positive integers. Taking an arbitrary point on the line linking the three balls and denoting it by *O*, the unit vector of Ball 1 pointing Ball 3 is denoted by ***i***. Establishing the linear coordinate system with Point *O* as the origin and the ***i***-direction as the positive direction of coordinate axis, let x be the coordinate.

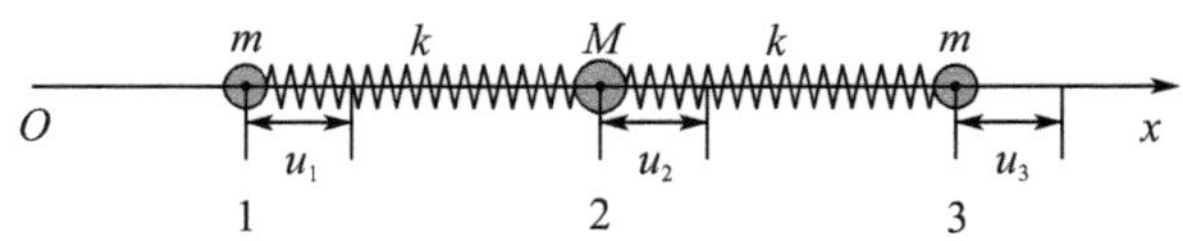

Figure 2

Taking *the three balls* as the object of study, respectively. Studying *the state at arbitrary Time t*. Considering ***Newton's second law*** (the ***i***-direction component equation):

$$\begin{cases} k\left[(x_2-x_1)-(x_{20}-x_{10})\right]=m\ddot{x}_1 & \text{(1a)} \\ -k\left[(x_2-x_1)-(x_{20}-x_{10})\right]+k\left[(x_3-x_2)-(x_{30}-x_{20})\right]=M\ddot{x}_2 & \text{(1b)} \\ -k\left[(x_3-x_2)-(x_{30}-x_{20})\right]=m\ddot{x}_3 & \text{(1c)} \end{cases}$$

Defining u_n (n = 1, 2, 3)—the ***i***-direction component of the displacement of each ball at Time t:

$$u_n = x_n - x_{n0} \qquad n=1,2,3 \tag{2}$$

$$\Rightarrow \quad \begin{cases} k(u_2 - u_1) = m\ddot{u}_1 & \text{(3a)} \\ -k(u_2 - u_1) + k(u_3 - u_2) = M\ddot{u}_2 & \text{(3b)} \\ -k(u_3 - u_2) = m\ddot{u}_3 & \text{(3c)} \end{cases}$$

Let the trial solution to u_n be:

$$u_n = \text{Re}\left[\tilde{A}_n \cos\omega t\right] \quad n = 1,2,3 \tag{4}$$

ω—the characteristic frequency of the system.

$\tilde{A}_n$ (n = 1, 2, 3)—the complex amplitude of the simple harmonic vibration of each ball.

$$\Rightarrow \quad \begin{cases} (k - m\omega^2)\tilde{A}_1 - k\tilde{A}_2 = 0 & \text{(5a)} \\ -k\tilde{A}_1 + (2k - M\omega^2)\tilde{A}_2 - k\tilde{A}_3 = 0 & \text{(5b)} \\ -k\tilde{A}_2 + (k - m\omega^2)\tilde{A}_3 = 0 & \text{(5c)} \end{cases}$$

The condition for the system of equations above to have non-trivial solutions is that its coefficient determinant is zero:

$$\Rightarrow \quad \begin{vmatrix} k - m\omega^2 & -k & 0 \\ -k & 2k - M\omega^2 & -k \\ 0 & -k & k - m\omega^2 \end{vmatrix} = 0 \tag{6}$$

$$\Rightarrow \quad \omega^2\left[Mm^2\omega^4 - 2km(M+m)\omega^2 + k^2(M+2m)\right] = 0 \tag{7}$$

$$\Rightarrow \quad \begin{cases} \omega_1 = \sqrt{\dfrac{k}{m}} & \text{(8a)} \\ \omega_2 = \sqrt{\dfrac{k}{\dfrac{Mm}{M+2m}}} & \text{(8b)} \\ \omega_3 = 0 & \text{(8c)} \end{cases}$$

It can be seen that the system has two vibration modes (the zero frequency ω_3 represents the rigid translation of the whole system, not the internal vibration mode).

Mode 1: $\omega = \omega_1 = \sqrt{\dfrac{k}{m}}$

(8a) $\rightarrow$ (5),

$$\Rightarrow \qquad \begin{cases} \tilde{A}_1 = -\tilde{A}_3 & (9a) \\ \tilde{A}_2 = 0 & (9b) \end{cases}$$

It can be seen that in this vibration mode, the middle ball remains stationary, the left and right balls perform the simple harmonic vibrations with the same amplitude and the phase difference π (that is, their vibration directions are opposite all the time). Based on this, we can draw the schematic diagram of Mode 1(see Fig. 3a).

Mode 2: $\omega = \omega_2 = \sqrt{k/\dfrac{Mm}{M+2m}}$

(8b) $\rightarrow$ (5),

$$\Rightarrow \qquad \begin{cases} \tilde{A}_1 = \tilde{A}_3 & (10a) \\ \tilde{A}_2 = -\dfrac{2m}{M}\tilde{A}_3 & (10b) \end{cases}$$

It can be seen that in this vibration mode, the left and right balls perform the simple harmonic vibration with the same amplitude and the same phase (that is, their vibration directions are the same all the time), and the phase difference between the middle ball and the balls on both sides is π (that is, their vibration directions is opposite all the time). Based on this, we can draw the schematic diagram of Mode 2 (see Fig. 3b).

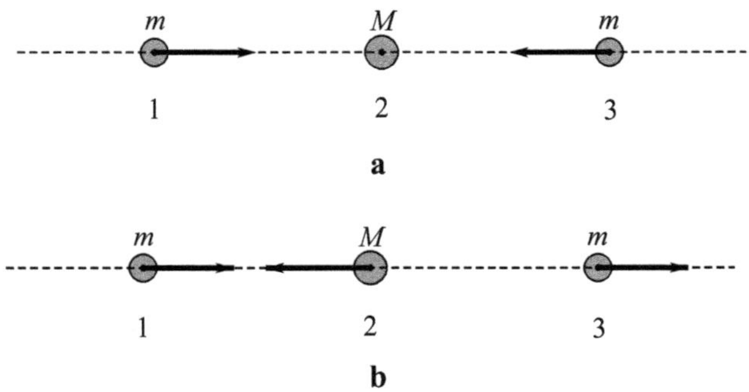

Figure 3a, b

4.2.4 Coupled Simple Pendulum System 1 (d, *C*)

Two light rigid strings of equal length l and two small balls of equal mass m are connected in sequence and hung on a ceiling. Find the modes of the tiny vibration of the system near the equilibrium position.

Solution

Time zero is arbitrarily set.

As shown in Fig. 1, the balls above and below are denoted by Ball 1 and Ball 2, respectively, the hanging point of the string above is denoted by O_1, the position of Ball 1 at arbitrary Time t is denoted by O_2. Establishing the Cartesian coordinate system with the frame $(O_1, \boldsymbol{i}, \boldsymbol{j})$ (let x, y be the Cartesian coordinates), the polar coordinate system 1 (let r_1, θ_1 be the polar coordinates) and the polar coordinate system 2 fixed to Ball 1 (let r_2, θ_2 be the polar coordinates).

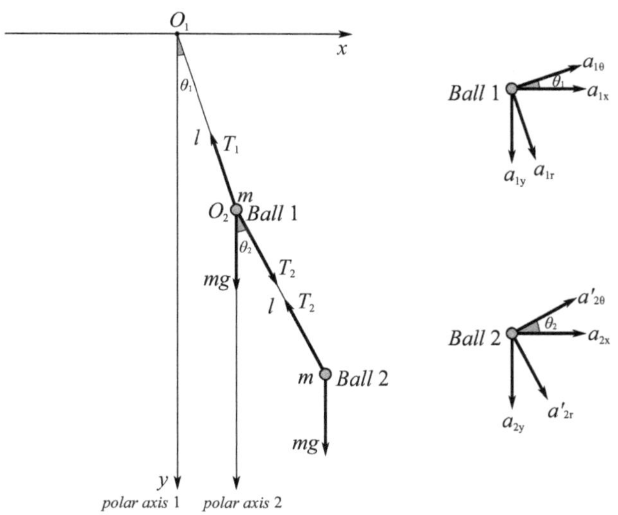

Figure 1

Taking *Ball 1* and *the system composed of Ball 1 and Ball 2* as the object of study, respectively. Studying *the state at arbitrary Time t*. Both considering ***Newton's second law*** (the component equations in the $(\boldsymbol{i}, \boldsymbol{j})$-directions):

$$\begin{cases} -T_1 \sin\theta_1 + T_2 \sin\theta_2 = ma_{1x} & \text{(1a)} \\ mg - T_1 \cos\theta_1 + T_2 \cos\theta_2 = ma_{1y} & \text{(1b)} \end{cases}$$

$$\begin{cases} -T_1 \sin\theta_1 = ma_{1x} + ma_{2x} & \text{(2a)} \\ 2mg - T_1 \cos\theta_1 = ma_{1y} + ma_{2y} & \text{(2b)} \end{cases}$$

T_1 and T_2—the magnitude of the tension in the string above and the string below at Time *t*.

(a_{1x}, a_{1y})—the components of the acceleration of Ball 1 in the $(\boldsymbol{i}, \boldsymbol{j})$-directions at Time *t*.

(a_{2x}, a_{2y})—the components of the acceleration of Ball 2 in the $(\boldsymbol{i}, \boldsymbol{j})$-directions at Time *t*.

Considering ***the components form of the acceleration of a particle in polar coordinate system*** [the constraint condition: the length of both strings remains constant]:

$$\begin{cases} a_{1r} = -l\dot{\theta}_1^2 & \text{(3a)} \\ a_{1\theta} = l\ddot{\theta}_1 & \text{(3b)} \end{cases}$$

$$\begin{cases} a'_{2r} = -l\dot{\theta}_2^2 & \text{(4a)} \\ a'_{2\theta} = l\ddot{\theta}_2 & \text{(4b)} \end{cases}$$

$(a_{1r}, a_{1\theta})$—the components of the acceleration of Ball 1 in the $(\boldsymbol{r}_1^0, \boldsymbol{\theta}_1^0)$-directions at arbitrary Time *t*.

$(a'_{2r}, a'_{2\theta})$—the components of the acceleration of Ball 2 relative to Ball 1 in the $(\boldsymbol{r}_2^0, \boldsymbol{\theta}_2^0)$-directions at arbitrary Time *t*.

Considering the transformation of the acceleration of Ball 1 from Base $(\boldsymbol{r}_1^0, \boldsymbol{\theta}_1^0)$ to Base $(\boldsymbol{i}, \boldsymbol{j})$, and the transformation of the acceleration of Ball 2 relative to Ball 1 from Base $(\boldsymbol{r}_2^0, \boldsymbol{\theta}_2^0)$ to Base $(\boldsymbol{i}, \boldsymbol{j})$:

$$\Rightarrow \begin{cases} \begin{pmatrix} a_{1x} \\ a_{1y} \end{pmatrix} = \begin{pmatrix} \cos\theta_1 & \sin\theta_1 \\ -\sin\theta_1 & \cos\theta_1 \end{pmatrix} \begin{pmatrix} a_{1\theta} \\ a_{1r} \end{pmatrix} = \begin{pmatrix} l\ddot{\theta}_1 \cos\theta_1 - l\dot{\theta}_1^2 \sin\theta_1 \\ -l\ddot{\theta}_1 \sin\theta_1 - l\dot{\theta}_1^2 \cos\theta_1 \end{pmatrix} & \begin{matrix} \text{(5a)} \\ \text{(5b)} \end{matrix} \\ \begin{pmatrix} a'_{2x} \\ a'_{2y} \end{pmatrix} = \begin{pmatrix} \cos\theta_2 & \sin\theta_2 \\ -\sin\theta_2 & \cos\theta_2 \end{pmatrix} \begin{pmatrix} a'_{2\theta} \\ a'_{2r} \end{pmatrix} = \begin{pmatrix} l\ddot{\theta}_2 \cos\theta_2 - l\dot{\theta}_2^2 \sin\theta_2 \\ -l\ddot{\theta}_2 \sin\theta_2 - l\dot{\theta}_2^2 \cos\theta_2 \end{pmatrix} & \begin{matrix} \text{(6a)} \\ \text{(6b)} \end{matrix} \end{cases}$$

(a'_{2x}, a'_{2y})—the components of the acceleration of Ball 2 relative to Ball 1 in the $(\boldsymbol{i}, \boldsymbol{j})$-directions at arbitrary Time *t*.

Taking *Ball 2* as the object of study. Studying *the state at arbitrary Time t*. Considering ***theorem for the composition of accelerations of a particle*** [taking the ground as the stationary reference frame, and the translational reference frame that is at rest with respect to Ball 1 as the moving reference frame] (the component equations in the $(\boldsymbol{i}, \boldsymbol{j})$-directions):

$$\begin{cases} a_{2x} = a_{1x} + a'_{2x} & (7a) \\ a_{2y} = a_{1y} + a'_{2y} & (7b) \end{cases}$$

θ_1 and θ_2 are small quantities, taking the following approximations:

$$\sin\theta_1 \approx \theta_1, \cos\theta_1 \approx 1, \sin\theta_2 \approx \theta_2, \cos\theta_2 \approx 1 \qquad (8a\text{–}d)$$

Combining (5) to (8),

$$\Rightarrow \begin{cases} \begin{cases} a_{1x} = l\ddot{\theta}_1 & (9a) \\ a_{1y} = 0 & (9b) \end{cases} \\ \begin{cases} a_{2x} = l(\ddot{\theta}_1 + \ddot{\theta}_2) & (10a) \\ a_{2y} = 0 & (10b) \end{cases} \end{cases}$$

(8) to (10) $\rightarrow$ (1) and (2),

$$\Rightarrow \begin{cases} \begin{cases} -T_1\theta_1 + T_2\theta_2 = ml\ddot{\theta}_1 & (11a) \\ mg - T_1 + T_2 = 0 & (11b) \end{cases} \\ \begin{cases} -T_1\theta_1 = ml\ddot{\theta}_1 + ml(\ddot{\theta}_1 + \ddot{\theta}_2) & (12a) \\ 2mg - T_1 = 0 & (12b) \end{cases} \end{cases}$$

$$\Rightarrow \begin{cases} \ddot{\theta}_1 + 2\dfrac{g}{l}\theta_1 - \dfrac{g}{l}\theta_2 = 0 & (13a) \\ \ddot{\theta}_2 + 2\dfrac{g}{l}\theta_2 - 2\dfrac{g}{l}\theta_1 = 0 & (13b) \end{cases}$$

Let the trial solutions of θ_1 and θ_2 be:

$$\begin{cases} \theta_1 = \mathrm{Re}\left[\tilde{\Theta}_1 \cos(\omega t)\right] & (14a) \\ \theta_2 = \mathrm{Re}\left[\tilde{\Theta}_2 \cos(\omega t)\right] & (14b) \end{cases}$$

ω—the characteristic frequency of the system.

$\tilde{\Theta}_1$ and $\tilde{\Theta}_2$—the complex amplitude (angular) of the simple harmonic vibration of Ball 1 and Ball 2.

$$\Rightarrow \quad \begin{cases} \left(\omega^2 - 2\dfrac{g}{l}\right)\tilde{\Theta}_1 + \dfrac{g}{l}\tilde{\Theta}_2 = 0 & (15a) \\ 2\dfrac{g}{l}\tilde{\Theta}_1 + \left(\omega^2 - 2\dfrac{g}{l}\right)\tilde{\Theta}_2 = 0 & (15b) \end{cases}$$

The condition for the system of equations above to have non-trivial solutions is that its coefficient determinant is zero:

$$\Rightarrow \quad \begin{vmatrix} \omega^2 - 2\dfrac{g}{l} & \dfrac{g}{l} \\ 2\dfrac{g}{l} & \omega^2 - 2\dfrac{g}{l} \end{vmatrix} = 0 \tag{16}$$

$$\Rightarrow \quad \left(\omega^2 - 2\frac{g}{l}\right)^2 - 2\left(\frac{g}{l}\right)^2 = 0 \tag{17}$$

$$\Rightarrow \quad \begin{cases} \omega_1 = \sqrt{(2+\sqrt{2})\dfrac{g}{l}} & (18a) \\ \omega_2 = \sqrt{(2-\sqrt{2})\dfrac{g}{l}} & (18b) \end{cases}$$

It can be seen that the system has two vibration modes.

Mode 1: $\omega = \omega_1 = \sqrt{(2+\sqrt{2})\dfrac{g}{l}}$

(18a) $\rightarrow$ (15),

$$\Rightarrow \quad \frac{\tilde{\Theta}_1}{\tilde{\Theta}_2} = \frac{\sqrt{2}}{2} \tag{19}$$

It can be seen that the rotation directions of the two strings are the same all the time in this vibration mode.

Based on this, we can draw the schematic diagram of Mode 1(see Fig. 2a).

Mode 2: $\omega = \omega_2 = \sqrt{(2-\sqrt{2})\dfrac{g}{l}}$

(18b) $\rightarrow$ (15),

$$\Rightarrow \quad \frac{\tilde{\Theta}_1}{\tilde{\Theta}_2} = -\frac{\sqrt{2}}{2} \tag{20}$$

It can be seen that the rotation directions of the two strings are opposite all the time in this vibration mode.

Based on this, we can draw the schematic diagram of Mode 2 (see Fig. 2b).

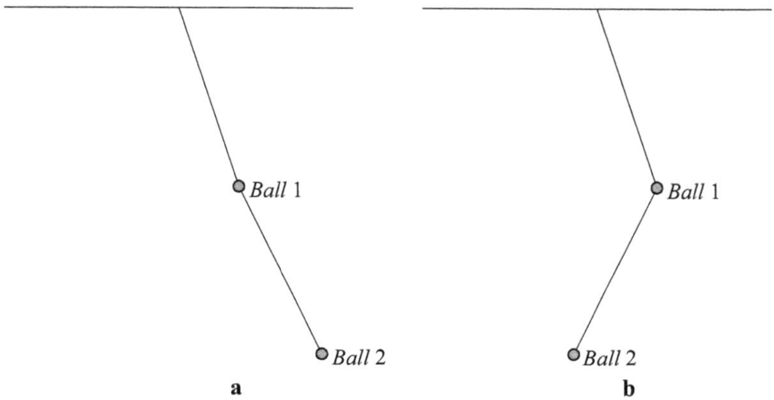

Figure 2a, b

4.2.5 Coupled Simple Pendulum System 2 (d, *C*)

Two simple pendulums of mass m and length l are hung at the same height (see Fig. 1). The two pendulum balls are connected by a light spring whose stiffness coefficient is k and the natural length is equal to the distance between the two hanging points.

(1) Find the kinematics equations of the tiny vibrations of the two pendulum balls near their equilibrium positions.

(2) Please design some special initial conditions and deduce the specific forms of the kinematics equations of the two pendulum balls.

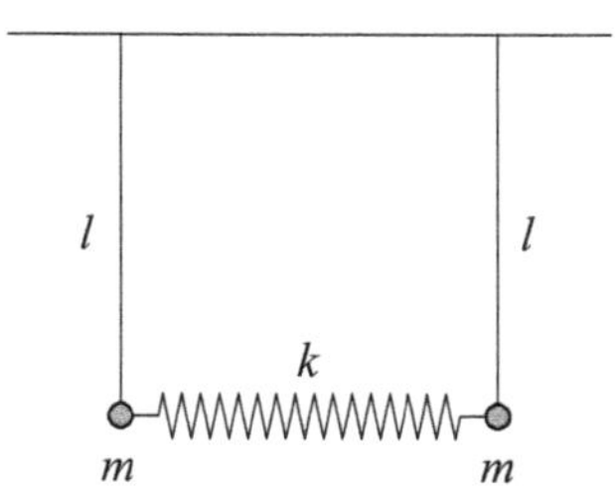

Figure 1

Solution to Problem (1)

Time zero is arbitrarily set.

The ball on the left is denoted by Ball 1, and the ball on the right is denoted by Ball 2.

As shown in Fig. 2, respectively establishing the linear coordinate system 1 (let x_1 be the coordinate) and the polar coordinate system 1 (let r_1, θ_1 be the polar coordinates) for describing Ball 1, and respectively establishing the linear coordinate system 2 (let x_2 be the coordinate) and the polar coordinate system 2 (let r_2, θ_2 be the polar coordinates) for describing Ball 2.

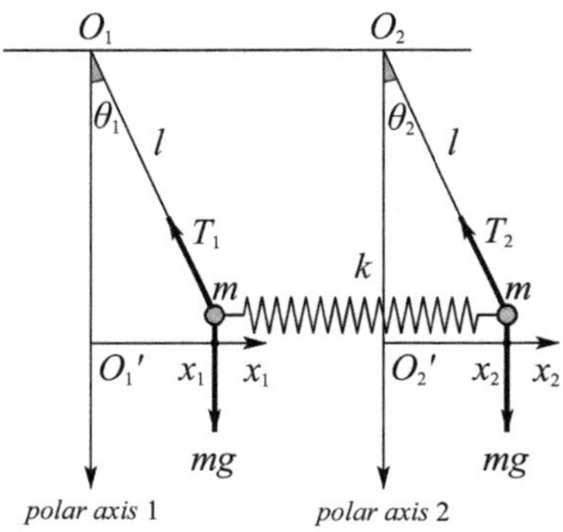

Figure 2

Taking *Ball 1 and Ball 2* as the object of study respectively. Studying *the state at arbitrary Time t*. Both considering ***Newton's second law*** (for Ball 1: the $\boldsymbol{i}$-direction and $-\boldsymbol{r}_1^0$-direction component equations, for Ball 2: the $\boldsymbol{i}$-direction and $-\boldsymbol{r}_2^0$-direction component equations):

$$\begin{cases} -T_1 \sin\theta_1 + k(x_2 - x_1) = m\ddot{x}_1 & \text{(1a)} \\ T_1 - mg\cos\theta_1 - k(x_2 - x_1)\sin\theta_1 \approx 0 & \text{(1b)} \end{cases}$$

$$\begin{cases} -T_2 \sin\theta_2 - k(x_2 - x_1) = m\ddot{x}_2 & \text{(2a)} \\ T_2 - mg\cos\theta_2 + k(x_2 - x_1)\sin\theta_2 \approx 0 & \text{(2b)} \end{cases}$$

T_1 and T_2—the magnitude of the tension in the strings acting on Ball 1 and Ball 2 at Time t.

Considering the following geometric relations in Fig. 2:

$$\begin{cases} x_1 = l\sin\theta_1 & (3a) \\ x_2 = l\sin\theta_2 & (3b) \end{cases}$$

θ_1 and θ_2 are small quantities, taking the following approximations:

$$\sin\theta_1 \approx \theta_1, \cos\theta_1 \approx 1, \sin\theta_2 \approx \theta_2, \cos\theta_2 \approx 1 \quad (4a\text{–}d)$$

$$\Rightarrow \begin{cases} T_1 = mg\cos\theta_1 + k(x_2 - x_1)\sin\theta_1 \approx mg & (5a) \\ T_2 = mg\cos\theta_2 - k(x_2 - x_1)\sin\theta_2 \approx mg & (5b) \end{cases}$$

Defining the constant ω_0:

$$\omega_0 = \sqrt{\frac{g}{l}} \quad (6)$$

(4) to (6) $\rightarrow$ (1a) and (2a),

$$\Rightarrow \begin{cases} -m\omega_0^2 x_1 + k(x_2 - x_1) = m\ddot{x}_1 & (7a) \\ -m\omega_0^2 x_2 - k(x_2 - x_1) = m\ddot{x}_2 & (7b) \end{cases}$$

$$\Rightarrow \begin{cases} (\ddot{x}_1 + \ddot{x}_2) + \omega_0^2(x_1 + x_2) = 0 & (8a) \\ (\ddot{x}_1 - \ddot{x}_2) + (\omega_0^2 + \dfrac{2k}{m})(x_1 - x_2) = 0 & (8b) \end{cases}$$

Defining the variables q_1 and q_2:

$$\begin{cases} q_1 = x_1 + x_2 & (9a) \\ q_2 = x_1 - x_2 & (9b) \end{cases}$$

$$\Rightarrow \begin{cases} \ddot{q}_1 + \omega_0^2 q_1 = 0 & (10a) \\ \ddot{q}_2 + \left(\omega_0^2 + \dfrac{2k}{m}\right) q_2 = 0 & (10b) \end{cases}$$

The motions described by q_1 and q_2 are simple harmonic vibrations, and their angular frequencies ω_1 and ω_2 are as follows:

$$\Rightarrow \begin{cases} \omega_1 = \omega_0 & (11a) \\ \omega_2 = \sqrt{\omega_0^2 + \dfrac{2k}{m}} & (11b) \end{cases}$$

The general solutions of (10a) and (10b) are as follows:

$$\begin{cases} q_1 = A_1' \cos(\omega_1 t + \varphi_1) & (12a) \\ q_2 = A_2' \cos(\omega_2 t + \varphi_2) & (12b) \end{cases}$$

A'_1 and φ_1—the amplitude and initial phase of the simple harmonic vibration described by q_1.

A'_2 and φ_2—the amplitude and initial phase of the simple harmonic vibration described by q_2.

(12) → (9),

$$\Rightarrow \begin{cases} x_1 = \frac{1}{2}\left[A_1' \cos(\omega_1 t + \varphi_1) + A_2' \cos(\omega_2 t + \varphi_2)\right] & (13a) \\ x_2 = \frac{1}{2}\left[A_1' \cos(\omega_1 t + \varphi_1) - A_2' \cos(\omega_2 t + \varphi_2)\right] & (13b) \end{cases}$$

Defining the constants A_1 and A_2:

$$\begin{cases} A_1 = \frac{1}{2} A_1' & (14a) \\ A_2 = \frac{1}{2} A_2' & (14b) \end{cases}$$

$$\Rightarrow \begin{cases} x_1 = A_1 \cos(\omega_1 t + \varphi_1) + A_2 \cos(\omega_2 t + \varphi_2) & (15a) \\ x_2 = A_1 \cos(\omega_1 t + \varphi_1) - A_2 \cos(\omega_2 t + \varphi_2) & (15b) \end{cases}$$

Equations (15a) and (15b) are the kinematics equations of the two pendulum balls performing the tiny vibrations near their equilibrium positions. Ball 1 and Ball 2 both simultaneously participate in the tiny simple harmonic vibrations with ω_1 and ω_2 as the characteristic frequencies, respectively.

Solution to Problem (2)

Special case (a): The initial state is shown in Fig. 3a:

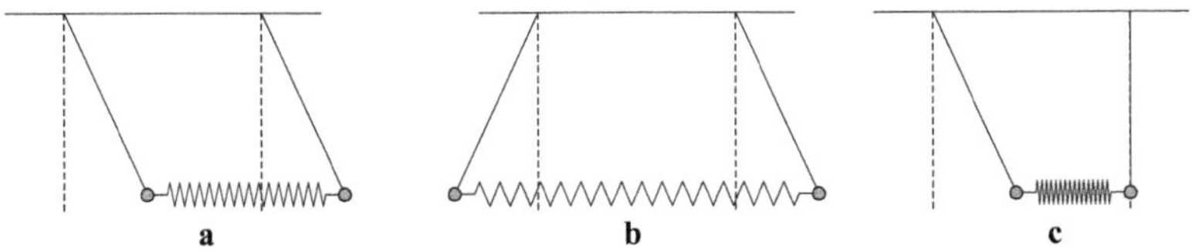

Figure 3a–c

$$x_1\big|_{t=0} = x_2\big|_{t=0} > 0 \tag{16a}$$

$$\dot{x}_1\big|_{t=0} = \dot{x}_2\big|_{t=0} = 0 \tag{16b}$$

(15) → (16),

$$\Rightarrow \quad A_2 = 0 \tag{17}$$

$$\Rightarrow \quad \begin{cases} x_1 = A_1\cos(\omega_1 t + \varphi_1) & (18a) \\ x_2 = A_1\cos(\omega_1 t + \varphi_1) & (18b) \end{cases}$$

In this case, the simple harmonic vibration with ω_2 as the characteristic frequency disappears, and Ball 1 and Ball 2 synchronously perform the simple harmonic vibration with ω_1 as the characteristic frequency.

Special case (b): The initial state is shown in Fig. 3b:

$$x_1\big|_{t=0} = -x_2\big|_{t=0} < 0 \tag{19a}$$

$$\dot{x}_1\big|_{t=0} = \dot{x}_2\big|_{t=0} = 0 \tag{19b}$$

(15) → (19),

$$\Rightarrow \quad A_1 = 0 \tag{20}$$

$$\Rightarrow \quad \begin{cases} x_1 = A_2\cos(\omega_2 t + \varphi_2) & (21a) \\ x_2 = -A_2\cos(\omega_2 t + \varphi_2) & (21b) \end{cases}$$

In this case, the simple harmonic vibration with ω_1 as the characteristic frequency disappears, and Ball 1 and Ball 2 perform the simple harmonic vibrations with the same characteristic frequency ω_2 and with opposite phase.

Special case (c): The initial state is shown in Fig. 3c:

$$x_1\big|_{t=0} = A \tag{22a}$$

$$x_2\big|_{t=0} = 0 \tag{22b}$$

$$\dot{x}_1\big|_{t=0} = \dot{x}_2\big|_{t=0} = 0 \tag{22c}$$

(15) → (22),

$$\Rightarrow \quad \begin{cases} \varphi_1 = \varphi_2 = 0 & (23a) \\ A_1 = A_2 \equiv \dfrac{A}{2} & (23b) \end{cases}$$

$$\Rightarrow \begin{cases} x_1 = \dfrac{A}{2}(\cos\omega_1 t + \cos\omega_2 t) \\ \quad = \left\{A\cos\left[\dfrac{1}{2}(\omega_2 - \omega_1)t\right]\right\}\cos\left[\dfrac{1}{2}(\omega_2 + \omega_1)t\right] & \text{(24a)} \\ x_2 = \dfrac{A}{2}(\cos\omega_1 t - \cos\omega_2 t) \\ \quad = \left\{A\sin\left[\dfrac{1}{2}(\omega_2 - \omega_1)t\right]\right\}\sin\left[\dfrac{1}{2}(\omega_2 + \omega_1)t\right] & \text{(24b)} \end{cases}$$

That is, the motions of Ball 1 and Ball 2 can be regarded as the simple harmonic vibrations with $\frac{1}{2}(\omega_2 + \omega_1)$ as the angular frequency and $A\cos\left[\frac{1}{2}(\omega_2 - \omega_1)t\right]$ and $A\sin\left[\frac{1}{2}(\omega_2 - \omega_1)t\right]$ as the time-varying amplitude, respectively.

4.2.6 The Tiny Vibration of a Homogeneous Rod Hung by a Light Rigid String (d, *C*)

A homogeneous rod of mass m and length $2l$ is hung on a ceiling by a light rigid string of length l (see Fig. 1). Find the vibration modes of the rod near its equilibrium position.

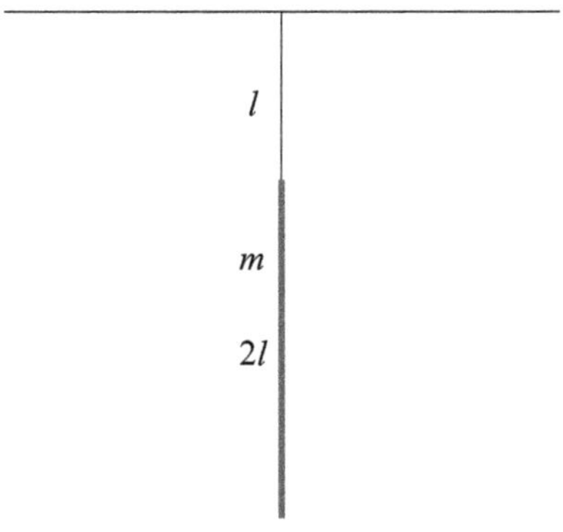

Figure 1

Solution

Once the rod is in mechanical equilibrium is defined as Time zero.

As shown in Fig. 2, the fixed point of the string is denoted by *O*, the upper and lower ends of the rod are denoted by *A* and *B*, respectively, the center of mass of the rod is denoted by *C*. Establishing the Cartesian coordinate system with the frame (*O*, ***i***, ***j***, ***k***) (let *x*, *y*, *z* be the Cartesian coordinates), the cylindrical coordinate system 1 (let *r*, *θ*, *z* be the cylindrical coordinates), and the cylindrical coordinate system 2 fixed to Point *A* (let *ρ*, *φ*, *z* be the cylindrical coordinates).

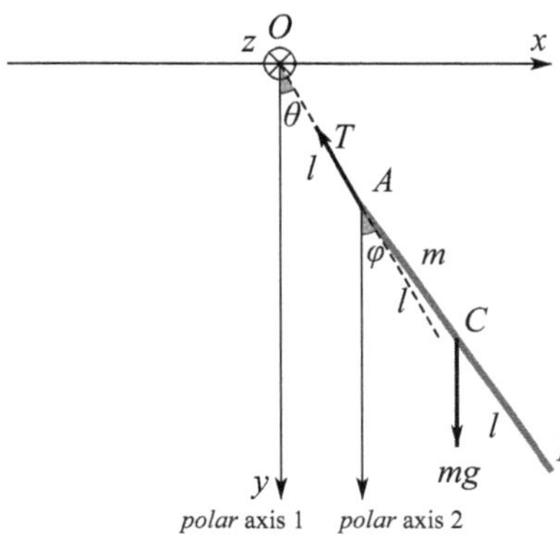

Figure 2

Taking *the rod* as the object of study. Studying *the state at arbitrary Time t*. Considering (a) ***theorem of motion of center of mass*** (the ***i***-direction and ***j***-direction component equations), (b) ***law of rotation*** [in its center-of-mass reference frame, taking its central axis parallel to the *z*-axis as the rotation axis] (the −***k***-direction component equation), (c) ***in the center-of-mass reference frame, the resultant torque of the inertia force acting on system of particles is zero***, (d) ***the form of moment of inertia of a homogeneous rod about one of its central axes***:

$$\begin{cases} -T\sin\theta = m\ddot{x}_C & \text{(1a)} \\ mg - T\cos\theta = m\ddot{y}_C & \text{(1b)} \end{cases}$$

$$T\sin\theta \cdot l\cos\varphi - T\cos\theta \cdot l\sin\varphi = I\ddot{\varphi} \tag{2}$$

$$I = \frac{1}{12}m(2l)^2 \tag{3}$$

T—the magnitude of the tension in the string acting on the rod at Time *t*.

I—the moment of inertia of the rod about its central axis parallel to the z-axis.

Considering the following geometric relations in Fig. 2:

$$\begin{cases} x_C = l\sin\theta + l\sin\varphi & (4a) \\ y_C = l\cos\theta + l\cos\varphi & (4b) \end{cases}$$

θ and φ are small quantities, taking the following approximations:

$$\sin\theta \approx \theta, \sin\varphi \approx \varphi, \cos\theta \approx 1, \cos\varphi \approx 1 \qquad (5a–d)$$

$$\Rightarrow \quad \begin{cases} x_C \approx l(\theta+\varphi) & (6a) \\ y_C \approx 2l & (6b) \end{cases}$$

(5) and (6) → (1) and (2):

$$\Rightarrow \quad \begin{cases} \begin{cases} -T\theta = ml(\ddot{\theta}+\ddot{\varphi}) & (7a) \\ mg - T = 0 & (7b) \end{cases} \\ Tl(\theta-\varphi) = I\ddot{\varphi} & (8) \end{cases}$$

$$\Rightarrow \quad \begin{cases} l(\ddot{\theta}+\ddot{\varphi}) + g\theta = 0 & (9a) \\ \dfrac{l}{3}\ddot{\varphi} + g(\varphi-\theta) = 0 & (9b) \end{cases}$$

Let the trial solutions of θ and φ be:

$$\begin{cases} \theta = \mathrm{Re}[\tilde{\Theta}\sin(\omega t)] & (10a) \\ \varphi = \mathrm{Re}[\tilde{\Phi}\sin(\omega t)] & (10b) \end{cases}$$

ω—the characteristic frequency of the system.

$\tilde{\Theta}$ and $\tilde{\Phi}$—the complex amplitude (angular) of the simple harmonic vibrations of the string and the rod.

$$\Rightarrow \quad \begin{cases} (g - l\omega^2)\tilde{\Theta} - l\omega^2\tilde{\Phi} = 0 & (11a) \\ g\tilde{\Theta} + \left(\dfrac{1}{3}l\omega^2 - g\right)\tilde{\Phi} = 0 & (11b) \end{cases}$$

The condition for the system of equations above to have non-trivial solutions is that its coefficient determinant is zero:

$$\Rightarrow \quad \begin{vmatrix} g - l\omega^2 & -l\omega^2 \\ g & \dfrac{1}{3}l\omega^2 - g \end{vmatrix} = 0 \qquad (12)$$

$$\Rightarrow \qquad (g-l\omega^2)\left(\frac{1}{3}l\omega^2-g\right)+gl\omega^2=0$$

$$\Rightarrow \qquad l^2\omega^4-7gl\omega^2+3g^2=0 \tag{13}$$

$$\Rightarrow \qquad \begin{cases} \omega_1=\sqrt{\dfrac{7+\sqrt{37}}{2}\dfrac{g}{l}} & \text{(14a)} \\ \omega_2=\sqrt{\dfrac{7-\sqrt{37}}{2}\dfrac{g}{l}} & \text{(14b)} \end{cases}$$

It can be seen that the system has two vibration modes.

Mode 1: $\omega=\omega_1=\sqrt{\dfrac{7+\sqrt{37}}{2}\dfrac{g}{l}}$

(14a) $\rightarrow$ (11a),

$$\Rightarrow \qquad \frac{\tilde{\Phi}}{\tilde{\Theta}}=\frac{g}{l\omega_1^2}-1=\frac{2}{7+\sqrt{37}}-1<0 \tag{15}$$

It can be seen that the rotation directions of the string and the rod are opposite all the time in this vibration mode. Based on this, we can draw the schematic diagram of Mode 1 (see Fig. 2a).

Mode 2: $\omega_2=\sqrt{\dfrac{7-\sqrt{37}}{2}\dfrac{g}{l}}$

(14b) $\rightarrow$ (11a),

$$\Rightarrow \qquad \frac{\tilde{\Phi}}{\tilde{\Theta}}=\frac{g}{l\omega_2^2}-1=\frac{2}{7-\sqrt{37}}-1>0 \tag{16}$$

Figure 2a, b

It can be seen that the rotation directions of the string and the rod are the same all the time in this vibration mode. Based on this, we can draw the schematic diagram of Mode 2 (see Fig. 2b).

4.2.7 Benzene Ring Model (*c*, *C*)[8]

Consider a benzene ring model as follows: six small beads of mass m are strung on a fixed smooth light ring, the beads are connected alternately by six springs, whose natural length are all one sixth of the circumference of the ring and the stiffness coefficients are all k (see Fig. 1).

(1) Determine all vibration modes (tiny vibration) of the system and draw corresponding schematic diagrams, use the arrow to indicate the complex amplitude of each bead, and paint the stationary beads black.

(2) Determine which vibration modes are related to the vibration modes of actual benzene ring.

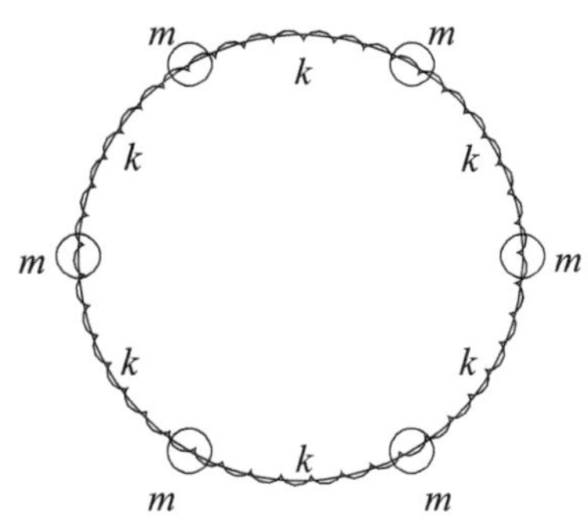

Figure 1

Solution to Problem (1)

As shown in Fig. 2, sorting the six beads with positive integers anticlockwise. The center of the ring is denoted by O. Taking the center O as the pole, the direction of the origin O pointing NO. 1 bead as the positive direction of polar axis, the anticlockwise direction as the positive direction of polar angle, establishing the polar coordinate system, let r, θ be the polar coordinate.

[8]This problem is a test question from Princeton University in 1977. It is also included in the following book: Yongde Zhang, et al. (1988). *American Physics Test Questions and Solutions* (University of Science and Technology of China Press).

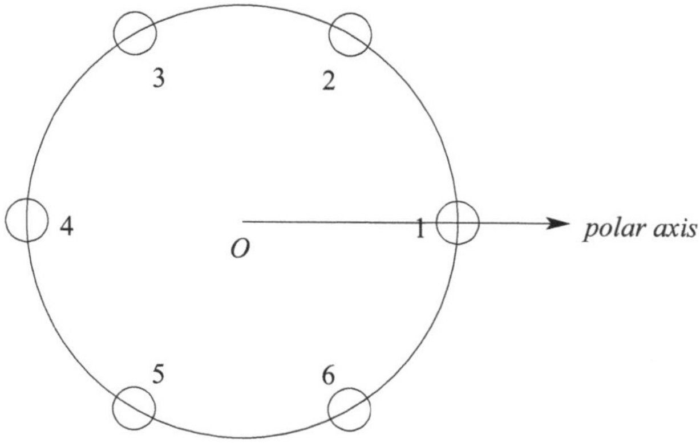

Figure 2

Once each bead is in the mechanical equilibrium position in the angular direction (all springs are the natural length) is defined as Time zero.

Taking *each bead* as the object of study, respectively. Studying *the state at arbitrary Time t*. Considering ***Newton's second law*** (the $\boldsymbol{\theta}^0$-direction component equation):

$$m\ddot{\theta}_i = k(\theta_{i+1} - \theta_i) - k(\theta_i - \theta_{i-1}), \quad i = 1,2,\cdots,6 \tag{1}$$

θ_i—the $\boldsymbol{\theta}^0$-direction component of the angular displacement of the No. i bead relative to its equilibrium position at Time t.

$$\Rightarrow \quad m\ddot{\theta}_i + k(-\theta_{i-1} + 2\theta_i - \theta_{i+1}) = 0, \quad i = 1,2,\cdots,6 \tag{2}$$

Let the trial solution to θ_i be:

$$\theta_i = \mathrm{Re}[\tilde{A}_i e^{i\omega t}], \quad i = 1,2,\cdots,6 \tag{3}$$

ω—the angular frequency of the tiny vibration of the system.

$\tilde{A}_i$—the complex amplitude (angular) of the No. i bead.

$$\Rightarrow \quad -m\omega^2 \tilde{A}_i + k(-\tilde{A}_{i-1} + 2\tilde{A}_i - \tilde{A}_{i+1}) = 0, \quad i = 1,2,\cdots,6 \tag{4}$$

Defining the constant ε:

$$\varepsilon = \frac{m\omega^2}{k} - 2 \tag{5}$$

$$\Rightarrow \quad \tilde{A}_{i-1} + \varepsilon \tilde{A}_i + \tilde{A}_{i+1} = 0, \quad i = 1,2,\cdots,6 \tag{6}$$

The condition for the system of equations above to have non-trivial solutions is that its coefficient determinant is zero:

$$\Rightarrow \quad \begin{vmatrix} \varepsilon & 1 & 0 & 0 & 0 & 1 \\ 1 & \varepsilon & 1 & 0 & 0 & 0 \\ 0 & 1 & \varepsilon & 1 & 0 & 0 \\ 0 & 0 & 1 & \varepsilon & 1 & 0 \\ 0 & 0 & 0 & 1 & \varepsilon & 1 \\ 1 & 0 & 0 & 0 & 1 & \varepsilon \end{vmatrix} = 0 \tag{7}$$

$$\Rightarrow \quad \varepsilon^6 - 6\varepsilon^4 + 9\varepsilon^2 - 4 = 0 \tag{8}$$

$$\Rightarrow \quad (\varepsilon+1)^2(\varepsilon-1)^2(\varepsilon+2)(\varepsilon-2) = 0$$

$$\Rightarrow \quad \begin{cases} \varepsilon_1 = 2, \quad \varepsilon_2 = -2, \quad \varepsilon_3 = 1 \\ \varepsilon_4 = 1, \quad \varepsilon_5 = -1, \quad \varepsilon_6 = 1 \end{cases} \tag{9a–f}$$

$$\Rightarrow \quad \begin{cases} \omega_1 = \sqrt{\dfrac{4k}{m}}, \quad \omega_2 = 0, \quad \omega_3 = \sqrt{\dfrac{3k}{m}} \\ \omega_4 = \sqrt{\dfrac{3k}{m}}, \quad \omega_5 = \sqrt{\dfrac{k}{m}}, \quad \omega_6 = \sqrt{\dfrac{k}{m}} \end{cases} \tag{10a–f}$$

It can be seen that the system has three vibration modes (the vibration models corresponding to ω_3 and ω_4 are degenerate, the vibration models corresponding to ω_5 and ω_6 are degenerate; the zero frequency ω_2 represents the rigid rotation of the whole system, not the internal vibration mode).

Using system of equations (6) to find out the relations between all the complex amplitudes corresponding to the angular frequencies calculated above can determine all the vibration modes, which is the most conventional method, but the calculation is very complicated. The following will try to find out some possible relations of the complex amplitudes according to the spatial symmetry and the conservation relations in the vibration process, and then use the Newton dynamic equation to verify and further solve them.

Considering the spatial symmetry and the conservation of angular momentum of the system, the following six vibration modes to be verified can be found (see Fig. 3 a–f), which are called vibration mode 1 to 6.

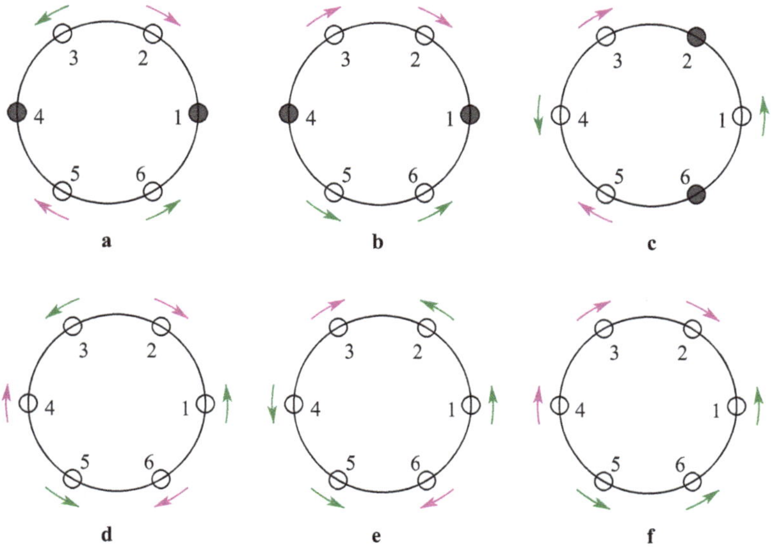

Figure 3a–f

Mode 1 to be verified

The relations between the complex amplitudes of the vibration mode 1 are as follows (see Fig. 3a):

$$\tilde{A}_1 = \tilde{A}_4 = 0, \quad \tilde{A}_3 = -\tilde{A}_2, \quad \tilde{A}_6 = -\tilde{A}_5 \tag{11a–c}$$

(11) → (6),

$$\Rightarrow \begin{cases} \tilde{A}_5 - \tilde{A}_2 = 0 & \text{(12a)} \\ -m\omega^2 \tilde{A}_2 + 3k\tilde{A}_2 = 0 & \text{(12b)} \\ m\omega^2 \tilde{A}_2 - 3k\tilde{A}_2 = 0 & \text{(12c)} \\ \tilde{A}_2 - \tilde{A}_5 = 0 & \text{(12d)} \\ -m\omega^2 \tilde{A}_5 + 3k\tilde{A}_5 = 0 & \text{(12e)} \\ m\omega^2 \tilde{A}_5 - 3k\tilde{A}_5 = 0 & \text{(12f)} \end{cases}$$

There are only three independent equations in the system of equations above.

$$\Rightarrow \begin{cases} \omega = \sqrt{\dfrac{3k}{m}} & \text{(13)} \\ \tilde{A}_2 = \tilde{A}_5 & \text{(14)} \end{cases}$$

This solution is reasonable, that is, the vibration mode 1 exists.

Mode 2 to be verified

The relations between the complex amplitudes of the vibration mode 2 are as follows (see Fig. 3b):

$$\tilde{A}_1 = \tilde{A}_4 = 0, \quad \tilde{A}_6 = -\tilde{A}_2, \quad \tilde{A}_5 = -\tilde{A}_3 \tag{15a–c}$$

(15) → (6):

$$\Rightarrow \begin{cases} \tilde{A}_2 - \tilde{A}_2 = 0 & \text{(16a)} \\ -m\omega^2\tilde{A}_2 + k(2\tilde{A}_2 - \tilde{A}_3) = 0 & \text{(16b)} \\ -m\omega^2\tilde{A}_3 + k(-\tilde{A}_2 + 2\tilde{A}_3) = 0 & \text{(16c)} \\ -\tilde{A}_3 + \tilde{A}_3 = 0 & \text{(16d)} \\ m\omega^2\tilde{A}_3 + k(-2\tilde{A}_3 + \tilde{A}_2) = 0 & \text{(16e)} \\ m\omega^2\tilde{A}_2 + k(\tilde{A}_3 - 2\tilde{A}_2) = 0 & \text{(16f)} \end{cases}$$

There are only two independent equations in the system of equations above, which are rearranged as follows:

$$\Rightarrow \begin{cases} (2k - m\omega^2)\tilde{A}_2 - k\tilde{A}_3 = 0 & \text{(17a)} \\ k\tilde{A}_2 + (m\omega^2 - 2k)\tilde{A}_3 = 0 & \text{(17b)} \end{cases}$$

The condition for the system of equations above to have non-trivial solutions is that its coefficient determinant is zero:

$$\Rightarrow \quad \begin{vmatrix} 2k - m\omega^2 & -k \\ k & m\omega^2 - 2k \end{vmatrix} = 0 \tag{18}$$

$$\Rightarrow \quad (m\omega^2 - 2k)^2 - k^2 = 0 \tag{19}$$

$$\Rightarrow \quad \omega_{(1)} = \sqrt{\frac{3k}{m}}, \quad \omega_{(2)} = \sqrt{\frac{k}{m}} \tag{20a, b}$$

(20a) → (17),

$$\Rightarrow \quad \tilde{A}_2 = -\tilde{A}_3 \tag{21}$$

That is, $\omega_{(1)} = \sqrt{\frac{3k}{m}}$ corresponds to the vibration mode 1.

(20b) → (17),

$$\Rightarrow \quad \tilde{A}_2 = \tilde{A}_3 \tag{22}$$

This solution is reasonable, that is, the vibration mode 2 exists.

Mode 3 to be verified

The relations between the complex amplitudes of the vibration mode 3 are as follows (see Fig. 3c):

$$\tilde{A}_2 = \tilde{A}_6 = 0, \quad \tilde{A}_3 = -\tilde{A}_1, \quad \tilde{A}_5 = -\tilde{A}_4 \tag{23a–c}$$

(23) → (6):

$$\Rightarrow \quad \begin{cases} -m\omega^2 \tilde{A}_1 + 2k\tilde{A}_1 = 0 & (24a) \\ -\tilde{A}_1 + \tilde{A}_1 = 0 & (24b) \\ -m\omega^2 \tilde{A}_1 + k(2\tilde{A}_1 + \tilde{A}_4) = 0 & (24c) \\ -m\omega^2 \tilde{A}_4 + k(\tilde{A}_1 + 3\tilde{A}_4) = 0 & (24d) \\ -m\omega^2 \tilde{A}_4 + 3k\tilde{A}_4 = 0 & (24e) \\ \tilde{A}_4 - \tilde{A}_1 = 0 & (24f) \end{cases}$$

The system of equations above only has zero solution, that is, the vibration mode 3 does not exist.

Mode 4 to be verified

The relations between the complex amplitudes of the vibration mode 4 are as follows (see Fig. 3d):

$$\tilde{A}_2 = -\tilde{A}_1, \quad \tilde{A}_4 = -\tilde{A}_3, \quad \tilde{A}_6 = -\tilde{A}_5 \tag{25a–c}$$

(25) → (6):

$$\Rightarrow \quad \begin{cases} -m\omega^2 \tilde{A}_1 + 3k\tilde{A}_1 + k\tilde{A}_5 = 0 & (26a) \\ -m\omega^2 \tilde{A}_1 + 3k\tilde{A}_1 + k\tilde{A}_3 = 0 & (26b) \\ -m\omega^2 \tilde{A}_3 + 3k\tilde{A}_3 + k\tilde{A}_1 = 0 & (26c) \\ -m\omega^2 \tilde{A}_3 + 3k\tilde{A}_3 + k\tilde{A}_5 = 0 & (26d) \\ -m\omega^2 \tilde{A}_5 + 3k\tilde{A}_5 + k\tilde{A}_3 = 0 & (26e) \\ -m\omega^2 \tilde{A}_5 + 3k\tilde{A}_5 + k\tilde{A}_1 = 0 & (26f) \end{cases}$$

$$\Rightarrow \quad \tilde{A}_1 = \tilde{A}_3 = \tilde{A}_5 \tag{27}$$

The following three independent equations can be obtained:

$$\Rightarrow \quad -m\omega^2 \tilde{A}_i + 4k\tilde{A}_i = 0, \quad i = 1,2,3 \tag{28}$$

$$\Rightarrow \quad \omega = \sqrt{\frac{4k}{m}} \tag{29}$$

This solution is reasonable, that is, the vibration mode 4 exists.

Mode 5 to be verified

The relations between the complex amplitudes of the vibration mode 5 are as follows (see Fig. 3e):

$$\tilde{A}_3 = -(\tilde{A}_2 + \tilde{A}_4), \quad \tilde{A}_6 = -(\tilde{A}_1 + \tilde{A}_5) \tag{30a, b}$$

(30) → (6):

$$\Rightarrow \quad \begin{cases} -m\omega^2 \tilde{A}_1 + k(3\tilde{A}_1 - \tilde{A}_2 + \tilde{A}_5) = 0 & \text{(31a)} \\ -m\omega^2 \tilde{A}_2 + k(3\tilde{A}_2 - \tilde{A}_1 + \tilde{A}_4) = 0 & \text{(31b)} \\ -m\omega^2 (\tilde{A}_2 + \tilde{A}_4) + 3k(\tilde{A}_2 + \tilde{A}_4) = 0 & \text{(31c)} \\ -m\omega^2 \tilde{A}_4 + k(3\tilde{A}_4 - \tilde{A}_5 + \tilde{A}_2) = 0 & \text{(31d)} \\ -m\omega^2 \tilde{A}_5 + k(3\tilde{A}_5 - \tilde{A}_4 + \tilde{A}_1) = 0 & \text{(31e)} \\ -m\omega^2 (\tilde{A}_1 + \tilde{A}_5) + 3k(\tilde{A}_1 + \tilde{A}_5) = 0 & \text{(31f)} \end{cases}$$

From (31c) or (31f):

$$\Rightarrow \quad \omega = \sqrt{\frac{3k}{m}} \tag{32}$$

(32) → (31a), (31e),

$$\Rightarrow \quad \begin{cases} \tilde{A}_2 = \tilde{A}_5 & \text{(33a)} \\ \tilde{A}_1 = \tilde{A}_4 & \text{(33b)} \end{cases}$$

This solution is reasonable, that is, the vibration mode 5 exists.

Mode 6 to be verified

The relations between the complex amplitudes of the vibration mode 6 are as follows (see Fig. 3f):

$$\tilde{A}_2 = -\tilde{A}_1, \quad \tilde{A}_3 = -\tilde{A}_6, \quad \tilde{A}_4 = -\tilde{A}_5 \tag{34a–c}$$

(34) → (6),

$$\Rightarrow \quad \begin{cases} -m\omega^2 \tilde{A}_1 + k(3\tilde{A}_1 - \tilde{A}_6) = 0 & \text{(35a)} \\ m\omega^2 \tilde{A}_1 + k(-3\tilde{A}_1 + \tilde{A}_6) = 0 & \text{(35b)} \\ m\omega^2 \tilde{A}_6 + k(\tilde{A}_1 + \tilde{A}_5 - 2\tilde{A}_6) = 0 & \text{(35c)} \\ m\omega^2 \tilde{A}_5 + k(-3\tilde{A}_5 + \tilde{A}_6) = 0 & \text{(35d)} \\ -m\omega^2 \tilde{A}_5 + k(3\tilde{A}_5 - \tilde{A}_6) = 0 & \text{(35e)} \\ -m\omega^2 \tilde{A}_6 + k(-\tilde{A}_1 - \tilde{A}_5 + 2\tilde{A}_6) = 0 & \text{(35f)} \end{cases}$$

(35b) × a – (35c) × b – (35d) × c,

$$\Rightarrow \quad -m\omega^2(a\tilde{A}_1 + b\tilde{A}_6 + c\tilde{A}_5) +$$

$$k[(3a-b)\tilde{A}_1 + (-a+2b-c)\tilde{A}_6 + (-b+3c)\tilde{A}_5] = 0 \tag{36}$$

Taking $a\theta_1 + b\theta_2 + c\theta_3$ as the normal coordinate, a, b and c shall satisfy the following system of equations:

$$\begin{cases} 3a - b = a & (37a) \\ -a + 2b - c = b & (37b) \\ -b + 3c = c & (37c) \end{cases}$$

$$\Rightarrow \quad \begin{cases} b = 2a & (38a) \\ c = a & (38b) \end{cases}$$

$$\Rightarrow \quad -m\omega^2(\tilde{A}_1 + 2\tilde{A}_6 + \tilde{A}_5) + k(\tilde{A}_1 + 2\tilde{A}_6 + \tilde{A}_5) = 0 \tag{39}$$

$$\Rightarrow \quad \omega = \sqrt{\frac{k}{m}} \tag{40}$$

(40) $\rightarrow$ (35),

$$\Rightarrow \quad \begin{cases} \tilde{A}_5 = \tilde{A}_1 & (41a) \\ \tilde{A}_6 = 2\tilde{A}_1 & (41b) \end{cases}$$

This solution is reasonable, that is, the vibration mode 6 exists.

Solution to Problem (2)

The actual benzene ring structure is a regular hexagon, similar to the above benzene ring model, but there is no fixed ring. The center of mass of the benzene ring should be stationary when there is no translational excitation, that is, only the center of mass is stationary among the vibration modes to be verified above, i.e., the vibration modes 1, 4, and 5 are related to the actual benzene ring.

4.3 The Vibration of Continuous System

The vibration of continuous system (with continuous distribution of mass and elastic properties) is highly similar to that of discrete system in physical analysis. From the mathematical point of view, continuum can be regarded as the degradation of discrete system, so the vibration problem of continuum can be transformed into the vibration problem of discrete system for research, which is a deeply enlightening and easy to apply research method.

As typical continuous systems, the vibration in the rod, chord and beam has been deeply studied.

In this section, some representative problems of vibration in continuum will be studied.

4.3.1 The Longitudinal Wave in a Homogeneous Isotropic Continuum and in a Light Springs-Balls System (d, *C*)

1. Find the propagation velocity of the longitudinal wave in a infinitely extended homogeneous isotropic continuum of density ρ and Young's modulus Y.
2. A great many of small balls of equal mass m are alternately connected by a great many of light springs of equal stiffness coefficient k and natural length l in collinear, placed on a smooth horizontal plane as a whole (see Fig. 1). Initially, all the balls are at rest, and all the springs are in the natural length. Apply an excitation with the angular frequency of ω in the direction of the connecting line of all the balls, $\omega << \sqrt{\dfrac{k}{m}}$.

   Find the propagation velocity of the longitudinal wave in this springs-balls system, then find its degeneration form in the situation of $l \to 0$.

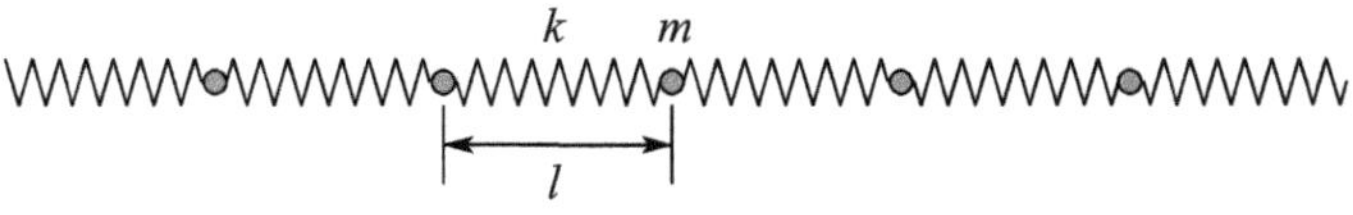

Figure 1

Solution to Problem (1)

Taking an arbitrary time point after the mechanical wave propagates all over the continuum and defining it as Time zero.

As shown in Fig. 2, taking an arbitrary point in the continuum and denoting it by O, the unit vector in the propagation direction of the longitudinal wave (horizontally to the right in the figure) is denoted by ***i***. Establishing the linear coordinate system with Point *O* as the origin and the ***i***-direction as the positive direction of coordinate axis, let *x* be the coordinate.

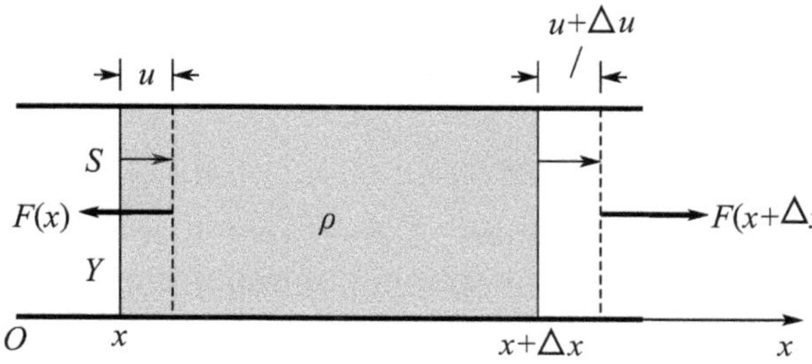

Figure 2

Taking *the cylindrical infinitesimal element of cross-sectional area S located between x and x + Δx before the vibration is excited [hereinafter referred to as element Δx]* as the object of study. Studying *the state at arbitrary Time t*. Considering ***Newton's second law*** (the ***i***-direction component equation):

$$F(x+\Delta x)-F(x)=\rho S\Delta x\frac{\partial^2 u}{\partial t^2} \tag{1}$$

F(*x*)—the −***i***-direction component of the elastic force acting on the left side of element Δ*x* at Time *t*.

F(*x* + Δ*x*)—the ***i***-direction component of the elastic force acting on the right side of element Δ*x* at Time *t*.

u—the ***i***-direction component of wave displacement at *x* at Time *t*.

Considering ***Hooke's law***:

$$\begin{cases} F(x)=YS\left.\dfrac{\partial u}{\partial x}\right|_x & \text{(2a)} \\ F(x+\Delta x)=YS\left.\dfrac{\partial u}{\partial x}\right|_{x+\Delta x} & \text{(2b)} \end{cases}$$

$$\Rightarrow \qquad \rho S\Delta x\frac{\partial^2 u}{\partial t^2}=YS\left.\frac{\partial u}{\partial x}\right|_{x+\Delta x}-YS\left.\frac{\partial u}{\partial x}\right|_x=YS\frac{\partial^2 u}{\partial x^2}\Delta x \tag{3}$$

$$\Rightarrow \qquad \frac{\partial^2 u}{\partial t^2} - \frac{Y}{\rho}\frac{\partial^2 u}{\partial x^2} = 0 \qquad (4)$$

Let the trial solution of Eq. (4) be:

$$u = A\cos\left[\omega\left(t - \frac{x}{v_{||}}\right)\right] \qquad (5)$$

A—the amplitude of element Δx.

$v_{||}$—the ***i***-direction component of the propagation velocity of the longitudinal wave in the homogeneous isotropic continuum.

$$\Rightarrow \qquad -\omega^2 A\cos\left[\omega\left(t - \frac{x}{v_{||}}\right)\right] = \frac{Y}{\rho}\left(-\frac{\omega^2}{v_{||}^2}\right)A\cos\left[\omega\left(t - \frac{x}{v_{||}}\right)\right] \qquad (6)$$

$$\Rightarrow \qquad v_{||} = \sqrt{\frac{Y}{\rho}} \qquad (7)$$

It can be seen that the propagation velocity of the longitudinal wave in homogeneous isotropic continuum is only determined by its inertia property and elasticity property.

Solution to Problem (2)

Taking an arbitrary time point after the mechanical wave propagates all over the system and defining it as Time zero.

As shown in Fig. 3, sorting all the balls from left to right with integers, setting the ball excited by the external force as No. 0 ball. The initial position of No. 0 ball is denoted by *O*, the unit vector in the direction of increasing sequence number is denoted by ***i***. Establishing the linear coordinate system with Point *O* as the origin and the ***i***-direction as the positive direction of coordinate axis, let *x* be the coordinate.

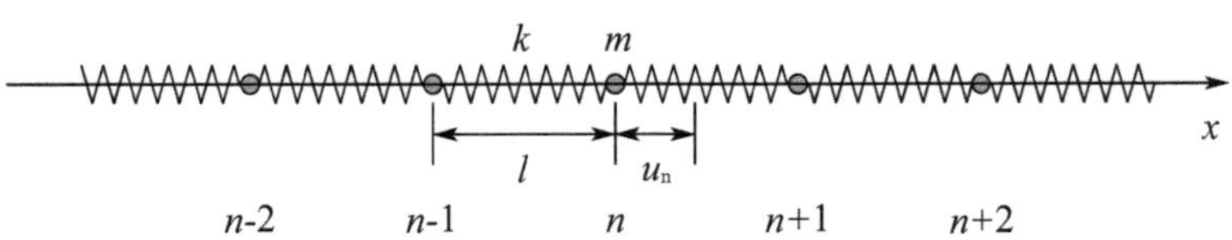

Figure 3

Taking *No. n ball* as the object of study. Studying *the state at arbitrary Time t*. Considering ***Newton's second law*** (the ***i***-direction component equation):

$$k(x_{n+1}-x_n)-k(x_n-x_{n-1})=m\ddot{x}_n \tag{8}$$

Defining u_n—the ***i***-direction component of the displacement of No. n ball at arbitrary Time t.

$$u_n=x_n-x_{n0}=x_n-nl \tag{9}$$

It can be seen from the symmetry that the angular frequencies and amplitudes of the simple harmonic vibrations of all balls are equal, and the phase differences between two small balls adjacent to each other at all positions are equal. According to this, let the trial solution to u_n be:

$$u_n=A\cos(\omega t-n\Delta\varphi) \tag{10}$$

A—the amplitude of the simple harmonic vibrations of all balls.

$\Delta\varphi$—the phase difference between two small balls adjacent to each other at all positions.

$$\begin{aligned}\Rightarrow\quad -m\omega^2 A\cos(\omega t-n\Delta\varphi) &= kA\{\cos[\omega t-(n+1)\Delta\varphi]+\cos[\omega t-(n-1)\Delta\varphi]\}- \\ &\quad 2kA\cos(\omega t-n\Delta\varphi) \\ &= 2kA\cos(\omega t-n\Delta\varphi)\cos\Delta\varphi-2kA\cos(\omega t-n\Delta\varphi) \\ &= -2kA\cos(\omega t-n\Delta\varphi)(1-\cos\Delta\varphi)\end{aligned} \tag{11}$$

$$\Rightarrow\qquad \cos\Delta\varphi=1-\frac{m\omega^2}{2k} \tag{12}$$

It can be seen from $\omega << \sqrt{\dfrac{k}{m}}$ that $\Delta\varphi$ is a small quantity, taking the second-order approximation of $\cos\Delta\varphi$:

$$\cos\Delta\varphi\approx 1-\frac{1}{2}(\Delta\varphi)^2 \tag{13}$$

$$\Rightarrow\qquad \Delta\varphi=\omega\sqrt{\frac{m}{k}} \tag{14}$$

Considering the definition of ***phase difference***:

$$\omega=\frac{\Delta\varphi}{\Delta t} \tag{15}$$

Δt—the time taken for the longitudinal wave to travel between two adjacent balls.

Defining $v_{||}$—the ***i***-direction component of the propagation velocity of the longitudinal wave between two adjacent balls. Considering the definition of ***wave velocity***:

$$v_{||} = \frac{l}{\Delta t} \tag{16}$$

Combining (14), (15) and (16),

$$\Rightarrow \quad v_{||} = l\sqrt{\frac{k}{m}} \tag{17}$$

It can be seen that the propagation velocity of the longitudinal wave in the springs-balls system is only determined by the inertia property and elasticity property, which is similar to the continuum.

Under the situation of $l \to 0$, the springs-balls system will degenerate to the homogeneous isotropic continuum. Defining its density ρ:

$$\rho = \frac{m}{Sl} \tag{18}$$

Defining Y—Young's modulus of the homogeneous isotropic continuum which can be regarded as being degenerated from the springs-balls system. Considering ***the relation between stiffness coefficient and Young's modulus***:

$$k = Y\frac{S}{l} \tag{19}$$

Combining (17), (18) and (19),

$$\Rightarrow \quad v_{||} = \sqrt{\frac{Y}{\rho}} \tag{20}$$

This solution is the same as the solution obtained in Problem (1).

4.3.2 Spring Vibrator Considering Spring Mass (*c*, *C*)[9]

There is a spring of mass m, stiffness coefficient k and natural length L, it has uniform mass distribution and elastic property when it

[9]This problem is a test question from Princeton University in 1980. It is also included in the following book: Yongde Zhang, et al. (1988). *American Physics Test Questions and Solutions* (University of Science and Technology of China Press).

in the natural length. The spring is placed on a smooth horizontal table with one end fixed and the other end tied with a block of mass M, forming a spring oscillator, $m << M$. Find the minimum normal frequency of the tiny vibration of the spring oscillator.

Solution [1]

Studying the spring as a homogeneous isotropic continuum.

Taking an arbitrary time point after the mechanical wave propagates all over the spring and defining it as Time zero.

As shown in Fig. 1, establishing the linear coordinate system, let x be the coordinate.

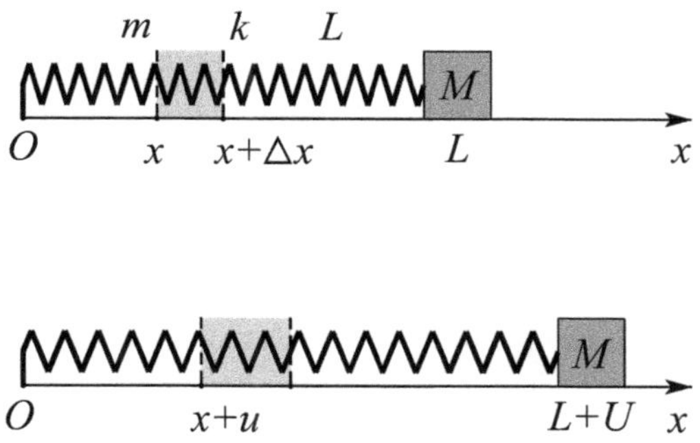

Figure 1

Taking *the mass infinitesimal element of the spring between x and x + Δx when the spring is in natural length (hereinafter referred to as element Δx)* as the object of study. Studying *the state at arbitrary Time t.* Considering ***the wave equation of longitudinal wave in homogeneous isotropic continuum***:

$$\frac{\partial^2 u}{\partial t^2} - \frac{Y}{\rho}\frac{\partial^2 u}{\partial x^2} = 0 \tag{1}$$

u—the ***i***-direction component of the displacement of element Δx at Time t.

Y—Young's modulus of the spring when is regarded as a homogeneous isotropic continuum.

ρ—the body mass density of the spring when is regarded as a homogeneous isotropic continuum.

Considering ***the relation between stiffness coefficient and Young's modulus***:

$$k = Y\frac{S}{L} \tag{2}$$

S—the cross-sectional area of the spring.

Considering the definition of ***body mass density***:

$$\rho = \frac{m}{SL} \tag{3}$$

(2) and (3) → (1),

$$\Rightarrow \qquad \frac{\partial^2 u}{\partial t^2} - \frac{k}{m}L^2\frac{\partial^2 u}{\partial x^2} = 0 \tag{4}$$

Equation (4) is the wave equation of mechanical wave in homogeneous isotropic continuum, and its wave velocity v is:

$$\Rightarrow \qquad v = \sqrt{\frac{k}{m}}L \tag{5}$$

Using the method of separation of variables to solve Eq. (4), u is decomposed into the product of function $X(x)$ which depends only on x and function $T(t)$ which depends only on t:

$$u(x,t) = X(x)T(t) \tag{6}$$

$$\Rightarrow \qquad X\frac{\mathrm{d}^2 T}{\mathrm{d}t^2} - \frac{k}{m}L^2 T\frac{\mathrm{d}^2 X}{\mathrm{d}t^2} = 0 \tag{7}$$

$$\Rightarrow \qquad \frac{1}{T}\frac{\mathrm{d}^2 T}{\mathrm{d}t^2} - \frac{k}{m}L^2\frac{1}{X}\frac{\mathrm{d}^2 X}{\mathrm{d}t^2} = 0 \tag{8}$$

The first term of the equation above is only related to t, and the second term is only related to x. The condition that the formula holds for arbitrary x and t is that both terms are equal to a constant, which is denoted by $-\mu$ (μ is positive):

$$\frac{1}{T}\frac{\mathrm{d}^2 T}{\mathrm{d}t^2} = \frac{k}{m}L^2\frac{1}{X}\frac{\mathrm{d}^2 X}{\mathrm{d}t^2} = -\mu \tag{9}$$

$$\Rightarrow \qquad \begin{cases} \dfrac{\mathrm{d}^2 X}{\mathrm{d}t^2} + \dfrac{\mu}{k/m}\dfrac{1}{L^2}X = 0 & \text{(10a)} \\[2ex] \dfrac{\mathrm{d}^2 T}{\mathrm{d}t^2} + \mu T = 0 & \text{(10b)} \end{cases}$$

Defining ω—the angular frequency:

$$\mu = \omega^2 \tag{11}$$

The general solution of Eq. (4) is as follows:

$$\Rightarrow \quad u(x,t) = \sum_{n=1}^{\infty}\left[A_n \sin\left(\frac{\omega_n}{\sqrt{k/m}}\frac{x}{L}\right) + B_n \cos\left(\frac{\omega_n}{\sqrt{k/m}}\frac{x}{L}\right)\right]\cos(\omega_n t + \varphi_n) \tag{12}$$

A_n, B_n, φ_n—the integral constants.

Considering the boundary condition:

$$u(x,t)\big|_{x=0} = 0 \tag{13}$$

$$\Rightarrow \qquad B_n = 0 \tag{14}$$

$$\Rightarrow \qquad u(x,t) = \sum_{n=1}^{\infty} A_n \sin\left(\frac{\omega_n}{\sqrt{k/m}}\frac{x}{L}\right)\cos(\omega_n t + \varphi_n) \tag{15}$$

Taking *the block* as the object of study. Studying *the state at arbitrary Time t*. Considering ***Newton's second law*** (the ***i***-direction component equation):

$$-SY\frac{\partial u}{\partial x}\bigg|_{x=L} = M\frac{\partial^2 u}{\partial t^2}\bigg|_{x=L} \tag{16}$$

$$\Rightarrow \qquad -kL\frac{\partial u}{\partial x}\bigg|_{x=L} = M\frac{\partial^2 u}{\partial t^2}\bigg|_{x=L} \tag{17}$$

(15) → (17),

$$\Rightarrow \qquad \frac{\omega_n}{\sqrt{k/m}}\tan\left(\frac{\omega_n}{\sqrt{k/m}}\right) = \frac{m}{M} \tag{18}$$

Defining the variable ξ, function $f_1(\xi)$ and $f_2(\xi)$:

$$\xi = \frac{\omega_n}{\sqrt{k/m}} \tag{19}$$

$$\begin{cases} f_1(\xi) = \tan(\xi) & \text{(20a)} \\ f_2(\xi) = \dfrac{m}{M}\dfrac{1}{\xi} & \text{(20b)} \end{cases}$$

Equation (18) can be transformed into the following form:

$$f_1(\xi) = f_2(\xi) \tag{21}$$

The solution to the equation above corresponds to the intersection point of the images of functions $f_1(\xi)$ and $f_2(\xi)$, (see Fig. 2).

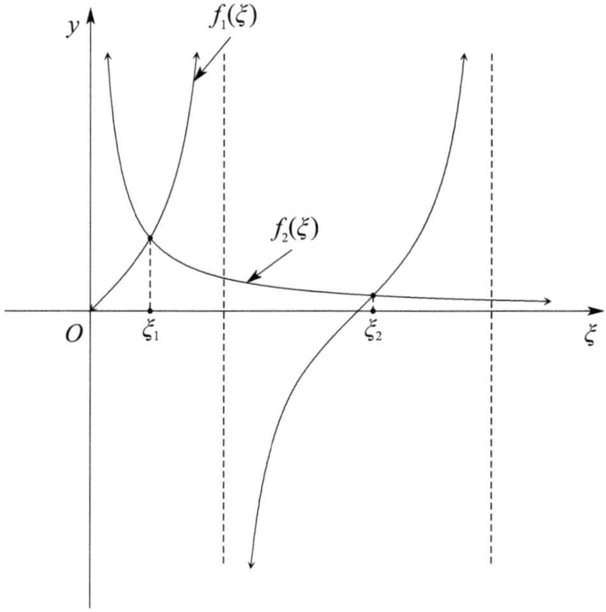

Figure 2

It can be seen from Fig. 2 that the minimum normal frequency is ω_1.

It can be seen from Eq. (18) and $m/M << 1$:

$$\Rightarrow \qquad \frac{\omega_1}{\sqrt{k/m}} \ll 1 \tag{22}$$

Taking the low order approximation of $\tan\left(\frac{\omega_1}{\sqrt{k/m}}\right)$:

$$\Rightarrow \qquad \tan\left(\frac{\omega_1}{\sqrt{k/m}}\right) \approx \frac{\omega_1}{\sqrt{k/m}} + \frac{1}{3}\left(\frac{\omega_1}{\sqrt{k/m}}\right)^3 \tag{23}$$

(23) → (18),

$$\Rightarrow \qquad \frac{1}{3}\frac{m^2}{k^2}\omega_1^4 + \frac{m}{k}\omega_1^2 - \frac{m}{M} = 0 \tag{24}$$

$$\Rightarrow \quad \omega_1^2 = 3k\frac{-1+\sqrt{1+\frac{4}{3}\frac{m}{M}}}{2m} \approx 3k\frac{-1+\left(1+\frac{2}{3}\frac{m}{M}-\frac{2}{9}\frac{m^2}{M^2}\right)}{2m}$$

$$= k\frac{3M-m}{3M^2} = \frac{k}{\dfrac{M}{1-\dfrac{m}{3M}}} \approx \frac{k}{M\left(1+\dfrac{m}{3M}\right)} = \frac{k}{M+\dfrac{m}{3}} \tag{25}$$

$$\Rightarrow \quad \omega_1 \approx \sqrt{\frac{k}{M+\dfrac{m}{3}}} \tag{26}$$

Defining λ—the wavelength of the longitudinal wave in the spring. Considering the definition of ***wavelength***, then place (5) and (26) in to obtain:

$$\Rightarrow \quad \lambda = \frac{2\pi}{\omega_1}v = \sqrt{\frac{M+\dfrac{m}{3}}{m}}\cdot 2\pi L \gg L \tag{27}$$

It can be seen that the wavelength of the longitudinal wave in the spring is much longer than the nature length of the spring (also much longer than the length of the spring at arbitrary Time t); therefore, all points on the spring perform the simple harmonic vibrations with approximately equal phase all the time.

Solution [2]

It is approximately assumed that all points on the spring perform the simple harmonic vibrations with approximately equal phase all the time, the mass of the spring should be approximately uniformly distributed all the time.

Defining U — the $\boldsymbol{i}$-direction component of the displacement of the block at arbitrary Time t. The mass of the spring is uniformly distributed all the time can be expressed as follows:

$$\frac{x+u}{x} \approx \frac{L+U}{L} \tag{28}$$

$$\Rightarrow \quad u = \frac{x}{L}U \tag{29}$$

$$\Rightarrow \quad \dot{u} = \frac{x}{L}\dot{U} \tag{30}$$

Defining E_{km}, E_{kM} and E_k—the kinetic energy of the spring, the block and the system composed of the spring and the block at arbitrary Time t. Considering the definition of ***kinetic energy***:

$$\Rightarrow \qquad E_{km} = \int_{\Gamma} \frac{1}{2}\left(\frac{m}{L}\mathrm{d}x\right)\dot{u}^2 = \frac{m\dot{U}^2}{2L^3}\int_0^L x^2\mathrm{d}x = \frac{1}{6}m\dot{U}^2 \tag{31}$$

$$E_{kM} = \frac{1}{2}M\dot{U}^2 \tag{32}$$

$$\Rightarrow \qquad E_k = E_{kM} + E_{km} = \frac{1}{2}\left(M + \frac{m}{3}\right)\dot{U}^2 \tag{33}$$

Defining E_p—the elastic potential energy of the spring at arbitrary Time t (taking the nature length of the spring as the potential energy zero). Considering ***theorem of elastic potential energy***:

$$E_p = \frac{1}{2}kU^2 \tag{34}$$

Defining E—the mechanical energy of the system composed of the spring and the block at arbitrary Time t.

$$\Rightarrow \qquad E = E_k + E_p = \frac{1}{2}\left(M + \frac{m}{3}\right)\dot{U}^2 + \frac{1}{2}kU^2 \tag{35}$$

Taking *the system composed of the block and the spring* as the object of study. Studying *the instantaneous process (the small time interval dt) at arbitrary Time t*. Considering ***law of conservation of mechanical energy*** [the work done by the external force and non-conservative internal force acting on the system is zero; therefore, the mechanical energy of the system is conserved]:

$$\frac{\mathrm{d}E}{\mathrm{d}t} = 0 \tag{36}$$

$$\Rightarrow \qquad \ddot{U} + \frac{k}{M + m/3}U = 0 \tag{37}$$

$$\Rightarrow \qquad \ddot{u} + \frac{k}{M + m/3}u = 0 \tag{38}$$

The two equations above are the differential equations of the block and an arbitrary mass infinitesimal element of the spring performing the simple harmonic vibrations along the spring, respectively, and their equal angular frequency ω is:

$$\Rightarrow \qquad \omega = \sqrt{\frac{k}{M + m/3}} \tag{39}$$

4.3.3 The Transverse Wave in a Homogeneous String and in a Light Strings-Beads System (d, *C*)

1. Consider a long soft string of line mass density η straightened under the tension T. Apply a small transverse disturbance to it, find the propagation velocity of the transverse wave in the string.
2. Consider a soft light string of length $L = (N + 1)a$ straightened under the tension T, with both ends fixed. N beads of mass M are strung on the string in equidistance. Find the characteristic frequencies and vibration modes of the tiny vibration of this strings-beads system, and find the propagation velocity of the transverse wave in this strings-beads system.

Solution to Problem 1

Taking an arbitrary time point after the mechanical wave propagates all over the string and defining it as Time zero.

As shown in the Fig. 1, when the string is at rest, taking an arbitrary point on it and denoting it by O. Establishing the Cartesian coordinate system with the frame $(O, \boldsymbol{i}, \boldsymbol{j})$, let x, y be the Cartesian coordinates.

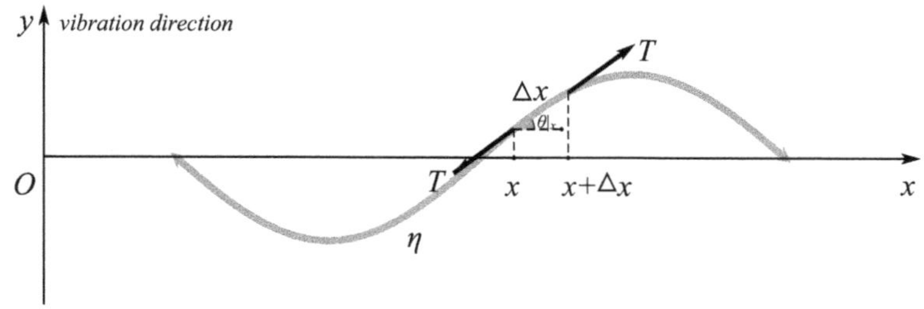

Figure 1

Taking *the line element of string between x and x + Δx before the vibration is excited [hereinafter referred to as element Δx]* as the object of study. Studying *the state at arbitrary Time t.* Considering ***Newton's second law*** (the $\boldsymbol{j}$-direction component equation):

$$T\sin\theta\big|_{x+\Delta x} - T\sin\theta\big|_{x} = \eta\Delta x\frac{\partial^2 y}{\partial t^2} \tag{1}$$

$\theta|_x$—the angle between the tangent direction of the string at x and the $\boldsymbol{i}$-direction at Time t.

$\theta|_x$ is a small quantity, taking the following approximation, and considering the geometric meaning of derivative:

$$\sin\theta\big|_x \approx \tan\theta\big|_x = \frac{\partial y}{\partial x}\bigg|_x \tag{2}$$

$$\Rightarrow \qquad \eta\Delta x\frac{\partial^2 y}{\partial t^2} = T\left(\frac{\partial y}{\partial x}\bigg|_{x+\Delta x} - \frac{\partial y}{\partial x}\bigg|_x\right) = T\frac{\partial^2 y}{\partial x^2}\Delta x \tag{3}$$

$$\Rightarrow \qquad \frac{\partial^2 y}{\partial t^2} - \frac{T}{\eta}\frac{\partial^2 y}{\partial x^2} = 0 \tag{4}$$

Equation (4) is the wave equation of mechanical wave in homogeneous isotropic continuum, and its wave velocity v is:

$$\Rightarrow \qquad v = \sqrt{\frac{T}{\eta}} \tag{5}$$

Solution to Problem 2

Taking an arbitrary time point after the mechanical wave propagates all over the system and defining it as Time zero

As shown in the Fig. 2, the left end of the string is denoted by O. Establishing the Cartesian coordinate system with the frame $(O, \boldsymbol{i}, \boldsymbol{j})$, let x, y be the Cartesian coordinates.

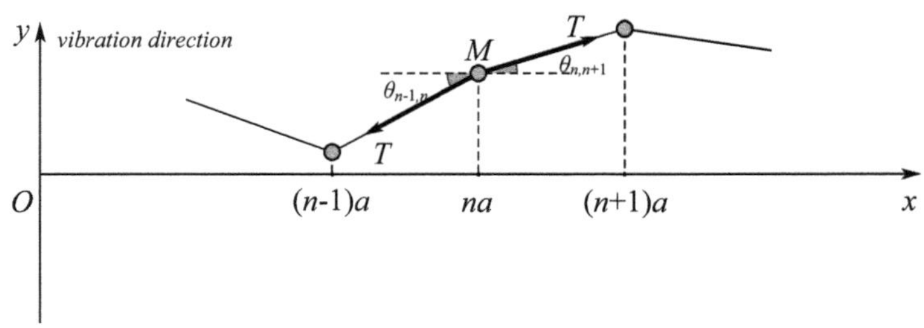

Figure 2

Sorting the beads on the string with positive integers from left to right.

Taking *No. n bead* as the object of study. Studying *the state at arbitrary Time t*. Considering ***Newton's second law*** (the ***j***-direction component equation):

$$T\sin\theta_{n,n+1} - T\sin\theta_{n-1,n} = M\frac{\mathrm{d}^2 y_n}{\mathrm{d}t^2} \tag{6}$$

$\theta|_{n,\,n+1}$—the angle between the direction of No. n bead pointing to No. $(n + 1)$ bead and the ***i***-direction at Time t.

$\theta|_{n,\,n+1}$ is a small quantity, taking the following approximation, and considering the following geometric relation in Fig. 2:

$$\sin\theta_{n,n+1} \approx \tan\theta_{n,n+1} = \frac{y_{n+1} - y_n}{a} \tag{7}$$

$$\Rightarrow \qquad \frac{\mathrm{d}^2 y_n}{\mathrm{d}t^2} + \frac{T}{Ma}[2y_n - (y_{n+1} + y_{n-1})] = 0 \tag{8}$$

Let the trial solution to y_n be:

$$y_n = A_n \cos(\omega t + \varphi) \tag{9}$$

ω, φ and A_n—the angular frequency, initial phase and amplitude of No. n bead.

A_n is related to the vibration mode, assuming that it has the following form:

$$A_n = A\sin(kna) + B\cos(kna) \tag{10}$$

k—the wave number related to the vibration mode.

Considering the boundary conditions:

$$\begin{cases} A_0 = 0 & \text{(11a)} \\ A_{N+1} = 0 & \text{(11b)} \end{cases}$$

$(10) \to (11a)$,

$$\Rightarrow \qquad B = 0 \tag{12}$$

$$\Rightarrow \qquad A_n = A\sin(kna) \tag{13}$$

$(13) \to (11b)$,

$$\Rightarrow \qquad \sin k(N+1)a = 0 \tag{14}$$

$$\Rightarrow \qquad k(N+1)a = m\pi, \qquad m = 1, 2, \cdots, N \tag{15}$$

Equation (15) only has the discrete solutions corresponding to the positive integer m one by one, let it be k_m:

$$\Rightarrow \qquad k_m = \frac{m}{N+1}\frac{\pi}{a} \tag{16}$$

$$\Rightarrow \qquad y_n = A\sin(k_m na)\cos(\omega_m t + \varphi_m) \tag{17}$$

(17) $\rightarrow$ (8),

$$\Rightarrow \quad 0 = -\omega_m^2 \sin k_m na$$

$$+\frac{T}{Ma}\{2\sin(k_m na) - \sin(k_m(n+1)a) - \sin[k_m(n-1)a]\}$$

$$= -\omega_m^2 \sin k_m na + \frac{T}{Ma}[2\sin(k_m na) - 2\sin(k_m na)\cos(k_m a)] \tag{18}$$

$$\Rightarrow \qquad \omega_m = \sqrt{\frac{2T}{Ma}\left[1-\cos(k_m a)\right]} = 2\sqrt{\frac{T}{Ma}}\sin\left(\frac{1}{2}k_m a\right)$$

$$= 2\sqrt{\frac{T}{Ma}}\sin\left(\frac{1}{2}\frac{m}{N+1}\pi\right) \tag{19}$$

From (17),

$$\Rightarrow \quad y_n = \frac{1}{2}A[\sin(\omega_m t + k_m na + \varphi_m) - \sin(\omega_m t - k_m na + \varphi_m)]$$

$$= \frac{1}{2}A[\sin(\omega_m t + k_m x_n + \varphi_m) - \sin(\omega_m t - k_m x_n + \varphi_m)] \tag{20}$$

In the equation above, the first term represents the mechanical wave that propagates toward the left, and the second term represents the mechanical wave that propagates toward the right, whose wave functions are defined as y_{n1} and y_{n2}, respectively:

$$\begin{cases} y_{n1} = \dfrac{1}{2}A\sin(\omega_m t + k_m x_n + \varphi_m) & \text{(21a)} \\ y_{n2} = -\dfrac{1}{2}A\sin(\omega_m t - k_m x_n + \varphi_m) & \text{(21b)} \end{cases}$$

Defining v_{n1} and v_{n2}—the wave velocity of the mechanical waves that propagates toward the left and toward the right, respectively.

Considering the defining of ***wave number***:

$$\Rightarrow \quad \begin{cases} v_{n1} = -\dfrac{\omega_m}{k_m} = -\dfrac{\sin\left(\dfrac{1}{2}\dfrac{m}{N+1}\pi\right)}{\dfrac{1}{2}\dfrac{m}{N+1}\pi}\sqrt{\dfrac{T}{M/a}} & \text{(22a)} \\ v_{n2} = \dfrac{\omega_m}{k_m} = \dfrac{\sin\left(\dfrac{1}{2}\dfrac{m}{N+1}\pi\right)}{\dfrac{1}{2}\dfrac{m}{N+1}\pi}\sqrt{\dfrac{T}{M/a}} & \text{(22b)} \end{cases}$$

Under the situation of $a \to 0$ (i.e., $N \to \infty$), the light springs-beads system will degenerate to the homogeneous string. Defining its line mass density η:

$$\eta = \frac{M}{a} \tag{23}$$

$$\Rightarrow \quad \omega_m \approx \sqrt{\frac{T}{Ma}}\frac{m}{N+1}\pi = \sqrt{\frac{T}{M/a}}\frac{m}{(N+1)a}\pi = \frac{m\pi}{L}\sqrt{\frac{T}{\eta}} \tag{24}$$

$$\Rightarrow \quad \begin{cases} v_{n1} \approx -\dfrac{\dfrac{1}{2}\dfrac{m}{N+1}\pi}{\dfrac{1}{2}\dfrac{m}{N+1}\pi}\sqrt{\dfrac{T}{\eta}} = -\sqrt{\dfrac{T}{\eta}} & \text{(25a)} \\ v_{n2} \approx \dfrac{\dfrac{1}{2}\dfrac{m}{N+1}\pi}{\dfrac{1}{2}\dfrac{m}{N+1}\pi}\sqrt{\dfrac{T}{\eta}} = \sqrt{\dfrac{T}{\eta}} & \text{(25b)} \end{cases}$$

4.3.4 The Reflection and Transmission of a Transverse Wave at a Small Bead Strung on a Homogeneous String (*c*, *C*)[10]

Consider a long soft string of line mass density η straightened under the tension T, a small bead of mass m is strung on it (see Fig. 1). A train of harmonic waves of angular frequency ω is incident along the string from the left, and are reflected and transmitted when reaching the bead. Find the energy reflection coefficient and the energy transmission coefficient in this string-bead system.

[10]This problem is a test question from MIT in 1980. It is also included in the following book: Yongde Zhang, et al. (1988). *American Physics Test Questions and Solutions* (University of Science and Technology of China Press).

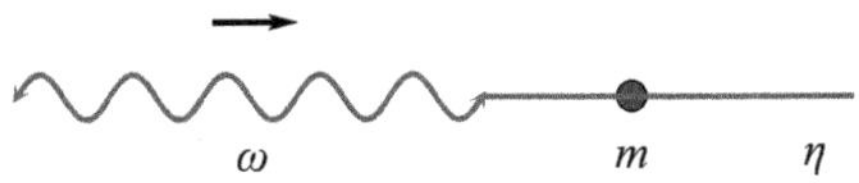

Figure 1

Solution

As shown in Figure 2, the position of the bead before the vibration is excited is denoted by O. Establishing the Cartesian coordinate system with the frame $(O, \boldsymbol{i}, \boldsymbol{j})$, let x, y be the Cartesian coordinates.

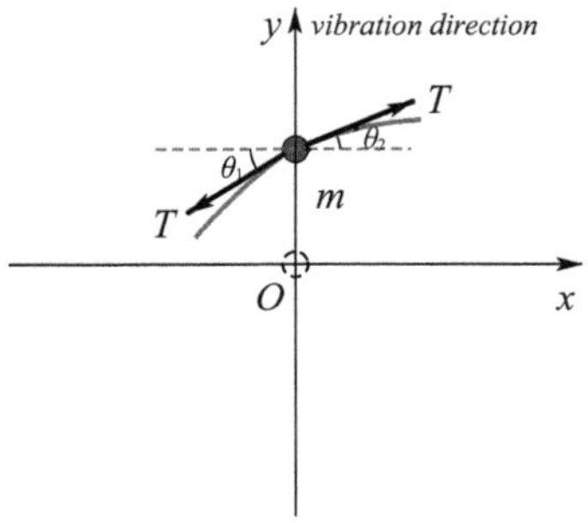

Figure 2

Once the wave displacement of the bead is the maximum displacement in the $\boldsymbol{j}$-direction is defined as Time zero.

Considering the first boundary condition: the wave displacement is continuous at $x = 0$.

$$y_{\text{bead}} = y_1\big|_{x\to 0^-} = y_2\big|_{x\to 0^+} \tag{1}$$

y_1 and y_2—the wave displacement at an arbitrary point of the negative x-axis and an arbitrary point of the positive x-axis at arbitrary Time t.

Defining y_i and y_r—the wave displacement of the incident wave and the reflected wave at an arbitrary point of the negative x-axis at arbitrary Time t. y_t—the wave displacement of the transmitted wave at an arbitrary point of the positive x-axis at arbitrary Time t:

$$\begin{cases} y_1 = y_i + y_r & \text{(2a)} \\ y_2 = y_t & \text{(2b)} \end{cases}$$

$$\Rightarrow \qquad y_i\big|_{x\to 0^-} + y_r\big|_{x\to 0^-} = y_t\big|_{x\to 0^+} \tag{3}$$

Taking *the bead* as the object of study. Studying *the state at arbitrary Time t*. Considering ***Newton's second law*** (the ***j***-direction component equation):

$$-T\sin\theta_1\big|_{x\to 0^-} + T\sin\theta_2\big|_{x\to 0^+} = m\frac{\partial^2 y_{\text{bead}}}{\partial t^2} \tag{4}$$

The equation above can be regarded as the second boundary condition at $x = 0$.

θ_1 and θ_2—the angle between the tangent direction of the string at an arbitrary point of the negative x-axis and the ***i***-direction and the angle between the tangent direction of the string at an arbitrary point of the positive x-axis and the ***i***-direction.

θ_1 and θ_2 are small quantities, taking the following approximations and considering the geometric meaning of derivative:

$$\begin{cases} \sin\theta_1\big|_{x\to 0^-} \approx \tan\theta_1\big|_{x\to 0^-} = \left.\dfrac{\partial y_1}{\partial x}\right|_{x\to 0^-} & \text{(5a)} \\ \sin\theta_2\big|_{x\to 0^+} \approx \tan\theta_2\big|_{x\to 0^+} = \left.\dfrac{\partial y_2}{\partial x}\right|_{x\to 0^+} & \text{(5b)} \end{cases}$$

Combining (1) to (5),

$$\Rightarrow \qquad -T\left.\frac{\partial(y_i + y_r)}{\partial x}\right|_{x\to 0^-} + T\left.\frac{\partial y_t}{\partial x}\right|_{x\to 0^+} = m\left.\frac{\partial^2 y_t}{\partial t^2}\right|_{x\to 0^+} \tag{6}$$

Let the wave functions of the incident wave, the reflected wave and the transmitted wave be:

$$\begin{cases} y_i = A\cos(\omega t - kx) & \text{(7a)} \\ y_r = B\cos(\omega t + kx + \varphi_r) & \text{(7b)} \\ y_t = C\cos(\omega t - kx + \varphi_t) & \text{(7c)} \end{cases}$$

φ_r and φ_t—the initial phase of the reflected wave and the transmitted wave.

k— the equal wave number of the three waves.

A, B and C—the amplitude of the incident wave, the reflected wave and the transmitted wave.

Considering the definition of ***wave number***:

$$k = \frac{\omega}{v} \tag{8}$$

v—the equal wave velocity of the three waves.

Considering ***the wave velocity form of plane mechanical wave in homogeneous isotropic continuum***:

$$v = \sqrt{\frac{T}{\eta}} \tag{9}$$

(7), (8) and (9) $\rightarrow$ (3) and (6),

$$\Rightarrow \begin{cases} A\cos\omega t + B\cos(\omega t + \varphi_r) = C\cos(\omega t + \varphi_t) & \\ -A\sin\omega t + B\sin(\omega t + \varphi_r) + C\sin(\omega t + \varphi_r) = & (10\text{a}) \\ -\dfrac{m\omega}{\sqrt{T\eta}} C\cos(\omega t + \varphi_r) & (10\text{b}) \end{cases}$$

$$\Rightarrow \begin{cases} A\cos\omega t + B\cos\omega t\cos\varphi_r - B\sin\omega t\sin\varphi_r & \\ = C\cos\omega t\cos\varphi_t - C\sin\omega t\sin\varphi_t & (11\text{a}) \\ -A\sin\omega t + B\sin\omega t\cos\varphi_r + B\cos\omega t\sin\varphi_r + C\sin\omega t\cos\varphi_t & \\ + C\cos\omega t\sin\varphi_t = -\dfrac{m\omega}{\sqrt{T\eta}} C\cos\omega t\cos\varphi_t + \dfrac{m\omega}{\sqrt{T\eta}} C\sin\omega t\sin\varphi_t & (11\text{b}) \end{cases}$$

Equations (11a) and (11b) hold for arbitrarily Time t, both can be understood as the identical equations for t, and the necessary condition for their holds are the coefficients of $\sin\omega t$ and $\cos\omega t$ on both sides are equal:

$$\Rightarrow \begin{cases} \begin{cases} A + B\cos\varphi_r = C\cos\varphi_t & (12\text{a}) \\ B\sin\varphi_r = C\sin\varphi_t & (12\text{b}) \end{cases} \\ \begin{cases} -A + B\cos\varphi_r + C\cos\varphi_t = \dfrac{m\omega}{\sqrt{T\eta}} C\sin\varphi_t & (12\text{c}) \\ B\sin\varphi_r + C\sin\varphi_t = -\dfrac{m\omega}{\sqrt{T\eta}} C\cos\varphi_t & (12\text{d}) \end{cases} \end{cases}$$

$$\Rightarrow \begin{cases} B\sin\varphi_r = C\sin\varphi_t = -\dfrac{\dfrac{m\omega}{2\sqrt{T\eta}}}{1+\left(\dfrac{m\omega}{2\sqrt{T\eta}}\right)^2}A & \text{(13a, b)} \\ B\cos\varphi_r = -\dfrac{\left(\dfrac{m\omega}{2\sqrt{T\eta}}\right)^2}{1+\left(\dfrac{m\omega}{2\sqrt{T\eta}}\right)^2}A & \text{(13c)} \\ C\cos\varphi_t = \dfrac{1}{1+\left(\dfrac{m\omega}{2\sqrt{T\eta}}\right)^2}A & \text{(13d)} \end{cases}$$

$$\Rightarrow \begin{cases} \tan\varphi_r = \dfrac{2\sqrt{T\eta}}{m\omega} & \text{(14a)} \\ \tan\varphi_t = -\dfrac{m\omega}{2\sqrt{T\eta}} & \text{(14b)} \end{cases}$$

A, *B* and *C* all take positive values, from (13a) and (13b),

$$\Rightarrow \begin{cases} \sin\varphi_r < 0 & \text{(15a)} \\ \sin\varphi_t < 0 & \text{(15b)} \end{cases}$$

Therefore, φ_r is in the third quadrant and φ_t is in the fourth quadrant.

$$\Rightarrow \begin{cases} \varphi_r = \pi + \arctan\dfrac{2\sqrt{T\eta}}{m\omega} & \text{(16a)} \\ \varphi_t = 2\pi - \arctan\dfrac{m\omega}{2\sqrt{T\eta}} & \text{(16b)} \end{cases}$$

(16) → (13),

$$\Rightarrow \quad \begin{cases} B = \dfrac{\dfrac{m\omega}{2\sqrt{T\eta}}}{\sqrt{1+\left(\dfrac{m\omega}{2\sqrt{T\eta}}\right)^2}} A & \text{(17a)} \\ C = \dfrac{1}{\sqrt{1+\left(\dfrac{m\omega}{2\sqrt{T\eta}}\right)^2}} A & \text{(17b)} \end{cases}$$

Defining the energy reflection coefficient R and the energy transmission coefficient T in this string-bead system:

$$\Rightarrow \quad \begin{cases} R = \left(\dfrac{B}{A}\right)^2 = \dfrac{\left(\dfrac{m\omega}{2\sqrt{T\eta}}\right)^2}{1+\left(\dfrac{m\omega}{2\sqrt{T\eta}}\right)^2} & \text{(18a)} \\ T = \left(\dfrac{C}{A}\right)^2 = \dfrac{1}{1+\left(\dfrac{m\omega}{2\sqrt{T\eta}}\right)^2} & \text{(18b)} \end{cases}$$

$$\Rightarrow \quad R^2 + T^2 = 1 \quad (19)$$

That is, the propagation process of the mechanical wave in this string-bead system is conserved in energy.

4.3.5 The Reflection and Transmission of Sound Wave at Water Surface (d, *C*)

A train of plane sound waves is incident vertically into the water from the air, and are reflected and transmitted on the water surface. Find the energy reflection coefficient and the energy transmission coefficient. Given the bulk modulus of elasticity of air and water are K_1 and K_2, respectively, and the density are ρ_1 and ρ_2, respectively.

Solution

Taking an arbitrary point on the water surface and denoting it by *O*, the unit vector in the vertical downward direction is denoted by ***i***. Establishing the linear coordinate system with Point *O* as the origin and the ***i***-direction as the positive direction of coordinate axis, let *x* be the coordinate. The following only studies the propagation of wave trains on the *x*-axis.

Once the wave displacement at the origin *O* is the maximum displacement in the ***i***-direction is defined as Time zero.

Considering the first boundary condition: the wave displacement is continuous at $x = 0$.

$$y_1\big|_{x\to 0^-} = y_2\big|_{x\to 0^+} \tag{1}$$

y_1 and y_2—the wave displacement at an arbitrary point in the air and an arbitrary point in the water at arbitrary Time *t*.

Defining y_i and y_r—the wave displacement of the incident wave and the reflected wave at an arbitrary point in the air at arbitrary Time *t*. y_t—the wave displacement of the transmitted wave at an arbitrary point in the water at arbitrary Time *t*.

$$\begin{cases} y_1 = y_i + y_r & (2a) \\ y_2 = y_t & (2b) \end{cases}$$

$$\Rightarrow \qquad y_i\big|_{x\to 0^-} + y_r\big|_{x\to 0^-} = y_t\big|_{x\to 0^+} \tag{3}$$

Considering the second boundary condition: the wave pressure is continuous at $x = 0$.

$$p_1\big|_{x\to 0^-} = p_2\big|_{x\to 0^+} \tag{4}$$

p_1 and p_2—the wave pressure at an arbitrary point in the air and an arbitrary point in the water at arbitrary Time *t*.

Considering ***Hooke's law***:

$$\begin{cases} p_1 = K_1 \dfrac{\partial y_1}{\partial x} & (5a) \\ p_2 = K_2 \dfrac{\partial y_2}{\partial x} & (5b) \end{cases}$$

Combining (2), (4) and (5),

$$\Rightarrow \qquad K_1 \frac{\partial y_i}{\partial x}\bigg|_{x\to 0^-} + K_1 \frac{\partial y_r}{\partial x}\bigg|_{x\to 0^-} = K_2 \frac{\partial y_t}{\partial x}\bigg|_{x\to 0^+} \tag{6}$$

Let the wave functions of the incident wave, the reflected wave and the transmitted wave be:

$$\begin{cases} y_i = A\cos(\omega t - k_1 x) & \text{(7a)} \\ y_r = B\cos(\omega t + k_1 x + \varphi_r) & \text{(7b)} \\ y_t = C\cos(\omega t - k_2 x + \varphi_t) & \text{(7c)} \end{cases}$$

ω—the angular frequency of the incident wave, the reflected wave and the transmitted wave.

φ_r and φ_t—the initial phase of the reflected wave and the transmitted wave.

k_1 and k_2—the wave number of the sound wave in the air and in the water.

A, B and C—the amplitude of the incident wave, the reflected wave and the transmitted wave (A is defined as positive, the signs of B and C are modulated by the value range of φ_r and φ_t, respectively).

Considering the definition of ***wave number***:

$$k_1 = \frac{\omega}{v_1}, \quad k_2 = \frac{\omega}{v_2} \tag{8a, b}$$

v_1 and v_2—the wave velocity of the sound wave in the air and in the water.

Considering ***the wave velocity form of plane mechanical wave in homogeneous isotropic continuum***:

$$v_1 = \sqrt{\frac{K_1}{\rho_1}}, \quad v_1 = \sqrt{\frac{K_2}{\rho_2}} \tag{9a, b}$$

(7), (8) and (9) $\rightarrow$ (3) and (6),

$$\Rightarrow \begin{cases} A\cos\omega t + B\cos(\omega t + \varphi_r) = C\cos(\omega t + \varphi_t) & \text{(10a)} \\ \sqrt{K_1\rho_1}\, A\sin\omega t - \sqrt{K_1\rho_1}\, B\sin(\omega t + \varphi_r) = \sqrt{K_2\rho_2}\,\sin(\omega t + \varphi_r) & \text{(10b)} \end{cases}$$

$$\Rightarrow \begin{cases} A\cos\omega t + B\cos\omega t\cos\varphi_r - B\sin\omega t\sin\varphi_r & \\ = C\cos\omega t\cos\varphi_t - C\sin\omega t\sin\varphi_t & \text{(11a)} \\ \sqrt{K_1\rho_1}\, A\sin\omega t - \sqrt{K_1\rho_1}\, B\sin\omega t\cos\varphi_r - \sqrt{K_1\rho_1}\, B\cos\omega t\sin\varphi_r & \\ = \sqrt{K_2\rho_2}\, C\sin\omega t\cos\varphi_t + \sqrt{K_2\rho_2}\,\cos\omega t\sin\varphi_t & \text{(11b)} \end{cases}$$

Equations (11a) and (11b) hold for arbitrarily Time t, both can be understood as the identical equations for t, and the necessary condition for their holds are the coefficients of $\sin\omega t$ and $\cos\omega t$ on both sides are equal:

$$\Rightarrow \quad \begin{cases} \begin{cases} A+B\cos\varphi_r = C\cos\varphi_t & (12a) \\ B\sin\varphi_r = C\sin\varphi_t & (12b) \end{cases} \\ \begin{cases} \sqrt{K_1\rho_1}(A-B\cos\varphi_r) = \sqrt{K_2\rho_2}C\cos\varphi_t & (12c) \\ -\sqrt{K_1\rho_1}B\sin\varphi_r = \sqrt{K_2\rho_2}C\sin\varphi_t & (12d) \end{cases} \end{cases}$$

Equations (12b) and (12d) are regarded as the homogeneous linear equations of $\sin\varphi_r$ and $\cos\varphi_t$, investigating its coefficient determinant:

$$\begin{vmatrix} B & -C \\ -\sqrt{K_1\rho_1}B & -\sqrt{K_2\rho_2}C \end{vmatrix} = \left(-\sqrt{K_2\rho_2} - \sqrt{K_1\rho_1}\right)BC \neq 0 \qquad (13)$$

That is, there is only zero solution to this system of equations.

$$\Rightarrow \qquad \sin\varphi_r = \sin\varphi_t = 0 \qquad (14)$$

Taking zero solution for both φ_r and φ_t:

$$\varphi_r = \varphi_t = 0 \qquad (15)$$

(15) $\rightarrow$ (12a) and (12c),

$$\Rightarrow \quad \begin{cases} A+B = C & (16a) \\ \sqrt{K_1\rho_1}(A-B) = \sqrt{K_2\rho_2}C & (16b) \end{cases}$$

$$\Rightarrow \quad \begin{cases} B = \dfrac{\sqrt{K_1\rho_1} - \sqrt{K_2\rho_2}}{\sqrt{K_1\rho_1} + \sqrt{K_2\rho_2}}A & (17a) \\ C = \dfrac{2\sqrt{K_1\rho_1}}{\sqrt{K_1\rho_1} + \sqrt{K_2\rho_2}}A & (17b) \end{cases}$$

It can be seen from (17a) that when $\sqrt{K_1\rho_1} > \sqrt{K_2\rho_2}$, $B > 0$, the reflected wave has no half-wave loss, when $\sqrt{K_1\rho_1} < \sqrt{K_2\rho_2}$, $B < 0$, the reflected wave has half-wave loss (B taking positive value corresponds to φ_r taking π).

Defining the energy reflection coefficient R and the energy transmission coefficient T,

$$\Rightarrow \begin{cases} R = \left(\dfrac{B}{A}\right)^2 = \left(\dfrac{\sqrt{K_1\rho_1} - \sqrt{K_2\rho_2}}{\sqrt{K_1\rho_1} + \sqrt{K_2\rho_2}}\right)^2 & \text{(18a)} \\ T = \left(\dfrac{C}{A}\right)^2 = \left(\dfrac{2\sqrt{K_1\rho_1}}{\sqrt{K_1\rho_1} + \sqrt{K_2\rho_2}}\right)^2 & \text{(18b)} \end{cases}$$

$$\Rightarrow \qquad R^2 + T^2 = 1 \qquad (19)$$

That is, the propagation process of the plane sound wave at the water surface is conserved in energy.

4.3.6 Shallow Water Wave (d, *C*)

Consider a shallow water body which is wide enough, given its depth is h. Find the wave velocity of the mechanical wave in the water under the weak excitation in a horizontal direction.

Solution

As shown in Fig. 1, taking an arbitrary point on the water bottom and denoting it by O, the unit vector of the propagation direction of the mechanical wave in the water (horizontally to the right in the figure) is denoted by $\boldsymbol{i}$, the unit vector in the vertical upward direction is denoted by $\boldsymbol{j}$, $\boldsymbol{i} \times \boldsymbol{j}$ is denoted by $\boldsymbol{k}$ (perpendicular to the plane of the figure outward). Establishing the Cartesian coordinate system with the frame $(O, \boldsymbol{i}, \boldsymbol{j}, \boldsymbol{k})$, let x, y, z be the Cartesian coordinates.

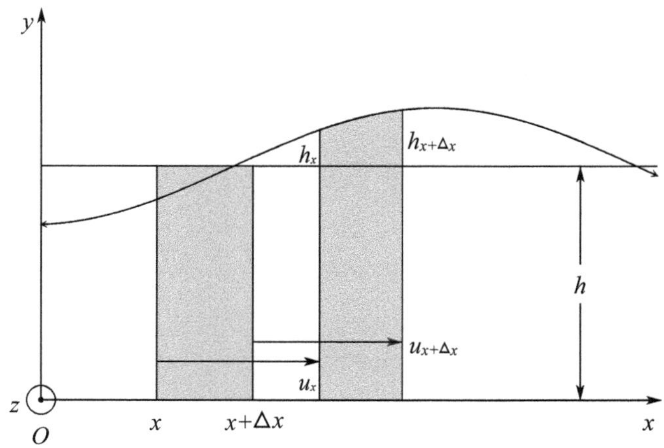

Figure 1

Before the mechanical wave is excited, taking the water infinitesimal element (of unit length in the ***k***-direction) between x and $x+\Delta x$, denoting it as element Δx.

Water is incompressible; therefore, in the process of the mechanical wave's propagating in the water in the horizontal direction, the vertical motion of the water is inevitable. That the volume of the element Δx is constant in its vibration process can be expressed as follows:

$$h\Delta x \approx (h+h_x)[\Delta x+(u_{x+\Delta x}-u_x)]=(h+h_x)\left(1+\frac{\partial u}{\partial x}\right)\Delta x \tag{1}$$

u_x—the ***i***-direction component of the wave displacement at x at arbitrary Time t.

h_x—the depth increment of element Δx at arbitrary Time t (approximately taking the depth increment of its left side).

Known that the mechanical wave in the water is weak, it can be seen that:

$$\left|\frac{\partial u}{\partial x}\right| \ll 1 \tag{2}$$

Taking the first-order approximation of Eq. (1),

$$\Rightarrow \quad h_x=-\frac{\partial u}{\partial x}\left(1+\frac{\partial u}{\partial x}\right)^{-1}h\approx-\frac{\partial u}{\partial x}\left(1-\frac{\partial u}{\partial x}\right)h\approx-\frac{\partial u}{\partial x}h \tag{3}$$

$$\Rightarrow \quad |h_x| \ll h \tag{4}$$

That is, the depth change of the water body at an arbitrary position is small enough. Therefore, the motion of the water body in the ***j***-direction can be ignored.

Taking *element* Δx as the object of study. Studying *the state at arbitrary Time t.* Considering ***theorem of motion of center of mass*** (the ***i***-direction component equation):

$$F_\rho=\rho h\Delta x\frac{\partial^2 u}{\partial t^2} \tag{5}$$

F_ρ—the ***i***-direction component of the resultant force exerted on element Δx by the water on both sides of it.

ρ—the density of the water.

Considering (a) the definition of ***pressure***, (b) ***basic equation of hydrostatic pressure*** (the motion of the water in the ***j***-direction can be ignored; therefore, the water pressure can approximately take the hydrostatic pressure):

$$F_\rho = \int_{\Sigma_x} \rho g y \mathrm{d}S_x - \int_{\Sigma_{x+\Delta x}} \rho g y \mathrm{d}S_{x+\Delta x}$$

$$= \frac{1}{2}\rho g(h+h_x)\cdot(h+h_x) - \frac{1}{2}\rho g(h+h_{x+\Delta x})\cdot(h+h_{x+\Delta x})$$

$$\approx \rho g h_x h - \rho g h_{x+\Delta x} h$$

$$= -\rho g h \frac{\partial h_x}{\partial x}\Delta x \qquad (6)$$

This operation has considered h_x and $h_{x+\Delta x}$ as small quantities and taken the first-order approximations.

$$\Rightarrow \qquad \frac{\partial^2 u}{\partial t^2} - gh\frac{\partial^2 u}{\partial x^2} = 0 \qquad (7)$$

Equation (7) is the wave equation of mechanical wave in homogeneous isotropic continuum, and its wave velocity v is:

$$\Rightarrow \qquad v = \sqrt{gh} \qquad (8)$$

The wave function is as follows:

$$\Rightarrow \qquad u(x,t) = A\cos\left[\omega\left(t - \frac{x}{v}\right)\right] = A\cos\left[\omega\left(t - \frac{x}{\sqrt{gh}}\right)\right] \qquad (9)$$

$$\Rightarrow \qquad h_x = -\omega\sqrt{\frac{h}{g}}A\sin\left[\omega\left(t - \frac{x}{\sqrt{gh}}\right)\right] \qquad (10)$$

It can be seen that the shape of the free surface of water is a sine surface.

4.3.7 Sound Velocity in the Air (d, *C*)

Determine the sound velocity in the air. Given the molar mass of the air is μ, the adiabatic coefficient is γ, and the temperature is T. It is assumed that the frequency of the sound wave is very high, and the air vibration is very weak.

Solution

When the sound wave propagates continuously in the air, taking an arbitrary time point and defining it as Time zero.

As shown in Fig. 1, taking an arbitrary point in the propagation path of the sound wave and denoting it by *O*, and the unit vector in the propagation direction (horizontally to the right in the figure) is denoted by ***i***. Establishing the linear coordinate system with Point *O* as the origin and the ***i***-direction as the positive direction of coordinate axis, let *x* be the coordinate.

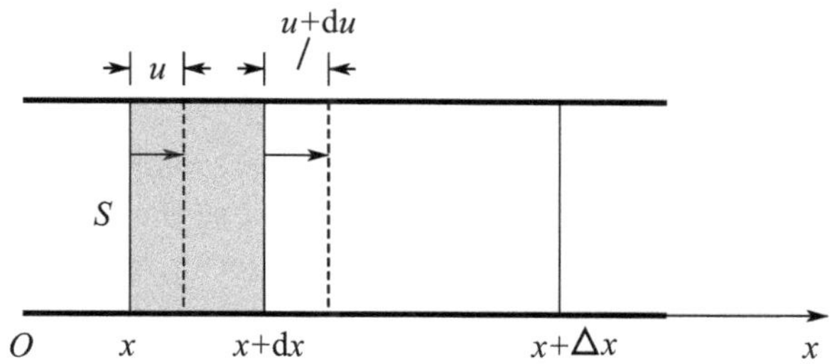

Figure 1

Taking *the cylindrical air infinitesimal element of cross-sectional area S between x and x + dx before the sound wave is excited [hereinafter referred to as element dx]* as the object of study. Studying *the process from the time point when it starts to vibrate to arbitrary Time t.* Considering ***the adiabatic equation of ideal gas*** [the frequency of the sound wave is high enough, that is, the deformation period of element d*x* is extremely short that there is no time for it to achieve the heat transfer with the surrounding air; therefore, an arbitrary process that it undergoes can be considered as an adiabatic process] (the pressure at each point of element d*x* can approximately take the pressure at its left side):

$$p_0(S\mathrm{d}x)^{\gamma} = p_x[S(\mathrm{d}x+\mathrm{d}u)]^{\gamma} = (p_0+\mathrm{d}p_x)[S(\mathrm{d}x+\mathrm{d}u)]^{\gamma} \tag{1}$$

p_0—the air pressure when the sound wave is not excited.

p_x—the pressure of element d*x* at Time *t*.

dp_x—the pressure increment of element d*x* in this process.

u—the ***i***-direction component of the wave displacement at *x* at Time *t*.

d*u*—the ***i***-direction component of the wave displacement at *x* + d*x* relative to that at *x* at Time *t*.

$$\Rightarrow \quad 1=\left(1+\frac{\mathrm{d}p_x}{p_0}\right)\left(1+\frac{\mathrm{d}u}{\mathrm{d}x}\right)^{\gamma}\approx\left(1+\frac{\mathrm{d}p_x}{p_0}\right)\left(1+\gamma\frac{\mathrm{d}u}{\mathrm{d}x}\right)\approx 1+\frac{\mathrm{d}p_x}{p_0}+\gamma\frac{\mathrm{d}u}{\mathrm{d}x} \tag{2}$$

$$\Rightarrow \quad \mathrm{d}p_x=-\gamma p_0\frac{\partial u}{\partial x} \tag{3}$$

$$\Rightarrow \quad p_x=p_0+\mathrm{d}p_x=p_0\left(1-\gamma\frac{\partial u}{\partial x}\right) \tag{4}$$

Taking *the cylindrical air element of cross-sectional area S between x and x + Δx before the sound wave is excited* as the object of study. Studying *the state at arbitrary Time t.* Considering ***Newton's second law*** (the ***i***-direction component equation):

$$p_xS-p_{x+\Delta x}S=\rho S\Delta x\frac{\partial^2u}{\partial t^2} \tag{5}$$

ρ—the air density when the sound wave is not excited.

$$\Rightarrow \quad -\frac{\partial p_x}{\partial x}\Delta xS=\rho S\Delta x\frac{\partial^2u}{\partial t^2} \tag{6}$$

$(4)\rightarrow(6)$,

$$\Rightarrow \quad \gamma p_0\frac{\partial^2u}{\partial x^2}\Delta xS=\rho S\Delta x\frac{\partial^2u}{\partial t^2} \tag{7}$$

$$\Rightarrow \quad \frac{\partial^2u}{\partial t^2}-\frac{\gamma p_0}{\rho}\frac{\partial^2u}{\partial x^2}=0 \tag{8}$$

Equation (8) is the wave equation of mechanical wave in homogeneous isotropic continuum, and its wave velocity v is:

$$\Rightarrow \quad v=\sqrt{\frac{\gamma p_0}{\rho}} \tag{9}$$

Taking *the air* as the object of study. Studying *the state when the sound wave is not excited.* Considering ***state equation of ideal gas*** (equivalent form):

$$\rho=\frac{p_0\mu}{RT} \tag{10}$$

$$\Rightarrow \quad v=\sqrt{\frac{\gamma RT}{\mu}} \tag{11}$$

4.3.8 The Propagation of Sound Wave in the Atmosphere with Temperature Gradient (*c*, *C*)[11]

At night, due to the radiation cooling of the Earth's surface, the air will have a vertical temperature gradient (temperature increases with altitude), leading to the air density's increase with altitude, which in turn causing the sound speed to increase with altitude. Taking an arbitrary point on the horizontal ground and denoting it by O, establishing the Cartesian coordinate system with the frame (O, $\boldsymbol{i}, \boldsymbol{j}$), where $\boldsymbol{j}$ is vertical upward, let x, y be the Cartesian coordinates. Given that the sound speed at an arbitrary point in space as a function of the y coordinate is as follows: $v = v_0(1 + \alpha^2 y^2)$, where v_0 is the sound speed at the ground and α is a constant. Consider a sound wave excited at the origin O propagating in the O-xy plane, the angle between whose excitation direction and the $\boldsymbol{j}$-direction is θ_0. It is assumed that within the range of sound wave propagation: $\alpha y \ll 1$. Find the trajectory of the sound wave propagation in space and find the position where the sound can be most clearly heard on the ground.

Solution

As shown in Fig. 1, the air is divided into a series of thin layers in the direction parallel to the ground, and the interfaces are sorted with positive integers from the bottom to the top. The air density in an arbitrary thin layer can be regarded as uniformly distributed, in which the sound speed is equal everywhere and the sound wave can propagate along a straight line. When the sound wave passes from one thin layer to another, it is refracted at the interface of the two thin layers.

Taking *the air in the propagation path of the sound wave* as the object of study. Studying *the refraction process of the sound wave at the No. (i + 1) interface*. Considering ***law of refraction***:

$$\frac{\sin\theta_i}{v_i} = \frac{\sin\theta_{i+1}}{v_{i+1}} \tag{1}$$

[11]This problem is a test question from Princeton University in 1981. It is also included in the following book: Yongde Zhang, et al. (1988). *American Physics Test Questions and Solutions* (University of Science and Technology of China Press).

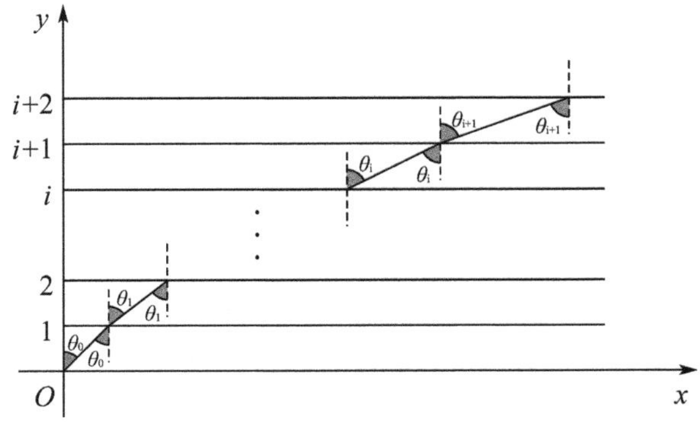

Figure 1

θ_i—the incident angle of the sound wave at the No. $(i + 1)$ interface (i.e., the refraction angle at the No. i interface)

v_i—the sound speed in the air thin layer between the No. i interface and the No. $(i + 1)$ interface.

Equation (1) is the recursion satisfied by θ_i:

$$\Rightarrow \qquad \frac{\sin\theta_i}{v_i} = \frac{\sin\theta_0}{v_0} \tag{2}$$

Let the thickness of all air thin layers tend to zero, θ_i will tend to the angle between the actual propagation direction of the sound wave and the ***j***-direction, which is denoted by θ:

$$\Rightarrow \qquad \frac{\sin\theta}{v} = \frac{\sin\theta_0}{v_0} \tag{3}$$

Known:

$$v = v_0\,(1 + a^2 y^2) \tag{4}$$

$$\Rightarrow \qquad \sin\theta = \sin\theta_0\,(1 + \alpha^2 y^2) \tag{5}$$

Considering ***the geometric meaning of derivative***, then place (5) in to obtain:

$$\Rightarrow \qquad \frac{\mathrm{d}y}{\mathrm{d}x} = \cot\theta = \sqrt{\csc^2\theta - 1} = \sqrt{\frac{1}{\sin^2\theta} - 1}$$

$$= \sqrt{\csc^2\theta_0(1+\alpha^2 y^2)^{-2} - 1}$$

$$\approx \sqrt{\csc^2\theta_0(1-2\alpha^2 y^2) - 1}$$

$$= \sqrt{\cot^2\theta_0 - 2\alpha^2\csc^2\theta_0\, y^2} \tag{6}$$

This operation has considered αy as a small quantity and taken the low order approximation.

$$\Rightarrow \qquad \frac{\mathrm{d}(\sqrt{2}\alpha\sec\theta_0 y)}{\sqrt{1-2\alpha^2\sec^2\theta_0 y^2}} = \sqrt{2}\alpha\csc\theta_0\mathrm{d}x$$

$$\Rightarrow \qquad \int_0^y \frac{\mathrm{d}(\sqrt{2}\alpha\sec\theta_0 y)}{\sqrt{1-2\alpha^2\sec^2\theta_0 y^2}} = \sqrt{2}\alpha\csc\theta_0\int_0^x \mathrm{d}x$$

$$\Rightarrow \qquad \arcsin(\sqrt{2}\alpha\sec\theta_0 y) = \sqrt{2}\alpha\csc\theta_0 x$$

$$\Rightarrow \qquad y = \frac{\sin(\sqrt{2}\alpha\csc\theta_0 x)}{\sqrt{2}\alpha\sec\theta_0} \tag{7}$$

Equation (7) is the trajectory equation of the sound wave propagation in space.

The position where the sound can be most clearly heard on the ground is the intersection of the sound wave propagation trajectory and the ground:

$$\Rightarrow \qquad x\big|_{y=0} = \begin{cases} 0 & \text{(8a)} \\ \dfrac{\pi}{\sqrt{2}\alpha}\sin\theta_0 & \text{(8b)} \end{cases}$$

Solution (8a) represents the position of the source of the sound wave, which is a trivial solution. Solution (8b) represents the position where the sound can be most clearly heard on the ground.

We can draw the schematic diagram of the trajectory of sound wave (see Fig. 2).

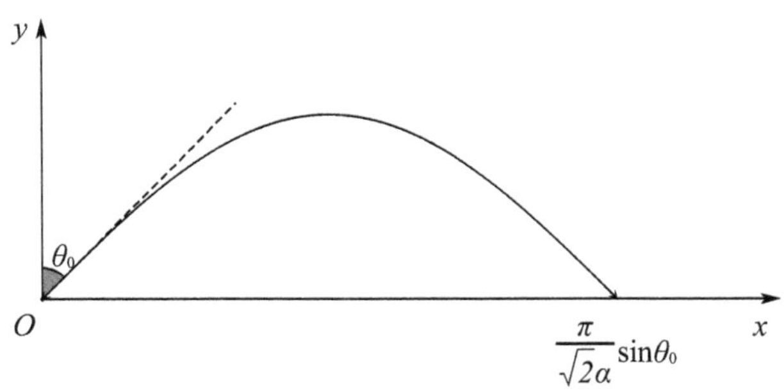

Figure 2

Chapter 5

Two-Body Model

5.1 Central Force Field

The Kepler model is a special two-body model, which simulates the dynamic phenomena of two particles under the action of the inverse-square force (represented by universal gravitation and Coulomb force) exerted by each other. In classical mechanics, the Kepler model is the only one in the N-body model that can accurately solve the orbit. In addition, for the bound state motion of the Kepler model, its orbit is closed. These particularities determine that the Kepler model has important research value and become a basic approximation method to study the N-body model.

Kepler model has perfect space-time symmetry (space translation invariance, time translation invariance, spatial isotropy, time-reversal invariance), which determines that all kinds of motion integration (conserved quantity) theories become the ideal theoretical tools to study it.

A two-body model can be transformed into a one-body model in the center-of-mass reference frame, so the center-of-mass reference frame naturally becomes the ideal reference frame for studying the Kepler model.

The problem of the motion of celestial bodies has been studied deeply and explained perfectly based on the Kepler model theory.

Solving Physics Problems: Exploring New Thinking Paradigms
Yipeng Gu

ISBN 978-981-4877-41-1 (Hardcover), 978-1-003-16254-4 (eBook)
www.jennystanford.com

There are many interesting and important applied research topics, such as interplanetary trajectory, orbit maneuver, spacecraft attitude, etc. In this section, we will study some Kepler problems which are mainly about the motion of celestial bodies.

5.1.1 Kepler Model (d, *B*)

Johannes Kepler summed up three laws of planetary motion based on the motion data of the planets in the solar system observed by Tycho Brahe. The contents are as follows (the common expressions):

1. Each planet moves around the sun in its respective elliptical orbit, with the sun at one of the two foci.
2. The area swept in unit time by the vector pointing from the sun to the same planet is constant.
3. The ratio of the cube of the semi-major axis of the elliptical orbit to the square of the revolutionary period is the same for all planets in the solar system.

The three laws above are called Kepler's first, second and third law of planetary motion respectively.

(1) Try to induce the law of universal gravitation according to Kepler's three laws of planetary motion and other known laws of physics.

(2) Try to prove Kepler's three laws of planetary motion according to the law of universal gravitation and other known laws of physics.

Solution to Problem (1)

The following studies the two-body model composed of an arbitrary planet and the sun. It is assumed that the effects of other planets on them can be ignored.

The solar reference frame can be approximately regarded as an ideal inertial reference frame; the following will be studied in the solar reference frame. The size of the sun and the planet is much smaller than the distance between them; therefore, they can be regarded as particles.

Time zero is arbitrarily set.

As shown in Fig. 1, the position of the sun is denoted by O, and the unit vector in the angular velocity direction of the planet's revolution around the sun is denoted by $\boldsymbol{k}$ (perpendicular to the plane of the figure outward). Taking Point O as the pole, the direction of the sun pointing to the perihelion of the planet (horizontally to the right in the figure) as the positive direction of the polar axis, the right-hand helix direction with regard to the $\boldsymbol{k}$-direction as the positive direction of polar angle, establishing the polar coordinate system, let r, θ be the polar coordinates.

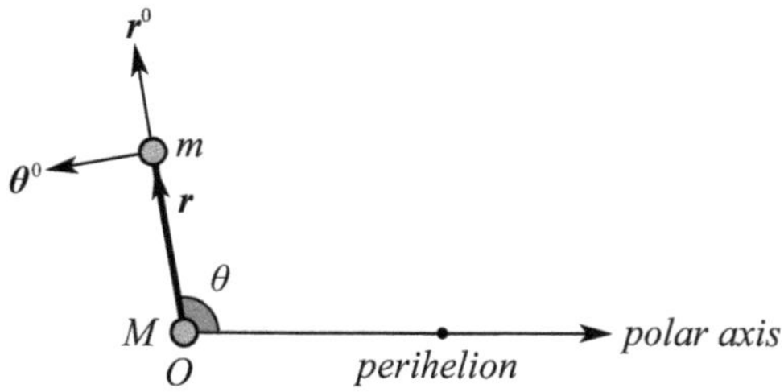

Figure 1

Taking *the planet* as the object of study. Studying *the state at arbitrary Time t*. Considering ***Newton's second law***:

$$\boldsymbol{F} = m\boldsymbol{a} \tag{1}$$

m—the mass of the planet.

$\boldsymbol{a}$—the acceleration of the planet at Time t.

$\boldsymbol{F}$—the force exerted on the planet by the sun at Time t.

Considering ***the components form of the acceleration of a particle in polar coordinate system***:

$$\boldsymbol{a} = (\ddot{r} - r\dot{\theta}^2)\boldsymbol{r}^0 + (2\dot{r}\dot{\theta} + r\ddot{\theta})\boldsymbol{\theta}^0 \tag{2}$$

Known that the areal velocity of the same planet (i.e., the area swept in unit time by the vector pointing from the sun to the same planet) is constant (Kepler's second law of planetary motion):

$$\frac{\mathrm{d}S}{\mathrm{d}t} = C \tag{3}$$

$\dfrac{\mathrm{d}S}{\mathrm{d}t}$—the areal velocity of the planet at arbitrary Time t.

As shown in Fig. 2, considering the area formula of triangle:

$$\mathrm{d}S \approx \frac{1}{2}\left|\boldsymbol{r} \times \mathrm{d}\boldsymbol{s}\right| \tag{4}$$

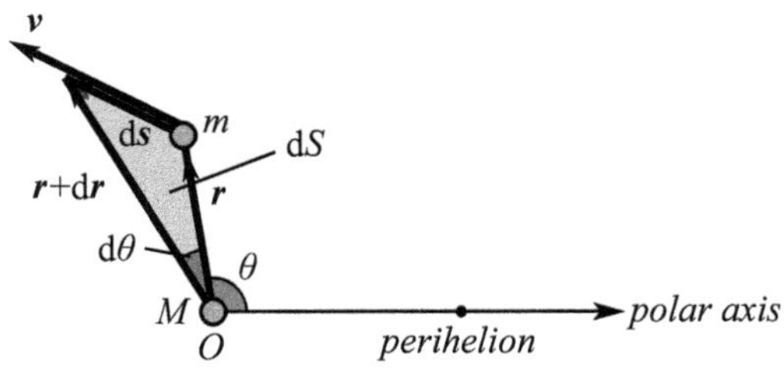

Figure 2

d$\boldsymbol{s}$—the displacement of the planet in small time period dt at arbitrary Time t.

Considering the definition of ***velocity***:

$$\boldsymbol{v} = \frac{\mathrm{d}\boldsymbol{s}}{\mathrm{d}t} \tag{5}$$

$$\Rightarrow \quad \frac{\mathrm{d}S}{\mathrm{d}t} = \frac{1}{2}\left|\boldsymbol{r} \times \boldsymbol{v}\right| \tag{6}$$

Considering ***the components form of the velocity of a particle in polar coordinate system***:

$$\boldsymbol{v} = \dot{r}\boldsymbol{r}^0 + r\dot{\theta}\boldsymbol{\theta}^0 \tag{7}$$

$$\Rightarrow \quad \frac{\mathrm{d}S}{\mathrm{d}t} = \frac{1}{2}r^2\dot{\theta} \tag{8}$$

$$\Rightarrow \quad \frac{1}{2}r^2\dot{\theta} = C \tag{9}$$

Taking the derivative of the terms on both sides of the equation above with respect to t:

$$\Rightarrow \quad \frac{1}{2}r(2\dot{r}\dot{\theta}+r\ddot{\theta}) = 0$$

$$\Rightarrow \quad 2\dot{r}\dot{\theta} + r\ddot{\theta} = 0 \tag{10}$$

(10) → (2),

$$\boldsymbol{a} = (\ddot{r} - r\dot{\theta}^2)\boldsymbol{r}^0 \tag{11}$$

That is, the acceleration of the planet has only radial component at arbitrary Time t.

Known that the obit of each planet is an ellipse with the sun at one of the two foci (Kepler's first law of planetary motion). Considering the curve equation of conic section in polar coordinate system:

$$r = \frac{p}{1+e\cos\theta} \tag{12}$$

e and p—the eccentricity and the semi-latus rectum of the orbit equation of the planet.

$$\Rightarrow \qquad p = r(1+e\cos\theta) \tag{13}$$

Taking the second derivative of the terms on both sides of the equation above with respect to t:

$$\Rightarrow \qquad 0 = \ddot{r}(1+e\cos\theta) - 2\dot{r}(e\dot{\theta}\sin\theta) - re(\ddot{\theta}\sin\theta + \dot{\theta}^2\cos\theta)$$

$$= \ddot{r} + (\ddot{r} - r\dot{\theta}^2)e\cos\theta - (2\dot{r}\dot{\theta} + r\ddot{\theta})e\sin\theta$$

$$= \ddot{r} + (\ddot{r} - r\dot{\theta}^2)e\cos\theta$$

$$= (\ddot{r} - r\dot{\theta}^2)(1+e\cos\theta) + r\dot{\theta}^2$$

$$= (\ddot{r} - r\dot{\theta}^2)\frac{p}{r} + r\dot{\theta}^2 \tag{14}$$

$$\Rightarrow \qquad \ddot{r} - r\dot{\theta}^2 = -r^2\dot{\theta}^2\frac{1}{p} \tag{15}$$

Defining T—the revolutionary period of the planet, S—the area of the planet's elliptical orbit.

$$\frac{\mathrm{d}S}{\mathrm{d}t} = \frac{S}{T} \tag{16}$$

Considering the area formula of ellipse:

$$\Rightarrow \qquad S = \pi ab \tag{17}$$

a and b—the length of the semi-major axis and semi-minor axis of the planet's elliptical orbit.
Combining (8), (16) and (17):

$$\Rightarrow \qquad \dot{\theta} = \frac{2\pi ab}{T}\frac{1}{r^2} \tag{18}$$

Considering the relations of elliptic parameters:

$$p = \frac{b^2}{a} \tag{19}$$

(18), (19) $\rightarrow$ (15),

$$\Rightarrow \qquad \ddot{r} - r\dot{\theta}^2 = -\frac{4\pi^2a^2b^2}{T^2}\frac{a}{b^2}\frac{1}{r^2} = -4\pi^2\frac{a^3}{T^2}\frac{1}{r^2} \tag{20}$$

Known that the ratio of the cube of the semi-major axis of the elliptical orbit to the square of the revolutionary period is the same for all planets (Kepler's and third law of planetary motion):

$$\frac{a^3}{T^2} = C_{\text{sun}} \tag{21}$$

C_{sun}—a constant related to the characteristics of the sun.

Combining (11), (20) and (21):

$$\Rightarrow \qquad \boldsymbol{a} = -4\pi^2 C_{\text{sun}} \frac{1}{r^2} \boldsymbol{r}^0 \tag{22}$$

$$\Rightarrow \qquad \boldsymbol{F} = -4\pi^2 C_{\text{sun}} \frac{m}{r^2} \boldsymbol{r}^0 \tag{23}$$

Defining $\boldsymbol{F}'$—the force exerted on the sun by the planet at arbitrary Time t, and infer that it should have the isomorphic form with $\boldsymbol{F}$:

$$\boldsymbol{F}' = -4\pi^2 C_{\text{planet}} \frac{M}{r^2} (-\boldsymbol{r}^0) \tag{24}$$

M—the mass of the sun.

C_{planet}—a constant related to the characteristics of the planet.

Considering ***Newton's third law***:

$$\boldsymbol{F} = -\boldsymbol{F}' \tag{25}$$

$$\Rightarrow \qquad C_{\text{sun}} m = C_{\text{planet}} M \tag{26}$$

Supposing that C_{sun} and C_{sun} are directly proportional to the mass of the sun and the planet, respectively, defining the proportionality coefficient as $G/4\pi^2$:

$$\begin{cases} C_{\text{sun}} = \dfrac{G}{4\pi^2} M & \text{(27a)} \\ C_{\text{planet}} = \dfrac{G}{4\pi^2} m & \text{(27b)} \end{cases}$$

$$\Rightarrow \qquad \begin{cases} \boldsymbol{F} = -G\dfrac{Mm}{r^2} \boldsymbol{r}^0 & \text{(28a)} \\ \boldsymbol{F}' = -G\dfrac{Mm}{r^2} (-\boldsymbol{r}^0) & \text{(28b)} \end{cases}$$

$\boldsymbol{F}$ and $\boldsymbol{F}'$ are called the universal gravity between the sun and the planet. The law is extended to any two particles, called ***law of universal gravitation***.

Solution to Problem (2)

Time zero is arbitrarily set.

Taking *the planet* as the object of study. Studying *the state at arbitrary Time t*. Considering ***Newton's second law***:

$$\boldsymbol{F} = m\frac{\mathrm{d}\boldsymbol{v}}{\mathrm{d}t} \tag{29}$$

m—the mass of the planet.

$\boldsymbol{v}$—the velocity of the planet at Time t.

$\boldsymbol{F}$—the force exerted on the planet by the sun at Time t.

Taking *the sun and the planet* as the object of study. Studying *the state at arbitrary Time t*. Considering ***law of universal gravitation***:

$$\boldsymbol{F} = -G\frac{Mm}{r^2}\boldsymbol{r}^0 \tag{30}$$

M—the mass of the sun.

Defining $\boldsymbol{M}$—the resultant torque acting on the planet with the pole O (the sun) as the reference point, considering the defining of ***torque***:

$$\Rightarrow \qquad \boldsymbol{M} = \boldsymbol{r}\times\boldsymbol{F} = -G\frac{Mm}{r^2}\boldsymbol{r}\times\boldsymbol{r}^0 = 0 \tag{31}$$

$$\Rightarrow \qquad \boldsymbol{r}\times\frac{\mathrm{d}\boldsymbol{v}}{\mathrm{d}t} = 0 \tag{32}$$

Considering ***product rule*** and the defining of ***velocity***:

$$\Rightarrow \quad \frac{\mathrm{d}}{\mathrm{d}t}(\boldsymbol{r}\times\boldsymbol{v}) = \frac{\mathrm{d}\boldsymbol{r}}{\mathrm{d}t}\times\boldsymbol{v} + \boldsymbol{r}\times\frac{\mathrm{d}\boldsymbol{v}}{\mathrm{d}t} = \boldsymbol{v}\times\boldsymbol{v} + \boldsymbol{r}\times\frac{\mathrm{d}\boldsymbol{v}}{\mathrm{d}t} = \boldsymbol{r}\times\frac{\mathrm{d}\boldsymbol{v}}{\mathrm{d}t} = 0 \tag{33}$$

That is, the direction of the plane determined by $\boldsymbol{r}$ and $\boldsymbol{v}$ does not change with time; therefore, the Planet move in the same plane all the time (this conclusion can also be obtained from the fact that the angular momentum of the planet with the pole O (the sun) as the reference point is conserved).

Repeating the analysis in the solution to Problem (1), we can also obtain:

$$\frac{\mathrm{d}S}{\mathrm{d}t} = \frac{1}{2}\left|\boldsymbol{r}\times\boldsymbol{v}\right| \tag{34}$$

$\frac{\mathrm{d}S}{\mathrm{d}t}$—the areal velocity of the planet at arbitrary Time t.

$$\Rightarrow \qquad \frac{\mathrm{d}S}{\mathrm{d}t} = C \tag{35}$$

That is, the areal velocity of the same planet (i.e., the area swept in unit time by the vector pointing from the sun to the same planet) is constant, thereupon Kepler's second law of planetary motion is proved.

Establishing the polar coordinate system in the same way as **Solution to Problem (1)**.

Taking *the system composed of the planet and the sun* as the object of study. Studying *the process from Time zero to arbitrary Time t*. Considering (a) ***law of conservation of mechanical energy*** [the work done by the external force and non-conservative internal force acting on the system is zero; therefore, the mechanical energy of the system is conserved] (taking the configuration in which the planet and the sun are infinitely apart as the zero point of universal gravitational potential energy), (b) ***law of conservation of angular momentum*** [taking the pole O (the sun) as the reference point, the resultant external torque acting on the system is zero; therefore, the angular momentum of the system is conserved] (the $\boldsymbol{k}$-direction component equation):

$$\begin{cases} \frac{1}{2}m(\dot{r}^2 + r^2\dot{\theta}^2) - G\frac{Mm}{r} = E & \text{(36a)} \\ mr^2\dot{\theta} = L & \text{(36b)} \end{cases}$$

E and L—the mechanical energy and the $\boldsymbol{k}$-direction component of the angular momentum of the system at Time zero.

$$\Rightarrow \qquad \dot{\theta} = \frac{L}{mr^2} \tag{37}$$

Taking the following differential transformation:

$$\Rightarrow \qquad \dot{r} = \frac{\mathrm{d}r}{\mathrm{d}\theta}\dot{\theta} = \frac{L}{mr^2}\cdot\frac{\mathrm{d}r}{\mathrm{d}\theta} \tag{38}$$

(37), (38) $\rightarrow$ (36a),

$$\Rightarrow \qquad \frac{1}{r^4}\left(\frac{\mathrm{d}r}{\mathrm{d}\theta}\right)^2 + \frac{1}{r^2} - \frac{2GMm^2}{L^2 r} = \frac{2mE}{L^2} \tag{39}$$

The equation above is the first-order nonlinear differential equation of r with θ as the variable.

Defining the variable u:

$$u = \frac{1}{r} \tag{40}$$

Taking the following differential transformation:

$$\Rightarrow \qquad \frac{\mathrm{d}r}{\mathrm{d}\theta} = \frac{\mathrm{d}r}{\mathrm{d}u}\frac{\mathrm{d}u}{\mathrm{d}\theta} = -\frac{1}{u^2}\frac{\mathrm{d}u}{\mathrm{d}\theta} \tag{41}$$

(40), (41) → (39),

$$\Rightarrow \qquad \left(\frac{\mathrm{d}u}{\mathrm{d}\theta}\right)^2 + u^2 - \frac{2GMm^2}{L^2}u = \frac{2mE}{L^2} \tag{42}$$

Taking the derivative of the terms on both sides of the equation above with respect to θ:

$$\Rightarrow \qquad 2\frac{\mathrm{d}u}{\mathrm{d}\theta}\cdot\frac{\mathrm{d}^2u}{\mathrm{d}\theta^2} + 2u\frac{\mathrm{d}u}{\mathrm{d}\theta} - \frac{2GMm^2}{L^2}\frac{\mathrm{d}u}{\mathrm{d}\theta} = 0$$

$$\Rightarrow \qquad \frac{\mathrm{d}u}{\mathrm{d}\theta}\left(\frac{\mathrm{d}^2u}{\mathrm{d}\theta^2} + u - \frac{GMm^2}{L^2}\right) = 0 \tag{43}$$

Firstly, we study the situation when $\frac{\mathrm{d}u}{\mathrm{d}\theta}$ is not zero:

$$\Rightarrow \qquad \frac{\mathrm{d}^2u}{\mathrm{d}\theta^2} + u = \frac{GMm^2}{L^2} \tag{44}$$

The general solution to the equation above is as follows:

$$\Rightarrow \qquad u = \frac{GMm^2}{L^2} + A\cos\theta + B\sin\theta \tag{45}$$

A and B—the integration constants.

Considering the initial condition:

$$r\big|_{\theta=0} = r_{\min} \tag{46}$$

$$\Rightarrow \qquad u\big|_{\theta=0} = u_{\max} \tag{47}$$

The necessary condition for u to take the extreme value is as follows:

$$\Rightarrow \qquad \left.\frac{\mathrm{d}u}{\mathrm{d}\theta}\right|_{\theta=0} = 0 \tag{48}$$

(45) → (48),

$$\Rightarrow \qquad B = 0 \qquad (49)$$

$$\Rightarrow \qquad u = \frac{GMm^2}{L^2} + A\cos\theta \qquad (50)$$

$$\Rightarrow \qquad \begin{cases} u_{\max} = \dfrac{GMm^2}{L^2} + A & (51a) \\ u_{\min} = \dfrac{GMm^2}{L^2} - A & (51b) \end{cases}$$

where A is regarded as greater than zero (if A is less than zero, $u_{\max}$ and $u_{\min}$ will be interchanged, which will not affect the later analysis).

Under the condition of Eq. (48), Eq. (42) degenerates to the following form:

$$\Rightarrow \qquad u_{\mathrm{m}}^2 - \frac{2GMm^2}{L^2}u_{\mathrm{m}} - \frac{2mE}{L^2} = 0 \qquad (52)$$

This is the equation satisfied by $u_{\max}$ and $u_{\min}$.

Considering Vieta theorem:

$$\Rightarrow \qquad u_{\max}u_{\min} = -\frac{2mE}{L^2} \qquad (53)$$

(51) → (53),

$$\Rightarrow \qquad A = \frac{GMm^2}{L^2}\sqrt{1 + \frac{2EL^2}{G^2M^2m^3}} \qquad (54)$$

$$\Rightarrow \qquad u = \frac{GMm^2}{L^2}\left(1 + \sqrt{1 + \frac{2EL^2}{G^2M^2m^3}}\cos\theta\right) \qquad (55)$$

$$\Rightarrow \qquad r = \frac{\dfrac{L^2}{GMm^2}}{1 + \sqrt{1 + \dfrac{2EL^2}{G^2M^2m^3}}\cos\theta} \qquad (56)$$

The equation above is a conic section equation, that is, the planet's orbit is a conic section, with the sun at one of the two foci.

Defining the eccentricity e and the semi-latus rectum p of the conic orbit above:

$$\begin{cases} p = \dfrac{L^2}{GMm^2} & (57a) \\ e = \sqrt{1 + \dfrac{2EL^2}{G^2M^2m^3}} & (57b) \end{cases}$$

$$\Rightarrow \qquad r = \frac{p}{1 + e\cos\theta} \qquad (58)$$

According to the range of E, the shapes of the conic orbits can be classified as follows:

$$\Rightarrow e \begin{cases} <1 & \text{when } E < 0 & \text{the orbit is an ellipse} \\ =1 & \text{when } E = 0 & \text{the orbit is a parabola} \\ >1 & \text{when } E > 0 & \text{the orbit is a branch of a hyperbola} \end{cases} \qquad (59)$$

Because the planet moves around the sun in a bound state, the mechanical energy of the system is less than zero (taking the configuration in which the planet and the sun are infinitely apart as the zero point of universal gravitational potential energy), that is, the planet's orbit is an ellipse, with the sun at one of the two foci; thereupon Kepler's first law of planetary motion is proved.

As shown in Fig. 3, introducing the natural frame ($\boldsymbol{\tau}$, $\boldsymbol{n}$) of the planet:

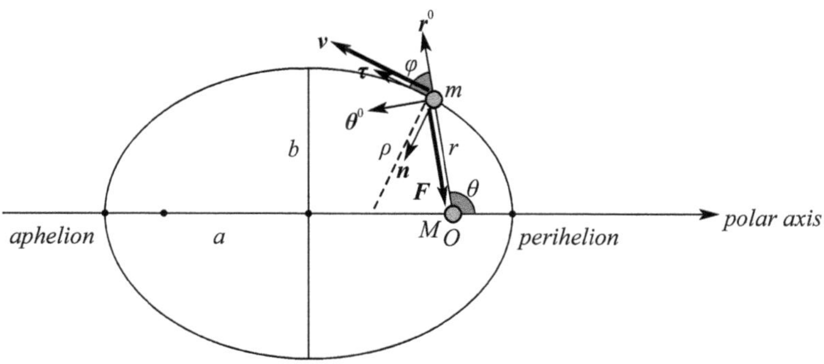

Figure 3

Taking *the planet* as the object of study. Studying *the state at arbitrary Time t*. Considering ***Newton's second law*** (the $\boldsymbol{n}$-direction component equation), then place (30) in to obtain:

$$ma_n = \boldsymbol{F} \cdot \boldsymbol{n} = G\frac{Mm}{r^2}(-\boldsymbol{r}^0 \cdot \boldsymbol{n}) = G\frac{Mm}{r^2}\sin\varphi \qquad (60)$$

a_n—the $\boldsymbol{n}$-direction component of the acceleration of the planet at Time t.

φ—the angle between $\boldsymbol{r}^0$ and $\boldsymbol{\tau}$ at Time t (i.e., the angle between $\boldsymbol{r}^0$ and $\boldsymbol{v}$).

Considering ***the normal component form of acceleration***:

$$a_n = \frac{v^2}{\rho} \tag{61}$$

ρ—the radius of curvature of the planet's orbit at the position where it is at arbitrary Time t.

Considering the form of the radius of curvature at an arbitrary point on the plane curve in polar coordinate system:

$$\rho = \frac{\left[r^2 + \left(\frac{\mathrm{d}r}{\mathrm{d}\theta}\right)^2\right]^{3/2}}{\left|r^2 + 2\left(\frac{\mathrm{d}r}{\mathrm{d}\theta}\right)^2 - r\frac{\mathrm{d}^2 r}{\mathrm{d}\theta^2}\right|} \tag{62}$$

(58) $\rightarrow$ (62),

$$\Rightarrow \qquad \rho = p\frac{(1+e^2+2e\cos\theta)^{3/2}}{(1+e\cos\theta)^3} \tag{63}$$

Considering the angle formula between two lines:

$$\tan\varphi = \frac{k_{\tau} - k_{r^0}}{1 + k_{\tau}k_{r^0}} \tag{64}$$

k_{τ} and k_{r^0} —the slope of $\boldsymbol{\tau}$ and $\boldsymbol{r}^0$.

Considering the geometric meaning of derivative:

$$k_{r^0} = \tan\theta \tag{65}$$

Considering the slope form of the tangent at an arbitrary point of an ellipse:

$$k_{\tau} = -\frac{b^2}{a^2}\frac{x}{y} = -\frac{e+\cos\theta}{\sin\theta} \tag{66}$$

(65), (66) $\rightarrow$ (64),

$$\tan\varphi = -\frac{1+e\cos\theta}{e\sin\theta} \tag{67}$$

$$\Rightarrow \qquad \sin\varphi = \pm\frac{\tan\varphi}{\sqrt{1+\tan^2\varphi}} = \frac{1+e\cos\theta}{\sqrt{1+2e\cos\theta+e^2}} \tag{68}$$

(61), (62) and (68) $\rightarrow$ (60),

$$\Rightarrow \qquad v = \sqrt{GM\frac{1+e^2+2e\cos\theta}{p}} \tag{69}$$

(58), (68) and (69) $\rightarrow$ (34),

$$\Rightarrow \qquad \frac{\mathrm{d}S}{\mathrm{d}t} = \frac{1}{2}|\boldsymbol{r}\times\boldsymbol{v}| = \frac{1}{2}rv\sin\varphi = \frac{1}{2}\sqrt{GMp} = \frac{1}{2}\sqrt{GM\frac{b^2}{a}} \tag{70}$$

Defining T—the revolutionary period of the planet, S—the area of the planet's elliptical orbit.

$$\frac{\mathrm{d}S}{\mathrm{d}t} = \frac{S}{T} \tag{71}$$

Considering the area formula of ellipse:

$$\Rightarrow \qquad S = \pi ab \tag{72}$$

a and b—the length of the semi-major axis and semi-minor axis of the planet's elliptical orbit.

$$\Rightarrow \qquad T = \frac{\pi ab}{\frac{\mathrm{d}S}{\mathrm{d}t}} = \frac{2\pi}{\sqrt{GM}}a^{3/2} \tag{73}$$

$$\Rightarrow \qquad \frac{a^3}{T^2} = \frac{GM}{4\pi^2} \tag{74}$$

That is, for the planet, the ratio of the cube of the semi-major axis of its elliptical orbit to the square of its revolutionary period is proportional to the mass of the sun, that is, the ratio of the cube of the semi-major axis of the elliptical orbit to the square of the revolutionary period is the same for all planets in the solar system, thereupon Kepler's third law of planetary motion is proved.

5.1.2 The Motion of Two Particles from Rest under the Universal Gravitation (d, *AC*)

Consider two particles A and B of mass m_A and m_B, they are only affected by each other's universal gravitation. Initially, the distance between the two particles is r_0, and both are at rest. Find the time taken for the two particles to meet.

Solution [1] (type A)

The time point when the two particles start to move is defined as Time zero.

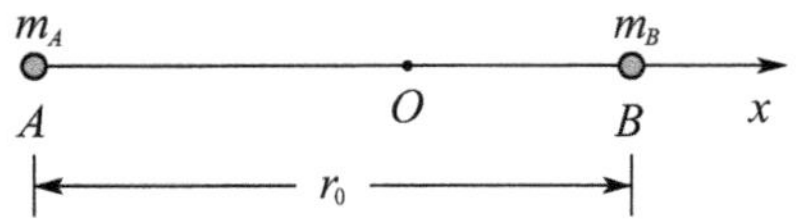

Figure 1

As shown in Fig. 1, the position of the center of mass of the two particles at Time zero is denoted by O, the unit vector of Particle A pointing to Particle B is denoted by $\boldsymbol{i}$. Establishing the linear coordinate system with Point O as the origin and the $\boldsymbol{i}$-direction as the positive direction of coordinate axis, let x be the coordinate.

Taking *the system composed of the two particles* as the object of study. Studying *the process from Time zero to arbitrary Time t.* Considering (a) ***law of conservation of momentum*** [the resultant external force acting on the system is zero; therefore, the momentum of the system is conserved], (b) ***the total momentum of the system of particles is equal to the center-of-mass momentum*** [integrating into (a)].

$$(m_A + m_B)\, \boldsymbol{v}_C = 0 \tag{1}$$

$\boldsymbol{v}_C$—the center-of-mass velocity of the system at Time t.

$$\Rightarrow \qquad \boldsymbol{v}_C = 0 \tag{2}$$

That is, the center of mass of the system is static all the time; therefore, the two particles will meet at the origin O.

Taking *the system composed of the two particles* as the object of study. Studying *the process from Time zero to arbitrary Time t.* Considering (a) ***law of conservation of momentum*** [the analysis method is the same as above] (the $\boldsymbol{i}$-direction component equation), (b) ***law of conservation of mechanical energy*** [the work done by the external force and non-conservative internal force acting on the system is zero; therefore, the mechanical energy of the system is conserved] (taking the configuration in which the two particles are infinitely apart as the zero point of universal gravitational potential energy):

$$\begin{cases} m_A v_A + m_B v_B = 0 & \text{(3a)} \\ -G\dfrac{m_A m_B}{r_0} = \left(\dfrac{1}{2}m_A v_A^2 + \dfrac{1}{2}m_B v_B^2\right) - G\dfrac{m_A m_B}{r} & \text{(3b)} \end{cases}$$

r—the distance between the two particles at Time t.

v_A and v_B—the $\boldsymbol{i}$-direction component of the velocity of Particle A and Particle B at Time t.

$$\Rightarrow \qquad v_A = \sqrt{2G\frac{m_B^2}{m_A + m_B}\left(\frac{1}{r} - \frac{1}{r_0}\right)} \tag{4}$$

Considering the definition of ***velocity***:

$$v_A = \frac{\mathrm{d}x_A}{\mathrm{d}t} \tag{5}$$

Considering the definition of ***center of mass***:

$$x_A = -\frac{m_B}{m_A + m_B} r \tag{6}$$

$$\Rightarrow \qquad v_A = -\frac{m_B}{m_A + m_B}\frac{\mathrm{d}r}{\mathrm{d}t} \tag{7}$$

Combining (4) and (7),

$$\Rightarrow \qquad -\frac{\mathrm{d}r}{\sqrt{\dfrac{1}{r} - \dfrac{1}{r_0}}} = \sqrt{2G(m_A + m_B)}\mathrm{d}t \tag{8}$$

Defining the variable θ:

$$\frac{1}{r} = \frac{1}{r_0}\sec^2\theta \tag{9}$$

$$\Rightarrow \qquad r = r_0\cos^2\theta \tag{10}$$

$$\Rightarrow \qquad \mathrm{d}r = -2r_0\sin\theta\cos\theta\,\mathrm{d}\theta \tag{11}$$

$$\Rightarrow \qquad \mathrm{d}t = \frac{2r_0^{3/2}}{\sqrt{2G(m_A + m_B)}}\cos^2\theta\,\mathrm{d}\theta \tag{12}$$

$$\Rightarrow t_{\text{meetion}} = \int_0^{t_{\text{meetion}}} \mathrm{d}t = \frac{r_0^{3/2}}{2\sqrt{2G(m_A + m_B)}} \int_0^{\pi/2} (1 + \cos 2\theta)\, \mathrm{d}(2\theta)$$

$$= \frac{\pi r_0^{3/2}}{2\sqrt{2G(m_A + m_B)}} \tag{13}$$

t_{meetion}—the time taken for the two particles to meet.

Solution [2] (type C)

As shown in Fig. 2, the universal gravitation exerted on Particle A by Particle B can be regarded as generated from an imaginary particle B' fixed at the origin O, and the motion of Particle A can be regarded as the elliptical motion around Particle B' (the eccentricity degenerates to 1, and the elliptical trajectory degenerates to a straight line).

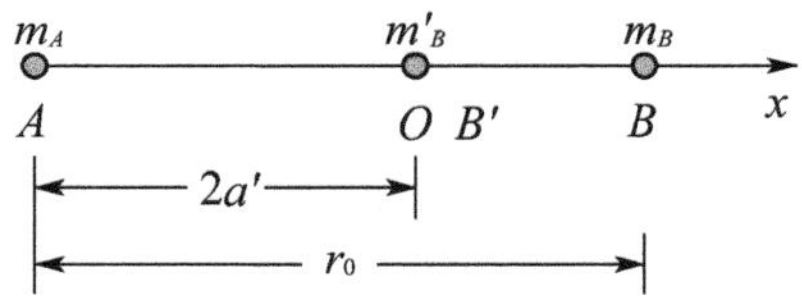

Figure 2

Considering ***the motion period form of the elliptical motion of a particle under the action of universal gravitation***:

$$T' = \frac{2\pi a'^{3/2}}{\sqrt{Gm'_B}} \tag{14}$$

m'_B—the mass of Particle B'.

a'—the length of the semi-major axis of the elliptical trajectory of Particle A around Particle B'.

T'—the motion period of the elliptical motion of Particle A around Particle B'.

From the geometric relation, we can know that:

$$a' = \frac{1}{2}\left(-x_A\big|_{t=0}\right) = \frac{1}{2}\frac{m_B}{m_A + m_B} r_0 \tag{15}$$

According to the assumption, at arbitrary Time t, the universal gravitation exerted on Particle A by Particle B is equal to that exerted on Particle A by Particle B':

$$G\frac{m_A m_B}{r^2} = G\frac{m_A m_B'}{x_A^2} \tag{16}$$

Combining (6) and (16),

$$\Rightarrow \qquad m_B' = \frac{m_B^3}{(m_A + m_B)^2} \tag{17}$$

(15) and (17) $\rightarrow$ (14),

$$\Rightarrow \qquad T' = \frac{\pi r_0^{3/2}}{\sqrt{2G(m_A + m_B)}} \tag{18}$$

When Particle A reaches the origin O, its trajectory is half of the complete elliptical trajectory around Particle B'; therefore, the time taken for the two particles to meet is half of the motion period of the elliptical motion of Particle A around Particle B':

$$\Rightarrow \qquad t_{\text{meetion}} = \frac{T'}{2} = \frac{\pi r_0^{3/2}}{2\sqrt{2G(m_A + m_B)}} \tag{19}$$

Solution [3] (type A)

As shown in Fig. 3, the translational reference frame that is at rest with respect to Particle B is denoted as Reference frame S'. In Reference frame S', the universal gravitation exerted on Particle A by Particle B can be regarded as generated from an imaginary particle B'' fixed at the position of Particle B, and the motion of Particle A can be regarded as the elliptical motion around Particle B'' (the eccentricity degenerates to 1, and the elliptical trajectory degenerates to a straight line).

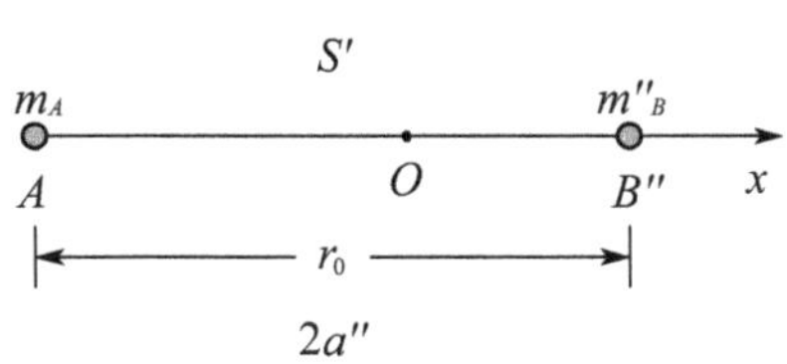

Figure 3

Considering ***the motion period form of the elliptical motion of a particle under the action of universal gravitation***:

$$T'' = \frac{2\pi a''^{3/2}}{\sqrt{Gm_B''}} \tag{20}$$

m''_B—the mass of Particle B''.

a''—the length of the semi-major axis of the elliptical trajectory of Particle A around Particle B''.

T''—the motion period of the elliptical motion of Particle A around Particle B''.

From the geometric relation, we can know that:

$$a'' = \frac{r_0}{2} \tag{21}$$

Taking *Particle B* as the object of study. Studying *the state at arbitrary Time t*. Considering ***Newton's second law*** (the ***i***-direction component equation):

$$-G\frac{m_A m_B}{r^2} = m_B a_B \tag{22}$$

a_B—the ***i***-direction component of the acceleration of Particle B at Time t.

$$\Rightarrow \qquad a_B = -G\frac{m_A}{r^2} \tag{23}$$

According to the assumption, in Reference frame S', at arbitrary Time t, the universal gravitation exerted on Particle A by Particle B is equal to that exerted on Particle A by Particle B'':

$$G\frac{m_A m_B}{r^2} - m_A a_B = G\frac{m_A m''_B}{r^2} \tag{24}$$

(23) → (24),

$$\Rightarrow \qquad m''_B = m_A + m_B \tag{25}$$

(21) and (25) → (20),

$$\Rightarrow \qquad T'' = \frac{\pi r_0^{3/2}}{\sqrt{2G(m_A + m_B)}} \tag{26}$$

When Particle A reaches the origin O, its trajectory is half of the complete elliptical trajectory around Particle B''; therefore, the time taken for the two particles to meet is half of the motion period of the elliptical motion of Particle A around Particle B'':

$$\Rightarrow \qquad t_{\text{meetion}} = \frac{T''}{2} = \frac{\pi r_0^{3/2}}{2\sqrt{2G(m_A + m_B)}} \tag{27}$$

5.1.3 An Explosion in a Binary Star System (*c, AC*)[1]

Consider a supernova explosion model as follows: A binary star system consists of two stars *A* and *B* of mass *M* and *m*. Initially, they both perform the uniform circular motion under each other's universal gravitation. Suddenly an explosion occurs on Star *A* with a mass loss of ΔM. During the explosion, the mass distribution and velocity distribution of the explosion part are spherically symmetric in the translational reference frame that is at rest with respect to Star *A*. The remnant of Star *A* is called Star *A′*. Supposing the effect of the explosion part on Star *A′* and Star *B* in the later motion process can be ignored. Find the condition for Star *A′* and Star *B* to perform bound state motion.

Solution (type C)

The states at the time points before and after the explosion are called the initial state and the final state, respectively. The system formed by Star *A* and Star *B* before the explosion is called the original system, and the system formed by Star *A′* and Star *B* after the explosion is called the new system. As shown in Figs. 1 and 2, the centers of mass of the original system and the new system are denoted by *O* and *C*, respectively. The center-of-mass reference frame of the original system and the new system are denoted by Reference frame *S* and Reference frame *S′*, respectively. The unit vector of the velocity direction of Star *B* at the initial state is denoted by ***i***. Establishing the linear coordinate system fixed to Reference frame *S* with Point *O* as the origin and the ***i***-direction as the positive direction of coordinate axis, let *x* be the coordinate.

Taking *Star A and Star B* as the object of study respectively. Studying *the initial state*. Considering ***Newton's second law*** [in Reference frame *S*] (for each star: the component equation in the radial inward direction):

$$\begin{cases} G\dfrac{Mm}{d^2} = M\dfrac{v_1^2}{r_1} & \text{(1a)} \\ G\dfrac{Mm}{d^2} = m\dfrac{v_2^2}{r_2} & \text{(1b)} \end{cases}$$

[1]This problem is a test question from MIT in 1982. It is also included in the following book: Yongde Zhang, et al. (1988). *American Physics Test Questions and Solutions* (University of Science and Technology of China Press).

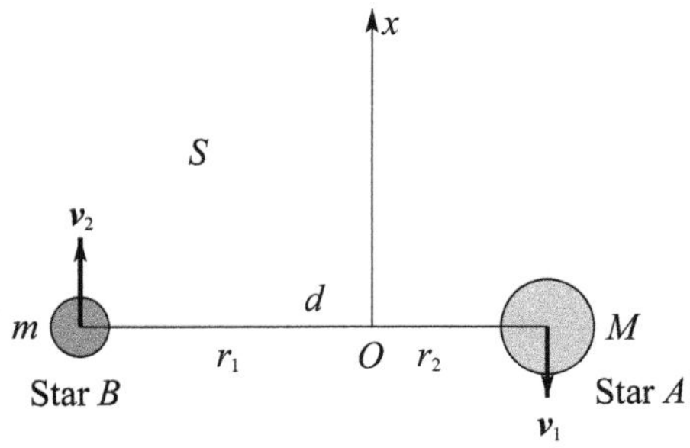

Figure 1

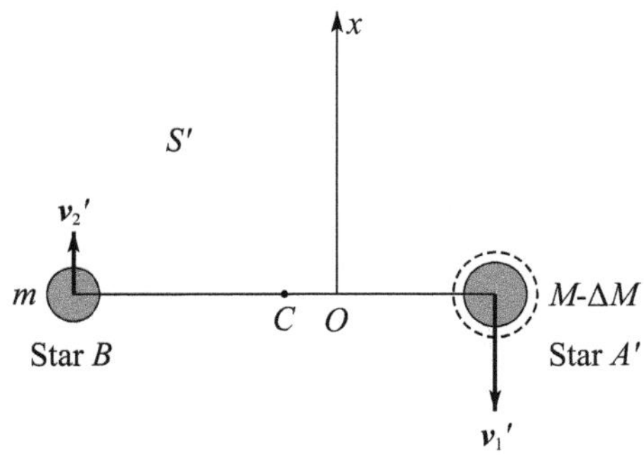

Figure 2

d—the distance between Star A and Star B at the initial state.

r_1 and r_2—the distance from Star A and Star B to the center of mass O at the initial state.

v_1 and v_2—the ***i***-direction component of the velocity of Star A and Star B in Reference frame S at the initial state.

Considering the definition of ***center of mass***:

$$\begin{cases} r_1 = \dfrac{m}{M+m}d & \text{(2a)} \\ r_2 = \dfrac{M}{M+m}d & \text{(2b)} \end{cases}$$

$$\Rightarrow \quad \begin{cases} v_1 = -\dfrac{m}{M+m}\sqrt{G\dfrac{M+m}{d}} & \text{(3a)} \\ v_2 = \dfrac{M}{M+m}\sqrt{G\dfrac{M+m}{d}} & \text{(3b)} \end{cases}$$

Considering ***theorem of center-of-mass velocity***:

$$\Rightarrow \quad v_C = \frac{(M-\Delta M)v_1 + mv_2}{M-\Delta M+m} = \sqrt{G\frac{M+m}{d^3}}\frac{\Delta Mm}{(M-\Delta M+m)(M+m)}d \quad (4)$$

v_C—the ***i***-direction component of the velocity of the center of mass C in Reference frame S at the final state.

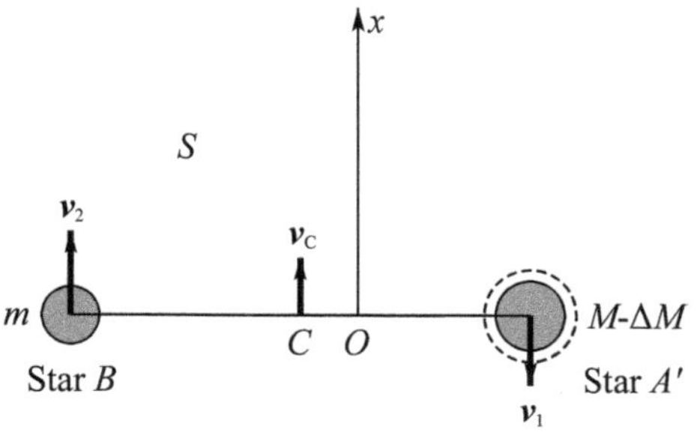

Figure 3

Taking *Star A′ and Star B* as the object of study respectively (see Fig. 3). Studying *the final state*. Considering ***theorem for composition of velocities of a particle*** [taking Reference frame S as the static reference frame, and Reference frame S' as the moving reference frame]:

$$\begin{cases} v_1 = v_C + v_1' & (5a) \\ v_2 = v_C + v_2' & (5b) \end{cases}$$

v'_1 and v'_2—the ***i***-direction component of the velocity of Star A' and Star B in Reference frame S' at the final state.

$$\Rightarrow \begin{cases} v_1' = -\sqrt{G\frac{M+m}{d^3}}\frac{m}{M-\Delta M+m}d & (6a) \\ v_2' = \sqrt{G\frac{M+m}{d^3}}\frac{M-\Delta M}{M-\Delta M+m}d & (6b) \end{cases}$$

Defining E'_k—the kinetic energy of the new system in Reference frame S' at the final state. Considering the definition of ***kinetic energy***, then place (6) in to obtain:

$$\Rightarrow \quad E_k' = \frac{1}{2}(M-\Delta M)v_1'^2 + \frac{1}{2}mv_2'^2 = \frac{1}{2}G\frac{(M+m)(M-\Delta M)m}{(M-\Delta M+m)d} \quad (7)$$

Defining E_p—the universal gravitational potential energy of the new system at the final state (taking the configuration in which the two stars are infinitely apart as the potential energy zero). Considering ***theorem of universal gravitational potential energy***:

$$E_p = -G\frac{(M-\Delta M)m}{d} \tag{8}$$

Defining E'—the mechanical energy of the new system in Reference frame S' at the final state.

$$\Rightarrow \qquad E' = E'_k + E_p = G\frac{(2\Delta M - M - m)(M-\Delta M)\,m}{2(M-\Delta M+m)d} \tag{9}$$

Considering ***the orbit-energy relation of the two-body model under the universal gravitation***, it can be inferred that the condition for Star A' and Star B to perform the bound state motion is that the mechanical energy of the new system in Reference frame S' is less than zero:

$$E' < 0 \tag{10}$$

$$\Rightarrow \qquad \Delta M < \frac{1}{2}(M+m) \tag{11}$$

Another method to get v_C (type A)

Taking *the original system* as the object of study. Studying *the instantaneous process of the explosion*. Considering (a) ***law of conservation of momentum*** [in Reference frame S, the resultant external force acting on the system is zero; therefore, the momentum of the system is conserved] (the ***i***-direction component equation), (b) ***the total momentum of the system of particles is equal to the center-of-mass momentum*** [integrating into (a)]:

$$0 = (M-\Delta M+m)v_C + \Delta M v_1 \tag{12}$$

v_C can be obtained by combining (3a) and (12).

Another method to get E'_k (type C)

Considering ***König theorem***:

$$E_k = E'_k + E_{kC} \tag{13}$$

E_k—the kinetic energy of the new system in Reference frame S at the final state.

E_{kC}—the translational kinetic energy of the new system in Reference frame S at the final state.

$$E_k = \frac{1}{2}(M - \Delta M)v_1^2 + \frac{1}{2}mv_2^2 \tag{14}$$

$$E_{kC} = \frac{1}{2}(M - \Delta M + m)v_C^2 \tag{15}$$

E'_k can be obtained by combining (13) to (15).

The third method to get E'_k (type A)

Considering ***the form of available energy in the center-of-mass reference frame***:

$$E'_k = \frac{1}{2}\frac{(M - \Delta M)m}{(M - \Delta M) + m}(v_1 - v_2)^2 \tag{16}$$

E'_k can be obtained by combining (3) and (16).

5.1.4 The Stability of a Particle's Circular Orbit in a Conservative Central Force Field (d, *C*)

A particle of mass m moves in a conservative central force field $f(r)$. Find the condition for the circular orbit of the particle to be stable (under small disturbance in radial direction), and find the angular frequency of the tiny vibration of the particle in radial direction under this condition.

Solution [1]

Time zero is arbitrarily set.

The center of the force field $f(r)$ is denoted by O, the unit vector of the angular momentum direction of the particle with Point O as the reference point is denoted by $\boldsymbol{k}$. Taking Point O as the pole, an arbitrary direction in the motion plane of the particle as the positive direction of the polar axis, the right-hand helix direction with regard to the $\boldsymbol{k}$-direction as the positive direction of polar angle, establishing the polar coordinate system, let r, θ be the polar coordinates.

Defining $V(r)$—the potential energy associated with $f(r)$:

$$\frac{\partial V(r)}{\partial r} = -f(r) \tag{1}$$

Taking *the system composed of the particle and center of force (the field source of the central force field)* as the object of study. Studying *the state at arbitrary Time t*. Considering (a) ***law of conservation of mechanical energy*** [the work done by the external force and non-conservative internal force acting on the system is zero; therefore, the mechanical energy of the system is conserved], (b) ***law of conservation of angular momentum*** [taking the pole O as the reference point, the resultant external torque acting on the system is zero; therefore, the angular momentum of the system is conserved] (the $\boldsymbol{k}$-direction component equation):

$$\begin{cases} E=\dfrac{1}{2}m\left(v_r^2+v_\theta^2\right)+V(r)=\text{const} & \text{(2a)} \\ L=mv_\theta r=\text{const} & \text{(2b)} \end{cases}$$

E—the mechanical energy of the system at Time t.

L—the $\boldsymbol{k}$-direction component of the angular momentum of the system at Time t [taking the pole O as the reference point].

v_r and v_θ—the $\boldsymbol{r}^0$-direction and $\boldsymbol{\theta}^0$-direction component of the velocity of the particle at Time t.

Considering ***the components form of the velocity of a particle in polar coordinate system***:

$$\begin{cases} v_r=\dot{r} & \text{(3a)} \\ v_\theta=r\dot{\theta} & \text{(3b)} \end{cases}$$

$$\Rightarrow \qquad E=\frac{1}{2}m\dot{r}^2+\frac{L^2}{2mr^2}+V(r) \tag{4}$$

The last two terms of Eq. (4) are the single-valued function of the polar radius r, defining $V_{\text{eff}}(r)$—the equivalent potential energy of the system composed of the particle and the field source of the central force field:

$$V_{\text{eff}}(r)=\frac{L^2}{2mr^2}+V(r) \tag{5}$$

$$\Rightarrow \qquad E=\frac{1}{2}m\dot{r}^2+V_{\text{eff}}(r) \tag{6}$$

The following condition shall be met at arbitrary Time t when the particle performs a uniform circular motion:

$$\left.\frac{\partial V_{\text{eff}}(r)}{\partial r}\right|_{r=r_0}=0 \tag{7}$$

r_0—the orbit radius of the circular orbit of the particle.

From (5),

$$\Rightarrow \quad \left.\frac{\partial V_{\text{eff}}(r)}{\partial r}\right|_{r=r_0}=\left.\frac{\partial V(r)}{\partial r}\right|_{r=r_0}-\frac{L^2}{mr_0^3} \tag{8}$$

$$\Rightarrow \quad \left.\frac{\partial V(r)}{\partial r}\right|_{r=r_0}=\frac{L^2}{mr_0^3} \tag{9}$$

From (5) and (9),

$$\Rightarrow \quad \left.\frac{\partial^2 V_{\text{eff}}(r)}{\partial r^2}\right|_{r=r_0}=\left.\frac{\partial^2 V(r)}{\partial r^2}\right|_{r=r_0}+\frac{3L^2}{mr_0^4}$$

$$=\left.\frac{\partial^2 V(r)}{\partial r^2}\right|_{r=r_0}+\frac{3}{r_0}\left.\frac{\partial V(r)}{\partial r}\right|_{r=r_0} \tag{10}$$

(1) $\rightarrow$ (10),

$$\Rightarrow \quad \left.\frac{\partial^2 V_{\text{eff}}(r)}{\partial r^2}\right|_{r=r_0}=-f'(r_0)-\frac{3}{r_0}f(r_0) \tag{11}$$

The condition for the orbit of the particle to be stable is as follows:

$$\left.\frac{\partial^2 V_{\text{eff}}(r)}{\partial r^2}\right|_{r=r_0}>0 \tag{12}$$

$$\Rightarrow \quad f'(r_0)+\frac{3}{r_0}f(r_0)<0 \tag{13}$$

In the following study, as a special case, $f(r)$ is taken as the following form:

$$f(r)=-\frac{k}{r^n} \tag{14}$$

k—non-negative constant.

$$\Rightarrow \qquad \frac{nk}{r_0^{n+1}} - \frac{3k}{r_0^{n+1}} < 0$$

$$\Rightarrow \qquad n < 3 \tag{15}$$

The Taylor series expansion of $V_{\text{eff}}(r)$ at $r = r_0$ is as follows:

$$V_{\text{eff}}(r) = V_{\text{eff}}(r_0) + \left.\frac{\partial V_{\text{eff}}(r)}{\partial r}\right|_{r=r_0}(r - r_0) + \frac{1}{2}\left.\frac{\partial^2 V_{\text{eff}}(r)}{\partial r^2}\right|_{r=r_0}(r - r_0)^2 + \cdots \tag{16}$$

Taking the second-order approximation of the Taylor series expansion when r approximates r_0, then place (7) in to obtain:

$$\Rightarrow \qquad V_{\text{eff}}(r) \approx V_{\text{eff}}(r_0) + \frac{1}{2}\left.\frac{\partial^2 V_{\text{eff}}(r)}{\partial r^2}\right|_{r=r_0}(r - r_0)^2 \tag{17}$$

(17) → (6),

$$\Rightarrow \qquad E = \frac{1}{2}m\dot{r}^2 + V_{\text{eff}}(r_0) + \frac{1}{2}\left.\frac{\partial^2 V_{\text{eff}}(r)}{\partial r^2}\right|_{r=r_0}(r - r_0)^2 \tag{18}$$

Taking the derivative of the terms on both sides of the equation above with respect to t:

$$\Rightarrow \qquad \ddot{r} + \frac{1}{m}\left.\frac{\partial^2 V_{\text{eff}}(r)}{\partial r^2}\right|_{r=r_0}(r - r_0) = \frac{1}{m}\frac{\mathrm{d}E}{\mathrm{d}t} = 0 \tag{19}$$

Defining δ—the increment of the particle's polar radius at arbitrary Time t (δ is a small quantity):

$$\delta = r - r_0 \tag{20}$$

$$\Rightarrow \qquad \ddot{\delta} + \frac{1}{m}\left.\frac{\partial^2 V_{\text{eff}}(r)}{\partial r^2}\right|_{r=r_0}\delta = 0 \tag{21}$$

Equation (21) is the differential equation of the particle performing the simple harmonic vibration in the radial direction of the circular orbit, and its angular frequency ω is as follows, then place (11) in to obtain:

$$\Rightarrow \qquad \omega = \sqrt{\frac{1}{m}\left.\frac{\partial^2 V_{\text{eff}}(r)}{\partial r^2}\right|_{r=r_0}} = \sqrt{-\left[\frac{3}{r_0}f(r_0) + f'(r_0)\right]/m} \tag{22}$$

Solution [2]

The time point when the particle is disturbed is defined as Time zero.

Taking *the particle* as the object of study. Studying *the state at arbitrary Time t*. Considering ***Newton's second law*** (the $\mathbf{r}^0$-direction component equation):

$$f(r) = ma_r \tag{23}$$

a_r—the $\mathbf{r}^0$-direction component of the acceleration of the particle at Time t.

Considering ***the radial component form of acceleration***:

$$a_r = \ddot{r} - r\dot{\theta}^2 \tag{24}$$

$$\Rightarrow \quad f(r) = m(\ddot{r} - r\dot{\theta}^2) \tag{25}$$

Known:

$$\ddot{r}\big|_{r=r_0} = 0 \tag{26}$$

$$\Rightarrow \quad f(r_0) = -mr_0\left(\dot{\theta}\big|_{r=r_0}\right)^2 \tag{27}$$

Taking *the particle* as the object of study. Studying *the process from Time zero to arbitrary Time t*. Considering ***law of conservation of angular momentum*** (the $\mathbf{k}$-direction component equation):

$$\Rightarrow \quad mr_0^2\,\dot{\theta}\big|_{r=r_0} = mr^2\dot{\theta} \tag{28}$$

Combining (27) and (28),

$$\Rightarrow \quad \dot{\theta} = \frac{r_0^2}{r^2}\sqrt{\frac{-f(r_0)}{mr_0}} \tag{29}$$

(29) $\rightarrow$ (25),

$$\Rightarrow \quad \ddot{r} = \frac{1}{m}\left[-\frac{r_0^3}{r^3}f(r_0) + f(r)\right] \tag{30}$$

Considering the following relation:

$$df(r) = f(r) - f(r_0) = f'(r_0)(r - r_0) = f'(r_0)\delta \tag{31}$$

$$\Rightarrow \quad \ddot{\delta} = \frac{1}{m}\left\{\left[1 - \frac{r_0^3}{(r_0+\delta)^3}\right]f(r_0) + f'(r_0)\delta\right\}$$

$$= \frac{1}{m}\left\{\left[1-\left(1+\frac{\delta}{r_0}\right)^{-3}\right]f(r_0)+f'(r_0)\delta\right\}$$

$$\approx \frac{1}{m}\left\{\left[1-\left(1-\frac{3\delta}{r_0}\right)\right]f(r_0)+f'(r_0)\delta\right\}$$

$$= \frac{\frac{3}{r_0}f(r_0)+f'(r_0)}{m}\delta \tag{32}$$

This operation has taken the first-order approximation of δ.

$$\Rightarrow \qquad \ddot{\delta}+\frac{-\left(\frac{3}{r_0}f(r_0)+f'(r_0)\right)}{m}\delta=0 \tag{33}$$

Equation (33) is the differential equation of the particle performing the simple harmonic vibration in the radial direction of the circular orbit, and its angular frequency ω is:

$$\Rightarrow \qquad \omega=\sqrt{-\left(\frac{3}{r_0}f(r_0)+f'(r_0)\right)/m} \tag{34}$$

5.1.5 The Precession of a Planetary Orbit in a Universal Gravitation Field (*c*, *C*)[2]

Consider a star of mass M, around which there is a large area of spherically symmetric cosmic dust of density ρ. In the range of the cosmic dust, there is a planet of mass m orbiting the star in a circular orbit. Given that the cosmic dust is very thin, its resistance acting on the planet can be ignored, but its universal gravitation acting on the planet must be considered. The angular momentum of the planet with the star as its reference point is denoted by $\boldsymbol{L}$. If the planet is slightly disturbed in the radial direction in the motion process, try to describe the motion of the planet.

Solution

Time zero is arbitrarily set.

[2]This problem is a test question of CUSPEA (China-U.S. Physics Examination and Application) in 1983.

As shown in Fig. 1, the position of the star is denoted by O, and the unit vector in the direction of $\boldsymbol{L}$ is denoted by $\boldsymbol{k}$. Taking Point O as the pole, an arbitrary direction in the motion plane of the planet as the positive direction of the polar axis, the right-hand helix direction with regard to the $\boldsymbol{k}$-direction as the positive direction of polar angle, establishing the polar coordinate system, let r, θ be the polar coordinates.

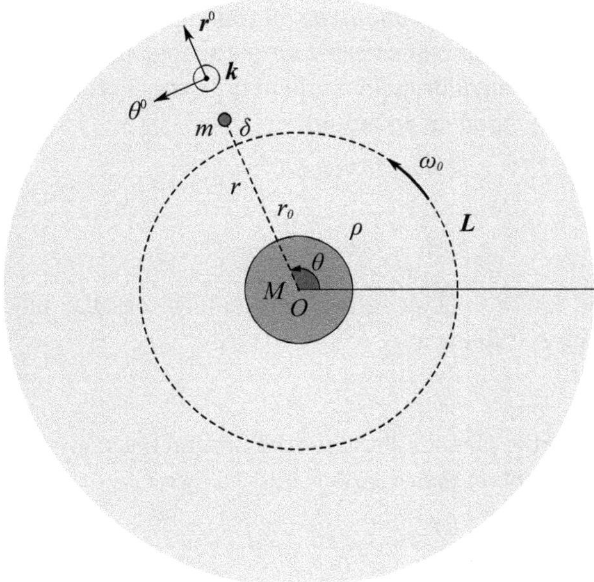

Figure 1

Taking *the planet* as the object of study. Studying *the state at arbitrary Time t*. Considering ***Newton's second law*** [we can imagine a sphere *SP*, of which the radius is same to that of planet's circular orbit, the resultant universal gravitation exerted on the planet by the cosmic dust outside the *SP* region is zero) (the $\boldsymbol{r}^0$-direction component equation), (b) ***the radial component form of acceleration*** [integrating into (a)]:

$$-G\frac{Mm}{r^2}-G\frac{\rho\cdot\frac{4}{3}\pi r^3 m}{r^2}=m(\ddot{r}-r\dot{\theta}^2) \tag{1}$$

Defining the constant k:

$$k = \frac{4}{3}\pi G\rho \tag{2}$$

$$\Rightarrow \qquad \ddot{r} = -G\frac{M}{r^2} - kr + mr\dot{\theta}^2 \tag{3}$$

Taking *the system composed of the star, the planet and the cosmic dust* as the object of study. Studying *an arbitrary process*. Considering ***law of conservation of angular momentum*** [taking the pole O as the reference point, the resultant external torque acting on the system is zero; therefore, the angular momentum of the system is conserved] (the $\boldsymbol{k}$-direction component equation):

$$L = mr^2\dot{\theta} = \text{const} \tag{4}$$

$$\Rightarrow \qquad \ddot{r} = -G\frac{M}{r^2} - kr + \frac{L^2}{m^2r^3} \tag{5}$$

The condition for the planet to perform uniform circular motion around the star is as follows:

$$\ddot{r}\Big|_{r=r_0} = 0 \tag{6}$$

r_0—the radius of the planet's circular orbit around the star.

(3) → (6), we obtain the equation satisfied by r_0:

$$\Rightarrow \qquad -G\frac{M}{r_0^2} - kr_0 + \frac{L^2}{m^2r_0^3} = 0 \tag{7}$$

Defining δ—the increment of the planet's polar radius at arbitrary Time t after it is slightly disturbed in the radial direction (δ is a small quantity):

$$\delta = r - r_0 \tag{8}$$

(8) → (5),

$$\begin{aligned}
\Rightarrow \qquad \ddot{\delta} &= -G\frac{M}{(r_0+\delta)^2} - k(r_0+\delta) + \frac{L^2}{m^2(r_0+\delta)^3} \\
&= -G\frac{M}{r_0^2}\left(1+\frac{\delta}{r_0}\right)^{-2} - k(r_0+\delta) + \frac{L^2}{m^2r_0^3}\left(1+\frac{\delta}{r_0}\right)^{-3} \\
&\approx -G\frac{M}{r_0^2}\left(1-\frac{2\delta}{r_0}\right) - k(r_0+\delta) + \frac{L^2}{m^2r_0^3}\left(1-\frac{3\delta}{r_0}\right)
\end{aligned}$$

$$=-\left(-2G\frac{M}{r_0^3}+kr_0^2+\frac{3L^2}{m^2r_0^4}\right)\delta \tag{9}$$

This operation has taken the first-order approximation of δ.

Combining (7) and (9),

$$\Rightarrow \qquad \ddot{\delta}+\left(\frac{L^2}{m^2r_0^4}+3k\right)\delta=0 \tag{10}$$

Equation (10) is the differential equation of the planet performing the simple harmonic vibration in the radial direction, and its angular frequency ω_r is:

$$\Rightarrow \qquad \omega_r=\sqrt{\frac{L^2}{m^2r_0^4}+3k} \tag{11}$$

Defining ω_0—the $\boldsymbol{k}$-direction component of the angular velocity of the uniform circular motion of the planet around the star in the orbit r_0.

Considering the definition of the ***angular velocity***:

$$\omega_0=\dot{\theta} \tag{12}$$

Combining (4) and (12),

$$\Rightarrow \qquad \omega_0=\frac{L}{mr_0^2} \tag{13}$$

$$\Rightarrow \qquad \omega_r=\sqrt{\omega_0^2+3k}=\omega_0\sqrt{1+\frac{3k}{\omega_0^2}}\approx\omega_0\left(1+\frac{3k}{2\omega_0^2}\right)=\omega_0+\frac{3k}{2\omega_0} \tag{14}$$

This operation has considered k as a small quantity and taken the first-order approximation.

If there is no cosmic dust, the planet will perform a stable elliptical motion around the star if it deviates from the circular orbit. Due to the presence of the dust, the angular frequency of the planet's simple harmonic vibration in the radial direction will be slightly greater than the average angular velocity of its motion around the star [since k and δ are small quantities, the average angular velocity of the planet after it deviates from the circular orbit is approximately equal to the angular velocity of the uniform circular motion]. Therefore the orbit of the planet around the star can be approximately regarded as an ellipse with very slow precession, and the direction of the precession angular velocity is opposite to the direction of the planet's revolution [in this problem, the direction of the precession angular velocity is clockwise (see Fig. 2)].

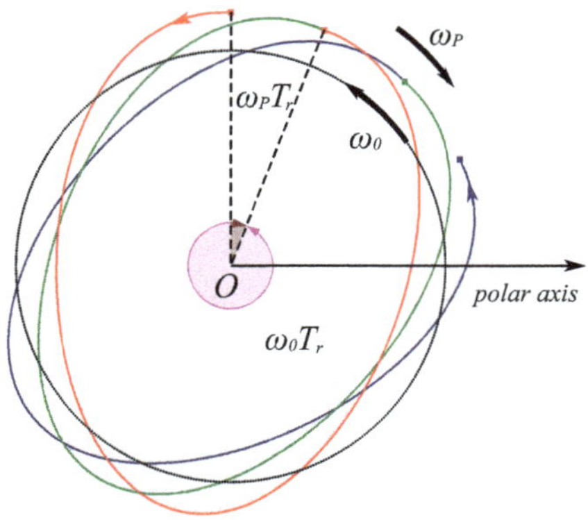

Figure 2

Defining T_r—the motion period of the simple harmonic vibration of the planet in the radial direction.

$$\omega_r T_r = 2\pi \tag{15}$$

Defining ω_P—the $-\boldsymbol{k}$-direction component of the precession angular velocity of the elliptic orbit.

Considering the following geometric relation in Fig. 2:

$$\omega_P T_r + \omega_0 T_r \approx 2\pi \tag{16}$$

$$\Rightarrow \qquad \omega_P = \omega_r - \omega_0 \tag{17}$$

(14) → (17), then place (2) and (13) in to obtain:

$$\Rightarrow \qquad \omega_P = \frac{3k}{2\omega_0} = 2\pi G\rho \frac{mr_0^2}{L} \tag{18}$$

Equations (13) and (18) retains r_0 determined by Eq. (7).

5.1.6 The Time a Comet Spent within the Range of the Earth Orbit (d, *C*)

The earth's orbit around the sun is approximately regarded as a circle. Consider a comet orbiting around the sun, whose orbit intersects the two ends of one diameter of the earth's orbit. Q: How many earth years does the comet travel in the area encircled by the earth's orbit?

Solution [1]

As shown in Fig. 1, the two intersections of the earth's orbit and the comet's orbit are denoted by *A* and *B*, respectively, establishing the polar coordinate system, let *r*, θbe the polar coordinates.

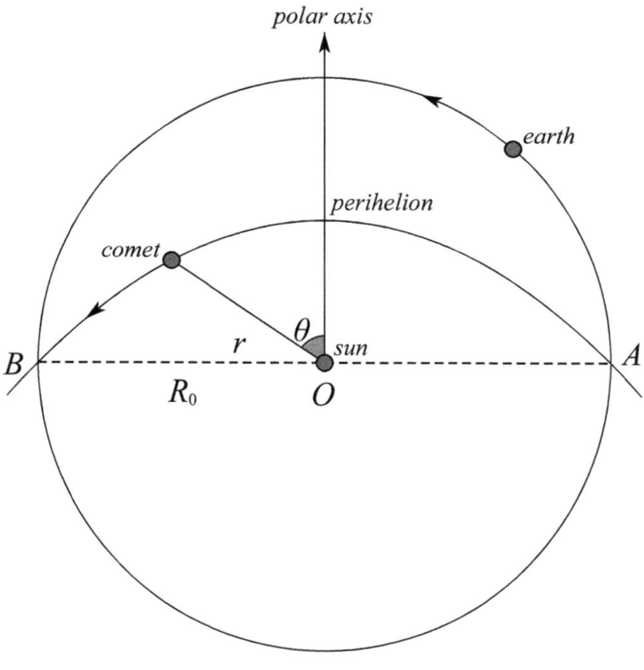

Figure 1

The time point when the comet reaches Point *A* is defined as Time zero.

Taking *the comet* as the object of study. Studying *the state at arbitrary Time t*. Considering ***the orbit equation of a particle in the inverse-square central force field*** [the comet's orbit is a parabola]:

$$r = \frac{p}{1+\cos\theta} \tag{1}$$

p—the semi-latus rectum of the comet's orbit.

Known that the comet's orbit intersects the two ends of one diameter of the earth's orbit:

$$r\Big|_{\theta=\frac{\pi}{2},\frac{3\pi}{2}} = R_0 \tag{2}$$

$$\Rightarrow \qquad p = R_0 \tag{3}$$

$$\Rightarrow \qquad r = \frac{R_0}{1+\cos\theta} \tag{4}$$

Taking *the comet* as the object of study. Studying *an arbitrary process of its motion in the universal gravitational field of the sun*. Considering ***law of conservation of angular momentum*** [taking the sun as the reference point, the external torque acting on the comet is zero; therefore, its angular momentum is conserved] (the component equation in the direction perpendicular to the plane of the figure outward):

$$L = mr^2\dot{\theta} = \text{const} \tag{5}$$

m—the mass of the comet.

L—at arbitrary Time t, the component of the angular momentum of the comet with the sun as its reference point in the direction perpendicular to the plane of the figure outward.

$$\Rightarrow \qquad \mathrm{d}t = \frac{m}{L} r^2 \mathrm{d}\theta \tag{6}$$

Considering ***the semi-latus rectum form of the orbit equation when the inverse-square central force field is the universal gravitational field***:

$$p = \frac{L^2}{GMm^2} \tag{7}$$

M—the mass of the sun.

Combining (3) and (7),

$$\Rightarrow \qquad L = \sqrt{GMm^2R_0} \tag{8}$$

$$\Rightarrow \qquad \mathrm{d}t = \frac{R_0^{3/2}}{\sqrt{GM}} \frac{1}{(1+\cos\theta)^2} \mathrm{d}\theta \tag{9}$$

$$\Rightarrow \qquad \tau = \int_0^{\tau} \mathrm{d}t = \frac{R_0^{3/2}}{\sqrt{GM}} \int_{-\frac{\pi}{2}}^{\frac{\pi}{2}} \frac{\mathrm{d}\theta}{(1+\cos\theta)^2} \tag{10}$$

τ—the time taken for the comet to travel in the area encircled by the earth's orbit.

Defining the definite integral I:

$$I = \int_{-\frac{\pi}{2}}^{\frac{\pi}{2}} \frac{d\theta}{(1+\cos\theta)^2} \tag{11}$$

Defining the integral variable φ:

$$\varphi = \frac{\theta}{2} \tag{12}$$

$$\Rightarrow \quad I = \frac{1}{2}\int_{-\frac{\pi}{4}}^{\frac{\pi}{4}} \frac{d\varphi}{\cos^4\varphi} = \frac{1}{2}\int_{-\frac{\pi}{4}}^{\frac{\pi}{4}} \sec^4\varphi \, d\varphi = \frac{1}{2}\int_{-\frac{\pi}{4}}^{\frac{\pi}{4}} (\tan^2\varphi + 1)\, d(\tan\varphi)$$

$$= \frac{1}{2}\tan\varphi\left(\frac{1}{3}\tan^2\varphi + 1\right)\Bigg|_{-\frac{\pi}{4}}^{\frac{\pi}{4}} = \frac{4}{3} \tag{13}$$

$$\Rightarrow \quad \tau = \frac{4}{3}\frac{R_0^{3/2}}{\sqrt{GM}} \tag{14}$$

Taking *the earth* as the object of study. Studying *the state at arbitrary Time t*. Considering ***Newton's second law*** (the $-\boldsymbol{r}^0$-direction component equation):

$$G\frac{Mm_e}{R_0^2} = m_e\omega_0^2 R_0 \tag{15}$$

m_e—the mass of the earth.

ω_0—the angular velocity magnitude of the earth performing the uniform circular motion around the sun.

$$\Rightarrow \quad \omega_0 = \sqrt{G\frac{M}{R_0^3}} \tag{16}$$

Defining T_0—the motion period of the earth performing the uniform circular motion around the sun (i.e., earth year):

$$\Rightarrow \quad T_0 = \frac{2\pi}{\omega_0} = 2\pi\frac{R_0^{3/2}}{\sqrt{GM}} \tag{17}$$

Combining (14) and (17):

$$\Rightarrow \qquad \tau = \frac{2}{3\pi} T_0 \tag{18}$$

Solution [2]

Taking *the comet* as the object of study. Studying *the state at arbitrary Time t*. Considering ***Kepler's second law***:

$$\frac{\mathrm{d}S}{\mathrm{d}t} = \frac{1}{2} r^2 \dot{\theta} = \frac{1}{2} |\boldsymbol{r} \times \boldsymbol{v}| = \text{const} \tag{19}$$

$\boldsymbol{r}$—the polar radius vector of the comet at Time t.

$\boldsymbol{v}$—the velocity of the comet at Time t.

$\frac{\mathrm{d}S}{\mathrm{d}t}$—the areal velocity of the comet orbiting the sun at Time t.

When the comet reaches the perihelion, the $\boldsymbol{r}^0$-direction component of its velocity is zero, and the form of its areal velocity degenerates into:

$$\Rightarrow \qquad \frac{\mathrm{d}S}{\mathrm{d}t} = \frac{1}{2} r\big|_{\theta=0} v\big|_{\theta=0} \tag{20}$$

From (4),

$$\Rightarrow \qquad r\big|_{\theta=0} = \frac{R_0}{2} \tag{21}$$

Considering ***the orbit-energy relation of the two-body model based on the universal gravitation*** [when the mechanical energy of the system is zero, the orbit is a parabola]:

$$\frac{1}{2} m (v\big|_{\theta=0})^2 - G \frac{Mm}{r\big|_{\theta=0}} = 0 \tag{22}$$

$$\Rightarrow \qquad v\big|_{\theta=0} = 2\sqrt{G \frac{M}{R_0}} \tag{23}$$

$$\Rightarrow \qquad \frac{\mathrm{d}S}{\mathrm{d}t} = \frac{1}{2} \sqrt{GMR_0} \tag{24}$$

(4) → (19),

$$\Rightarrow \qquad \mathrm{d}S = \frac{1}{2} r^2 \mathrm{d}\theta = \frac{1}{2} \frac{R_0^2}{(1+\cos\theta)^2} \mathrm{d}\theta \tag{25}$$

$$\Rightarrow \qquad S = \int_0^S \mathrm{d}S = \frac{1}{2}R_0^2 \int_{-\frac{\pi}{2}}^{\frac{\pi}{2}} \frac{1}{(1+\cos\theta)^2} \mathrm{d}\theta = \frac{2}{3}R_0^2 \qquad (26)$$

S—the area swept by the comet's polar radius during its motion in the area encircled by the earth's orbit.

$$\Rightarrow \qquad \tau = \frac{S}{\frac{\mathrm{d}S}{\mathrm{d}t}} = \frac{4}{3}\frac{R_0^{3/2}}{\sqrt{GM}} \qquad (27)$$

5.1.7 Tides Caused by the Moon and the Sun (d, *C*)

It is approximately assumed that the equatorial plane coincides with the lunar orbit plane and the ecliptic plane. Estimate the height ratio of the tides caused by the moon and the sun on the equator of the earth.

Solution

First, we estimate the tidal height caused by the moon at the earth's equator.

Time zero is arbitrarily set.

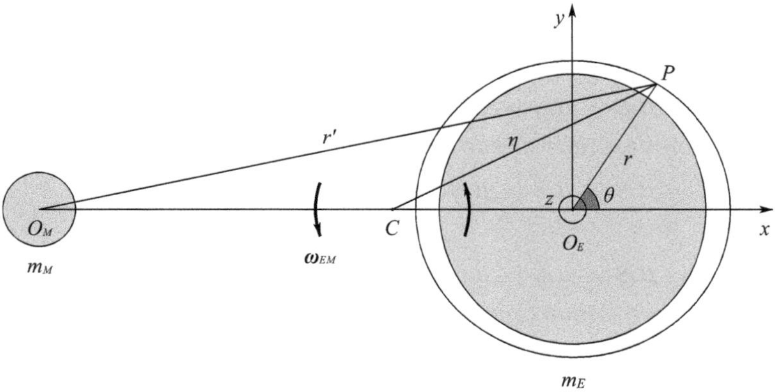

Figure 1

As shown in Fig. 1, the center of mass of the earth, the moon and Earth-Moon system are denoted by O_E, O_M and C, respectively, taking an arbitrary point on the water surface at the earth's equator and denoting it by P. The center-of-mass reference frame of the Earth-

Moon system is called the Earth-Moon center-of-mass reference frame, which can be approximately considered as an inertial reference frame. The rotating reference frame with the center of mass C as the center and the rotating angular velocity of $O_M O_E$ as the rotating angular velocity is called the Earth-Moon rotating reference frame.

The unit vector in the direction of $\overrightarrow{O_M O_E}$ is denoted by $\boldsymbol{i}$, the unit vector in the direction of rotating angular velocity of $O_M O_E$ is denoted by $\boldsymbol{k}$, $\boldsymbol{k} \times \boldsymbol{i}$ is denoted by $\boldsymbol{j}$. Establishing the Cartesian coordinate system with the frame (O_E, $\boldsymbol{i}$, $\boldsymbol{j}$, $\boldsymbol{k}$) fixed to Earth-Moon rotating reference frame, let x, y, z be the Cartesian coordinates. Taking Point O_E as the pole, the positive x-axis as the polar axis, the right-hand helix direction with regard to the $\boldsymbol{k}$-direction as the positive direction of polar angle, establishing the polar coordinate system fixed to Earth-Moon rotating reference frame, let r, θ be the polar coordinates.

Considering the definition of ***center of mass***:

$$r_E = \frac{m_M}{m_E + m_M} r_{EM} \tag{1}$$

m_E and m_M—the mass of the earth and the moon.

r_{EM}—the distance between Point O_E and Point O_M.

r_E—the distance between Point O_E and the center of mass C.

Taking *the earth* as the object of study. Studying *the state at arbitrary Time t*. Considering ***Newton's second law*** [in Earth-Moon center-of-mass reference frame] ($-\boldsymbol{i}$-direction component equation):

$$G\frac{m_E m_M}{r_{EM}^2} = m_E \omega_{EM}^2 r_E \tag{2}$$

ω_{EM}—the $\boldsymbol{k}$-direction component of the rotating angular velocity of $O_M O_E$ in Earth-Moon center-of-mass reference frame.

$$\Rightarrow \qquad \omega_{EM} = \sqrt{\frac{G(m_E + m_M)}{r_{EM}^3}} \tag{3}$$

Considering that $\Delta O_E P O_M$ and $\Delta O_E P C$ satisfy cosine theorem (see Fig. 1):

$$\begin{cases} r' = (r_{EM}^2 + r^2 + 2r_{EM} r\cos\theta)^{1/2} & \text{(4a)} \\ \eta = (r_E^2 + r^2 + 2r_E r\cos\theta)^{1/2} & \text{(4b)} \end{cases}$$

r'—the distance between Point P and Point O_M.

η—the distance between Point P and the center of mass C.

Defining V_E—the universal gravitational potential of the earth at Point P (taking infinity as potential zero), V_M—the universal gravitational potential of the moon at Point P (taking infinity as potential zero), $V_{\text{centrifugal}}$—the inertial centrifugal force potential at Point P in Earth-Moon rotating reference frame (taking the center of mass C as potential zero), V—the total potential at Point P in Earth-Moon rotating reference frame.

Considering ***the form of universal gravitational potential of a sphere with a spherically symmetric mass at an arbitrary point in space***:

$$V_E = -\frac{Gm_E}{r} \tag{5}$$

$$\Rightarrow \quad V_M = -\frac{Gm_M}{r'} = -\frac{Gm_M}{r_{EM}}\left[1+\left(\frac{r}{r_{EM}}\right)^2 + 2\frac{r}{r_{EM}}\cos\theta\right]^{-1/2}$$

$$\approx -\frac{Gm_M}{r_{EM}}\left[1-\frac{r}{r_{EM}}\cos\theta + \frac{1}{2}\left(\frac{r}{r_{EM}}\right)^2(3\cos^2\theta - 1)\right] \tag{6}$$

This operation has considered r/r_{EM} as a small quantity and taken the second-order approximation.

Considering ***theorem of potential energy of inertial centrifugal force***:

$$\Rightarrow V_{\text{centrifugal}} = -\frac{1}{2}\omega_{EM}^2\eta^2 = -\frac{1}{2}\frac{G(m_E+m_M)}{r_{EM}^3}(r_E^2 + r^2 + 2r_E r\cos\theta)$$

$$= -\frac{Gm_M}{r_{EM}}\left[\frac{1}{2}\frac{m_M}{m_E+m_M} + \frac{1}{2}\frac{m_E+m_M}{m_M}\left(\frac{r}{r_{EM}}\right)^2 + \frac{r}{r_{EM}}\cos\theta\right] \tag{7}$$

$$\Rightarrow \quad V = V_E + V_M + V_{\text{centrifugal}}$$

$$= -\frac{Gm_E}{r} - \frac{Gm_M}{r_{EM}}\left[\frac{2m_E+3m_M}{2(m_E+m_M)} + \frac{m_E+3m_M\cos^2\theta}{2m_M}\left(\frac{r}{r_{EM}}\right)^2\right] \tag{8}$$

Considering ***principle of minimum potential energy***, we can know that ***the surface of the water on the earth is an equipotential surface***:

$$V = C \tag{9}$$

$$\Rightarrow \quad -\frac{Gm_E}{r} - \frac{Gm_M}{r_{EM}}\left[\frac{2m_E + 3m_M}{2(m_E + m_M)} + \frac{m_E + 3m_M\cos^2\theta}{2m_M}\left(\frac{r}{r_{EM}}\right)^2\right] = C \tag{10}$$

Defining r_0—the radius of the solid part of the earth, h—the water depth at an arbitrary point:

$$r = r_0 + h \tag{11}$$

$$\begin{aligned}
\Rightarrow \quad C &= -\frac{Gm_E}{r_0}\left(1 + \frac{h}{r_0}\right)^{-1} - \\
&\quad \frac{Gm_M}{r_{EM}}\left[\frac{2m_E + 3m_M}{2(m_E + m_M)} + \frac{m_E + 3m_M\cos^2\theta}{2m_M}\left(\frac{r_0}{r_{EM}}\right)^2\left(1 + \frac{h}{r_0}\right)^2\right] \\
&\approx -\frac{Gm_E}{r_0}\left(1 - \frac{h}{r_0}\right) - \\
&\quad \frac{Gm_M}{r_{EM}}\left[\frac{2m_E + 3m_M}{2(m_E + m_M)} + \frac{m_E + 3m_M\cos^2\theta}{2m_M}\left(\frac{r_0}{r_{EM}}\right)^2\left(1 + \frac{2h}{r_0}\right)\right] \\
&\approx -\frac{Gm_E}{r_0}\left(1 - \frac{h}{r_0}\right) - \frac{Gm_M}{r_{EM}}\left[\begin{array}{l}\dfrac{2m_E + 3m_M}{2(m_E + m_M)} + \\ \dfrac{m_E + 3m_M\cos^2\theta}{2m_M}\left(\dfrac{r_0}{r_{EM}}\right)^2\end{array}\right] \\
&= \frac{Gm_E}{r_0^2}h - \frac{3Gm_M r_0^2}{2r_{EM}^3}\cos^2\theta - \\
&\quad \frac{Gm_E}{r_0} - \frac{Gm_M}{r_{EM}}\left[\frac{2m_E + 3m_M}{2(m_E + m_M)} + \frac{m_E}{2m_M}\left(\frac{r_0}{r_{EM}}\right)^2\right]
\end{aligned} \tag{12}$$

This operation has considered h and r/r_{EM} as small quantities and taken the first-order approximation and second-order approximation respectively.

$$\Rightarrow \qquad \frac{Gm_E}{r_0^2}h - \frac{3Gm_M r_0^2}{2r_{EM}^3}\cos^2\theta = C'$$

$$\Rightarrow \qquad h = \frac{3m_M r_0^4 \cos^2\theta}{2m_E r_{EM}^3} + C'' \tag{13}$$

$$\Rightarrow \qquad \begin{cases} h_{\max} = h\big|_{\theta=0} = \dfrac{3m_M r_0^4}{2m_E r_{EM}^3} + C'' & (14a) \\ h_{\min} = h\big|_{\theta=\pi/2} = C'' & (14b) \end{cases}$$

Defining $(\Delta h)_M$—the tidal height caused by the moon on the earth's equator:

$$\Rightarrow \qquad (\Delta h)_M = h_{\max} - h_{\min} = \frac{3m_M r_0^4}{2m_E r_{EM}^3} \tag{15}$$

Defining $(\Delta h)_S$—the tidal height caused by the sun on the earth's equator, in the same way, we can obtain:

$$\Rightarrow \qquad (\Delta h)_S = \frac{3m_S r_0^4}{2m_E r_{ES}^3} \tag{16}$$

m_S—the mass of the sun.

r_{ES}—the distance between the earth and the sun.

$$\Rightarrow \qquad \frac{(\Delta h)_S}{(\Delta h)_M} = \frac{m_S r_{EM}^3}{m_M r_{ES}^3} \approx \frac{2.0\times 10^{30}\times(3.8\times 10^8)^3}{7.3\times 10^{22}\times(1.5\times 10^{11})^3} \approx 0.44 \tag{17}$$

It can be seen from the estimated results that the main contribution of the tide on the earth's surface is the moon.

5.1.8 Dual-Impulse Hohmann Transfer and Triple-Pulse Hohmann Transfer (d, *C*)

A spacecraft orbiting the earth plans to transfer from the initial circular orbit (denoted by Orbit ①) to the target circular orbit (denoted by Orbit ②) coaxial with it (see Fig. 1), two schemes called "Dual-impulse Hohmann transfer" and "Triple-pulse Hohmann transfer" respectively are considered:

The Dual-impulse Hohmann transfer scheme is as follows: an elliptical transfer orbit (denoted by Orbit i) is coaxial with Orbits

① and ②, and Orbits ① and ② are tangent to Orbit i at Points A and B, respectively. In the first step, the spacecraft moves along Orbit ①, changes its velocity at Point A, and enters Orbit i. In the second step, the spacecraft changes its velocity at Point B, and enters Orbit ②.

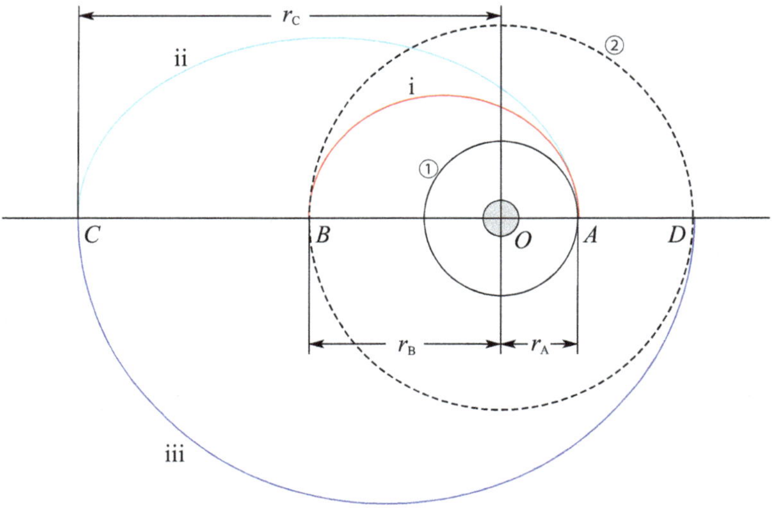

Figure 1

The Triple-pulse Hohmann transfer scheme is as follows: two elliptical transfer orbits (denoted by Orbit ii and Orbit iii respectively) are coaxial with Orbits ① and ②, and Orbits ① is tangent to Orbit ii at Point A, Orbit ii is tangent to Orbit iii at Point C, Orbit iii is tangent to Orbit ② at Point D. In the first step, the spacecraft moves along Orbit ①, changes its velocity at Point A, and enters Orbit ii. In the second step, the spacecraft changes its velocity at Point C, and enters Orbit iii. In the third step, the spacecraft changes its velocity at Point D, and enters Orbit ②.

Given $\qquad r_B = \alpha r_A, r_C = \gamma r_A,\quad$ where $\gamma > \alpha > 1$.

The spacecraft can change its velocity by impulse maneuver, the mass of each jet is far less than that of the spacecraft, and the time of each jet can be ignored. The energy consumption during each jet is positively related to the speed change caused by the impulse maneuver. Try to compare the energy consumption of Dual-impulse Hohmann transfer and Triple-pulse Hohmann transfer.

Solution

The energy consumption of the two schemes can be compared by comparing the sums of the speed changes caused by all impulse maneuvers respectively in the two schemes.

For the motion of the spacecraft around the earth in an arbitrary elliptical orbit, considering ***the mechanical energy form of two-body model performing the bound state motion based on universal gravitation***:

$$\frac{1}{2}mv^2 - G\frac{Mm}{r} = -G\frac{Mm}{2a} \tag{1}$$

M and m—the mass of the earth and the spacecraft.

a—the length of the semi-major axis of the elliptical orbit of the spacecraft around the earth.

r—the distance from the spacecraft to the earth at arbitrary time.

v—the speed of the spacecraft at arbitrary time.

$$\Rightarrow \qquad v = \sqrt{2 - \frac{r}{a}}\sqrt{\frac{GM}{r}} \tag{2}$$

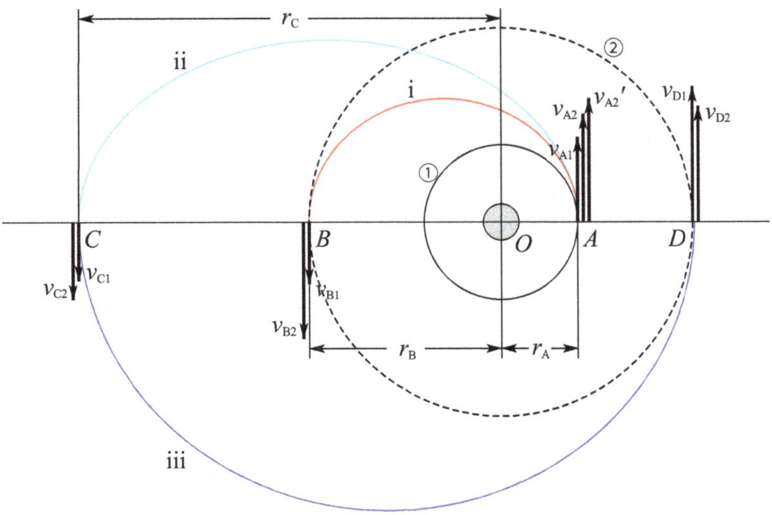

Figure 2

For Dual-impulse Hohmann transfer, defining v_{A1} and v_{A2}—the speed of the spacecraft at the time point before and after the jet at Point A. v_{B1} and v_{B2}—the speed of the spacecraft at the time point before and after the jet at Point B. Δv_A and Δv_B—the speed change of the spacecraft respectively caused by the impulse maneuver at Points A and B. $\Delta v_{\text{Dual-impulse}}$—the sum of speed changes caused by the two impulse maneuvers of the spacecraft (see Fig. 2).

$$\Rightarrow \quad \Delta v_{\text{Dual-impulse}}$$

$$= \Delta v_A + \Delta v_B = \left|v_{A2} - v_{A1}\right| + \left|v_{B2} - v_{B1}\right|$$

$$= \left| v\big|_{r=r_A,\,2a=r_A+r_B} - v\big|_{r=r_A,\,a=r_A} \right| + \left| v\big|_{r=r_B,\,a=r_B} - v\big|_{r=r_B,\,2a=r_A+r_B} \right|$$

$$= \left| \sqrt{\frac{2\alpha}{1+\alpha}}\sqrt{\frac{GM}{r_A}} - \sqrt{\frac{GM}{r_A}} \right| + \left| \sqrt{\frac{1}{\alpha}}\sqrt{\frac{GM}{r_A}} - \sqrt{\frac{1}{\alpha}\frac{2}{1+\alpha}}\sqrt{\frac{GM}{r_A}} \right|$$

$$= \left(\sqrt{\frac{2\alpha}{1+\alpha}} - 1 \right)\sqrt{\frac{GM}{r_A}} + \sqrt{\frac{1}{\alpha}}\left(1 - \sqrt{\frac{2}{1+\alpha}} \right)\sqrt{\frac{GM}{r_A}}$$

$$= \left[\sqrt{\frac{1}{\alpha}} + \frac{\sqrt{2}(\alpha-1)}{\sqrt{\alpha(1+\alpha)}} - 1 \right]\sqrt{\frac{GM}{r_A}} \tag{3}$$

For Triple-pulse Hohmann transfer, defining v'_{A2} — the speed of the spacecraft at the time point after the jet at Point A. v_{C1} and v_{C2}—the speed of the spacecraft at the time point before and after the jet at Point C. v_{D1} and v_{D2}—the speed of the spacecraft at the time point before and after the jet at Point D. $\Delta v'_A$, Δv_C and Δv_D—the speed change of the spacecraft respectively caused by the impulse maneuver at Points A, C and D. $\Delta v_{\text{Triple-pulse}}$—the sum of speed changes caused by the three impulse maneuvers of the spacecraft (see Fig. 2).

$$\Rightarrow \quad \Delta v_{\text{Triple-impulse}}$$

$$= \Delta v'_A + \Delta v_C + \Delta v_D = \left|v'_{A2} - v_{A1}\right| + \left|v_{C2} - v_{C1}\right| + \left|v_{D2} - v_{D1}\right|$$

$$= \left| v\big|_{r=r_A,\,2a=r_A+r_C} - v\big|_{r=r_A,\,a=r_A} \right| + \left| v\big|_{r=r_C,\,2a=r_B+r_C} - v\big|_{r=r_C,\,2a=r_A+r_C} \right| +$$
$$\left| v\big|_{r=r_B,\,a=r_B} - v\big|_{r=r_B,\,2a=r_B+r_C} \right|$$

$$=\left|\sqrt{\frac{2\gamma}{1+\gamma}}\sqrt{\frac{GM}{r_A}}-\sqrt{\frac{GM}{r_A}}\right|+$$

$$\left|\sqrt{\frac{1}{\gamma}\frac{2\alpha}{\alpha+\gamma}}\sqrt{\frac{GM}{r_A}}-\sqrt{\frac{1}{\gamma}\frac{2}{1+\gamma}}\sqrt{\frac{GM}{r_A}}\right|+$$

$$\left|\sqrt{\frac{1}{\alpha}}\sqrt{\frac{GM}{r_A}}-\sqrt{\frac{1}{\alpha}\left(\frac{2\gamma}{\alpha+\gamma}\right)}\sqrt{\frac{GM}{r_A}}\right|$$

$$=\left(\sqrt{\frac{2\gamma}{1+\gamma}}-1\right)\sqrt{\frac{GM}{r_A}}+\sqrt{\frac{2}{\gamma}}\left(\sqrt{\frac{\alpha}{\alpha+\gamma}}-\sqrt{\frac{1}{1+\gamma}}\right)\sqrt{\frac{GM}{r_A}}+$$

$$\sqrt{\frac{1}{\alpha}}\left(\sqrt{\frac{2\gamma}{\alpha+\gamma}}-1\right)\sqrt{\frac{GM}{r_A}}$$

$$=\left[\sqrt{\frac{2(\alpha+\gamma)}{\alpha\gamma}}+\sqrt{\frac{2}{\gamma(1+\gamma)}}(\gamma-1)-\sqrt{\frac{1}{\alpha}}-1\right]\sqrt{\frac{GM}{r_A}} \tag{4}$$

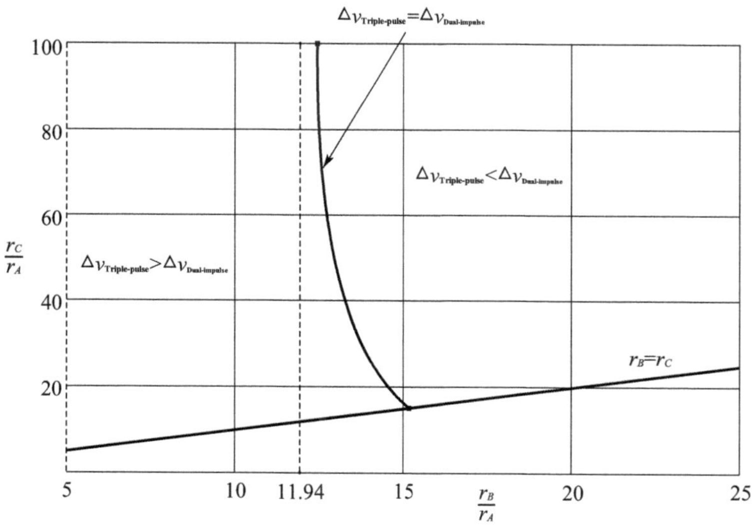

Figure 3

Based on the above results we can get a graphical representation of the relation between $\Delta v_{\text{Dual-impulse}}$ and $\Delta v_{\text{Triple-pulse}}$ (see Fig. 3). It can be seen that if the radius of the target circular orbit is smaller than 11.94 times of the radius of the initial circular orbit, the Dual-impulse Hohmann transfer scheme will be more energy-saving; if the radius of the target circular orbit is larger than about 15 times of the radius of the initial circular orbit, the Triple-pulse Hohmann transfer scheme will be more energy-saving.

5.1.9 The Docking of a Space Station with Its Launched Spacecraft (d, *C*)

Consider a space station of mass M with a spacecraft of mass m on it orbiting the earth in a circular orbit, whose orbit radius is n ($n = 1.25$) times the radius of the earth. At one point, the space station launches the spacecraft in the motion direction, the mass of jet during the launch can be ignored. After the launch, the spacecraft and the space station will respectively move along their new elliptical orbit. The distance between the apogee of the spacecraft's elliptical orbit and the earth's center is $8n$ times of the earth's radius. Q: In order to ensure that the spacecraft can just meet the space station after one circle around the earth, what should the value of M/m be?

Solution

The time point before the space station launches the spacecraft is defined as Time zero.

As shown in Fig. 1, the earth's center is denoted by O, the launch position of the spacecraft is denoted by A, the perigee of the space station's elliptical orbit is denoted by B, and the apogee of the spacecraft's elliptical orbit is denoted by C. Taking Point O as the pole, the direction of $\overrightarrow{OA}$ as the positive direction of the polar axis, anticlockwise direction as the positive direction of polar angle, establishing the polar coordinate system, let r, θ be the polar coordinates.

Taking *the system composed of the space station and the spacecraft (which can be regarded as a particle)* as the object of study. Studying *the state at Time zero*. Considering ***Newton's second law*** (the $-\mathbf{r}^0$-direction component equation):

$$G\frac{M_e(M+m)}{(nR)^2} = (M+m)\frac{v_{M+m}^2}{nR} \tag{1}$$

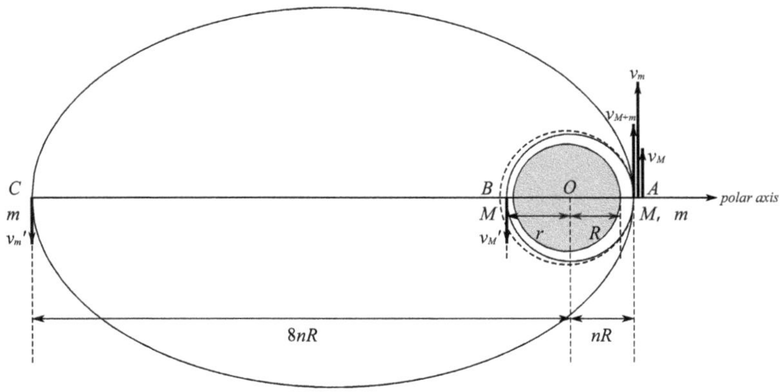

Figure 1

M_e—the mass of the earth.

v_{M+m}—the $\boldsymbol{\theta}^0$-direction component of the velocity of the space station and the spacecraft at Time zero.

$$\Rightarrow \qquad v_{M+m} = \sqrt{\frac{GM_e}{nR}} \tag{2}$$

Taking *the system composed of the space station and the spacecraft* as the object of study. Studying *the instantaneous process of the space station's launching the spacecraft.* Considering ***law of conservation of momentum*** [the mass of the jet during the launch can be ignored, and their gravity is a finite quantity, acting on an extremely short process, its impulse is an infinitely small quantity, which can be ignored; therefore, the momentum of the system is conserved] (the $\boldsymbol{\theta}^0$-direction component equation):

$$(M+m)\,v_{M+m} = Mv_M + mv_m \tag{3}$$

v_M and v_m—the $\boldsymbol{\theta}^0$-direction component of the velocity of the space station and the spacecraft at the time point after the space station launches the spacecraft.

Taking *the system composed of the space station and the earth* as the object of study. Studying *the process of the space station's moving from Point A to Point B.* Considering (a) ***law of conservation of mechanical energy*** [the work done by the external force and non-conservative internal force acting on the system is zero; therefore, the mechanical energy of the system is conserved] (b)

law of conservation of angular momentum [taking Point O as the reference point, the resultant external torque acting on the system is zero; therefore, the angular momentum of the system is conserved] (the component equation in the direction perpendicular to the plane of the figure outward):

$$\begin{cases} \frac{1}{2}Mv_M^2 - G\frac{M_eM}{nR} = \frac{1}{2}Mv_M'^2 - G\frac{M_eM}{r} & \text{(4a)} \\ Mv_M nR = Mv_M' r & \text{(4b)} \end{cases}$$

r—the polar radius of Point B.

v'_M—the $\boldsymbol{\theta}^0$-direction component of the velocity of the space station at Point B.

$$\Rightarrow \qquad v_M = \sqrt{\frac{2r}{r+nR}} \cdot \sqrt{\frac{GM_e}{nR}} \tag{5}$$

Taking *the system composed of the spacecraft and the earth* as the object of study. Studying *the process of the spacecraft's moving from Point A to Point C*. Considering (a) ***law of conservation of mechanical energy***, (b) ***law of conservation of angular momentum*** (the component equation in the direction perpendicular to the plane of the figure outward) [the analyses method are the same as the above analyses on the space station-earth system]:

$$\begin{cases} \frac{1}{2}mv_m^2 - G\frac{M_em}{nR} = \frac{1}{2}mv_m'^2 - G\frac{M_em}{8nR} & \text{(6a)} \\ mv_m nR = mv_m' \cdot 8nR & \text{(6b)} \end{cases}$$

v'_m—the $\boldsymbol{\theta}^0$-direction component of the velocity of the spacecraft at Point C.

$$\Rightarrow \qquad v_m = \frac{4}{3}\sqrt{\frac{GM_e}{nR}} \tag{7}$$

(5) and (7) $\rightarrow$ (3):

$$\Rightarrow \qquad \frac{m}{M} = 3\left(1 - \sqrt{\frac{2r}{r+nR}}\right) \tag{8}$$

The condition that the spacecraft can just meet the space station after one circle around the earth is that the motion period of the spacecraft is a positive integral times of the motion period of the space station:

$$T_m = kT_M \quad k \in \mathbf{N}_+ \tag{9}$$

T_m and T_M—the motion period of the spacecraft and the space station.

Considering ***Kepler's third law***:

$$\frac{T_m}{T_M} = \left(\frac{9nR}{r+nR}\right)^{3/2} \tag{10}$$

$$\Rightarrow \qquad r = \frac{9-k^{2/3}}{k^{2/3}}nR \tag{11}$$

(11) → (8):

$$\Rightarrow \qquad \frac{m}{M} = 3-\sqrt{18-2k^{2/3}} \tag{12}$$

Considering (a) the mass of the space station and the spacecraft are both positive, (b) the spacecraft is not allowed to hit the earth:

$$\begin{cases} \dfrac{m}{M} > 0 \Rightarrow k > \left(\dfrac{9}{2}\right)^{3/2} \approx 9.55 & \text{(13a)} \\ r > R \Rightarrow k < \left(\dfrac{9n}{n+1}\right)^{3/2} \approx 11.18 & \text{(13b)} \end{cases}$$

$$\Rightarrow \qquad 9.55 < k < 11.18 \tag{14}$$

$$\Rightarrow \qquad k = 10,11 \tag{15}$$

(15) → (12):

$$\Rightarrow \qquad \frac{m}{M} = \begin{cases} 0.048 & k=10 \\ 0.153 & k=11 \end{cases} \tag{16}$$

5.1.10 The Multiple Orbital Transfers of a Spacecraft (d, *C*)

Consider a spacecraft orbiting the earth in a circular orbit of radius r_0 (see Fig. 1). The spacecraft can change its orbit by impulse maneuver, the mass of each jet is far less than that of the spacecraft, and the time of each jet can be ignored. Given the mass of the earth is *M* and the radius is *R*.

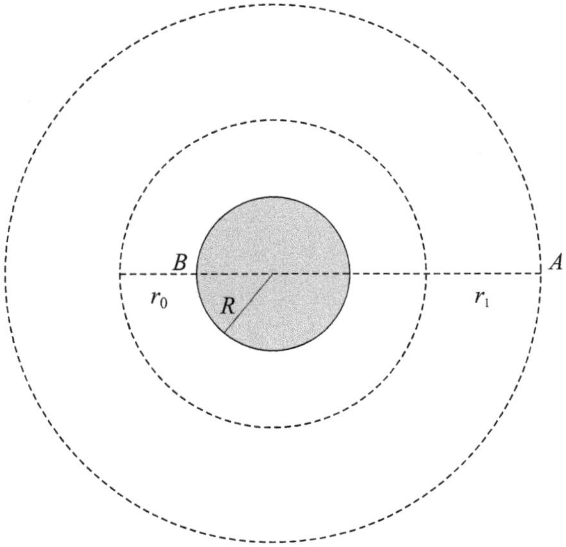

Figure 1

(1) In order to make the spacecraft escape the bond of the earth from Orbit r_0, the spacecraft performs the first impulse maneuver. What is the minimum velocity increment required? In what direction should the spacecraft jet?

(2) During the flight to escape from the earth as required above, the spacecraft performs a second impulse maneuver which makes it fly in a higher circular orbit with the radius of r_1. What is the minimum velocity increment required? In what direction should the spacecraft jet?

(3) In order to return to the earth from the Orbit r_1, the spacecraft performs a third impulse maneuver. Requiring the spacecraft to reach the certain Point B along the tangential direction of the earth's surface. There are two different schemes: (i) Jetting forward at Point A which is collinear with Point B and geocenter. (ii) Jetting radially outward at an appropriate point. Find the velocity increment required by the two schemes of impulse maneuver respectively.

Solution to Problem (1)

As shown in Fig. 2, taking *the spacecraft* as the object of study. Studying *the state at arbitrary Time t during its orbiting in Orbit r_0*.

Considering ***Newton's second law*** (the radial inward component equation):

$$G\frac{Mm}{r_0^2} = m\frac{v_0^2}{r_0} \tag{1}$$

m—the mass of the spacecraft.

v_0—the speed of the spacecraft at arbitrary Time t during its orbiting in Orbit r_0.

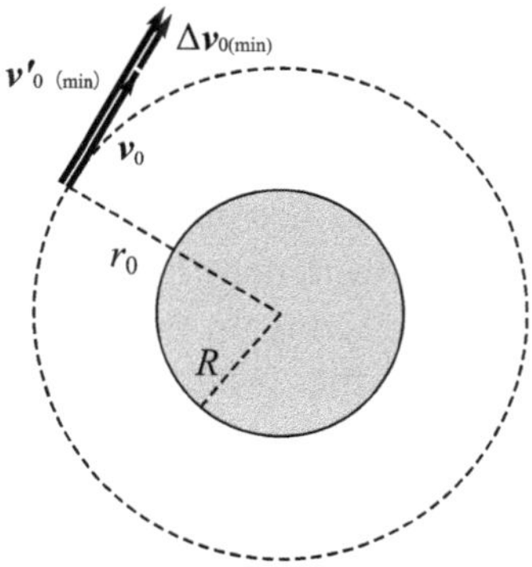

Figure 2

$$\Rightarrow \qquad v_0 = \sqrt{\frac{GM}{r_0}} \tag{2}$$

Considering ***the orbit-energy relation of the two-body model under the universal gravitation***, it can be inferred that the condition for the spacecraft to escape the bond of the earth is that the mechanical energy of the system composed of the spacecraft and the earth is greater than or equal to zero:

$$\frac{1}{2}mv_0'^2 - G\frac{Mm}{r_0} \geq 0 \tag{3}$$

v'_0—the speed at which the spacecraft can escape the bond of the earth after the first impulse maneuver in Orbit r_0.

$$\Rightarrow \qquad v'_{0(\min)} = \sqrt{\frac{2GM}{r_0}} \tag{4}$$

Defining $\Delta \boldsymbol{v}_{0(\min)}$—the minimum velocity increment required for the first impulse maneuver of the spacecraft.

It is easy to prove that when the directions of $\boldsymbol{v}'_{0(\min)}$ and $\boldsymbol{v}_0$ are the same, the speed increment required for the first impulse maneuver is the smallest, therefore the spacecraft should jet backward:

$$\Rightarrow \qquad \Delta v_{0(\min)} = v'_{0(\min)} - v_0 \tag{5}$$

(2) and (4) → (5):

$$\Rightarrow \qquad \Delta v_{0(\min)} = (\sqrt{2}-1)\sqrt{\frac{GM}{r_0}} \tag{6}$$

Solution to Problem (2)

Taking *the system composed of the spacecraft and the earth* as the object of study (see Fig. 3). Studying *the process of the spacecraft's moving from Orbit r_0 to Orbit r_1*. Considering (a) ***law of conservation of mechanical energy*** [the work done by the external force and non-conservative internal force acting on the system is zero; therefore, the mechanical energy of the system is conserved] (b) ***law of conservation of angular momentum*** [taking the geocenter as the reference point, the resultant external torque acting on the system is zero; therefore, the angular momentum of the system is conserved] (the component equation in the direction perpendicular to the plane of the figure inward):

$$\begin{cases} \frac{1}{2}mv'^{2}_{0(\min)} - G\frac{Mm}{r_0} = \frac{1}{2}mv_1^2 - G\frac{Mm}{r_1} = 0 & (7a) \\ mv'_{0(\min)}r_0 = mv_1 r_1 \cos\varphi & (7b) \end{cases}$$

v_1—the speed of the spacecraft when it reaches Orbit r_1.

φ—the angle between the velocity direction of the spacecraft when it reaches Orbit r_1 and the angular direction.

(4) → (7),

$$\Rightarrow \qquad \begin{cases} v_1 = \sqrt{\frac{2GM}{r_1}} & (8a) \\ \cos\varphi = \sqrt{\frac{r_0}{r_1}} & (8b) \end{cases}$$

Taking *the spacecraft* as the object of study. Studying *the state at arbitrary Time t during its orbiting in Orbit r_1*. Considering ***Newton's second law*** (the radial inward component equation):

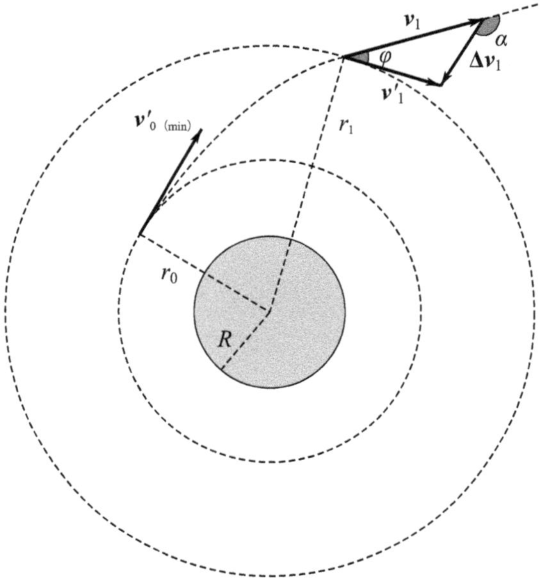

Figure 3

$$G\frac{Mm}{r_1^2} = m\frac{v_1'^2}{r_1} \tag{9}$$

v'_1—the speed of the spacecraft at arbitrary Time t during its orbiting in Orbit r_1.

$$\Rightarrow \qquad v_1' = \sqrt{\frac{GM}{r_1}} \tag{10}$$

Defining $\Delta \boldsymbol{v}_1$—the velocity increment required for the second impulse maneuver of the spacecraft. Considering the vector property of ***velocity***:

$$\boldsymbol{v}_1' = \boldsymbol{v}_1 + \Delta \boldsymbol{v}_1 \tag{11}$$

Defining α—the angle between $\Delta \boldsymbol{v}_1$ and $\boldsymbol{v}_1$.
Considering the following geometric relations in Fig. 3:

$$\begin{cases} \Delta v_1 = \sqrt{v_1'^2 + v_1^2 - 2v_1' v_1 \cos\varphi} & \text{(12a)} \\ \dfrac{v_1'}{\sin\alpha} = \dfrac{\Delta v_1}{\sin\varphi} & \text{(12b)} \end{cases}$$

(8a) and (10) → (12),

$$\Rightarrow \quad \begin{cases} \Delta v_1 = \sqrt{\dfrac{GM}{r_1}\left(3-2\sqrt{\dfrac{2r_0}{r_1}}\right)} & (13a) \\ \sin\alpha = \dfrac{\sqrt{1-\dfrac{r_0}{r_1}}}{\sqrt{3-2\sqrt{\dfrac{2r_0}{r_1}}}} & (13b) \end{cases}$$

Solution to Problem (3)

(i) The first scheme to jet

Taking *the system composed of the spacecraft and the earth* as the object of study (see Fig. 4). Studying *the process of the spacecraft's returning from Orbit r_1 to the earth with the first scheme.* Considering (a) ***law of conservation of mechanical energy***, (b) ***law of conservation of angular momentum*** (the component equation in the direction perpendicular to the plane of the figure inward):

$$\begin{cases} \dfrac{1}{2}mv''^2_{1(\text{Scheme 1})} - G\dfrac{Mm}{r_1} = \dfrac{1}{2}mv^2_{2(\text{Scheme 1})} - G\dfrac{Mm}{R} & (14a) \\ mv''_{1(\text{Scheme 1})}r_1 = mv_{2(\text{Scheme 1})}R & (14b) \end{cases}$$

$v''_{1(\text{Scheme 1})}$—the speed of the spacecraft at the time point after jetting forward at Point *A*.

$v_{2(\text{Scheme 1})}$—the speed of the spacecraft at the time point when it reaches Point *B*.

$$\Rightarrow \quad v''_{1(\text{Scheme 1})} = \sqrt{\frac{2GMR}{r_1(r_1+R)}} \tag{15}$$

Defining $\Delta v'_{1(\text{Scheme 1})}$—the velocity increment of the spacecraft in the process of jetting forward at Point *A*.

$$\Delta v'_{1(\text{Scheme 1})} = v'_1 - v''_{1(\text{Scheme 1})} \tag{16}$$

(10) and (15) → (16):

$$\Rightarrow \quad \Delta v'_{1(\text{Scheme 1})} = \sqrt{\frac{GM}{r_1}}\left(1-\sqrt{\frac{2R}{r_1+R}}\right) \tag{17}$$

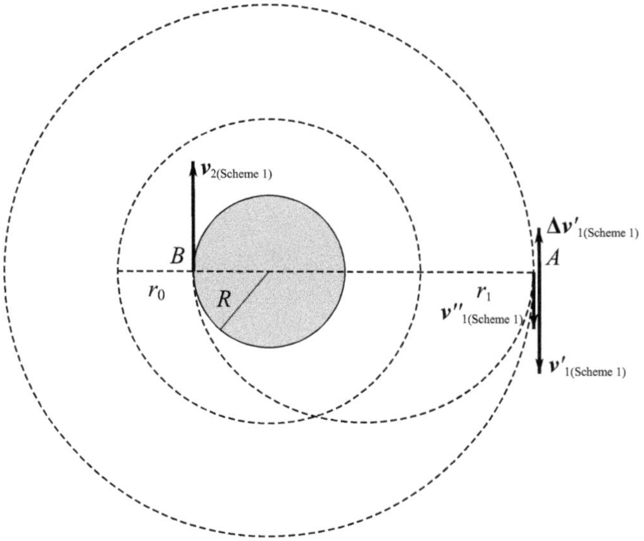

Figure 4

(ii) The second scheme to jet

As shown in Fig. 5, the position of the impulse maneuver of the spacecraft in Orbit r_1 is denoted by A'.

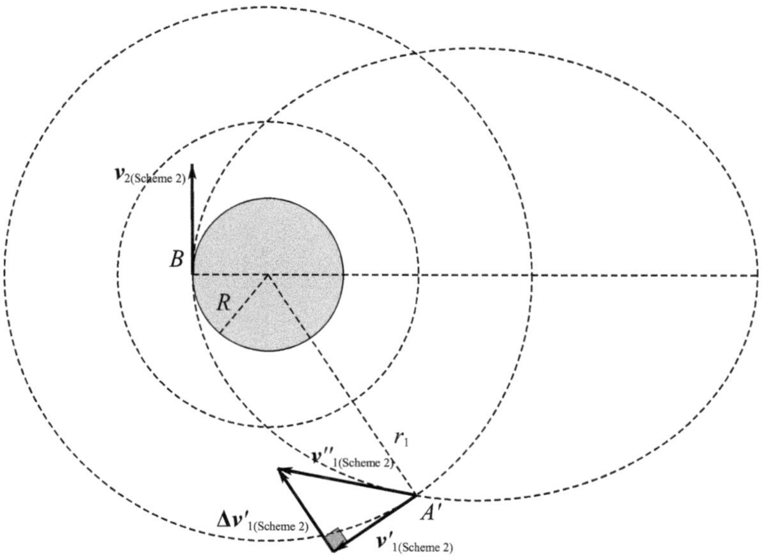

Figure 5

Taking *the system composed of the spacecraft and the earth* as the object of study. Studying *the process of the spacecraft's returning from Orbit r_1 to the earth with the second scheme.* Considering (a) ***law of conservation of mechanical energy***, (b) ***law of conservation of angular momentum*** (the component equation in the direction perpendicular to the plane of the figure inward):

$$\begin{cases} \frac{1}{2}mv''^2_{1(\text{Scheme 2})} - G\frac{Mm}{r_1} = \frac{1}{2}mv^2_{2(\text{Scheme 2})} - G\frac{Mm}{R} & (18a) \\ mv'_1 r_1 = mv_{2(\text{Scheme 2})}R & (18b) \end{cases}$$

$v''_{1(\text{Scheme 2})}$—the speed of the spacecraft at the time point after jetting radially outward at Point A'.

$v_{2(\text{Scheme 2})}$—the speed of the spacecraft at the time point when it reaches Point B.

$$\Rightarrow \qquad v''_{1(\text{Scheme 2})} = \sqrt{GM\left[\frac{r_1}{R^2} - 2\left(\frac{1}{R} - \frac{1}{r_1}\right)\right]} \qquad (19)$$

Defining $\Delta v'_{1(\text{Scheme 2})}$—the velocity increment of the spacecraft in the process of jetting radially outward at Point A'. Considering the vector property of ***velocity***:

$$\boldsymbol{v}''_{1(\text{Scheme 2})} = \boldsymbol{v}'_{1(\text{Scheme 2})} + \Delta\boldsymbol{v}'_{1(\text{Scheme 2})} \qquad (20)$$

Considering the following geometric relation in Fig. 5:

$$\Delta v'_{1(\text{Scheme 2})} = \sqrt{v''^2_{1(\text{Scheme 2})} - v'^2_1} \qquad (21)$$

(10) and (19) → (21),

$$\Rightarrow \qquad \Delta v'_{1(\text{Scheme 2})} = \sqrt{\frac{GM}{r_1}}\left(\frac{r_1}{R} - 1\right) \qquad (22)$$

5.1.11 The Orbit Change of a Synchronous Satellite due to an Accidental Maneuver (*c*, *C*)[3]

A geostationary satellite is orbiting stably, and suddenly a wrong control command causes its apogee engine (be used for providing the thrust needed to reach the final orbit) to be fired again and jet

[3]This problem is a test question of the 36th IPhO (International Physics Olympiad) in 2005.

radially outward. Despite the quick reaction of the ground crew to shut the engine off, the satellite inevitably produced a non-negligible velocity variation of the magnitude Δv and the direction pointing to the earth. Given the radius of the earth is R, the gravitational acceleration is g, and rotation period of the earth is T_0.

(1) Find the orbit of the satellite after the jet.

(2) Find:

(i) The condition for the satellite to perform an elliptical motion around the earth after the jet, as well as the perigee and apogee of the elliptical orbit and the motion period in this case.

(ii) The condition for the satellite to perform a parabolic motion after the jet, as well as the perigee of the parabolic orbit in this case.

(iii) The condition for the satellite to perform a hyperbolic motion after the jet, as well as the escape velocity and the impact parameter in this case.

Solution to Problem (1)

Taking *the satellite* as the object of study. Studying *an arbitrary state during it orbiting the earth before the jet.* Considering ***Newton's second law*** (the radial inward component equation):

$$G\frac{Mm}{r_0^2} = m\frac{v_0^2}{r_0} \tag{1}$$

M and m—the mass of the earth and the satellite.

r_0—the radius of the circular orbit of the satellite before the jet.

v_0—the speed of the satellite at an arbitrary state during it orbiting the earth before the jet.

Considering the definition of ***gravitational acceleration***:

$$g = \frac{GM}{R^2} \tag{2}$$

Considering the revolution period of the geostationary satellite is equal to the rotation period of the earth:

$$T_0 = \frac{2\pi r_0}{v_0} \tag{3}$$

Combining (1) to (3),

$$\Rightarrow \quad \begin{cases} r_0 = \left(\dfrac{gR^2T_0^2}{4\pi^2}\right)^{1/3} & \text{(4a)} \\ v_0 = \left(\dfrac{2\pi gR^2}{T_0}\right)^{1/3} & \text{(4b)} \end{cases}$$

The time point after the jet is denoted by Time zero.

As shown in Fig. 1, the geocenter is denoted by O, the unit vector perpendicular to the plane of the figure inward is denoted by $\boldsymbol{k}$. taking the geocenter as the pole, the direction of the earth pointing to the perigee of the satellite's new orbit as the positive direction of the polar axis, the right-hand helix direction with regard to the $\boldsymbol{k}$-direction as the positive direction of polar angle, establishing the polar coordinate system, let r, θ be the polar coordinates.

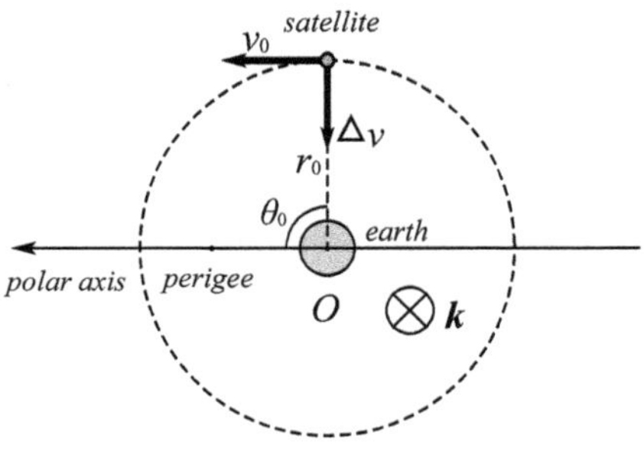

Figure 1

Taking *the satellite* as the object of study. Studying *the state at arbitrary Time t*. Considering ***the orbit equation of a particle in the inverse-square central force field***:

$$r = \frac{p}{1 + e\cos\theta} \tag{5}$$

p and e—the semi-latus rectum and eccentricity of the orbit equation of the satellite after the jet.

Considering ***the form of the semi-latus rectum and eccentricity of the orbit equation when the inverse-square central force field is the universal gravitation field***:

$$\begin{cases} p = \dfrac{L^2}{GMm^2} & (6a) \\ e = \sqrt{1 + \dfrac{2EL^2}{G^2M^2m^3}} & (6b) \end{cases}$$

E and L—the mechanical energy [taking the configuration in which the satellite and the earth are infinitely apart as the zero point of potential energy of universal gravitation] and the angular momentum magnitude [taking the pole O as the reference point] of the system composed of the earth and the satellite.

Considering the definition of ***mechanical energy*** and ***angular momentum***:

$$\begin{cases} E = E\big|_{\substack{r=r_0 \\ \theta=\theta_0}} = \dfrac{1}{2}m\left[v_0^2 + (\Delta v)^2\right] - G\dfrac{Mm}{r_0} & (7a) \\ L = L\big|_{\substack{r=r_0 \\ \theta=\theta_0}} = mv_0 r_0 & (7b) \end{cases}$$

θ_0—the polar angle of the satellite at Time zero.

Δv—the $-\boldsymbol{r}^0$-direction component of the velocity increment of the satellite obtained during the jet.

Defining the constant α:

$$\alpha = \frac{\Delta v}{v_0} \tag{8}$$

$$\Rightarrow \qquad E = \frac{1}{2}mv_0^2(\alpha^2 - 1) \tag{9}$$

(7b) and (9) $\rightarrow$ (6),

$$\Rightarrow \qquad \begin{cases} p = r_0 & (10a) \\ e = \alpha & (10b) \end{cases}$$

$$\Rightarrow \qquad r = \frac{r_0}{1 + \alpha\cos\theta} \tag{11}$$

Considering the initial condition:

$$r\big|_{\theta=\theta_0} = r_0 \tag{12}$$

$$\Rightarrow \qquad \cos\theta_0 = 0$$

$$\Rightarrow \qquad \theta_0 = \frac{\pi}{2} \tag{13}$$

The schematic diagram is as shown in Fig. 1.

Solution to Problem (2)

(i) Considering ***the orbit-energy relation of the two-body model under the universal gravitation***, it can be inferred that the condition for the satellite to perform an elliptical motion around the earth is that the mechanical energy of the system composed of the satellite and the earth is less than zero:

$$E < 0 \tag{14}$$

(9) → (14),

$$\Rightarrow \qquad \alpha < 1 \tag{15}$$

From (11),

$$\Rightarrow \qquad \begin{cases} r_{\min}\big|_{\theta=0} = \dfrac{r_0}{1+\alpha} & \text{(16a)} \\ r_{\max}\big|_{\theta=\pi} = \dfrac{r_0}{1-\alpha} & \text{(16b)} \end{cases}$$

From the geometric relation in Fig. 2, we can know that:

$$\Rightarrow \qquad a = \frac{1}{2}(r_{\max} + r_{\min}) = \frac{r_0}{1-\alpha^2} \tag{17}$$

a—the length of the semi-major axis of the satellite's elliptical orbit.

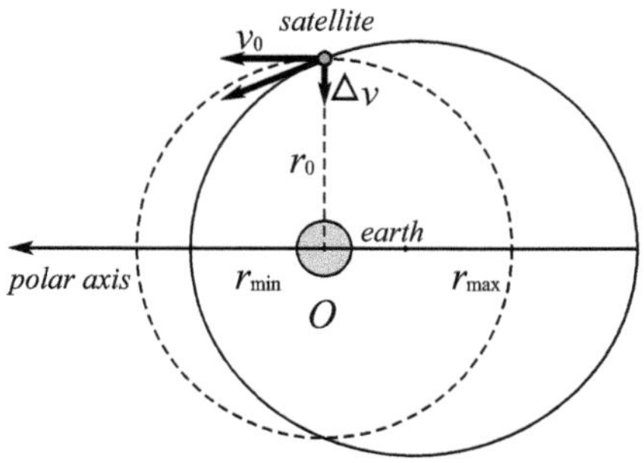

Figure 2

Considering ***Kepler's third law***:

$$\frac{T^2}{a^3} = \frac{T_0^2}{r_0^3} \tag{18}$$

T—the motion period of the elliptical motion of the satellite.

$$\Rightarrow \qquad T = \frac{T_0}{(1-\alpha^2)^{3/2}} \tag{19}$$

(ii) Considering ***the orbit-energy relation of the two-body model under the universal gravitation***, it can be inferred that the condition for the satellite to perform a parabolic motion is that the mechanical energy of the system composed of the satellite and the earth is equal to zero:

$$E = 0 \tag{20}$$

$$\Rightarrow \qquad \alpha = 1 \tag{21}$$

Combining (11) and (21),

$$\Rightarrow \qquad r_{\min}\big|_{\theta=0} = \frac{r_0}{2} \tag{22}$$

The schematic diagram is as shown in Fig. 3.

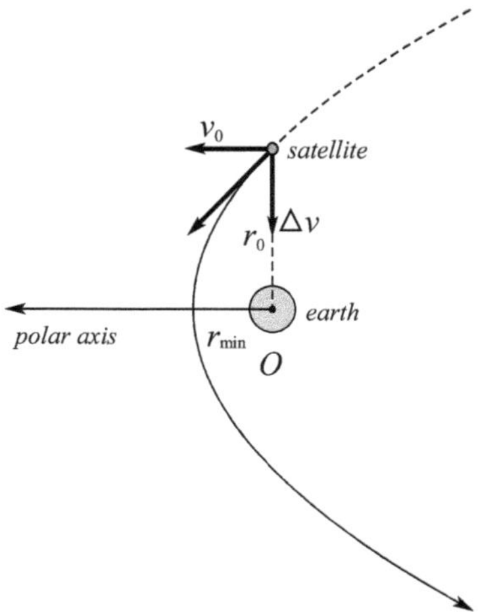

Figure 3

(iii) Considering *the* ***orbit-energy relation of the two-body model under the universal gravitation***, it can be inferred that the condition for the satellite to perform a hyperbolic motion is that the mechanical energy of the system composed of the satellite and the earth is greater than zero:

$$E > 0 \tag{23}$$

$$\Rightarrow \qquad \alpha > 1 \tag{24}$$

Considering the definition of ***mechanical energy*** and ***angular momentum*** (see Fig. 4):

$$\begin{cases} E = E\big|_{r\to\infty} = \dfrac{1}{2}mv_\infty^2 & \text{(25a)} \\ L = L\big|_{r\to\infty} = mv_\infty b & \text{(25b)} \end{cases}$$

v_∞—the escape velocity of the satellite.

b—the distance from earth to the asymptote of satellite's hyperbolic orbit, i.e., the length of the semi-imaginary axis of the satellite's hyperbolic orbit.

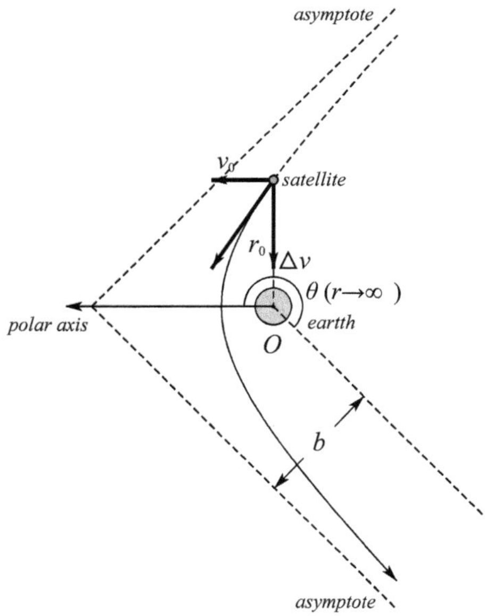

Figure 4

(25) → (7b) and (9),

$$\Rightarrow \quad \begin{cases} v_\infty = v_0\sqrt{\alpha^2-1} & (26a) \\ b = \dfrac{r_0}{\sqrt{\alpha^2-1}} & (26b) \end{cases}$$

From (11),

$$\Rightarrow \quad \cos\theta\Big|_{r\to\infty} = -\frac{1}{\alpha} \quad (27)$$

$\theta|_{r\to\infty}$—the angle between the escape direction of the satellite and the polar axis.

$$\Rightarrow \quad \theta\Big|_{r\to\infty} = \pi + \arccos\frac{1}{\alpha} \quad (28)$$

The schematic diagram is as shown in Fig. 4.

5.1.12 Escaping from the Solar System via "Gravitation Slingshot" (*c, C*)[4]

There are two schemes for launching a spacecraft from the earth outside the Solar system. Scheme 1: launch the spacecraft with a velocity large enough to escape from the Solar system directly. Scheme 2: let the spacecraft first approach Mars, and with its help (gravitational field) change its motion direction and reach the velocity necessary to escape from the Solar system. For these two schemes, find the minimum launching speed and the launching direction of the spacecraft relative to the earth respectively.

Given the speed of the earth's revolution around the sun is 30 km/s, the ratio of the distance from the earth to the sun and the distance from Mars to the sun is 2:3, the radius of the earth is 6400 km, the acceleration of gravity at the earth's surface is 9.80 m/s^2.

Solution

Scheme 1:
Taking *the system composed of the spacecraft and the earth* as the object of study. Studying *the process of the spacecraft's escaping from the earth (defined as escaping from the sphere of effect of the earth,*

[4]This problem is a test question of the 16th IPhO (International Physics Olympiad) in 1985.

the displacement of the spacecraft relative to the sun can be ignored in this process). Considering ***law of conservation of mechanical energy*** [in the earth reference frame] [the work done by the external force and non-conservative internal force acting on the system is zero (the effect of the sun can be ignored); therefore, the mechanical energy of the system is conserved]:

$$\frac{1}{2}mv_{AE}^2 - G\frac{M_E m_A}{R_E} \approx \frac{1}{2}mv_{AE}'^2 \qquad (1)$$

M_E and m_A—the mass of the earth and the spacecraft.

v_{AE}—the speed of the spacecraft relative to the earth at the time point when it is launched.

v'_{AE}—the speed of the spacecraft relative to the earth at the time point when it escapes from the earth.

Taking *the spacecraft* as the object of study. Studying *the state at the time point when it escapes from the earth*. Considering ***theorem for composition of velocities of a particle*** [taking the sun as the static reference frame and the earth as the moving reference frame]:

$$\boldsymbol{v}_{AS} = \boldsymbol{v}_E + \boldsymbol{v}'_{AE} \qquad (2)$$

$\boldsymbol{v}_E$, $\boldsymbol{v}'_{AE}$ and $\boldsymbol{v}_{AS}$—at the time point when the spacecraft escapes from the earth, the velocity of the earth relative to the sun, the velocity of the spacecraft relative to the earth, and the velocity of the spacecraft relative to the sun.

$$\Rightarrow \qquad v_{AS}^2 = v_E^2 + v_{AE}'^2 + 2v_E v_{AE}'^2 \cos\theta(\boldsymbol{v}_E, \boldsymbol{v}'_{AE}) \qquad (3)$$

θ ($\boldsymbol{v}_E$, $\boldsymbol{v}'_{AE}$)—the angle between $\boldsymbol{v}_E$ and $\boldsymbol{v}'_{AE}$.

Considering ***the orbit-energy relation of the two-body model under the universal gravitation***, it can be inferred that the condition for the spacecraft to escape from the bond of the sun is that the mechanical energy of the system composed of the spacecraft and the sun is greater than or equal to zero:

$$\frac{1}{2}mv_{AS}^2 - G\frac{M_S m_A}{R_{ES}} \geq 0 \qquad (4)$$

M_S—the mass of the sun.

$$\Rightarrow \qquad v_{AS(\min)} = \sqrt{\frac{2GM_S}{R_{ES}}} \qquad (5)$$

Taking *the earth* as the object of study. Studying *the state at arbitrary time*. Considering ***Newton's second law*** (the component equation in the direction of the earth pointing to the sun):

$$G\frac{M_S M_E}{R_{ES}^2} = M_E \frac{v_E^2}{R_{ES}} \tag{6}$$

$$\Rightarrow \qquad v_{AS(\min)} = \sqrt{2}v_E \tag{7}$$

Combining (3) and (7),

$$\Rightarrow \qquad v'_{AE(\min)}\Big|_{\substack{v_{AS}=v_{AS(\min)} \\ \theta(v_E, v'_{AE})=0}} = v_{AS(\min)} - v_E = (\sqrt{2}-1)v_E \tag{8}$$

Combining (1) and (8),

$$\Rightarrow \qquad v_{AE(\min)}\Big|_{v'_{AE}=v'_{AE(\min)}} = \sqrt{(\sqrt{2}-1)^2 v_E^2 + 2G\frac{M_E}{R_E}} \tag{9}$$

$$g = G\frac{M_E}{R_E^2} \tag{10}$$

$$\Rightarrow \qquad v_{AE(\min)} = \sqrt{(\sqrt{2}-1)^2 v_E^2 + 2R_E g} = 16.4\text{km/s} \tag{11}$$

Scheme 2:

Taking *the system composed of the spacecraft and the earth* as the object of study. Studying *the process of the spacecraft's escaping from the earth*. Considering ***law of conservation of mechanical energy*** [the analysis method is the same as that in Scheme 1]:

$$\frac{1}{2}m_A v_{AE*}^2 - G\frac{M_E m_A}{R_E} \approx \frac{1}{2}m_A v'^2_{AE*} \tag{12}$$

v_{AE*}—the speed of the spacecraft relative to the earth at the time point when it is launched.

v'_{AE*}—the speed of the spacecraft relative to the earth at the time point when it escapes from the earth.

Taking *the spacecraft* as the object of study (see Fig. 1). Studying *the state at the time point when it escapes from the earth*. Considering ***theorem for composition of velocities of a particle*** [taking the sun as the static reference frame and the earth as the moving reference frame]:

$$\boldsymbol{v}_{AS*} = \boldsymbol{v}_E + \boldsymbol{v}'_{AE*} \tag{13}$$

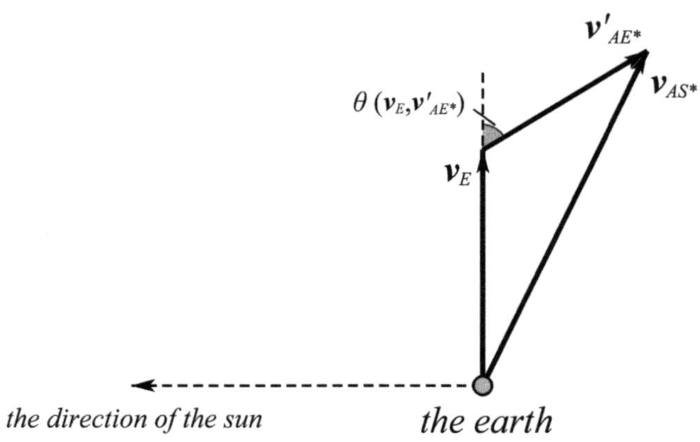

Figure 1

$\boldsymbol{v}'_{AE*}$ and $\boldsymbol{v}_{AS*}$—the velocity of the spacecraft relative to the earth and the sun at the time point when it escapes from the earth.

$$\Rightarrow \qquad v_{AS*}^2 = v_E^2 + v_{AE*}'^2 + 2v_E v_{AE*}'^2 \cos\theta(\boldsymbol{v}_E, \boldsymbol{v}'_{AE*}) \tag{14}$$

$\theta\,(\boldsymbol{v}_E, \boldsymbol{v}'_{AE*})$—the angle between $\boldsymbol{v}_E$ and $\boldsymbol{v}'_{AE*}$.

Taking *the system composed of the spacecraft and the sun* as the object of study. Studying *the process of the spacecraft moving from the earth orbit to Mars orbit*. Considering (a) ***law of conservation of mechanical energy*** [in the sun reference frame] [the work done by the external force and non-conservative internal force acting on the system is zero; therefore, the mechanical energy of the system is conserved] (b) ***law of conservation of angular momentum*** [in the sun reference frame] [taking the sun as the reference point, the resultant external torque acting on the system is zero; therefore, the angular momentum of the system is conserved]:

$$\begin{cases} \dfrac{1}{2}m_A v_{AS*}^2 - G\dfrac{M_S m_A}{R_{ES}} = \dfrac{1}{2}m_A v_{AS}'^2 - G\dfrac{M_S m_A}{R_{MS}} & \text{(15a)} \\ m_A \boldsymbol{R}_{ES} \times \boldsymbol{v}_{AS*} = m_A \boldsymbol{R}_{MS} \times \boldsymbol{v}'_{AS} & \text{(15b)} \end{cases}$$

$\boldsymbol{v}'_{AS}$—the velocity of the spacecraft relative to the sun at the time point when it enters the gravitational field of Mars.

Defining $v'_{AS||}$ and $v'_{AS\perp}$—the components of $\boldsymbol{v}'_{AS}$ in the tangential and normal directions of Mars orbit.

$$\Rightarrow \qquad m_A R_{ES} v_{AS*(\min)} = m_A R_{MS} v'_{AS||} \tag{16}$$

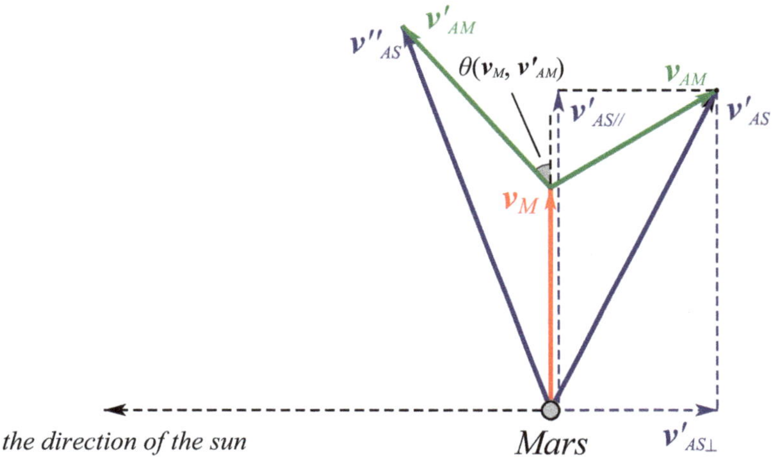

Figure 2

Taking *the spacecraft* as the object of study (see Fig. 2). Studying *the state at the time point when it enters the gravitational field of Mars.* Considering ***theorem for composition of velocities of a particle*** [taking the sun as the static reference frame and Mars as the moving reference frame]:

$$\boldsymbol{v}'_{AS} = \boldsymbol{v}_M + \boldsymbol{v}_{AM} \tag{17}$$

$\boldsymbol{v}_M$ and $\boldsymbol{v}_{AM}$—the velocity of Mars relative to the sun and the velocity of the spacecraft relative to Mars at the time point when the spacecraft enters the gravitational field of Mars.

$$\Rightarrow \qquad v_{AM}^2 = (v'_{AS||} - v_M)^2 + v'^2_{AS\perp} \tag{18}$$

Taking *Mars* as the object of study. Studying *the state at arbitrary time*. Considering ***Newton's second law*** (the component equation in the direction of Mars pointing to the sun):

$$G\frac{M_S M_M}{R_{MS}^2} = M_M \frac{v_M^2}{R_{MS}} \tag{19}$$

M_M—the mass of Mars.

$$\Rightarrow \qquad v_M = \sqrt{\frac{GM_S}{R_{MS}}} \tag{20}$$

$$\Rightarrow \qquad v_M = \sqrt{\frac{R_{ES}}{R_{MS}}} v_E \tag{21}$$

Taking *the system composed of the spacecraft and Mars* as the object of study. Studying *the process of the spacecraft's crossing the Mars's sphere of effect.* Considering ***law of conservation of mechanical energy*** [in the Mars reference frame] [the work done by the external force and non-conservative internal force acting on the system is zero (the effect of the sun can be ignored in that process); therefore, the mechanical energy of the system is conserved]:

$$\frac{1}{2} m_A v_{AM}^2 \approx \frac{1}{2} m_A v_{AM}'^2 \tag{22}$$

v'_{AM}—the speed of the spacecraft relative to Mars at the time point when it escapes from the gravitational field of Mars.

$$\Rightarrow \qquad v_{AM} = v'_{AM} \tag{23}$$

Taking *the spacecraft* as the object of study. Studying *the state at the time point when it escapes from Mars.* Considering ***theorem for composition of velocities of a particle*** [taking the sun as the static reference frame and Mars as the moving reference frame]:

$$\boldsymbol{v}''_{AS} = \boldsymbol{v}_M + \boldsymbol{v}'_{AM} \tag{24}$$

$\boldsymbol{v}'_{AM}$ and $\boldsymbol{v}''_{AS}$—the velocity of the spacecraft relative to Mars and the sun at the time point when it escapes from Mars.

$$\Rightarrow \qquad v_{AS}''^2 = v_M^2 + v_{AM}'^2 + 2 v_M v'_{AM} \cos\theta\,(\boldsymbol{v}_M, \boldsymbol{v}'_{AM}) \tag{25}$$

$\theta\,(\boldsymbol{v}_M, \boldsymbol{v}'_{AM})$—the angle between $\boldsymbol{v}_M$ and $\boldsymbol{v}'_{AM}$.

Considering ***the orbit-energy relation of the two-body model under the universal gravitation***, it can be inferred that the condition for the spacecraft can further escape from the sun after escaping from Mars is that the mechanical energy of the system composed of the spacecraft and the sun is greater than or equal to zero:

$$\frac{1}{2} m v_{AS}''^2 - G\frac{M_S m_A}{R_{MS}} \geq 0 \tag{26}$$

$$\Rightarrow \qquad v''_{AS(\min)} = \sqrt{\frac{2GM_S}{R_{MS}}} = \sqrt{\frac{2R_{ES}}{R_{MS}}} v_E \qquad (27)$$

From (25),

$$\Rightarrow \qquad v'_{AM(\min)}\Big|_{\substack{v''_{AS}=v''_{AS(\min)} \\ \theta(\mathbf{v}_M,\mathbf{v}'_{AM})=0}} = v''_{AS(\min)} - v_M = (\sqrt{2}-1)\sqrt{\frac{R_{ES}}{R_{MS}}} v_E \qquad (28)$$

From (23),

$$\Rightarrow \qquad v_{AM(\min)} = (\sqrt{2}-1)\sqrt{\frac{R_{ES}}{R_{MS}}} v_E \qquad (29)$$

(21) and (29) → (16) and (18):

$$\Rightarrow \qquad v^2_{AS^*(\min)} - 2\left(\frac{R_{ES}}{R_{MS}}\right)^{3/2} v_E v_{AS^*(\min)} + \left(2\sqrt{2}\frac{R_{ES}}{R_{MS}} - 2\right) v_E^2 = 0 \qquad (30)$$

$$\Rightarrow \qquad v_{AS^*(\min)} = v_E\left[\left(\frac{R_{ES}}{R_{MS}}\right)^{3/2} + \sqrt{\left(\frac{R_{ES}}{R_{MS}}\right)^3 - \left(2\sqrt{2}\frac{R_{ES}}{R_{MS}} - 2\right)}\right] \qquad (31)$$

From (14),

$$\Rightarrow \quad v'_{AE^*(\min)}\Big|_{\substack{v_{AS^*}=v_{AS^*(\min)} \\ \theta(\mathbf{v}_E,\mathbf{v}'_{AE})=0}} = v_{AS^*(\min)} - v_E$$

$$= v_E\left[\left(\frac{R_{ES}}{R_{MS}}\right)^{3/2} + \sqrt{\left(\frac{R_{ES}}{R_{MS}}\right)^3 - \left(2\sqrt{2}\frac{R_{ES}}{R_{MS}} - 2\right)} - 1\right] \qquad (32)$$

From (12),

$$\Rightarrow \quad v_{AE^*(\min)}\Big|_{v'_{AE^*}=v'_{AE(\min)}}$$

$$= \left\{ v_E^2\left[\left(\frac{R_{ES}}{R_{MS}}\right)^{3/2} + \sqrt{\left(\frac{R_{ES}}{R_{MS}}\right)^3 - \left(2\sqrt{2}\frac{R_{ES}}{R_{MS}} - 2\right)} - 1\right]^2 + 2gR_E \right\}^{1/2}$$

$$= 12.5 \text{ km/s} \qquad (33)$$

Through comparison, it can be seen that Scheme 2 is more energy-saving.

We can draw the trajectory of the spacecraft escaping from the solar system via "gravitation slingshot" as follows (see Fig. 3):

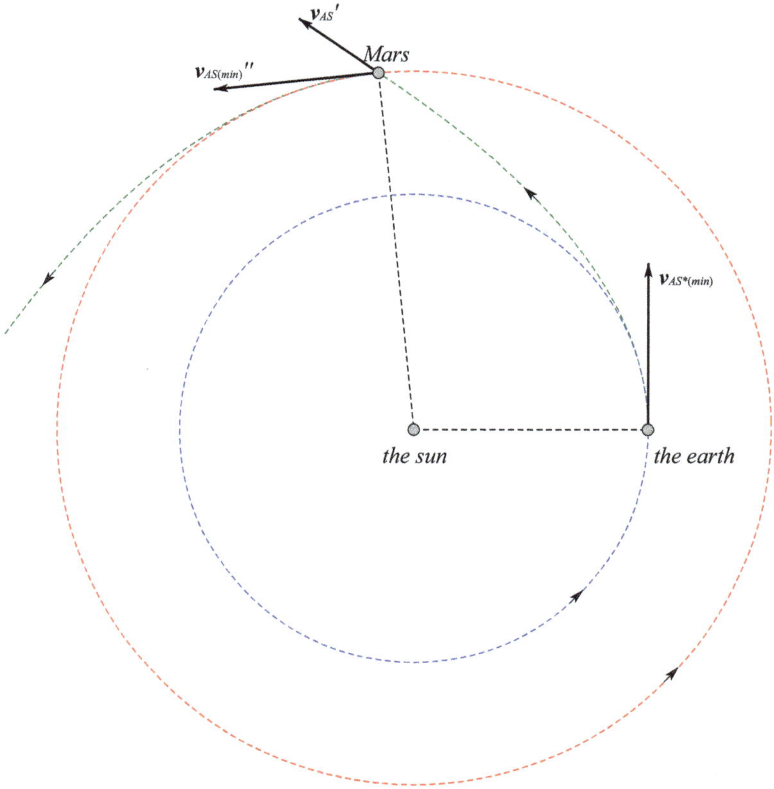

Figure 3

5.1.13 Gravitational Capture (*b, B*)

A spacecraft is scheduled to perform the following space mission: entering the gravitational field of Mars at a large enough relative speed parallel to the motion direction of Mars (the orbit relative to Mars is a hyperbola), and entering the target elliptical orbit around Mars (coplanar with the original orbit) through an impulse maneuver when it reaches the perigee. Given the mass of Mars is M, the spacecraft enters the gravitational field of Mars at the relative speed v_∞, the eccentricity of the target elliptical orbit is e. The distance between Mars and the asymptote of the spacecraft's hyperbolic orbit is defined as the aiming distance of the spacecraft relative Mars, try to determine the minimum impulse maneuvering speed change needed for the spacecraft to perform this mission and the corresponding aiming distance.

Solution

Time zero is arbitrarily set.

The translational reference frame that is at rest with respect to Mars can be approximately regarded as an ideal inertial reference frame; the following will be studied in this reference frame.

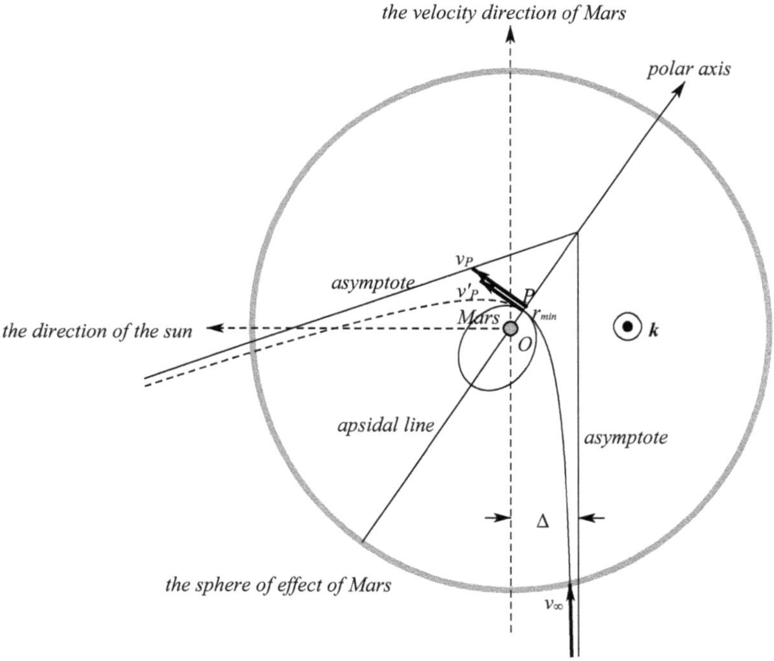

Figure 1

As shown in Fig. 1, the position of Mars is denoted by *O*, the perigee of the spacecraft's hyperbolic orbit is denoted by *P*, the unit vector perpendicular to the plane of the figure outward is denoted by $\boldsymbol{k}$. Taking Point *O* as the pole, the direction of Point *O* pointing to Point *P* as the positive direction of the polar axis, the right-hand helix direction with regard to the $\boldsymbol{k}$-direction as the positive direction of polar angle, establishing the polar coordinate system, let r, θ be the polar coordinates.

Taking *the system composed of Mars and the spacecraft* as the object of study. Studying *the process from the spacecraft entering the gravitational field of Mars to the spacecraft reaching the perigee P*. Considering (a) ***law of conservation of mechanical energy*** [the work done by the external force and non-conservative internal force acting on the system is zero; therefore, the mechanical energy of the system is conserved] (taking the configuration in which the spacecraft and Mars are infinitely apart as the zero point of potential energy of universal gravitation), (b) ***law of conservation of angular momentum*** [taking the pole O as the reference point, the resultant external torque acting on the system is zero; therefore, the angular momentum of the system is conserved] (the $\boldsymbol{k}$-direction component equation):

$$\begin{cases} \frac{1}{2}mv_\infty^2 = \frac{1}{2}mv_p^2 - G\frac{Mm}{r_{\text{min}}} & \text{(1a)} \\ mv_\infty \Delta = mv_p r_{\text{min}} & \text{(1b)} \end{cases}$$

m—the mass of the spacecraft.

v_p—the $\boldsymbol{\theta}^0$-direction component of the velocity of the spacecraft when it reaches the perigee P.

r_{min}—the perigee distance of the hyperbolic orbit.

Δ—the aiming distance of the spacecraft relative Mars.

$$\Rightarrow \qquad r_{\text{min}}^2 + \frac{2GM}{v_\infty^2} r_{\text{min}} - \Delta^2 = 0 \tag{2}$$

$$\Rightarrow \qquad r_{\text{min}} = \sqrt{\frac{G^2M^2}{v_\infty^4} + \Delta^2} - \frac{GM}{v_\infty^2} \tag{3}$$

(3) → (1b),

$$\Rightarrow \qquad v_p = v_\infty \frac{\Delta}{\sqrt{\frac{G^2M^2}{v_\infty^4} + \Delta^2} - \frac{GM}{v_\infty^2}} \tag{4}$$

Taking *the spacecraft* as the object of study. Studying *the state at the time point after the impulse maneuver*. Considering ***the orbit equation of a particle in the inverse-square central force field***:

$$r' = \frac{p}{1 + e\cos\theta'} \tag{5}$$

p—the semi-latus rectum of the elliptical orbit equation of the spacecraft.

The energy of the system decreases after the impulse maneuver; therefore, the impulse maneuver is a deceleration process. In order to minimize the speed change obtained by the impulse maneuver, the position of the impulse maneuver should be the point with the maximum speed of the elliptical orbit, and the spacecraft should perform the impulse maneuver by jetting fuel forward. That is, the perigee of the hyperbolic orbit should coincide with the perigee of the elliptical orbit, and the two orbits are tangent at this point.

$$\Rightarrow \qquad r_{\min} = r'_{\min} = r'\big|_{\theta'=0} = \frac{p}{1+e} \tag{6}$$

Considering ***the form of the semi-latus rectum of the orbit equation when the inverse-square central force field is the universal gravitation field***:

$$p = \frac{L^2}{GMm^2} \tag{7}$$

L—the $\boldsymbol{k}$-direction component of the angular momentum of the spacecraft (taking the pole O as the reference point) at the time point after the impulse maneuver at the perigee P.

Considering the definition of ***angular momentum***:

$$L = mv'_p r_{\min} \tag{8}$$

v'_p—the $\boldsymbol{\theta}^0$-direction component of the velocity of the spacecraft relative to Mars at the time point after the impulse maneuver at the perigee P.

$$\Rightarrow \qquad v'_p = \sqrt{\frac{GM}{r_{\min}}(1+e)} \tag{9}$$

Let's take $r_{\min}$ as the independent variable to study the extremum problem.

From (1),

$$\Rightarrow \qquad v_p = \sqrt{v_\infty^2 + \frac{2GM}{r_{\min}}} \tag{10}$$

Defining $|\Delta v_p|$—the speed change of the spacecraft obtained by the impulse maneuver.

$$\Rightarrow \qquad \left|\Delta v_p\right| = v_p - v_p' = \sqrt{v_\infty^2 + \frac{2GM}{r_{\min}}} - \sqrt{\frac{GM}{r_{\min}}(1+e)} \tag{11}$$

$$\Rightarrow \qquad \frac{\mathrm{d}\left|\Delta v_p\right|}{\mathrm{d}r_{\min}} = \frac{1}{r_{\min}^{3/2}}\left[-\frac{GM}{\sqrt{r_{\min}v_\infty^2 + 2GM}} + \frac{1}{2}\sqrt{GM(1+e)}\right] \tag{12}$$

$$\Rightarrow \qquad \frac{\mathrm{d}^2\left|\Delta v_p\right|}{\mathrm{d}r_{\min}^2} = \frac{1}{r_{\min}^{5/2}}\left[\frac{GM(2r_{\min}v_\infty^2 + 3GM)}{(r_{\min}v_\infty^2 + 2GM)^{3/2}} - \frac{3}{4}\sqrt{GM(1+e)}\right] \tag{13}$$

The necessary condition for $|\Delta v_p|$ to take the minimum value is as follows:

$$\frac{\mathrm{d}\left|\Delta v_p\right|}{\mathrm{d}r_{\min}} = 0 \tag{14}$$

$$\Rightarrow \qquad r_{\min} = 2\frac{1-e}{1+e}\frac{GM}{v_\infty^2} \tag{15}$$

$$\Rightarrow \qquad \left.\frac{\mathrm{d}^2\left|\Delta v_p\right|}{\mathrm{d}r_{\min}^2}\right|_{r_{\min}=2\frac{1-e}{1+e}\frac{GM}{v_\infty^2}} = \frac{\sqrt{2}(1+e)^3}{64(1-e)^{3/2}}\frac{v_\infty^5}{(GM)^2} > 0 \tag{16}$$

That is, $|\Delta v_p|$ takes the minimum value when $r_{\min}$ takes the form of Eq. (15).

(15) → (3),

$$\Rightarrow \qquad \Delta = \frac{2\sqrt{2}\sqrt{1-e}}{1+e}\frac{GM}{v_\infty^2} \tag{17}$$

(15) → (11),

$$\Rightarrow \qquad \left|\Delta v_p\right|_{\min} = \sqrt{\frac{1-e}{2}}v_\infty \tag{18}$$

5.2 Collision

Collision model is a special two-body model. Although the research on this model started in the macro world, it has played a greater value in people's access to micro-world information.

In the study of the collision between two macro objects (including objects that can be regarded as particles), the following approximate scheme is often adopted: avoiding the specific information of the interaction between the two objects in the collision, and it is considered that the interaction occurs in an extremely short period of time, and the motion states of the two objects have a sudden-change in the collision process, that is, the motion state of each objects changes discontinuously with time. This approximate study scheme can provide computational convenience, but due to the loss of interaction information, it is necessary to propose additional settings for the related physical quantities (such as the velocity of the two objects in the collision, the conversion between the mechanical energy and internal energy of the system, etc.) to form a problem that can be solved perfectly.

The study on the collision between two microparticles often adopts the strict scheme: it is necessary to make clear the specific information of the interaction between the two objects in the collision. The motion states of two micro particles are regarded as continuously changing with time, and their motion trajectories are smooth (except for frontal collision). This strict study scheme can make more efficient simulation on the objective facts, and the scattering theory has fully developed it. As a special case, the problem of the collision (scattering) under the inverse-square repulsive force can also be classified as the Kepler problem.

The study difficulty of a collision problem mainly depends on the degrees of freedom of the objects involved in the collision and the form of their interaction. Therefore, it is more difficult to study the collision problems involving rotation (such as the general collision involving rigid bodies) than that involving only translation.

In this section, we will study some interesting collision problems.

5.2.1 Free Collision Model of Two Particles (d, *B*)

Consider two particles 1 and 2 of mass m_1 and m_2, respectively. A completely elastic collision between the two particles is as follows: before the collision the reference frame that is at rest with regard to Particle 2 is defined as the laboratory reference frame, in the laboratory reference frame, Particle 1 collides with Particle 2 at the velocity $\boldsymbol{v}_{10} = v_{10}\boldsymbol{k}$, $v_{10} > 0$, the unit vector of the velocity direction of Particle 1 in the center-of-mass reference frame at the time point after the collision is $\boldsymbol{k}'$.

(1) Find the velocity of the two particles in the center-of-mass reference frame at the time point after the collision.
(2) Find the velocity of the two particles in the laboratory reference frame at the time point after the collision.

Solution to Problem (1)

Considering ***theorem of center-of-mass velocity***:

$$\boldsymbol{v}_{CO} = \frac{m_1}{m_1 + m_2}\boldsymbol{v}_{10} = \frac{m_1}{m_1 + m_2}v_{10}\boldsymbol{k} \tag{1}$$

$\boldsymbol{v}_{CO}$—the center-of-mass velocity of the system composed of the two particles at the time point before the collision.

Taking *the two particles* as the object of study respectively. Studying *the state at the time point before the collision*. Considering ***theorem for composition of velocities of a particle*** [taking the laboratory reference frame as the static reference frame, and the center-of-mass reference frame as the moving reference frame], then place Eq. (1) in to obtain:

$$\Rightarrow \begin{cases} \boldsymbol{v}'_{10} = \boldsymbol{v}_{10} - \boldsymbol{v}_{CO} = \dfrac{m_2}{m_1}\boldsymbol{v}_{CO} = \dfrac{m_2}{m_1 + m_2}v_{10}\boldsymbol{k} & \text{(2a)} \\ \boldsymbol{v}'_{20} = 0 - \boldsymbol{v}_{CO} = -\boldsymbol{v}_{CO} = -\dfrac{m_1}{m_1 + m_2}v_{10}\boldsymbol{k} & \text{(2b)} \end{cases}$$

$\boldsymbol{v}'_{10}$ and $\boldsymbol{v}'_{20}$—the velocity of Particle 1 and Particle 2 in the center-of-mass reference frame at the time point before the collision.

The vector diagram corresponding to Eq. (2) is as follows (see Fig. 1):

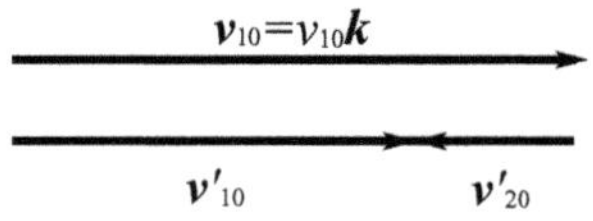

Figure 1

Taking *the system composed of the two particles* as the object of study. Studying *the instantaneous process of the collision between the two*. Considering (a) ***law of conservation of momentum*** [in the center-of-mass reference frame, the resultant external force acting on the system is zero; therefore, the momentum of the system is conserved], (b) ***law of conservation of mechanical energy*** [the collision between the two particles is completely elastic; therefore, the mechanical energy of the system in the center-of-mass reference frame is conserved], (c) ***the total momentum of system of particles in center-of-mass reference frame is zero*** [integrating into (a)]:

$$\begin{cases} m_1\mathbf{v}'_{10} + m_2\mathbf{v}'_{20} = m_1\mathbf{v}'_1 + m_2\mathbf{v}'_2 = 0 & (3a) \\ \frac{1}{2}m_1 v'^2_{10} + \frac{1}{2}m_2 v'^2_{20} = \frac{1}{2}m_1 v'^2_1 + \frac{1}{2}m_2 v'^2_2 & (3b) \end{cases}$$

$\mathbf{v}'_1$ and $\mathbf{v}'_2$ — the velocity of Particle 1 and Particle 2 in the center-of-mass reference frame at the time point after the collision.

The system of equations above can only take the following solutions:

$$\Rightarrow \begin{cases} v'_1 = v'_{10} = \dfrac{m_2}{m_1 + m_2} v_{10} & (4a) \\ v'_2 = v'_{20} = \dfrac{m_1}{m_1 + m_2} v_{10} & (4b) \end{cases}$$

That is, in the center-of-mass reference frame, the velocity magnitude of the two particles does not change, only the velocity direction changes.

Defining θ_1 and θ_2—the angle between $\mathbf{v}'_1$ and $\boldsymbol{k}$, and the angle between $\mathbf{v}'_2$ and $\boldsymbol{k}$. The vector diagram corresponding to Eq. (3a) is as shown in Fig. 2.

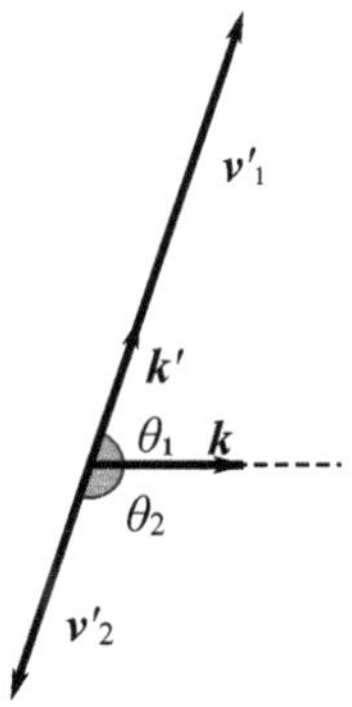

Figure 2

Considering the following geometric relation in Fig. 2:

$$\theta_1 + \theta_2 = \pi \tag{5}$$

θ_1 is the angle between $\boldsymbol{k}'$ and $\boldsymbol{k}$:

$$\Rightarrow \qquad \cos\theta_1 = \boldsymbol{k}\cdot\boldsymbol{k}' \ \Rightarrow \ \theta_1 = \arccos \boldsymbol{k}\cdot\boldsymbol{k}' \tag{6}$$

From the fact that the direction of $\boldsymbol{k}'$ is arbitrary, it can be seen that the range of θ_1 and θ_2 are as follows:

$$\Rightarrow \qquad \theta_1, \theta_2 \in [0, \pi] \tag{7}$$

Solution to Problem (2)

Taking *the system composed of the two particles* as the object of study. Studying *the instantaneous process of the collision between the two.* Considering (a) ***law of conservation of momentum***, (b) ***the total momentum of system of particles is equal to the center-of-mass momentum*** [integrating into (a)]:

$$(m_1 + m_2)\boldsymbol{v}_C = (m_1 + m_2)\boldsymbol{v}_{C0} \tag{8}$$

$\boldsymbol{v}_C$—the center-of-mass velocity of the system composed of the two particles at the time point after the collision.

$$\Rightarrow \qquad \boldsymbol{v}_C = \boldsymbol{v}_{C0} = \frac{m_1}{m_1 + m_2} v_{10}\boldsymbol{k} \tag{9}$$

Taking *the two particles* as the object of study respectively. Studying *the state at the time point after the collision*. Considering ***theorem for composition of velocities of a particle*** [taking the laboratory reference frame as the static reference frame and the center-of-mass reference frame as the moving reference frame]:

$$\begin{cases} \boldsymbol{v}_1 = \boldsymbol{v}_C + \boldsymbol{v}_1' & (10a) \\ \boldsymbol{v}_2 = \boldsymbol{v}_C + \boldsymbol{v}_2' & (10b) \end{cases}$$

$\boldsymbol{v}_1$ and $\boldsymbol{v}_2$—the velocity of Particle 1 and Particle 2 in the laboratory reference frame at the time point after the collision.

The vector diagram corresponding to Eq. (10) is as follows (see Fig. 3):

Defining Θ_1 and Θ_2—the angle between $\boldsymbol{v}_1$ and $\boldsymbol{k}$, and the angle between $\boldsymbol{v}_2$ and $\boldsymbol{k}$ (called the scattering angles of Particle 1 and Particle 2 in the laboratory reference frame, respectively).

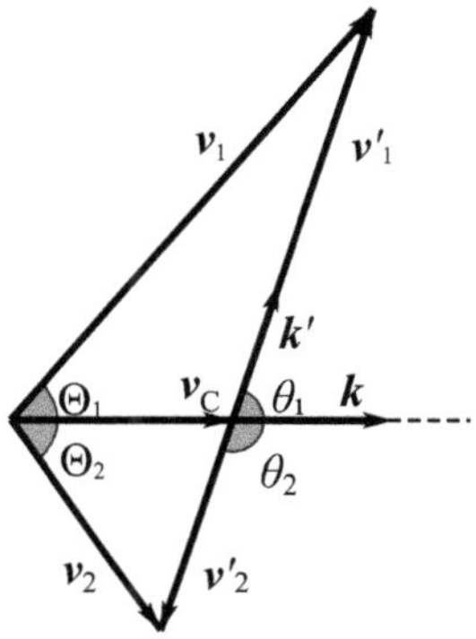

Figure 3

Considering the following geometric relation in Fig. 3, then place (4) and (9) in to obtain:

$$\Rightarrow \begin{cases} \begin{cases} v_1 = \sqrt{v_C^2 + v_1'^2 + 2v_C v_1' \cos\theta_1} = \dfrac{\sqrt{m_1^2 + m_2^2 + 2m_1 m_2 \cos\theta_1}}{m_1 + m_2} v_{10} & \text{(11a)} \\ v_2 = \sqrt{v_C^2 + v_2'^2 - 2v_C v_2' \cos\theta_1} = \dfrac{2m_1}{m_1 + m_2} v_{10} \sin\dfrac{\theta_1}{2} & \text{(11b)} \end{cases} \\ \begin{cases} \tan\Theta_1 = \dfrac{v_1' \sin\theta_1}{v_C + v_1' \cos\theta_1} = \dfrac{m_2 \sin\theta_1}{m_1 + m_2 \cos\theta_1} & \text{(11c)} \\ \tan\Theta_2 = \dfrac{v_2' \sin\theta_1}{v_C - v_2' \cos\theta_1} = \dfrac{\sin\theta_1}{1 - \cos\theta_1} = \tan\left(\dfrac{\pi}{2} - \dfrac{\theta_1}{2}\right) = \tan\dfrac{\theta_2}{2} & \text{(11d)} \end{cases} \end{cases}$$

From (11c), we obtain the expression of θ_1 based on Θ_1:

$$\Rightarrow \qquad \cos\theta_1 = -\frac{m_1}{m_2}\sin^2\Theta_1 \pm \cos\Theta_1\sqrt{1 - \frac{m_1^2}{m_2^2}\sin^2\Theta_1} \qquad (12)$$

In order to make Eq. (12) meaningful, the range of Θ_1 will be limited as follows:

When $m_1 < m_2$, the range of Θ_1 is as follows:

$$\Rightarrow \qquad \Theta_1 \in [0, \pi] \qquad (13)$$

When $m_1 \geq m_2$, the range of Θ_1 is as follows:

$$\Rightarrow \qquad \sin\Theta_{1\max} = \frac{m_2}{m_1} \Rightarrow \Theta_1 \in \left[0, \arcsin\frac{m_2}{m_1}\right] \qquad (14)$$

The value range of Θ_1 can also be easily obtained through vector diagram, Drawing the vector diagram corresponding to Eq. (10a), and drawing the circle with $|v'_1|$ as the radius, in three cases (see Fig. 4a–c). It can be seen that the range of Θ_1 will be limited when $|v_C| \geq |v'_1|$, and will not be limited when $|v_C| < |v'_1|$. By using the value of $\boldsymbol{v}_C$ and $\boldsymbol{v}'_1$ obtained above and combining the geometric relations in the Figures, the same results as above can be obtained.

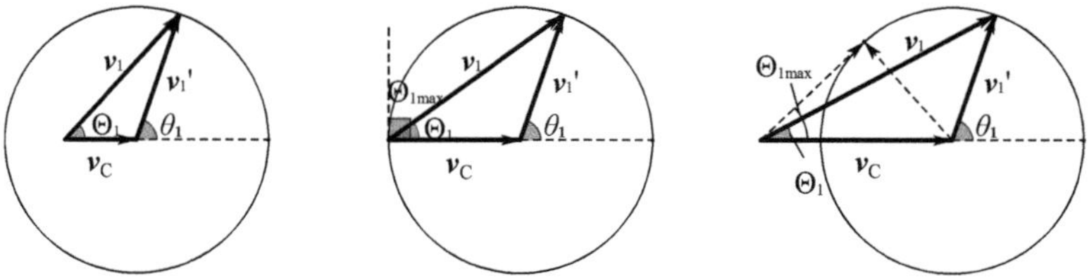

a $|v_C| < |v'_1|$ **b** $|v_C| = |v'_1|$ **c** $|v_C| > |v'_1|$

Figure 4 a–c

In the situation of $\theta_1 = \pi$ (frontal collision), Eqs. (11a–d) degenerates to:

$$\Rightarrow \left\{\begin{array}{l} \left\{\begin{array}{ll} v_1 = \dfrac{|m_1 - m_2|}{m_1 + m_2} v_{10} = \begin{cases} \dfrac{m_1 - m_2}{m_1 + m_2} v_{10} & m_1 \geq m_2 \\ \dfrac{m_2 - m_1}{m_1 + m_2} v_{10} & m_1 < m_2 \end{cases} & (15a) \\ v_2 = \dfrac{2m_1}{m_1 + m_2} v_{10} & (15b) \end{array}\right. \\ \left\{\begin{array}{ll} \tan\Theta_1 = 0 \Rightarrow \begin{cases} \Theta_1 = 0 & m_1 \geq m_2 \\ \Theta_1 = \pi & m_1 < m_2 \end{cases} & (15c) \\ \tan\Theta_2 = 0 \Rightarrow \Theta_2 = \begin{cases} \Theta_2 = 0 & m_1 \geq m_2 \\ \Theta_2 = 0 & m_1 < m_2 \end{cases} & (15d) \end{array}\right. \end{array}\right.$$

The verification of *Newton-collision-formula*

In the laboratory reference frame, the initial velocity of Particle 2 is zero; therefore, its final velocity direction is the direction of the line of the impact, whose unit vector is denoted by $\boldsymbol{k}''$ (see Fig. 5).

Calculating the ratio of relative speeds after and before the collision, then place (2) and (4) in to obtain:

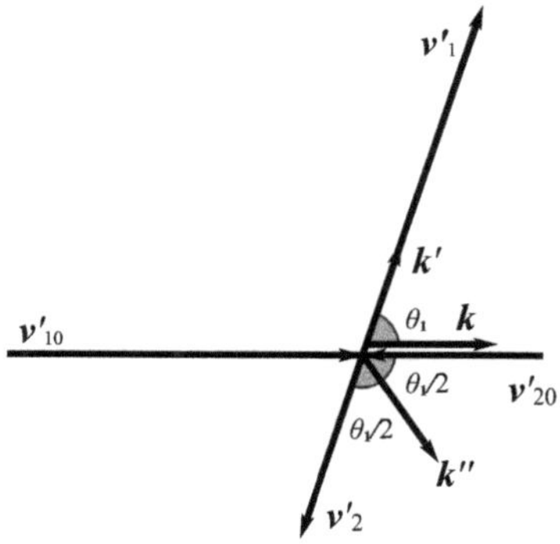

Figure 5

$$\frac{(\boldsymbol{v}_2' - \boldsymbol{v}_1') \cdot \boldsymbol{k}''}{(\boldsymbol{v}_{10}' - \boldsymbol{v}_{20}') \cdot \boldsymbol{k}''} = \frac{(v_1' + v_2')(-\boldsymbol{k}' \cdot \boldsymbol{k}'')}{(v_{10}' + v_{20}')(\boldsymbol{k} \cdot \boldsymbol{k}'')} = 1 \tag{16}$$

That is, ***Newton-collision-formula*** holds for this problem.

5.2.2 The Collision between a Small Ball and an Inclined Plane That Can Slide Freely (*b*, *A*)

An inclined body of mass M and inclination angle θ is statically placed on a smooth horizontal ground (see Fig. 1). A small ball of mass m falls on the inclined body at the initial velocity of the magnitude v_0 and the direction vertical downward. The collision between the ball and the inclined body and the collision between the inclined body and the ground happen successively in an extremely short time. The collision between the inclined body and the ground is completely inelastic (the separation velocity is zero). Given that the collision coefficient of restitution between the ball and the inclined body is e for the whole process composed of these two collision processes, the friction between the ball and the inclined body can be ignored. Find the velocity of the ball and the inclined body after the collision and find the reflection angle of the ball's trajectory about the normal of the inclined body (respectively solve in the ground reference frame and the inclined body reference frame).

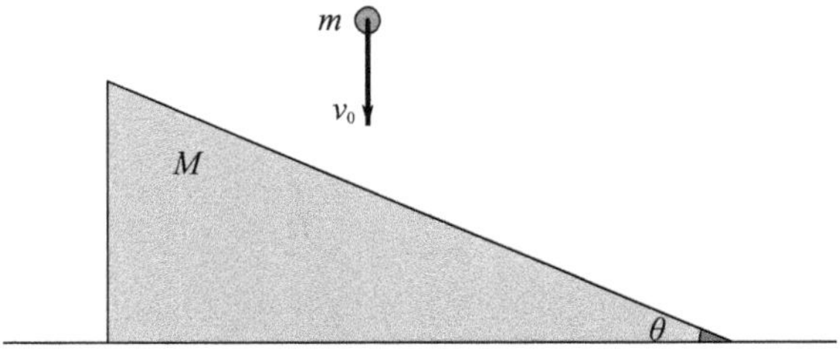

Figure 1

Solution [1]

As shown in Fig. 2, establishing the Cartesian coordinate system with the frame (O, $\boldsymbol{i}$, $\boldsymbol{j}$), let x, y be the Cartesian coordinates, introducing the tangential unit vector $\boldsymbol{\tau}$ and normal unit vector $\boldsymbol{n}$ of the inclined plane at the collision point.

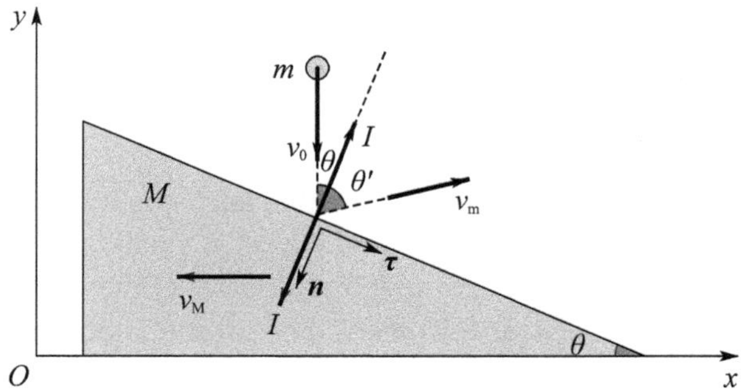

Figure 2

Taking *the ball and the inclined body* as the object of study respectively. Studying *the whole collision process*. Considering ***momentum theorem*** [the gravity of the ball is a finite quantity, acting on an extremely short process, its impulse is an infinitely small quantity, which can be ignored] (for the ball: the $\boldsymbol{i}$-direction and $\boldsymbol{j}$-direction component equations, for the inclined body: the $-\boldsymbol{i}$-direction component equation):

$$\begin{cases} \begin{cases} I\sin\theta = mv_m\sin(\theta+\theta') & \text{(1a)} \\ I\cos\theta = mv_m\cos(\theta+\theta') - (-mv_0) & \text{(1b)} \end{cases} \\ I\sin\theta = Mv_M & \text{(1c)} \end{cases}$$

v_m and v_M—the speed of the ball and the inclined body at the time point after the collision.

θ and θ'—the incident angle and reflection angle of the ball's trajectory in the ground reference frame about the normal of the inclined plane.

I—the impulse magnitude of the impact force between the ball and the inclined body in the collision process between them.
Considering ***Newton-collision-formula***:

$$e = \frac{v'_{Mn} - v'_{mn}}{v_{mn} - v_{Mn}} = \frac{v_M \sin\theta - (-v_m \cos\theta')}{v_0 \cos\theta - 0} \tag{2}$$

v_{mn} and v'_{mn}—the ***n***-direction component of the velocity of the ball at the time point before and after the collision.

v_{Mn} and v'_{Mn}—the ***n***-direction component of the velocity of the inclined body at the time point before and after the collision.

Defining the constant α:

$$\alpha = \frac{e - \frac{m}{M}\sin^2\theta}{1 + \frac{m}{M}\sin^2\theta} \tag{3}$$

Combining (1) to (3),

$$\Rightarrow \begin{cases} v_M = (1+\alpha)\dfrac{m\sin\theta\cos\theta}{M}v_0 & \text{(4a)} \\ v_m = (\sin^2\theta + \alpha^2\cos^2\theta)^{1/2} v_0 & \text{(4b)} \\ \tan\theta' = \dfrac{1}{\alpha}\tan\theta \Rightarrow \theta' = \arctan\left(\dfrac{1}{\alpha}\tan\theta\right) & \text{(4c)} \end{cases}$$

The range of e is as follows:

$$e \le 1 \tag{5}$$

$(5) \rightarrow (3)$,

$$\alpha \le \frac{1 - \frac{m}{M}\sin^2\theta}{1 + \frac{m}{M}\sin^2\theta} < 1 \tag{6}$$

$$\Rightarrow \quad \theta' > \theta \tag{7}$$

Defining θ''—the reflection angle of the ball's trajectory in the inclined body reference frame about the normal of the inclined plane.

Considering the definition of ***reflection angle***:

$$\tan\theta'' = \frac{v_m \sin\theta' - (-v_M \cos\theta)}{v_m \cos\theta' - (-v_M \sin\theta)} \tag{8}$$

(4) → (8),

$$\Rightarrow \quad \tan\theta'' = \frac{(\sin^2\theta + \alpha^2\cos^2\theta)^{1/2}\sin\theta' + (1+\alpha)\frac{m\sin\theta\cos\theta}{M}\cos\theta}{(\sin^2\theta + \alpha^2\cos^2\theta)^{1/2}\cos\theta' + (1+\alpha)\frac{m\sin\theta\cos\theta}{M}\sin\theta}$$

$$= \frac{(\sin^2\theta + \alpha^2\cos^2\theta)^{1/2}\frac{\tan\theta'}{\sqrt{1+\tan^2\theta'}} + (1+\alpha)\frac{m\sin\theta\cos\theta}{M}\cos\theta}{(\sin^2\theta + \alpha^2\cos^2\theta)^{1/2}\frac{1}{\sqrt{1+\tan^2\theta'}} + (1+\alpha)\frac{m\sin\theta\cos\theta}{M}\sin\theta}$$

$$= \frac{\tan\theta + (1+\alpha)\frac{m}{M}\sin\theta\cos\theta}{\alpha + (1+\alpha)\frac{m}{M}\sin^2\theta} \tag{9}$$

(3) → (9),

$$\Rightarrow \qquad \tan\theta'' = \frac{1}{e}\tan\theta + \left(1+\frac{1}{e}\right)\frac{\frac{m}{M}\sin\theta\cos\theta}{1+\frac{m}{M}\sin^2\theta}$$

$$\Rightarrow \qquad \theta'' = \arctan\left(\frac{1}{e}\tan\theta + \left(1+\frac{1}{e}\right)\frac{\frac{m}{M}\sin\theta\cos\theta}{1+\frac{m}{M}\sin^2\theta}\right) > \theta \tag{10}$$

Solution 2

Taking *the ball* as the object of study. Studying *the whole collision process*. Considering ***law of conservation of momentum*** [the gravity of the ball is a finite quantity, acting on an extremely short process, its impulse is an infinitely small quantity, which can be ignored] (the $\boldsymbol{\tau}$-direction component equation):

$$mv_0 \sin\theta = mv_m \sin\theta' \tag{11}$$

Taking *the system composed of the ball and the inclined body* as the object of study. Studying *the whole collision process*. Considering ***law of conservation of momentum*** [the ***i***-direction component of the resultant external force acting on the system is zero; therefore, the ***i***-direction component of the momentum of the system is conserved] (the ***i***-direction component equation):

$$mv_m \sin(\theta+\theta') - Mv_M = 0 \tag{12}$$

The results of v_m, v_M and θ' can be obtained by combining Eqs. (2), (11) and (12).

5.2.3 The Repeated Collisions between a Box and an Object in the Box (*b*, *A*)

A cuboid box of mass M and length L is resting on a smooth horizontal ground, in which there is a block of mass m (which can be regarded as a particle) resting next to its left wall (see Fig. 1), $M/m = a$. The block is subjected to an instantaneous impact and obtains an initial velocity of the magnitude v_0 and the direction horizontal right. Given that the friction between the block and the bottom of the box can be ignored, and the collision coefficient of restitution between them is e.

(1) Find the total kinetic energy loss of the block and the box after infinite collisions.

(2) In order to make the total kinetic energy loss of the block and the box not exceed $\dfrac{a-1}{a+1}$ times of the initial kinetic energy, find the upper limit of the number of collisions satisfies this condition.

(3) Find the time taken from the block starting to move to the completion of the above number of collisions.

All the above results need to be calculated when: $a = 10$, $e = \sqrt[4]{\dfrac{1}{2}}$.

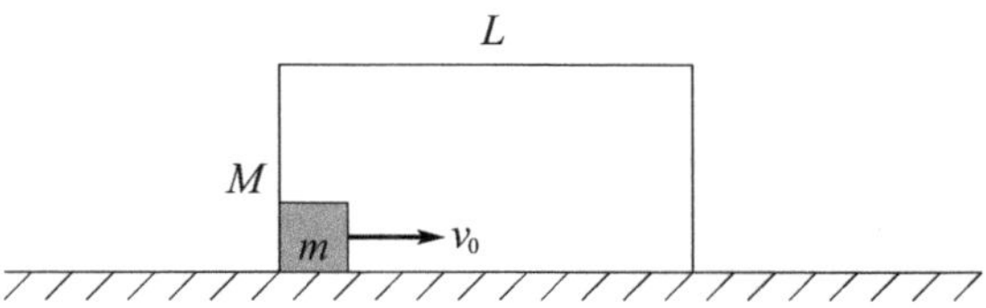

Figure 1

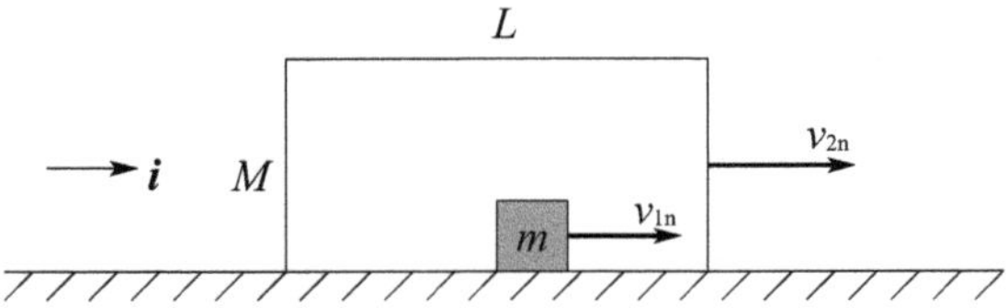

Figure 2

Solution to Problem (1)

The time point when the block is impacted is defined as Time zero.

As shown in Fig. 2, the unit vector in the horizontal right direction is denoted by ***i***.

Considering ***the total momentum of the system of particles in the center-of-mass reference frame is zero***:

$$mv'_{1n} + Mv'_{2n} = 0 \tag{1}$$

v'_{1n} and v'_{2n}—the ***i***-direction component of the velocity of the block and the box at the time point after the n-th collision between them.

Considering ***Newton-collision-formula***:

$$e = \frac{v'_{2i} - v'_{1i}}{v'_{1(i-1)} - v'_{2(i-1)}} \qquad i = 1,2,3,\cdots \tag{2}$$

$$\Rightarrow \qquad \frac{v'_{2n} - v'_{1n}}{v_0} = (-1)^{n-1} \prod_{i=1}^{n} \frac{v'_{2i} - v'_{1i}}{v'_{1(i-1)} - v'_{2(i-1)}} = (-1)^{n-1} e^n \tag{3}$$

$$\Rightarrow \qquad v'_{2n} - v'_{1n} = (-1)^{n-1} e^n v_0 \tag{4}$$

Combining (1) and (4),

$$\Rightarrow \qquad \begin{cases} v'_{1n} = -\dfrac{M}{M+m}(-1)^{n-1} e^n v_0 & \text{(5a)} \\ v'_{2n} = \dfrac{m}{M+m}(-1)^{n-1} e^n v_0 & \text{(5b)} \end{cases}$$

Taking *the system composed of the block and the box* as the object of study. Studying *the process from Time zero to arbitrary Time t.* Considering (a) ***law of conservation of momentum*** [the resultant external force acting on the system is zero; therefore, the momentum of the system is conserved] (the ***i***-direction component equation),

(b) ***the total momentum of the system of particles is equal to the center-of-mass momentum*** [integrating into (a)]:

$$mv_0 = (M+m)v_C \tag{6}$$

v_C—the ***i***-direction component of the center-of-mass velocity of the system at Time *t*.

$$\Rightarrow \qquad v_C = \frac{m}{M+m}v_0 \tag{7}$$

Defining E_{k0}—the initial kinetic energy of the system, E_{kn}—the kinetic energy of the system at the time point after the *n*-th collision.

Considering (a) the definition of ***kinetic energy of particle***, (b) ***König theorem***:

$$E_{k0} = \frac{1}{2}mv_0^2 \tag{8}$$

$$E_{kn} = \frac{1}{2}(M+m)v_C^2 + \left(\frac{1}{2}mv_{1n}'^2 + \frac{1}{2}Mv_{2n}'^2\right) \tag{9}$$

$$\Rightarrow \qquad E_{kn} = \frac{1}{2}\frac{m^2}{M+m}v_0^2 + \frac{1}{2}\frac{Mm}{M+m}e^{2n}v_0^2 \tag{10}$$

Defining ΔE_{kn}—the total kinetic energy loss of the system after *n* collisions.

$$\Rightarrow \qquad \Delta E_{kn} = E_{k0} - E_{kn} = \frac{1}{2}\frac{Mm}{M+m}(1-e^{2n})v_0^2$$

$$= \frac{1}{2}mv_0^2\frac{M}{M+m}(1-e^{2n}) = E_{k0}\frac{a}{a+1}(1-e^{2n}) \tag{11}$$

$$\Rightarrow \qquad \frac{\Delta E_{kn}}{E_{k0}} = \frac{a}{a+1}(1-e^{2n}) \tag{12}$$

$$\Rightarrow \qquad \left.\frac{\Delta E_{kn}}{E_{k0}}\right|_{n\to\infty} = \lim_{n\to\infty}\frac{a}{a+1}(1-e^{2n}) = \frac{a}{a+1}\underset{a=10}{=}\frac{10}{11} \tag{13}$$

Another solution for v'_{1n} and v'_{2n}

Considering ***the total momentum of the system of particles in the center-of-mass reference frame is zero***:

$$mv_{1i}' + Mv_{2i}' = 0 \tag{14}$$

Combining (2) and (14),

$$\Rightarrow \begin{cases} \dfrac{v'_{1i}}{v'_{1(i-1)}} = -e & (15a) \\ \dfrac{v'_{2i}}{v'_{2(i-1)}} = -e & (15b) \end{cases}$$

$$\Rightarrow \begin{cases} \dfrac{v'_{1n}}{v'_{10}} = \sum_{i=1}^{n} \dfrac{v'_{1i}}{v'_{1(i-1)}} = (-1)^n e^n & (16a) \\ \dfrac{v'_{2n}}{v'_{20}} = \sum_{i=1}^{n} \dfrac{v'_{2i}}{v'_{2(i-1)}} = (-1)^n e^n & (16b) \end{cases}$$

Taking *the block and the box* as the object of study respectively. Studying *the state at Time zero*. Considering ***theorem for composition of velocities of a particle*** [taking the ground as the static reference frame and center-of-mass reference frame of the system as the moving reference frame], then place (7) in to obtain:

$$\Rightarrow \begin{cases} v'_{10} = v_0 - v_C = \dfrac{M}{M+m} v_0 & (17a) \\ v'_{20} = 0 - v_C = -\dfrac{m}{M+m} v_0 & (17b) \end{cases}$$

$$\Rightarrow \begin{cases} v'_{1n} = \dfrac{M}{M+m} (-1)^n e^n v_0 & (18a) \\ v'_{2n} = -\dfrac{m}{M+m} (-1)^n e^n v_0 & (18b) \end{cases}$$

Another solution for the kinetic energy of the system in the center-of-mass reference frame after an arbitrary collision.

Defining E'_{ki}—the kinetic energy of the system in the center-of-mass reference frame after the *i*-th collision. Considering ***energy loss form in frontal impact of two particles based on collision coefficient of restitution***:

$$\frac{E'_{ki} - E'_{k(i+1)}}{E'_{ki}} = 1 - e^2 \quad (19)$$

$$\Rightarrow \quad \frac{E'_{k(i+1)}}{E'_{ki}} = e^2 \quad (20)$$

$$\Rightarrow \qquad \frac{E'_{kn}}{E'_{k0}} = \sum_{i=0}^{n-1} \frac{E'_{k(i+1)}}{E'_{ki}} = e^{2n} \qquad (21)$$

Considering ***the form of available energy in the center-of-mass reference frame***:

$$E'_{k0} = \frac{1}{2}\frac{Mm}{M+m}v_0^2 \qquad (22)$$

$$\Rightarrow \qquad E'_{kn} = \frac{1}{2}\frac{Mm}{M+m}e^{2n}v_0^2 \qquad (23)$$

Solution to Problem (2)

$$\frac{\Delta E_k}{E_{k0}} \le \frac{a-1}{a+1} \qquad (24)$$

$$\Rightarrow \qquad e^{2n} \ge \frac{1}{a} \qquad (25)$$

Substituting $a = 10$, $e = \sqrt[4]{\frac{1}{2}}$ into the inequation above, we obtain:

$$\Rightarrow \qquad n \le \frac{2}{\lg 2} \approx 6.6 \qquad (26)$$

Defining N—the upper limit of the number of collisions that ensures the total kinetic energy loss of the block and the box not exceed $\frac{a-1}{a+1}$ times of the initial kinetic energy.

$$\Rightarrow \qquad N = 6 \qquad (27)$$

Solution to Problem (3)

Defining t_n—the time taken from the (n-1)-th collision to the n-th collision. Considering ***the displacement-time relation of uniform linear motion***, then place (5) in to obtain:

$$t_n = \frac{L}{\left|v'_{2(n-1)} - v'_{1(n-1)}\right|} = \frac{L}{e^{n-1}v_0} \qquad (28)$$

Defining T—the time taken from Time zero to the 6-th collision.

$$\Rightarrow \ T=\sum_{n=1}^{6}t_n=\sum_{n=1}^{6}\frac{L}{e^{n-1}v_0}=\frac{L}{v_0}\sum_{n=1}^{6}\frac{1}{e^{n-1}}=\frac{L}{v_0}\frac{e^6-1}{e^6-e^5}\underset{e=\sqrt[4]{\frac{1}{2}}}{=}\frac{L}{v_0}\frac{1-2^{3/2}}{1-2^{1/4}} \tag{29}$$

5.2.4 Total Number of Collisions Occurred in a Double Blocks-Wall System (*c*, *B*)[5]

Consider two blocks A and B of mass M and m resting on a smooth horizontal ground, on the left side of Block B there is a vertical wall (see Fig. 1). Block A is subject to an instant impact and obtains an initial velocity pointing to Block B. The collision between Blocks A and B and the collision between Block B and the wall are both completely elastic. Find the total number of collisions that can occur between the two blocks and between Block B and the wall, and find the approximate solution under the condition of $M \gg m$.

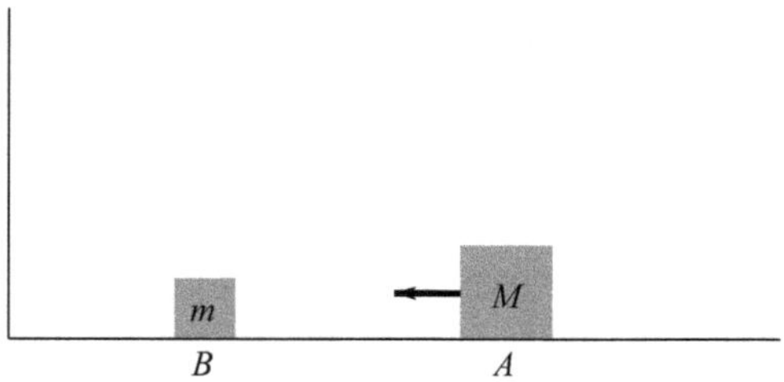

Figure 1

Solution

As shown in Fig. 2, the intersection of the line passing through the two blocks and the wall is denoted by O, the unit vector in the horizontal right direction is denoted by $\boldsymbol{i}$. Establishing the linear coordinate system with Point O as the origin and the $\boldsymbol{i}$-direction as the positive direction of coordinate axis, let x be the coordinate.

Taking *the system composed of the two blocks* as the object of study. Studying *the instantaneous process of the i-th collision between*

[5]G. Galperin. Playing pool with π (the number π from a billiard point of view), *Regular & Chaotic Dynamics [J]*, 2003, 8(4): 375–394.

them. Considering (a) ***law of conservation of momentum*** [the resultant external force acting on the system is zero; therefore, the momentum of the system is conserved] (the ***i***-direction component equation), (b) ***law of conservation of mechanical energy*** [the collision between the two blocks is completely elastic; therefore, the mechanical energy of the system is conserved]:

$$\begin{cases} MV_i + mv_i' = MV_{i+1} + mv_{i+1} & (1a) \\ \dfrac{1}{2}MV_i^2 + \dfrac{1}{2}mv_i'^2 = \dfrac{1}{2}MV_{i+1}^2 + \dfrac{1}{2}mv_{i+1}^2 & (1b) \end{cases}$$

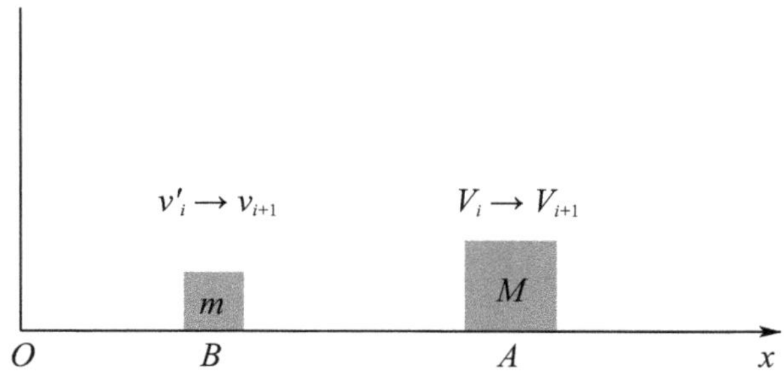

Figure 2

V_i and v_i—the ***i***-direction component of the velocity of Blocks *A* and *B* at the time point after the *i*-th collision between them.

v'_i—the ***i***-direction component of the velocity of Block *B* at the time point after it collides with the wall for the *i*-th time (i.e., the ***i***-direction component of the velocity of Block *B* at the time point before it collides with Block *A* for the $(i + 1)$-th time).

Taking *the system composed of Block B and the wall* as the object of study. Studying *the instantaneous process of the i-th collision between them*. Considering ***Newton-collision-formula for completely elastic collision*** [for Block *B*, its collision with the wall can be regarded as the completely elastic collision with the object of infinite mass]:

$$e = \frac{0 - v_i'}{v_i - 0} = 1 \quad (2)$$

$$\Rightarrow \quad v_i' = -v_i \quad (3)$$

Defining *N*—the mass ratio of Blocks *A* and *B*:

$$N = \frac{M}{m} \tag{4}$$

(3) and (4) → (1),

$$\Rightarrow \quad \begin{cases} N(V_{i+1}+V_i)(V_{i+1}-V_i) = -(v_{i+1}+v_i)(v_{i+1}-v_i) & \text{(5a)} \\ N(V_{i+1}-V_i) = -(v_{i+1}+v_i) & \text{(5b)} \end{cases}$$

The terms on both sides of Eq. (5a) are divided by the terms on both sides of Eq. (5b):

$$\Rightarrow \quad V_{i+1}+V_i = v_{i+1}-v_i \tag{6}$$

Equations (5b) and (6) are rearranged as follows:

$$\Rightarrow \quad \begin{cases} -NV_i + v_i = -NV_{i+1} - v_{i+1} & \text{(7a)} \\ V_i + v_i = -V_{i+1} + v_{i+1} & \text{(7b)} \end{cases}$$

$$\Rightarrow \quad \begin{pmatrix} -N & 1 \\ 1 & 1 \end{pmatrix}\begin{pmatrix} V_i \\ v_i \end{pmatrix} = \begin{pmatrix} -N & -1 \\ -1 & 1 \end{pmatrix}\begin{pmatrix} V_{i+1} \\ v_{i+1} \end{pmatrix} \tag{8}$$

$$\Rightarrow \quad \begin{pmatrix} V_{i+1} \\ v_{i+1} \end{pmatrix} = \begin{pmatrix} -N & -1 \\ -1 & 1 \end{pmatrix}^{-1}\begin{pmatrix} -N & 1 \\ 1 & 1 \end{pmatrix}\begin{pmatrix} v_i \\ V_i \end{pmatrix} = \begin{pmatrix} \dfrac{N-1}{N+1} & -\dfrac{2}{N+1} \\ \dfrac{2N}{N+1} & \dfrac{N-1}{N+1} \end{pmatrix}\begin{pmatrix} V_i \\ v_i \end{pmatrix} \tag{9}$$

Defining E_0—the initial mechanical energy of the system.

From (1b),

$$\frac{1}{2}MV_i^2 + \frac{1}{2}mv_i^2 = \frac{1}{2}MV_0^2 = E_0 \tag{10}$$

$$\Rightarrow \quad \frac{V_i^2}{\dfrac{2E_0}{Nm}} + \frac{v_i^2}{\dfrac{2E_0}{m}} = 1 \tag{11}$$

Defining the constant a:

$$a = \frac{2E_0}{Nm} \tag{12}$$

$$\Rightarrow \quad \frac{V_i^2}{a} + \frac{v_i^2}{Na} = 1 \tag{13}$$

It can be seen that, the array (V_i, v_i) satisfies the elliptic equation. That is, in the velocity space (O, V, v) of Blocks A and B, the points described by the arrays are distributed on the ellipse described by Eq. (13) (see Fig. 3).

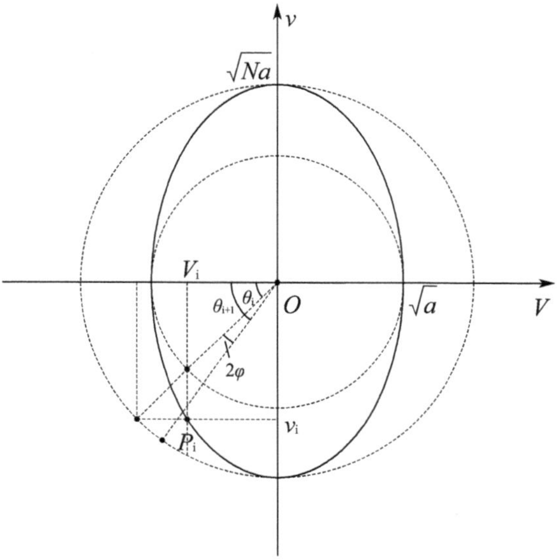

Figure 3

Defining the variable θ_i:

$$\begin{cases} V_i = -\sqrt{a}\cos\theta_i & \text{(14a)} \\ v_i = -\sqrt{Na}\sin\theta_i & \text{(14b)} \end{cases}$$

In the velocity space (O, V, v), Point (V_i, v_i) is denoted by P_i.
$(14) \to (9)$,

$$\Rightarrow \begin{cases} V_{i+1} = -\sqrt{a}\left(\dfrac{N-1}{N+1}\cos\theta_i - \dfrac{2\sqrt{N}}{N+1}\sin\theta_i\right) & \text{(15a)} \\ v_{i+1} = -\sqrt{Na}\left(\dfrac{2\sqrt{N}}{N+1}\cos\theta_i + \dfrac{N-1}{N+1}\sin\theta_i\right) & \text{(15b)} \end{cases}$$

Defining the constant φ:

$$\tan\varphi = \frac{1}{\sqrt{N}} \tag{16}$$

$$\Rightarrow \begin{cases} \dfrac{2\sqrt{N}}{N+1} = \dfrac{2\cot\varphi}{\cot^2\varphi+1} = \sin 2\varphi & (17a) \\ \dfrac{N-1}{N+1} = \dfrac{\cot^2\varphi-1}{\cot^2\varphi+1} = \cos 2\varphi & (17b) \end{cases}$$

$$\Rightarrow \begin{cases} V_{i+1} = -\sqrt{a}(\cos 2\varphi\cos\theta_i - \sin 2\varphi\sin\theta_i) = -\sqrt{a}\cos(\theta_i + 2\varphi) & (18a) \\ v_{i+1} = -\sqrt{Na}(\sin 2\varphi\cos\theta_i + \cos 2\varphi\sin\theta_i) = -\sqrt{Na}\sin(\theta_i + 2\varphi) & (18b) \end{cases}$$

Combining (14) and (18),

$$\Rightarrow \qquad \theta_{i+1} = \theta_i + 2\varphi \qquad (19)$$

Considering the initial conditions:

$$\begin{cases} V_0 = -\sqrt{\dfrac{2E_0}{Nm}} = -\sqrt{a} & (20a) \\ v_0 = 0 & (20b) \end{cases}$$

(14) $\rightarrow$ (20),

$$\Rightarrow \qquad \theta_0 = 0 \qquad (21)$$

$$\Rightarrow \qquad \theta_i = 2i\varphi \qquad (22)$$

$$\Rightarrow \begin{cases} V_i = -\sqrt{a}\cos(2i\varphi) & (23a) \\ v_i = -\sqrt{Na}\sin(2i\varphi) & (23b) \end{cases}$$

$$\Rightarrow \qquad v_i' = \sqrt{Na}\sin(2i\varphi) \qquad (24)$$

There are two possible situations:

Situation 1: After the last collision between the two blocks, Block *B* and the wall will not collide again.

In this situation, the number of the last collision between the two blocks is denoted by *j*; the following relations should be satisfied:

$$\begin{cases} B \text{ can collide with } A \text{ for the } j \text{ th time:} \\ \qquad v'_{j-1} > V_{j-1} & (25a) \\ B \text{ can't collide with the wall for the } j \text{ th time, that is} \\ \text{after the } j \text{ th collision with } A, B \text{ moves away from the wall:} \\ \qquad v_j \geq 0 & (25b) \end{cases}$$

(23) and (24) → (25),

$$\Rightarrow \qquad \begin{cases} \sqrt{Na}\sin[2(j-1)\varphi] > -\sqrt{a}\cos[2(j-1)\varphi] & \text{(26a)} \\ -\sqrt{Na}\sin(2j\varphi) \geq 0 & \text{(26b)} \end{cases}$$

(16) → (26),

$$\Rightarrow \qquad \begin{cases} \sin[2(j-1)\varphi+\varphi] > 0 \Rightarrow (2j-1)\varphi < \pi & \text{(27a)} \\ \sin(2j\varphi) \leq 0 \Rightarrow 2j\varphi \geq \pi & \text{(27b)} \end{cases}$$

$$\Rightarrow \qquad \frac{\pi}{2\varphi} \leq j < \frac{\pi}{2\varphi} + \frac{1}{2} \tag{28}$$

The condition that *j* has a reasonable integer solution is as follows:

$$\text{Celiing}\left(\frac{\pi}{2\varphi}\right) = \text{Floor}\left(\frac{\pi}{2\varphi} + \frac{1}{2}\right) \tag{29}$$

where function Ceiling (x) represents the minimum integer not smaller than the variable x, function Floor (x) represents the maximum integer not greater than the variable x.

When this condition can be satisfied, defining n_1—the total number of collisions that can occur between the two blocks and between Block *B* and the wall in Situation 1. In Situation 1, the number of collisions between the two Blocks is one more than that between Block *B* and the wall:

$$\Rightarrow \qquad n_1 = 2(j-1) + 1 \tag{30}$$

$$\Rightarrow \qquad \frac{\pi}{\varphi} - 1 \leq n_1 < \frac{\pi}{\varphi} \tag{31}$$

$$\Rightarrow \qquad n_1 = \text{Floor}\left(\frac{\pi}{\varphi}\right) = \text{Floor}\left(\frac{\pi}{\arctan\frac{1}{\sqrt{N}}}\right) \underset{N\to\infty}{\to} \text{Floor}(\sqrt{N}\pi) \tag{32}$$

In addition, we can draw the schematic diagram of Situation 1 in the velocity space (see Fig. 4):

Situation 2: After the last collision of the two blocks, Block *B* collides with the wall again.

In this situation, the number of the last collision between the two blocks is denoted by *j*; the following relations should be satisfied:

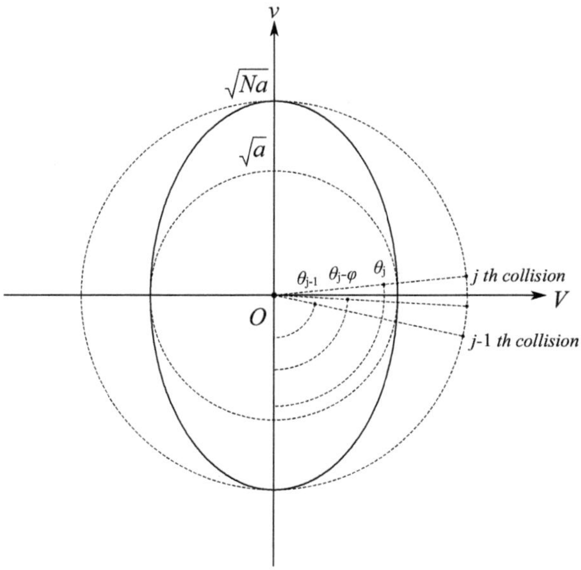

Figure 4

$$\begin{cases} B \text{ can collide with the wall for the } j \text{ th time, that is} \\ \text{after the } j \text{ th collision with } A, B \text{ moves towards the wall:} \\ \qquad v_j < 0 & \text{(33a)} \\ B \text{ can't collide with } A \text{ for the } (j+1) \text{ th time:} \\ \qquad v'_j \le V_j & \text{(33b)} \end{cases}$$

(23) and (24) $\rightarrow$ (33),

$$\Rightarrow \quad \begin{cases} -\sqrt{Na}\sin(2j\varphi) < 0 & \text{(34a)} \\ \sqrt{Na}\sin(2j\varphi) \le -\sqrt{a}\cos(2j\varphi) & \text{(34b)} \end{cases}$$

(16) $\rightarrow$ (34),

$$\Rightarrow \quad \begin{cases} \sin(2j\varphi) > 0 \Rightarrow 2j\varphi < \pi & \text{(35a)} \\ \sin[(2j+1)\varphi] \le 0 \Rightarrow (2j+1)\varphi \ge \pi & \text{(35b)} \end{cases}$$

$$\Rightarrow \quad \frac{\pi}{2\varphi} - \frac{1}{2} \le j < \frac{\pi}{2\varphi} \qquad (36)$$

The condition that *j* has a reasonable integer solution is as follows:

$$\text{Celiing}\left(\frac{\pi}{2\varphi}-\frac{1}{2}\right)=\text{Floor}\left(\frac{\pi}{2\varphi}\right) \quad (37)$$

When this condition can be satisfied, defining n_2—the total number of collisions that can occur between the two blocks and between Block *B* and the wall in Situation 2. In Situation 2, the number of collisions between the two Blocks is equal to that between Block *B* and the wall.

$$\Rightarrow \quad n_2 = 2j \quad (38)$$

$$\Rightarrow \quad \frac{\pi}{\varphi}-1\le n_2<\frac{\pi}{\varphi} \quad (39)$$

$$\Rightarrow \quad n_2=\text{Floor}\left(\frac{\pi}{\varphi}\right)=\text{Floor}\left(\frac{\pi}{\arctan\frac{1}{\sqrt{N}}}\right)\underset{N\to\infty}{\to}\text{Floor}(\sqrt{N}\pi) \quad (40)$$

It is easy to prove that for an arbitrary *N*, one of the conditions (29) and (37) must hold.

In addition, we can draw the schematic diagram of Situation 2 in the velocity space (see Fig. 5):

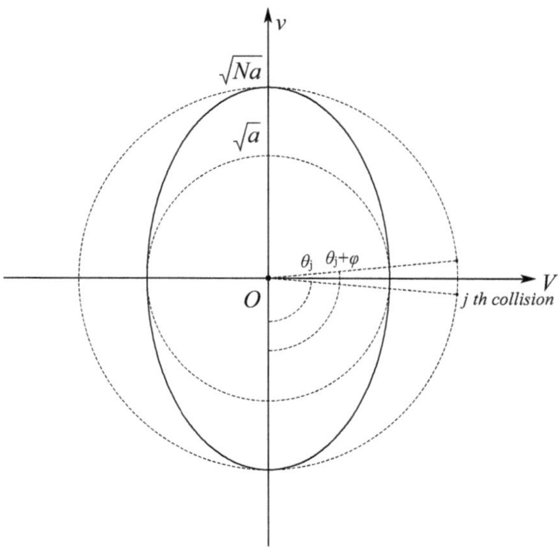

Figure 5

In both situations, the form of the total number of collisions is the same, which can be uniformly expressed as n:

$$\Rightarrow \qquad n = \text{Floor}\left(\frac{\pi}{\arctan\frac{1}{\sqrt{N}}}\right) \underset{N\to\infty}{\to} \text{Floor}(\sqrt{N}\pi) \tag{41}$$

Considering the following special cases:

$$N = 10^{2k} \qquad k \text{ is a large positive integer} \tag{42}$$

$$\Rightarrow \qquad n = \text{Floor}\,(10^k \pi) \tag{43}$$

That is, the total number of collisions will reflect the numerical law of π.

5.2.5 The Collision of Two Balls through a Light Rigid String (d, *C*)

As shown in Fig. 1, there is a small ball of mass m_1 (called Ball 1) placed on a smooth horizontal ground and another small ball of mass m_2 (called Ball 2) placed in a smooth straight thin groove on the horizontal ground. Respectively tie the two balls to the two ends of a light rigid string of length l. Initially, the two balls are both at rest, the line segment linking them is perpendicular to the groove and with distance kl from each other, $0 < k < 1$. Ball 1 is subject to an instant impact and obtains an initial velocity of the magnitude v_0 and the direction parallel to the groove.

(1) Find the velocity of the two balls when Ball 2 starts to move.
(2) Find the tension in the string when Ball 2 starts to move.

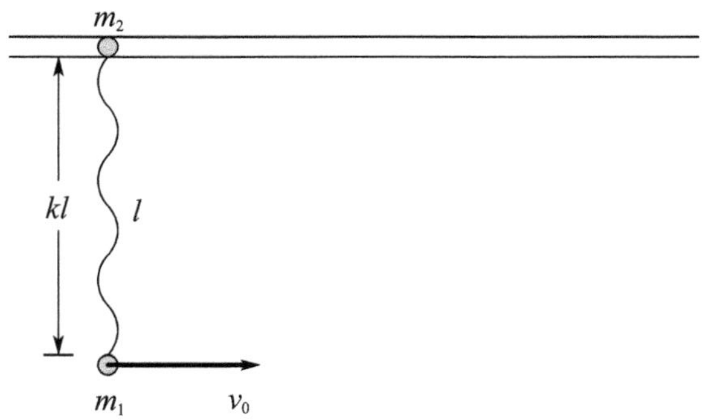

Figure 1

Solution [1] to Problem (1)

The instantaneous process of the string being straightened is a series of collision processes that occur in the system composed of the two balls and all the particles that compose the string.

As shown in Fig. 2, the initial position of Ball 1 is denoted by O, the unit vector of the initial velocity direction of Ball 1 is denoted by $\boldsymbol{i}$ (horizontally to the right in the figure), the unit vector of Ball 1 pointing to Ball 2 at the initial time is denoted by $\boldsymbol{j}$, $\boldsymbol{i} \times \boldsymbol{j}$ is denoted by $\boldsymbol{k}$. Establishing the Cartesian coordinate system with the frame (O, $\boldsymbol{i}$, $\boldsymbol{j}$, $\boldsymbol{k}$), let x, y, z be the Cartesian coordinates.

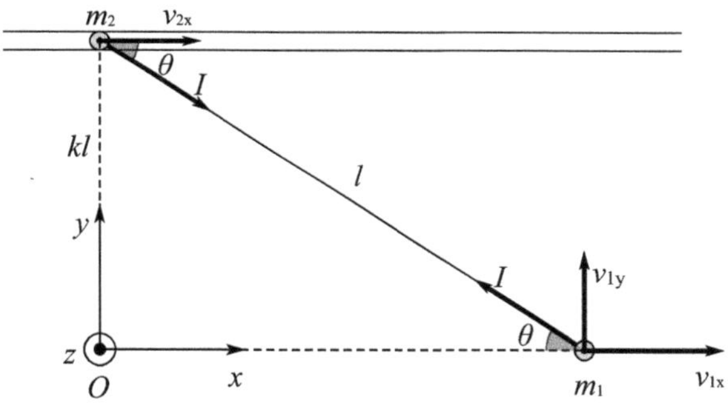

Figure 2

Defining θ—the angle between the string and the groove at the time point after the collision. Considering the following geometric relation in Fig. 2:

$$\sin\theta = \frac{kl}{l} = k \tag{1}$$

Taking *Ball 1 and Ball 2* as the object of study respectively. Studying *the instantaneous process of the collision*. Considering ***momentum theorem*** [the average value of the tension in the string with respect to time is an infinitely large quantity, acting in an extremely short process, its impulse is an finite quantity, which cannot be ignored] (for Ball 1: the $\boldsymbol{i}$-direction and $\boldsymbol{j}$-direction component equations, for Ball 2: the $\boldsymbol{i}$-direction component equation):

$$\begin{cases} \begin{cases} -I\cos\theta = m_1 v_{1x} - m_1 v_0 & \text{(2a)} \\ I\sin\theta = m_1 v_{1y} & \text{(2b)} \end{cases} \\ I\cos\theta = m_2 v_{2x} & \text{(2c)} \end{cases}$$

I—the impulse magnitude of the tension in the string in the instantaneous process of the collision.

v_{1x} and v_{1y}—the ***i***-direction and ***j***-direction component of the velocity of Ball 1 at the time point after the collision.

v_{2x}—the ***i***-direction component of the velocity of Ball 2 at the time point after the collision.

Considering the constraint condition: the length of the string remains unchanged, it can be seen that the components of the velocity of the two balls are equal in the tangential direction of the string at the time point after the collision:

$$v_{1x}\cos\theta - v_{1y}\sin\theta = v_{2x}\cos\theta \quad (3)$$

Combining (1), (2) and (3),

$$\Rightarrow \begin{cases} I = \dfrac{m_1 m_2 \cos\theta}{m_1\cos^2\theta + m_2} v_0 & (4) \\ \begin{cases} v_{1x} = \dfrac{m_1\cos^2\theta + m_2\sin^2\theta}{m_1\cos^2\theta + m_2} v_0 & (5a) \\ v_{1y} = \dfrac{m_2\sin\theta\cos\theta}{m_1\cos^2\theta + m_2} v_0 & (5b) \\ v_{2x} = \dfrac{m_1\cos^2\theta}{m_1\cos^2\theta + m_2} v_0 & (5c) \end{cases} \end{cases}$$

Solution [2] to Problem (1)

Taking *the system composed of the two balls* as the object of study. Studying *the instantaneous process of the collision*. Considering (a) ***law of conservation of momentum*** [the ***i***-direction component of the resultant external force acting on the system is zero; therefore, the ***i***-direction component of the momentum of the system is conserved] (the ***i***-direction component equation), (b) ***law of conservation of angular momentum*** [taking the space point that coincides with Ball 2 as the reference point, the resultant external torque acting on the system is zero; therefore, the angular momentum of the system is conserved] (the ***k***-direction component equation):

$$\begin{cases} m_1 v_0 = m_1 v_{1x} + m_2 v_2 & (6a) \\ m_1 v_0 l\sin\theta = m_1 v_{1x} l\sin\theta + m_1 v_{1y} l\cos\theta & (6b) \end{cases}$$

Equation (5) can be obtained by combining Eqs. (3) and (6).

Solution to Problem (2)

As shown in Fig. 3, the unit vector of Ball 1 pointing to Ball 2 at the time point after the collision is denoted by $\boldsymbol{\tau}$.

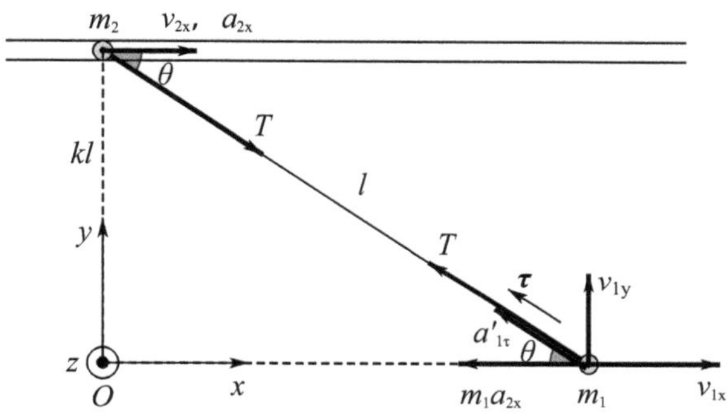

Figure 3

Taking *Ball 1 and Ball 2* as the object of study respectively. Studying *the state at the time point after the collision*. Considering ***Newton's second law*** [for Ball 1: in the reference frame of Ball 2, for Ball 2: in the ground reference frame] (for Ball 1: the $\boldsymbol{\tau}$-direction component equation, for Ball 2: the $\boldsymbol{i}$-direction component equation):

$$\begin{cases} T + m_1 a_{2x}\cos\theta = m_1 a'_{1\tau} & (7a) \\ T\cos\theta = m_2 a_{2x} & (7b) \end{cases}$$

T—the magnitude of the tension in the string at the time point after the collision.

$a'_{1\tau}$—the $\boldsymbol{\tau}$-direction component of the acceleration of Ball 1 relative to Ball 2 at the time point after the collision.

a_{2x}—the $\boldsymbol{i}$-direction component of the acceleration of Ball 2 at the time point after the collision.

Considering ***the radial component form of acceleration*** [the constraint condition: the length of the string remains unchanged]:

$$a'_{1\tau} = \frac{[(v_{1x}\sin\theta + v_{1y}\cos\theta) - v_{2x}\sin\theta]^2}{l} \tag{8}$$

$[(5) \to (8)] \to (7)$,

$$\Rightarrow \qquad T = \frac{m_1 v^2 \sin^2\theta}{\left(1 + \frac{m_1}{m_2}\cos^2\theta\right)^3 l} \qquad (9)$$

5.2.6 The Collisions between a Small Ball and the Inner Wall of a Free Homogeneous Ring (d, *B*)

A homogeneous ring of mass M with a small gap on it is statically placed on a smooth horizontal ground (see Fig. 1). A small ball of mass m enters the ring through the gap with the initial velocity of the magnitude v_0. Given that the inner wall of the ring is smooth, and the collision between the ball and the ring is completely elastic. It is assumed that the small ball can leave the ring through the gap after N collisions with the ring.

(1) Determine the initial velocity direction of the ball.
(2) Find the center-of-mass velocity of the ring and the velocity of the ball when the ball leaves the ring through the gap.

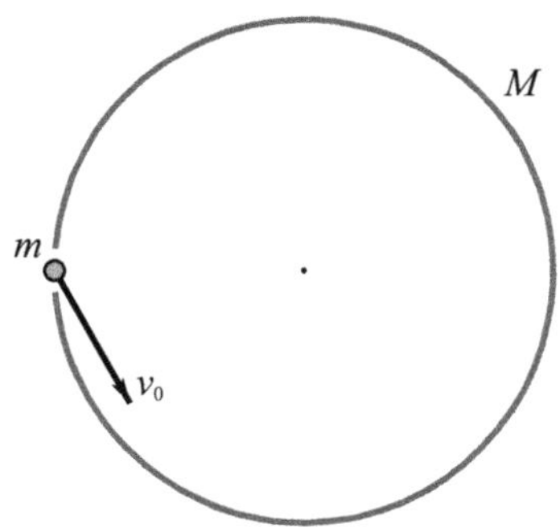

Figure 1

Solution to Problem (1)

The time point when the ball enters the ring is defined as Time zero.

As shown in Fig. 2, the center of the ring is is denoted by O, at Time zero, the point on the ground that coincides with the center O is denoted by O', the unit vector that the gap pointing to the center O is denoted by $\boldsymbol{i}$ (horizontally to the right in the figure), the unit vector in the vertical upward direction is denoted by $\boldsymbol{k}$, $\boldsymbol{k} \times \boldsymbol{i}$ is denoted by $\boldsymbol{j}$. Establishing the Cartesian coordinate system with the frame (O', $\boldsymbol{i}$, $\boldsymbol{j}$, $\boldsymbol{k}$), let x, y, z be the Cartesian coordinates.

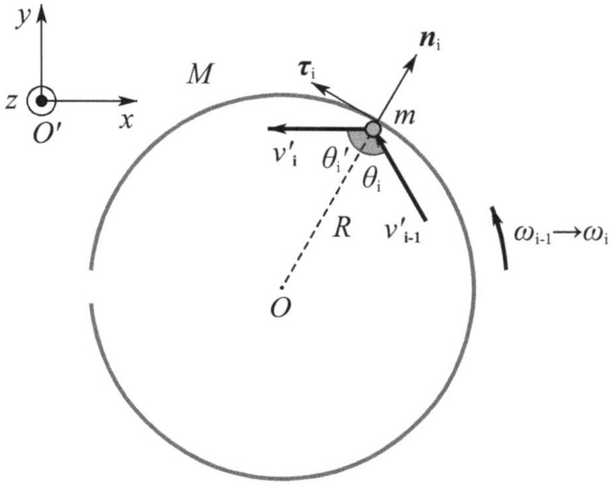

Figure 2

Taking *the ring* as the object of study. Studying *the instantaneous process of the n-th collision*. Considering (a) ***law of conservation of angular momentum*** [in its center-of-mass reference frame, taking the center O as the reference point, the resultant external torque acting on the ring is zero; therefore, the angular momentum of the ring is conserved] (the $\boldsymbol{k}$-direction component equation), (b) ***the form of angular momentum of a rigid body performing fixed-axis rotation*** [integrating into (a)], (c) ***the form of moment of inertia of a homogeneous ring about its central axis perpendicular to the plane of the ring***:

$$I\omega_{i-1} = I\omega_i \tag{1}$$

$$I = MR^2 \tag{2}$$

I—the moment of inertia of the ring about its central axis perpendicular to the plane of the ring.

ω_i—the $\boldsymbol{k}$-direction component of the rotating angular velocity of the ring at the time point after the i-th collision.

$\Rightarrow$

$$\omega_{i-1} = \omega_i \tag{3}$$

$$\Rightarrow \qquad \omega_i = \omega_0 = 0 \tag{4}$$

That is, the ring never rotates.

Introducing the tangential unit vector $\boldsymbol{\tau}_i$ and normal unit vector $\boldsymbol{n}_i$ of the ring at the collision point of the i-th collision.

Taking *the system composed of the ball and the ring* as the object of study. Studying *the instantaneous process of the i-th collision.* Considering (a) ***law of conservation of momentum*** [the resultant external force acting on the system is zero; therefore, the momentum of the system is conserved] (the $\boldsymbol{n}_i$-direction and $\boldsymbol{\tau}_i$-direction component equations), (b) ***law of conservation of mechanical energy*** [the collision between the ball and the ring is free and completely elastic; therefore, the mechanical energy of the system is conserved], (c) ***the total momentum of the system of particles is equal to the center-of-mass momentum*** [integrating into (a)], (d) ***König theorem*** [integrating into (b)], (e) ***the form of kinetic energy of a rigid body performing fixed-axis rotation*** [integrating into (d)].

$$\begin{cases} Mv_{M(i-1)n_i} + mv_{m(i-1)n_i} = Mv_{M(i)n_i} + mv_{m(i)n_i} & \text{(5a)} \\ Mv_{M(i-1)\tau_i} + mv_{m(i-1)\tau_i} = Mv_{M(i)\tau_i} + mv_{m(i)\tau_i} & \text{(5b)} \end{cases}$$

$$\left[\frac{1}{2}M(v_{M(i-1)n_i}^2 + v_{M(i-1)\tau_i}^2) + \frac{1}{2}I\omega_{i-1}^2\right] + \frac{1}{2}m(v_{m(i-1)n_i}^2 + v_{m(i-1)\tau_i}^2)$$

$$= \left[\frac{1}{2}M(v_{M(i)n_i}^2 + v_{M(i)\tau_i}^2) + \frac{1}{2}I\omega_i^2\right] + \frac{1}{2}m(v_{m(i)n_i}^2 + v_{m(i)\tau_i}^2) \tag{6}$$

$v_{M(i)ni}$ and $v_{M(i)\tau i}$—the $\boldsymbol{n}_i$-direction and $\boldsymbol{\tau}_i$-direction component of the center-of-mass velocity of the ring at the time point after the i-th collision.

$v_{m(i)ni}$ and $v_{m(i)\tau i}$—the $\boldsymbol{n}_i$-direction and $\boldsymbol{\tau}_i$-direction component of the velocity of the ball at the time point after the i-th collision.

Taking *the ball* as the object of study. Studying *the instantaneous process of the i-th collision.* Considering ***law of conservation of momentum*** [the $\boldsymbol{\tau}_i$-direction component of the resultant external force acting on the ball is zero; therefore, the $\boldsymbol{\tau}_i$-direction component of the momentum of the ball is conserved] (the $\boldsymbol{\tau}_i$-direction component equation):

$$mv_{m(i-1)\tau_i} = mv_{m(i)\tau_i} \tag{7}$$

Combining (4) to (7),

$$\Rightarrow \begin{cases} v_{M(i-1)\tau_i} = v_{M(i)\tau_i} & \text{(8a)} \\ v_{m(i-1)\tau_i} = v_{m(i)\tau_i} & \text{(8b)} \\ v_{m(i-1)n_i} - v_{M(i-1)n_i} = v_{M(i)n_i} - v_{m(i)n_i} & \text{(8c)} \end{cases}$$

Taking *the ball* as the object of study. Studying the state at the time point after the i-th collision. Considering ***theorem for composition of velocities of a particle*** [taking the ground as the static reference frame, and the center-of-mass reference frame of the ring as the moving reference frame] (the $\boldsymbol{n}_i$-direction and $\boldsymbol{\tau}_i$-direction component equations):

$$\begin{cases} v_{m(i)n_i} = v_{M(i)n_i} + v'_{m(i)n_i} & \text{(9a)} \\ v_{m(i)\tau_i} = v_{M(i)\tau_i} + v'_{m(i)\tau_i} & \text{(9b)} \end{cases}$$

$v'_{m(i)ni}$ and $v'_{m(i)\tau i}$—the $\boldsymbol{n}_i$-direction and $\boldsymbol{\tau}_i$-direction component of the velocity of the ball relative to the center of mass of the ring at the time point after the i-th collision.

Combining (8) and (9),

$$\begin{cases} v'_{m(i-1)n_i} = -v'_{m(i)n_i} & \text{(10a)} \\ v'_{m(i-1)\tau_i} = v'_{m(i)\tau_i} & \text{(10b)} \end{cases}$$

It can be seen that the speed of the ball in the center-of-mass reference frame of the ring remains constant.

$$v'_{m(i-1)} = v'_{m(i)} \tag{11}$$

$$\Rightarrow \quad v'_{m(i)} = v_0 \tag{12}$$

Defining θ_i and θ_i'—the incident angle and reflection angle of the ball's trajectory in the center-of-mass reference frame of the ring in the process of the i-th collision.

$$\begin{cases} \tan\theta_i = \dfrac{v'_{m(i-1)\tau}}{v'_{m(i-1)n}} & \text{(13a)} \\ \tan\theta_i' = \dfrac{v'_{m(i)\tau}}{-v'_{m(i)n}} & \text{(13b)} \end{cases}$$

Combining (10) and (13),

$$\Rightarrow \qquad \theta_i = \theta_i' \tag{14}$$

That is, the incident angle and reflection angle of the ball's trajectory in the center-of-mass reference frame of the ring are equal for an arbitrary collision process.

From the geometric relation, we can know that:

$$\Rightarrow \qquad \theta_{i-1} = \theta_i \tag{15}$$

$$\Rightarrow \qquad \theta_i = \theta_0 \tag{16}$$

There are N collisions between the ball and the ring, that is, the trajectory of the ball in the center-of-mass reference frame of the ring is a $(N + 1)$ - regular polygon. From the geometric relation, we can obtain:

$$\Rightarrow \qquad \theta_0 = \frac{\pi}{2} - \frac{\pi}{N+1} \tag{15}$$

Solution to Problem (2)

As shown in Fig. 3, taking *the system composed of the ball and the ring* as the object of study. Studying *the whole process from the ball entering the ring to the ball leaving the ring*. Considering ***law of conservation of momentum*** [the resultant external force acting on the system is zero; therefore, the momentum of the system is conserved] (the ***i***-direction and ***j***-direction component equations):

$$\begin{cases} mv_0 \cos\theta_0 = Mv_{Mx} + mv_{mx} & \text{(16a)} \\ -mv_0 \sin\theta_0 = Mv_{My} + mv_{my} & \text{(16b)} \end{cases}$$

v_{Mx} and v_{My}—the ***i***-direction and ***j***-direction component of the center-of-mass velocity of the ring at the time point when ball leaves the ring.

v_{mx} and v_{my}—the ***i***-direction and ***j***-direction component of the velocity of the ball at the time point when the ball leaves the ring.

Taking *the ball* as the object of study. Studying *the state at the time point when the ball leaves the ring*. Considering ***theorem for composition of velocities of a particle*** [taking the ground as the static reference frame, and the center-of-mass reference frame of the ring as the moving reference frame] (the ***i***-direction and ***j***-direction component equations):

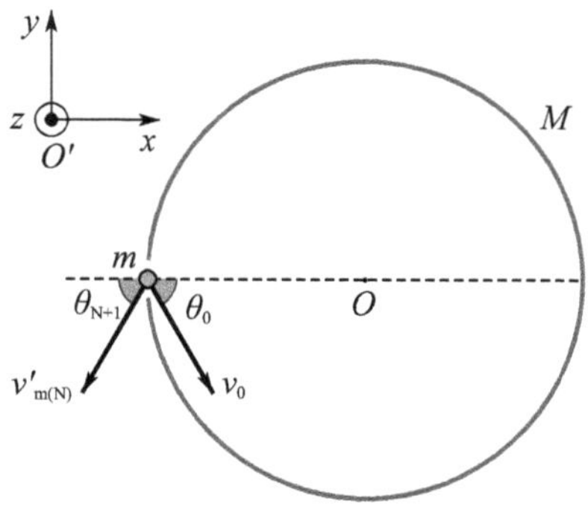

Figure 3

$$\Rightarrow \quad \begin{cases} v_{mx} = v_{Mx} - v'_{m(N)} \cos\theta_{N+1} = v_{Mx} - v_0 \cos\theta_0 & (17a) \\ v_{my} = v_{My} - v'_{m(N)} \sin\theta_{N+1} = v_{My} - v_0 \sin\theta_0 & (17b) \end{cases}$$

Combining (16) and (17),

$$\Rightarrow \quad \begin{cases} v_{Mx} = \dfrac{2m}{M+m} v_0 \sin\dfrac{\pi}{N+1} & (18a) \\ v_{My} = 0 & (18b) \\ v_{mx} = \dfrac{m-M}{M+m} v_0 \sin\dfrac{\pi}{N+1} & (18c) \\ v_{my} = -v_0 \cos\dfrac{\pi}{N+1} & (18d) \end{cases}$$

5.2.7 The Collision between Two Homogeneous Rods (c, C)[6]

Two homogeneous rods A and B of mass m_A and m_B and equal length l are placed on a smooth horizontal ground (see Fig. 1). Establishing a Cartesian coordinate system with the frame $(O, \mathbf{i}, \mathbf{j}, \mathbf{k})$, let x, y, z be the Cartesian coordinates. Initially, Rod A is at rest in the region $(-l + \varepsilon, \varepsilon)$ on the x-axis, ε is a small quantity, Rod B is parallel to x-axis and located in the region $(0, l)$, performing the translational motion with the center-of-mass velocity $v_0\mathbf{j}$. There is a completely elastic collision between the left end of Rod B and the right end of Rod A when Rod B reaches the x-axis. Find the center-of-mass velocity and rotating

[6]This problem is a test question of PRE-CUSPEA.

angular velocity of Rods A and B at the time point after the collision respectively.

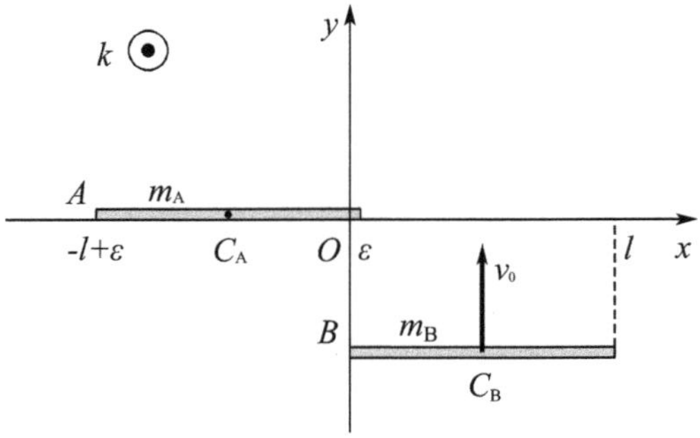

Figure 1

Solution

As shown in Fig. 2, taking *the system composed of Rods A and B* as the object of study. Studying *the instantaneous process of the collision*. Considering (a) ***law of conservation of momentum*** [the resultant external force acting on the system is zero; therefore, the momentum of the system is conserved] (the ***j***-direction component equation), (b) ***law of conservation of mechanical energy*** [the collision between Rod A and Rod B is a free and completely elastic collision; therefore, the mechanical energy of the system is conserved], (c) ***law of conservation of angular momentum*** [taking the origin O as the reference point, the resultant external torque acting on the system is zero; therefore, the angular momentum of the system is conserved] (the ***k***-direction component equation), (d) ***König theorem*** [integrating into (b)], (e) ***the form of kinetic energy of a rigid body performing fixed-axis rotation*** [integrating into (d)], (f) ***theorem for the composition of angular momentums based on center-of-mass reference frame*** [integrating into (c)], (g) ***the form of angular momentum of a rigid body performing fixed-axis rotation [integrating into*** (f)] :

$$\begin{cases} m_B v_0 = m_A v_A + m_B v_B & \text{(1a)} \\ \dfrac{1}{2} m_B v_0^2 = \left(\dfrac{1}{2} m_A v_A^2 + \dfrac{1}{2} I_A \omega_A^2\right) + \left(\dfrac{1}{2} m_B v_B^2 + \dfrac{1}{2} I_B \omega_B^2\right) & \text{(1b)} \\ m_B v_0 \dfrac{l}{2} = \left(-m_A v_A \dfrac{l}{2} + I_A \omega_A\right) + \left(m_B v_B \dfrac{l}{2} + I_B \omega_B\right) & \text{(1c)} \end{cases}$$

v_A and v_B—the ***j***-component of the center-of-mass velocity of Rods *A* and *B* at the time point after the collision.

ω_A and ω_B—the ***k***-component of the rotating angular velocity of Rods *A* and *B* at the time point after the collision.

I_A and I_B—the moment of inertia of Rods *A* and *B* about one of their central axes.

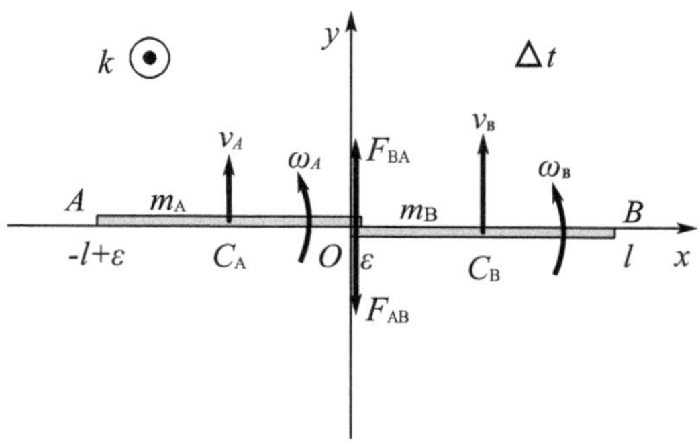

Figure 2

Taking *Rod A and Rod B* as the object of study respectively. Studying *the instantaneous process of the collision*. Considering (a) ***theorem of angular momentum*** [for each rod: in its own center-of-mass reference frame, taking its own center of mass as the reference point] (the ***k***-direction component equation), (b) ***the form of angular momentum of a rigid body performing fixed-axis rotation*** [integrating into (a)], (c) ***in the center-of-mass reference frame, the resultant torque of the inertia force acting on system of particles is zero***:

$$\begin{cases} \int_0^{\Delta t} F_{BA} \dfrac{l}{2} \mathrm{d}t = I_A \omega_A & \text{(2a)} \\ \int_0^{\Delta t} F_{AB} \dfrac{l}{2} \mathrm{d}t = I_B \omega_B & \text{(2b)} \end{cases}$$

Δt—the time taken of the collision process.

F_{BA}—the ***j***-direction component of the impact force exerted on Rod *A* by Rod *B* at arbitrary Time *t* in the collision process.

F_{AB}—the −***j***-direction component of the impact force exerted on Rod *B* by Rod *A* at arbitrary Time *t* in the collision process.

Considering ***Newton's third law***:

$$F_{BA} = F_{AB} \tag{3}$$

Considering ***the form of moment of inertia of a homogeneous rod about one of its central axes***:

$$\begin{cases} I_A = \dfrac{1}{12} m_A l^2 & \text{(4a)} \\ I_B = \dfrac{1}{12} m_B l^2 & \text{(4b)} \end{cases}$$

Combining (1) to (4),

$$\Rightarrow \begin{cases} \begin{cases} v_A = \dfrac{m_B}{2(m_A + m_B)} v_0 & \text{(5a)} \\ \omega_A = \dfrac{3m_B}{m_A + m_B} \dfrac{v_0}{l} & \text{(5b)} \end{cases} \\ \begin{cases} v_B = \dfrac{m_A + 2m_B}{2(m_A + m_B)} v_0 & \text{(6a)} \\ \omega_B = \dfrac{3m_A}{m_A + m_B} \dfrac{v_0}{l} & \text{(6b)} \end{cases} \end{cases}$$

5.2.8 The Collision between Two Homogeneous Disks (*c*, *C*)[7]

Two homogeneous disks A and B of equal mass and radius R_A and R_B are located on a smooth horizontal plane (see Fig. 1). Establishing a Cartesian coordinate system with the frame $(O, \boldsymbol{i}, \boldsymbol{j}, \boldsymbol{k})$, let x, y, z be the Cartesian coordinates. Initially, Disk B is at rest with its center C_B located at the origin O, Disk A performs translational motion with the center-of-mass velocity $v_0\boldsymbol{i}$, and the distance from its center C_A to the x-axis is b. There is a collision between the two disks. Given that in the normal direction of the collision point, the separation velocity magnitude of the two disks' center-of-mass is equal to the approaching velocity magnitude; the coefficient of friction between the two disks is large enough that the two disks will stop sliding relative to each other before the end of the collision. Find the center-of-mass velocity of the two disks after the collision.

[7]This problem is a test question of the 24th IPhO (International Physics Olympiad) in 1994.

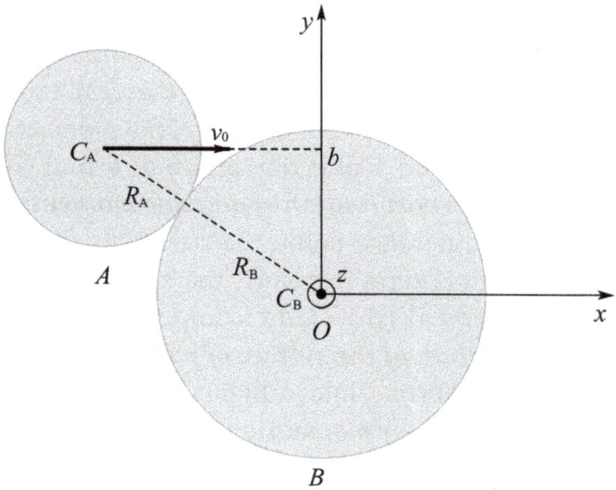

Figure 1

Solution

As shown in Fig. 2, when Disks A and B collide, the angle between $\overrightarrow{C_A C_B}$ and $\boldsymbol{i}$ is denoted by φ. The space point where Disks A and B collides is denoted by P, and the points on Disk A and Disk B coincident with Point P are denoted by P_A and P_B, respectively. Introducing the tangential unit vector $\boldsymbol{\tau}$ and normal unit vector $\boldsymbol{n}$ of Disk A at Point P.

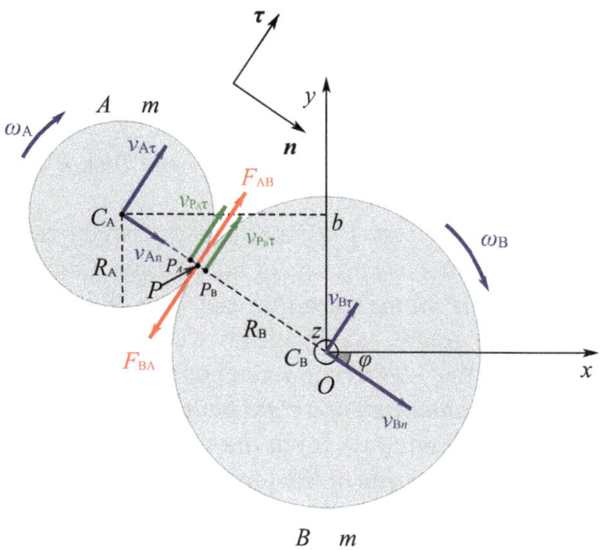

Figure 2

Taking *the system composed of Disks A and B* as the object of study. Studying *the instantaneous process of the collision*. Considering (a) ***law of conservation of momentum*** [the resultant external force acting on the system is zero; therefore, the momentum of the system is conserved] (the $\boldsymbol{\tau}$-direction and $\boldsymbol{n}$-direction component equations), (b) ***law of conservation of angular momentum*** [taking the origin O as the reference point, the resultant external torque acting on the system is zero; therefore, the angular momentum of the system is conserved] (the $-\boldsymbol{k}$-direction component equation), (c) ***the total momentum of the system of particles is equal to the center-of-mass momentum*** [integrating into (a)], (d) ***theorem for the composition of angular momentums based on center-of-mass reference frame*** [integrating into (b)], (e) ***the form of angular momentum of a rigid body performing fixed-axis rotation*** [integrating into (d)]:

$$\begin{cases} mv_0 \sin\varphi = mv_{A\tau} + mv_{B\tau} & (1a) \\ mv_0 \cos\varphi = mv_{An} + mv_{Bn} & (1b) \end{cases}$$

$$mv_0 b = [mv_{A\tau}(R_A + R_B) + I_A\omega_A] + I_B\omega_B \quad (2)$$

m—the mass of Disk A and Disk B.

$v_{A\tau}$ and v_{An}—the $\boldsymbol{\tau}$-direction and $\boldsymbol{n}$-direction component of the center-of-mass velocity of Disk A at the time point after the collision.

$v_{B\tau}$ and v_{Bn}—the $\boldsymbol{\tau}$-direction and $\boldsymbol{n}$-direction component of the center-of-mass velocity of Disk B at the time point after the collision.

ω_A and ω_B—the $-\boldsymbol{k}$-direction component of the rotating angular velocity of Disk A and Disk B at the time point after the collision.

I_A and I_B—the moment of inertia of Disk A and Disk B about their central axes parallel to the z-axis.

Taking *Disk A and Disk B* as the object of study respectively. Studying *the instantaneous process of the collision*. Considering (a) ***theorem of angular momentum*** [for each disk: in its own center-of-mass reference frame, taking its own center of mass as the reference point] (the $-\boldsymbol{k}$-direction component equation), (b) ***the form of angular momentum of a rigid body performing fixed-axis rotation*** [integrating into (a)], (c) ***in the center-of-mass reference frame, the resultant torque of the inertia force acting on system of particles is zero***:

$$\begin{cases} \int_0^{\Delta t} F_{BA}R_A \mathrm{d}t = I_A\omega_A & \text{(3a)} \\ \int_0^{\Delta t} F_{AB}R_B \mathrm{d}t = I_B\omega_B & \text{(3b)} \end{cases}$$

Δt—the time taken of the collision process.

F_{BA}—the $-\boldsymbol{\tau}$-direction component of the friction exerted on Disk *A* by Disk *B* at arbitrary Time *t* in the collision process.

F_{AB}—the $\boldsymbol{\tau}$-direction component of the friction exerted on Disk *B* by Disk *A* at arbitrary Time *t* in the collision process.

Considering ***Newton's third law***:

$$F_{BA} = F_{AB} \tag{4}$$

Considering ***the form of moment of inertia of a homogeneous disk about its central axis perpendicular to the plane of the disk***:

$$\begin{cases} I_A = \frac{1}{2}mR_A^2 & \text{(5a)} \\ I_B = \frac{1}{2}mR_B^2 & \text{(5b)} \end{cases}$$

Known that in the normal direction of the collision point, the separation velocity magnitude of the two disks' center-of-mass is equal to the approaching velocity magnitude:

$$v_{Bn} - v_{An} = v_0 \cos\varphi \tag{6}$$

Known that the two disks will stop sliding relative to each other before the end of the collision:

$$v_{P_A\tau} = v_{P_B\tau} \tag{7}$$

$v_{P_A\tau}$ and $v_{P_B\tau}$—the $\boldsymbol{\tau}$-direction component of the velocity of Point P_A and Point P_B at the time point after the collision.

Taking *Point P_A and Point P_B* as the object of study respectively. Studying *the state at the time point after the collision*. Considering (a) ***theorem for composition of velocities of a particle*** [for Point P_A: taking the ground as the static reference frame, and center-of-mass reference frame of Disk *A* as the moving reference frame; for Point P_B: taking the ground as the static reference frame, and the center-of-mass reference frame of Disk *B* as the moving reference frame] (the $\boldsymbol{\tau}$-direction component equation), (b) ***the transverse component form of velocity*** [integrating into (a)]:

$$v_{P_A\tau} = v_{A\tau} - \omega_A R_A \tag{8a}$$

$$v_{P_B\tau} = v_{B\tau} + \omega_B R_B \tag{8b}$$

$$\Rightarrow \quad v_{A\tau} - \omega_A R_A = v_{B\tau} + \omega_B R_B \tag{9}$$

Combining (1) to (5), (6) and (9),

$$\Rightarrow \quad \begin{cases} \begin{cases} v_{A\tau} = \dfrac{5}{6} v_0 \sin\varphi & (10a) \\ v_{An} = 0 & (10b) \end{cases} \\ \begin{cases} v_{B\tau} = \dfrac{1}{6} v_0 \sin\varphi & (11c) \\ v_{Bn} = v\cos\varphi & (11d) \end{cases} \end{cases}$$

Chapter 6

Fluid

6.1 Hydrostatics

The pressure distribution of static fluid in the potential field of mass forces can be described by the Euler equilibrium equation, and its important inference is Pascal's law. The problems of the force of the static fluid on the object immersed in it, and the shape of the isobaric surface of the static fluid (especially the free surface) are very basic and have practical value.

In this section, some classical hydrostatic problems will be studied.

6.1.1 The Ideal Gas in a Rotating Device (*c*, *C*)[1]

Consider a pipe device composed of a thin pipe and a thick pipe perpendicular to each other (see Fig. 1). Immerse the open end of the thin pipe vertically in a liquid, and seal off the open end of the thick pipe. Make the pipe device rotate at a constant angular velocity $\boldsymbol{\omega}$ with the thin pipe as the rotation axis. Given the atmospheric density is ρ_a, the liquid density is ρ_L, the length of the thick pipe is L, the system can keep constant temperature approximately. Find the rising height of liquid in the thin pipe when the system reaches stability.

[1]Yongde Zhang, et al. (1988). *American Physics Test Questions and Solutions* (University of Science and Technology of China Press).

Solving Physics Problems: Exploring New Thinking Paradigms
Yipeng Gu

ISBN 978-981-4877-41-1 (Hardcover), 978-1-003-16254-4 (eBook)
www.jennystanford.com

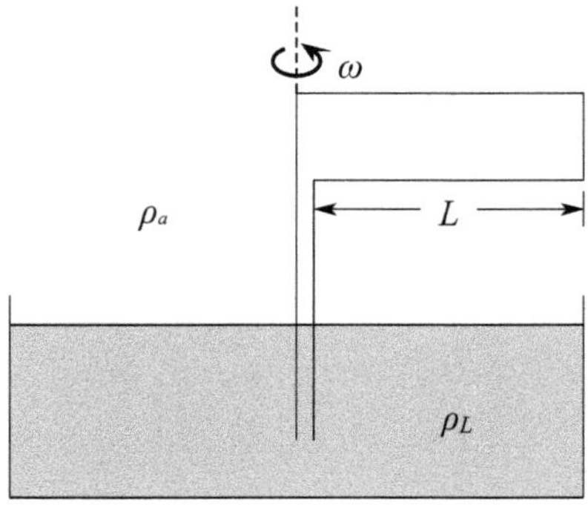

Figure 1

Solution

The time point when the system reaches stability is defined as Time zero.

As shown in Fig. 2, establishing the cylindrical coordinate system, let *r*, *φ*, *z* be the cylindrical coordinates.

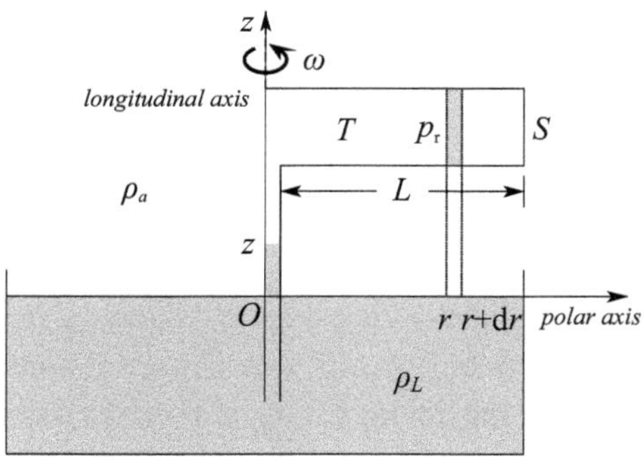

Figure 2

Taking *the disk-shaped air infinitesimal element between r and r + dr in the thick pipe (hereinafter referred to as element dr)* as the object of study. Studying *the state at arbitrary Time t.* Considering ***Newton's second law*** [in the rotating reference frame fixed to the pipe device] (the $\mathbf{r}^0$-direction component equation):

$$\rho_r S\mathrm{d}r \cdot \omega^2 r + p_r S - (p_r + \mathrm{d}p_r)S = 0 \tag{1}$$

ρ_r—the air density of element d*r* at Time *t*.

p_r—the air pressure of element d*r* at Time *t*.

S—the cross-sectional area of the thick pipe.

Taking *the external atmosphere and element dr* as the object of study respectively. Studying *the state at arbitrary Time t*. Considering ***the state equation of ideal gas***:

$$\begin{cases} p_a = \dfrac{\rho_a}{m} kT & \text{(2a)} \\ p_r = \dfrac{\rho_r}{m} kT & \text{(2b)} \end{cases}$$

m—the average molecular mass of the air.

T—the constant temperature of the system.

(2b) → (1),

$$\Rightarrow \quad \frac{dp_r}{p_r} = \frac{m\omega^2}{kT} r dr \tag{3}$$

$$\Rightarrow \quad \int_{p_0}^{p_r} \frac{dp}{p} = \frac{m\omega^2}{kT} \int_0^r r dr$$

$$\Rightarrow \quad \ln \frac{p_r}{p_0} = \frac{m\omega^2}{2kT} r^2$$

$$\Rightarrow \quad p_r = p_0 e^{\frac{m\omega^2}{2kT} r^2} \tag{4}$$

Taking *the air in the pipe device* as the object of study. Studying *the process from the initial state to the state when the system reaches stability*. Considering ***law of conservation of mass*** [the air mass in the thin pipe can be ignored]:

$$\rho_a SL = \int_0^L \rho_r S dr \tag{5}$$

(2) → (5),

$$\Rightarrow \quad p_a L = \int_0^L p_r dr \tag{6}$$

$$\Rightarrow p_a L = p_0 \int_0^L e^{\frac{m\omega^2}{2kT} r^2} dr \approx p_0 \int_0^L \left(1 + \frac{m\omega^2}{2kT} r^2\right) dr = p_0 L \left(1 + \frac{m\omega^2 L^2}{6kT}\right) \tag{7}$$

$$\Rightarrow \quad p_0 = \frac{p_a}{1 + \dfrac{m\omega^2 L^2}{6kT}} \approx p_a \left(1 - \frac{m\omega^2 L^2}{6kT}\right) \tag{8}$$

The operations in Eqs. (7) and (8) have considered $\frac{m\omega^2 L^2}{kT}$ as a small quantity and taken the first-order approximation.

Taking *the liquid* as the object of study. Studying *the state at arbitrary Time t*. Considering ***basic equation of hydrostatic pressure***:

$$p_a = p_0 + \rho_L g z \tag{9}$$

z—the z-coordinate of the liquid level in the thin pipe when the system reaches stability.

$$\Rightarrow \qquad z = \frac{p_a}{6\rho_L g} \cdot \frac{m\omega^2 L^2}{kT} \tag{10}$$

(2a) → (10),

$$\Rightarrow \qquad z = \frac{\rho_a \omega^2 L^2}{6\rho_L g} \tag{11}$$

6.1.2 The Liquid in a Rotating Glass (d, *C*)

A cylindrical glass of cross-sectional radius R is placed vertically, and a homogeneous liquid of height h is placed inside. The glass is rotated around its central axis at a constant angular velocity of the magnitude ω and the direction vertical upward. After a long enough time, the liquid reaches a steady state. Given that the glass is high enough that the liquid will not reach the rim of the glass when it reaches stability.

(1) Find the shape of the free surface of the liquid when it reaches the steady state (given that the bottom of the glass will not be exposed at this time).

(2) Find the condition for the bottom of the glass not to be exposed when the liquid reaches the steady state.

Solution [1] to Problem (1)

According to the symmetry, when the liquid reaches the steady state, its free surface is rotationally symmetric with respect to the central axis of the glass.

The time point when the liquid reaches the steady state is defined as Time zero.

As shown in Fig. 1, when the liquid reaches the steady state, the lowest point of its free surface is denoted by *O*, the unit vector in the vertical upward direction is denoted by $\boldsymbol{k}$. Establishing the cylindrical coordinate system (fixed to the glass reference frame) with Point *O* as the pole and the $\boldsymbol{k}$-direction as the positive direction of the longitudinal axis, let r, φ, z be the cylindrical coordinates. Taking an arbitrary point on the free surface of the liquid except the pole *O* and denoting it by *P*, and the tangent unit vector of the free surface of the liquid at that point is denoted by $\boldsymbol{\tau}$ (in the vertical plane and away from the pole *O*).

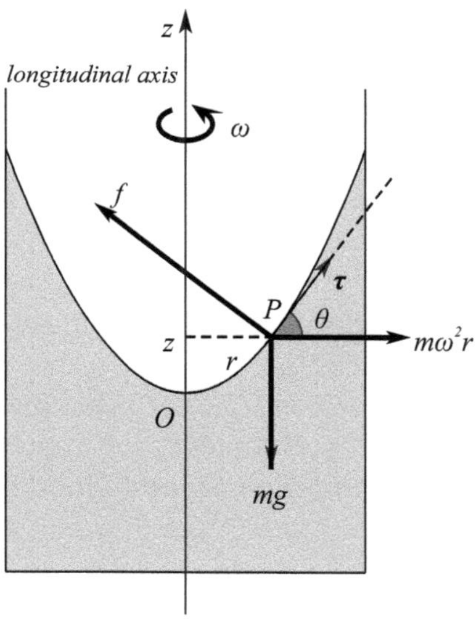

Figure 1

Taking *the liquid infinitesimal element of mass m at Point P* as the object of study. Studying *the state at arbitrary Time t*. Considering ***Newton's second law*** [in the glass reference frame] (the $\boldsymbol{\tau}$-direction component equation):

$$m\omega^2 r\cos\theta - mg\sin\theta = 0 \tag{1}$$

θ—the angle between $\boldsymbol{\tau}$ and $\boldsymbol{r}^0$ at Time t.

$$\Rightarrow \qquad \tan\theta = \frac{\omega^2 r}{g} \tag{2}$$

Considering ***the geometric meaning of derivative***:

$$\tan\theta = \frac{\mathrm{d}z}{\mathrm{d}r} \tag{3}$$

$$\Rightarrow \qquad \frac{\mathrm{d}z}{\mathrm{d}r} = \frac{\omega^2 r}{g} \tag{4}$$

$$\Rightarrow \qquad \mathrm{d}z = \frac{\omega^2 r}{g}\mathrm{d}r$$

$$\Rightarrow \qquad \int_0^z \mathrm{d}z = \frac{\omega^2}{g}\int_0^r r\mathrm{d}r$$

$$\Rightarrow \qquad z = \frac{\omega^2}{2g}r^2 \tag{5}$$

That is, when the liquid reaches the steady state, the shape of its free surface is a paraboloid of revolution.

Solution [2] to Problem (1)

When the liquid reaches the steady state, taking an arbitrary point in the liquid and denoting it by Q.

Taking *the liquid infinitesimal element of unit volume at Point Q* as the object of study (see Fig. 2). Studying *the state at arbitrary Time t*. Considering ***Euler equilibrium equation*** [in the glass reference frame]:

$$\boldsymbol{f} = \nabla p \tag{6}$$

$\boldsymbol{f}$—in the glass reference frame, the force density acting on the liquid at Point Q at Time t.

p—the liquid pressure at Point Q at Time t.

$$\boldsymbol{f} = \rho\omega^2 r\boldsymbol{r}^0 - \rho g\boldsymbol{k} \tag{7}$$

ρ—the density of the liquid.

Considering ***the gradient form in the cylindrical coordinate system***:

$$\nabla p = \frac{\partial p}{\partial r}\boldsymbol{r}^0 + \frac{1}{r}\frac{\partial p}{\partial \varphi}\boldsymbol{\varphi}^0 + \frac{\partial p}{\partial z}\boldsymbol{k} \tag{8}$$

$$\Rightarrow \qquad \rho\omega^2 r\boldsymbol{r}^0 - \rho g\boldsymbol{k} = \frac{\partial p}{\partial r}\boldsymbol{r}^0 + \frac{1}{r}\frac{\partial p}{\partial \varphi}\boldsymbol{\varphi}^0 + \frac{\partial p}{\partial z}\boldsymbol{k} \tag{9}$$

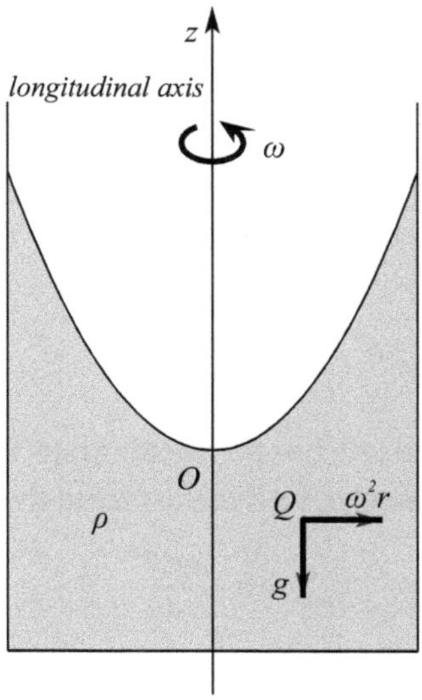

Figure 2

$$\Rightarrow \quad \begin{cases} \dfrac{\partial p}{\partial r} = \rho\omega^2 r & (10a) \\ \dfrac{\partial p}{\partial \varphi} = 0 & (10b) \\ \dfrac{\partial p}{\partial z} = -\rho g & (10c) \end{cases}$$

$$\Rightarrow \quad \mathrm{d}p = \frac{\partial p}{\partial r}\mathrm{d}r + \frac{1}{r}\frac{\partial p}{\partial \varphi}\mathrm{d}\varphi + \frac{\partial p}{\partial z}\mathrm{d}z = \rho\omega^2 r\mathrm{d}r - \rho g\mathrm{d}z \qquad (11)$$

$$\Rightarrow \quad \int_{p_0}^{p} \mathrm{d}p = \int_0^r \rho\omega^2 r\mathrm{d}r - \rho g\int_0^z \mathrm{d}z$$

$$\Rightarrow \quad p = p_0 + \frac{1}{2}\rho\omega^2 r^2 - \rho g z \qquad (12)$$

p_0—the atmospheric pressure at the free surface of the liquid.

Equation (12) is the curved surface equation of the isobaric surface with pressure p in the liquid.

$$\Rightarrow \quad z\big|_{p=p_0} = \frac{\omega^2}{2g}\left(r\big|_{p=p_0}\right)^2 \qquad (13)$$

Equation (13) is the surface equation of the liquid free surface, which is a paraboloid of revolution, the same as Eq. (5).

Solution to Problem (2)

Considering ***law of conservation of mass*** [the liquid is incompressible; therefore its volume is conserved] (see Fig. 3):

$$\pi R^2 h = \pi R^2 h_{\min} + \int_0^{\frac{\omega^2}{2g}R^2} (\pi R^2 - \pi r^2)\mathrm{d}z$$

$$= \pi R^2 h_{\min} + \frac{\pi \omega^2 R^4}{4g} \tag{14}$$

$h_{\min}$—when the liquid reaches the steady state, the distance from the lowest point (the center) of its free surface to the bottom of the glass.

$$\Rightarrow \qquad h_{\min} = h - \frac{\omega^2 R^2}{4g} \tag{15}$$

The condition for the bottom of the glass not to be exposed when the liquid reaches the steady state is as follows:

$$h_{\min} > 0 \tag{16}$$

$$\Rightarrow \qquad \omega < \frac{2\sqrt{gh}}{R} \tag{17}$$

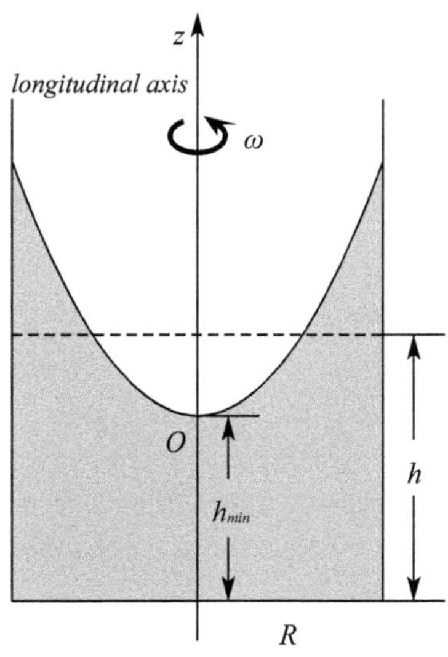

Figure 3

6.1.3 A Rotating Liquid Planet (d, *C*)

Consider a planet performing rotation about its central axis at a very small angular speed. In order to determine the surface shape of the planet, the planet can be roughly regarded as a planet made of a homogeneous incompressible fluid. Given its mass is M, density is ρ, and rotating angular speed is ω. When the planet reaches stability, the radius of its equatorial great-circle is R_e, and the distance between its two poles and its center is R_p, given that $\omega \ll \sqrt{GM/R_e^3}$, $R_e - R_p \ll R_e$.

(1) Find the surface equation of the planet and find the difference between R_e and R_p.
(2) Find the fluid pressure at an arbitrary point near the planet's surface.

Solution [1] to Problem (1)

Time zero is arbitrarily set.

It can be seen from the symmetry that the planet's surface is a rotating surface with its rotation axis as the axis of symmetry.

The rotating reference frame fixed to the planet is denoted by S'-system.

As shown in Fig. 1, the center of the planet is denoted by O, taking an arbitrary point on the equatorial great-circle and denoting it by N, the unit vector in $\overrightarrow{ON}$ direction is denoted by $\boldsymbol{i}$, the unit vector in the planet's rotating angular velocity direction is denoted by $\boldsymbol{j}$. Establishing the Cartesian coordinate system with the frame $(O, \boldsymbol{i}, \boldsymbol{j})$ fixed to S'-system, let x, y be the Cartesian coordinates.

In the O-xy plane, taking an arbitrary point on the planet's surface and denoting it by P, the tangential direction of the planet's surface away from the x-axis at that point is denoted by $\boldsymbol{\tau}$.

Taking a point near Point P in the planet and denoting it by P', defining $\boldsymbol{F}_G$ and $\boldsymbol{F}_c$—the universal gravitation and inertial centrifugal force (in S'-system) acting on the fluid infinitesimal element of unit mass at Point P' at Time t. Considering (a) ***law of universal gravitation***, (b) ***theorem of inertial centrifugal force***:

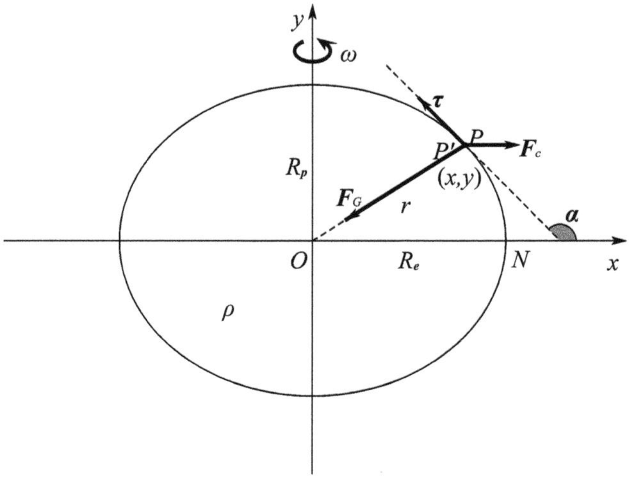

Figure 1

$$\begin{cases} \boldsymbol{F}_G \approx -G\dfrac{M}{r^2}\dfrac{\boldsymbol{r}}{r} = -G\dfrac{Mx}{r^3}\boldsymbol{i} - G\dfrac{My}{r^3}\boldsymbol{j} & \text{(1a)} \\ \boldsymbol{F}_c = \omega^2 x\boldsymbol{i} & \text{(1b)} \end{cases}$$

Taking *the fluid infinitesimal element of unit mass at Point P′* as the object of study. Studying *the state at arbitrary Time t.* Considering ***Newton's second law*** [in S'-system] (the $\boldsymbol{\tau}$-direction component equation):

$$(F_{Gx} + F_c)\cos\alpha + F_{Gy}\sin(\pi - \alpha) = 0 \tag{2}$$

α—the angle between $\boldsymbol{\tau}$ and $\boldsymbol{i}$.

$$\Rightarrow \qquad \tan\alpha = -\frac{F_{Gx} + F_c}{F_{Gy}} = -\left(1 - \frac{\omega^2 r^3}{GM}\right)\frac{x}{y} \tag{3}$$

Considering the geometric meaning of derivative:

$$\tan\alpha = \frac{\mathrm{d}y}{\mathrm{d}x} \tag{4}$$

$$\Rightarrow \qquad \left(1 - \frac{\omega^2 r^3}{GM}\right)x\mathrm{d}x + y\mathrm{d}y = 0 \tag{5}$$

Integrating the equation above under the following approximation:

$$r \approx R_e \tag{6}$$

$$\Rightarrow \qquad \int\left(1-\frac{\omega^2 R_e^3}{GM}\right)x\mathrm{d}x+\int y\mathrm{d}y=0$$

$$\Rightarrow \qquad \left(1-\frac{\omega^2 R_e^3}{GM}\right)x^2+y^2=C \tag{7}$$

C—the integral constant.

Considering the boundary conditions:

$$\begin{cases} y\big|_{x=0}=R_p & \text{(8a)} \\ x\big|_{y=0}=R_e & \text{(8b)} \end{cases}$$

(7) $\rightarrow$ (8b),

$$C=\left(1-\frac{\omega^2 R_e^3}{GM}\right)R_e^2 \tag{9}$$

$$\Rightarrow \qquad \frac{x^2}{R_e^2}+\frac{y^2}{\left(1-\frac{\omega^2 R_e^3}{GM}\right)R_e^2}=1 \tag{10}$$

That is, the section of the *O-xy* plane on the planet is an ellipse with a small eccentricity; therefore, the planet's surface is a rotating ellipsoid with a small eccentricity.

(10) $\rightarrow$ (8a),

$$\Rightarrow \qquad R_p^2=\left(1-\frac{\omega^2 R_e^3}{GM}\right)R_e^2 \tag{11}$$

Defining ε—the difference between R_e and R_p.

$$\varepsilon=R_e-R_p \tag{12}$$

$$\Rightarrow \qquad 1=\left(1-\frac{\omega^2 R_e^3}{GM}\right)\frac{R_e^2}{(R_e-\varepsilon)^2}=\left(1-\frac{\omega^2 R_e^3}{GM}\right)\left(1-\frac{\varepsilon}{R_e}\right)^{-2}$$

$$\approx\left(1-\frac{\omega^2 R_e^3}{GM}\right)\left(1+\frac{2\varepsilon}{R_e}\right)\approx 1-\frac{\omega^2 R_e^3}{GM}+\frac{2\varepsilon}{R_e} \tag{13}$$

This operation has considered $\omega \ll \sqrt{GM/R_e^3}$ and $\varepsilon \ll R_e$, and taken the first-order approximation.

$$\Rightarrow \qquad \varepsilon = \frac{\omega^2 R_e^4}{2GM} \ll R_e \tag{14}$$

Solution [2] to Problem (1)

For the fluid infinitesimal element of unit mass at Point P' at Time t, defining $U_G(x, y)$—the universal gravitational potential energy (taking infinity as the potential energy zero), $U_c(x, y)$—the potential energy of inertial centrifugal force (in S'-system, taking the origin O as the potential energy zero) and $U(x, y)$—the total potential energy.

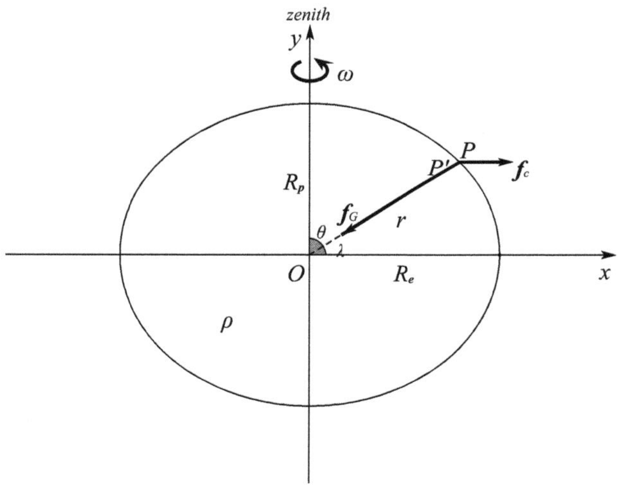

Figure 2

Considering ***the form of universal gravitational potential of a homogeneous sphere at an arbitrary point in space*** [the planet can be approximated as a homogeneous sphere]:

$$U_G(x,y) \approx -G\frac{M}{r} \tag{15}$$

Considering the definition of ***potential energy***:

$$U_c(x,y) = \int \boldsymbol{F}_c \cdot \mathrm{d}\boldsymbol{r} = \omega^2 \int_x^0 x\mathrm{d}x = -\frac{1}{2}\omega^2 x^2 \tag{16}$$

$$\Rightarrow \qquad U(x,y) = U_G(x,y) + U_c(x,y) = -G\frac{M}{r} - \frac{1}{2}\omega^2 x^2 \tag{17}$$

Considering ***principle of minimum potential energy***, we can know that the planet's surface is an equipotential surface:

$$U(x,y)=\text{const} \tag{18}$$

$$\Rightarrow \qquad -G\frac{M}{r}-\frac{1}{2}\omega^2x^2=\text{const} \tag{19}$$

(19) → (8),

$$\Rightarrow \qquad -G\frac{M}{R_p}=-G\frac{M}{R_e}-\frac{1}{2}\omega^2R_e^2 \tag{20}$$

(12) → (20),

$$\Rightarrow \qquad 0=\frac{1}{R_e-\varepsilon}-\frac{1}{R_e}-\frac{\omega^2R_e^2}{2GM}=\frac{1}{R_e}\left[\left(1-\frac{\varepsilon}{R_e}\right)^{-1}-1\right]-\frac{\omega^2R_e^2}{2GM}$$

$$\approx\frac{1}{R_e}\left[\left(1+\frac{\varepsilon}{R_e}\right)-1\right]-\frac{\omega^2R_e^2}{2GM}=\frac{\varepsilon}{R_e^2}-\frac{\omega^2R_e^2}{2GM} \tag{21}$$

This operation has considered $\varepsilon \ll R_e$ and taken the first-order approximation.

$$\Rightarrow \qquad \varepsilon=\frac{\omega^2R_e^4}{2GM} \tag{22}$$

Solution to Problem (2)

Taking the ***j***-direction as the zenith direction, the ***i***-direction as the reference direction on the reference plane, establishing the spherical coordinate system, let r, θ, φ be the spherical coordinates.

Taking *the fluid infinitesimal element of unit volume at Point P′* as the object of study. Studying *the state at arbitrary Time t.* Considering ***Euler equilibrium equation*** [in S'-system]:

$$\boldsymbol{f}=\nabla p \tag{23}$$

$\boldsymbol{f}$—in S'-system, the force density acting on the fluid at Point P' at Time t.

p—the fluid pressure at Point P' at Time t.

Defining $\boldsymbol{f}_{\mathrm{G}}$ and $\boldsymbol{f}_{\mathrm{c}}$—the force density of universal gravitation and inertial centrifugal force acting on the fluid at Point P' at Time t.

From (1),

$$\begin{cases} \boldsymbol{f}_G \approx -G\dfrac{M\rho}{r^2}\boldsymbol{r}^0 & (24\text{a}) \\ \boldsymbol{f}_c = \rho\omega^2 r\cos\lambda(\cos\lambda\boldsymbol{r}^0 + \sin\lambda\boldsymbol{\theta}^0) & (24\text{b}) \end{cases}$$

λ—the latitude of Point P.

$$\Rightarrow \quad \boldsymbol{f} = \boldsymbol{f}_G + \boldsymbol{f}_c = \rho\left(-G\frac{M}{r^2} + \omega^2 r\cos^2\lambda\right)\boldsymbol{r}^0 + \rho\omega^2 r\sin\lambda\cos\lambda\boldsymbol{\theta} \qquad (25)$$

Considering the gradient form in the spherical coordinate system:

$$\nabla p = \frac{\partial p}{\partial r}\boldsymbol{r}^0 + \frac{1}{r}\frac{\partial p}{\partial \theta}\boldsymbol{\theta}^0 + \frac{1}{r\sin\theta}\frac{\partial p}{\partial \varphi}\boldsymbol{\varphi}^0 \qquad (26)$$

$$\Rightarrow \quad \rho\left(-G\frac{M}{r^2} + \omega^2 r\cos^2\lambda\right)\boldsymbol{r}^0 + \rho\omega^2 r\sin\lambda\cos\lambda\boldsymbol{\theta}^0$$

$$= \frac{\partial p}{\partial r}\boldsymbol{r}^0 + \frac{1}{r}\frac{\partial p}{\partial \theta}\boldsymbol{\theta}^0 + \frac{1}{r\sin\theta}\frac{\partial p}{\partial \varphi}\boldsymbol{\varphi}^0 \qquad (27)$$

$$\Rightarrow \quad \begin{cases} \dfrac{\partial p}{\partial r} = \rho\left(-G\dfrac{M}{r^2} + \omega^2 r\cos^2\lambda\right) & (28\text{a}) \\ \dfrac{\partial p}{\partial \lambda} = -\dfrac{\partial p}{\partial \theta} = -\rho\omega^2 r\sin\lambda\cos\lambda & (28\text{b}) \\ \dfrac{\partial p}{\partial \varphi} = 0 & (28\text{c}) \end{cases}$$

$$\Rightarrow \quad \mathrm{d}p = \frac{\partial p}{\partial r}\mathrm{d}r + \frac{\partial p}{\partial \lambda}\mathrm{d}\lambda + \frac{\partial p}{\partial \varphi}\mathrm{d}\varphi$$

$$= \rho\left(-G\frac{M}{r^2} + \omega^2 r\cos^2\lambda\right)\mathrm{d}r - \rho\omega^2 r\sin\lambda\cos\lambda\mathrm{d}\lambda \qquad (29)$$

Considering the following differential relation:

$$\frac{\partial}{\partial \lambda}\left[\rho\left(-G\frac{M}{r^2} + \omega^2 r\cos^2\lambda\right)\right] = \frac{\partial}{\partial r}(-\rho\omega^2 r\sin\lambda\cos\lambda) \qquad (30)$$

That is, Eq. (29) is an exact equation.

$$\Rightarrow \quad \mathrm{d}p = \mathrm{d}\left(G\frac{\rho M}{r} + \frac{1}{2}\rho\omega^2 r^2\cos^2\lambda\right) \qquad (31)$$

$$\Rightarrow \qquad p = G\frac{\rho M}{r} + \frac{1}{2}\rho\omega^2 r^2 \cos^2\lambda + C_1 \tag{32}$$

C_1—the integral constant.

Considering the boundary condition:

$$p\big|_{r=R_e,\lambda=0} = 0 \tag{33}$$

$$\Rightarrow \qquad C_1 = -G\frac{\rho M}{R_e} - \frac{1}{2}\rho\omega^2 R_e^2 \tag{34}$$

$$\Rightarrow \qquad p = G\rho M\left(\frac{1}{r} - \frac{1}{R_e}\right) + \frac{1}{2}\rho\omega^2(r^2\cos^2\lambda - R_e^2) \tag{35}$$

Defining h—the depth of an arbitrary point near Point P.

$$r = R_e - h \tag{36}$$

$$\Rightarrow \; p = G\rho M\frac{1}{R_e}\left[\left(1 - \frac{h}{R_e}\right)^{-1} - 1\right] + \frac{1}{2}\rho\omega^2 R_e^2\left[\left(1 - \frac{h}{R_e}\right)^2 \cos^2\lambda - 1\right]$$

$$\approx G\rho M\frac{1}{R_e}\left[\left(1 + \frac{h}{R_e}\right) - 1\right] + \frac{1}{2}\rho\omega^2 R_e^2\left[\left(1 - \frac{2h}{R_e}\right)\cos^2\lambda - 1\right]$$

$$= G\rho M\frac{h}{R_e^2} + \frac{1}{2}\rho\omega^2 R_e^2\left[\left(1 - \frac{2h}{R_e}\right)\cos^2\lambda - 1\right] \tag{37}$$

This operation has considered $h \ll R_e$ and taken the first-order approximation.

6.1.4 A Soap Water Film between Two Rings (*c*, *C*)[2]

Two rings of equal radius R are coaxially placed, with the distance d. Between the two rings there is a soap water film. Find the shape of the soap water film when it is in mechanical equilibrium.

Solution

The mechanical equilibrium state of the soap water film is called the equilibrium state for short.

As shown in Fig. 1, the two rings are denoted by Ring A and Ring B, respectively, and their centers are denoted by O_A and O_B, respectively.

[2]This problem is a test question from MIT in 1983. It is also included in the following book: Yongde Zhang, et al. (1988). *American Physics Test Questions and Solutions* (University of Science and Technology of China Press).

The midpoint of O_AO_B is denoted by O. Establishing the cylindrical coordinate system, let r, φ, z be the cylindrical coordinates.

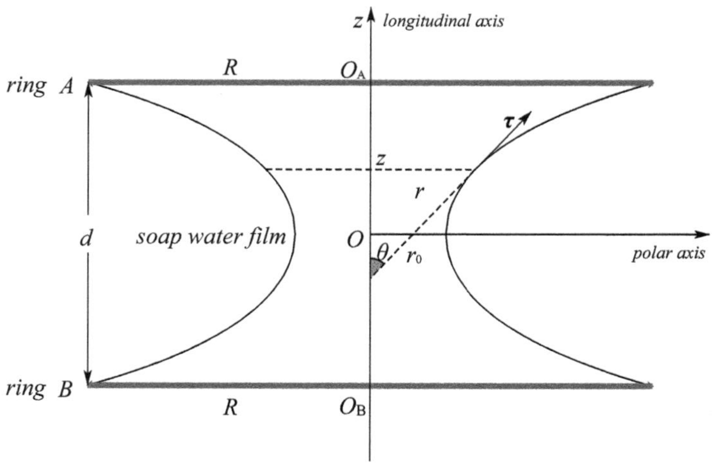

Figure 1

It can be seen from the symmetry that in the equilibrium state, the intersecting line of the soap water film and an arbitrary plane perpendicular to the z-axis is a circle, which is denoted by the z-circle.

Taking *the annulus-form infinitesimal element of the soap water film between the z-circle and the z + dz-circle* as the object of study. Studying *the equilibrium state*. Considering ***Newton's second law*** [the $\boldsymbol{k}$-direction component equation]:

$$f_{z+\mathrm{d}z} - f_z = 0 \tag{1}$$

f_z—the $\boldsymbol{k}$-direction component of the surface tension exerted on the soap water below the z-circle by the soap water above the z-circle.

Taking *the soap water on both sides of an arbitrary* ***dl*** *line element on the z-circle [hereinafter referred to as element dl]* as the object of study. Studying *the equilibrium state*. Considering ***surface tension theorem*** (need to consider both the inner and outer surfaces):

$$\mathrm{d}\boldsymbol{f} = 2\sigma \mathrm{d}l\,\boldsymbol{\tau} \tag{2}$$

$\mathrm{d}\boldsymbol{f}$—the surface tension exerted on the soap water below element $\mathrm{d}\boldsymbol{l}$ by the soap water above element $\mathrm{d}\boldsymbol{l}$.

$\boldsymbol{\tau}$—the unit vector in the tangent direction of the soap water film at element d$\boldsymbol{l}$ (on the $\boldsymbol{k}$ side, perpendicular to the element d$\boldsymbol{l}$).

$$\Rightarrow \quad f_z = \oint_L \mathrm{d}\boldsymbol{f} \cdot \boldsymbol{k} = 2\sigma\cos\theta \oint_L \mathrm{d}l = 2\sigma\cos\theta \cdot 2\pi r = 4\pi\sigma r\cos\theta \tag{3}$$

θ—the angle between $\boldsymbol{\tau}$ and $\boldsymbol{k}$.

r—the radius of the z-circle.

(3) → (1),

$$\Rightarrow \quad 0 = (r\cos\theta)\big|_{z+\mathrm{d}z} - (r\cos\theta)\big|_z = \frac{\partial(r\cos\theta)}{\partial z}\mathrm{d}z \tag{4}$$

$$\Rightarrow \quad \frac{\partial(r\cos\theta)}{\partial z} = 0 \tag{5}$$

$$\Rightarrow \quad r\cos\theta = \text{const} \tag{6}$$

From the symmetry we can see that:

$$\theta\big|_{z=0} = 0 \tag{7}$$

Defining r_0—the radius of the 0-circle.

$$\Rightarrow \quad r\cos\theta = r\big|_{z=0} = r_0 \tag{8}$$

Considering the geometric meaning of derivative:

$$\frac{\mathrm{d}r}{\mathrm{d}z} = \tan\theta \tag{9}$$

Combining (8) and (9),

$$\Rightarrow \quad \frac{r}{\sqrt{1+\left(\frac{\mathrm{d}r}{\mathrm{d}z}\right)^2}} = r_0 \tag{10}$$

It is assumed that the shape of the soap water film is the thinnest in the middle and thicker on both sides, that is:

$$\left.\frac{\mathrm{d}r}{\mathrm{d}z}\right|_{z>0} > 0 \tag{11}$$

In the region $z > 0$, Eq. (10) can be transformed into:

$$\Rightarrow \quad \mathrm{d}z = \frac{\mathrm{d}r}{\sqrt{\frac{r^2}{r_0^2} - 1}} \tag{12}$$

$$\Rightarrow \quad z=\int_0^z \mathrm{d}z=\int_{r_0}^{r}\frac{\mathrm{d}r}{\sqrt{\frac{r^2}{r_0^2}-1}}=r_0\ln\left|\frac{r}{r_0}+\sqrt{\frac{r}{r_0}-1}\right|=r_0\operatorname{arccosh}\frac{r}{r_0} \tag{13}$$

$$\Rightarrow \quad r=r_0\cosh\frac{z}{r_0} \tag{14}$$

Equation (14) is r as a function of z with r_0 as the parameter. Considering the boundary condition:

$$r\big|_{z=d/2}=R \tag{15}$$

$$\Rightarrow \quad r_0\cosh\frac{d}{2r_0}=R \tag{16}$$

Equation (16) is the transcendental equation satisfied by r_0.

From (14),

$$\Rightarrow \quad \left.\frac{\mathrm{d}r}{\mathrm{d}z}\right|_{z>0}=\left.\left(\sinh\frac{z}{r_0}\right)\right|_{z>0}>0 \tag{17}$$

That is, the assumption above holds.

6.1.5 The Contraction Process of a Soap Bubble (*a*, *A*)

Make a spherical soap bubble, in its initial state, its radius is R_0, and the pressure of its internal gas and the external atmosphere are both P_0. The surface tension coefficient of the soap water forming the soap bubble is α, $\alpha \ll P_0R_0$. The soap bubble will eventually reach the mechanical equilibrium state, and it is approximately assumed that the gas inside the soap bubble undergoes an isothermal change during this process. Find the heat absorbed by the gas inside the soap bubble from the outside in the whole process.

Solution [1]

Taking *the infinitesimal element of unit area at an arbitrary point on the soap bubble in the final state* as the object of study. Studying *the final state*. Considering ***Newton's second law*** (the component equation in the radial outward direction):

$$p\big|_{R=R_0+\Delta R} - (p_0 + p_\alpha\big|_{R=R_0+\Delta R}) = 0 \tag{1}$$

ΔR—the radius increment of the soap bubble in the whole process, $\Delta R > 0$ corresponds to the expansion of the soap bubble; $\Delta R < 0$ corresponds to the contraction of the soap bubble.

$p\,|_{R=R_0+\Delta R}$ —the pressure of the gas inside the soap bubble at the final state.

$p_\alpha|_{R=R_0+\Delta R}$ —the additional pressure of the surface tension of the soap bubble at the final state.

Considering ***Laplace's law*** (need to consider both the inner and outer surfaces):

$$p_\alpha\big|_R = 2\cdot\alpha\left(\frac{1}{R}+\frac{1}{R}\right) \tag{2}$$

(2) → (1),

$$\Rightarrow \qquad p\big|_{R=R_0+\Delta R} = p_0 + \frac{4\alpha}{R_0+\Delta R} \tag{3}$$

Taking *the gas inside the soap bubble* as the object of study. Studying *the whole process*. Considering ***Boyle–Mariotte's law***:

$$p_0\cdot\frac{4}{3}\pi R_0^3 = p\,|_{R=R_0+\Delta R}\cdot\frac{4}{3}\pi(R_0+\Delta R)^3 \tag{4}$$

(3) → (4),

$$\Rightarrow \qquad p_0R_0^3 = \left(p_0 + \frac{4\alpha}{R_0+\Delta R}\right)(R_0+\Delta R)^3$$

$$= p_0(R_0+\Delta R)^3 + 4\alpha(R_0+\Delta R)^2$$

$$= p_0R_0^3\left(1+\frac{\Delta R}{R_0}\right)^3 + 4\alpha R_0^2\left(1+\frac{\Delta R}{R_0}\right)^2$$

$$\approx p_0R_0^3\left(1+3\frac{\Delta R}{R_0}\right) + 4\alpha R_0^2\left(1+2\frac{\Delta R}{R_0}\right)$$

$$= p_0R_0^3 + 4\alpha R_0^2 + (3p_0R_0^2 + 8\alpha R_0)\Delta R \tag{5}$$

$$\Rightarrow \qquad \Delta R = -\frac{4\alpha R_0}{3p_0R_0 + 8\alpha} = -\frac{4\alpha}{3p_0\left(1+\dfrac{8\alpha}{3p_0R_0}\right)}$$

$$\approx -\frac{4\alpha}{3p_0}\left(1-\frac{8\alpha}{3p_0R_0}\right) \approx -\frac{4\alpha}{3p_0} \ll R \tag{6}$$

Taking *the system composed the soap bubble and the gas inside it* as the object of study. Studying *the whole process.* Considering ***principle of energy-work*** [the energy contained in the system are the internal energy of the gas and the surface energy of the soap bubble, the work done by the external atmosphere and the heat absorbed by the system from the outside change the energy of the system (the constraint condition: the gas undergoes an isothermal change; therefore, its internal energy is constant)]:

$$p_0\left[\frac{4}{3}\pi R_0^3 - \frac{4}{3}\pi(R_0+\Delta R)^3\right] + Q = E_{\alpha 2} - E_{\alpha 1} \tag{7}$$

Q—the heat absorbed by the system composed of the soap bubble and the gas inside it from the outside.

$E_{\alpha 1}$ and $E_{\alpha 2}$—the surface energy of the soap bubble at the initial state and final state.

Considering ***surface energy theorem*** (need to consider both the inner and outer surfaces):

$$\begin{cases} E_{\alpha 1} = 2\cdot\alpha\cdot 4\pi R_0^2 & \text{(8a)} \\ E_{\alpha 2} = 2\cdot\alpha\cdot 4\pi (R_0+\Delta R)^2 & \text{(8b)} \end{cases}$$

Combining (3), (4), (6) to (8),

$$\Rightarrow \qquad Q = 8\pi\alpha(R_0+\Delta R)^2 - 8\pi\alpha R_0^2 - \frac{4\alpha}{R_0+\Delta R}\cdot\frac{4}{3}\pi(R_0+\Delta R)^3$$

$$= \frac{8}{3}\pi\alpha\left(R_0+\Delta R\right)^2 - 8\pi\alpha R_0^2$$

$$= \frac{8}{3}\pi\alpha R_0^2\left(1+\frac{\Delta R}{R_0}\right)^2 - 8\pi\alpha R_0^2$$

$$\approx \frac{8}{3}\pi\alpha R_0^2\left(1+2\frac{\Delta R}{R_0}\right) - 8\pi\alpha R_0^2$$

$$= -\frac{16}{3}\pi\alpha R_0^2 + \frac{16}{3}\pi\alpha R_0 \Delta R$$

$$\approx -\frac{16}{3}\pi\alpha R_0^2 \tag{9}$$

Solution [2]

Taking *the gas inside the soap bubble* as the object of study. Studying *the whole process.* Considering ***first law of thermodynamics***:

$$Q = W \tag{10}$$

W—the work done by the gas inside the soap bubble to the outside in the whole process.

Considering the definition of ***work***:

$$W = \int p \mathrm{d}V \tag{11}$$

p—the pressure of the gas inside the soap bubble at an arbitrary state.

V—the volume of the soap bubble at an arbitrary state.

Taking *the gas inside the soap bubble* as the object of study. Studying *the whole process.* Considering ***Boyle–Mariotte's law***:

$$p_0 V_0 = pV \tag{12}$$

V_0—the volume of the soap bubble at the initial state.

(6) → [(12) → (11)],

$$\Rightarrow \ W = p_0 V_0 \int_{V_0}^{V|_{R=R_0+\Delta R}} \frac{\mathrm{d}V}{V} = p_0 V_0 \ln \frac{V|_{R=R_0+\Delta R}}{V_0}$$

$$= p_0 \cdot \frac{4}{3}\pi R_0^3 \ln \frac{\frac{4}{3}\pi (R_0 + \Delta R)^3}{\frac{4}{3}\pi R_0^3} = p_0 \cdot \frac{4}{3}\pi R_0^3 \ln \left(1 + \frac{\Delta R}{R_0}\right)^3$$

$$= 4\pi p_0 R_0^3 \ln\left(1 + \frac{\Delta R}{R_0}\right) \approx 4\pi p_0 R_0^3 \cdot \frac{\Delta R}{R_0} = 4\pi p_0 R_0^2 \Delta R = -\frac{16}{3}\pi\alpha R_0^2 \tag{13}$$

(13) → (10),

$$\Rightarrow \qquad Q = -\frac{16}{3}\pi\alpha R_0^2 \qquad (14)$$

Solution [3]

Taking *the soap bubble* as the object of study. Studying *the whole process*. Considering ***principle of energy-work*** [the work done by the external atmosphere and the gas inside the soap bubble change the surface energy of the soap bubble]:

$$p_0\left[\frac{4}{3}\pi R_0^3 - \frac{4}{3}\pi(R_0 + \Delta R)^3\right] + W = E_{\alpha 2} - E_{\alpha 1} \qquad (15)$$

Q can be obtained by combining Eqs. (6), (8), (10) and (15).

Operations (5), (6), (9) and (13) have considered $\alpha \ll P_0 R_0$ and taken the first-order approximation.

6.1.6 The Rising of a Liquid along a Vertical Flat Plate (*b, AC*)

An infinite flat plate is partially and vertically immersed in a liquid with which it has a wetting effect. Given the contact angle between them is θ, the liquid's density is ρ, and the surface tension coefficient is α.

(1) Find the height of the liquid rising along this flat plate.
(2) Find the surface shape of the rising part of the liquid.

Solution to Problem (1) (type C)

As shown in Fig. 1, establishing the Cartesian coordinate system with the frame $(O, \boldsymbol{i}, \boldsymbol{j}, \boldsymbol{k})$, let x, y, z be the Cartesian coordinates. It can be seen from the symmetry that when the liquid reaches mechanical equilibrium, the shape of the liquid surface is translationally invariant with respect to the z-coordinate. Taking an arbitrary point on the liquid surface and denoting it by P.

Taking *the liquid infinitesimal element of unit area at Point P* as the object of study. Studying *the state when the liquid reaches the mechanical equilibrium*. Considering ***Newton's second law*** (the component equation in the normal outward direction of the liquid surface at Point P):

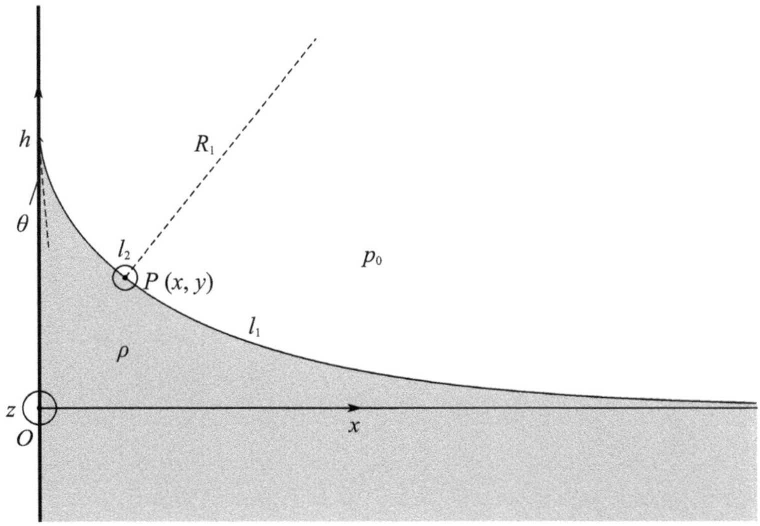

Figure 1

$$0=(p_0-\rho g y)-p_0+p_\alpha=p_\alpha-\rho g y \tag{1}$$

p_0—the atmospheric pressure.

p_α—the additional pressure of the liquid surface tension.

Considering ***Laplace's law*** [taking the directions perpendicular to the z-axis and parallel to the z-axis as the tangential directions of the two orthogonal curves, respectively, the two curves are denoted by l_1 and l_2, respectively]:

$$p_\alpha=\alpha\left(\frac{1}{R_1}+\frac{1}{R_2}\right) \tag{2}$$

R_1 and R_2—the radius of curvature of the curves l_1 and l_2.

Considering ***the form of the radius of curvature at an arbitrary point on the plane curve in Cartesian coordinate system***:

$$\begin{cases} R_1=\dfrac{\left[1+\left(\dfrac{\mathrm{d}y}{\mathrm{d}x}\right)^2\right]^{3/2}}{\dfrac{\mathrm{d}^2y}{\mathrm{d}x^2}} & \text{(3a)} \\ R_2=\infty & \text{(3b)} \end{cases}$$

Combining (1) to (3),

$$\Rightarrow \qquad \frac{\dfrac{d^2y}{dx^2}}{\left[1+\left(\dfrac{dy}{dx}\right)^2\right]^{3/2}}=\frac{\rho g}{\alpha}y \tag{4}$$

Taking the following differential transformation:

$$\frac{d^2y}{dx^2}=\frac{d}{dx}\left(\frac{dy}{dx}\right)=\frac{d}{dy}\left(\frac{dy}{dx}\right)\frac{dy}{dx} \tag{5}$$

Defining the variable u:

$$u=\frac{dy}{dx} \tag{6}$$

(5) and (6) → (4),

$$\frac{udu}{(1+u^2)^{3/2}}=\frac{\rho g}{\alpha}ydy \tag{7}$$

$$\Rightarrow \qquad \int_0^u \frac{udu}{(1+u^2)^{3/2}}=\frac{\rho g}{\alpha}\int_0^y ydy$$

$$\Rightarrow \qquad 1-\frac{1}{\sqrt{1+u^2}}=\frac{\rho g}{2\alpha}y^2$$

$$\Rightarrow \qquad u=-\sqrt{\frac{1}{\left(1-\dfrac{\rho g}{2\alpha}y^2\right)^2}-1}$$

$$\Rightarrow \qquad \frac{dy}{dx}=-\sqrt{\frac{1}{\left(1-\dfrac{\rho g}{2\alpha}y^2\right)^2}-1} \tag{8}$$

Defining h—the height of liquid rising along this flat plate.

Considering the boundary condition:

$$\left.\frac{dy}{dx}\right|_{y=h}=-\cot\theta \tag{9}$$

(8) → (9),

$$\Rightarrow \qquad h=\sqrt{\frac{2\alpha}{\rho g}(1-\sin\theta)} \tag{10}$$

Solution to Problem (2) (type A)

From (8),

$$\Rightarrow \qquad \mathrm{d}x = -\frac{\mathrm{d}y}{\sqrt{\dfrac{1}{\left(1-\dfrac{\rho g}{2\alpha}y^2\right)^2}-1}} \tag{11}$$

$$\Rightarrow \qquad \int_0^x \mathrm{d}x = -\int_h^y \frac{\mathrm{d}y}{\sqrt{\dfrac{1}{\left(1-\dfrac{\rho g}{2\alpha}y^2\right)^2}-1}} \tag{12}$$

Defining the variable φ:

$$1-\frac{\rho g}{2\alpha}y^2 = \cos\varphi \tag{13}$$

$$\Rightarrow \qquad \frac{\rho g}{2\alpha}y^2 = 1-\cos\varphi = 2\sin^2\frac{\varphi}{2}$$

$$\Rightarrow \qquad y = 2\sqrt{\frac{\alpha}{\rho g}}\sin\frac{\varphi}{2}$$

$$\Rightarrow \qquad \mathrm{d}y = \sqrt{\frac{\alpha}{\rho g}}\cos\frac{\varphi}{2}\mathrm{d}\varphi \tag{14}$$

(13) and (14) $\rightarrow$ (12),

$$\Rightarrow \qquad x = -\sqrt{\frac{\alpha}{\rho g}}\int_{\varphi(h)}^{\varphi(y)} \frac{\cos\frac{\varphi}{2}\mathrm{d}\varphi}{\tan\varphi}$$

$$= -\frac{1}{2}\sqrt{\frac{\alpha}{\rho g}}\int_{\varphi(h)}^{\varphi(y)}\left(\csc\frac{\varphi}{2}-2\sin\frac{\varphi}{2}\right)\mathrm{d}\varphi$$

$$= \sqrt{\frac{\alpha}{\rho g}}\left(\ln\left|\csc\frac{\varphi}{2}+\cot\frac{\varphi}{2}\right|-2\cos\frac{\varphi}{2}\right)\Bigg|_{\varphi(h)}^{\varphi(y)}$$

$$= \sqrt{\frac{\alpha}{\rho g}}\left(\ln\left|\csc\frac{\varphi}{2}+\sqrt{\csc^2\frac{\varphi}{2}-1}\right|-2\cos\frac{\varphi}{2}\right)\Bigg|_{\varphi(h)}^{\varphi(y)}$$

$$= \sqrt{\frac{\alpha}{\rho g}} \left(\operatorname{arccosh} \csc \frac{\varphi}{2} - 2\cos\frac{\varphi}{2} \right) \Bigg|_{\varphi(h)}^{\varphi(y)}$$

$$= \sqrt{\frac{\alpha}{\rho g}} \left[\begin{array}{l} \operatorname{arccosh} \csc \dfrac{\varphi(y)}{2} - \operatorname{arccosh} \csc \dfrac{\varphi(h)}{2} \\ -2\left(\cos\dfrac{\varphi(y)}{2} - \cos\dfrac{\varphi(h)}{2} \right) \end{array} \right]$$

$$= \sqrt{\frac{\alpha}{\rho g}} \left[\begin{array}{l} \operatorname{arccosh}\left(2\sqrt{\dfrac{\alpha}{\rho g}}\dfrac{1}{y} \right) - \operatorname{arccosh}\left(2\sqrt{\dfrac{\alpha}{\rho g}}\dfrac{1}{h} \right) \\ -2\left(\sqrt{1-\dfrac{1}{4}\dfrac{\rho g}{\alpha}y^2} - \sqrt{1-\dfrac{1}{4}\dfrac{\rho g}{\alpha}h^2} \right) \end{array} \right] \tag{15}$$

6.1.7 The Quasi-Static Motion of an Inverted Bucket in Water (*c*, *A*)[3]

Consider a bucket of mass m, height l_0, sectional area S and self volume V_0. Invert it at the bottom of a wide pool of depth H_0, with air of height l filled in it (see Fig. 1). Given the atmospheric pressure is p_0, the density of the water is ρ.

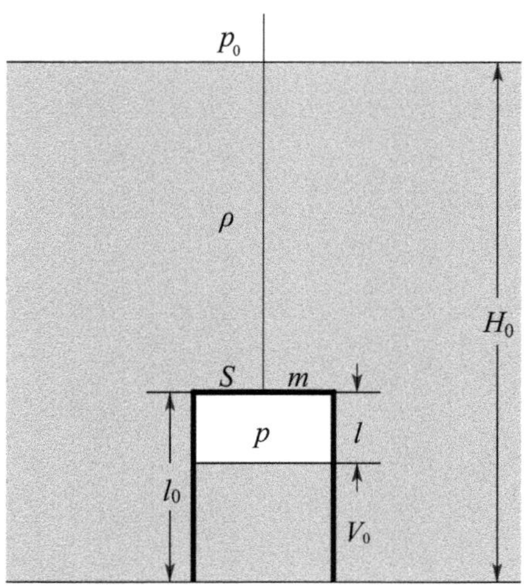

Figure 1

[3]This problem is a test question of the 7th CphO (China Physics Olympiads) in 1990.

Use a sling to lift the bucket slowly (which can be regarded as quasi-static process), so that the bucket reaches the water surface. Given that H_0 is large enough, when the bucket rises to a certain height, the magnitude of the tension in the sling will decrease to zero, and then the bucket will automatically float up.

Ignore the resistance of the water and the influence of saturated vapor pressure. The water temperature is uniform and constant. Find the variation of the mechanical energy of the bucket and the water (including the pool water and the water inside the bucket) in the process from the beginning of pulling the sling until the magnitude of the tension in the sling decreases to zero (only the facts needed to solve this problem need to be given, without calculation results).

Solution

The time point when start pulling the sling is denoted by Time zero.

As shown in Fig. 2, taking an arbitrary point on the pool bottom and denoting it by O, the unit vector in the vertical upward direction is denoted by $\boldsymbol{i}$. Establishing the linear coordinate system with Point O as the origin and the $\boldsymbol{i}$-direction as the positive direction of coordinate axis, let x be the coordinate.

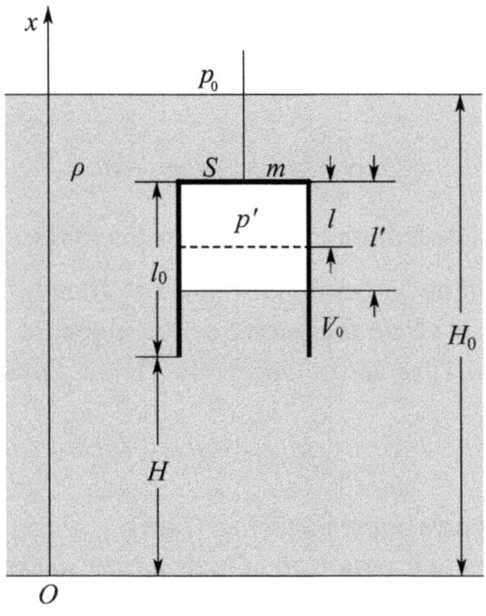

Figure 2

Taking *the bucket* as the object of study. Studying *the state at the time point when stop pulling the sling*. Considering ***theorem of motion of center of mass*** [the bucket is in mechanical equilibrium] (the ***i***-direction component equation):

$$\rho(V_0 + l'S)g - mg = 0 \tag{1}$$

l'—the height of the air inside the bucket at the time point when stop pulling the sling.

Taking *the air inside the bucket* as the object of study. Studying *the whole process of pulling the sling*. Considering ***Boyle–Mariotte's law***:

$$plS = p'l'S \tag{2}$$

p and p'—the pressure of the air inside the bucket at the time point when start pulling the sling and stop pulling the sling.

Considering ***basic equation of hydrostatic pressure***:

$$\begin{cases} p = p_0 + \rho g\left[H_0 - (l_0 - l)\right] & \text{(3a)} \\ p' = p_0 + \rho g\left[H_0 - H - (l_0 - l')\right] & \text{(3b)} \end{cases}$$

H — the rising height of the bucket in the whole process.

Taking *the bucket* as the object of study. Studying *the state at arbitrary Time t*. Considering ***theorem of motion of center of mass*** [the bucket is in mechanical equilibrium] (the ***i***-direction component equation):

$$F_x + \rho(V_0 + l_x S)g - mg = 0 \tag{4}$$

F_x—the ***i***-direction component of the tension in the sling at Time t.

l_x—the height of the air inside the bucket at Time t.

Taking *the air inside the bucket* as the object of study. Studying *the process from Time zero to arbitrary Time t*. Considering ***Boyle–Mariotte's law***:

$$plS = p_x l_x S \tag{5}$$

p_x—the air pressure in the bucket at Time t.

Considering ***basic equation of hydrostatic pressure***:

$$p_x = p_0 + \rho g\left[H_0 - x - (l_0 - l_x)\right] \tag{6}$$

Taking *the system composed of the bucket and the water (including the pool water and the water inside the bucket) and the earth* as the object of study. Studying *the whole process of pulling of sling*. Considering ***theorem of mechanical energy*** [the free surface of the pool water is regarded as infinity; therefore, the atmospheric pressure acting on the pool water surface can be regarded as an infinitely large quantity, in the whole process of pulling of sling, the rising height of the pool water surface can be regarded as an infinitely small quantity. The work done by the infinite atmospheric pressure acting on the infinitesimal displacement is a finite quantity, which cannot be ignored]:

$$W_F + W_{p_0} + W_p = \Delta E \tag{7}$$

ΔE—the gravitational potential energy increment of the bucket and the water (including the pool water and the water inside the bucket) in the whole process.

W_F—the work done by the sling to the bucket in the whole process.

W_{P_0} —the work done by atmosphere to the pool water in the whole process.

W_P—the work done by the air inside the bucket to the bucket and the water inside the bucket in the whole process.
Considering the definition of ***work***:

$$W_F = \int_0^H F_x \mathrm{d}x \tag{8}$$

$$W_{p_0} = -p_0 \Delta V = -p_0(l' - l)S \tag{9}$$

ΔV—in the whole process of pulling of sling, the volume of the space where the pool water surface rises (equal to the volume increment of the air inside the bucket).

Considering ***the form of work done by ideal gas in isothermal process***:

$$W_p = pV \ln \frac{V'}{V} = plS \ln \frac{l'S}{lS} = plS \ln \frac{l'}{l} \tag{10}$$

V and V'—the volume of the air inside the bucket at the time point when start and stop pulling the sling.

ΔE can be obtained by combining the above equations.

6.2 Ideal Fluid

In some practical problems, the real fluid can be simulated approximately by the ideal fluid, such as the airflow around a streamlined airfoil, the airflow in the airway of a jet engine, etc. The steady flow of ideal fluid in the potential field of mass forces is the simplest hydrodynamic model, its dynamic law can be described by the Bernoulli equation, as a degradation of theorem of mechanical energy, this theoretical form is very simple and practical.

In this section, some representative ideal fluid problems will be studied.

6.2.1 The Flowing of the Liquid in a Tank Out of a Drain Hole/Drain Pipe (d, *C*)

1. A cylindrical tank of sectional diameter D is placed horizontally, at the lowest of its side wall there is a circular drain hole of sectional diameter d, $d \ll D$ (see Fig. 1). Initially, the drain hole is closed by a valve, fill a homogeneous incompressible liquid to the tank, the height of which is H. If the valve is removed, find the time taken for the liquid to be completely drained.

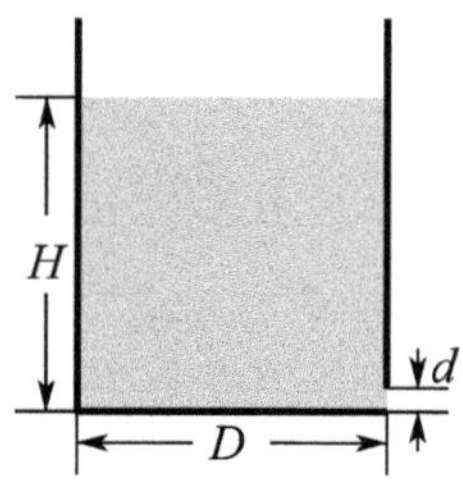

Figure 1

2. At the drain hole of the tank mentioned in the previous question, a circular drain pipe of sectional diameter d and length L is connected horizontally (see Fig. 2). Initially, the open end of the drain pipe is closed by a valve, fill a homogeneous incompressible liquid to the tank, the height of which is H. If the valve is removed, find the function relation between the drain speed and the height of the liquid in the initial stage.

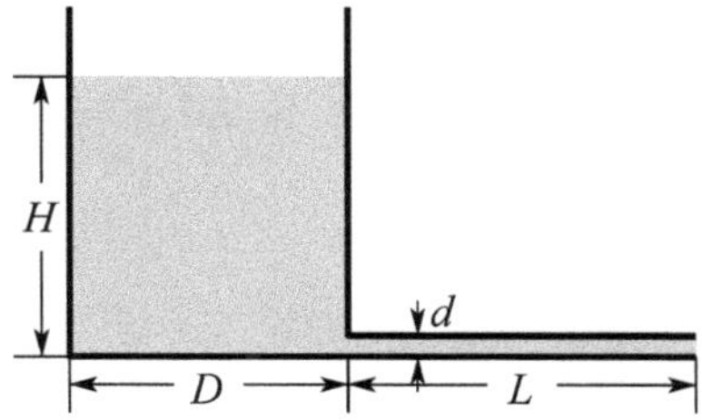

Figure 2

Solution to Problem (1)

Because $d \ll D$, the drain process of the liquid is quite slow, which can be regarded as the steady flow.

The time point when the valve is removed is defined as Time zero.

As shown in Fig. 3, establishing the linear coordinate system, let z be the coordinate.

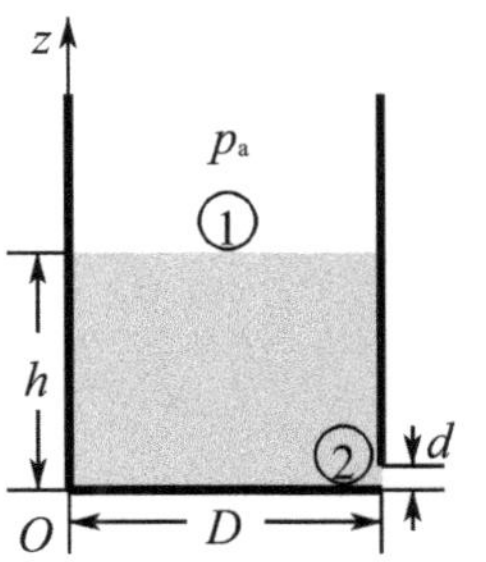

Figure 3

At arbitrary Time t, the free surface of the liquid in the tank is denoted by Section ①, and the drain hole is denoted by Section ②.

Taking *the liquid flowing along an arbitrary streamline between Section ① and Section ② at arbitrary Time t* as the object of study. Studying *the state at Time t*. Considering ***Bernoulli's equation for steady incompressible fluid***:

$$p_1 + \rho g z_1 + \frac{1}{2}\rho v_1^2 = p_2 + \rho g z_2 + \frac{1}{2}\rho v_2^2 \tag{1}$$

ρ—the density of the liquid.

p_1 and p_2—the liquid pressure at Section ① and Section ② at Time *t*.

v_1 and v_2—the velocity of flow of the liquid at Section ① and Section ② at Time *t*.

Known:

$$p_1 = p_2 = p_a \tag{2}$$

$$z_1 - z_2 = h \tag{3}$$

$$v_1 \approx 0 \tag{4}$$

p_a—the atmospheric pressure.

h—the height of the liquid at arbitrary Time *t*.

$$\Rightarrow \quad v_2 \approx \sqrt{2gh} \tag{5}$$

Taking *the liquid in the tank at arbitrary Time t* as the object of study. Studying *the instantaneous process (the small time interval dt) at Time t*. Considering ***law of conservation of mass*** (in an arbitrary process, the mass of the liquid flowing out of the drain hole is equal to the mass decrement of the liquid in the tank):

$$\frac{1}{4}\pi d^2 \cdot v_2 \mathrm{d}t = -\mathrm{d}\left(\frac{1}{4}\pi D^2 h\right) \tag{6}$$

$$\Rightarrow \quad \mathrm{d}t = -\left(\frac{D}{d}\right)^2 \frac{\mathrm{d}h}{\sqrt{2gh}} \tag{7}$$

$$\Rightarrow \quad T = \int_0^T \mathrm{d}t = -\left(\frac{D}{d}\right)^2 \frac{1}{\sqrt{2g}} \int_H^0 \frac{\mathrm{d}h}{\sqrt{h}} = \left(\frac{D}{d}\right)^2 \sqrt{\frac{2}{g}} \sqrt{H} \tag{8}$$

T—the time taken for the liquid to be completely drained.

Solution to Problem (2)

In the initial extremely short stage, the flow of the liquid is the non-steady flow.

The time point when the valve is removed is defined as Time zero.

As shown in Fig. 4, establishing the linear coordinate system, let z be the coordinate.

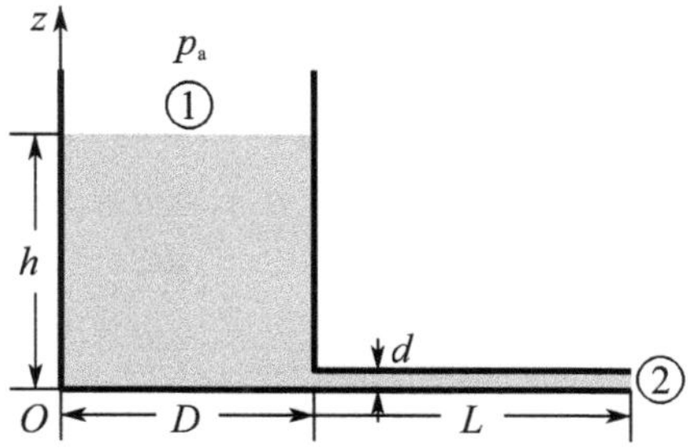

Figure 4

At arbitrary Time t, the free surface of the liquid in the tank is denoted by Section ①, and the open end of the drain pipe is denoted by Section ②.

Taking *the liquid flowing along an arbitrary streamline between Section ① and Section ② at arbitrary Time t in the initial stage* as the object of study. Studying *the state at Time t*. Considering ***Bernoulli's equation for non-steady incompressible fluid***:

$$p_1 + \rho g z_1 + \frac{1}{2}\rho v_1^2 = p_2 + \rho g z_2 + \frac{1}{2}\rho v_2^2 + \rho\int_1^2 \frac{\partial v}{\partial t}\mathrm{d}s \tag{1}$$

ρ—the density of the liquid.

p_1 and p_2—the liquid pressure at Section ① and Section ② at Time t.

v_1 and v_2—the velocity of flow of the liquid at Section ① and Section ② at Time t.

v—the velocity of flow of the liquid at an arbitrary position of the streamline at Time t.

Known:

$$p_1 = p_2 = p_a \tag{2}$$

$$z_1 - z_2 = h \tag{3}$$

$$v_1 \approx 0 \tag{4}$$

p_a—the atmospheric pressure.

h—the height of the liquid at arbitrary Time t in the initial stage.

$$\Rightarrow \qquad gh \approx \frac{1}{2}v_2^2 + \int_1^2 \frac{\partial v}{\partial t}\mathrm{d}s \tag{5}$$

The main contribution of the integral term on the right side of Eq. (5) comes from the flow of the liquid in the long pipe, and the velocity of flow of the liquid in the long pipe depends only on time:

$$\Rightarrow \qquad \int_1^2 \frac{\partial v}{\partial t} ds \approx \frac{dv_2}{dt} \int_0^L ds = L\frac{dv_2}{dt} \tag{6}$$

$$\Rightarrow \qquad dt = \frac{2L dv_2}{2gh - v_2^2} \tag{7}$$

$$\Rightarrow \qquad t = \int_0^t dt = \int_0^{v_2} \frac{2L dv_2}{2gh - v_2^2} = \frac{\sqrt{2}L}{\sqrt{gh}} \operatorname{arctanh}\left(\frac{v_2}{\sqrt{2gh}}\right) \tag{8}$$

$$\Rightarrow \qquad v_2 = \sqrt{2gh} \tanh\left(\frac{\sqrt{gh}}{\sqrt{2}L} t\right) \tag{9}$$

Considering the properties of tanh x:

$$\tanh x \to \begin{cases} x & \text{when } x \to 0 & (10a) \\ 1 & \text{when } x \to \infty & (10b) \end{cases}$$

$$\Rightarrow \qquad v_2 \approx \begin{cases} \sqrt{2gh} \cdot \dfrac{\sqrt{gh}}{\sqrt{2}L} t = \dfrac{gh}{L} t & \text{when } t \ll \dfrac{L}{\sqrt{gh}} & (11a) \\ \sqrt{2gh} \cdot 1 = \sqrt{2gh} & \text{when } t \gg \dfrac{L}{\sqrt{gh}} & (11b) \end{cases}$$

Comparing this result with the result of Problem (1), it can be seen that when the drain time is long enough, the liquid flow will tend to be steady.

6.2.2 A Semi Cylindrical Hangar in Wind (d, *C*)

Consider a semi cylindrical hangar of sectional radius R and length l exposed to the wind (see Fig. 1), the direction of the wind velocity at an arbitrary point in space is perpendicular to its central axis. The wind velocity at an arbitrary point around the hangar can be determined by the velocity potential, whose functional form in the polar coordinate system provided in the figure is as follows: $\varphi = -v_0(r + \frac{R^2}{r}) \cos \theta$, where v_0 is the wind speed in the distance.

Given the air density is ρ. There is a small door at Point (r, π) (denoted by A) of the hangar, if the door is opened and the air pressure inside the hangar reaches evenly distributed, find the force exerted on the hangar by the air.

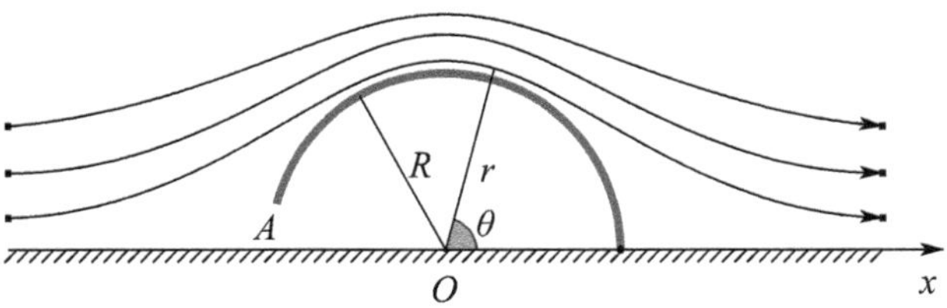

Figure 1

Solution

Known:

$$\varphi = -v_0\left(r + \frac{R^2}{r}\right)\cos\theta \tag{1}$$

Considering the definition of ***velocity potential***:

$$\Rightarrow \begin{cases} v_r = -\left.\dfrac{\partial\varphi}{\partial r}\right|_{r=R} = v_0\left(1 - \dfrac{R^2}{r^2}\right)\cos\theta\Big|_{r=R} = 0 & \text{(2a)} \\ v_\theta = -\left.\dfrac{1}{r}\dfrac{\partial\varphi}{\partial\theta}\right|_{r=R} = -v_0\left(1 + \dfrac{R^2}{r^2}\right)\sin\theta\Big|_{r=R} = -2v_0\sin\theta & \text{(2b)} \end{cases}$$

(v_r, v_θ)—the $(\boldsymbol{r}^0, \boldsymbol{\theta}^0)$-directions components of the wind velocity at the point (r, θ) near the hangar.

Taking *the air flowing along the specific streamline between an arbitrary point in the distance and an arbitrary point near the hangar at arbitrary Time t* as the object of study. Studying *the state at Time t*. Considering ***Bernoulli's equation for steady incompressible fluid***:

$$p_0 + \frac{1}{2}\rho v_0^2 = p_\theta + \frac{1}{2}\rho v_\theta^2 \tag{3}$$

p_θ—the air pressure at the point (r, θ) near the hangar.

$$\Rightarrow \qquad p_\theta = p_0 + \frac{1}{2}\rho v_0^2(1 - 4\sin^2\theta) \tag{4}$$

Defining p_A—the air pressure at the door.

$$\Rightarrow \qquad p_A = p_\theta\big|_{\theta=\pi} = p_0 + \frac{1}{2}\rho v_0^2 \tag{5}$$

The air pressure inside the hangar is evenly distributed, equal to the air pressure at the door.

It can be seen from the symmetry that the direction of the resultant force exerted on the hangar by the air is vertical. Defining F—the vertical up component of the resultant force exerted on the hangar by the air. Considering the definition of ***pressure***, then place (4) and (5) in to obtain:

$$\begin{aligned}\Rightarrow \qquad F &= \int_\Sigma (p_A - p_\theta)\mathrm{d}S \cdot \sin\theta \\ &= \int_0^\pi (p_A - p_\theta) lR\mathrm{d}\theta \cdot \sin\theta \\ &= 2\rho v_0^2 lR \int_0^\pi \sin^3\theta \mathrm{d}\theta \\ &= 2\rho v_0^2 lR \int_0^\pi (\cos^2\theta - 1)\mathrm{d}(\cos\theta) \\ &= 2\rho v_0^2 lR \left(\frac{1}{3}\cos^3\theta - \cos\theta\right)\Bigg|_0^\pi \\ &= \frac{8}{3}\rho v_0^2 lR \end{aligned} \tag{6}$$

6.2.3 The Flow of the Liquid Squeezed by a Pair of Coaxial Parallel Disks (d, *C*)

As shown in Fig. 1, two disks of equal radius R are placed coaxially in the horizontal direction, the lower disk is fixed, and an incompressible liquid of density ρ is filled between the two disks. An external force is applied to the upper disk to make it move downward at the speed u, the liquid between the two disks flows outward because of the squeeze. The distance between the two disks at arbitrary time is denoted by h. Find the external force acting on the upper disk at arbitrary time.

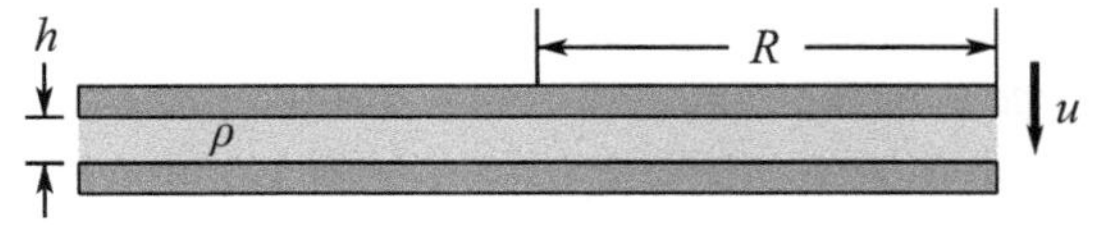

Figure 1

Solution

As shown in Fig. 2, establishing the cylindrical coordinate system, let r, θ, z be the cylindrical coordinates.

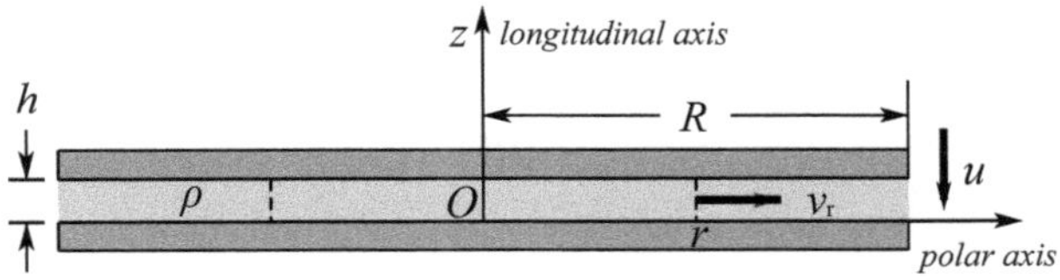

Figure 2

Between the two disks, the cylindrical surface with the longitudinal axis as the central axis, whose sectional radius tends to be zero is called the central cylindrical surface.

Taking *the liquid flowing along an arbitrary radial streamline between the central cylindrical surface and the cylindrical surface of sectional radius r at arbitrary Time t* as the object of study. Studying *the state at Time t*. Considering ***Bernoulli's equation for non-steady incompressible fluid***:

$$p_0 + \frac{1}{2}\rho v_0^2 = p_r + \frac{1}{2}\rho v_r^2 + \rho\int_0^r \frac{\partial v_r}{\partial t}\mathrm{d}r \tag{1}$$

p_0 and p_r—the liquid pressure at the central cylindrical surface and the cylindrical surface of sectional radius r at Time t.

v_0 and v_r—the $\boldsymbol{r}^0$-direction component of velocity of flow of the liquid at the central cylindrical surface and the cylindrical surface of sectional radius r at Time t.

The velocity of flow of the liquid approximates to zero at the position tends to the central axis:

$$v_0 \approx 0 \tag{2}$$

Taking *the liquid inside the cylindrical surface of sectional radius r at arbitrary Time t* as the object of study. Studying *the instantaneous process (the small time interval dt) at Time t*. Considering ***law of conservation of mass*** (in an arbitrary process, the mass of the liquid flowing out of the cylindrical surface of sectional radius r is equal to the mass decrement of the liquid inside it):

$$2\pi rh\cdot v_r\mathrm{d}t=-\mathrm{d}(\pi r^2 h) \tag{3}$$

$$\Rightarrow \qquad v_r=-\frac{r}{2h}\frac{\mathrm{d}h}{\mathrm{d}t} \tag{4}$$

Considering the definition of ***velocity***:

$$u=-\frac{\mathrm{d}h}{\mathrm{d}t} \tag{5}$$

$$\Rightarrow \qquad v_r=u\frac{r}{2h} \tag{6}$$

$$\Rightarrow \qquad \frac{\partial v_r}{\partial t}=-u\frac{r}{2h^2}\frac{\mathrm{d}h}{\mathrm{d}t}=\frac{u^2 r}{2h^2} \tag{7}$$

$$\Rightarrow \qquad \int_0^r\frac{\partial v_r}{\partial t}\mathrm{d}r=\frac{u^2}{2h^2}\int_0^r r\mathrm{d}r=\frac{u^2 r^2}{4h^2} \tag{8}$$

(2), (6) and (8) $\rightarrow$ (1),

$$\Rightarrow \qquad p_r=p_0-\frac{3}{8}\frac{\rho u^2 r^2}{h^2} \tag{9}$$

Considering the boundary condition:

$$p_r\big|_{r=R}=p_a \tag{10}$$

p_a—the atmospheric pressure.

$$\Rightarrow \qquad p_0=p_a+\frac{3}{8}\frac{\rho u^2 R^2}{h^2} \tag{11}$$

$$\Rightarrow \qquad p_r=p_a+\frac{3}{8}\frac{\rho u^2}{h^2}(R^2-r^2) \tag{12}$$

Taking *the upper disk* as the object of study. Studying *the state at arbitrary Time t*. Considering ***theorem of motion of center of mass*** [the upper disk is in mechanical equilibrium] (the $-\boldsymbol{k}$-direction component equation):

$$F + \int_0^R p_a \cdot 2\pi r dr - \int_0^R p_r \cdot 2\pi r dr = 0 \tag{13}$$

F—the $-\boldsymbol{k}$-direction component of the external force acting on the upper disk at Time t.

(12) → (13),

$$\Rightarrow \quad F = \int_0^R (p_r - p_a) \cdot 2\pi r dr$$

$$= \frac{3\pi}{4} \frac{\rho u^2}{h^2} \int_0^R (R^2 - r^2) r dr$$

$$= \frac{3\pi}{16} \frac{\rho u^2 R^4}{h^2} \tag{14}$$

6.2.4 Wind Turbine (d, *C*)

Consider a propeller type wind wheel whose central axis is fixed (see Fig. 1). A disk-shaped control volume is used to encircle the wind wheel, whose cross-sectional area is A. An incompressible airflow of density ρ flows to the wind wheel at the speed v_1 from the distance, the speed decreases to v_2 when it reaches the distance through the wind wheel. Find the energy transfer power from the airflow to the wind wheel.

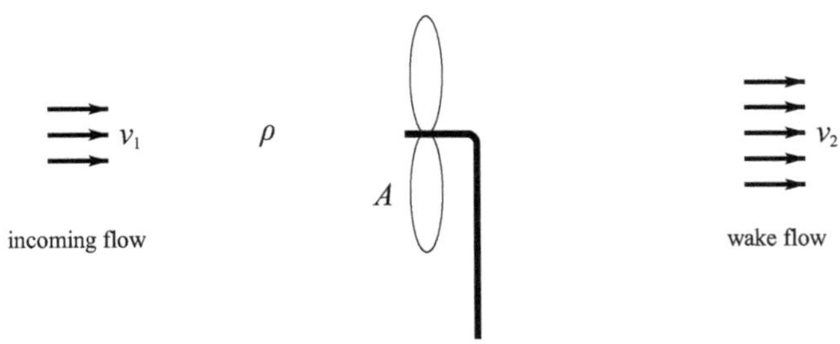

Figure 1

Solution

As shown in Fig. 2, the stream tube encircling the airflow flowing through the wind wheel is denoted by Σ, whose cross sections far away from the wind wheel in the incoming flow direction and in the wake flow direction are denoted by ① and ② respectively, and the cross sections close to the wind wheel in the incoming flow direction and in the wake flow direction are denoted by ⓐ and ⓑ respectively.

The unit vector in the direction of the airflow velocity is denoted by ***i***.

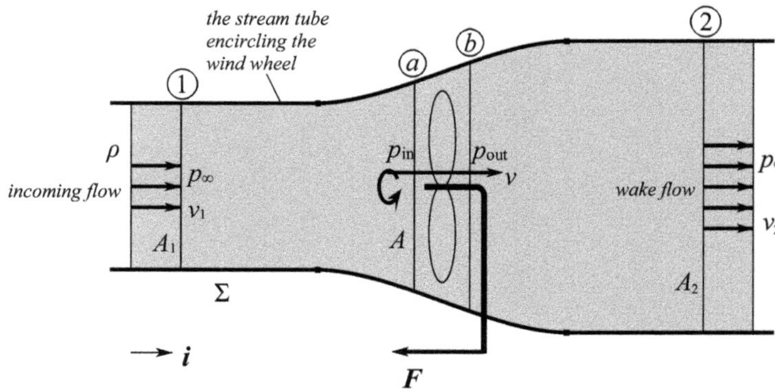

Figure 2

Taking *the airflow flowing in the stream tube Σ at arbitrary Time t* as the object of study. Studying *the state at Time t*. Considering ***flow conservation***:

$$\rho A_1 v_1 = \rho A_2 v_2 = \rho A v \tag{1}$$

A_1 and A_2—the area of Section ① and Section ②.

v—the ***i***-direction component of the velocity of the airflow when it reaches the wind wheel.

Taking *the system composed of three parts: (1) the airflow between Section ① and Section ② at arbitrary Time t, (2) the airflow infinitesimal element entering Section ① in the instantaneous process (the small time interval dt) at Time t, (3) the wind wheel* as the object of study. Studying *that instantaneous process*. Considering ***momentum theorem*** (the ***i***-direction component equation):

$$-F\mathrm{d}t = \rho A_2 v_2 \mathrm{d}t \cdot v_2 - \rho A_1 v_1 \mathrm{d}t \cdot v_1 \tag{2}$$

F—the −***i***-direction component of the external force to fix the wind wheel at Time t.

$$\Rightarrow \quad F = \rho A v (v_1 - v_2) \tag{3}$$

Taking *the wind wheel* as the object of study. Studying *the state at arbitrary Time t*. Considering ***theorem of motion of center of mass*** [the center of mass of the wind wheel remains stationary] (the ***i***-direction component equation):

$$-F+(p_{\text{in}}-p_{\text{out}})A=0 \tag{4}$$

p_{in} and p_{out}—the pressure of the airflow at Section ⓐ and Section ⓑ at Time *t*.

$$\Rightarrow \quad p_{\text{in}}-p_{\text{out}}=\rho v(v_1-v_2) \tag{5}$$

Taking *the airflow flowing along an arbitrary streamline between Section ① and Section ⓐ at arbitrary Time t, and the airflow flowing along an arbitrary streamline between Section ⓑ and Section ② at arbitrary Time t* as the object of study respectively. Studying *the state at Time t*. Considering ***Bernoulli's equation for steady incompressible fluid***:

$$\begin{cases} p_{\infty}+\dfrac{1}{2}\rho v_1^2=p_{\text{in}}+\dfrac{1}{2}\rho v^2 & \text{(6a)} \\ p_{\text{out}}+\dfrac{1}{2}\rho v^2=p_{\infty}+\dfrac{1}{2}\rho v_2^2 & \text{(6b)} \end{cases}$$

p_{∞}—the pressure of the airflow far away from the wind wheel.

$$\Rightarrow \quad p_{\text{in}}-p_{\text{out}}=\frac{1}{2}\rho(v_1^2-v_2^2) \tag{7}$$

Combining (5) and (7),

$$\Rightarrow \quad v=\frac{1}{2}(v_1+v_2) \tag{8}$$

Defining *P*—the energy transfer power from the airflow to the wind wheel at arbitrary Time *t*.

Considering the definition of ***power***:

$$P=(p_{\text{in}}-p_{\text{out}})A\cdot v \tag{9}$$

(7) and (8) → (9),

$$\Rightarrow \quad P=\frac{1}{4}\rho A(v_1+v_2)(v_1^2-v_2^2) \tag{10}$$

6.2.5 The Fluid Force on the Wind Grids of an Axial-Flow Turbine (d, *C*)

As shown in Fig. 1, in an axial-flow turbine, the blades are equidistant installed on the circumference of the rotor, and the fluid flowing through the turbine flows along the cylindrical surface coaxial with the rotor (the fluid velocity only has axial and circumferential components, and has no radial component). The blades are densely distributed, the adjacent blades can be regarded as approximately parallel to each other from the view at an arbitrary point between them, and both ends of each blade can be regarded as extending to infinity. Given the distance between the adjacent blades is a (based on the above understanding), the fluid flowing through the turbine is an ideal incompressible fluid, whose density is ρ. The circumferential and axial components of the flow velocity and wake velocity are, respectively, (u_1, v_1) and (u_2, v_2). Find the force exerted on an arbitrary blade by the fluid at arbitrary time.

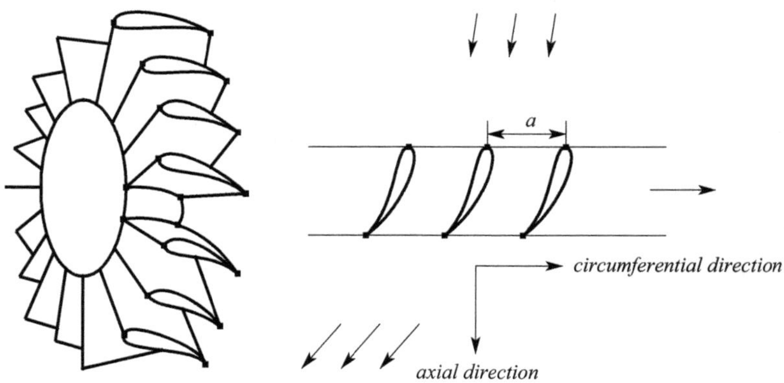

Figure 1

Solution

Time zero is arbitrarily set.

As shown in Fig. 2, sorting all blades with positive integers, the following studies Blade i, whose center is denoted by O, the unit vector in Point O's velocity direction at arbitrary Time t is denoted by $\boldsymbol{i}$, and the unit vector in the rotor's axial direction (the direction of the fluid passing through the turbine) is denoted by $\boldsymbol{j}$, establishing the Cartesian coordinate system with the frame $(O, \mathbf{i}, \mathbf{j})$, let x, y be the Cartesian coordinates.

At Time t, the space area where the fluid flows is divided into a series of stream tubes with the same shape, each blade is in the middle of one of the stream tubes, the stream tube which encloses Blade i is denoted by Stream Tube i. Taking a closed curve $ABCDA$ in the O-xy plane, where BC and AD are located at the boundary of the Stream Tube i, AB and CD are located in the distance and parallel to the x-axis.

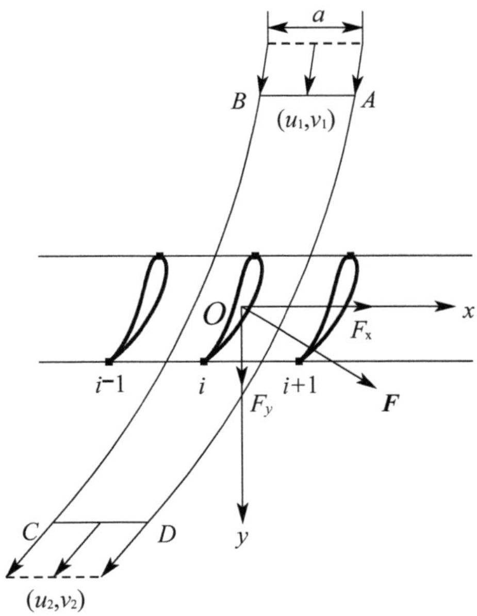

Figure 2

Taking *the fluid in Stream Tube i between AB and CD at Time t* as the object of study. Studying *the state at Time t*. Considering ***flow conservation***:

$$\rho a v_1 = \rho a v_2 \tag{1}$$

$$\Rightarrow \quad v_1 = v_2 \tag{2}$$

Defining v:

$$v = v_1 = v_2 \tag{3}$$

Taking *the system composed of the fluid in Stream Tube i between AB and CD at Time t and the fluid infinitesimal element passing through the section where AB is in the instantaneous process (the small time interval dt) at Time t* as the object of study. Studying *that instantaneous process*. Considering ***momentum theorem*** (the ***i***-direction and ***j***-direction component equations):

$$\begin{cases} -F_x l dt = \rho lavdt \cdot u_2 - \rho lavdt \cdot u_1 & \text{(4a)} \\ \left[-F_y + (p_1 - p_2)a\right] l dt = \rho lavdt \cdot v_2 - \rho lavdt \cdot v_1 & \text{(4b)} \end{cases}$$

l—the length of each blade.

(F_x, F_y)—the ***i***-direction and ***j***-direction components of the force exerted on Blade *i* of unit length perpendicular to the *O-xy* plane by the fluid at Time *t*.

p_1 and p_2—the fluid pressure at *AB* and *CD* at Time *t*.

$$\Rightarrow \quad \begin{cases} F_x = \rho av(u_1 - u_2) & \text{(5a)} \\ F_y = (p_1 - p_2)a & \text{(5b)} \end{cases}$$

Taking *the fluid flowing along an arbitrary streamline between AB and CD at Time t* as the object of study. Studying *the state at Time t*. Considering ***Bernoulli's Equation for steady incompressible fluid***:

$$p_1 + \frac{1}{2}\rho(u_1^2 + v_1^2) = p_2 + \frac{1}{2}\rho(u_2^2 + v_2^2) \tag{6}$$

$$\Rightarrow \quad F_y = \frac{1}{2}\rho a(u_2^2 - u_1^2) \tag{7}$$

Defining Γ_{AB}, Γ_{BC}, Γ_{CD} and Γ_{DA}—the velocity integrals of the fluid along the paths *AB, BC, CD* and *DA*. Γ—the velocity circulation along the closed curve *ABCDA*.

$$\Gamma = \Gamma_{AB} + \Gamma_{BC} + \Gamma_{CD} + \Gamma_{DA} \tag{8}$$

It can be seen from the symmetry that the velocity integrals of the fluid along the paths *BC* and *DA* are equal in magnitude and opposite in sign.

$$\Gamma_{BC} + \Gamma_{DA} = 0 \tag{9}$$

$$\Rightarrow \quad \Gamma = \Gamma_{AB} + \Gamma_{CD} \tag{10}$$

$$\begin{cases} \Gamma_{AB} = -u_1 a & \text{(11a)} \\ \Gamma_{CD} = u_2 a & \text{(11b)} \end{cases}$$

$$\Rightarrow \quad \Gamma = (u_2 - u_1)a \tag{12}$$

(12) → (5a) and (7),

$$\Rightarrow \quad \begin{cases} F_x = -\rho v \Gamma & (13a) \\ F_y = \dfrac{1}{2}\rho(u_1 + u_2)\Gamma & (13b) \end{cases}$$

Defining $\bar{u}$ —the average value of the ***i***-direction component of the velocity of the incoming flow and wake flow of the fluid:

$$\bar{u} = \frac{1}{2}(u_1 + u_2) \tag{14}$$

$$\Rightarrow \quad F_y = \rho \bar{u} \Gamma \tag{15}$$

Defining ***F***—the force exerted on Blade *i* of unit length perpendicular to the *O*-*xy* plane by the fluid.

$$\Rightarrow \quad \boldsymbol{F} = F_x \boldsymbol{i} + F_y \boldsymbol{j} = -\rho v \Gamma \boldsymbol{i} + \rho \bar{u} \Gamma \boldsymbol{j} \tag{16}$$

$$\Rightarrow \quad |\boldsymbol{F}| = \sqrt{F_x^2 + F_y^2} = \rho \sqrt{\bar{u}^2 + v^2}\,\Gamma \tag{17}$$

Defining $\bar{\boldsymbol{v}}$ —the average value of the velocity of flow of the incoming flow and wake flow of the fluid:

$$\bar{\boldsymbol{v}} = \bar{u}\boldsymbol{i} + v\boldsymbol{j} \tag{18}$$

$$\Rightarrow \quad |\boldsymbol{F}| = \rho |\bar{\boldsymbol{v}}| \Gamma \tag{19}$$

Combining (16) and (18),

$$\Rightarrow \quad \boldsymbol{F} \cdot \bar{\boldsymbol{v}} = F_x \bar{u} + F_y v = 0 \tag{20}$$

That is, ***F*** is perpendicular to $\bar{\boldsymbol{v}}$.

Equations (16) and (19) retain Γ determined by (12) and $\bar{\boldsymbol{v}}$ determined by (14) and (18).

6.3 Viscous Fluid

In most cases, the viscosity of the fluid cannot be ignored, such as the fall of dust in the air, the slow motion of bubbles in the liquid, and the flow of fluid near the solid wall. Due to the complexity of viscous stress, the dynamics of viscous fluid is more difficult to study than the dynamics of ideal fluid.

In this section, some simple viscous fluid problems (limited to laminar flow) will be studied.

6.3.1 Poiseuille Flow (d, *C*)

Consider a viscous fluid of viscosity coefficient η flowing in a straight circular pipe of sectional radius *r* and length *l*. The pressure on both sides of the straight circular pipe is kept at p_1 and p_2 respectively, $p_1 > p_2$. After a long enough time, the fluid in the pipe can reach the steady flow, the velocity of flow at the pipe wall is approximately zero, and the velocity of flow at the central axis is maximum.

(1) Find the velocity distribution of the fluid in the pipe when it reaches the steady flow.

(2) Find the volume rate of flow of the fluid in the pipe when it reaches the steady flow.

Solution to Problem (1)

The time point when the fluid reaches the steady flow is defined as Time zero.

As shown in Fig. 1, establishing the cylindrical coordinate, let *r*, θ, *x* be the cylindrical coordinates.

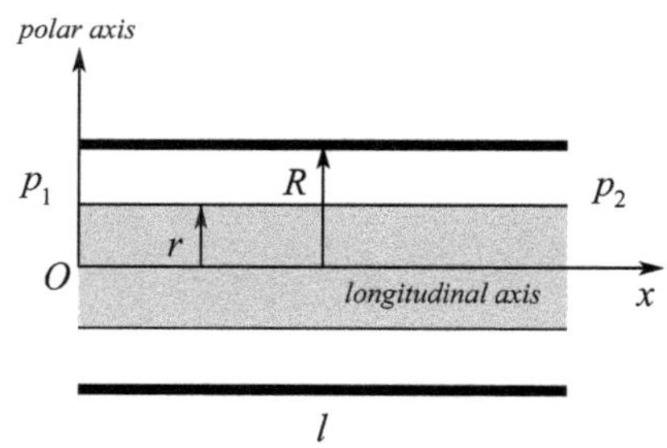

Figure 1

Considering (a) ***flow conservation***, (b) ***spatial symmetry***, it can be seen that the $\boldsymbol{r}^0$-direction and $\boldsymbol{\theta}^0$-direction component of the velocity of flow at an arbitrary point in the pipe is zero.

Defining Δp—the pressure difference on both sides of the pipe:

$$\Delta p = p_1 - p_2 \tag{1}$$

Taking *the fluid inside the cylindrical surface of sectional radius r* as the object of study. Studying *the state at arbitrary Time t*. Considering ***Newton's second law*** [the fluid is in mechanical equilibrium] (the $\boldsymbol{i}$-direction component equation):

$$\Delta p \cdot \pi r^2 + F_{\eta r} = 0 \tag{2}$$

$F_{\eta r}$—for the cylindrical surface of radius r, the $\boldsymbol{i}$-direction component of the viscous force exerted on the inner liquid by the outer liquid at Time t.

Considering ***theorem of viscous force***:

$$F_{\eta r} = \int_{\Sigma} \eta \mathrm{d}S \frac{\mathrm{d}v_r}{\mathrm{d}r} = \eta \frac{\mathrm{d}v_r}{\mathrm{d}r} \int_{\Sigma} \mathrm{d}S = \eta \frac{\mathrm{d}v_r}{\mathrm{d}r} \cdot 2\pi r l \tag{3}$$

v_r—the $\boldsymbol{i}$-direction component of the velocity of flow at the polar radius r at arbitrary Time t.

$$\Rightarrow \qquad \mathrm{d}v_r = -\frac{\Delta p}{2\eta l} r\mathrm{d}r \tag{4}$$

Known that the velocity of flow at the pipe wall is zero:

$$v_r\big|_{r=R} = 0 \tag{5}$$

$$\Rightarrow \qquad \int_0^{v_r} \mathrm{d}v_r = -\frac{\Delta p}{2\eta l} \int_R^r r\mathrm{d}r \tag{6}$$

$$\Rightarrow \qquad v_r = \frac{\Delta p}{4\eta l}(R^2 - r^2) \tag{7}$$

This is the velocity distribution of the fluid in the pipe when it reaches the steady flow. Based on this, we can draw the schematic diagram of the distribution of the velocity of flow (see Fig. 2).

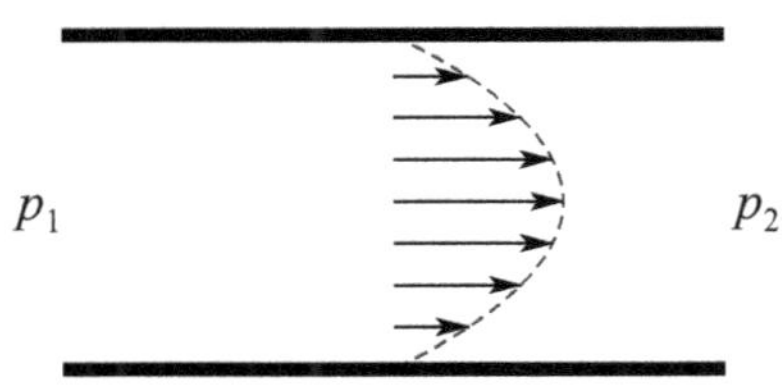

Figure 2

Solution to Problem (2)

Defining Q—the volume rate of flow of the fluid in the pipe when it reaches the steady flow.

Considering the definition of ***volume rate of flow***, then place (7) in to obtain:

$$\Rightarrow \qquad Q=\int v_r\cdot 2\pi r\mathrm{d}r=\frac{\pi\Delta p}{2\eta l}\int_0^R (R^2-r^2)r\mathrm{d}r=\frac{\pi\Delta pR^4}{8\eta l} \tag{8}$$

This conclusion is called ***Poiseuille's law***.

Another method to get Eq. (4)

As shown in Fig. 3, taking *the fluid infinitesimal element between the two cylindrical surfaces of sectional radius r and r + dr* as the object of study. Studying *the state at arbitrary Time t*. Considering ***Newton's second law*** [the fluid infinitesimal element is in mechanical equilibrium] (the ***i***-direction component equation):

$$0=\Delta p\cdot 2\pi r\mathrm{d}r+F_{\eta(r+\mathrm{d}r)}-F_{\eta r}=\Delta p\cdot 2\pi r\mathrm{d}r+\mathrm{d}F_{\eta r} \tag{9}$$

(3) → (9),

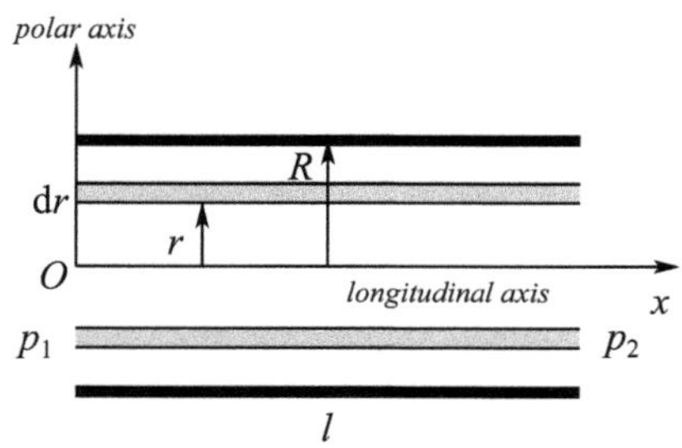

Figure 3

$$\Rightarrow \qquad \Delta pr\mathrm{d}r+\eta l\mathrm{d}\left(r\frac{\mathrm{d}v_r}{\mathrm{d}r}\right)=0 \tag{10}$$

Known that the velocity of flow at the central axis is maximum. The necessary condition for v_r to take the maximum value is as follows:

$$\left.\frac{\mathrm{d}v_r}{\mathrm{d}r}\right|_{r=0}=0 \tag{11}$$

$$\Rightarrow \qquad \Delta p\int_0^r r\mathrm{d}r+\eta l\int_0^{r\frac{\mathrm{d}v_r}{\mathrm{d}r}}\mathrm{d}\left(r\frac{\mathrm{d}v_r}{\mathrm{d}r}\right)=0 \tag{12}$$

$$\Rightarrow \qquad \frac{1}{2}\Delta pr^2+\eta lr\frac{\mathrm{d}v_r}{\mathrm{d}r}=0 \tag{13}$$

$$\Rightarrow \qquad \mathrm{d}v_r = -\frac{\Delta p}{2\eta l} r\mathrm{d}r \qquad (14)$$

6.3.2 An Oil Lamp without Wick (*a*, *A*)

A simple model of an oil lamp without wick is as follows: as shown in Fig. 1, a container of sectional area S contains a fluid fuel of density ρ and viscosity coefficient η. Initially, the height of the fuel in the container is H_0, and a thin circular pipe of sectional radius R and length l is vertically inserted into the container to the bottom, $l > H_0$, $\pi R^2 \ll S$. Fill the pipe with the same kind of fuel and ignite the fuel at the top. The fuel in the container can flow into the pipe slowly, in the process of the fuel's burning, the air pressure p near the top of the pipe approximately satisfies the following equation: $p = p_0 - a \frac{\mathrm{d}m}{\mathrm{d}t}$, where p_0 is the atmospheric pressure, $\frac{\mathrm{d}m}{\mathrm{d}t}$ is the mass of the fuel burned in unit time, and a is a constant related to the properties of the fuel, $a > {8\eta l}/{\pi R^4 \rho}$. It is assumed that the fuel in the thin pipe can quickly reach the steady flow (the velocity of flow at the pipe wall is approximately zero). Given that the capillarity can be ignored. Find the height of the remaining fuel in the container at arbitrary time and the time taken for the fuel to be burnt out.

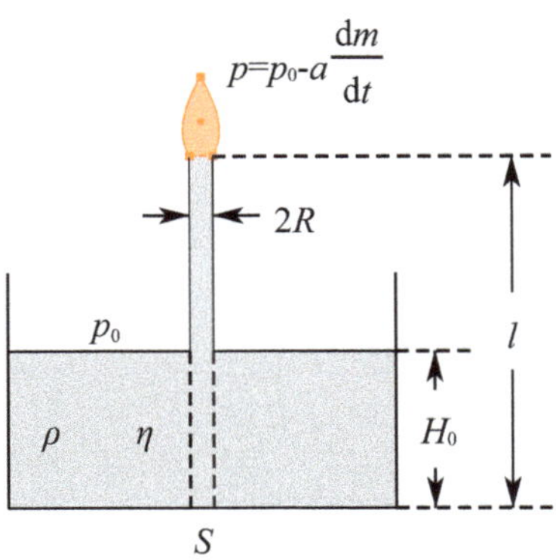

Figure 1

Solution

The time point when the fuel is ignited is defined as Time zero. Taking *all the fuel in the container and the pipe at arbitrary Time t* as the object of study. Studying *the instantaneous process (the small time interval dt) at Time t*. Considering ***law of conservation of mass***:

$$-\frac{\mathrm{d}}{\mathrm{d}t}(\rho SH)=\rho Q \tag{1}$$

H—the height of the fuel in the container at Time t.

Q—the volume rate of flow of the fuel in the pipe at Time t.

Taking *the fuel in the pipe at arbitrary Time t* as the object of study. Studying *the state at Time t*. Considering ***Poiseuille's law***:

$$\Rightarrow \qquad Q=\frac{\pi[(p_{\text{bottom}}-p_{\text{top}})-\rho gl]R^4}{8\eta l} \tag{2}$$

p_{bottom} and p_{top}—the pressure at the bottom and top of the pipe at Time t.

Taking *the fuel in the container at arbitrary Time t* as the object of study. Studying *the state at Time t*. Considering ***basic equation of hydrostatic pressure*** [the fuel in the container flows into the pipe slowly; therefore, it can be regarded as static fluid]:

$$p_{\text{bottom}}=p_0+\rho gH \tag{3}$$

Known:

$$p_{\text{top}}=p_0-a\frac{\mathrm{d}m}{\mathrm{d}t}=p_0+a\frac{\mathrm{d}}{\mathrm{d}t}(\rho SH) \tag{4}$$

Combining (1) to (4),

$$\Rightarrow \qquad \frac{\mathrm{d}H}{H-l}=\frac{\pi R^4\rho g}{(\pi R^4\rho a-8\eta l)S}\mathrm{d}t \tag{5}$$

Known:

$$l>H \tag{6}$$

$$a>8\eta l/\pi R^4\rho \tag{7}$$

$$\Rightarrow \qquad \mathrm{d}H<0 \tag{8}$$

It can be seen that the fuel in the container falls continuously, that is, the fuel can burn continuously.

Defining the constant α:

$$\alpha = \frac{\pi R^4 \rho g}{(\pi R^4 \rho a - 8\eta l)S} \tag{9}$$

$$\Rightarrow \qquad \frac{\mathrm{d}H}{H-l} = \alpha \mathrm{d}t \tag{10}$$

$$\int_{H_0}^{H} \frac{\mathrm{d}H}{H-l} = \alpha \int_0^t \mathrm{d}t$$

$$\Rightarrow \qquad \ln\frac{H-l}{H_0-l} = \alpha t$$

$$\Rightarrow \qquad H = H_0 e^{\alpha t} + l(1-e^{\alpha t}) \tag{11}$$

$$\Rightarrow \qquad t\big|_{H=0} = \frac{1}{\alpha}\ln\frac{l}{l-H_0} \tag{12}$$

6.3.3 Measurement of Viscosity Coefficient of Gas by Means of Exhaust Device (*b*, *A*)

A container of volume V_0 contains an ideal gas of molar mass μ, the inside of the container is connected with the outside through a thin circular pipe of cross-sectional radius r and length L, and the opening of the pipe is upward. There is a valve at the joint of the thin pipe and the container. Initially, the valve is closed, the gas pressure in the container is p_1, and the atmospheric pressure is p_a, $p_1 > p_a$. Open the valve, the gas starts to flow out of the container from the pipe. Given that r is quite small, and the viscosity coefficient of the gas is quite large that the flow of the gas is quite slow. The whole system keeps the constant temperature at T_0. Given that the time taken for the gas pressure in the container to decrease to p_2 is T. Find the viscosity coefficient of the gas.

Solution

The time point when the valve is opened is defined as Time zero.

As shown in Fig. 1, the joint between the thin pipe and the container is denoted by O, and the unit vector in the vertical upward direction is denoted by $\boldsymbol{i}$. Establishing the linear coordinate system with Point O as the origin and the $\boldsymbol{i}$-direction as the positive direction of coordinate axis, let x be the coordinate.

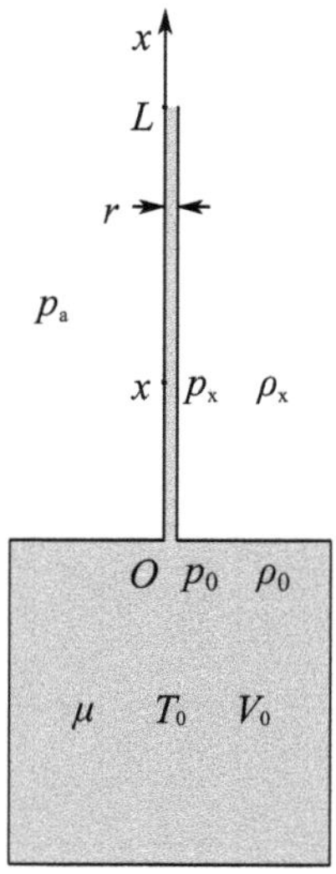

Figure 1

Taking *the gas in the pipe at arbitrary Time t* as the object of study. Studying *the instantaneous process at Time t.* Considering ***law of conservation of mass*** [the flow of the gas in the pipe is slow enough; therefore, the gas in the pipe can be regarded as performing the steady flow approximately, that is, the mass flow of the gas is approximately equal everywhere]:

$$\frac{\partial}{\partial x}\left(\rho_x \left.\frac{\mathrm{d}V}{\mathrm{d}t}\right|_x\right)=0 \tag{1}$$

ρ_x—the gas density at x in the pipe at Time t.

Taking *the gas infinitesimal element between x and x + dx in the pipe at arbitrary Time t* as the object of study. Studying *the state at Time t.* Considering ***Poiseuille's law***:

$$\left.\frac{\mathrm{d}V}{\mathrm{d}t}\right|_x=\frac{\pi r^4}{8\eta}\frac{p_x-p_{x+\mathrm{d}x}-\rho_x g\mathrm{d}x}{\mathrm{d}x}=-\frac{\pi r^4}{8\eta}\left(\frac{\mathrm{d}p_x}{\mathrm{d}x}+\rho_x g\right) \tag{2}$$

η—the viscosity coefficient of the gas.

p_x—the gas pressure at x in the pipe at arbitrary Time t.

Considering ***the state equation of ideal gas***:

$$\rho_x=\frac{p_x\mu}{RT_0} \tag{3}$$

$$\Rightarrow \qquad \frac{\partial}{\partial x}\left[p_x\left(\frac{\mathrm{d}p_x}{\mathrm{d}x}+\frac{\mu g}{RT_0}p_x\right)\right]=0 \tag{4}$$

Defining the constant α:

$$\alpha=\frac{\mu g}{RT_0} \tag{5}$$

$$\Rightarrow \qquad p_x\left(\frac{\mathrm{d}p_x}{\mathrm{d}x}+\alpha p_x\right)=C(t) \tag{6}$$

$C(t)$—the time dependent integral constant.

Defining the variable u:

$$u=p_x^2 \tag{7}$$

$$\Rightarrow \qquad \frac{\mathrm{d}u}{\alpha u-C(t)}=-2\mathrm{d}x \tag{8}$$

$$\Rightarrow \qquad \int_{p_0^2}^{p_x^2}\frac{\mathrm{d}u}{\alpha u-C(t)}=-2\int_0^x \mathrm{d}x$$

$$\Rightarrow \qquad \ln\frac{\alpha p_x^2-C(t)}{\alpha p_0^2-C(t)}=-2\alpha x$$

$$\Rightarrow \qquad p_x=\sqrt{\frac{1}{\alpha}\left[(\alpha p_0^2-C(t))e^{-2\alpha x}+C(t)\right]} \tag{9}$$

Considering the boundary condition:

$$p_L=p_a \tag{10}$$

$$\Rightarrow \qquad C(t)=\alpha\frac{p_a^2-p_0^2e^{-2\alpha L}}{1-e^{-2\alpha L}} \tag{11}$$

$$p_x=\sqrt{\frac{(p_0^2-p_a^2)e^{-2\alpha x}+p_a^2-p_0^2e^{-2\alpha L}}{1-e^{-2\alpha L}}} \tag{12}$$

This is the pressure distribution in the pipe at arbitrary Time t.

Taking *the gas in the container at arbitrary Time t* as the object of study. Studying *the instantaneous process (the small time interval dt) at Time t*. Considering ***law of conservation of mass***:

$$\rho_0\left.\frac{\mathrm{d}V}{\mathrm{d}t}\right|_{x=0}\mathrm{d}t=-\mathrm{d}(\rho_0V_0) \tag{13}$$

(12) → (2),

$$\Rightarrow \quad \left.\frac{\mathrm{d}V}{\mathrm{d}t}\right|_{x=0} = -\frac{\pi r^4}{8\eta}\left.\left(\frac{\mathrm{d}p_x}{\mathrm{d}x} + \alpha p_x\right)\right|_{x=0} = \frac{\alpha}{(1-e^{-2\alpha L})}\frac{\pi r^4}{8\eta}\frac{p_0^2 e^{-2\alpha L} - p_a^2}{p_0} \tag{14}$$

(14) → (13),

$$\Rightarrow \quad \mathrm{d}t = -\frac{(1-e^{-2\alpha L})e^{2\alpha L}}{\alpha}\frac{8\eta V_0}{\pi r^4}\frac{\mathrm{d}p_0}{p_0^2 - p_a^2 e^{2\alpha L}} \tag{15}$$

$$\Rightarrow \quad T = \int_0^T \mathrm{d}t = -\frac{(1-e^{-2\alpha L})e^{2\alpha L}}{\alpha}\frac{8\eta V_0}{\pi r^4}\int_{p_1}^{p_2}\frac{\mathrm{d}p_0}{p_0^2 - p_a^2 e^{2\alpha L}}$$

$$= \frac{(1-e^{-2\alpha L})e^{\alpha L}}{\alpha}\frac{4\eta V_0}{\pi r^4 p_a}\ln\frac{(p_2 + p_a e^{\alpha L})(p_1 - p_a e^{\alpha L})}{(p_2 - p_a e^{\alpha L})(p_1 + p_a e^{\alpha L})} \tag{16}$$

$$\Rightarrow \quad \eta = \frac{\alpha}{(1-e^{-2\alpha L})e^{\alpha L}}\frac{\pi r^4 p_a T}{4V_0 \ln\dfrac{(p_2 + p_a e^{\alpha L})(p_1 - p_a e^{\alpha L})}{(p_2 - p_a e^{\alpha L})(p_1 + p_a e^{\alpha L})}} \tag{17}$$

This result retains the constant α defined by Eq. (5).

6.3.4 Measurement of Viscosity Coefficient of Liquid by Means of Rotating Device (d, *B*)

As shown in Fig. 1, two cylindrical shell of radius r_1 and r_2 and height l are coaxially placed, between which a liquid of density ρ is filled. The inner cylindrical shell is fixed, and the outer cylindrical shell is rotated at a constant angular velocity $\boldsymbol{\omega}_0$ relative to the central axis under the external force. After a sufficiently long time, the liquid reaches the steady flow, the thin liquid layer in contact with the outer cylindrical shell rotates synchronously with the outer cylindrical shell, the thin liquid layer in contact with the inner cylindrical shell remains stationary. Given the liquid pressure at the inner wall of the outer cylindrical shell is p_0, the torque magnitude of the viscous force exerted on the inner cylindrical shell by the liquid is M_{r1} (with the central axis of the cylindrical shells as the rotation axis), the gravity of the liquid can be ignored.

(1) Find the radial distribution of the rotating angular velocity of the liquid.
(2) Find the volume rate of flow of the liquid at arbitrary vertical section between the two cylindrical shells.
(3) Find the viscosity coefficient of the liquid.

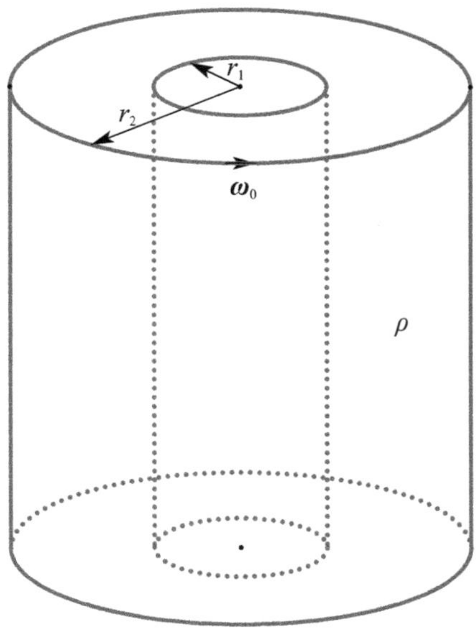

Figure 1

Solution to Problem (1)

The time point when the fluid reaches the steady flow is defined as Time zero.

As shown in Fig. 2 (the top view), the center of the bottom of the rotating device is denoted by O, the unit vector in the vertical up direction is denoted by $\boldsymbol{k}$. Taking Point O as the pole, an arbitrary direction in the horizontal plane as the positive direction of the pole axis, the $\boldsymbol{k}$-direction as the positive direction of the longitudinal axis, establishing the cylindrical coordinate system, let r, θ, z be the cylindrical coordinates.

Taking *the liquid infinitesimal element between the two cylindrical surfaces of sectional radius* r *and* $r + dr$ as the object of study. Studying *the state at arbitrary Time t.* Considering ***law of rotation*** [the liquid infinitesimal element can be regarded as a rigid body in the state of mechanical equilibrium] [taking the z-axis as the rotation axis] (the $\boldsymbol{k}$-direction component equation):

$$M_{r+\mathrm{d}r} - M_r = 0 \tag{1}$$

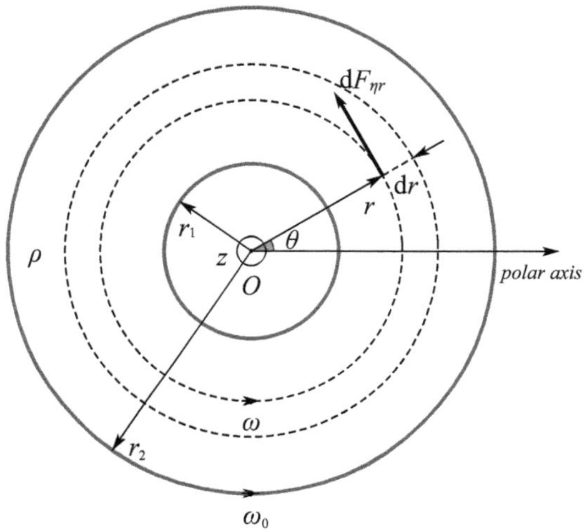

Figure 2

M_r—for the cylindrical surface of sectional radius r, the $\boldsymbol{k}$-direction component of the torque of the viscous force exerted on the inner liquid by the outer liquid at Time t.

$$M_{r+\mathrm{d}r} = M_r + \frac{\mathrm{d}M_r}{\mathrm{d}r}\mathrm{d}r \tag{2}$$

$$\Rightarrow \qquad \frac{\mathrm{d}M_r}{\mathrm{d}r} = 0 \tag{3}$$

Considering the definition of ***torque***:

$$M_r = \int_\Sigma \mathrm{d}F_{\eta r} \cdot r \tag{4}$$

$\mathrm{d}F_{\eta r}$—for the arbitrary surface element on the cylindrical surface of sectional radius r, the $\boldsymbol{\theta}^0$-direction component of the viscous force exerted on the inner liquid by the outer liquid at Time t.

Considering ***theorem of viscous force***:

$$dF_{\eta r} = \eta dS \cdot r \frac{d\omega}{dr} \tag{5}$$

η—the viscosity coefficient of the liquid.

dS—the area of the surface element on the cylindrical surface of sectional radius r.

$$\Rightarrow \quad M_r = \eta r^2 \frac{d\omega}{dr} \int_{\Sigma} dS = \eta r^2 \frac{d\omega}{dr} \cdot 2\pi r l \tag{6}$$

(6) → (3),

$$\Rightarrow \quad \frac{d^2\omega}{dr^2} = -\frac{3}{r} \frac{d\omega}{dr} \tag{7}$$

Defining the variable u:

$$u = \frac{d\omega}{dr} \tag{8}$$

$$\Rightarrow \quad \frac{du}{dr} = -\frac{3}{r} u \tag{9}$$

$$\Rightarrow \quad \frac{du}{u} = -3 \frac{dr}{r}$$

$$\Rightarrow \quad \int \frac{du}{u} = -3 \int \frac{dr}{r}$$

$$\Rightarrow \quad u = \frac{C_1}{r^3}$$

$$\Rightarrow \quad d\omega = \frac{C_1}{r^3} dr$$

$$\Rightarrow \quad \int d\omega = C_1 \int \frac{dr}{r^3}$$

$$\Rightarrow \quad \omega = -\frac{1}{2} \frac{C_1}{r^2} + C_2 \tag{10}$$

Considering the boundary conditions:

$$\begin{cases} \omega|_{r=r_1} = 0 & (11a) \\ \omega|_{r=r_2} = \omega_0 & (11b) \end{cases}$$

$$\Rightarrow \quad \begin{cases} C_1 = \omega_0 \dfrac{2r_1^2 r_2^2}{r_2^2 - r_1^2} & \text{(12a)} \\ C_2 = \omega_0 \dfrac{r_2^2}{r_2^2 - r_1^2} & \text{(12b)} \end{cases}$$

$$\Rightarrow \quad \omega = \omega_0 \frac{r_2^2}{r_2^2 - r_1^2}\left(1 - \frac{r_1^2}{r^2}\right) \tag{13}$$

Solution to Problem (2)

Defining Q—the volume rate of flow of the liquid at an arbitrary vertical section between the two cylindrical shells. Considering the definition of ***volume rate of flow***:

$$Q = \int \omega r \cdot l \mathrm{d}r \tag{14}$$

(13) → (14),

$$\Rightarrow \quad Q = \omega_0 l \frac{r_2^2}{r_2^2 - r_1^2}\int_{r_1}^{r_2}\left(1 - \frac{r_1^2}{r^2}\right) r\mathrm{d}r = \omega_0 l\left(\frac{1}{2}r_2^2 - \frac{r_1^2 r_2^2}{r_2^2 - r_1^2}\ln\frac{r_2}{r_1}\right) \tag{15}$$

Solution to Problem (3)

From (13),

$$\Rightarrow \quad \frac{\mathrm{d}\omega}{\mathrm{d}r} = \omega_0 \frac{2r_1^2 r_2^2}{r_2^2 - r_1^2}\frac{1}{r^3} \tag{16}$$

(16) → (6),

$$\Rightarrow \quad M_{r_1} = \eta r_1^2 \cdot \left.\frac{\mathrm{d}\omega}{\mathrm{d}r}\right|_{r=r_1} \cdot 2\pi r_1 l = 4\pi\eta l\omega_0 \frac{r_1^2 r_2^2}{r_2^2 - r_1^2} \tag{17}$$

$$\Rightarrow \quad \eta = \frac{M_{r_1}}{4\pi l\omega_0}\frac{r_2^2 - r_1^2}{r_1^2 r_2^2} \tag{18}$$

Index

For Product Safety Concerns and Information please contact our EU representative GPSR@taylorandfrancis.com
Taylor & Francis Verlag GmbH, Kaufingerstraße 24, 80331 München, Germany

www.ingramcontent.com/pod-product-compliance
Ingram Content Group UK Ltd.
Pitfield, Milton Keynes, MK11 3LW, UK
UKHW022200190926
475702UK00003B/129
* 9 7 8 9 8 1 4 8 7 7 4 1 1 *